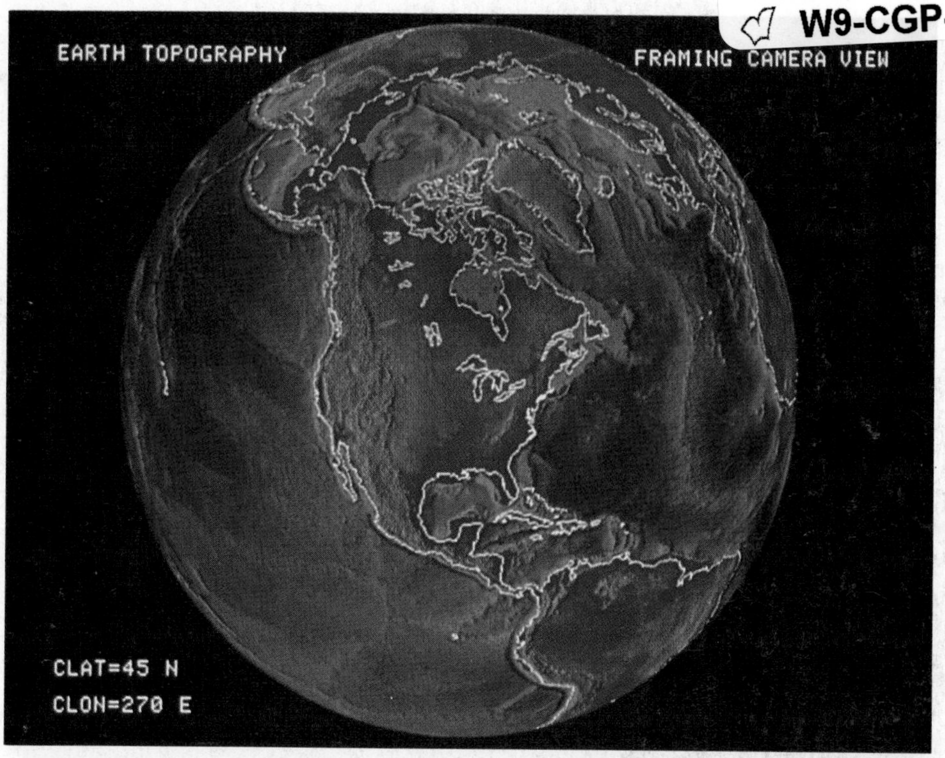

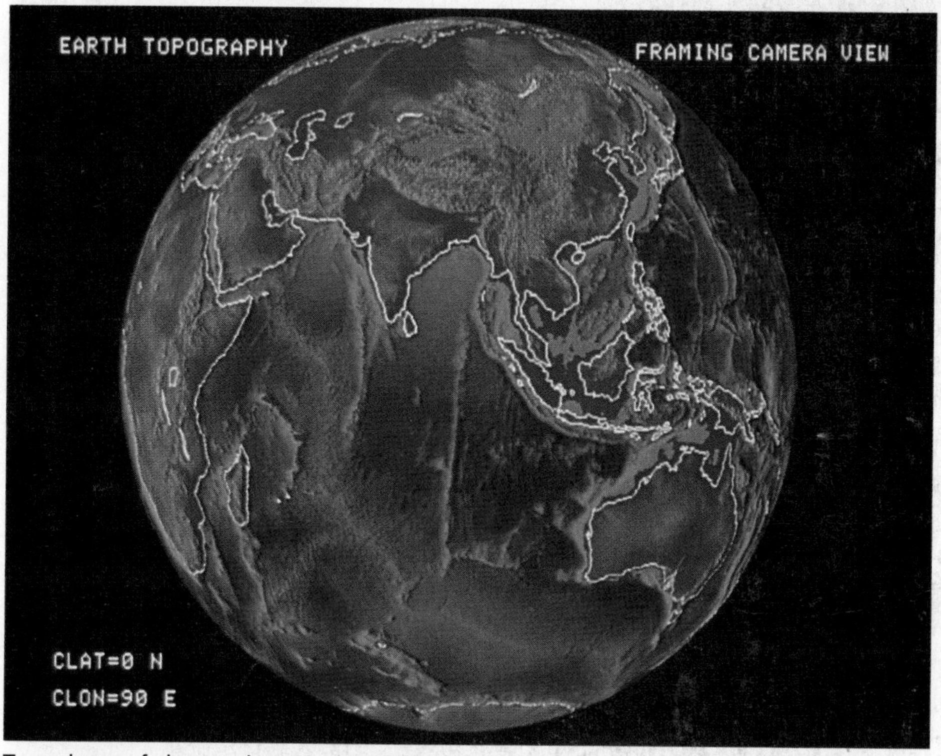

Two views of the earth as a planet. They depict the earth as it would appear if photographed by a spacecraft about 500,000 km above the earth. Topography of continents and ocean basins are shown by shaded relief and color. (Courtesy of M. Edwards and R. Arvidson, NASA Regional Planetary Image Facility, Washington University.)

Contemporary Physical Geology

Third Edition

Harold L. Levin

Washington University
St. Louis

Saunders College Publishing

Philadelphia Fort Worth Chicago
San Francisco Montreal Toronto
London Sydney Tokyo

Text Typeface: Century Book
Compositor: York Graphic Services
Acquisitions Editor: John Vondeling
Developmental Editor: Margaret Mary Anderson
Managing Editor: Carol Field
Project Editor: Margaret Mary Anderson
Copy Editor: Sarah Fitz-Hugh
Manager of Art and Design: Carol Bleistine
Art Director: Carol Bleistine
Art and Design Coordinator: Doris Bruey
Text Designer: Bill Boehm
Cover Designer: Lawrence R. Didona
Text Artwork: J & R Technical Services
Director of EDP: Tim Frelick
Production Manager: Bob Butler

Cover Credit: Front: Sandstone forms, Salt Point State Park, California. © DAVID MUENCH, 1989.
Back: Otter Cliffs, Acadia National Park, Maine. © JOHN G. WIDMAIER, JR., 1989.
Title page, Preface, and Contents photographs: Peter L. Kresan Photography and John G. Widmaier, Jr.

Printed in the United States of America
CONTEMPORARY PHYSICAL GEOLOGY, Third Edition

0-03-031139-X

Library of Congress Catalog Card Number: 89-043107

0123 015 987654321

For Emily and Natalie

Preface

While reflecting on the progress of science in his time, Charles Lyell concluded that never before had geology revealed "so many novel and unexpected truths and overcome so many preconceived opinions." The great nineteenth-century geologist had in mind the changing view of the earth from that of a static planet unaltered since its origin, to a "theatre of reiterated change." Lyell believed that he lived during a time of revolution in the geological sciences. Today, we are experiencing another such revolution, and Lyell's remarks about the progress of geology are as appropriate as they were over a century ago. As a result of bold new discoveries, geologists are now able to recognize the deeper and more fundamental causes for the many changes the earth has undergone. They have synthesized older interpretations of geologic events into a grand scenario of a dynamic planet on which ocean floors are in constant movement, continents break apart and admit the ocean, and geomagnetic poles switch about. These provocative ideas, together with the still important traditional geologic topics, are important components of the body of knowledge likely to enrich any student's liberal education.

Changes in the Third Edition

In preparing this third edition of *Contemporary Physical Geology,* it has been my foremost objective to present the more important aspects of geology as non-technically as the nature of the material will permit. I have kept in mind the needs of non-science majors seeking knowledge about their planetary home, while simultaneously fulfilling a science requirement. To this end, discussions throughout the text have been refashioned or augmented for greater clarity, as well as for the inclusion of information about recent developments, such as the earthquake in Armenia. Short "take-home messages" have been added to emphasize or summarize important points. The book includes only a few, simple chemical equations and almost no mathematics. It does contain hundreds of colorful illustrations. For a text intended for liberal arts majors, the adage about one picture being worth ten thousand words is particularly valid. As further aids, there is a comprehensive glossary, lists of terms to understand, and questions for home study or discussion in the classroom.

Organization of the Third Edition

The chapter sequence follows the traditional pattern of most introductory geology courses. At the same time, an effort has been made to instill enough background information into each chapter to permit instructors to adjust the order of topics according to their own course design. Chapter 1 is an introduction to the science, but also contains a brief overview of plate tectonics so that this important concept can be immediately integrated into subsequent discussions. Background information about geologic time, structures, and basic seismology enhances a student's understanding of sea floor spreading, and therefore these topics precede the full treatment of plate tectonics in Chapter 11. Chapters 12 through 18 will familiarize the student with agents and processes that operate at the earth's surface, whereas Chapter 19 deals with sources of energy, ore deposits, and other earth materials that constitute natural resources. The final chapter on "Planets and Moons" is written as a more or less independent unit that can be inserted in a topic sequence of the instructor's choosing. The chapter reminds the student that the earth is but one of nine planets, and shares many aspects of its origin, composition, and properties with neighboring bodies in space.

Acknowledgments

Most of the changes in this edition resulted from the suggestions of a particularly insightful and diligent group of reviewers and colleagues. In my own department at Washington University, I have benefited immensely from discussions with Sam A. Bowring and Ian Duncan. I happily also thank Thomas W. Broadhead (University of Tennessee), Larry E. Davis (Washington State University), Bryan Gregor (Wright State University), Bryce M. Hand (Syracuse University), Terence T. Kato (California State University, Chico), Peter L. Kresan (University of Arizona), Albert M. Kudo (University of New Mexico), Lawrence L. Malinconico (Southern Illinois University, Carbondale), William K. Steele (Eastern Washington University), Daniel A. Sundeen (University of Southern Mississippi), and Harry A. Wagner (Victoria College).

The assistance and direction of my friends at Saunders College Publishing were indispensible to this revision. Associate Publisher John Vondeling provided the stimulus and encouragement that moved the project forward on schedule. Carol Bleistine made certain the book would be attractive and easy to use by assiduously overseeing the illustration and design programs. The editor who guided the work from manuscript to galleys, to page proofs, and to finished product was Margaret Mary Anderson. I am deeply grateful for her professionalism, patience, and absolute commitment to quality.

My thanks conclude with a special note of gratitude to my wife Kay, who has cheerfully endured my preoccupation with the revision.

Contents Overview

Contents

Contemporary Physical Geology
Third Edition

The Scope and Substance of Geology

1

The Red Sea, shown here in a photograph taken from the Gemini spacecraft, was born about 30 million years ago, when the Arabian Peninsula broke away from Africa. The seaway continues to widen today. Hot solutions beneath the central part of the Red Sea are now creating ore deposits there. (Courtesy of NASA.)

EXAM I: Ch. 1, 2, 3, 4, 5, 11, 17

The light in the world comes principally from two sources—the sun, and the student's lamp.

Bovee, 1842

Introduction

In a dim corner of eastern Africa about 3 million years ago, an ancestral member of the human family stooped to pick up a sharp-edged rock and discovered that it could be used as a tool. This simple act was the beginning of humanity's utilization of the many and varied resources of the earth. Unlike the antelope and bison on the horizon, these early members of our own lineage were not content to passively accept what the earth had to offer. They, and their descendents, chose rather to exploit the environment and shape it to their needs. The resulting interaction between *Homo sapiens* and the earth has produced both benefits and problems. To understand the benefits, to gain knowledge useful in solving the problems, and to enhance our aesthetic appreciation of this extraordinary planet are sufficient reasons for undertaking the study of geology.

Geology: Study of the Earth and Earthlike Planets

In a deep tunnel of a South African diamond mine, miners glance inquisitively at a young woman as she uses a geology pick to break loose a piece of hard, black, kimberlite rock. She seems not to notice the miners as she slowly turns the piece of rock in the light of her head lamp. A slight smile crosses her face as she recognizes nodules of iron-rich minerals that confirm a theory that the rock had been carried upward as a melt from depths far beyond the reach of the deepest drill hole.

On a quiet evening in the Mediterranean Sea, two scientists clad in shorts and tennis shoes shift their gaze slowly along the length of a cylindric core of sediment that had been brought up from the sea floor only moments before. Suddenly, one exclaims, "There it is!" "It" is a layer of gypsum that provides the two researchers with evidence that 6 million years ago the blue Mediterranean was a dry, hot, desert basin.

Tiny particles of ice decorate the beard of a man taking a sample of rock cuttings that flow from a pipe near the base of a thundering rotary drilling rig located in Alaska. He carries his sample into a trailer beside the huge derrick, washes the rock chips carefully, and examines them with the aid of a microscope. On the well record, next to the appropriate depth, he enters the words "sandstone, well sorted, fine-grained, *oil stained.*" He holds a chip of rock to his nose and sniffs the faint odor of petroleum. Another sample of the rock is placed in a box containing a fluorescent lamp. The observer notes the vivid bluish hues that indicate the presence of crude oil. "Maybe," he thinks, "just maybe, this one will be an oil well."

The researcher in the diamond mine, the investigators aboard the oceanographic vessel, and the explorer for petroleum are all engaged in the practice of a science devoted to the study of the earth—its origin, history, composition, properties, and resources. That science is called *geology,* a word derived from the Greek *geo,* meaning earth, and *logos* meaning discourse. Today, a new breed of earth scientists has expanded the definition of geology to include geologic investigations of other planets in our solar system. This expansion is appropriate, since geologic knowledge is employed in such extraterrestrial tasks as interpreting photographs of moons of other planets (Fig. 1–1), in estimating the power of volcanoes on Jupiter's moons, and in identifying the minerals of moon rocks.

As a discipline, geology draws heavily on all other sciences and frequently merges into them. Many geologists consider themselves chemists or physicists who concentrate on the earth. Like chemists and physicists, geologists often set up carefully controlled experiments in the laboratory. Others collect data in the field. Their laboratory is

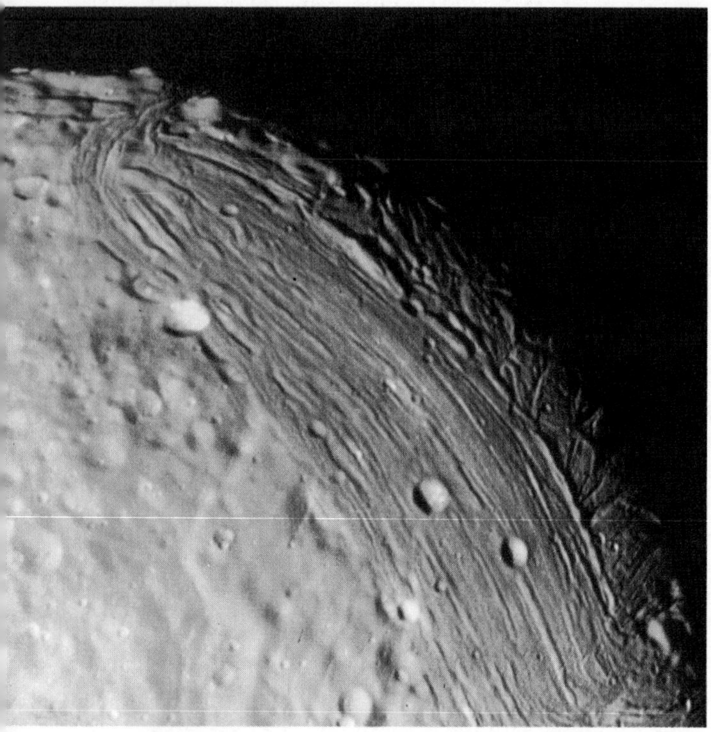

Figure 1–1 Voyager 2 spacecraft encountered distant Uranus, the seventh planet from the sun, in January of 1986. Ten moons of Uranus were discovered, including Miranda, shown here. This computer mosaic of Miranda images shows the moon's varied landscapes and indicates a complex geologic history. (Courtesy of NASA.)

an entire planet affected by a multitude of complex interacting processes (Fig. 1–2). Geologists are required to reach conclusions based on studies of the observed *results* of events whose *causes* may be traced back billions of years. An understanding of how earth materials originate, and the interpretation of the forces that produce mountain chains and volcanoes, must often be inferred from limited evidence or by analogy with events that can still be observed today.

Geology is largely concerned with rocks and minerals and the record they preserve of changes produced by processes that have long been at work on the surface and deep within the earth.

The Scientific Method in Geology

As they strive to learn all they can about the earth, geologists employ the same procedures used by scientists in other disciplines. The procedures are sometimes rather formally referred to as the scientific method. The scientific method is merely a general scheme for attacking problems, and is certainly not a series of steps that scientists strictly and consciously follow. As we look at the progression of scientific work, however, we see that it often begins with the formulation of questions, proceeds to the collection of observations or data, and is followed by the development of an explanation or hypothesis that can be subjected to testing and scrutiny by other scientists.

Scientific methodology usually begins with the collection of data needed to answer a question. A hypothesis that is supported by the data can then be developed and tested for its validity.

As an example of scientific methodology, consider an investigation by Anita Harris, Jack Epstein, and Leonard Harris of the United States Geological Survey. These geologists were intrigued by the observation that a group of fossils called conodonts exhibited a range of colors from pale yellow to black. Conodonts (Fig. 1–3) are the microscopic hard parts of organisms that lived on earth from about 560 to about 200 million years ago. They are composed of apatite—the same mineral of which bone is made—and they are abundant at many localities in all kinds of sedimentary rocks. The earth scientists at the U.S. Geological Survey asked the question: "Why did conodonts of the same age, but from different parts of a region, have different colors?" They then set about collecting the laboratory and field data needed to provide an answer. For their laboratory observations they selected pale yellow conodonts that came from a limestone formation that had never been deeply buried. The conodonts from the limestone were heated in an electric furnace. The investigators increased the temperature from 300°C to 600°C in 50° steps over a period of 10 to 50 days. To their de-

Figure 1–2 Geologist on U.S.G.S. survey team measuring dimensions of a huge mudflow resulting from the eruption of Mount St. Helens. Mount St. Helens is in the background. Skamania County, Washington. (Courtesy of U.S. Geological Survey.)

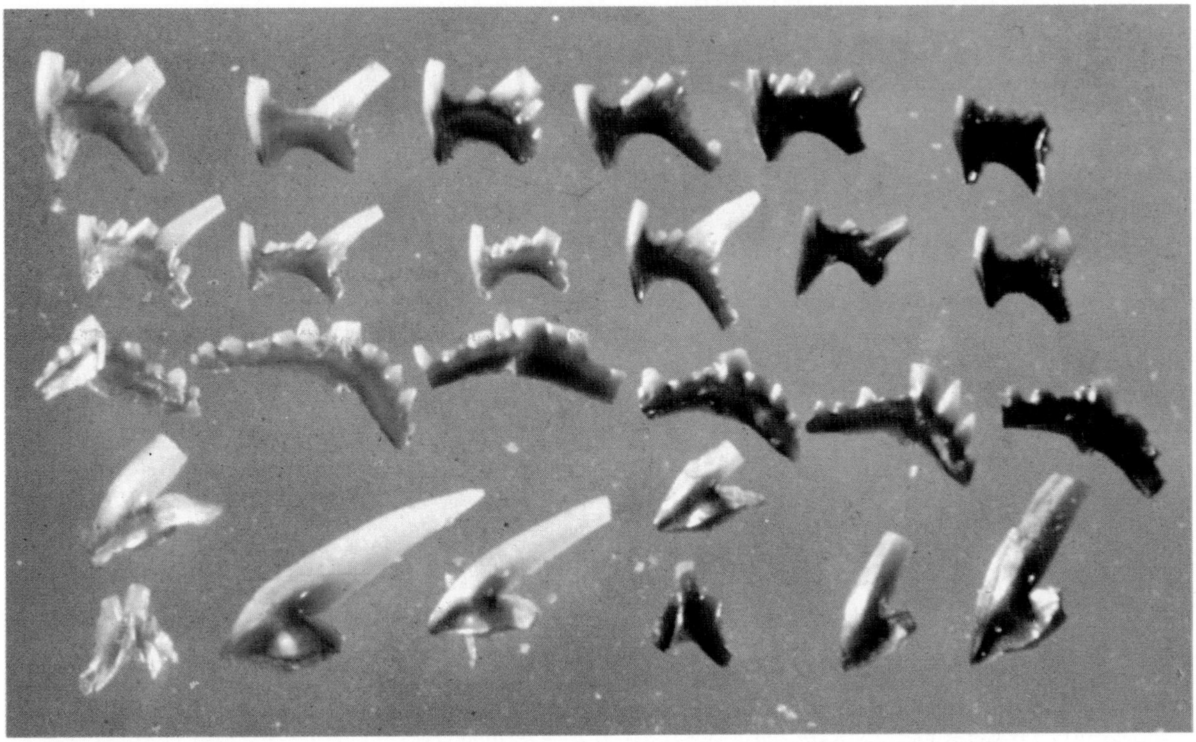

Figure 1–3 Conodonts showing varying degree of color alteration resulting from deep burial and attendant increases in temperature. (Courtesy U.S. Geological Survey).

light, they observed that the conodonts changed in color through five phases from pale yellow to black. On further heating, so as to achieve conditions that change sedimentary rocks into metamorphic rocks, black conodonts changed sequentially to gray, opaque white, and finally crystal clear. The scientists then compared the color changes produced experimentally with the colors of untreated conodonts collected in the Appalachian Mountains from rocks that had been subjected to various depths and durations of burial. When plotted on maps, their data dramatically revealed that light-colored conodonts are found in the western Appalachians where burial was least, and dark-colored conodonts occur along the eastern side of the Appalachians where burial was greatest (Fig. 1–4).

The original question, "Why did conodonts of the same age, but from different parts of a region, differ in color?" could now be answered with a hypothesis stating that depth of burial and attendant increase in temperature are the dominant factors in conodont color alteration. Further, conodont color can be used as an index to assess the heating history of ancient rocks.

In the course of their investigation, another interesting fact was noted. It became apparent that rocks containing black conodonts, having experienced high temperatures, are less likely to yield

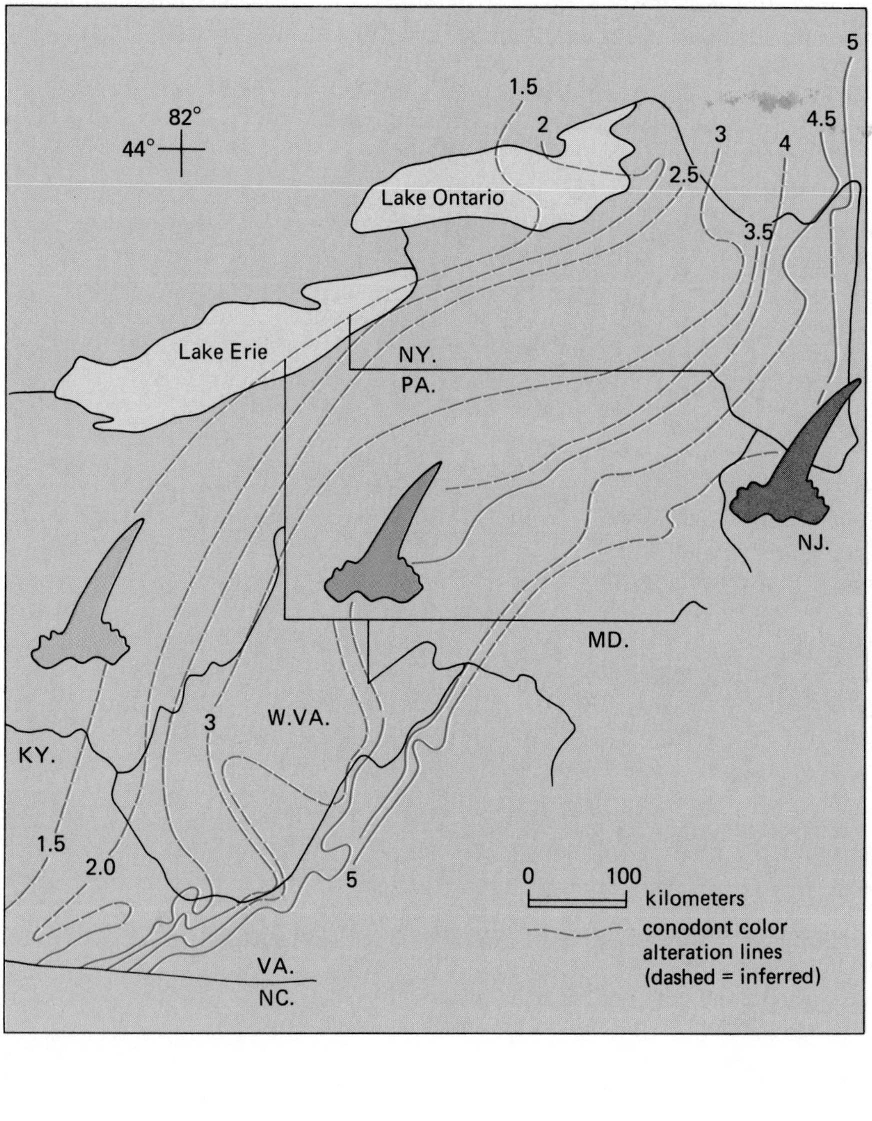

Figure 1–4 Map showing the west-to-east gradation in conodont color resulting from eastward increases in the temperatures to which the conodont-bearing strata were subjected. In general, the darker-colored conodonts indicate higher temperatures. Conodont alteration colors range from black (numbered 5), through dark brown (numbered 3), to pale brown (numbered 1.5). (Simplified from Epstein, A.G., Epstein, J.B., and Harris, L.D., *Conodont Color Alteration—an Index to Organic Metamorphism,* U.S. Geological Survey Professional Paper 995, 1977; conodont drawings added to illustrate concept.)

commercial quantities of oil and gas. This index of alteration is now used routinely by many oil companies in their search for petroleum.

Time and Geology

Time is important in geology. It is involved in every change that occurs on and within the earth. The time during which geologic agencies act upon the crust influences the characteristics of continents and the configurations of landscapes. Time is involved in the cooling of molten masses of the rocks and has a bearing on the characteristics of rocks that ultimately form. Time is an ingredient in organic evolution and provides the framework for the history of animals and plants. Indeed, every landscape on earth contains the tangible imprint of the passage of time.

According to recent estimates, the earth is approximately 4700 million years old. The conception of so vast a span of years helps us to comprehend the fact that mountains are really only transitory forms and that processes that appear to be working imperceptibly slowly have nevertheless had ample time to change the surface of the earth. With geologic knowledge, each of us can expand his time window to better understand the earth. We need not be limited by what we perceive only in the span of our lives. In Chapter 8, we will describe how geologists deal with time. We will learn how they determine the ages of rocks and minerals and are thereby able to discern when momentous events occurred in the past. In this preparatory chapter, however, it will be useful to introduce the geologic time scale (Fig. 1–5) to which these events and past conditions can be referred.

The time scale was originally based on the superpositional relationships of fossil-bearing strata (oldest at the bottom, youngest at the top), and on the stage of evolutionary progress shown by fossilized organisms. Subsequently, in the early 1900s, radiometric methods were developed for determining the precise age of rocks in actual numbers of years. These actual or absolute ages were then added to the geologic time scale. As is indicated in Figure 1–5, the largest divisions of the time scale are the Archean, Proterozoic, and Phanerozoic **Eons**. Approximately 87 per cent of all geologic

The Geologic Time Scale

Time Units of the Geologic Time Scale*
(Numbers are absolute dates in millions of years before present)

Eon	Eras	Periods	Epochs	Date
Phanerozoic	Cenozoic	Quaternary	Recent or Holocene	
			Pleistocene	2.0
		Tertiary	Pliocene	5
			Miocene	24
			Oligocene	37
			Eocene	58
			Paleocene	66
	Mesozoic	Cretaceous		144
		Jurassic		208
		Triassic		245
	Paleozoic	Permian		286
		Pennsylvanian (Upper Carboniferous)		320
		Mississippian (Lower Carboniferous)		360
		Devonian		408
		Silurian		438
		Ordovician		505
		Cambrian		570
Proterozoic Eon			Precambrian	2500
Archean Eon			Precambrian	3800
No record			Precambrian	

Figure 1–5 The geologic time scale.

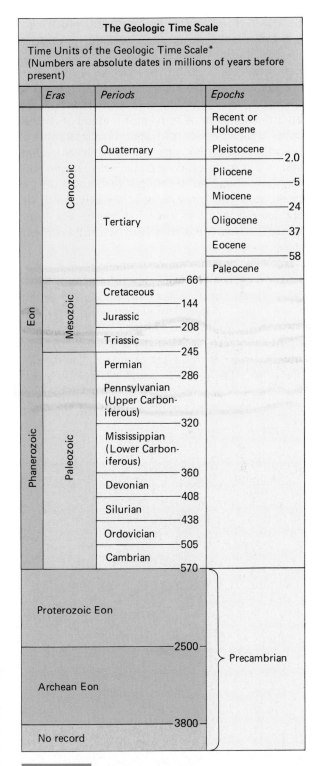

time is encompassed in the Archean and Proterozoic Eons, which are sometimes simply referred to as "Precambrian." To a geologist, the phrase "in the beginning" alludes to the Archean, which began about 4.6 billion years ago. By Archean time the earth had amassed most of its materials, possibly from some far older turbulent cloud of cosmic dust. Archean rocks reveal the characteristics of the earth during its early stages of development, and, in a few places, contain the microscopic, unicellular remains of primordial life. Rocks of the Proterozoic are generally less severely deformed and altered than those of the Archean. The fossil remains of life forms, however, are still scarce in these rocks.

The Phanerozoic Eon is divided into three subdivisions called **eras**. The oldest of these, the Paleozoic (from the Greek *Palaeos*, "ancient," plus *zoon*, "life"), lasted 325 million years and witnessed the appearance of fish, amphibians, and reptiles on earth. The Paleozoic was followed by the Mesozoic (from the Greek *mesos*, "medieval," plus *zoon*, "life"), which lasted 179 million years. Birds and mammals appeared in the Mesozoic, which is also popularly dubbed the "age of dinosaurs." The most recent era, the Cenozoic (*kainos*, "recent," plus *zoon*, "life") began 66 million years ago.

Geologic eras are divided into shorter time intervals called **periods**, and periods are divisible into still smaller units of time called **epochs**. We live in the Recent or Holocene Epoch.

The geologic time scale is the chronological standard to which events in the 4.6 billion year history of the earth can be referred.

An Overview of the Third Planet

Imagine that you have taken a journey by spacecraft to a location high above our solar system. From your vantage point in space, and provided you had suitable tools, you might look downward and study the entire solar system. You would see at its center a rather ordinary star, our sun. Circling the sun would be the nine planets (Fig. 1–6). On

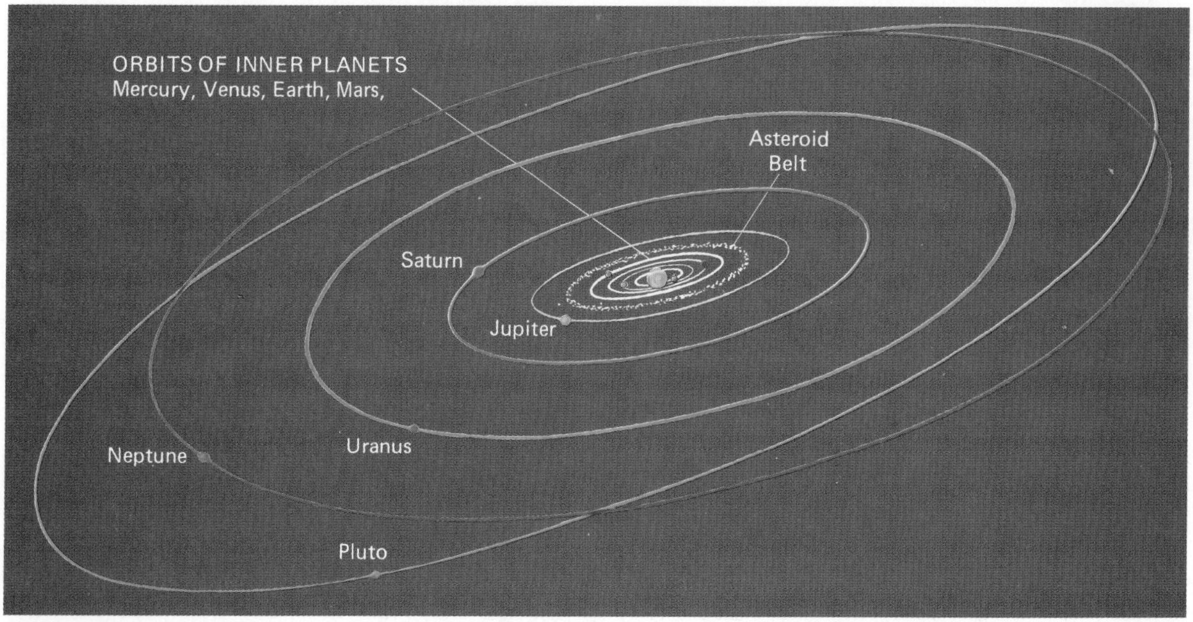

Figure 1–6 Orbits of the nine planets in the solar system.

close examination you would discern that the four planets nearest the sun are relatively small and that the next four are bigger and more widely spaced apart. Finally, the tiny planet Pluto would be seen at a distance from the sun of over 40 times that which separates the earth from the sun.

From your distant observation point, you would see little about the earth that is distinctive. It is the third planet from the sun and the largest of the inner group of rocky or terrestrial planets. Its moon is larger relative to the size of the earth than the moon of any other planet. Otherwise it would appear only as one planet among many, below average in size, and a mere speck when compared with the dimensions of the sun.

With patience and mathematical skills, those aboard the spacecraft might next begin to work out the paths of the planets. It would soon be evident that the orbital routes of all the planets lie in approximately the same plane. Thus, the solar system is like a disc with the sun at its center. Looking down on the solar system from a point high above the North Pole, you would observe an interesting similarity of movements. The revolution of the earth about the sun, of the moon about the earth, and the rotation of all three bodies on their individual axes would be similar, that is, in a *counterclockwise* direction. All of these movements play a role in our daily lives. The sun, besides being our source of energy and light, is a reference object for the keeping of time. A calendar year, for example, is defined as the time required for the earth to make one entire orbit around the sun. During this time, the earth rotates 365.25 times about an imaginary line known as its *axis*. The axis of rotation makes an angle of 66.5° with the orbital plane (or 23.5° from a line perpendicular to that plane), and this is the reason why the earth experiences seasonal changes (Fig. 1–7). Tides are caused largely by the gravitational attraction of the moon and sun for the earth.

All planets of the solar system revolve around the sun in the same counterclockwise direction. All except Venus have the same sense of rotational direction, and all orbit the sun in approximately the same plane.

Earth Dimensions and Density

From classical times it has been known that the earth is roughly spherical in shape. Actually, the

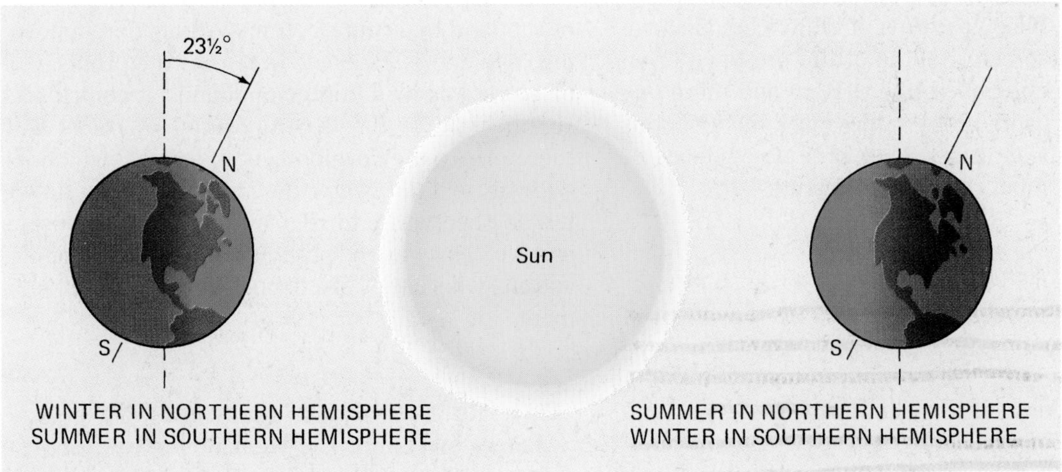

Figure 1–7 Seasons occur because a planet's axis is inclined with respect to the plane of its revolution around the sun. As shown here, when the earth's northern hemisphere is tilted toward the sun, it has its summer. At the same time the southern hemisphere is experiencing winter. (From Pasachoff, Jay M., *Contemporary Astronomy*, 4th ed., Philadelphia, Saunders College Publishing, 1989.)

planet is shaped more like a slightly flattened ball whose polar radius is about 21 km shorter than its equatorial radius. The average radius is 6371 km. The earth's overall density is about 5.5 grams per cubic centimeter (5.5 g/cm^3). While the earth's density is greater than that of any other planet in the solar system, it is not greatly different from that of Mercury, Venus, or Mars, the other three *inner* planets of the solar system. Because the average density of surface rocks is only about 2.7 g/cm^3, the material deep within the earth must have a density well in excess of the 5.5 g/cm^3 average. Very likely the material at the earth's center has a density of over 13.0 g/cm^3. The way this information is derived, and its importance, are discussed in Chapter 10.

The 5.5 g/cm^3 density of the earth is more than twice that of the density of surface rocks. To account for this difference, exceptionally dense materials must exist in the earth's central region.

Air, Water, and Rock

The splendid photographs of earth taken from space (Fig. 1–8) by the Apollo astronauts remind us that our planet is more than a rocky globe orbiting the sun. Wispy patterns of white clouds above the azure blue color tell us of the presence of an atmosphere and hydrosphere. Here and there one can discern patches of tan that mark the locations of continents. Greenish hues provide evidence of the planet's most remarkable feature—*life.*

The Atmosphere

We live beneath a thin but vital envelope of gases called the atmosphere. We refer to these gases as *air.* Dry air is composed mainly of nitrogen (78.03 per cent by volume) and oxygen (20.99 per cent). The remaining 0.98 per cent of air is made of argon, carbon dioxide, and very small quantities of other gases. One of these other components found mostly in the upper atmosphere is a form of oxygen called ozone. Ozone absorbs much of the sun's lethal ultraviolet radiation and is thus of critical importance to organisms on the surface of the earth. Air also contains 0.1 to 5.0 per cent water vapor. However, because this moisture content is so variable, it is not usually included in the list of atmosphere components.

Each day the earth receives radiation from the sun. At the surface of the earth, this energy is radiated back as heat and some is absorbed by water vapor, carbon dioxide, and ozone. This absorbed energy heats the atmosphere and drives the winds. It is a major contributor to weather and climate, both of which influence the nature of weathering and erosion at the earth's surface.

The Hydrosphere

The discontinuous envelope of water that covers 71 per cent of the earth's surface is called the **hydrosphere**. It includes not only the ocean, but the water contained in streams and lakes, water frozen in glaciers, water that occurs underground in pores and cavities of rocks, and water vapor in the air. If surface irregularities such as continents and deep oceanic basins and trenches were smoothed out, water would completely cover the earth to a depth of more than 2 km.

The process of sculpturing our landscapes is primarily dependent on water (Fig. 1–9). Glaciers composed of water in its solid form alter the shape of the land by scouring, transporting, and depositing rock debris. Because water has the property of dissolving many natural compounds, it contributes significantly to the decomposition of rocks and, therefore, to the development of soils on which we depend for food. Water moving relentlessly downhill as sheetwash, in rills, and in streams loosens and carries away the particles of rock to lower elevations where they are deposited as layers of sediment.

Water is important in many geologic processes, including the weathering and erosion of the earth's surface, and the transport and deposition of sediment.

The ocean basins hold most of the water in the hydrosphere. These basins are of enormous inter-

Figure 1–8 The earth from space, photographed by Apollo astronauts. (Courtesy of NASA.)

est to geologists, who have discovered that they are not permanent and immobile as once believed, but rather are dynamic and ever changing. The ocean basins grow and shrink, and their evolution has a direct relation to the formation of mountains, chains of volcanoes, deep sea trenches, and mid-oceanic ridges. The layers of sediment deposited on the ocean floor contain clues that allow geologists to decipher the earth's history. Here also one finds mineral resources and clues to the location of ore deposits elsewhere on the planet. The ocean provides part of our food supply and has a far-reaching influence on the climate we experience.

Figure 1–9 Water contained in streams is an effective agent in sculpturing our landscapes. Here, the erosive power of running water has carved the interesting goosenecks of the San Juan River, Utah. (Courtesy W.B. Hamilton.)

The Lithosphere and the "Spheres" Beneath

Somewhat like the concentric shells of an onion, the earth is composed of a series of layers (Fig.

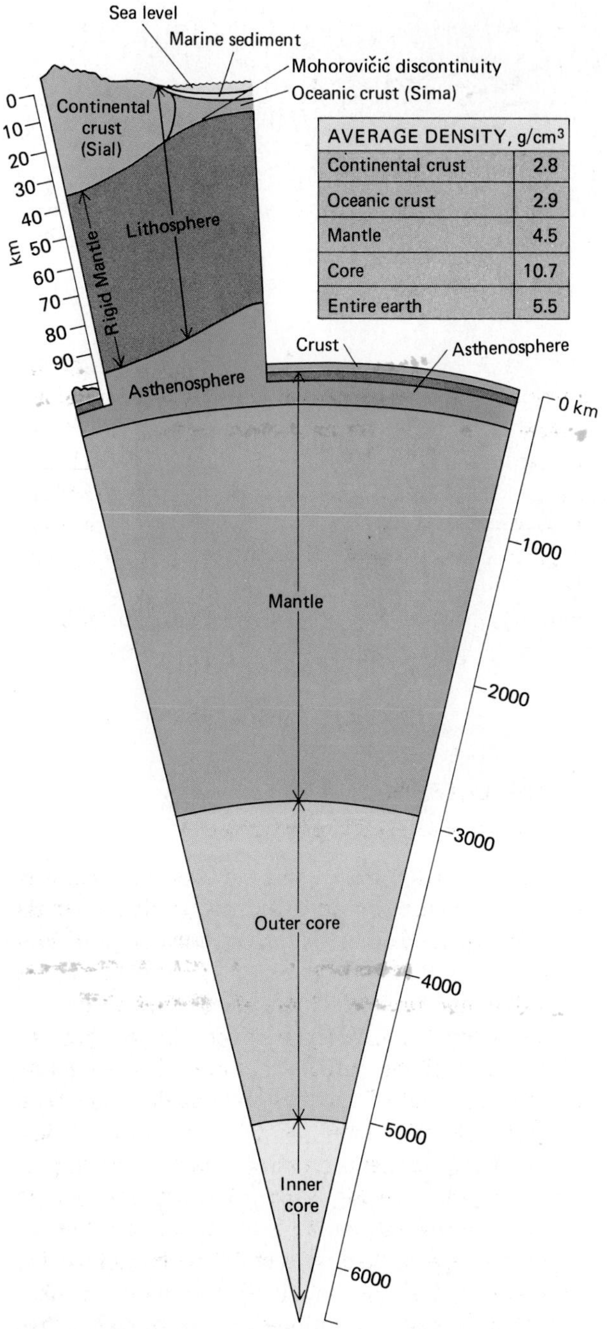

AVERAGE DENSITY, g/cm³	
Continental crust	2.8
Oceanic crust	2.9
Mantle	4.5
Core	10.7
Entire earth	5.5

Figure 1–10 Internal structure of the earth. Notice that the mantle extends from the base of the crust to the top of the core. The lithosphere includes both the crust and the layer of mantle that overlies the asthenosphere.

1–10). As will be described in a later chapter, the existence of these layers has been deduced from the study of earthquake waves that have passed through the earth. At the surface is the thin outer shell known as the crust. The base of the crust is defined by a surface below which the velocity of certain earthquake waves is significantly greater than in the rocks above. A plane of this kind is called a seismic discontinuity, and the seismic discontinuity at the bottom of the crust has been named the **Mohorovičić discontinuity** (or simply the "Moho") after its discoverer.

The crust of the earth really consists of two kinds of rock, each with its own distinctive general composition, thickness, and density. The continental crust has a composition somewhat similar to granite's, has a relatively low density, and ranges in thickness between 35 and 60 km. The crust comprising the ocean basins is somewhat denser, rarely exceeds 5 km in thickness, and has an average composition of basalt.

The layer beneath the earth's crust is called the **mantle**. The mantle has not yet been penetrated by drilling, but earthquake data indicate that it extends from the base of the crust to a depth of about 2900 km. The mantle has two upper parts that have the same composition but different physical characteristics. The uppermost of these two parts lies immediately below the crust to a depth of about 120 km. It is solid and, together with the crust, is referred to as the **lithosphere**. The lithosphere rides over a weak plastic second layer of the mantle known as the **asthenosphere**, which extends to a depth of about 250 km.

At the base of the mantle is yet another discontinuity that serves as the boundary between the mantle and the core. Geophysicists have recognized two parts of the core: a liquid outer zone and a solid inner core. Both parts are believed to be composed mainly of iron and nickel.

The earth's crust and the outer part of the mantle compose the lithosphere, which lies above a weak zone of the mantle called the asthenosphere.

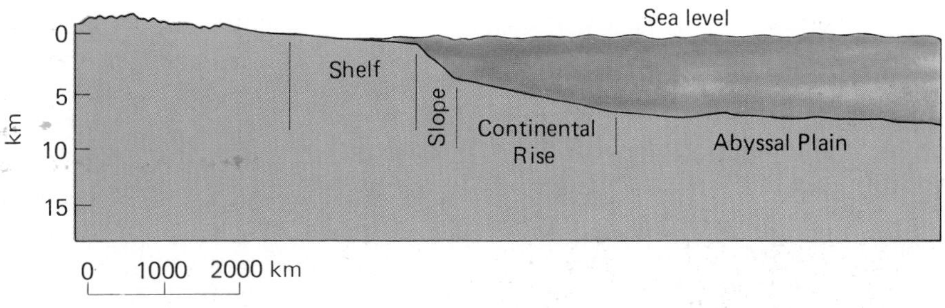

Figure 1–11 Profile showing the relationships of the continental shelf, continental slope, and continental rise.

Major Features of the Continents and Ocean

It requires only a quick glance at a map of the world to become convinced that the most conspicuous elements of the earth's surface are continents and ocean basins. Both continents and ocean basins contain distinctive geologic features that have developed in response to particular geologic processes. For example, the major features of continents are stable regions and orogenic belts. As is suggested by the name, stable regions are parts of the continents that are no longer disturbed by the kind of geologic forces that tend to distort rock layers and raise mountains. Plains and plateaus are characteristic of stable regions. **Orogeny** means mountain-building, and thus orogenic belts are zones in which great thicknesses of layered rocks have been strongly compressed, altered, and raised into lofty mountain chains.

A century ago, the ocean floors were considered to be rather featureless plains. Actually, they exhibit a variety of major features. Around the edges of the oceans are the submerged margins of the continents called continental shelves. The shelves are of enormous importance because they contain many off-shore oil traps, as well as deposits of sand, gravel, oyster shells, and diamonds. They are bounded seaward by the steeper continental slopes, which in turn drop off into less steep continental rises and eventually into the abyssal plains (Fig. 1–11). Rising above the floors of the abyssal plains in the Atlantic, Pacific, and Indian oceans are perhaps the most impressive features of the ocean basins. These features, called mid-oceanic ridges (Fig. 1–12), tower over 3500 meters above the sea floor. In contrast to the ridges, the ocean floor contains long, narrow, earthquake-prone depressions known as **deep-sea trenches**. The Aleutian trench (Fig. 1–13) extends 3500 km from Yakutat Bay in Alaska to the western tip of the Aleutian Islands. It has a maximal depth of more than 7500 m. Even greater depths are attained by the Marianas trench off the Philippine Islands in the western Pacific. Here the ocean floor descends to the awesome depth of over 11,000 meters.

Plate Tectonics— A Preliminary Overview

The world-encircling mid-oceanic ridges, the deep-sea trenches, and even our great mountain ranges are neither haphazardly formed nor randomly located. These features are manifestations of a dynamic process called **plate tectonics**.

The central idea of plate tectonics is that the lithosphere of the earth is composed of seven or eight large "plates" and several smaller ones (see Fig. 11–11). The plates, which are 75 to 120 km thick, consist of both crust and part of the uppermost mantle (see Fig. 1–10). Lithospheric plates rest on a weak plastic layer of the mantle that we defined as the asthenosphere. Two characteristics of plates are that they move relative to one another and that volcanic eruptions and earthquakes frequently occur along their borders. Geologists are uncertain about what causes them to move. Most believe the plates are conveyed rather like the scum on boiling jam by the drag of slowly moving convection currents in the underlying mantle.

(Text continues on p. 14.)

Figure 1–12 The physical features of the ocean floors and the continents. (Copyright © Marie Tharp.)

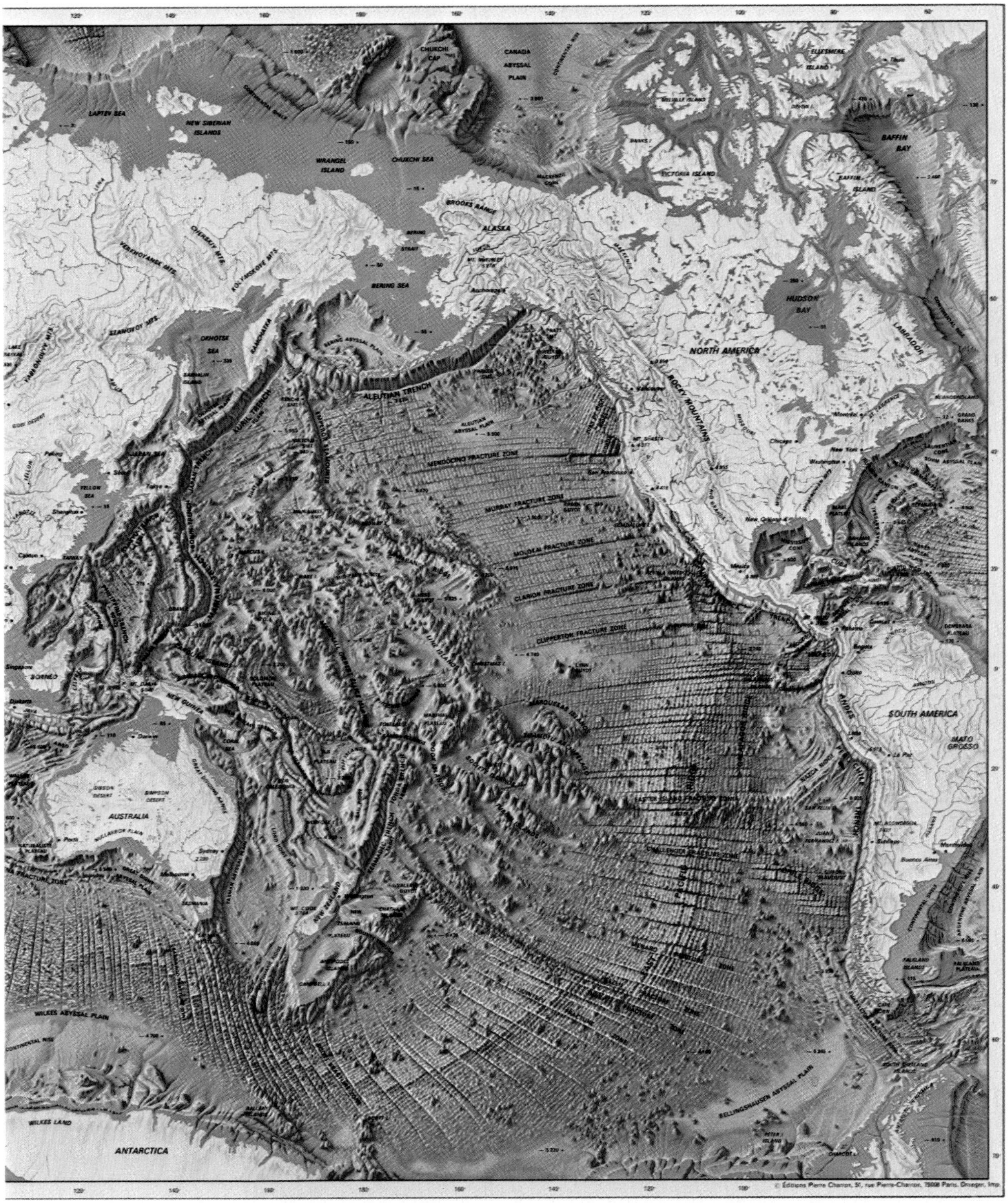
LAPTEV SEA
NEW SIBERIAN ISLANDS
CHUKCHI CAP
CANADA ABYSSAL PLAIN
WRANGEL ISLAND
CHUKCHI SEA
ELLESMERE ISLAND
BAFFIN BAY
VICTORIA ISLAND
BAFFIN ISLAND
BROOKS RANGE
ALASKA
BERING STRAIT
BERING SEA
HUDSON BAY
OKHOTSK SEA
LABRADOR
NORTH AMERICA
ALEUTIAN TRENCH
ROCKY MOUNTAINS
KURIL TRENCH
ALEUTIAN ABYSSAL PLAIN
JAPAN SEA
MENDOCINO FRACTURE ZONE
YELLOW SEA
San Francisco
MURRAY FRACTURE ZONE
MOLOKAI FRACTURE ZONE
CLARION FRACTURE ZONE
CLIPPERTON FRACTURE ZONE
BORNEO
NEW GUINEA
SOUTH AMERICA
MATO GROSSO
GIBSON DESERT
SIMPSON DESERT
AUSTRALIA
WILKES ABYSSAL PLAIN
CONTINENTAL RISE
BELLINGSHAUSEN ABYSSAL PLAIN
WILKES LAND
ANTARCTICA
© Éditions Pierre Charron, 51, rue Pierre-Charron, 75008 Paris. Draeger, Imp.

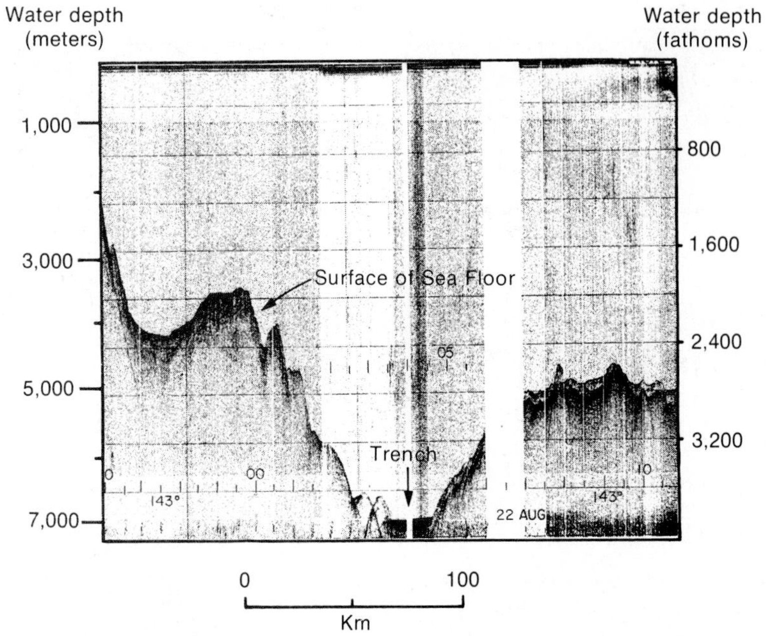

Figure 1–13 A continuous seismic profile across the Aleutian Trench. (Courtesy of Lamont-Doherty Geological Observatory, R.V. Conrad Cruise #12, 1969.)

The earth's lithosphere consists of several great plates that move over the underlying weak asthenosphere.

If one visualizes a globe sheathed in moving lithospheric plates, it becomes apparent that the boundaries between plates will differ according to the directions of plate movement. For example, zones along which plates move away from one another are called divergent boundaries. Along these boundaries, black lavas pour from volcanoes and fill the area between the diverging plates. The result is the great linear chains of mostly subsea volcanoes that constitute the mid-oceanic ridges (see Fig. 1–12). Iceland and the Azores represent small segments of the mid-Atlantic Ridge that have become emergent.

On a sphere such as the earth, it is not possible for lithospheric plates to move apart from one another in some places, unless they converge or slide past one another along other boundaries. Convergent boundaries are those along which plates move toward one another. They are also places where part of the crust is consumed and thereby provides room for new crust produced at divergent boundaries. This consumption may occur when one of the converging plates (usually the denser oceanic plate) sinks beneath the other plate to be melted and recycled in the underlying mantle (Fig. 1–14). The process is called **subduction**, and the tops of the subduction zones are recognized by the presence of deep-sea trenches. Convergent boundaries are also zones of intense volcanic and earthquake activity.

The third possibility for movement along a plate border is neither convergent nor divergent, but rather lateral, as two plates slide past each other. Such zones of horizontal movement are called shear or transform boundaries. They are located where crustal material is being neither added nor consumed. Shear boundaries are marked by great faults. Movements along these faults generate frequent and often intense earthquakes. The San Andreas fault zone in California marks a shear boundary.

A discussion of the historical development of plate tectonics theory and an explanation of the many lines of supporting evidence is presented in Chapter 11.

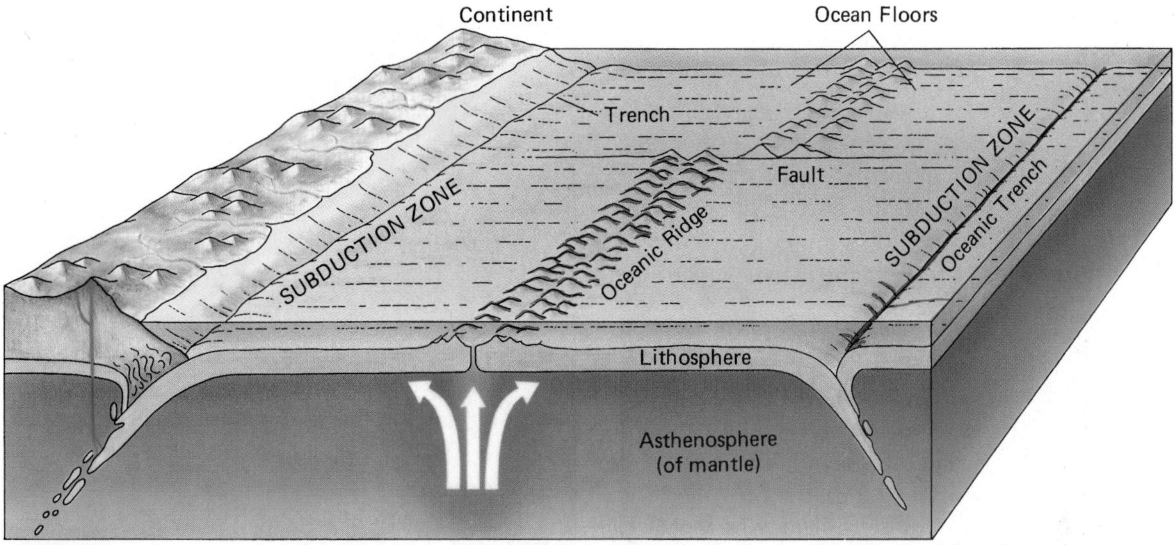

Figure 1–14 Lithospheric plate motions at divergent and convergent boundaries. The plates move somewhat like conveyor belts from a divergent boundary (oceanic ridge) to a convergent boundary where subduction occurs.

Boundaries between lithospheric plates may be divergent (plates move apart as at mid-oceanic spreading centers), convergent (boundaries associated with collision of plates), and shear boundaries (plates slide laterally past one another).

A Delicate Balance

In speculating about the origin of such features as deep-sea trenches, mid-oceanic ridges, and mountain ranges, one becomes quickly aware of the mobility of the lithosphere. The old notion embodied in the phrase "good old terra firma" is no longer valid. Crustal movements are continuous and may range from those that are imperceptibly slow to those that are rapid and violently destructive (Fig. 1–15). They may affect only small local areas or disturb a large region of a continent or ocean basin. The earth's instability is apparent to anyone who has been jolted by an earthquake or who has witnessed a volcano erupt. To the geologist, the continuum of unrest can be traced through

Figure 1–15 Damage caused by the 1985 Mexico City earthquake. (Courtesy St. Louis Science Center.)

folded strata (Fig. 1–16), layers of lava, and great displacements of crustal rocks. These features are manifestations of powerful inner forces capable of uplifting entire continents and changing the dimensions of the sea floors. However, just as there are powerful constructional forces on the earth, there are less dramatic, but equally significant, destructive forces. Geologic agents such as streams, glaciers, and wind erode the lands and carry the products of erosion to the ocean. One might surmise that the floors of the ocean would be the final resting place for these land-derived sediments, but this is not likely. As suggested earlier, the sea floors are not static but move along slowly like gargantuan conveyor belts. Sediment accumulating in the ocean basins is carried along on the moving floor to subduction zones, and there it may be scraped off or pulled downward into the asthenosphere and melted. Thus, the earth's surface is rather like a constantly changing battleground where internal forces raise and produce new land areas, where gravity and geologic agents operating near the surface work to wear the lands away, and where earth materials are continuously being changed and recycled.

Figure 1–17 Large quartz crystal (height 14 cm.)

Figure 1–16 These strata in the Scapegoat Mountains of Montana were originally laid down in horizontal layers. They have been folded and raised by compression of the earth's crust. Such constructive forces counterbalance the destructive processes that reduce the level of the land. (Courtesy U.S. Geological Survey.)

The Recycling of Earth Materials

We have only to look about us to see that the crust of the earth is composed of rocks and minerals. **Minerals** are naturally formed chemical elements or compounds having a definite range in chemical composition and usually a regular crystal structure (Fig. 1–17). Rocks are aggregates of one or more minerals that constitute an appreciable part of the earth's crust. Rocks can be identified according to their physical properties and mineralogic composition. The properties that provide clues to the origin of the rock are of greatest importance. By selecting properties having the greatest genetic significance, geologists have been able to group rocks into three great families according to the way in which they formed. **Igneous rocks** are those that have cooled

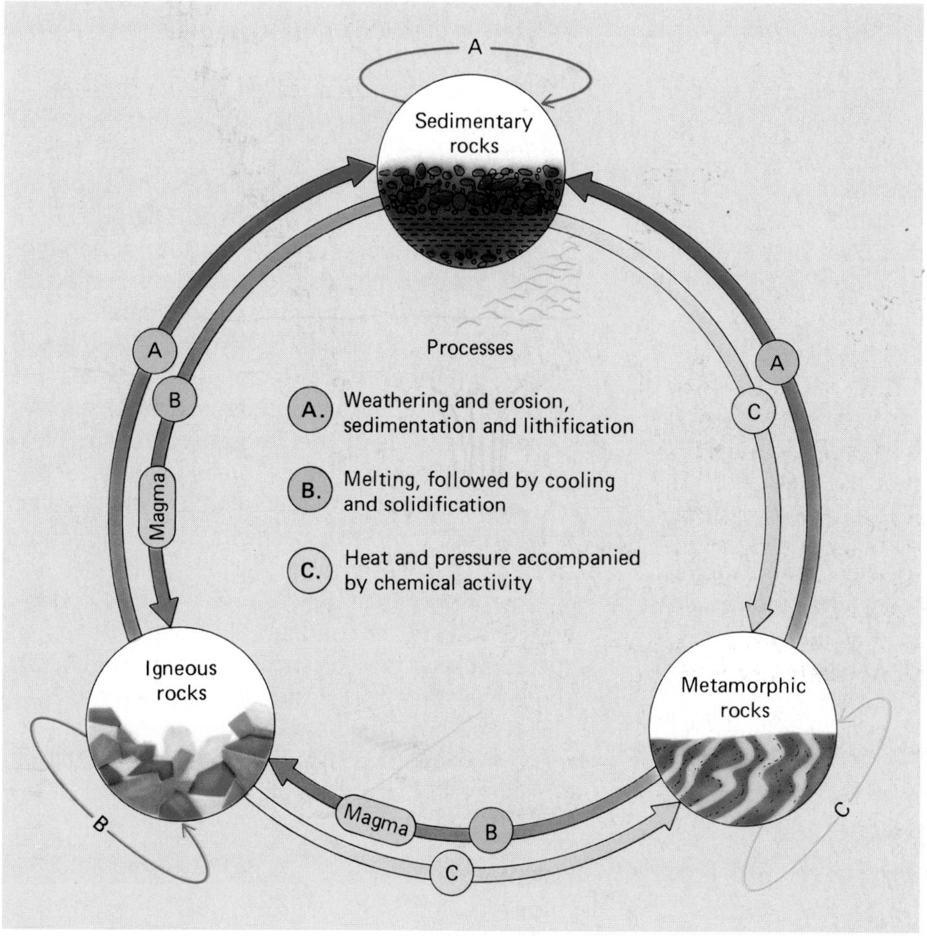

Figure 1–18 Geologic processes act continuously on earth to change one type of rock into another.

from a molten state. **Sedimentary rocks** are formed by the accumulation and consolidation of sediment. **Metamorphic rocks** are those that have formed in the solid state from preexisting rocks of any kind in response to heat, pressure, and associated chemical activity.

In thinking about the three major groupings of rocks, it is important to remember that rocks and minerals are not immune to change (Fig. 1–18). The earth's crust is dynamic and ever-changing. Any sedimentary or metamorphic rock may be melted to produce igneous rocks, and a previously existing rock of any category can be affected by the pressure and heat accompanying mountain-building so as to produce metamorphic rocks. The weathered and eroded residue of any family of rocks can be observed today being transported by rivers to the sea for deposition and eventual conversion into sedimentary rocks.

The three major categories of rocks are igneous, sedimentary, and metamorphic. Igneous rocks form from a hot molten mass (magma); sedimentary rocks form when particles or dissolved substances from preexisting rocks are transported and deposited; and metamorphic rocks form when preexisting rocks are changed in response to high temperature, high pressure, and accompanying chemical activity.

Summary

Geology is a division of science devoted to the study of the earth and similar planetary bodies. An important task of geology is to examine earth materials and structures and to discover how they come into existence. Geologists have in the world about them the tangible result of past events, and they must work backward in time to discover the causes of those events. In their work they use the usual procedures of science—namely, the formulation of questions about some phenomenon or condition, the collection of observations that relate to the questions, the development of a hypothesis to answer the questions, and the testing of the hypothesis.

An overview of the earth shows that it is the largest of the four inner planets, that it rotates around its axis once each day and revolves around the sun every 365.25 days. It is not perfectly spherical, but rather has the shape of an ellipsoid, with its equatorial radius 21 km longer than its polar radius. The earth's overall density is 5.5 g/cm^3.

The major components of the earth include the gaseous envelope of air called the *atmosphere,* the oceans and other water bodies that comprise the *hydrosphere,* and the solid earth itself. The latter is zoned internally and consists of a *core* surrounded by a *mantle* and *crust.* The outer stony crust and upper mantle constitute the *lithosphere,* which is believed to override the *asthenosphere*—a weak plastic zone of the mantle.

The surface of the lithosphere includes the continents, with their stable regions and orogenic belts, as well as the ocean basins. Developed on the oceanic tracts of the lithosphere are such features as mid-oceanic ridges, great fracture zones, and deep-sea trenches—all of which have a direct relationship to modern concepts of moving lithospheric plates, called *plate tectonics.*

Time and change are recurrent themes in geology. The lands are continuously being eroded, sporadically deformed, and uplifted anew. Rocks change to other rocks, mountains change to plains, sea floors migrate, and continents fragment. If nothing else, geology teaches us that we inhabit a lively and dynamic planet.

Terms to Understand

asthenosphere
atmosphere
core
crust
deep-sea trenches
eon
era
geology
hydrosphere
igneous rocks
lithosphere
mantle
metamorphic rocks
mineral
Mohorovičić discontinuity (Moho)
orogeny
period
plate tectonics
rock cycle
sedimentary rock
subduction

Questions for Review

1. Imagine a planet that is like the earth in every way, except that its axis is perpendicular to the plane of orbit. What characteristics of the earth would be lacking on such a planet?

2. Imagine that a space traveler from another planet makes a cautionary stop on the moon before landing on the earth. What features of earth might be discerned from that vantage point that would suggest life existed on earth?

3. Describe the general procedures employed by scientists as they attempt to solve particular problems. Does scientific thinking occur in areas of human endeavor other than science? Provide an example.

4. What are the two most abundant gases in air? What is the importance of ozone in the atmosphere?

5. Where is the earth's mantle located with respect to other major components of the planet? How do geophysicists determine the location of the top of the mantle?

6. What is the difference between a mineral and a rock? What are the three major categories of rocks?

7. What is the most obvious difference between the inner core and the outer core? Between the lithosphere and the asthenosphere?

8. Given the 4.7 billion years of geologic time, why have the destructive forces of weathering and erosion not worn the continents down to low-lying plains?

Minerals

2

Three common non-silicate minerals. At the left is an octohedral crystal of fluorite, a calcium fluoride mineral widely used as a source of fluorine in the chemical industry and as a flux in steelmaking. Pyrite (right) is an attractive iron sulfide mineral. Calcite (center) is a calcium carbonate mineral that is widespread among sedimentary and some metamorphic rocks. The transparent variety shown here is called Iceland spar. (Courtesy of C.D. Winters.)

A casual glance at crystals may lead to the idea that they were pure sports of nature, but this is simply an elegant way of declaring one's ignorance. With a thoughtful examination of them, we discover laws of arrangement.

Abbé René Just Haüy, 1801

Introduction

The quotation beneath the title of this chapter is a translation from the French of an article by René Just Haüy. It was published in the first year of the 19th century. Haüy, who had originally trained for the church and entered the priesthood, became interested in minerals when he noticed how large pieces of calcite (Fig. 2–1) can be broken into smaller pieces, all having a shape similar to that of the original piece. From his measurements of these and other solids displaying crystal form, he surmised that crystals are composed of small structural units that are of identical shape in all crystals of the same form. Today, mineralogists have demonstrated that Haüy's building blocks are actually groupings of still smaller particles called atoms. In this chapter, we will discuss the characteristics of atoms, the manner in which atoms are systematically located and repeated in three dimensions within minerals, and the way in which combinations of atoms affect the properties of all earth materials. Before beginning, however, it will be useful to recall that *minerals* are naturally formed solids having three-dimensional orderly internal arrangements of atoms and a definite chemical composition or range of composition. The phrase "range of composition" is necessary because some atoms are similar in size and properties to one another and, to a limited degree, can replace one another in the internal atomic structure of minerals.

Atoms

The Architecture of the Atom

If one were to take apart the constituents of a mineral, one might then be left with certain components that could not be separated further by chemical and physical methods. These basic ingredients, of which no further separation is possible, are called **elements.** Hydrogen is an element. No matter what we do to a sample of hydrogen gas, we cannot decompose it into other substances. The fundamental unit of an element is an **atom.** There are, at last count, 108 elements known; 13 of these are man-made and 4 do not occur on earth. Geology students, however, need not be overly concerned about the rather large number of elements, because only eight occur abundantly in the earth's surficial rocks. These eight abundant elements are oxygen, silicon, aluminum, iron, calcium, sodium, potassium, and magnesium. They constitute at least 98 per cent of the earth's crust by weight and over 95 per cent of the elements in the entire earth from core to crust (Table 2–1).

The idea that earth materials are composed of the atoms of elements was conceived long before the present atomic age. More than 25 centuries ago, the Hindu philosopher Kanadu proposed that matter was made of tiny, invisible "eternal particles." This same concept was proposed in classical Greek time by Democritus, who taught that atoms were small indivisible particles capable of produc-

Figure 2–1 Calcite showing characteristic rhombohedral shapes.

Table 2–1 Abundances of Chemical Elements in the Earth's Crust

Element and Symbol	*Percentage by Weight*	*Percentage by Number of Atoms*	*Percentage by Volume*
Oxygen (O)	46.6	62.6	93.8*
Silicon (Si)	27.7	21.2	0.9
Aluminum (Al)	8.1	6.5	0.5
Iron (Fe)	5.0	1.9	0.4
Calcium (Ca)	3.6	1.9	1.0
Sodium (Na)	2.8	2.6	1.3
Potassium (K)	2.6	1.4	1.8
Magnesium (Mg)	2.1	1.9	0.3
All other elements	1.5		
	100.0	100.0†	100.0†

*Note the high percentage of oxygen in the earth's crust.
†Includes only the first eight elements.
(Based on B. Mason, *Principles of Geochemistry,* New York, John Wiley & Sons, Inc., 1966.)

ing the properties of the substances they formed. Democritus, however, formulated his concept of the atom purely on speculation and intuition. He had no idea about the structure of the atom or that it might contain even smaller constituents.

Knowledge of some of the smaller constituents of atoms and their characteristics began to evolve during the late 1800s and early 1900s, mostly as a result of experiments carried out with radioactive substances. The atom was conceived of as having an extremely small but heavy **nucleus** surrounded by a cloud of rapidly moving particles with negative electrical charge called **electrons**. Electrons whirl around the nucleus at speeds so great that if permitted to do so, they would circle the earth in less than 1 second. They do not revolve in set paths as do planets around the sun, but swirl around the nucleus rather like a swarm of tiny insects around a light. Thus, it would be incorrect to think of an atom as a solid sphere because in reality atoms consist mostly of empty space. The electrons move so rapidly that they effectively fill that space and thereby give size to the atom.

Located in the nucleus of the atom are closely compacted subatomic particles called **protons**, each of which carries a charge of positive electricity equal to the charge of negative electricity carried by the electron.

Associated with the protons in the nucleus are electrically neutral particles having the same mass as protons. These particles are called **neutrons**. Modern atomic physics has provided evidence for the existence of other subatomic particles, but their contribution to the chemical behavior of elements is minimal.

Atomic Number and Atomic Mass

The number of protons within a nucleus determines the type of element to which an atom belongs. For example, if six protons occur in the nucleus of an atom, the element is carbon; if the number is eight, the element is oxygen; 14, silicon, and so on. Thus, all atoms of any one element have the same number of protons. The number of protons in the atom of an element defines that element's **atomic number**. In an electrically neutral atom, there are equal numbers of protons and electrons. The number of protons in naturally occurring elements ranges from 1 in hydrogen (Table 2–2) to 92 in uranium.

The **atomic mass** of an atom is approximately equal to the sum of the masses of its protons and neutrons. The mass of an electron is so small that it need not be considered in determining the atomic mass. Indeed, it would require 1837 electrons to equal the atomic mass of a single proton. By convention, atomic mass is noted as a superscript preceding the chemical symbol of an element, and the atomic number is placed beneath it as a subscript. Thus n$^{40}_{20}$Ca is translated as the element calcium having an atomic number of 20 and an atomic mass of 40.

Atoms are composed of protons, neutrons, and electrons. Protons and neutrons are located in the atomic nucleus. The atomic mass of an element is the sum of its protons and neutrons.

Isotopes

Although the atoms of any specific element are essentially identical in the manner in which they

Table 2–2 Structure of Some Geologically Important Elements

Element and Symbol	Atomic Number (Number of Protons in Nucleus)	Number of Neutrons in Nucleus	Atomic Mass	Electrons in Various Levels	Total Number of Electrons
Hydrogen (H)	1	0	1	1	(1)
Helium (He)	2	2	4	2	(2)
Carbon 12 (C)*	6	6	12	2–4	(6)
Carbon 14 (C)	6	8	14	2–4	(6)
Oxygen (O)	8	8	16	2–6	(8)
Sodium (Na)	11	12	23	2–8–1	(11)
Magnesium (Mg)	12	13	25	2–8–2	(12)
Aluminum (Al)	13	14	27	2–8–3	(13)
Silicon (Si)	14	14	28	2–8–4	(14)
Chlorine 35 (Cl)*	17	18	35	2–8–7	(17)
Chlorine 37 (Cl)	17	20	37	2–8–7	(17)
Potassium (K)	19	20	39	2–8–8–1	(19)
Calcium (Ca)	20	20	40	2–8–8–2	(20)
Iron (Fe)	26	30	56	2–8–14–2	(26)
Barium (Ba)	56	82	138	2–8–18–18–8–2	(56)
Lead 208 (Pb)*	82	126	208	2–8–18–32–18–4	(82)
Lead 206 (Pb)	82	124	206	2–8–18–32–18–4	(82)
Radium (Ra)	88	138	226	2–8–18–32–18–8–2	(88)
Uranium 238 (U)	92	146	238	2–8–18–32–18–12–2	(92)

*When two isotopes of an element are given, the most abundant is starred; for other elements, only the most abundant isotope is given. Note carefully that ordinary chemical atomic weights are not given; these are mixtures of isotopes and are therefore not whole numbers.

behave chemically, they are generally not identical in the number of neutrons they contain. Atoms of a particular element that differ in the number of neutrons they contain, and therefore also differ in atomic mass, are called **isotopes**. As an example, there are at least three different isotopes of oxygen (Table 2–3). All three have the same atomic number (8) but have mass numbers of 16, 17, and 18, respectively. Of these, oxygen 16 is the most abundant in nature. Isotopes are very useful in geologic studies. The isotopes of oxygen are used to determine the temperature of the oceans hundreds of millions of years ago, and isotopes of uranium, potassium, rubidium, and other elements are used to determine the age of rocks.

As suggested above, the chemical properties of isotopes of the same element are nearly identical. This is because the chemical behavior of atoms is determined not by protons or neutrons, but rather by the electrons. The number of electrons does not vary among the isotopes of an element.

Most isotopes have a high degree of stability and tend to remain unchanged over long spans of time. Others, however, are called unstable isotopes because they tend to break down into other isotopes.

Table 2–3 Stable Isotopes of Oxygen*

Isotope	Number of Protons	Number of Neutrons	Atomic Mass
^{16}O	8	8	16
^{17}O	8	9	17
^{18}O	8	10	18

*Isotopes of oxygen—oxygen 16, oxygen 17, and oxygen 18—are written ^{16}O, ^{17}O, ^{18}O, respectively. The superscript designates the sum of the protons and neutrons.

The unstable isotopes are the kind most useful in determining the age of rocks.

Although any atom of a single element will have a fixed number of protons and electrons, they may have different numbers of neutrons. Variation in the number of neutrons produces isotopes of the same element.

Electron Shells

Because of their negative charge, electrons are attracted to the positively charged nucleus of the atom but are prevented from collapsing inward to the nucleus by the centrifugal force associated with their rapid orbital spin. Although electrons may spin along any path, *on the average*, a certain quantity will remain within a particular *shell.* The shells are really zones in which there is a high probability of finding electrons (Fig. 2–2).

It is not possible to know the position of a particular electron at any designated moment. However, scientists have deduced the probable arrangement of the shells from X-ray studies, from atomic spectra, and from the atom's behavior. Such deductions provide a working hypothesis for the number of electron shells and the probable number of electrons in each shell (see Table 2–2). As the simplest of all the elements, hydrogen has only one proton and one electron. The helium atom has two electrons close to the nucleus—a condition also present in the innermost shell of all elements heavier than helium.

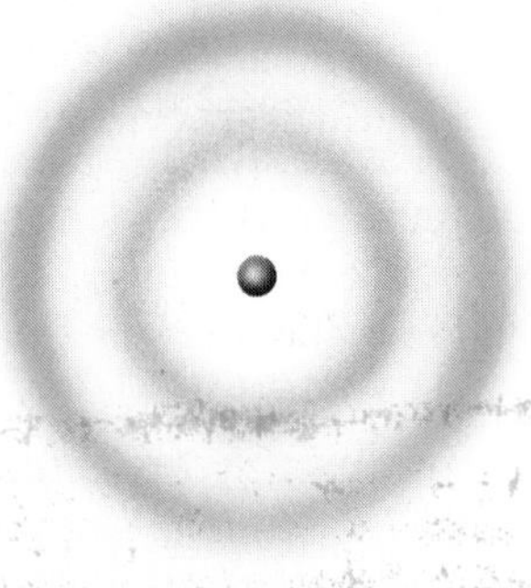

Figure 2–2 An atom is not a solid sphere but is composed mostly of space occupied by negatively charged electrons that whirl around the positively charged nucleus at such enormous speeds as to effectively "fill" the space. Within the cloud of rapidly orbiting electrons, there is a high probability of a specific number of electrons occupying particular zones surrounding the nucleus. Such a zone, in which the probability for finding electrons is high, is called an electron shell.

The greatest number of electron shells in even the most complex of atoms is seven. Each of the first four shells may contain no more than the following number of electrons: 2 in the first shell, 8 in the second, 18 in the third, and 32 in the fourth. The shells are always filled outward, so that an inner shell must be full before the next shell can have any electrons. In addition, electrons are distributed in such a way that the outermost shell contains eight or fewer electrons. Indeed, once an atom has acquired the eight outer-shell electrons, it becomes chemically quite stable. The chemical properties of atoms are thus dependent on their tendency to obtain eight electrons in this all-important outermost (or valence) shell. For example, if the outer shell contains exactly eight electrons (or two in helium, which has only two electrons), the atom will show little tendency to chemically interact. Thus, atoms like neon, argon, and krypton are called noble or inert gases because their outer shells contain the maximum quota of eight electrons. They are, in a figurative sense, "satisfied."

For those elements whose atoms have an outermost shell containing fewer than eight electrons, there is a tendency to share, gain, or lose outer-shell electrons so as to assume the stable configuration in which eight electrons occur in the outermost shell. For example, the conceptual two-dimensional diagram of the chlorine atom, shown in Figure 2–3, exhibits only seven electrons in the outer shell. Chlorine is "inclined" to borrow one electron. Sodium, which has only one electron in its outermost shell, is inclined to give up an electron. Therefore, by combining, both sodium and chlorine achieve a stable configuration. The number of electrons that an atom loses or receives, or the number of pairs it shares, is called its **valence**. Loss of an electron leaves an atom with a surplus of positive charges. Hence, it is said to have a *positive valence.* Likewise, a gain of electrons will produce a negative charge on the atom and result in

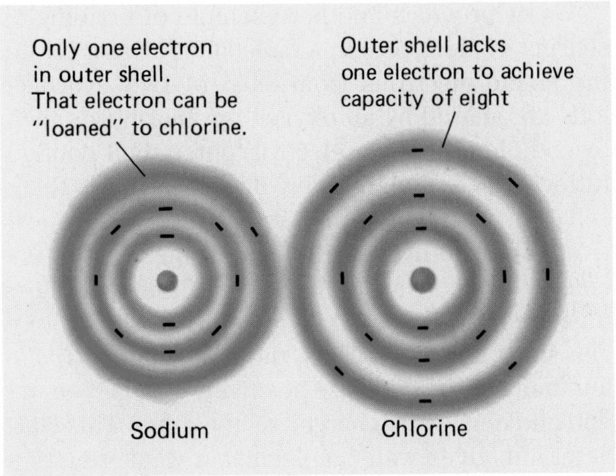
Only one electron
in outer shell.
That electron can be
"loaned" to chlorine.
Outer shell lacks
one electron to achieve
capacity of eight
Sodium
Chlorine

Figure 2–3 Diagrammatic representation of the formation of sodium chloride from sodium and chlorine.

negative valence. The electrons gained or lost are called valence electrons and are indicated by small plus or minus symbols, as in O^{-2} for oxygen or Ca^{+2} for calcium.

The chemical properties of atoms are related to their tendency to gain, share, or lose electrons so as to achieve a stable configuration of eight electrons in their outer electron shell.

Combining Atoms

As described above, the guiding rule in chemical combination is that when two or more atoms react with one another, they do so in a manner that will promote the attainment of eight electrons in the outer shell. The process is accomplished by accepting and donating electrons or by sharing electrons. Furthermore, atoms tend to achieve a complete outer shell by undergoing the smallest possible change. Thus, an element having fewer than four electrons in its outer shell will tend to donate those electrons, while an element having more than four will tend to borrow electrons to complete its outermost shell. Once an atom has either borrowed or donated an electron, it is no longer electrically neutral. It is now called an **ion**.

Examples of Chemical Bonding

Ionic Bonding

The reaction between an atom of sodium (Na^{+1}) and chlorine (Cl^{-1}) can again be used to illustrate one mechanism by which atoms may bond together. In the combination of these two elements, it is easier for sodium to lend the single electron in its outer shell (see Fig. 2–3) than to borrow seven. Upon lending this electron, the sodium atom becomes positively charged. A chlorine atom, on the other hand, carries 17 electrons: 2 in the first shell, 8 in the second shell, and the remaining 7 in the third, or outermost shell. Thus chlorine's outer shell is lacking one electron. For chlorine, it is far easier to acquire one electron than to donate seven. In the reaction between the two elements, each sodium atom gives its one outer-shell electron to one chlorine atom, with the result that each has the complete outer shell structure of a noble gas. The acquisition of the extra electron by the chlorine atom converts it into a negative ion. Because unlike charges of electricity attract, the negatively charged chlorine ion is drawn tightly to the positively charged sodium ion to form sodium chloride. Studies of halite, the mineral form of sodium chloride, reveal that each sodium ion is surrounded by exactly six larger chlorine ions, and similarly, each chlorine atom is in contact with six sodium ions. The cubic crystal form (Fig. 2–4) is the result of the manner in which the ions are packed, their electrical properties, and the sizes (atomic radii) of the participating ions (Fig. 2–5).

Covalent Bonding, a Sharing Situation

The completion of the outer shell of atoms by *sharing* pairs of electrons is another way atoms achieve the noble gas configuration of stability. This manner of combination is termed **covalent bonding.** In covalent bonding, the electrons of each atom orient themselves so that their influence is extended equally between the participating atoms. The shared electrons become indistinguishable as to their atom of origin and serve both nuclei simultaneously.

Examples of covalent bonding are the combining of two hydrogen atoms to form the H_2 mole-

Figure 2–4 An aggregate of cubic crystals of halite.

cule, two oxygen atoms to form O_2, or the combination of four hydrogen atoms and one carbon atom to form methane. Diamonds, which consist only of carbon atoms, are also bonded covalently. In a diamond, each carbon atom is surrounded by four other carbon atoms to form a tetrahedron. Within the tetrahedron, every atom shares an electron pair with each of its neighboring atoms (Fig. 2–6).

Water provides another example of bonding by sharing of electrons. In a molecule of water (H_2O), the single electrons from each of two hydrogen ions are shared by an oxygen ion that needs them to complete its eight-electron outer-shell configuration. The ions, however, do not arrange themselves in a straight line but rather the two tiny positively charged hydrogen ions become fixed at one end of the molecule, whereas two nonbonding or unshared electrons occur at the opposite or oxygen side (Fig. 2–7). Thus, the molecule is *dipolar* and behaves as if it were positively charged on one end and negatively charged on the other. This characteristic gives water molecules a weak electrical attraction for other ions. If these other ions are at the surface of mineral grains with which water comes into contact, the attraction may pull ions away from the mineral. In fact, the mineral is being dissolved. Such solution work by water molecules is an important aspect of the weathering of rocks at the earth's surface.

Ionic and covalent bonding are the most common kinds of bonds found in minerals. They are not, however, mutually exclusive. Both types of bonding may occur within certain minerals, and in many instances bonding may be intermediate or mixed in character.

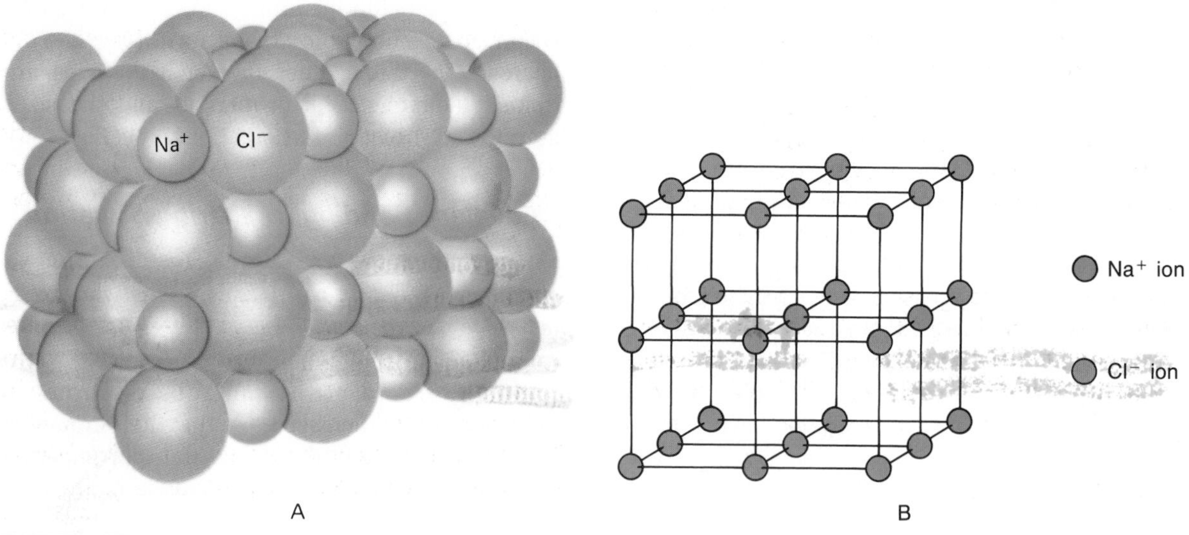

Figure 2–5 Structure of sodium chloride crystal. (A) Model showing relative sizes of the ions. (B) Ball-and-stick model showing cubic symmetry. (From Johnston, D.O., Netterville, J.T., Wood, J.L., and Jones, M.M., *Chemistry and the Environment,* Philadelphia, W.B. Saunders Company, 1973.)

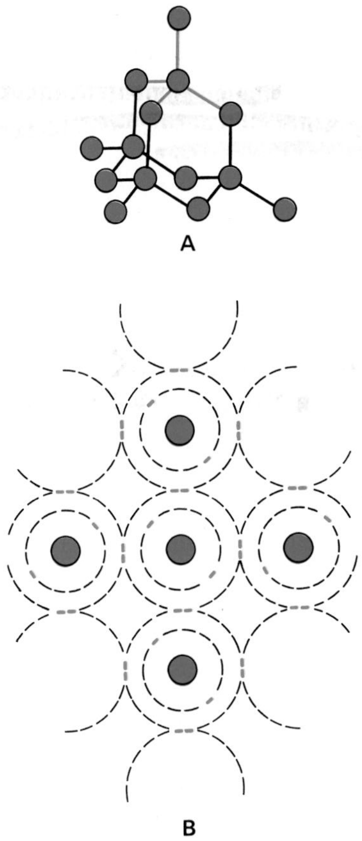

Figure 2–6 (A) The carbon atoms in a diamond are arranged in tetrahedrons consisting of four carbon atoms surrounding a central carbon atom. (B) A two-dimensional model of the atoms in a diamond. Note that each carbon atom, which individually has four electrons in its outer shell, shares one outer electron with each of its four neighbors. A stable outer shell of eight electrons is achieved by the sharing of pairs of electrons; this type of bonding is termed covalent.

Bonding is the attachment of one atom to one or more adjacent atoms. Ionic bonding (electron transfer) and covalent bonding (electron sharing) are the two most common kinds of bonds found in minerals.

Metallic Bonding

Metals can be visualized as closely packed aggregates of positive ions afloat in a "sea" of electrons. The valence electrons roam about independently of their atoms of origin. Rather than being

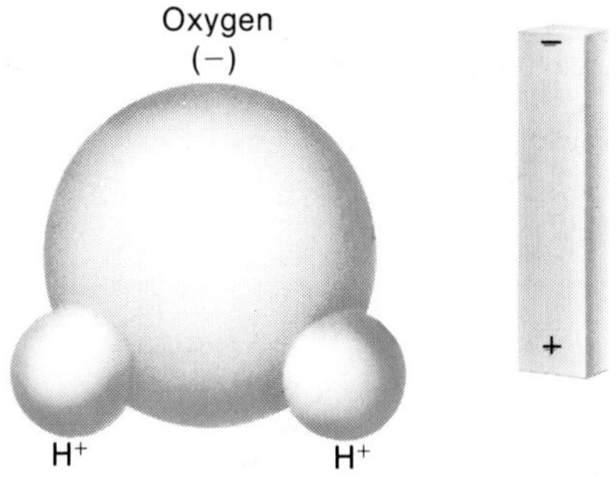

Figure 2–7 Model of a water molecule, H_2O. The negative charges are crowded close to the oxygen end of the molecule, while the side with the two hydrogen nuclei contains more positive charges. Thus, the molecule has polarity—somewhat like a bar magnet.

associated with only one or two atoms, these electrons are shared by the entire metal ion aggregate. The atoms forming the metal become positive ions by giving their outer-shell electrons to be shared among all other ions. The freedom of those electrons in being able to move about and not be retained by a particular atom accounts for such metallic properties as electrical conductivity, thermal conductivity, softness, reflectance, and malleability.

Atomic and Ionic Sizes

Another factor that influences which elements will combine to form minerals is the size of the ions and atoms (Fig. 2–8). The ions in minerals tend to be efficiently packed into tight geometric packages that leave little opportunity for any given ion to have so much surplus space that it can wobble about (Fig. 2–9). Combinations of ions of certain sizes can produce only specific kinds of atomic structures. In the sodium chloride structure (see Fig. 2–5), sodium ions are distinctly smaller than chlorine ions, and the difference in sizes of the two components permits each sodium ion to be snugly surrounded by exactly six chlorine ions. The result is the cubic structure characteristic of the sodium chloride mineral halite.

The relative sizes of ions influence what ions can be built into the atomic structure of a mineral.

Relative Sizes of Ions	Ions	Radius (in Å)
	Silicon	0.40
	Oxygen	1.40
	Potassium	1.42
	Calcium	0.99
	Sodium	0.97
	Iron (+2)	0.74
	Magnesium	0.66
	Aluminum	0.49

Figure 2–8 Sizes of some geologically important ions (one angstrom unit equals a length of 10^{-10} meter).

The Space Lattice

Crystals are solid bodies bounded by natural plane surfaces. The ions and atoms in crystals are arranged in three-dimensional geometric patterns that we call **space lattices**. The concept of an orderly internal arrangement of the constituents of crystals was accepted by scientists long before the formulation of atomic theory. In the 17th century, Nicolaus Steno observed that similar crystal faces found on different crystals of the same kind of mineral always met at the same angle, regardless of the size and distortions of shape in the crystal (Fig. 2–10). Steno's observation has since been called the **Law of Constancy of Interfacial Angles** and indicates that the internal arrangement of constituent units (later termed ions and molecules) must be orderly and that the components must be present in definite proportions. If this were not the case, the relationships of the crystal faces would not be consistent.

Unfortunately, Nicolaus Steno (and later, Abbé René Haüy) could only infer the existence of inter-

Figure 2–9 Diagram illustrating how the relative sizes of two ions can influence the packing arrangement.

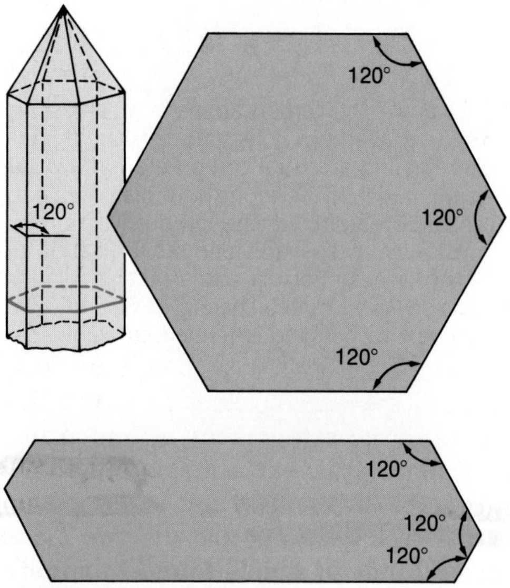

Figure 2–10 Cross sections of crystals of the same mineral having six-sided (hexagonal) symmetry. Although the lower crystal is misshapen, the angles between comparable surfaces (faces) of both crystals are 120°. This is because the internal atomic structure is identical in both crystals, and there is no internal distortion of that structure in the misshapen crystal.

nal ordering from exterior form. More direct evidence eluded scientists until 1912. At that time a method was developed that used X-rays to confirm Steno's theory that the constituent units (ions) that build a crystal are arranged in a symmetric three-dimensional array.

As indicated by their crystal form, constancy of interfacial angles on crystals, and their ability to produce consistent X-ray patterns, minerals have an orderly internal arrangement of atoms.

Crystal Form

A mineral grows or enlarges by addition of ions to its surfaces from a surrounding solution, melt, or gas. If the growing surfaces do not come into contact with other mineral grains, and provided that the temperature, pressure, and the concentration of ions in the liquid or gas are suitable, then the mineral will acquire a characteristic crystal form. The faces of the crystals will be parallel to uniform sheets of atoms within the crystal.

Except when crystals are growing into an open cavity (Fig. 2–11), they come into contact with each other and with other mineral grains, and therefore perfect crystals are not produced. Indeed, perfect crystals are relatively uncommon, and the more frequent pattern is for growing crystals to interfere with one another during growth to form an interlocking system of crystals. Only occasional crystal faces are seen in such **crystalline aggregates** (Fig. 2–12), but within individual mineral grains the ions and atoms are nevertheless arranged in a uniform geometric pattern.

When perfect or nearly perfect crystals of minerals do develop, their external form is determined not only by composition but also by such conditions as impurities, temperature, and pressure. The importance of temperature and pressure is clearly evident when one considers the crystal forms of carbon. Under high pressures and temperatures, carbon crystallizes in the octahedral form of diamond. At lower temperature and pressure conditions, soft, platy, six-sided crystals of graphite develop.

Figure 2–11 Well-developed crystals of the quartz mineral amethyst. These crystals grew within a cavity. Amethyst is prized as a semiprecious gemstone, as indicated by the faceted and polished specimens. Note that the crystal faces evident in the large unmodified specimen are natural planes of crystal growth, whereas the facets of gemstones are artificially produced.

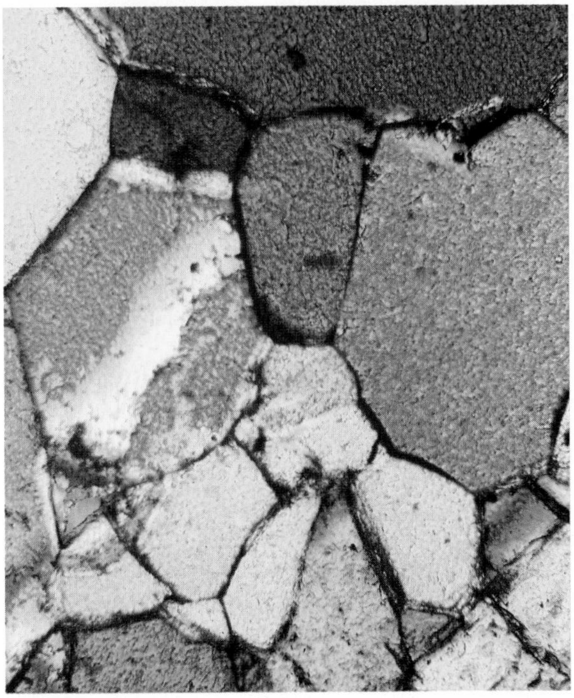

Figure 2–12 A thin slice (thin section) of a crystalline aggregate of the mineral olivine. Actual width of field is 3 mm.

A few common crystal forms are illustrated in Figure 2–13. Sometimes recognition of crystal form can be useful in identifying minerals in hand specimens. More often, however, the mineral occurs as a crystalline aggregate, or it may simply exhibit poor crystal development even though it has a regular atomic structure. Some minerals, like opal, are not crystalline at all.

A crystal is a solid body bounded by natural plane surfaces. Crystalline aggregates do not develop complete crystal form because of mutual encounters at the contacts between growing crystals.

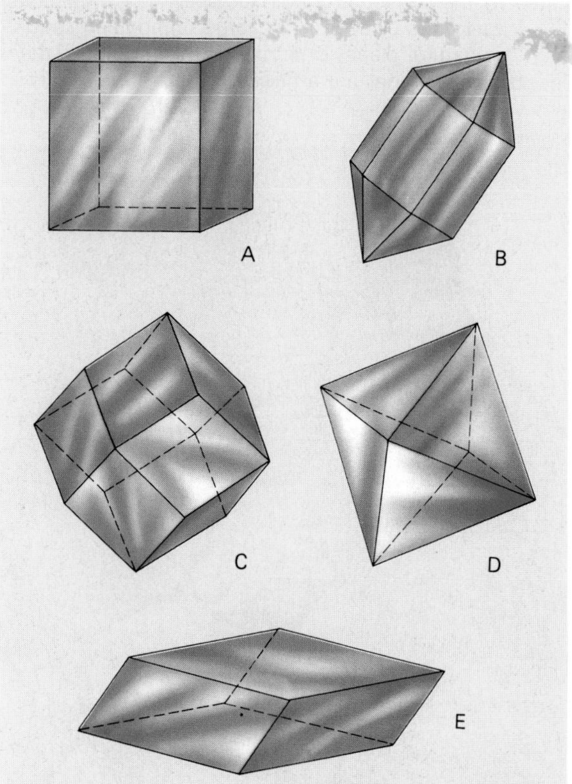

Figure 2–13 Examples of mineral crystals. (A) Cube (as in halite, galena, and pyrite). (B) Prism and pyramids (as in quartz). (C) Dodecahedron (as in garnet). (D) Octahedron (as in diamond). (E) Rhombohedron.

Common Rock-Forming Minerals

The rocks that form the earth's crust are composed of aggregates of minerals. Although over 3000 minerals have been discovered and scientifically described, most of these comprise only a small percentage of the crust and are rarely encountered. For our present purposes, it is important to consider only those minerals that compose the bulk of common rocks or that are particularly useful in making interpretations about the earth's history.

Properties Useful in Identification of Minerals

There are a large number of chemical and physical properties used by geologists in the identification of minerals. Some of these properties can be recognized only with the use of microscopes, X-ray equipment, or complex chemical tests. Fortunately, however, most of the common minerals can be identified with a knowledge of only a few easily recognized properties. Color, cleavage, hardness, crystal form, luster, specific gravity, and magnetism are among the easily used clues for the identification of minerals (see Appendix B).

Visual Properties

Color is the most obvious physical characteristic of a mineral. Like many other mineral properties, it is related to the kinds of atoms and ions that make up a mineral. Color in a mineral results from the absorption of certain wavelengths of light energy that constitute white light. The white light to which we are accustomed actually consists of energy of a great number of wavelengths. (We see these wavelengths as colors when white light is dispersed to form a rainbow.) Nearly complete absorption of white light would cause the mineral to be dark or black in color. Conversely, a mineral that absorbs few or no wavelengths would be very light colored or white.

Color can be a very useful property in mineral identification, especially for those minerals with colors that are constant. Malachite, for example, is always green (Fig. 2–14), lapis lazuli is blue, and

Figure 2–14 The two larger crystals shown here are malachite, $Cu_2CO_3(OH)_2$, a mineral that is invariably green and that often assumes the rounded (botryoidal) form seen here. The smaller specimen is azurite $CU_3(OH)_2(Co_3)_2$, another copper mineral that is nearly always dark blue in color.

almandite garnet is red (Fig. 2–15). Color, however, can vary in other minerals. Often the variation is caused by impurities. Quartz and fluorite, for example, may be colorless, white, pink (Fig. 2–16), purple, green, or blue. The mineral corundum includes gem varieties (Fig. 2–17) that are red (ruby) and blue (sapphire). Because of such variations, color should be used cautiously for identification, and should always be used in combination with other physical properties.

In order to avoid errors in identification caused by superficial color variations, it is often useful to observe the mineral's **streak** or color in powdered form. This can be quickly accomplished by rubbing the mineral across the surface of an unglazed porcelain "streak plate" and noting the color of the streak of powder that results.

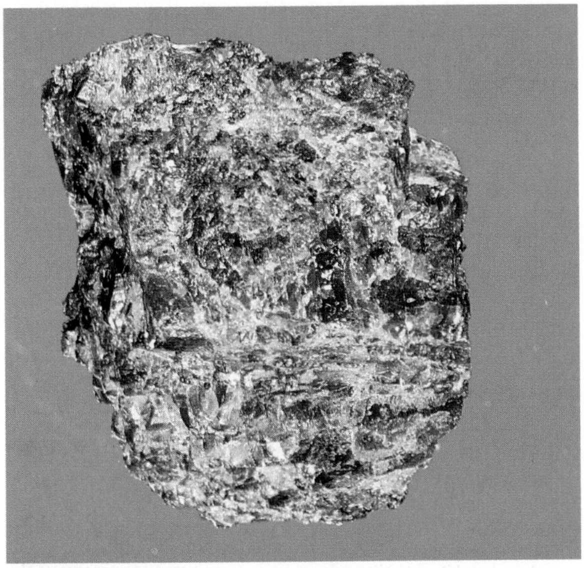

Figure 2–15 Among the distinctive physical properties of almandine garnet, shown here, is its wine-red color.

Figure 2–16 Quartz is a mineral whose color is not constant. It may be colorless, white, purple, green, blue, or pink, as shown in this specimen of rose quartz. Rose quartz is commonly used in jewelry.

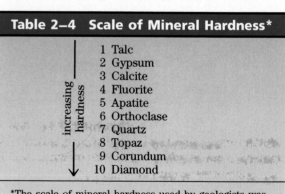

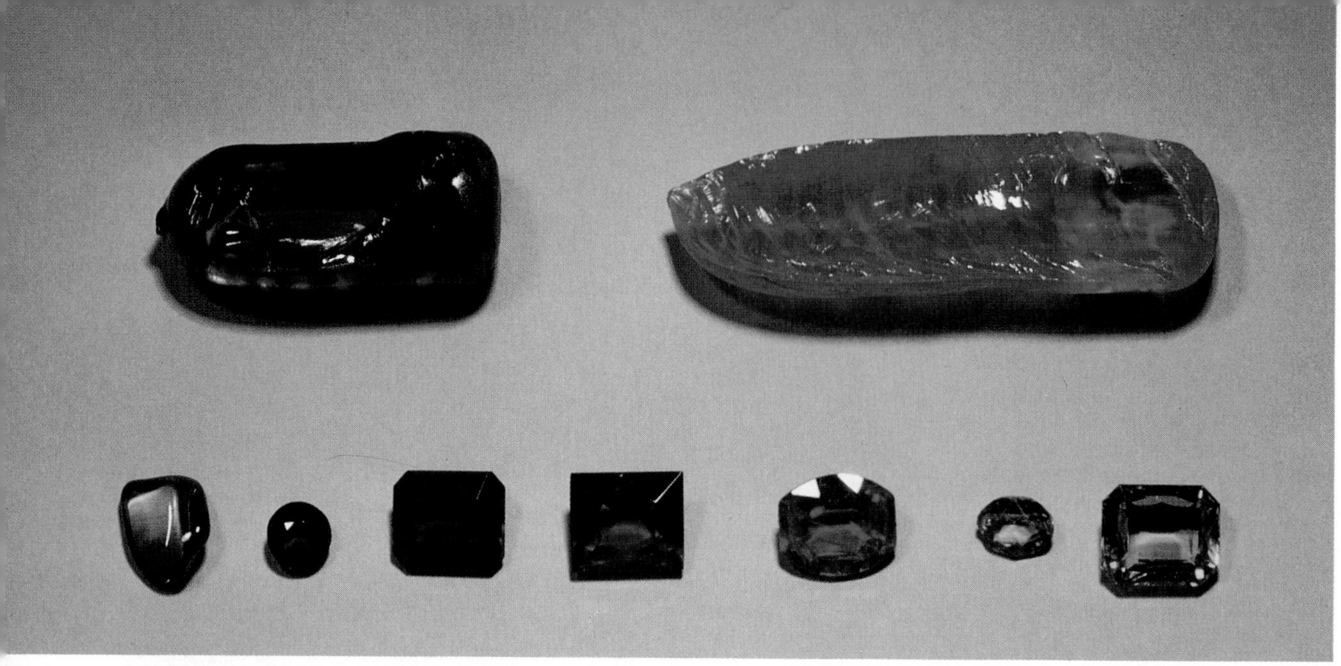

Figure 2–17 Varieties of corundum (Al_2O_3). The blue gems are sapphires and the red are rubies. At the top are broken pieces of synthetically produced ruby and sapphire.

The manner in which a mineral shines in reflected light is its **luster**. Those minerals that reflect light like metals are said to have *metallic luster*, whereas all others have nonmetallic luster. If the luster of a nonmetallic mineral is hard and brilliant like that of a diamond, it is said to have an *adamantine luster*. A glassy appearance is termed *vitreous luster*, whereas minerals with very low reflectance may simply be called *dull* or, if appropriate, *earthy*.

Hardness

The **hardness** of a mineral refers to its resistance to scratching by another substance of known hardness. To ensure uniformity in measuring hardness, geologists employ a scale formulated by the German mineralogist Frederich Mohs (1773 to 1839). Mohs arranged ten relatively common minerals in order of increasing hardness (Table 2–4). If an unknown mineral can be scratched by a mineral in the scale, it is softer than that mineral; if it cannot be scratched, it is harder. Mohs' scale is only a relative scale of hardness and does not provide exact mathematical relationships. Minerals have relatively small differences in hardness from 1 to 9, but increase 40 × in hardness between 9 (corundum) and 10 (diamond). In the field it is often expedient to make hardness tests with a penny (hardness of 3), fingernail (hardness of 2 to 2.5), or a steel nail (hardness of 5).

A mineral's hardness depends largely on the strength of bonds between the atoms or ions of its space lattice. For example, in diamonds, every carbon atom forms sturdy single bonds with four other carbon atoms to produce a strong three-dimensional lattice. In contrast, carbon atoms in graphite crystals are arranged in layers, and the forces between layers are quite weak (Fig. 2–18).

Table 2–4 Scale of Mineral Hardness*

increasing hardness ↓	
	1 Talc
	2 Gypsum
	3 Calcite
	4 Fluorite
	5 Apatite
	6 Orthoclase
	7 Quartz
	8 Topaz
	9 Corundum
	10 Diamond

*The scale of mineral hardness used by geologists was formulated in 1822 by Frederich Mohs. The scale begins with talc, a very soft mineral, and continues to diamond, the hardest of all minerals.

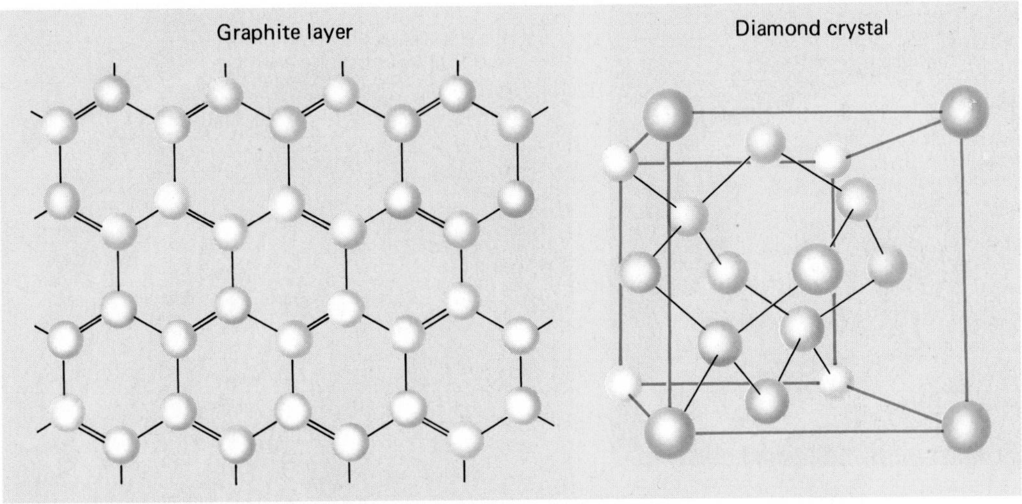

Figure 2–18 Lattice structures of diamond and graphite. In diamonds, each carbon atom is at the center of a tetrahedron, on each corner of which is another carbon atom. In graphite, the carbon atoms are linked in planes of hexagons, the layers of which are loosely bonded. As a result, the layers may slide along the weakly bonded parallel planes of atoms; this characteristic accounts for graphite's greasy feel. (From Masterton, W.L., Slowinski, E.J., and Stanitski, C.L., *Chemical Principles,* 6th ed., Philadelphia, Saunders College Publishing, 1985.)

Graphite has a hardness of only 1 to 2 on Mohs scale, whereas diamond has a value of 10. The hardness of some minerals is not uniform in all directions. Bond strengths may be greater along some planes in the lattice than along others, causing slight variations in hardness.

Cleavage and Fracture

We have discussed how ions and atoms in a crystal are arranged in definite layers or planes, all of which have a definite angular relationship to one another. If the bond strengths or number of bonds in one set of planes is greater than that of other planes, then it is likely that the structure will separate more readily along the directions of weakest bonding or where the spacing between planes is relatively wider. The tendency of a mineral to break smoothly along certain directions parallel to those planes of ions across which bonding is weakest is called **cleavage**. The surface along which the mineral breaks is the cleavage surface. Cleavage is usually described as poor, fair, or good, according to how smooth a cleavage surface is produced.

The number and direction of cleavage planes are important aids to identification. Minerals like mica and graphite, for example, have good cleavage in one direction (Fig. 2–19). If a crystal of halite is broken it will commonly produce three smooth lustrous surfaces at right angles to each other (Fig. 2–20). Thus halite is a mineral having

Figure 2–19 Mica, showing characteristic cleavage along one directional plane. "Direction," as used here, refers to a set of parallel planes. (Courtesy of Wards Natural Science Establishment, Inc., Rochester, New York.)

Figure 2–20 Piece of halite broken from large crystal and exhibiting good cleavage in three mutually perpendicular directions.

three good cleavage directions. Cleavage planes may be smooth and resemble crystal faces. However, cleavage planes are formed by breaking the mineral along planes of weakness. Crystal faces differ in that they are planes along which ions were added to a growing crystal in an orderly manner. Some minerals, such as quartz, may develop excellent crystal faces (see Fig. 1–16), yet do not exhibit cleavage at all.

Cleavage is the tendency of a mineral to split, with greater or lesser perfection, along planes parallel to possible crystal faces.

A surface of breakage in a mineral other than a cleavage plane is termed a **fracture** (Fig. 2–21). Fractures tend to be irregular and randomly oriented. They are described by such terms as *splintery, fibrous, uneven,* or *conchoidal* (see Fig. 4–15). Opal and glass frequently exhibit the shell-like arcs of conchoidal fractures. A *hackly* fracture is a ragged surface with sharp points and edges resembling the broken surfaces of cast iron.

Specific Gravity and Density

We are all familiar with the fact that some minerals seem heavier than others and that this can help us to identify them. To assure uniformity in determining whether one mineral is heavier than another, geologists utilize a property called **specific gravity**. Specific gravity is the number of times a mineral is as heavy as an equal volume of water. The iron sulphide mineral pyrite (Fig. 2–22), sometimes called fool's gold, has a specific gravity of 5.0, and is therefore five times as heavy as an equal volume of water. Real gold has a specific gravity of 19.3, and is thus readily distinguished from pyrite on the basis of specific gravity.

The **density** of a mineral or other body is its mass divided by the volume it occupies. Unlike specific gravity, which (like other ratios) is not expressed in units, density must always be expressed in units such as grams per cubic centimeter. Because 1 cm^3 of water at 25°C has a mass of 1 g, the density of a substance as expressed in the metric system has the same numerical value as its specific gravity. For example, quartz with a specific gravity of 2.65 has a density of 2.65 g/cm^3.

Figure 2–21 The splintery fracture characteristic of the mineral actinolite is a result of that mineral's chain structure. (Courtesy of Wards Natural Science Establishment, Inc., Rochester, New York.)

Figure 2–22 Crystalline pyrite from Leadville, Colorado. This iron sulfide mineral, commonly referred to as fool's gold, is known by its metallic luster, light-yellow color, brittleness, and lack of cleavage. Gold is softer than pyrite, and malleable. (Courtesy of Wards Natural Science Establishment, Inc., Rochester, New York.)

As is true for all other physical properties of minerals, density and specific gravity are dependent on the composition and the manner in which atoms and ions are arranged in their space lattice. The more tightly packed the atoms, and the more elements of greater mass, the greater also will be a mineral's specific gravity and density. As an example, both diamond and graphite are composed of carbon atoms. Graphite has a specific gravity of 2.2, whereas the specific gravity of diamond is 3.5 because it has more carbon atoms in a given volume.

Magnetism and Other Properties

In addition to the above characteristics by which a mineral may be recognized, certain minerals have other rather distinctive properties. The salty taste of halite, the malleability of gold, the flexibility of mica, the soapy feeling of talc, and the earthy odor of certain clays when moistened are all useful properties. A few minerals are attracted by a magnet, and, if magnetized, the mineral magnetite acts as a magnet itself (Fig. 2–23). Calcite is easily recognized by the way it effervesces when a drop of dilute hydrochloric acid is applied to its surface.

Such properties as hardness, color, density, cleavage, magnetism, and attributes peculiar to particular minerals can be used to identify many common minerals.

Silicates

Atomic Structure of Silicates

About 75 per cent by weight of the earth's crust is composed of the two elements oxygen and silicon (see Table 2–1). For the most part, oxygen and silicon occur in combination with other abundant elements such as aluminum, iron, calcium, sodium, potassium, and magnesium to form an important

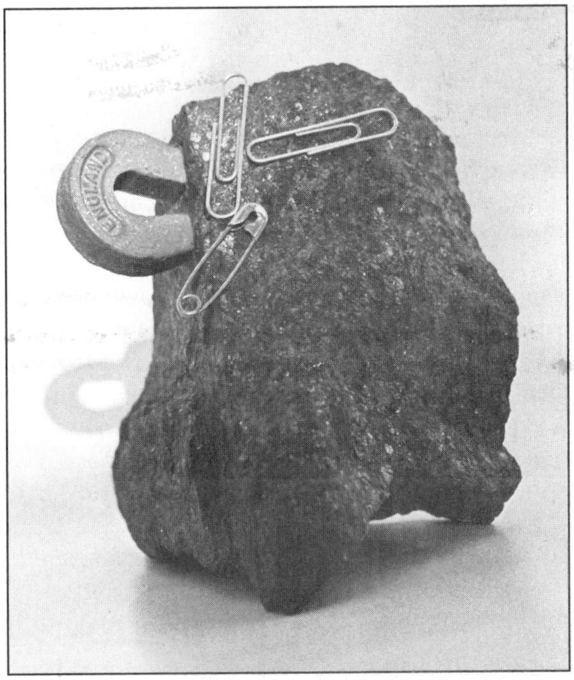

Figure 2–23 Magnetite, a mineral that is a natural magnet and attracts iron objects.

Figure 2–24 Veins of white quartz in dark igneous rock.

group of minerals called the **silicates**. A single family of silicate minerals, the **feldspars**, comprises about one-half of the material of the crust, and a single mineral species called **quartz** (Fig. 2–24 and see Fig. 1–22) represents a sizable portion of the remainder. The fundamental unit in the crystal structure of silicates is called a **silica tetrahedron** (Fig. 2–25). It consists of a compact tetrahedral arrangement of four oxygen ions around a central silicon ion. This tetrahedral shape is a consequence of the very small size and high charge of the silicon ion and of the relatively large size of the oxygen ions. The silica tetrahedron, however, is not an electrically neutral unit. The combining of four oxygen ions (each with two negative charges) and one silicon ion (with four positive charges) leaves the tetrahedron with four unpaired electrons. Because of the resulting charge, the tetrahedral unit must either bond to one or more additional positive ions (such as magnesium or iron) or covalently share oxygen atoms with neighboring tetrahedra. Among the abundant silicates, olivine and garnet are minerals that have all of their tetrahedra linked by metallic ions. In the other abundant silicates, the tetrahedra are strongly linked by sharing oxygens. The pattern developed by the connected tetrahedra not only forms the atomic structure of the mineral but also determines many of its properties, such as crystal form, cleavage, and specific gravity. The various silicate structural types are as follows:

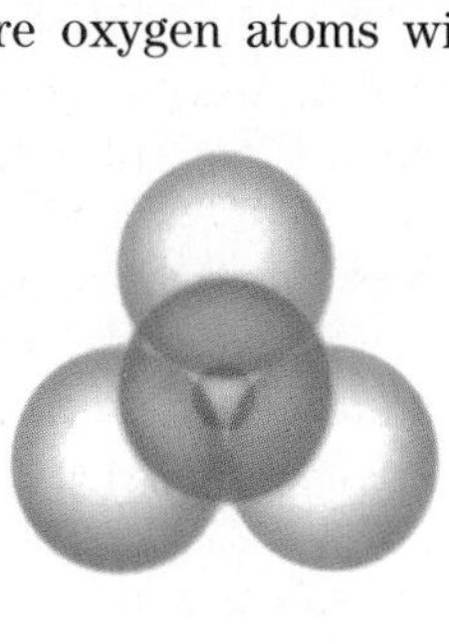

Figure 2–25 Model of the silicon-oxygen tetrahedron $(SiO_4)^{4-}$ viewed from above.

1. *Isolated Tetrahedral Structure.* These minerals are constructed of individual tetrahedra which are linked by positive iron and magnesium ions. Olivine, a mineral common to the lavas of the Hawaiian Islands, has this structure.

2. *Single- or Double-Chain Structure.* In single-chain structure, each tetrahedron shares two corner oxygen atoms of its base with two other tetrahedra; in the double chains, two single chains are combined by further sharing of oxygens (Fig. 2–26). Either kind of chain is bonded to adjacent chains by positive metallic ions. Because the bonds holding adjacent chains together are weaker than the silicon to oxygen bonds within each chain, cleavage develops parallel to the chains. The fibrous nature of amphibole asbestos results from chain structure (see Fig. 19–35).

3. *Sheet Structure.* In minerals having sheet structure, each tetrahedron shares three corners of its base with three other tetrahedra to form continuous sheets (Fig. 2–27). A good example of a mineral having sheet structure is mica. In mica, the bonding between pairs of sheets is weak, and this facilitates the ready sheet-by-sheet cleavage that characterizes this mineral.

4. *Framework Structure.* Each tetrahedron shares all four corners with other tetrahedra to form a continuous three-dimensional framework. Quartz is a mineral having a framework structure. Because the bonds are so strong in every direction in quartz, the mineral is exceptionally hard and has no tendency to break along preferred directions.

The silicate minerals of igneous rocks are constructed of silicon-oxygen tetrahedra in various combinations with other elements.

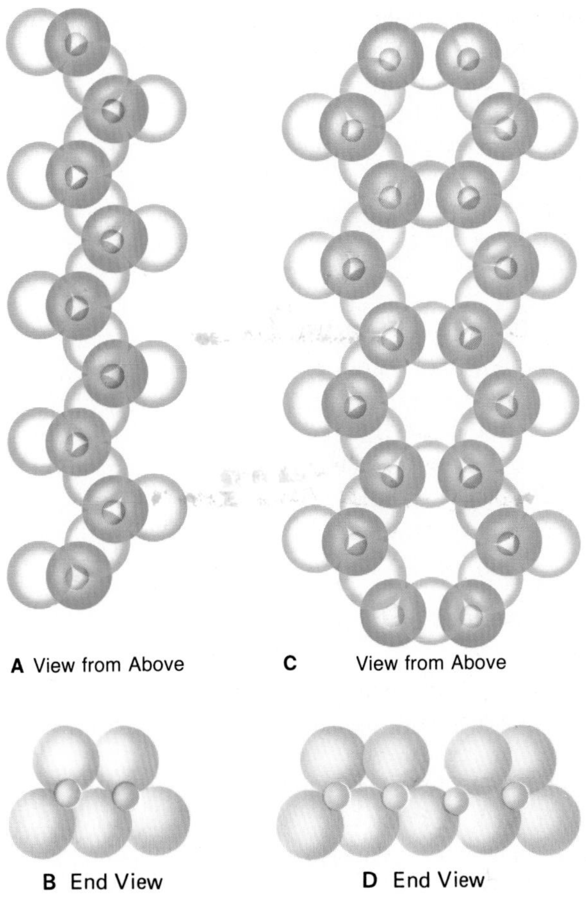

Figure 2–26 Single-chain tetrahedra viewed from above (A) and from an end (B). At right (C) and (D) are similar views of double-chain structures. Individual chains are bonded to adjacent chains by cations. In the overhead views, the small silicon ions are located beneath the uppermost oxygen ions.

Common Rock-Forming Silicates

The most abundant rock-forming silicates are included in one or more of the groups described below and in Table 2–5.

Quartz

The mineral *quartz* (SiO_2) is one of the most familiar and important of all the silicate minerals (Fig. 2–28). It is common in many different families of rocks. As mentioned earlier, quartz represents the ultimate in cross-linkage of silica tetrahedra; it therefore will not break along smooth planes. In quartz, the tetrahedra are joined only at the corners and in a relatively open arrangement. It is thus not a dense mineral, but it is quite hard because of the strong bonding in its framework structure. When quartz crystals are permitted to grow in an open cavity filled with water and dissolved silica, they may develop the hexagonal prisms topped by pyramids that are prized by crystal collectors (see Fig. 1–22). More frequently, the crystal faces cannot be discerned because the orderly addition of atoms had been interrupted by contact with other growing crystals.

Such minerals as chert, flint, jasper, and agate are varieties of a form of quartz called chalcedony. *Chalcedony* is composed of extremely small fibrous crystals of quartz. The crystals are so tiny that their study often requires the use of an electron microscope. Spaces between the small crystals are usually occupied by water molecules. Among the varieties of chalcedony, *chert* is exceptionally abundant in many sedimentary rock units. It is a dense, hard, usually white mineral or rock. *Flint* is the popular name for the dark gray or black variety of chalcedony much used by stone-age humans for making tools. *Jasper* is recognized by its opaque appearance and red or yellow color derived from iron oxide impurities. The term *agate* (Fig. 2–29) is used for chalcedony that exhibits bands of differing color or texture. There are many other varieties of quartz minerals than those briefly mentioned here. Properties useful in their identification are provided in Appendix B of this text.

The Feldspars

Feldspars are the most abundant constituents of rocks, composing about 60 per cent of the total weight of the earth's crust. There are two major families of feldspars: the *orthoclase* or *potassium feldspar* group (Fig. 2–30), which are the potassium aluminosilicates, and the *plagioclase* group (Fig. 2–31), which are the aluminosilicates of sodium and calcium. Members of the plagioclase group exhibit a wide range in composition—from a calcium-rich end member called *anorthite* ($CaAl_2Si_2O_8$) to a sodium-rich end member called *albite* ($NaAlSi_3O_8$). Between these two extremes, plagioclase minerals containing both sodium and calcium occur. The substitution of sodium for calcium, however, is not random but rather is governed by the temperature and composition of the

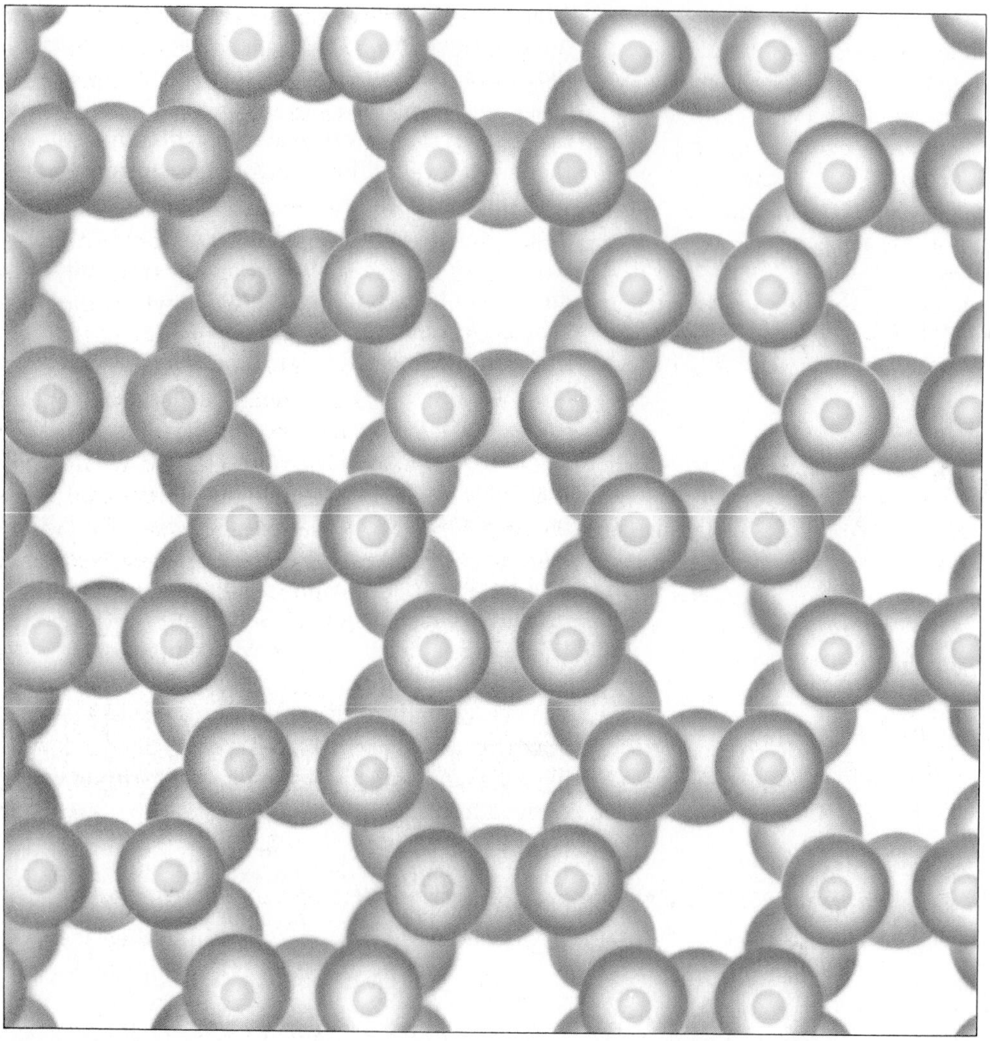

Figure 2–27 Silicate sheet structure viewed from above. The small spheres represent silicon ions that lie beneath the uppermost oxygen ions.

parent material. Thus, by examining the feldspar content of a once molten rock it is possible to infer the physical and chemical conditions under which it originated. Feldspars are nearly as hard as quartz and range in color from white or pink to bluish-gray. Silica tetrahedra in the feldspars are joined in a strong three-dimensional lattice that is characterized by planes of weaker bonding in two directions at (or nearly at) right angles to each other. Because of this bonding, the feldspars have good *cleavage* (break along smooth planes) in two directions. The resulting rectangular cleavage surfaces and a hardness of 6 are properties useful in the identification of feldspars.

The plagioclase feldspars provide an example of the manner in which ions can be interchanged in a mineral group. A chemical analysis of specimens of plagioclase taken from several different rocks would very probably reveal that the proportions of calcium, sodium, aluminum, and silicon (the principal cations in plagioclase) would differ among the specimens. This variability occurs because some ions resemble each other in size and electrical properties and are thus interchangeable in a given crystal. As indicated in Figure 2–8, calcium and sodium ions are large and nearly identical in size. Both aluminum and silicon are small ions and not greatly different in size. Thus, calcium ions might substitute for sodium ions freely if size alone were the only requirement. However, the electrical neutrality of the crystal must also be maintained. The electrical charge of the calcium ion is +2, whereas that of the sodium ion is +1. To counteract the surplus positive charge, an aluminum ion (+3) may substitute for a silicon ion (+4) to maintain electrical neutrality. Thus, $Ca^{2+} + Al^{3+}$ can interchange with Na^{1+} and Si^{4+}.

The Mica Group

As noted earlier, mica is a silicate mineral having sheet structure and is easily recognized by its perfect and conspicuous cleavage in one plane. The two chief varieties are the colorless or pale-colored **muscovite** mica (Fig. 2–32), which is a hydrous potassium aluminum silicate and the dark-colored **biotite** mica (Fig. 2–33), which also contains iron and magnesium.

Identification of large specimens of mica is rarely a problem because of its perfect planar cleavage and the way cleavage flakes snap back into place when they are bent and suddenly released. The micas are common constituents of igneous and metamorphic rocks, in which they can be recognized by their shiny surfaces and the ease with which they can be plucked loose with a pin or penknife. Before the manufacture of glass, one of the chief uses of muscovite mica was as window panes. This clear mica was quarried in Muscovy (an early name for Russia), and thus came to be known as "Muscovy glass," and eventually muscovite. Today, mica is used in the manufacture of electrical insulators and as a filler in plaster, roofing products, and rubber.

(Text continues on p. 42.)

Table 2–5 Common Rock-Forming Silicate Minerals

Silicate Mineral	*Composition*	*Physical Properties*
Quartz	Silicon dioxide (silica, SiO_2)	Hardness of 7 (on scale of 1 to 10); will not cleave (fractures unevenly); specific gravity: 2.65
Potassium feldspar group	Aluminosilicates of potassium	Hardness of 6.0–6.5; cleaves well in two directions; pink or white; specific gravity: 2.5–2.6
Plagioclase feldspar group	Aluminosilicates of sodium and calcium	Hardness of 6.0–6.5; cleaves well in two directions; white or gray; may show striations on cleavage planes; specific gravity: 2.6–2.7
Muscovite mica	Aluminosilicates of potassium with water	Hardness of 2–3; cleaves perfectly in one direction, yielding flexible thin plates; colorless; transparent in thin sheets; specific gravity: 2.8–3.0
Biotite mica	Aluminosilicates of magnesium, iron, potassium, with water	Hardness of 2.5–3.0; cleaves perfectly in one direction, yielding flexible thin plates; black to dark brown; specific gravity: 2.7–3.2
Pyroxene group	Silicates of aluminum, calcium, magnesium, and iron	Hardness of 5–6; cleaves in two directions at 90° black to dark green; specific gravity: 3.1–3.5
Amphibole group	Silicates of aluminum, calcium, magnesium, and iron	Hardness of 5–6; cleaves in two directions at 56° and 124°; black to dark green; specific gravity: 3.0–3.3
Olivine	Silicate of magnesium and iron	Hardness of 6.5–7.0; light green transparent to translucent; specific gravity: 3.2–3.6
Garnet group	Aluminosilicates iron, calcium, magnesium, and manganese	Hardness of 6.5–7.5; uneven fracture, red, brown, or yellow; specific gravity: 3.5–4.3

Figure 2–28 Quartz crystals that have grown around the walls of a cavity in rock. Height of specimen is 8 mm.

A

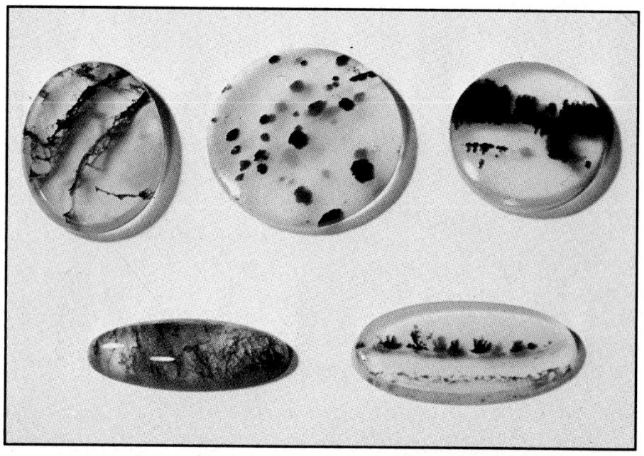

B

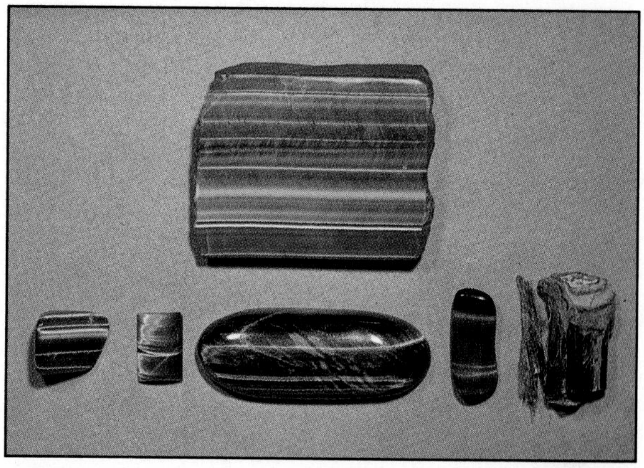

C

Figure 2–29 Three semiprecious gems belonging to the quartz family of silicate minerals. (A) Agate, a variety of quartz known as chalcedony, that exhibits bands of differing color or texture. (B) A variety of chalcedony termed "moss agate" because of the mosslike impurities of manganese oxide it contains. (C) Tiger's-eye quartz. The tiger's-eye effect results from light that is reflected off the fibrous structure of the mineral. The fibers are relics of asbestos fibers that have been replaced by silica.

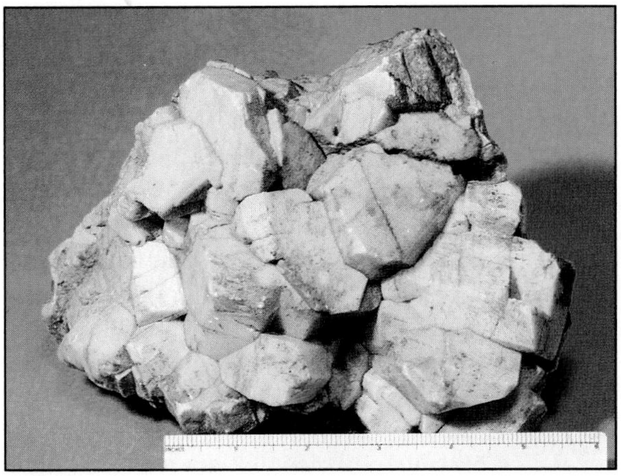

A

B

Figure 2–30 (A) Crystals of orthoclase feldspar. (B) Cleavage fragment of orthoclase feldspar.

Figure 2–31 The variety of plagioclase feldspar known as labradorite. The mineral often displays beautiful blue and gold reflections, as well as fine striations on cleavage planes. (Courtesy of Wards Natural Science Establishment, Inc., Rochester, New York.)

Figure 2–32 Muscovite mica.

Figure 2–33 Biotite mica.

Hornblende

Hornblende (Fig. 2–34) is a vitreous black or very dark green mineral. It is the most common member of a larger family of minerals called **amphiboles**, which have generally similar properties. Because of its content of iron and magnesium, hornblende (along with biotite, augite, and olivine) is designated a *ferromagnesian* or *mafic mineral.* Crystals of hornblende tend to be long and narrow. They have a double-chain silicate structure. Two good cleavages are developed parallel to the long axis and intersect each other at angles of 56° and 124° (Fig. 2–35A). The cleavage is a reflection of the location of planes of weaker bonds.

Figure 2–34 Hornblende. The black scaley flakes on the surface are biotite mica. Specimen is 6 cm long.

Augite

Just as hornblende is only one member of a family of minerals called amphiboles, **augite** is an important member of the **pyroxene** family, in which many other mineral species also occur. Like hornblende, it is a ferromagnesian mineral and thus dark colored. An augite crystal (Fig. 2–35B) is typically rather stumpy in shape, with good cleavages developed along two planes that are nearly at right angles. Thus, the cross section of a crystal appears nearly square rather than rhombic, as in hornblende. Unlike hornblende, which has a double-chain silicate structure, augite is constructed of single chains, and this accounts for its having differently shaped cleavage fragments.

Olivine

Olivine, another ferromagnesian mineral, was mentioned earlier as having isolated silicon-oxygen tetrahedra bonded together by iron or magnesium ions or both. Olivine contains variable proportions of iron and magnesium, and the substitution of these ions for each other is facilitated by their having similar ionic radii and electrical properties. The ions in olivine are so strongly held by ionic bonding that the mineral has a hardness of 6.5 to 7 on Mohs' scale. As you might guess from its name, this glassy-looking mineral often has a green color. Frequently, it occurs as masses of small sugary grains (Fig. 2–36) or as tiny vitreous crystals in black lavas. It is also an important constituent of stony meteorites. If large unblemished crystals of magnesium-rich olivine are found, they may be cut and polished into attractive gemstones called *peridots* (Fig. 2–37).

Garnet

Garnet is the name used for a group of mineral species closely related in composition, crystal form, and physical properties. In general composition, these minerals are silicates of aluminum with iron, calcium, or magnesium. The iron-rich variety,

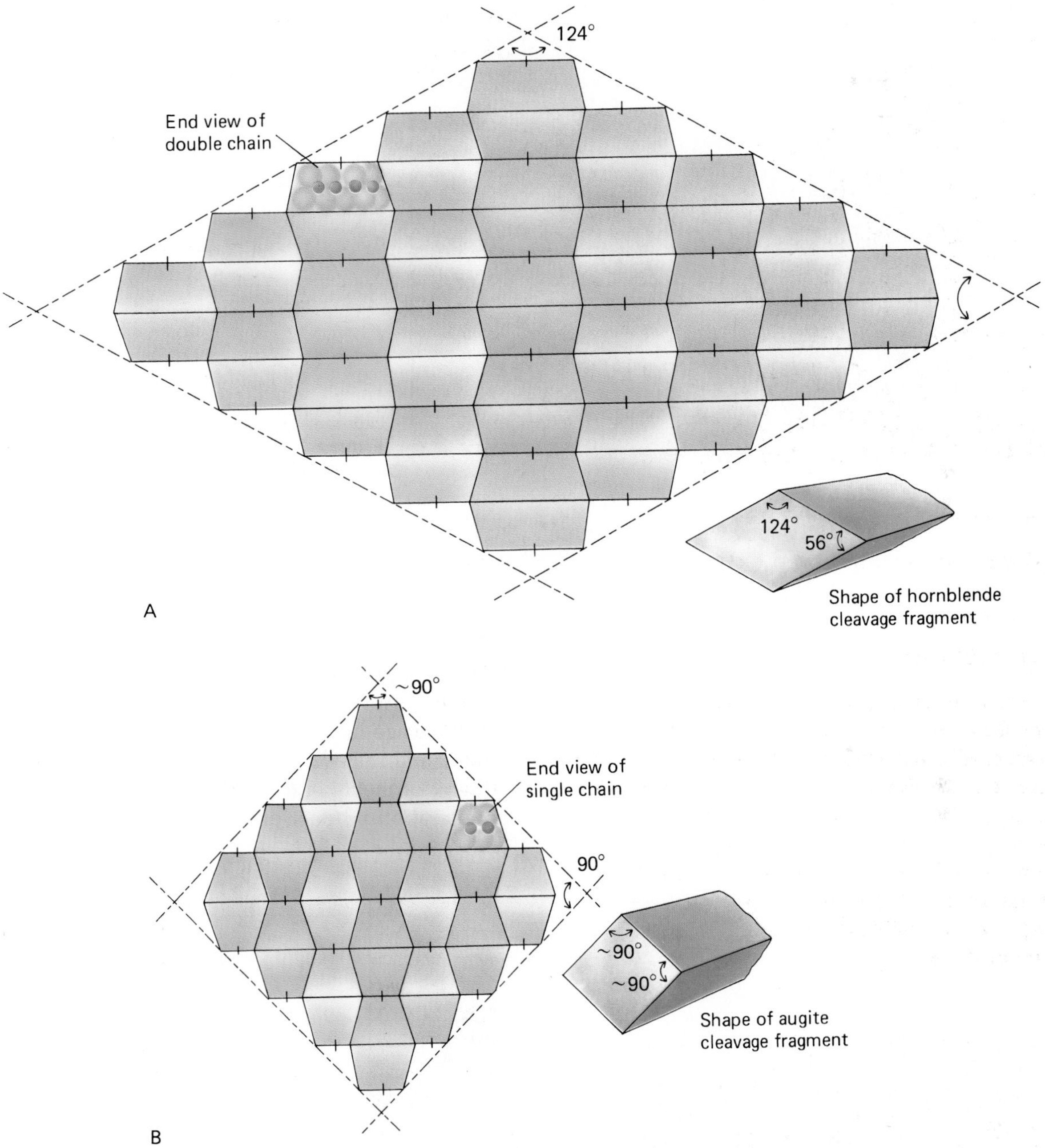

Figure 2–35 (A) Arrangement of double chains in the mineral hornblende, and (B) single chains in augite, and the relation of these arrangements to cleavage angles. The direction of strongest bonding is indicated by the short vertical lines.

Figure 2–36 Olivine.

almandite, is widely used in jewelry because of its rich dark red color. However, depending on their composition, garnets are found in all colors except blue. Garnets can usually be identified by their well-formed equidimensional crystals (Fig. 2–38) and their resinous-to-vitreous luster. Crystals will not cleave but rather break with a conchoidal fracture. Because of the sharp edges produced on fracturing and its hardness (7 to 7.5 on Mohs' scale), garnet is widely mined and sold as an abrasive material.

Chlorite

The **chlorites** are a group of greenish minerals having a platy form similar to that of micas. In composition, the minerals of the chlorite group are hydrous aluminum silicates of iron and magnesium. Chlorite can usually be identified by its perfect cleavage in one direction, greenish color and streak, and relative softness. It is a common constituent of metamorphic rocks and also occurs as an alteration product of ferromagnesian minerals.

Figure 2–37 Peridot, the gem variety of olivine.

Figure 2–38 Small crystals of almandite garnet collected from metamorphic rocks near Amity, New York. The crystals average about 4 mm in diameter.

Clay Minerals

The clay minerals are sheet-structure silicates of hydrogen and aluminum with additions of magnesium, iron, and potassium. At one time, clays were assumed to be amorphous mixtures of metal oxides and hydroxides. Through the use of X-ray equipment, however, it was demonstrated that clays are composed of crystalline units that, like mica, have sheet structure. The individual flakes are extremely small, but can be seen with the magnification provided by the electron microscope (Fig. 2–39). Sheets of tetrahedra in clays are weakly bonded, and the potassium, sodium, and magnesium ions may enter and leave exchangeable locations within the lattice structure. Clays are formed through the decomposition of other aluminum silicate minerals, especially the feldspars. Clay minerals are characteristically very soft, have low density, occur in earthy masses, and become rather plastic when wet. These characteristics permit one to recognize clay, but in order to identify the individual species of clay minerals it is necessary to use X-ray diffraction techniques.

Quartz, feldspar, mica, hornblende, augite, olivine, garnet, and clay are among the most common silicate minerals.

The Common Nonsilicate Minerals

Approximately 8 per cent of the earth's crust is composed of nonsilicate minerals. These include a host of **carbonates**, sulfides, sulfates, chlorides, and oxides. Among these groups, the carbonates such as **calcite** and **dolomite** are the most important. Calcite ($CaCO_3$), which is the main constituent of limestone and marble, forms in many ways. It is secreted as skeletal material by certain invertebrate animals, precipitated directly from sea water, or formed as dripstone in caverns. Calcite is easily recognized by its rhombohedron-shaped cleaved fragments and by the fact that an application of hydrochloric acid on its surface will cause effervescence. Clear cleavage rhombs of calcite, called iceland spar, exhibit a property known as **double refraction** (Fig. 2–40). For example, if a transparent piece of calcite is held over a dot that has been placed on white paper, two dots will appear when the calcite is held in certain positions.

Figure 2–39 Electron micrograph of the clay mineral kaolinite. The flaky, stack-of-cards character of the clay crystals is a manifestation of their silicate sheet structure. Magnified about 2000 times. (Courtesy of Kevex Corporation.)

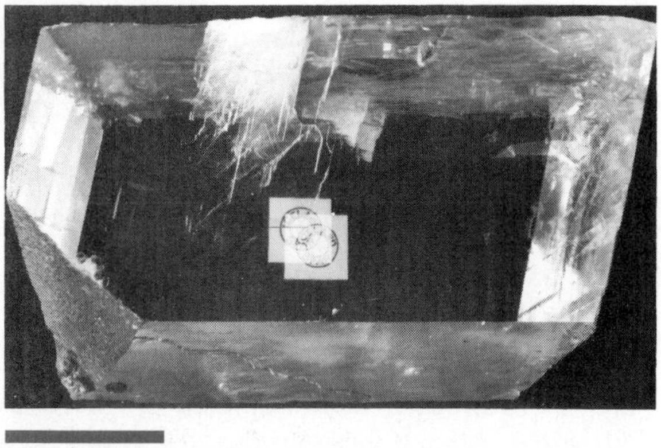

Figure 2–40 Cleavage fragment of exceptionally transparent calcite. The mineral's ability to cause double refraction is clearly exhibited. (Courtesy of the Institute of Geological Sciences, London.)

As the mineral is slowly rotated, one dot will appear to rotate around the other. The explanation for this bit of "magic" is that the light entering the specimen is bent or *refracted* and divided into two polarized rays, one of which is bent more than the other (Fig. 2–41). The image from the more refracted ray appears to move around the image of the less refracted ray as the piece of calcite is rotated.

Dolomite, $CaMg(CO_3)_2$, is a carbonate of calcium and magnesium that occurs as extensive layers of a rock called *dolostone*. In the field, geologists often distinguish dolomite by the fact that it will not effervesce in dilute hydrochloric acid unless it is powdered. Many ancient dolostones are thought to have been formed by recrystallization of still older limestones. Because calcium and magnesium ions are nearly the same size, it is possible for the magnesium to replace calcium in the atomic structure.

Other common nonsilicate minerals include common rock salt, or *halite* (NaCl), and gypsum. Halite is easily identified from its salty taste and perfect cubic cleavage. *Gypsum* is a soft (2 on Mohs' scale) hydrous calcium sulfate ($CaSO_4 \cdot 2H_2O$). The variety of gypsum called *satinspar* has a fine fibrous structure, whereas *selenite* (Fig. 2–42) will split into thin plates. The finely crystalline massive variety known as *alabaster* is widely used in carvings and sculpture (Fig. 2–43) because of its uniform texture and softness.

Iron Oxide Minerals

The chief iron oxide minerals are **magnetite** (Fe_3O_4), **hematite** (Fe_2O_3), and **goethite** ($HFeO_2$). Magnetite is a lustrous black mineral that is found in igneous rocks as well as some metamorphic rocks and clastic sedimentary rocks. It is an important ore of iron. Magnetite has a black streak, but its most distinctive property is related to magnetism. Pieces of magnetite can be picked up by an ordinary steel magnet, and the lodestone variety of magnetite is itself a natural magnet that will attract steel objects (see Fig. 2–23).

Hematite is even more important than magnetite as an ore of iron. It occurs as iron ore in thick and extensive beds of sedimentary origin that have been subsequently altered and enriched by solu-

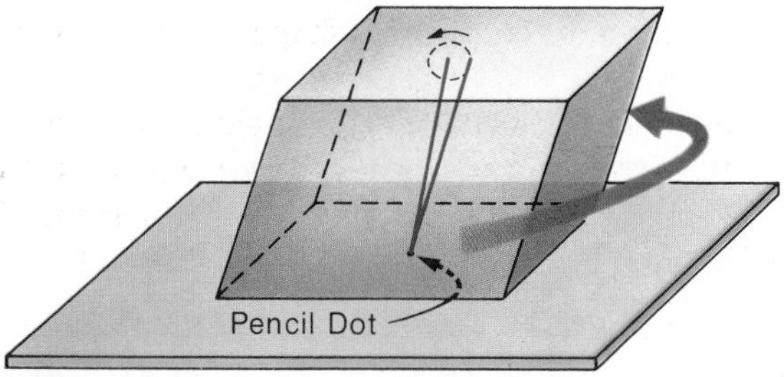

Figure 2–41 If a piece of clear calcite is placed as shown over a pencil dot, two dots will be viewed from a point above the specimen. If one then rotates the calcite as indicated by the arrow, the image of the dot from the more refracted ray will appear to move in a circular path around the other dot.

Figure 2–42 Selenite gypsum. This mineral crystallizes in the monoclinic crystal system. (Courtesy of the Institute of Geological Sciences, London.)

tions. Hematite is very common in a variety of sedimentary rocks, where it often provides a reddish or brownish coloration. The mineral varies in color from steel gray to brownish red but always exhibits a distinctive reddish streak.

Goethite generally forms as a weathering product of other iron-bearing minerals. It is the principal mineral in the amorphous hydrous iron oxide mineral known as limonite ($Fe_2O_3 \cdot 2H_2O$). Goethite has a yellow-brown streak and is generally yellow or brown in color.

The most important non-silicate mineral groups are carbonates, oxides of iron, such as hematite and magnetite, sulfates such as gypsum, and the sodium chloride mineral halite.

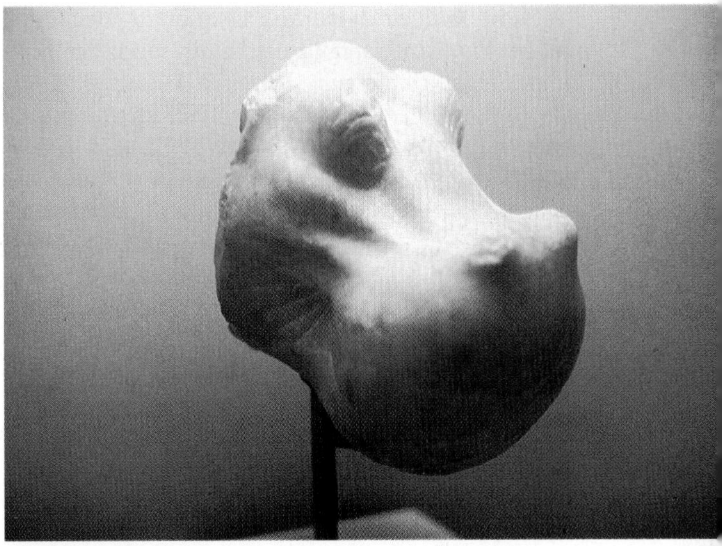

Figure 2–43 Head of a hippopotamus that was carved in alabaster gypsum by an ancient Egyptian sculptor (663 BCE, 26th Dynasty).

Mineralogic Clues to Geologic Events

Minerals are studied by geologists not as mere constituents of the crust but as clues to the history of the rocks in which they occur. Some minerals, such as halite, develop exclusively in ocean water and provide the geologist with evidence that a particular sediment was deposited in the sea rather than in a freshwater lake. A thick bed of halite indicates aridity and evaporation so extreme that the brine had become ten times saltier than ordinary sea water. The magnetic properties of magnetite can, in certain situations, provide clues to the position of continents relative to the earth's magnetic poles. Certain minerals form within a narrow range of conditions and can therefore be used to diagnose the pressures and temperatures involved in the formation of crustal rocks and mountains. Diamonds, for example, form only at high temperatures and extremely high pressures. Like diamond, graphite is a crystalline variety of carbon. However, it forms at lower temperatures and pressures. Other minerals contain radioactive isotopes that permit us to know the age of the parent rocks. By their size, crystals of feldspar give the geologist information about the rate at which ancient molten mass congealed. Even ordinary grains of quartz may hold clues to their history since being eroded from some ancient granitic terrain. The ancient buried mineral products of weathering may also provide information about past climatic conditions in a region.

Minerals as Gems

Some occurrences of the minerals introduced in this chapter are so attractive, durable, and rare that they are highly valued as gems. Their beauty is dependent on color, luster, and brilliance—properties that are enhanced when the "rough stones" are cut, faceted, or shaped and polished. Gems are sold by weight, and the basic unit of weight is called a gem carat. One gem carat is equal to ⅕ g. (The gem carat is not to be confused with the gold carat, which indicates the proportion of gold in an alloy and is not a unit of weight.)

To be considered a gem, a mineral must have beauty, rarity, and durability. Gems are sold by weight, and the basic international unit of gem weight is the gem carat, which is one-fifth of a gram.

Such familiar minerals as quartz, corundum, feldspar, olivine, and garnet occur sporadically as materials of gem quality. Gems in the quartz family are particularly well known and include the purple variety known as amethyst (see Fig. 2–11), yellow quartz or citrine, rose quartz, and the banded microcrystalline quartz called agate, and the mysterious tiger's-eye quartz (see Fig. 2–29), which results from the replacement of asbestos by microcrystalline quartz. Jasper is a variety of microcrystalline quartz that is rich in iron oxide.

The two most famous gems in the corundum family are blue sapphire and red ruby (see Fig. 2–17). The color differences in these corundum minerals result from small traces of metal oxides. For example, ruby is colored by chromium oxide and sapphire by oxides of iron and titanium. These and other impurities also produce exquisite hues of green, purple, and gold in sapphires. Feldspar gems include the attractive white opalescent moonstone and the green orthoclase called amazonite. Peridot is the gem variety of olivine. In the garnet family, almandite is prized for its clear wine-red color. There are, however, many other beautifully colored garnets, including pink, orange, and green varieties. Topaz (Fig. 2–44), known to us as the third hardest mineral in Mohs' scale, may occur as a lovely yellow gem, although stones colored pale blue and light green are also prized.

Diamond, the most highly treasured of gems, is a nonsilicate composed only of carbon. As mentioned earlier, diamonds (Fig. 2–45) are well known for their brilliance and hardness. They are derived from dark, dense igneous rocks called kimberlites. The principal factors involved in determining the value of a cut and faceted diamond is its weight, color, and degree of perfection. A so-called perfect diamond is one that shows no conspicuous inclusions or flaws when viewed at a magnification of 10 × by an experienced person. The weight of a diamond is usually expressed in a unit called a *point,* which is 1/100 of a carat. Thus, a 25-point diamond would weigh 0.25 carat.

Figure 2–44 Large topaz on display at the Field Museum in Chicago. Known as "Chalmer's Topaz," this Brazilian gem has a total weight of 5,890 carats.

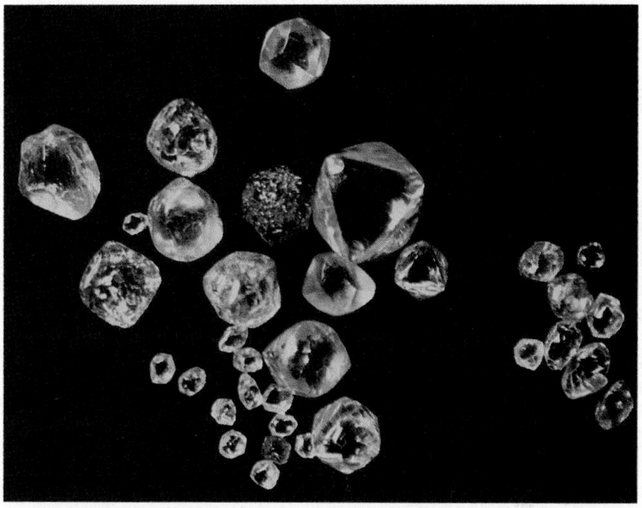

Figure 2–45 Raw diamonds obtained from alluvial deposits in Borneo. (Courtesy of S.C. Bergman.)

Summary

Minerals and rocks are the "documents" that contain the evidence of all the forces and processes that have affected the earth in the past. They are the basis for nearly all geologic studies.

Centuries ago, a Greek philosopher named Democritus speculated that water and rocks were made of invisibly tiny pieces of matter. Democritus coined the term *"atom"* to describe these particles. The present understanding of atoms is that they have a central nucleus containing positively charged *protons* and electrically neutral *neutrons.* Circling the nucleus are negatively charged *electrons.* The arrangement of electrons into shells, the number of atomic particles, the size of the atoms, and their electrical characteristics determine how the atoms will combine into molecules and ultimately into minerals.

Minerals are naturally occurring compounds that exist as crystals or crystalline solids, have either a definite chemical composition or restricted range of chemical compositions, and possess a systematic three-dimensional atomic order. Knowledge of their crystallinity is now well established as a result of X-ray diffraction studies. X-ray diffraction is also useful in the identification of minerals, but usually the more common minerals can be recognized without the aid of special instruments by their characteristic physical properties. The most commonly observed physical properties are *color, luster, cleavage, crystal form, hardness, specific gravity,* and *magnetism.*

Because the elements oxygen and silicon make up nearly 75 per cent of the earth's surficial rocks, the silicates are the most important of the rock-forming minerals. The fundamental unit in silicate mineral structures is a tetrahedral arrangement of four oxygen ions around a central silicon ion. The tetrahedra are joined by chemical bonds to form chains, sheets, or three-dimensional frameworks; these structures are the underlying causes for many of the specific properties of a mineral.

The principal rock-forming silicate mineral families are relatively few in number. They include the *quartz* family, potassium and plagioclase *feldspars,* muscovite and biotite *micas, hornblende, augite, olivine, garnet, chlorite,* and *clay.* Those silicate minerals such as biotite, augite, hornblende, and olivine that are rich in iron and magnesium (and usually black or green in color) are designated *ferromagnesian minerals.* Common nonsilicate rock-forming minerals are the *carbonates* (calcite and dolomite), evaporites (gypsum and halite), and iron oxide minerals (hematite, limonite, and magnetite).

Certain relatively rare specimens of otherwise common minerals are so beautiful that they have been used since ancient times as ornamental gems. The beauty of these gem varieties results from their color, luster, and the manner in which they transmit or reflect rays of light. Gems should also be relatively hard, otherwise wear will destroy their beauty. Their value is also influenced by their rarity. Minerals that combine the highest degree of beauty, hardness, and rarity include diamond, sapphire, ruby, emerald, and topaz. Somewhat more common gems include peridot (olivine), garnet, moonstone (a feldspar), and such members of the quartz family as amethyst, citrine, opal, tiger's-eye, and agate.

Terms to Understand

adamantine (luster)
amphibole
atom
atomic mass
atomic number
augite
biotite
calcite
carbonate
chert
chlorite
clay
cleavage
covalent bond
crystal
crystalline aggregate
crystallographic axes
density
dipolar
dolomite
double refraction
electron
elements
feldspars
ferromagnesian mineral
flint
fracture
garnet
hematite
hornblende
igneous
isometric crystal
ion
ionic bond
isotope
Law of Constancy of Interfacial Angles
limonite
luster
magnetite
mineral
molecule
muscovite
neutron
nucleus
olivine
orthoclase
plagioclase
proton
pyroxene
quartz
sedimentary rock
shell, electron
silicates
silica tetrahedron
space lattice (atomic)
specific gravity
texture (in rocks)
valence
vitreous (luster)

Questions for Review

1. What is a crystal? Why are complete and perfectly formed mineral crystals rare in nature?

2. In terms of atomic structure, how does a crystalline rock (like granite) differ from an amorphous rock (like volcanic glass)?

3. What is the difference between an atom and an ion?

4. In terms of atomic structure, explain why amphibole asbestos is fibrous and mica is platy, and why the mineral quartz does not break into fragments with smooth surfaces.

5. What are the eight most abundant elements composing the earth's crust? Of these, which has the largest ionic radius?

6. Explain how covalent bonding differs from ionic bonding.

7. Name four common rock-forming silicate minerals rich in iron and magnesium. Which of these have good cleavage in two directions? Which is the softest? Which is constructed of individual tetrahedra linked at their corners by metallic ions?

8. The minerals halite and calcite look somewhat alike. Without having to taste the specimens, how would you identify each?

9. Both diamond and graphite are minerals composed of carbon. Why are they so different in specific gravity, color, and hardness?

10. Provide an example of substitution of elements among a family of common rock-forming silicates. Under what limiting conditions can one element substitute for another in a mineral?

11. If you were a stone-age human and had to choose between calcite and chert for a stone axhead, which would you choose? Explain your choice.

12. What is the weight, in grams, of a 5-carat ruby? What would be the weight in carats of a gem described by your jeweler as a 75-point diamond?

Supplemental Readings and References

Blackburn, W.H., and Dennen, W.H., 1988. *Principles of Mineralogy,* Dubuque, Iowa, Wm. C. Brown, Publishers.

Dietrich, R.V., and Skinner, B., 1979. *Rocks and Minerals.* New York, John Wiley & Sons, Inc.

Ernst, W.G., 1969. *Earth Materials.* Englewood Cliffs, N.J., Prentice-Hall, Inc.

Pough, F.H., 1960. *A Field Guide to Rocks and Minerals.* Cambridge, Mass., Houghton Mifflin Co.

Shaub, B.M., 1975. *Treasures from the Earth.* New York, Crown Publishers, Inc.

Igneous Rocks and Processes

3

Thin section of olivine crystals viewed in polarized light. The brightly colored grains are all olivine in this crystalline aggregate. Width of field 22 mm.

Granites have been formed at considerable depth in the earth, and have cooled and crystallized slowly under great pressure.

Charles Lyell, 1882, Elements of Geology

Introduction

As described in Chapter 1, geology recognizes three major divisions of rock according to their mode of origin. These divisions are igneous, metamorphic, and sedimentary. In this chapter, we will examine igneous rocks and the processes responsible for their origin. **Igneous rocks** are those that have been formed by the solidification of molten rock material. The molten matter from which they are formed is termed **magma**. Magma may exist for long periods of time beneath the surface of the earth, intruding here and there into overlying rock yet never reaching the surface. When congealed, these magmas form the so-called plutonic or **intrusive igneous rocks**. When molten material is able to penetrate to the earth's surface and flow out onto the ground, it is called **lava**. Lavas and solid particles that have congealed from lavas constitute **extrusive igneous rocks**.

Intrusive Igneous Activity

The Nature of Magmas

A consideration of magmas is clearly essential to the study of igneous rocks. Magmas, after all, are the parent brew from which lavas and igneous rocks are derived. Lava is magma that has poured out through volcanic vents and is therefore accessible for us to observe. Much magma, however, solidifies far beneath the surface of the earth, and we cannot observe it directly. The composition and history of such magma must be ascertained from the study of the resultant igneous rock's mineral composition and texture.

Composition

Analyses of igneous rocks show that only the elements oxygen, silicon, aluminum, magnesium, calcium, sodium, potassium, and titanium are abundant. Of these elements, oxygen and silicon are essential. These two elements either may combine to form quartz or bond with other elements in specific ways to form the feldspars, ferromagnesian minerals, and micas that are the major constituents of igneous rocks.

Although the average abundance of silicon in igneous rocks of the crust is 27.72 per cent, that proportion has a *range* of 24 to 34 per cent. Furthermore, petrologists have noted that the number of other ions in igneous rocks has a definite relationship to silicon content. For example, as the amount of silicon increases, the amounts of iron, magnesium, and calcium necessarily decrease. Thus, rocks rich in ferromagnesian minerals are less likely to contain the mineral quartz, since there would be insufficient silica "left over" to form quartz after the ferromagnesian minerals had crystallized. The amount of potassium, on the other hand, is usually proportional to the amount of silica. Rocks rich in potassium (contained in such minerals as orthoclase and muscovite) are likely also to contain quartz.

Igneous rocks composed largely of quartz, orthoclase, sodic plagioclase, and muscovite generally are derived from melts rich in silicon. Those rich in calcium feldspar, hornblende, augite, and olivine are derived from melts with relatively low silicon content.

Sequence of Crystallization of Magma

The solidification of a magma does not occur at a definite temperature, as does, for example, the solidification of lead. It is a drawn-out affair in which different minerals form at different temperatures. As an example, consider a magma having the composition of the rock basalt, a dark-colored igneous rock composed largely of ferromagnesian minerals and plagioclase. As the magma begins to cool, olivine, pyroxene, and calcium plagioclase are among the earliest minerals to crystallize. Often these first-formed crystals are larger and more perfect than those formed later because, when they formed, there was ample space and an abundance of elements required for growth. Those minerals that crystallize later must fit into the remaining spaces and thus tend to be smaller and less perfect. Also, minerals enclosed in other minerals must have formed before the enclosing mineral. By using these observations of size and spatial relationships of minerals as viewed in thin sections of the rock, it is sometimes possible to make a reliable estimate of the order in which minerals crystallize in a magma.

During the first half of the present century, an eminent petrologist named N. L. Bowen studied the crystallization of artificial silicate melts in the laboratory and provided further evidence of the approximate order of crystallization in gradually cooling magmas. Bowen used melts of basaltic composition for his experiments and was able to ascertain the specific temperature range over which particular silicate minerals would begin to form. Olivine, for example, would crystallize first and at the highest temperature. Pyroxenes and the

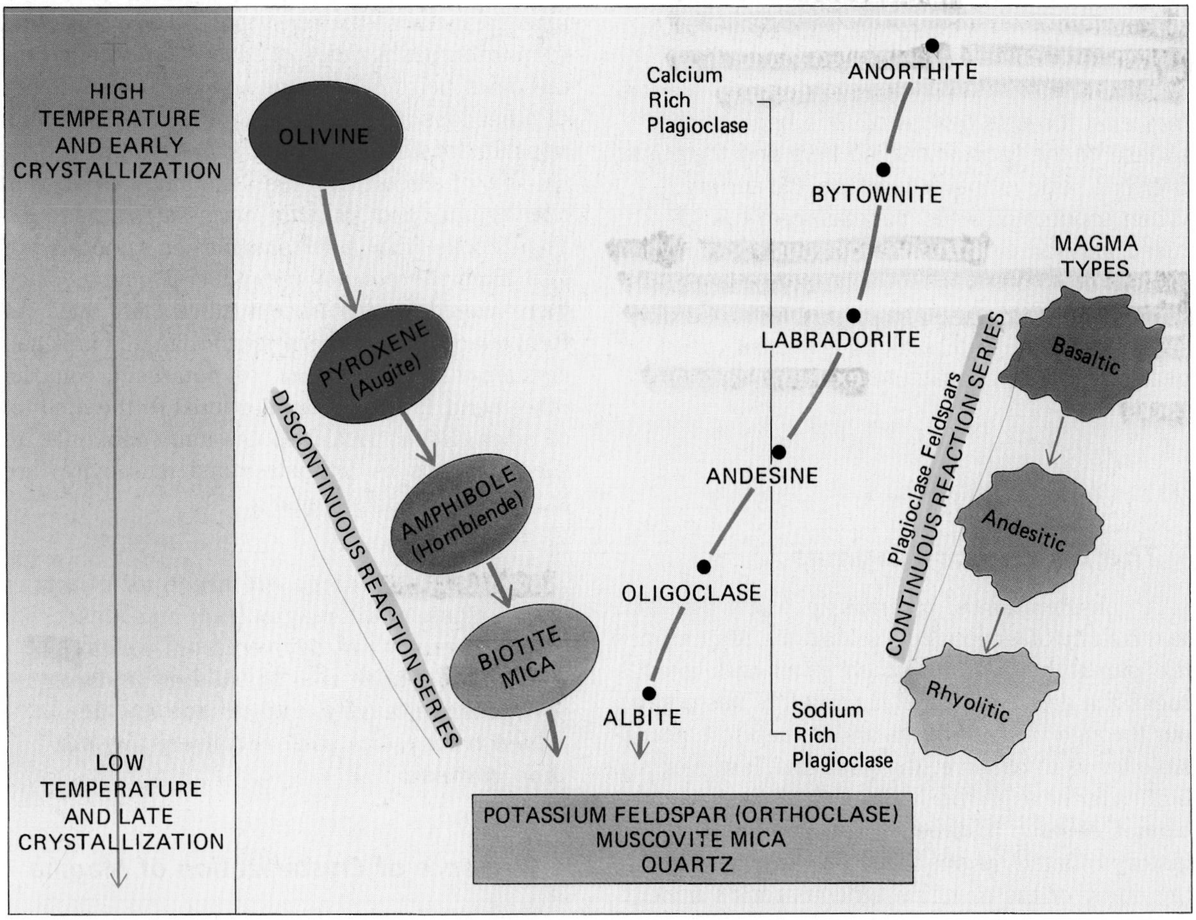

Figure 3–1 Bowen Reaction Series. Note that the earliest minerals to crystallize from a cooling magma are olivine and calcium-rich plagioclase minerals like bytownite. As crystallization proceeds, olivine reacts with the melt to form pyroxene, pyroxene reacts to form amphibole, etc. The plagioclase minerals also react with the remaining liquid, but rather than forming new minerals, change continuously in composition.

more calcium-rich feldspars would form next, followed by hornblende, biotite, and the more sodic feldspars. The order of crystallization, called the **Bowen Reaction Series**, is shown in Figure 3–1. It is called a *reaction series* because early formed crystals react with silica in the remaining liquid and change to the next lower mineral in the series. For example, the silica in the magma would react with olivine crystals by simultaneously dissolving and precipitating the dissolved components so as to form pyroxene (augite), and pyroxene crystals would begin to grow in spaces between earlier formed crystals. At lower temperatures, the silica in the melt would similarly react with pyroxene to form hornblende, and subsequently hornblende would react with the liquid to form biotite.

As these reactions proceed, ions are being continuously extracted from the magma and incorporated into newly forming minerals. When about half of the magma has crystallized, the remaining melt will be impoverished in iron, magnesium, and calcium. These ions now reside within the ferromagnesian minerals and plagioclase formed earlier. Conversely, the magma liquid has become enriched in such elements as sodium and potassium, as well as silicon. These elements will be incorporated into minerals that crystallize later. Such minerals are indicated at the bottom of the Bowen chart.

On the right branch of Bowen's chart, one can trace the changes that occur as calcium-rich plagioclase reacts with the magma to produce varieties of plagioclase that are successively richer in sodium. Because the plagioclase minerals maintain the same space lattice but change continuously in their content of calcium and sodium, the right side of the diagram was named the *continuous series*. The ferromagnesian or left side of the diagram depicts reactions that result in minerals with distinctly different structure. It was therefore termed a *discontinuous series*.

By examining thin sections of igneous rocks with the aid of a microscope, one can often see evidence of reactions like those described above. It is not uncommon, for example, to find plagioclase crystals in which the innermost layers or zones are more calcium-rich and the outer zones more sodium-rich (Fig. 3–2.) Similarly, reaction rims of pyroxene can be observed around olivine and rims of hornblende around pyroxene.

Figure 3–2 Plagioclase crystals exhibiting reaction rims. The core of the crystals consists of calcium-rich plagioclase and myriads of fine inclusions. The clear outer rims consist of sodium-rich plagioclase. The rock is an olivine andesite from Colima volcano in Mexico. (Courtesy J. Luhr.)

The continuous reaction series of Bowen's chart indicates continuous changes in plagioclase as calcium ions are exchanged for sodium ions. Entirely new minerals of differing crystal structure appear in succession in the discontinuous series part of the chart.

Factors Affecting Crystallization

As is evident from the Bowen Reaction Series, magmas are the product of heating. However, other factors are also involved in the formation of magmas and in their subsequent solidification. The two most important of these are volatiles and pressure.

All magmas contain volatiles. These consist primarily of water vapor and various gases, such as carbon dioxide, oxygen, hydrogen sulphide, and sulphur dioxide. Water vapor is the most important volatile because of its strong influence on melting temperatures. As a general rule, water lowers the melting temperature of a silicate rock. The effect is illustrated in Figure 3–3. At higher water vapor pressure (5000 atmospheres), sodium plagioclase melts at a temperature of about 765°C. When the water vapor pressure is only 1000 atmospheres, a temperature of over 900°C is required for melting.

Water continues to play a role in the magma after initial melting has occurred. The water dissolved in the melt interferes with the linkage (polymerization) of the silica tetrahedra. Such linkages are required to increase the viscosity of magma. Conversely, by inhibiting linkages, the viscosity of the magma is decreased. It becomes more fluid and mobile, and its progression through surrounding rock is enhanced.

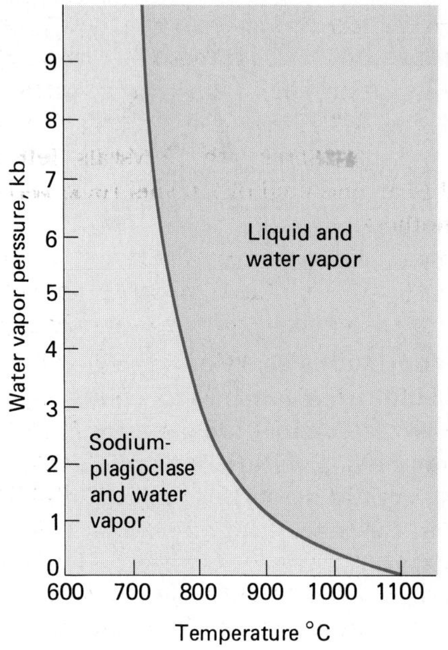

Figure 3–3 This curve shows that an increase in water pressure has the effect of lowering the temperature at which sodium plagioclase will melt. (After Burnham, C.W. and Davis, N.F., 1974.)

Considerable water exists in rocks of the upper mantle where most magmas are believed to originate. It finds its way there by way of subduction zones that convey wet oceanic sediments downward into the asthenosphere.

The effect of water on an igneous rock mass being heated is to decrease the melting point of constituent minerals.

Whereas water tends to lower the melting temperatures of rocks, pressure has the opposite effect. Melting occurs because increasing temperatures cause the atoms in a solid to vibrate increasingly faster, until they finally break their bonds and flow freely as a liquid. Pressure favors a denser, solid condition, and inhibits this change. As a result, the melting temperatures of materials at depth are considerably higher than they would be near the surface of the earth because of the great confining pressures of overlying rocks. The effect of pressure is illustrated in Figure 3–4. The black curve plotted on the graph indicates the estimated temperature at which basalt would become molten. As pressure increases with depth, the melting temperature of basalt also increases. Note that the temperature gradient curve and the melting curve approach closely in one area. A slight decrease in pressure in this area might result in melting.

The effect of pressure on an igneous rock is to raise the melting points of constituent minerals.

Causes of Diversity Among Intrusive Igneous Rocks

In formulating his reaction series, Bowen experimented with a melt of basaltic composition and then theorized that the different kinds of igneous rocks might be formed from a single basaltic magma. As we have noted, this process requires the sequential removal of earlier formed minerals as the magma gradually cooled. Bowen termed the process **fractional crystallization**. If the early formed crystals were not removed, a basaltic magma would yield only basaltic rocks. By remov-

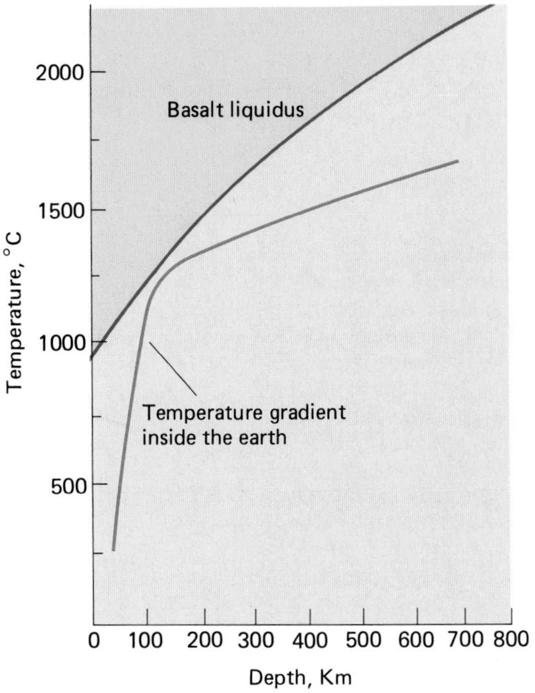

Figure 3–4 The melting temperature of material at depth is considerably increased by the confining pressures of great thicknesses of overlying rocks. The temperature of complete melting is known as the *liquidus*. (After Verhoogen, 1964.)

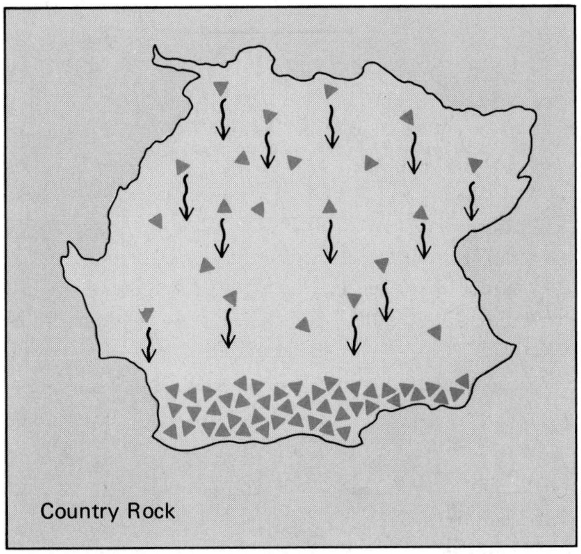

Figure 3–5 Fractional crystallization and accumulation of early formed crystals by gravity settling.

ing the earlier iron- and magnesium-rich minerals, the magma is gradually changed toward a more rhyolitic composition.

The removal of early formed minerals may be accomplished in various ways. In one process, early formed crystals having a specific gravity greater than that of the remaining magma may sink toward the bottom of the chamber. The mechanism has been termed **gravity settling** (Fig. 3–5). Gravity settling of minerals at the top of the reaction series diagram (see Fig. 3–1) would leave the remaining magma poorer in iron, magnesium, and calcium. This altered melt would be likely to form a rock composed of hornblende and plagioclases that are somewhat richer in sodium. If these minerals were then removed, yet another kind of rock would be formed.

Another mechanism for differentiating minerals in a magma has been called **filter pressing**. At some point in the crystallization of a magma, it exists as a mixture of solid crystals and hot liquid, a sort of mineral "slush." As the crystals pile up in the magma chamber, their accumulating weight exerts pressure on the remaining liquid, forcing it into spaces in the surrounding rock (Fig. 3–6). Should the magma chamber be compressed, this movement of fluid into adjacent rock would be enhanced. Ultimately, the crystals left behind would form one kind of igneous rock, and the liquid another.

Although fractional crystallization of a basaltic melt can indeed produce different kinds of igneous rocks, the process does not adequately explain all igneous bodies, nor can it account for the large amount of silica-rich (felsic) rock, which is the basic ingredient of continents. Only 10 per cent of an original basaltic melt can theoretically be converted to granite by fractional crystallization—hardly enough to provide the vast amount of granitic rock in continents. Processes other than fractional crystallization must operate to produce the rich variety of igneous rocks found at the earth's surface. One such process is partial melting (Fig. 3–7).

The term **partial melting** refers to a process whereby pre-existing rock is incompletely melted, and the molten fraction is separated from the still-

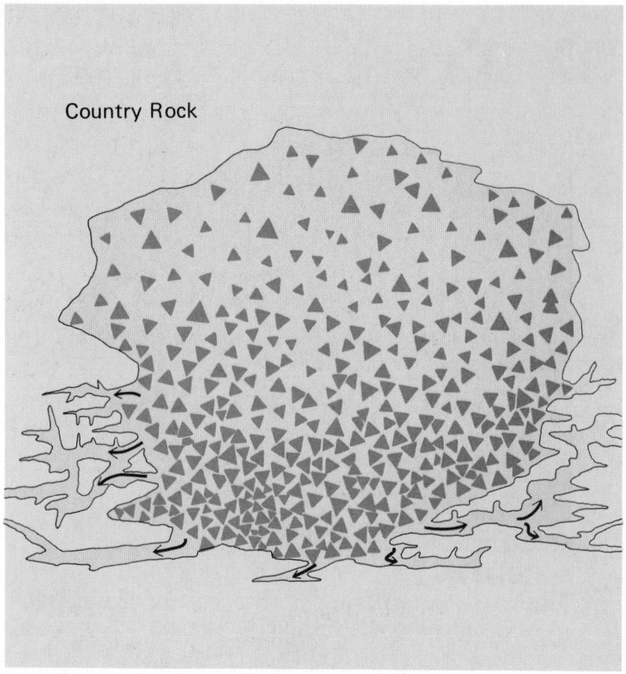

Figure 3–6 The compressional weight of the growing accumulation of crystals near the late stages of solidification of a magma may cause the remaining liquid to be squeezed into spaces in the surrounding rock. There it can crystallize to form a different kind of rock than that formed in the main body of magma. The process is called *filter pressing.*

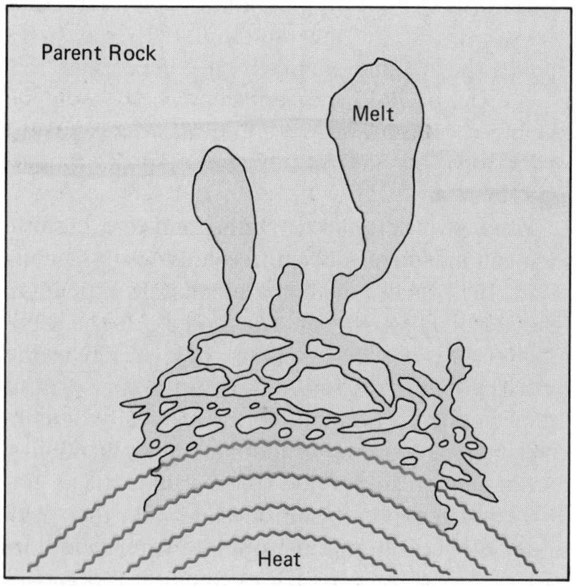

Figure 3–7 Conceptualized diagram of partial melting. The first portion of parent rock to be melted separates because of density difference, leaving residue of parent rock depleted of the separated chemical components.

solid portion. The melting may occur either because of decompression of ascending material from the upper mantle (as along mid-ocean ridges), or by heating of descending material (as along subduction zones). Like fractional crystallization, partial melting can produce rocks of granite composition but in a reverse manner. When a rock is heated and its components begin to melt, the minerals that crystallized last in the reaction series will be the first to melt. This process is illustrated in Figure 3–8 for a granitic rock containing water. The left side of the diagram bordered by the heavy black line represents solid rock, the central part a liquid-crystal "slush," and the right side represents melt containing water. By scanning the graph along the dotted line representing 3 kb pressure, we see that quartz and orthoclase begin to melt at 700°C, biotite at about 840°C, plagioclase at 900°C, and hornblende at nearly 1000°C. The "slush" persists over a wide range of temperatures.

Many other conditions and events can affect the nature of igneous rocks produced by a magma. Certainly the starting materials, be they midoceanic basalts or continental granites, provide limits on the kinds of melts that can be formed. Variety among igneous rocks also results from the mixing

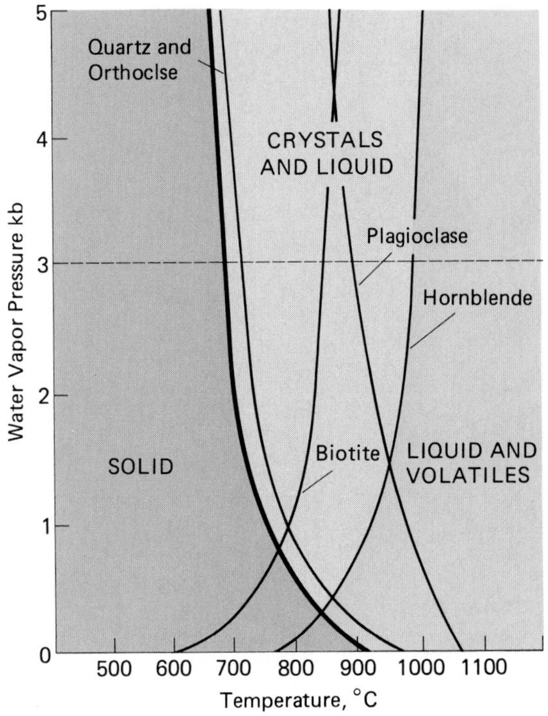

Figure 3–8 Order of melting of minerals in a water-rich granitic rock (granodiorite). (After Wyllie, P., 1971.)

of melts of different composition, and from a process known as **assimilation**. Assimilation is a general term used to describe the many reactions that occur between a rising body of magma and the country rock with which it comes into contact. Unlike either fractional crystallization or partial melting, which begin with a single parent material, both assimilation and magma mixing involve two source materials. The two sources combine to form rocks of intermediate composition.

Fractional crystallization, partial melting, assimilation, and mixing of magmas are important processes that account for diversity among igneous rocks.

The Significance of Texture and Color

Igneous rocks are an excellent example of a group of natural objects that can be classified and named on the basis of descriptive properties that have genetic significance. In the preceding section on the crystallization of magmas, we noted that the mineral composition of an intrusive rock can provide a clue to its temperature of formation as well as the chemistry of the melt from which it solidified. Rocks that are light colored (white, light gray, pink) have a high proportion of such light-colored minerals as quartz, orthoclase, sodium-rich plagioclase, and muscovite. They must have been derived from a melt having a high silica content, perhaps as high as 65 to 75 per cent. Dark-colored rocks are derived from magmas having a lower silica content (40 to 55 per cent) and are typically composed of a predominance of the dark-colored (gray, black, green) ferromagnesian minerals, as well as calcium plagioclases.

The color of igneous rocks is an indicator of their general composition. Rocks dark in color (black, gray, green) tend to be rich in ferromagnesian minerals and calcium plagioclase. Rocks light in color (white, light gray, pink), tend to contain abundant quartz, orthoclase, sodic feldspar, and muscovite.

The texture of an igneous rock can provide clues to that rock's cooling history. Texture refers to the size, shape, and arrangement of the component mineral grains in a rock. A rock having **aphanitic** texture is so finely crystalline that the individual minerals cannot be seen without the aid of a magnifying lens or microscope. In general, an aphanitic rock is considered to have experienced cooling so rapid that there was not sufficient time for the growth of large crystals. Rapid cooling, of course, characterizes volcanic rocks, but some intrusive rocks may also be cooled quickly if they are near the surface or have a broad thin shape and hence a large surface area from which heat may be quickly lost to the surrounding country rock. The extreme product of rapid chilling is represented by a natural glass in which few, if any, crystals have had time to form.

Most large bodies of magma are well insulated by the rocks that surround them and therefore retain their heat for long periods of time. This slow cooling favors the development of relatively fewer seed crystals and favors the growth of large crystals. In lava flows that cool quickly, many crystal nuclei quickly form, leaving fewer ions for each of the nuclei. The result is a rock composed of multitudes of tiny crystals.

The more coarsely crystalline igneous rocks are said to have **phaneritic** texture. Crystals in phaneritic rocks are clearly visible without magnification (Fig. 3–9). Magmas capable of retaining heat for long periods of time so as to develop phaneritic textures are characteristically those that are deeply buried and have abundant water. Water inhibits crystal nucleation by weakening the bonds between silica tetrahedra. At the same time, it lowers viscosity, which permits a rapid movement of ions to the relatively few nuclei. The coarsely crystalline rocks that form as a result of these conditions are known as **pegmatites** (Figs. 3–10 and 3–11).

In general, finely crystalline igneous rocks (aphanitic) cooled rapidly, coarsely crystalline rocks (phaneritic) cooled slowly, and often form magmas with a high water content.

As can be readily imagined, the cooling rates of particular magmas will vary. A deep-seated magma may begin its history by cooling slowly and developing a few large crystals. Then, while only partly crystallized, the magma may move into another environment in which cooling is more rapid and the remaining melt congeals quickly. The result is a rock having larger earlier formed crystals that are surrounded by a later formed ground-mass of smaller crystals. This type of texture is called **porphyritic** (Fig. 3–12), and the large crystals are called **phenocrysts**.

Figure 3–9 Hand specimen of granite, an intrusive igneous rock having phaneritic texture.

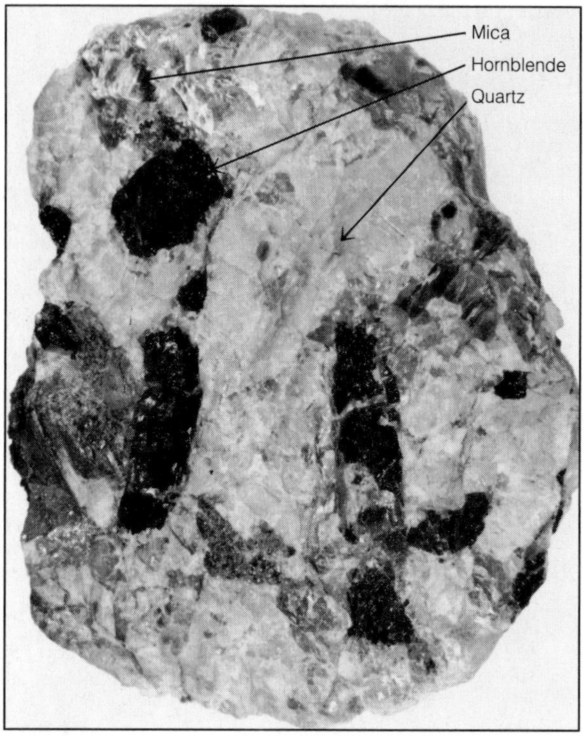

Figure 3–10 Pegmatite. Large crystals of hornblende (black), scaly masses of muscovite mica (upper left), and large intergrown crystals of gray glassy-looking quartz. (Courtesy of the Institute of Geological Sciences, London.)

Identification of Igneous Rocks

Texture and mineral composition are the most useful properties for inferring the origin of igneous rocks and are the basis for classification and identification. Although there are scores of minor igneous rock types, these represent variations of the relatively few major groups shown in Figure 3–13. In a widely used simple scheme, there are ten groups of crystalline igneous rocks: granite, grano-

Figure 3–12 Porphyritic texture in polished ornamental posts encircling the entrance to St. Paul's Cathedral in London. Large phenocrysts are orthoclase immersed in a more finely crystalline granite groundmass.

Figure 3–11 Pegmatite dikes in dark metamorphic rock. Front Range, near Parkdale, Colorado.

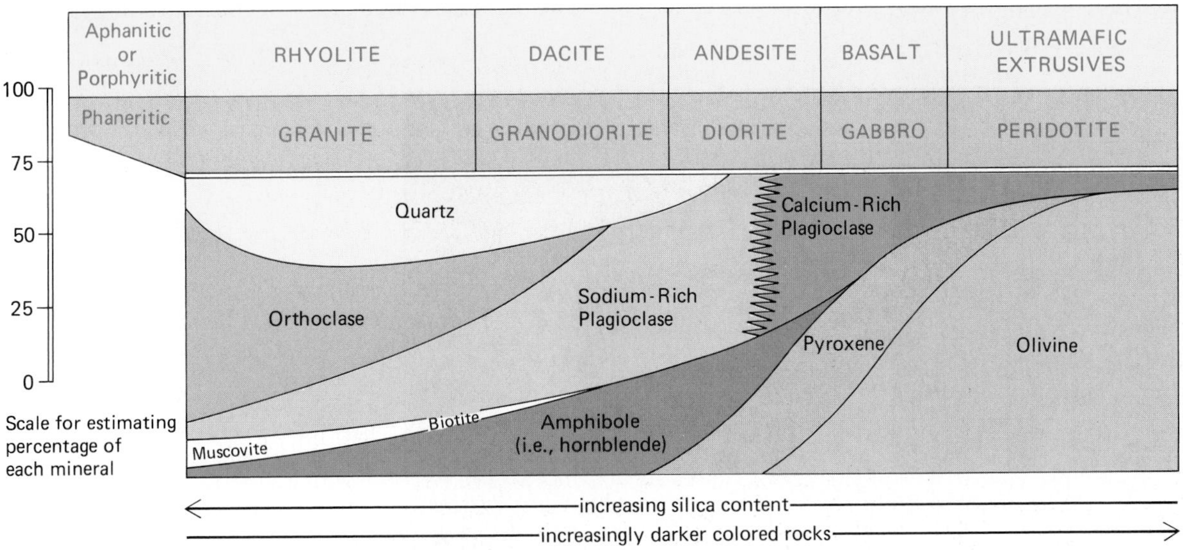

Figure 3–13 Diagram illustrating the mineralogic composition and textures of eight common igneous rocks. In order to ascertain the composition of a rock, estimate the percentages of each mineral beneath its name by reference to the percentage scale. (Adapted from Dietrich, R.V., *Virginia Minerals and Rocks,* Virginia Polytechnical Institute.)

diorite, diorite, gabbro, peridotite, rhyolite, dacite, andesite, basalt, and ultramafic extrusives.

In order to identify rocks in each of these groups, it is necessary to recognize texture and a few of the mineral constituents as they appear in the rock itself. To better identify the more common minerals, hold the rock in strong light and turn it back and forth so that you can see reflections from cleavage planes and crystal faces. Quartz can be recognized by its glassy appearance and mostly irregular reflecting surfaces. Feldspars show cleavage faces, may be white, gray, or pink and often have rectangular outlines. One can often identify the feldspar as plagioclase by thin parallel striations that develop on one cleavage face. Ferromagnesians are black or green. Among ferromagnesians, hornblende is more shiny and elongated than augite, and biotite may be distinguished by its smooth gleaming cleavage surfaces.

Granitic Rocks

The light-colored igneous rocks are, as we have already noted, derived from high-silica magmas. When such magmas crystallize, they yield a high proportion of orthoclase feldspar and quartz. Granite (see Figs. 3–9 and 3–14) is a familiar representative of this group and is recognized by its light color, phaneritic texture, and the presence of about 25 per cent quartz grains along with larger quantities of potassium feldspar (orthoclase) and sodium plagioclase. Muscovite and biotite are present, as are small quantities of hornblende. **Granite**, of course, is well known to most of us because it has been used for centuries in the construction of buildings, statues, and monuments. It is not, however, the most abundant granitic rock. **Granodiorite**, a somewhat less siliceous granitic rock, is more abundant. The igneous rocks of the Sierra Nevada are composed largely of granodiorite. As shown in Figure 3–13, the fine-grained equivalent of granodiorite is **dacite**.

The more finely crystalline but compositional equivalent of granite is **rhyolite** (see Fig. 4–21). Actually, the basic aphanitic texture of rhyolite may range into the glassy condition. Many rhyolites tend to be porphyritic, with small phenocrysts of orthoclase, sodic plagioclase, quartz, and mica. Some appear to be welded tuffs (a rock formed of small, hot volcanic particles that have been indurated by heat). In color, rhyolites tend to be white, light gray, pink, or orange.

Dioritic Rocks

Igneous rocks of the diorite clan include diorite itself and its aphanitic equivalent known as andesite. **Diorite** (Fig. 3–15) is a coarse- to fine-grained

Figure 3–14 Thin section of granite viewed with polarized light. "Scotch plaid" grains are orthoclase; brown or brightly colored grains are biotite; and gray-to-clear grains are quartz. Width of field 9 mm.

Figure 3–15 Diorite.

phaneritic rock containing less silica than granite or granodiorite but more than gabbro. Thus, in the gradations of composition across the rock chart, it is an intermediate rock type. The most abundant minerals are plagioclases, and these are compositionally in the mid range and contain both calcium and sodium. Quartz and orthoclase are very rare or absent in diorites. Among the ferromagnesian constituents, hornblende occurs in amounts about equal to plagioclase, and biotite mica is a common accessory mineral. The color of diorite is controlled by its compositional makeup of nearly equal amounts of black ferromagnesians and grayish plagioclase. Thus, it tends to be rather drab gray or greenish gray in color.

The somber color of diorite prevails also in **andesite**—its aphanitic equivalent. Andesites, which were first identified in the summit volcanoes of the Andes Mountains of South America, are mostly found along the more mountainous continental borders and island arcs that surround the Pacific Ocean. Andesites are intermediate between basalts and rhyolites in composition and also intermediate in relative abundance on the earth's surface.

Gabbroic Rocks

In **gabbro** (Fig. 3–16), the chief mineral is calcium plagioclase, which is darker and more calcium-rich than the plagioclase found in diorite. The typical ferromagnesian constituents are pyroxenes and olivine. Gabbros are typically coarsely crystalline plutonic rocks (Fig. 3–17). One variety called anorthosite is composed mostly of intergrown calcium-rich plagioclase crystals. Anorthosite is characteristic of the lighter-colored crustal areas of the moon. Varieties of gabbro are frequently polished and used as ornamental stone because of their rich dark hue and the iridescent play of colors reflected by the plagioclase crystals.

Basalt is the aphanitic equivalent of gabbro and is the most abundant extrusive igneous rock. Basalts are mostly black or very dark gray and there-

Figure 3–16 Thin section of gabbro viewed with polarized light. Gray crystals are plagioclase, colored crystals are ferromagnesian minerals. Width of field 30 mm.

Figure 3–17 Coarsely crystalline hand specimen of gabbro. Finer-grained examples are often darker in color.

fore darker than andesites. In thin section, they are found to contain myriads of elongate lathlike crystals of calcic plagioclase interspersed with pyroxene and olivine. Biotite and hornblende, which are common minerals in andesite, are sparse or absent in basalt.

Ultramafic Rocks

Ultramafic rocks are characterized by high density (3.3 g/cm^3), low percentage of silica (45 per cent), and a major mineral composition of pyroxenes and olivine. **Peridotite**, in fact, is an ultramafic rock composed of 70 to 90 per cent olivine, while **pyroxenite** is almost entirely an intergrowth of pyroxene crystals. The ultramafic rocks are not common near or on the surface of the earth. They are occasionally encountered in the lower parts of magmatic bodies where dense ferromagnesian minerals accumulated by fractional crystallization and gravity settling. At several locations around the world, peridotites have pushed their way upward from great depths and penetrated near-surface rocks. Geologists are keenly interested in these peridotites, for they may represent samples of the earth's mantle.

The progression from light-colored (high silica) to dark-colored (low silica) phaneritic igneous rocks is granite, diorite, gabbro, and peridotite.

Kinds of Intrusive Rock Bodies

Bodies of intrusive igneous rocks that have solidified from magmas located deep within the crust are called **plutons**. The shape, size, and relation of plutons to surrounding rock is quite varied. Some represent injections of once molten rock parallel to preexisting strata. Such plutons are said to be **concordant**. In contrast, a pluton that cuts across layering is spoken of as **discordant** (Fig. 3–18).

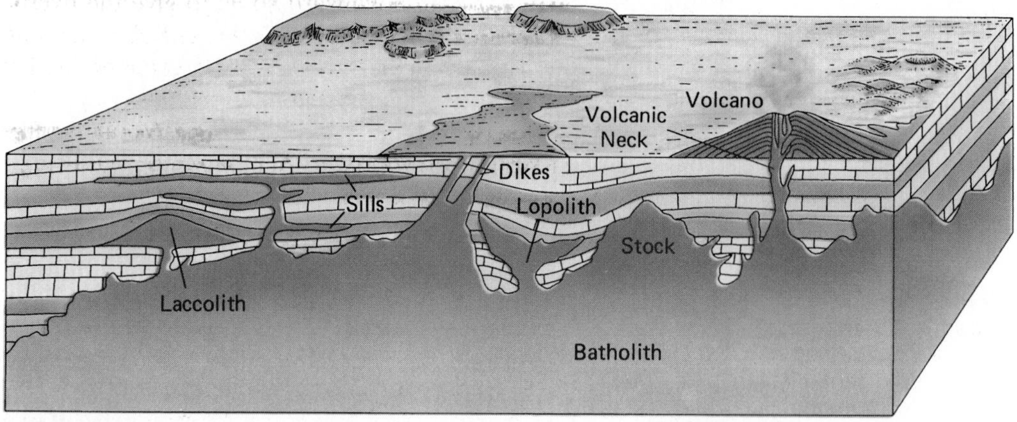

Figure 3–18 Nomenclature for bodies of intrusive igneous rocks. The discordant bodies include dikes, necks, stocks, and batholiths. Concordant bodies include sills, laccoliths, and lopoliths.

Figure 3–19 A dark-colored basaltic sill of Precambrian age on Banks Island, Northwest Territories. To make room for itself, the magma forced overlying rock upward and expanded a fracture where it changed level. (Courtesy of Geological Survey of Canada.)

Tabular Plutons: Sills and Dikes

A tabular body of rock is one shaped like a book or board, that is, having small thickness relative to its length and width. There are several kinds of tabular plutons. The concordant ones, which have been injected along bedding planes of sediment or between older lava flows, include **sills** (Fig. 3–19) and **lopoliths**. Sills may be either horizontal or tilted depending on the attitude of the enclosing beds. They range in thickness from only a centimeter to hundreds of meters. The massive cliff called the Palisades (Fig. 3–20), which faces New York City on the west side of the Hudson River, is a 300-m-thick sill. The molten rock that formed the sill was intruded parallel to layers of sandstone, so that both the upper and lower contacts cooled rapidly, forming aphanitic chilled zones. The thick molten layer between the chilled margins cooled more slowly, allowing time for the formation of larger crystals. The first of these to form was olivine, which by gravity settling, sank through the melt to accumulate as a layer above the lower chilled zone. A zone of plagioclase and pyroxene formed above the olivine layer, followed by an upper zone composed mostly of plagioclase.

Because both sills and lava flows are sheetlike and parallel to adjacent beds, the distinction between a flow that has been buried by later strata and a sill can be difficult to make. There are, however, a few characteristics that are useful in distinguishing between the two features (Fig. 3–21). Flows bake only the stratum beneath them, may fill the cracks only of the underlying bed, and in their upper portions may contain vesicles (small cavities made by gas bubbles in the lava). In contrast, sills bake the stratum above as well as the one below, penetrate cracks in both underlying and overlying beds, and rarely have vesicular borders.

Lopoliths are tabular plutons in which the roof and floor sag downward so as to give the overall shape of a bowl. An example is the Duluth lopolith which is estimated to have a diameter of 250 km and a thickness of 15 km.

Dikes (Figs. 3–22 and 3–23) are tabular bodies that are *discordant* in that they cut across preexisting rock layers. Dikes may be many kilometers long and range in thickness from paper-thin to tens of meters. Many dikes branch and cross one another so as to form a complex system called a *dike swarm.* They also occur in circular patterns called ring dikes. Ring dikes are thought to develop when lava rises into concentric fractures created around an area where overlying rocks sink into a magma chamber below. Although dikes may be composed of nearly any kind of igneous rock, most

(Text continues on p. 68.)

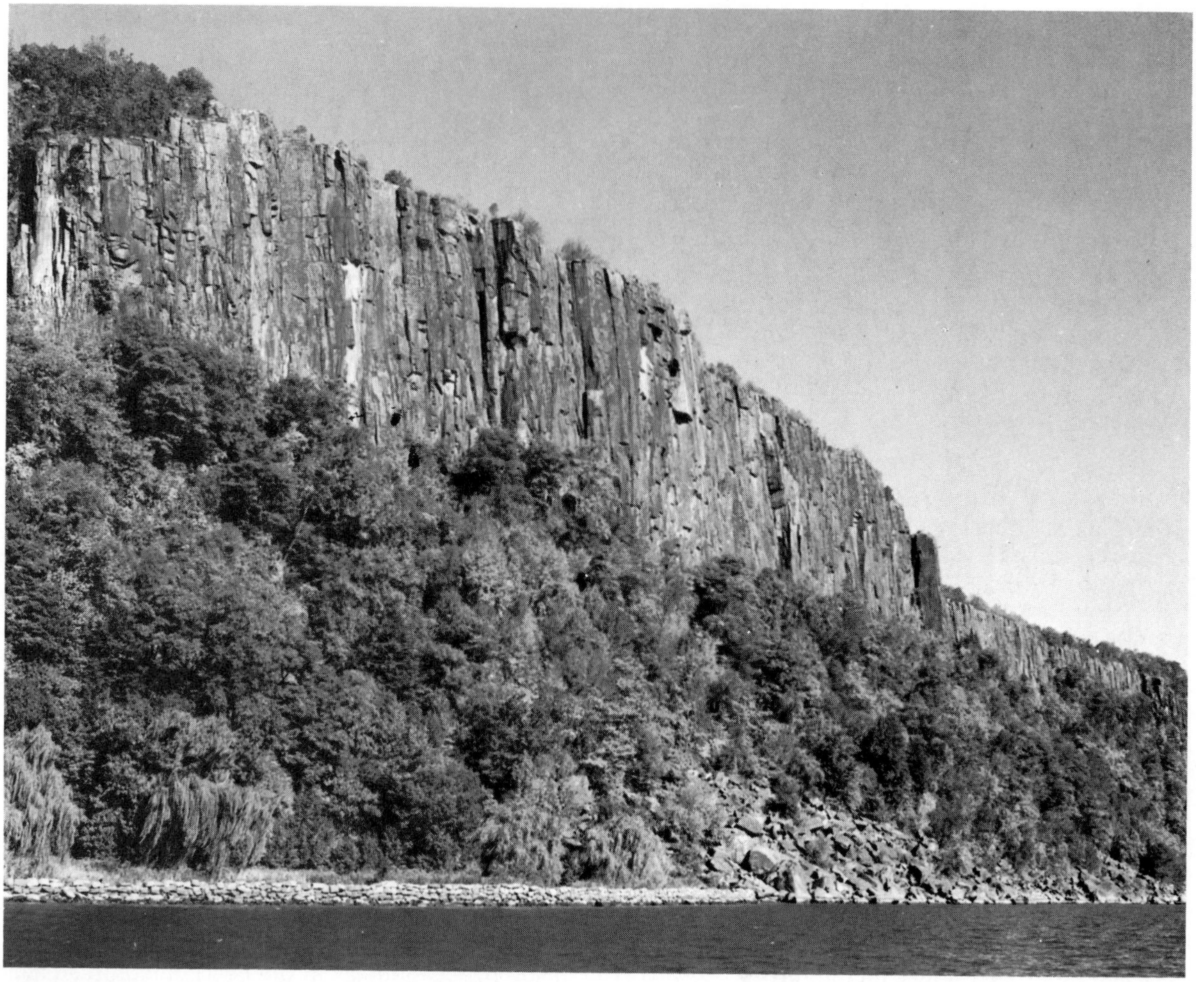

Figure 3–20 The Palisades Sill exposed along the banks of the Hudson River in New York. Note the well-developed vertical columnar jointing. (Courtesy of Palisades Interstate Park Commission.)

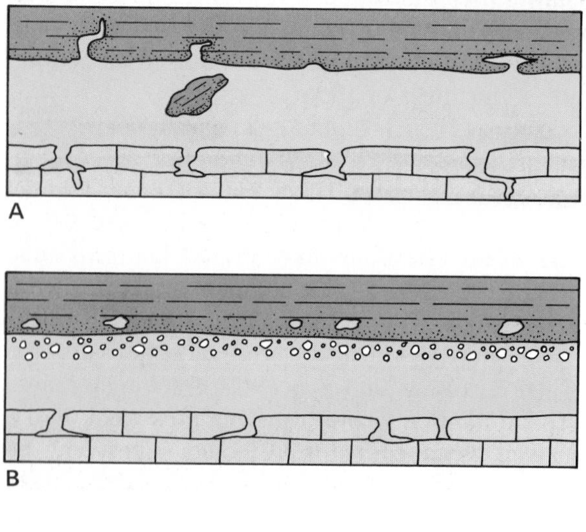

Figure 3–21 Differences between a sill (A) and a lava flow (B). Sills contain few if any vesicles, bake overlying and underlying rock, may contain inclusions of overlying rock, and send extensions of once-molten rock into beds above and below. Flows are vesicular, especially in their upper portions. They bake only the underlying rocks, and often pieces eroded from the flow will be found in the layer deposited on top of it.

Figure 3–22 A dark-colored basaltic dike cuts through lighter-colored rhyolitic rock (southeastern Missouri).

Figure 3–23 Basaltic dikes of early Proterozoic age, Coronation Gulf, Northwest Territories, Canada. (Photograph by Paul Hoffman.)

are basaltic in composition and were derived from magmas of low viscosity.

Batholiths

In many areas of former intrusive igneous activity, great discordant nontabular masses of intrusive rock occur. The largest of such bodies form the cores of great mountain ranges. They are known as **batholiths** (Fig. 3–24). Among the better known examples are the Idaho batholith, the Coast Range batholith of British Columbia, and the Sierra Nevada batholith of California. Each of these immense bodies of igneous rock actually represents the coalescence of several smaller intrusions of granites and granodiorites.

To be termed a batholith, the intrusive body must have an areal extent greater than 100 km^2. In shape, batholiths may be irregular or roughly cylindric. It is difficult to determine the depth to the base of existing batholiths, for at no place on earth is the bottom of a batholith exposed for study. However, investigations based on gravity measurements and earthquake data suggest that batholiths begin to form at depths of about 30 km. As intrusive masses, batholiths were once deeply covered by pre-existing rocks. We see batholithic rocks at the surface of the earth today only because great thicknesses of covering rocks have been removed by erosion. Batholiths are usually located in present or former mountainous belts, and such regions are subject to frequent episodes of uplift, which in

Figure 3–24 The massive cliffs in the Yosemite Valley, California, are composed of granitic rocks (granodiorite) that are part of the great Sierra Nevada batholith. El Capitan is the cliff on the left. On the right one can see Bridal Veil Falls.

turn greatly increase the rates of erosion needed to expose the batholiths.

Batholiths intrude into **country rock** (pre-existing rock) in a variety of ways. In some places there is evidence that the melt has been forcefully injected. Elsewhere, the heat of the magma appears to have been sufficient to melt surrounding country rock so that the magma melted its way upward. Such a process would, of course, change the composition of the magma. **Stoping**, a process by which magma moves upward as blocks of country rock are wedged loose and fall into the magma chamber, provides another mechanism for batholiths to work their way upward. The upward migration of magma may also be aided by density differences. Magma that is less dense than the enclosing rock tends to rise buoyantly toward the surface.

As the magma makes its way upward, pieces of the enclosing country rock may spall off and fall into the melt. Some of these fragments may melt and be assimilated by the magma. Included fragments of unmelted country rock are called **xenoliths**, meaning "stranger rocks" (Figs. 3–25, 3–26). Xenoliths also occur in lavas when pieces of rock are torn from the sides of volcanic vents and fissures during eruptions.

Laccoliths (see Fig. 3–18) are massive concordant plutons that have a mushroom shape. Like sills, they are formed by magma that has been injected into bedding planes. However, unlike sills, they dome up the overlying rock layers. Some laccoliths represent bulges on the surface of a sill, whereas others are self-contained structures fed by a vent from an underlying magma.

Concordant igneous bodies conform to the bedding planes of the country rock they have intruded. Sills, lopoliths, and laccoliths are examples. Igneous bodies that cut across bedding planes in country rock are discordant, and include dikes and batholiths.

Figure 3–25 Well developed foliation resulting from parallel alignment of biotite grains. The rock type is biotite schist and occurs as xenolith in granite (the surrounding lighter-colored rock). (From an exposure at Wadi al'Arabah, Jordan, courtesy of F. Bender and U.S. Geological Survey.)

Figure 3–26 Large xenoliths of the metamorphic rock schist engulfed in the intrusive igneous rock granite. The schist xenoliths can be recognized by their smooth, slab-like appearance. (Courtesy of L.E. Davis.)

The Global Distribution of Granitic and Basaltic Rocks

One of the most interesting observations about the global occurrence of igneous rocks is that granitic plutons occur within or on the edges (or former edges) of continents and that the crust beneath the ocean basins is practically devoid of granites and kindred intrusive rocks. This indicates that the crust of the continents and ocean basins originates by fundamentally different processes. The igneous rocks of the ocean basins are predominantly basalts that have apparently been generated by partial melting within the lower crust or upper mantle, as indicated by the high temperatures at which these basaltic melts are extruded and by their relatively rich ferromagnesian composition. The process of building the oceanic crust is going on continuously as prodigious amounts of basalt pour out of fissures along the world-encircling network of mid-oceanic ridges. These ridges are developed where tectonic plates separate and permit magma to well up and form new oceanic crust, which then moves away from the ridges at the rate of a few centimeters per year.

Because of the abundance of silica and aluminum and relative deficiency of iron and magnesium in continental crust, it could not have been derived directly from the upper mantle. Continental crust, however, can be formed from the melting of mixtures of ocean basalts and sediments. Weathering processes tend to remove iron and magnesium and concentrate silica and aluminum in sediments, and if the sedimentary materials can be melted, they may generate silica- and aluminum-rich granitic rocks. A likely place for this process to occur is along subduction zones where a lithospheric plate moves downward, carrying with it not only basalts of the ocean floor but water-saturated marine sediments as well. When these descending slabs reach the asthenosphere, partial melting occurs, and magmas of the kind that form granodiorites and andesites are produced.

The oceanic crust is composed of basalt formed by partial melting within the lower crust or upper mantle. The granitic character of the continental crust is often a consequence of partial melting of mixtures of sediment and basalt within subduction zones.

Summary

Igneous rocks are those that have cooled from a molten state. The melts from which igneous rocks are derived are called magmas while still beneath the earth's surface and lavas when they find their way to the surface through fissures and cylindric openings. Such openings, and the resulting buildup of extruded materials, form volcanoes. Rocks that solidify from lavas are frequently called *extrusive,* whereas those that crystallize at depth comprise the *intrusive* or plutonic family of igneous rocks.

Intrusive or plutonic rocks have a coarser texture than extrusive rocks. They solidified deep beneath the earth's surface in variously shaped masses called plutons. The coarser texture of plutonic rocks is a consequence of their having cooled slowly. We are able to examine plutons today, only because uplift and subsequent deep erosion have uncovered some of them.

The magma from which igneous rocks are formed consists mainly of the elements oxygen, silicon, aluminum, iron, calcium, sodium, potassium, magnesium, and titanium, as well as dissolved gases and water. From these basic ingredients are derived the common minerals of igneous rocks. These include feldspars, quartz, micas, hornblende, augite, and olivine. Quartz, feldspars, and ferromagnesian minerals are especially useful in the classification of intrusive rocks. For example, a coarsely crystalline *(phaneritic)* rock composed of quartz, orthoclase, sodic plagioclase, and a ferromagnesian mineral like hornblende or biotite would be granite. A phaneritic rock lacking quartz in which ferromagnesian minerals and calcic-plagioclase predominated would be a *gabbro.* Rocks having a composition intermediate between gabbro and granite might be classified as *diorite.*

The deep-seated igneous processes that produce plutons are far more difficult to study than igneous activity revealed to us directly by volcanic action. However, geophysical and field studies, procedures involving experimental melting and crystallization of rock materials in the laboratory, and trace element analyses have increased our understanding of magmas and the causes of diversity among igneous rocks. Among the mechanisms that provide for a variety of rock types are *fractional crystallization, partial melting, gravity settling,* and *assimilation*. Fractional crystallization is a process whereby minerals forming in a cooling magma are successively removed at progressively lower temperatures. The process by which a preexisting rock is incompletely melted and molten fractions are removed is called partial melting. Assimilation involved modification of the chemistry of a magma as a result of melting and assimilation of surrounding rocks.

Igneous rocks of the continents are basically different from those of the ocean floors. The dominant rock type of the ocean basins is basaltic, whereas that of the continents is granitic. *Oceanic basalts* are believed to be derived from magmas generated by partial melting within the asthenosphere. Massive extrusion of basaltic rocks occur along midoceanic ridge systems. *Granitic rocks* are apparently formed from magmas generated by the melting or partial melting of quartz-rich sedimentary and metamorphic rocks that have been deeply buried and strongly compressed during mountain-building activity. Granitic rocks like granodiorite may also result from the partial melting of a combination of basaltic crust and sediment as oceanic lithospheric plates plunge downward into the mantle at subduction zones.

Terms to Understand

andesite
aphanitic texture
assimilation
basalt
batholith
Bowen Reaction Series
concordant pluton
country rock
dacite
dike
dunite
discordant pluton
extrusive igneous rock
ferromagnesian mineral
filter pressing
fractional crystallization
gabbro
granite
granodiorite
gravity settling
igneous rocks
intrusive igneous rocks
laccolith
lava
lopolith
magma
partial melting
pegmatite
peridotite
phaneritic texture
pluton
pyroxenite
rhyolite
sill
stoping
ultramafic igneous rock
volatiles
xenolith

Questions for Review

1. What is the sequence of minerals in the Bowen Reaction Series? If an igneous rock contains quartz, what other minerals would you expect it to contain by reference to the Bowen Reaction Series? If it contained pyroxene, what companion minerals might be expected?

2. How may the mineral composition of an igneous rock reveal whether it was derived from a high-silica or low-silica magma?

3. For each of the following rocks, what inferences can be made as to cooling history and type of magma from which it is derived:
   a. Rock consisting of crystals of pyroxene in a dark fine-grained groundmass.
   b. Rock consisting of crystals of feldspar in a reddish fine-grained groundmass; no ferro-magnesian minerals visible.
   c. Rock having a coarsely crystalline texture, consisting of feldspar, quartz, and mica.

4. What is the difference between basalt and gabbro? What minerals account for the dark color in these rocks?

5. Upon what two factors does the classification of igneous rocks depend? Which of these factors are most important in distinguishing andesite from basalt? Granite from rhyolite?

6. Distinguish between a dike and a sill and between a sill and a lava flow.

7. Describe a theory that would account for the existence of large granitic batholiths such as the Sierra Nevada batholith.

8. Assuming magma forms by local melting of the earth's crust or upper mantle, why does it tend to rise toward the surface?

9. Summarize several possible mechanisms that would permit different kinds of igneous rocks to form from an initially basaltic melt.

10. How would you explain the olivine-rich layer that commonly occurs near the base of thick sills?

11. Draw and label a geologic cross section showing part of a batholith that is associated with the following features:
    a. Xenolith.
    b. Rhyolite dike that is older than the batholith.

12. In general, what is the effect of dissolved volatiles on the melting temperatures of silicate minerals? What is the effect of pressure?

Supplemental Readings and References

Barker, D.S., 1983. *Igneous Rocks.* Englewood Cliffs, N.J., Prentice Hall, Inc.

Best, M.G., 1982. *Igneous and Metamorphic Petrology.* San Francisco, W.H. Freeman & Co.

Ernst, W.G., 1969. *Earth Materials.* Englewood Cliffs, N.J., Prentice-Hall, Inc.

Maaloe, S., 1988. *Principles of Igneous Petrology.* New York, Springer-Verlag, Inc.

Tennissen, A.C., 1974. *The Nature of Earth materials.* Englewood Cliffs, N.J., Prentice Hall, Inc.

Volcanic Activity

4

Majestic Mount St. Helens, one year prior to its devastating 1980 eruption. (Courtesy of L.E. Davis.)

The Montagne Pelée presents no more danger to the inhabitants of Saint Pierre, than does Vesuvius to those of Naples.

Professor Landes of St. Pierre College the day before he and 30,000 residents of the town perished in the eruption of Mount Pelée

Introduction

Volcanoes are our most spectacular proof that there is sufficient heat in the earth beneath us to melt rocks. Nearly everyone would recognize a volcano if he or she saw one, yet volcanoes are not easily defined. A **volcano** is a vent in the earth's crust through which molten rock, steam, gas, and ash are expelled. The fragments expelled and the solidified molten rock usually form a mound or peak around the vent. For geologists, a volcano is a dynamic, usually accessible, natural laboratory where they can obtain valuable information about the earth's interior and the way in which igneous melts solidify to form rocks.

Volcanic activity (Figs. 4–1 and 4–2) has been an important process on this planet throughout the long span of geologic time. In the earliest stages of the earth's history, volcanoes contributed their products to form parts of the earth's crust. The gases and vapors emitted by the blistering array of primordial volcanoes provided the

Figure 4–1 Spectacular arching lava fountain, Kilauea's east rift eruption of 1983. Hawaiian Volcanoes National Park. (Courtesy of the U.S. Geological Survey; photograph by J.D. Griggs.)

Figure 4–2 Lightning strikes during the eruption of Galunggung volcano, West Java, Indonesia, December 3, 1982. (Photo by Ruska Hadian, volcanology survey of Ruska Hadian, volcanology survey of Indonesia. Courtesy of the U.S. Geological Survey.)

substance for the earlier atmosphere and the oceans. Perhaps volcanic eruptions are less frequent today than in those primordial times, yet nearly 800 volcanoes are active today or are known to have been active in historical times. We view some of the present-day eruptions on television and witness in safety the devastation they cause. Volcanoes evoke not only awe and fear, but also respect for the enormous forces that operate uneasily beneath the frail crust on which we live. Our planet is not a great lifeless blob of cold rock making an annual trip around the sun. It is constantly changing and vibrant with igneous activity. During 1979 alone, the Scientific Event Alert Network at the Smithsonian Institution in Washington, D.C., provided daily observations of eruptions in Japan (Usu volcano), the Philippines (Mayon volcano, Fig. 4–3), Guatemala (Fuego volcano), the Aleutians (Shishaldin and Westdahl volcanoes), and Mount St. Helens.

Figure 4–3 Mayon volcano in the Philippines. The summit of this composite volcano rises nearly 8000 feet above sea level. (Courtesy of L.E. Davis.)

Observations and measurements of volcanoes provide a basis for hypotheses about plate tectonics and the origin of the hydrosphere and atmosphere.

Two Notable Eruptions

Civilized humans have lived on this planet for only a few thousand years. During this time, many noteworthy volcanic eruptions have occurred. Nearly all of the recent eruptions and a few of the older ones have been documented and are a part of recorded history. Many of the most ancient accounts of volcanic activity are associated with superstition and mythology. Some, like Plato's story of the eruption of Santorini, contain just enough information to be intriguing.

Santorini

The islands of Santoria in the Aegean Sea mark the location of an eruption of particular interest to historians. This eruption, which occurred about 1470 B.C., may have destroyed a highly developed Minoan culture that had spread from Crete. Because the eruption resulted in explosive destruction of most of the island of Santorini, it may very well have given rise to the ancient legends of Atlantis, the "lost continent." The only known account of this event was written by Plato in 355 B.C. He described a city called Atlantis that was built on a circular plan about 24 km in diameter. The city was on an island that sank beneath the sea during a war with the Athenians. During this war, wrote Plato, "there occurred portentious earthquakes and floods, and one grievous day and night befell them when the whole body . . . of warriors was swallowed up by the earth, and the island of Atlantis in like manner was swallowed up by the sea and vanished."

Today, the name Santoria applies to a small group of islands that enclose a nearly circular bay about 10 km in diameter (Fig. 4–4). The islands are remnants of the large volcanic cone that was the probable source for Plato's story, and the bay is a submerged area that formed by collapse of the central part of the volcano. Careful investigations by archaeologists working in this area have revealed a story of volcanic and earthquake activity to rival that of Pompeii. The major eruptions were apparently preceded by earthquakes. These tremors so terrified the local inhabitants that they gathered their possessions and fled to safer ground. This exodus probably accounts for the absence of human skeletons in the excavations of the ruins of the principal town, Akrotira. The eruption itself was of enormous magnitude. Towns and villages were completely buried in ash, and great oceanic waves called **tsunami**, generated by earthquakes accompanying the eruption, ravaged towns all around Crete. Many believe these catastrophic events contributed to the eventual demise of the Minoan civilization.

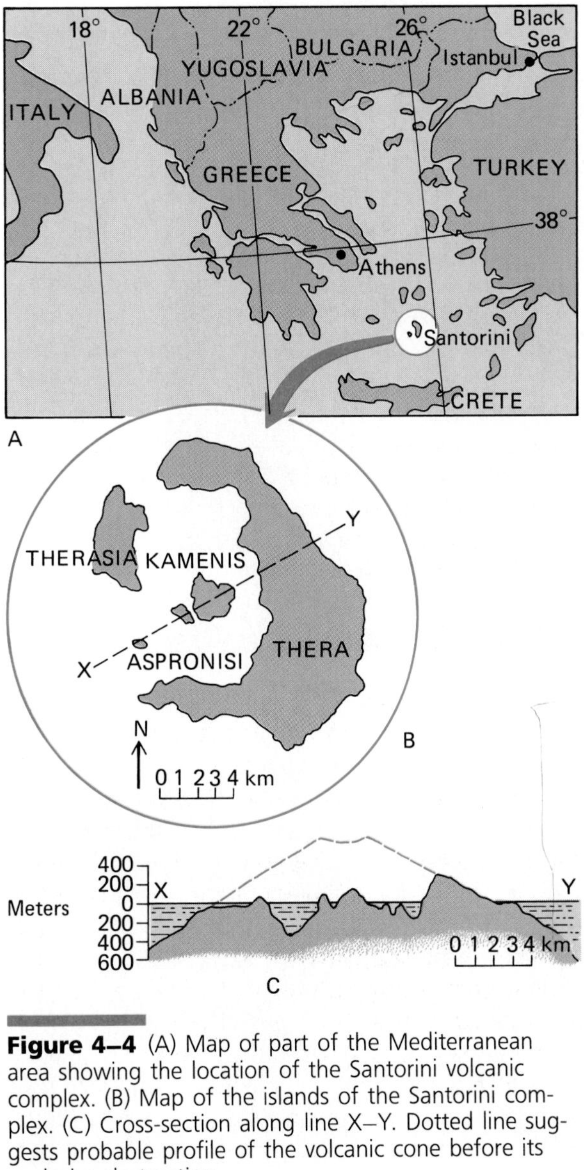

Figure 4–4 (A) Map of part of the Mediterranean area showing the location of the Santorini volcanic complex. (B) Map of the islands of the Santorini complex. (C) Cross-section along line X–Y. Dotted line suggests probable profile of the volcanic cone before its explosive destruction.

Mount St. Helens

Until recently many mainland Americans preferred to believe that volcanic eruptions were hazardous geologic events that occurred mostly elsewhere. This belief was shattered at 8:31 A.M. on May 18, 1980, when Mount St. Helens in Washington (Fig. 4–5) exploded with the energy equivalent of 50 million tons of TNT and a roar that was heard 300 km away from the once scenic mountain (Figs. 4–6A–E). A great turbulent cloud of hot gases, steam, pulverized rock, and ash burst later-

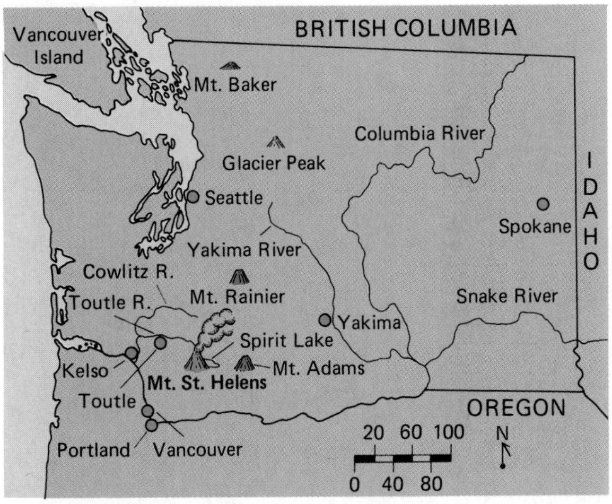

Figure 4–5 Location map, Mount St. Helens volcano in southeastern Washington State.

A

B

C

D

E

ally from the north side of the mountain, followed by a vertical column of gas and ash that rose 20,000 meters into the atmosphere, and began to drift slowly toward the east. An estimated 1 km^3 of airborne ash and other rock debris from the explosion blocked out the sun's light and caused automatic street lights to switch on in towns hundreds of kilometers downwind from the rumbling volcano. Turbulent hot masses of gas and ash surged down the mountain sides, flattening millions of trees in scorched irregular rows. Meanwhile, the ash fell like a dismal gray snow, blanketing streets, dangerously burdening the roofs of buildings, choking the engines of vehicles, and covering the leaves of trees and crops. Hot gas and ash from the volcano melted part of St. Helens' snow and ice cap, and the resulting meltwater mixed with ash and formed mudflows, which surged down the mountain slopes at speeds as great as 80 km per hour. The steaming turbulent mass of debris crashed through the once beautiful valley of the Toutle River and demolished 123 homes in the nearby town of Toutle. People camping or working near the volcano were killed by heat, gases, or burial under the downpour of ash. The May 18 eruption was clearly a major volcanic event. The fact that it was actually rather puny when compared with the famous eruptions of Krakatoa (an 1883 eruption in Indonesia) or Santorini (described above) was no consolation to the citizens of Washington as they viewed the devastated gray landscape, the hundreds of square kilometers of valuable trees toppled like matchsticks, buried roads, harbors choked with ash sludge, and smothered fields of wheat.

Mount St. Helens had been dormant since 1857. Its period of rest ended on March 27, 1980, when the peak emitted the first puffs of ash and gas. It seemed a warning signal. Intermittent activity of a relatively mild nature occurred between March 27, 1980, and the massive explosion of May 18 that obliterated the upper 400 meters of the once beautifully symmetric summit (see Fig. 4–6E). In the wake of the May 18 explosion, there were smaller but nevertheless impressive eruptions of ash.

The recent activity at Mount St. Helens and the older eruptions that gave us the volcanic peaks of the Cascades are surface manifestations of an ongoing collision between two of the earth's crustal plates. One of these is the American Plate, and the other is the small Gorda Plate or Juan de Fuca Plate of the eastern Pacific, which moves eastward on a collision course toward the coasts of Oregon and Washington. The Gorda Plate plunges beneath the coastline, and molten rock generated as the plate moves downward rises to supply the lava of the volcanoes (Fig. 4–7).

Volcanoes and Plate Tectonics

The majority of volcanoes that have been active within recorded history occur along somewhat discontinuous linear zones that extend for thousands of kilometers around the globe. The longest of these encircles the basin of the Pacific Ocean, and has been described as the "ring of fire." On the east side of the Pacific, the ring of fire includes such familiar volcanoes as Katmai in Alaska; Lassen, Hood (Fig. 4–8), Shasta, and Rainier in the United States; Popacatepetl and Paricutin in Mexico; and

Figure 4–6 Mount St. Helens erupts. (A) An aerial view of ash-laden Mount St. Helens on March 27, 1980. The craters visible at this time were destroyed during the subsequent eruption of May 18, 1980. (B) Photograph taken on April 10, 1980, showing the volcano from the east at a time when several steam and ash eruptions were in progress. This activity occurred about two weeks after the initial eruption of March 27 and five weeks before the great blast of May 18. (C) The massive May 18, 1980, eruption, during which clouds of steam and ash ascended to an altitude of over 20,000 meters. (D) A view from the northwest photographed nine hours after the initial explosion shows the destruction of part of the rim of the crater. (E) The new crater of Mount St. Helens as it appeared on May 23, five days after the great May 18 eruption. By projecting the slopes diagonally upward, one can obtain a rough idea of the amount of the mountain's northern side that was obliterated. (Photographs A–D, Austin Post, U.S. Geological Survey; E, Keith and Dorothy Stoffel, Washington State Division of Geology and Earth Resources.)

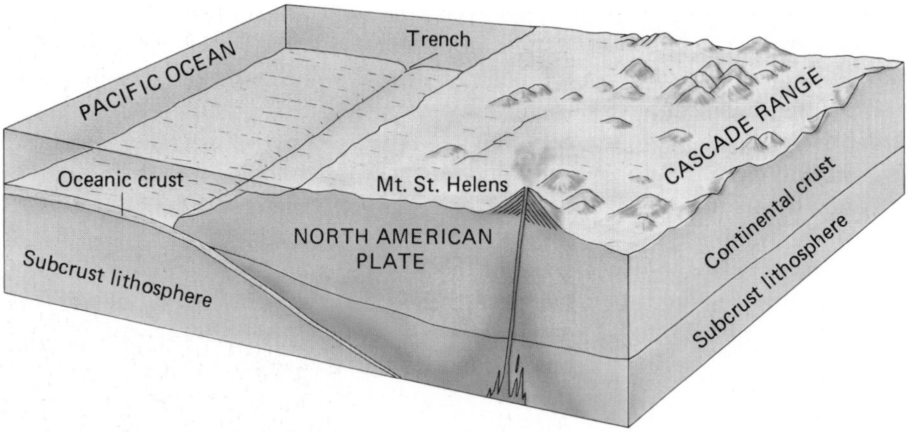

Figure 4–7 As the small Gorda Plate plunges beneath Oregon and Washington, molten rock is generated, which supplies the volcanoes of the Cascade Range.

Fuego in Guatemala. In addition to these highly visible volcanoes, however, there is another group that erupt unseen beneath the ocean along the great mid-oceanic ridges. Although most of these volcanoes are below sea level, a few have built themselves upward above the sea surface to form islands. Surtsey, south of Iceland is such an island that first appeared at the ocean surface in 1963. Iceland has a total of 22 similar volcanoes. Farther to the south in the Atlantic, the Azores have a similar origin.

Figure 4–8 Mt. Hood in Oregon is termed a *composite volcano* because it is composed of interbedded lava flows and layers of ash and cinders. Many volcanoes known for their symmetry and beauty are composite cones such as this one. The summit of Mt. Hood is at 11,245 feet above sea level.

Finally, there are a lesser number of volcanoes that occur between, rather than along, the linear zones. These so-called intraplate volcanoes are apparently related to rising plumes of mantle material beneath the crust.

The explanation for these global patterns of volcanoes is provided by plate tectonics, to which you were introduced in Chapter 1. The majority of volcanoes are located along the margins of tectonic plates (Fig. 4–9). The two kinds of plate margins that account for that volcanic activity are convergent boundaries that include subduction zones and divergent boundaries where sea-floor spreading occurs. As conditions differ in each of these kinds of plate boundaries, so too does the nature of volcanic activity.

Convergent Boundary Volcanism

As noted above, convergent boundaries are the sites of subduction zones. Here lie the oceanic trenches and adjacent volcanic island arcs or volcanically active continental margins. Along the subduction zones, slabs of oceanic crust bend and move downward into the upper mantle. When the slab has descended to depths of about 125 km, the

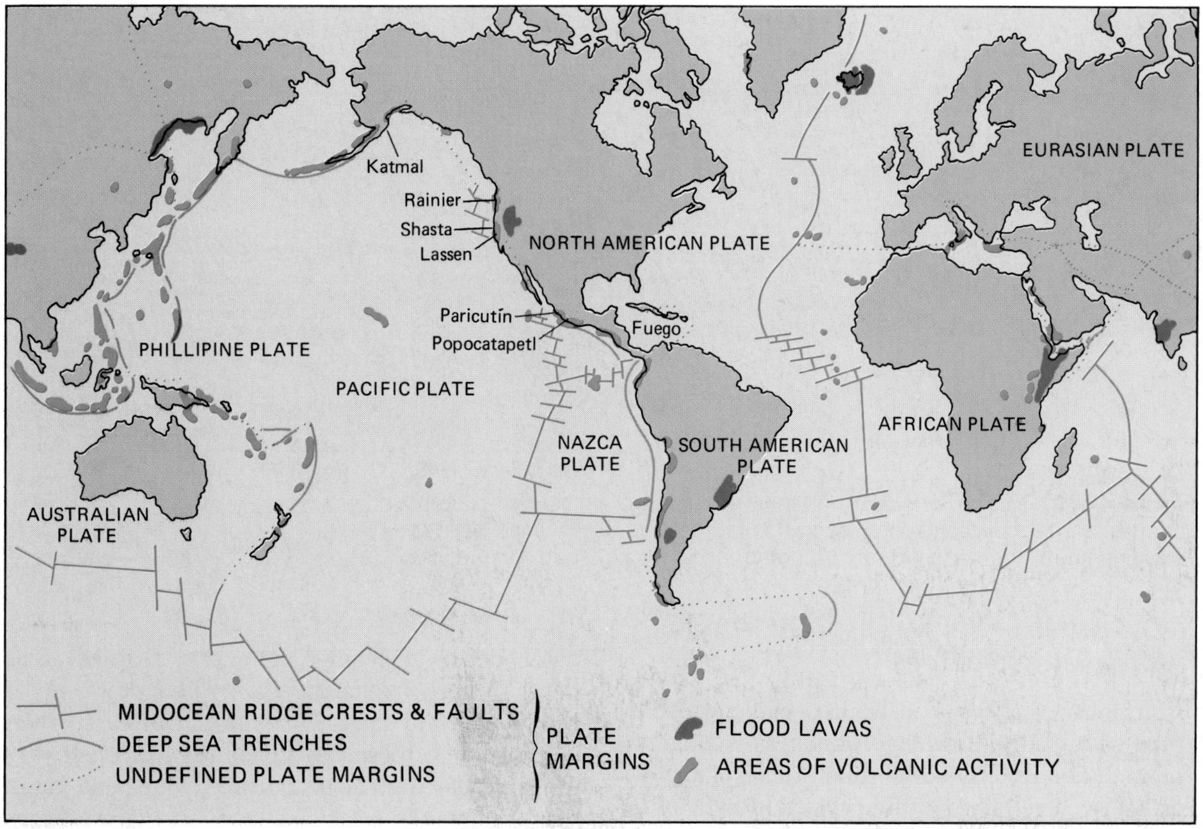

Figure 4–9 Tectonic plate boundaries and the distribution of active volcanoes on the earth. Note that volcanoes are not randomly distributed but rather tend to be associated with plate boundaries.

basalt slab with its covering of wet sediment undergoes partial melting (Fig. 4–10). Magmas, usually of andesitic composition, form and begin to work their way upward through overlying rocks. These ultimately provide the melts that are extruded at the surface.

At subduction zones, melting of the descending oceanic slab with its cover of sediment, provides the magmas required for volcanic activity.

Divergent Boundary Volcanism

The movement apart of tectonic plates along divergent boundaries provides an opportunity for underlying material to rise and spread laterally. The rising mantle material is primarily of peridotitic composition. As the rigid lithosphere above this material is pulled apart, pressure on the underlying rock is reduced. The reduction of pressure lowers the melting temperature of rocks. As a result, partial melting of the peridotitic material occurs, and melts of basaltic composition rise to fill the fissures formed by the lateral migration of plates. The outpourings of basaltic lavas produce the flows and submarine volcanoes of the midoceanic ridges.

At divergent plate boundaries, basaltic lavas rise to fill the space between separating plates.

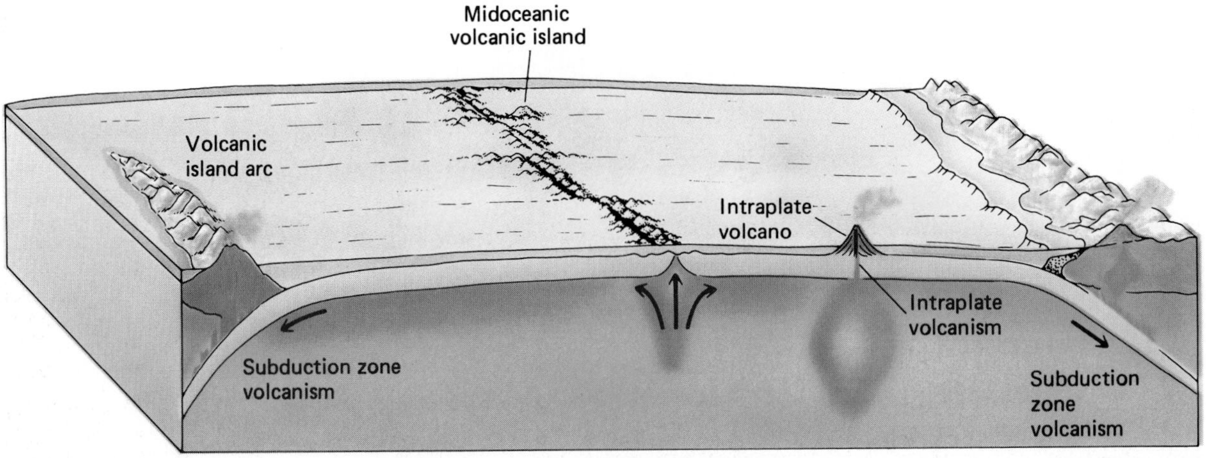

Figure 4–10 Most of the earth's volcanoes occur along convergent plate boundaries and associated subduction zones, along the mid-oceanic ridges where sea-floor spreading and plate divergence are occurring, or above hot spots in intraplate locations.

Intraplate Volcanoes

Intraplate volcanoes occur on oceanic crust, as is the case for the Hawaiian Islands, as well as on continental crust. The latter kind of interplate volcanism occurred during three cycles of activity in the Yellowstone region from about 600,000 to 2 million years ago. Each eruption produced tremendous quantities of rhyolitic ash and pumice. As will be described in Chapter 11, intraplate volcanoes are thought to develop over "hot spots," or rising plumes of hot mantle material. As this material rises and melts, there is an opportunity for fractional crystallization and assimilation of surrounding country rock so as to form silica-rich melts that produce rocks of a more rhyolitic composition.

Volcanoes located in central regions of tectonic plates develop over hot plumes of magma.

The Products of Volcanism

Volatiles

Whereas volcanic ash, glass, and rocks are the solid products of volcanism, the predominant volatiles are water and carbon dioxide. (Carbon monoxide, sulfur dioxide, hydrogen sulfide, and other gases occur in smaller quantities.) At the great depths where magmas are generated, they tend to hold water and gases in solution. However, when circumstances permit magmas to work their way upward near the earth's surface, water and gases may come out of solution somewhat in the manner of the bubbles in a carbonated soft drink. If the released gases and vapors become trapped as by a congealed plug of lava, pressures may build up underground until a volcanic explosion occurs and the pressures are released. Vast quantities of steam, gases, and fragmented rock may be thrown skyward, as was the case during the eruptions of Mt. St. Helens, the 1883 eruption of Krakatoa in Indonesia, and the 1902 eruption of Mt. Pelée on the island of Martinique in the West Indies.

In the eruption of Mt. Pelée, a great purplish cloud of gas and ash flowed rapidly down the slopes of the volcano and swept through the town of St. Pierre. Within the next two minutes all but two of the inhabitants of the town were killed—presumably by the searing heat (Fig. 4–11). One of the survivors was a murderer imprisoned in the town's dungeon, and the other a shoemaker whose escape seems to defy explanation. In the wake of the hot gases, the entire town was left in flames.

A cloud of incandescent dust, ash, and gases such as that which flowed down the slope of Mount Pelée is known as a **nuée ardente** (fiery

Figure 4–11 The smouldering ruins of St. Pierre as it appeared on May 14, 1902, six days following the May 8 eruption of Mount Pelée. (Courtesy of the Institute of Geological Sciences, London.)

cloud). A nuée ardente may take a high toll in lives whenever it passes through a heavily populated area. The extreme mobility of this fiery cloud is caused by the expanding hot gases, which contain cinders and ash. The solid particles give the cloud the density required to keep it close to the slope. Cooler air, trapped within the front of the hot moving mass, expands and increases the turbulence. It is this storm of turbulence that provides the destructive force of nuée ardente.

A nuée ardente is a turbulent cloud composed of gas and immense volumes of ash accompanying a volcanic eruption.

Pyroclastics

Even relatively quiet volcanic eruptions produce large quantities of ash, cinders, congealed blobs of lava, and solid fragments torn from the walls of vents. Such ejected materials are collectively termed **pyroclastics** (also termed **tephra**). They are classified according to their size and shape. Most *bombs* (Fig. 4–12), for example, are irregular, rounded, or ellipsoid masses ranging from fist to football size. They may display a twisted appearance resulting from the flow of still molten material during ejection, but the majority are simply unsymmetrical vesicular lumps. *Blocks* are larger solid angular pieces of rock. They consist of pieces of the crustal layers beneath the volcano or older lavas broken from the walls of the vent or edge of the crater. Blocks weighing hundreds of tons have been thrown several kilometers during particularly explosive eruptions.

In contrast to bombs and blocks, smaller pyroclastics (Fig. 4–13) include *lapilli*, which are frag-

Figure 4–12 Volcanic bombs. The specimen in the upper left part of the photograph is about 30 cm long. These objects are ejected during eruption as incandescent clots of liquid lava and largely consolidated during flight. (Courtesy of the Institute of Geological Sciences, London.)

Figure 4–13 Eruption of Paricutin Volcano, Michoacan, Mexico. This eruption was accompanied by great outpourings of tephra, including lapilli, ash, and bombs. There were also thin streams of lava, the red trails of which can be seen here on the flanks of the volcano. (Courtesy of the U.S. Geological Survey.)

ments between 64 and 2 mm, and *ash* (Fig. 4–14), which consists of still finer particles less than 2 mm in diameter. All of these fragments of volcanic ejecta can be consolidated or lithified to form the so-called pyroclastic rocks. If round in shape, the larger fragments form **agglomerates** and, if angular, **volcanic breccias**. The rocks from cinders and ash are called **volcanic tuffs**. **Welded tuffs** or **ignimbrites** are developed from ash and cinders that were part of a nuée ardente. As noted earlier, a nuée ardente accompanied the eruption of Mount Pelée. Because of the heat characteristic of a nuée ardente, the fine pyroclastics are still hot and soft when they are deposited and are welded together even with only slight compaction. Ignimbrites derived from the pyroclastics of continental volcanoes may attain thicknesses of hundreds of meters and extend over tens of thousands of square kilometers.

Pyroclastics may be consolidated to form agglomerates (having rounded fragments), volcanic breccias (having angular fragments), volcanic tuffs, and welded tuffs (ignimbrites).

Accumulations of volcanic ash have been known to bury entire towns and villages. When mixed with water, cinders can also contribute to the formation of mudflows called **lahars**. In tropical areas with heavy seasonal rainfall, such as Indonesia and the Philippines, lahars develop on the sides of active volcanoes and may flow downward at velocities of over 80 km per hour, causing death and destruction to everything in their paths. Lahars sometimes form when an eruption blasts up through the rim of a crater lake, expelling a lethal avalanche of boiling water, mud, and boulders.

Figure 4–14 Layers of ash and pumice exposed along the crater walls of Mount St. Helens. (Courtesy of L.E. Davis.)

Glasses

Most volcanic rocks, even though they cool rapidly from lavas, are crystalline solids. However, in such rocks, the individual mineral grains are usually too small to be seen without a microscope. But what about those lavas that have been chilled so quickly that there has not been time for the atoms in the melt to order themselves into the structures necessary to form crystals? The atoms and silica tetrahedra are frozen into the random positions they held in the liquid. The result is a noncrystalline mixture of silicates and silica called glass. **Obsidian** (Fig. 4–15) is a natural glass which, if it had cooled more slowly, would have developed crystals of feldspar, quartz, and ferromagnesians. Obsidian is usually colored black as a result of its content of impurities and its light-absorbing properties. It is easily recognized by its glassy appearance and distinctive conchoidal fracture.

Obsidian forms from lavas that have lost most of their dissolved gases. Like a bottle of beer that has been left open too long, such lavas have gone "flat." However, if the lava still contains abundant gas, bubbles may continue to escape while the melt cools. Such bubbles are preserved in the rock as cavities called **vesicles** (Fig. 4–16). **Pumice** (Fig. 4–17) is an extremely vesicular glass. In a very real sense, it is frozen silicate froth. Usually whitish or gray, pumice is light in weight and, because of its many sealed cavities, can sometimes actually float on water.

Obsidian has passed quickly from liquid to solid state, leaving no time for ions in the melt to arrange themselves in ordered ranks as crystals.

Fine-Grained Volcanic Rocks

Basalt

By far the most abundant volcanic rock is **basalt** (Fig. 4–18A&B). It underlies the oceans and is the rock from which Iceland and the Hawaiian Islands are constructed. The lavas from which basalts solidify have relatively low viscosity or resist-

Figure 4–15 Obsidian. Note conchoidal fracture, especially apparent at the top of the specimen.

Figure 4–17 Pumice, a type of glass produced by silica-rich lavas in which the gas content is so great as to cause the lava to "froth" as its rises in the chimney of the volcano and experiences rapid decrease in pressure. Some pieces of pumice will float on water because of the many air spaces formed by expanding gases.

Figure 4–16 Large and small vesicles in a solidified lava flow. Craters of the Moon National Monument. (Courtesy of L.E. Davis.)

ance to flow. As a result, they may flow over great distances before they congeal. The basalt flows of the Columbia Plateau (Fig. 4–19), the Deccan Plateau of India, and the maria of the moon cover tens of thousands of square kilometers.

Basalt is a dense dark gray or black rock composed of pyroxene, calcium-rich plagioclase feldspar, and minor amounts of magnetite and olivine. Basaltic lava pours from vents and fissures at temperatures of about 1200°C (Fig. 4–20) and becomes solid by the time the lava has cooled to 750°C. Crystallization is relatively rapid, so that a very finely crystalline texture is produced in which individual minerals cannot be discerned with the unaided eye. Thus, the rock is called aphanitic. However, one can view the uniformly sized crystals under the microscope by preparing a *thin section* of the rock. With the aid of a diamond saw, a rectangular piece of the basalt is cut and cemented onto a microscope slide. The mounted piece is then ground down until it is so thin that light can be transmitted through it. A cover glass is added to

A B

Figure 4–18 (A) Hand specimen of basalt. (B) Thin section of basalt viewed with polarized light. Tabular crystals are plagioclase and brightly colored crystals are ferromagnesian minerals. Width of field 2 mm.

Figure 4–19 Basalt lava flows exposed as stepped vertical cliffs by downcutting of the Columbia River (in foreground). These layered Columbia River basalts are part of the flood basalts that compose the Columbia plateau. (Courtesy of L.E. Davis.)

Figure 4–20 Geologists prepare to measure temperature of lava with a thermocouple probe. Hawaii Volcanoes National Park. (Photograph by T. Dugan, courtesy of the U.S. Geological Survey.)

complete the slide, and the myriads of feldspar and pyroxene grains can then be examined with the petrographic microscope.

Basalt is the most abundant extrusive igneous rock on earth. It is rich in calcium plagioclase and ferromagnesian minerals.

In order to understand the origin of basalt, it is necessary to refer briefly to a model of the earth's interior based on the study of earthquake waves. The model depicts the earth's basaltic crust as a thin zone about 10 km thick and overlying the mantle of denser olivine- and pyroxene-rich rocks. The boundary between the crust and the mantle is recognized by an abrupt change in the velocity of earthquake waves as they travel downward into the earth. For many years geologists believed that basaltic lavas originated from the lower part of the basaltic crust.

However, several recent lines of evidence suggest that the basaltic lavas may have come from molten pockets of upper-mantle material. For example, present-day volcanic activity is closely associated with deep earthquakes that occur within the mantle far beneath the crust. It is quite likely that fractures produced by these earthquakes serve as passages for the escape of molten material to the surface. A detailed study of earthquake shocks from particular volcanic eruptions in Hawaii indicates that the erupting lavas were derived from pockets of molten material within the upper mantle at depths of 65 to 100 km. However, mantle material (as indicated by many lines of evidence) is denser, richer in iron and magnesium, and more deficient in silica than are oceanic basalts.

How then can the melting of mantle material give rise to basaltic rocks that are less dense and contain more silicon? A process that accounts for this disparity is partial melting, which was mentioned earlier in the discussion of the origin of intrusive rocks. Partial melting occurs when a rock body (in this case upper-mantle rock) is partly melted and the liquid portion allowed to move to another location. The molten fraction, of course, forms from silicate minerals in the parent body that melt at relatively lower temperatures. This melt is usually less dense than the solids from which it was derived and thus tends to separate from the parent mass and work its way toward the surface. In this way melts of basaltic composition separate from denser rocks of the upper mantle and eventually make their way to the surface to form volcanoes.

Most basalts are derived from partial melts of magnetic bodies located in the upper mantle.

Many complex and interrelated factors control where in the mantle partial melting may occur or even if it will occur at all. Generally, heat in the 1100°C-to-1200°C-range is required. As noted in the previous chapter, however, the precise temperature for melting is also influenced by pressure and the water content of the rock. As pressure increases, the temperature at which particular minerals melt also rises. Thus a rock that would melt at 1000°C near the surface will not melt in deeper zones of higher pressure until it reaches far greater temperatures. Water has an effect opposite to that of pressure, for its presence will allow a rock to start melting at lower temperatures and shallower depths than it would have otherwise. Laboratory experiments indicate that the melting of "dry" mantle rock can occur at depths of about 350 km but that the presence of only a little water can cause partial melting and yield basaltic liquid from depths as shallow as 100 km.

The lavas from which basalts are formed consist chemically of about 50 per cent silica (SiO_2), with lesser percentages of oxides of aluminum, iron, calcium, magnesium, sodium, and potassium. Nonbasaltic lavas differ in their silica content, and this difference provides the basis for recognition of such other volcanic rocks as **rhyolite** and **andesite**, as well as many other transitional rock types.

Figure 4–21 Rhyolite porphyry. Phenocrysts are primarily quartz and orthoclase.

Rhyolite

Rhyolite is a gray or pink aphanitic rock derived from a highly siliceous lava. For example, rhyolitic lavas may consist of as much as 75 per cent silica. Quartz, potassium feldspar, sodium plagioclase, and lesser amounts of ferromagnesian minerals occur in rhyolite (Fig. 4–21). It is estimated that rhyolitic lavas erupt at temperatures of 800° to 1000°C. Usually they are far more viscous than basaltic lavas and flow very sluggishly, if at all.

Rhyolites are the products of continental volcanism rather than volcanoes that form on the ocean floor. This suggests that at least some rhyolites are formed by remelting of parts of the silica-rich continental crust. Another theory proposes that rhyolites are formed by the melting of sediments that had been eroded from the continents and deposited along continental margins. Highly silicic sediment, if caught in the vise of converging tectonic plates and melted, could produce magmas of rhyolitic composition.

Rhyolite is an aphanitic or porphyritic, light-colored, extrusive igneous rock rich in quartz, orthoclase, and sodic plagioclase.

It may also be possible to produce rhyolitic lavas by fractional crystallization. Imagine a deep reservoir of magma in which ferromagnesian minerals have formed early in the cooling period and settled to the bottom of the magma chamber. Early crystallization of these minerals leaves the original liquid with the higher proportion of silica needed for the formation of rhyolite.

Andesite

Andesite (Fig. 4–22) is a fine-grained light-gray rock intermediate in density and silica content (about 60 per cent) between basalt and rhyolite. Andesitic lavas are also intermediate in viscosity and thus form thicker flows than do basaltic lavas. Volcanoes of the andesitic type are common around the edges of continents bordering the Pacific Ocean and are generally more explosive than mid-oceanic volcanoes like those of Hawaii.

Because andesites are transitional in composition between rhyolites and basalts, it would appear that at least some andesitic lavas are the products of mixing of silica-rich and silica-poor melts. For example, a basaltic magma might work its way into pockets of rhyolitic magma and then blend with the more siliceous liquid. As another possibility, a basaltic magma might engulf, melt, and assimilate surrounding silica-rich rhyolitic rocks. Yet another method for infusing silica into basaltic materials to

Figure 4–22 Andesite.

produce andesite is provided by plate tectonics. In this method, the required amount of silica is derived from the thin layer of oceanic sediment that rests upon the basalt of an oceanic plate as it descends along a subduction zone (Fig. 4–23). As this plate plunges diagonally downward, masses of basalt and wet sediment mix and undergo partial melting to produce magmas of andesitic composition. These melts rise buoyantly and erupt to form the immense volumes of andesite often associated with subduction zones.

The association of andesites with subduction zones indicates an origin from melting oceanic crust enriched with silica from wet, ocean-floor sediment.

Features of Volcanic Rocks

As lavas cool and begin to congeal, they may develop a variety of textural features. We have already noted how gas bubbles produce vesicles. Vesicles are the most characteristic feature of the volcanic rock called **scoria** (Fig. 4–24), a dark-colored clinker-like rock that is partly crystalline and partly glassy.

The surfaces of solidified lavas may also develop distinctive appearances. **Pahoehoe** (pronounced pa-hoy-hoy) is ropy lava (Fig. 4–25). The ropy texture is produced in the still-plastic surface scum of lava by the drag of more rapidly flowing liquid lava below (Fig. 4–26). **Aa** is a blocky fragmented lava (Fig. 4–27). It is derived from a more viscous and slower-moving liquid. As this thicker lava flows down a slope, the upper layer hardens and is carried along in conveyor-belt fashion. On reaching the front of the lava flow, the brittle top layer is broken into a chaotic jumble of jagged fragments. If you were to walk upon this material in

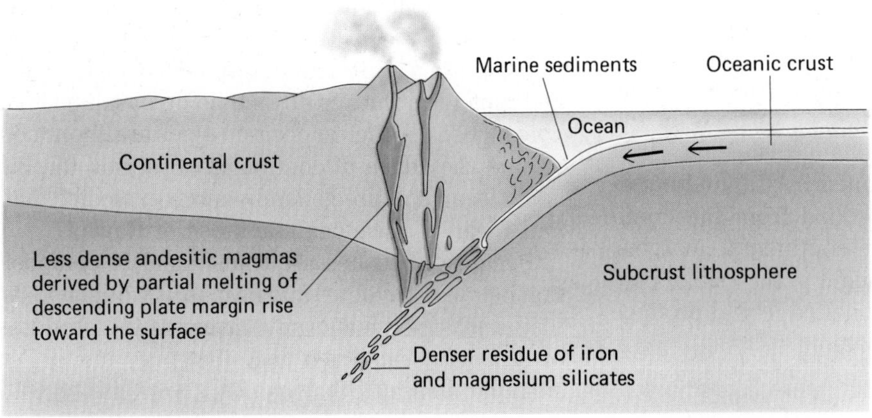

Figure 4–23 Subduction of sediment-bearing leading edge of an oceanic plate with partial melting at depth to produce andesitic magmas. The sketch depicts the mechanism responsible for the development of the Andes Mountains.

Figure 4–24 Scoria.

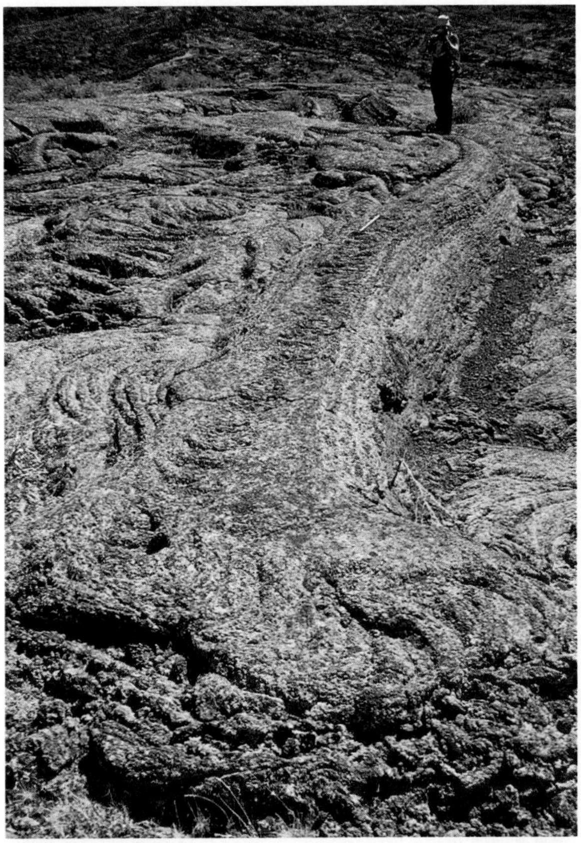

Figure 4–25 Basalt "lava stream" with pahoehoe surface, eastern Snake River Plains, Idaho. These basalts were extruded about 2000 years ago. They appear remarkably unaltered, partly because weathering is slow in this region of dry, cold winters. (Courtesy of G. Ross.)

Figure 4–26 Lobes of pahoehoe lava advance over black sand. Kilauea East Rift at Kaena Point, Hawaii Volcanoes National Park. (Courtesy of the U.S. Geological Survey.)

your bare feet, you might easily guess why native Hawaiians named it aa.

Lava that is extruded under water bulges outward into bulbous pillow-shaped masses that average about a meter in diameter (Figs. 4–28 and 4–29). Scuba-diving geologists have filmed these so-called pillows as they formed from tongues of

Figure 4–27 Scientists move away from advancing aa lava flow. Royal Gardens subdivision, Hawaii, 1983. (Photograph by J.D. Griggs, courtesy of the U.S. Geological Survey.)

Figure 4–28 Pillow structures formed by lavas being extruded on the floor of the Pacific Ocean along the Juan de Fuca ridge, a spreading center west of Oregon. Photo covers 5 m. (Courtesy of W.R. Normark and the U.S. Geological Survey.)

Figure 4–29 Ancient pillow lavas. These pillow lavas outcrop near Lake of the Woods, Ontario, Canada. They were extruded on the floor of an ocean basin over 2½ billion years ago. Glaciers moved across northern North America during the relatively recent Great Ice Age, beveled the surface of the pillows, and left behind the telltale scratches (glacial striae.) (Courtesy of K. Schultz.)

lava that were intermittently extruded from submarine fissures. They observed how the outer covering was initially somewhat plastic and molded itself into the rounded periphery. On further cooling, the surface became glassy and developed radial cracks. Insulated by the congealed skin of lava, the interior of the pillow cooled somewhat more slowly and became finely crystalline. Pillow lavas are useful indicators of underwater volcanic eruptions and also provide a means of locating the original top of deformed sequences of layered rocks that contain lava flows. The convex surfaces of the pillows indicate which way was "up" in the sequence.

Lava that is extruded beneath the surface of the sea forms lobate masses resembling a pile of bed pillows. These pillow lavas are clues to submarine eruptions.

Lava tunnels (Fig. 4–30), tubes, and caverns are among the most frequently visited tourist attractions on Hawaii and Craters of the Moon National Monument in Idaho. They are formed as the outer zone of large lava flows solidify, while an inner zone is still molten and flowing. After a period of time, the lava will drain out of the flow, so that only empty space in the form of a long tube remains.

Lava fountains (Fig. 4–31) are spectacular eruptions of lava shot into the air by the hydro-

Figure 4–30 Lava tunnel, Lava Beds National Monument, California. (Courtesy of J.C. Brice.)

static pressure on the liquid and the expansion of gases within it. They form along fissures or from ruptures in the roof rocks of tunnels and tubes. These fountains of incandescent lava may jet hundreds of meters into the air. As the still-liquid material falls to the ground, it builds **spatter cones** (Fig. 4–32) up to several meters high.

Perhaps the most distinctive feature formed in cooling lava flows are **columnar joints**. Joints are shrinkage cracks that form in rather regular patterns when lava cools and contracts. Columnar joints are perpendicular to the cooling surface and divide the rock into polygonal prisms or columns (Figs. 4–33 and 4–34). From a distance, the columns resemble bunches of giant fence posts. When viewed from above, they exhibit a pattern not unlike polygonal bathroom tiles. The columnar jointing of the Giant's Causeway in Ireland and the Devil's Post Pile in California are well-known examples.

Tensional stresses produced when lava cools and contracts produce columnar joints in lava flows, sills, and dikes.

Landforms Produced by Volcanoes

The viscosity of a magma is a critical factor in determining the eruptive style and ultimate shape of a volcano. The most important factors that influence viscosity are composition (including volatiles), pressure, and temperature. In regard to com-

Figure 4–31 Lava fountain being observed by geologists, Pu'u O'o eruption, Hawaii, 1986. The fountaining lava attained heights of 160 to 200 meters. (Courtesy of U.S. Dept. of the Interior, U.S. Geological Survey, Johnston Cascades Volcano Observatory. Photo by Lyn Topinka.)

Figure 4–32 Spatter cone. Spatter cones are mounds of lava spatter built up by lava fountains along a fissure or vent. This spatter cone located in Craters of the Moon National Monument, is nearly 10 m in height. (Courtesy of L.E. Davis.)

position, the generalization can be made that the more silica there is in a magma the more resistant it will be to flow. This is because, with an abundance of silica in the melt, tetrahedra readily form and begin to group into network structures that resist flow. An increase in pressure within the

Figure 4–33 Columnar jointing in Columbia River basalts, Columbia Plateau near Kendrick, Idaho. (Courtesy of L.E. Davis.)

Figure 4–34 Devil's Tower in northwestern Wyoming. The columnar jointing in this large volcanic conduit is conspicuous. The column of rock rises more than 240 meters above the surrounding area.

magma chamber also tends to increase viscosity. Heating has the reverse effect on viscosity. Other things being equal, a rise in temperature results in a decrease in the viscosity of the magma.

An increase in silica content or pressure increases the viscosity of magma, whereas an increase in temperature will decrease viscosity.

As an illustration of the effect of silica content on viscosity, one can compare the basaltic lavas of Kilauea, Hawaii, with those of Mount Pelée. Kilauean lavas contained about 50 per cent SiO_2 and flowed readily. Flow rates of 400 meters per hour were recorded for lavas moving over slopes as gentle as 3°. In contrast, Peléan lavas contained 70 per cent silica. In the 1903 eruption, lava was so viscous that it would hardly flow at all, but rather rose slowly from several vents so as to form grotesque pinnacles that then broke and rolled down the sides of the mountain.

In addition to the silica content of a melt, dissolved gases and water can also affect viscosity. In general, water tends to reduce viscosity by inhibiting linkage of tetrahedra in the melt. Carbon dioxide has the opposite effect, and therefore the ratio of this gas to water must be considered in assessing the overall effect of volatiles on viscosity.

Landforms Associated with Fissure Eruptions

In the canyon walls of the Columbia and Snake rivers of Washington, one can see scores of layers of flat-lying basaltic lava flows (see Fig. 4–19). They are termed **flood basalts** or **plateau basalts**. The lavas extruded were of low viscosity and therefore were able to spread widely before solidifying. Characteristically, flood basalts occur in areas where lavas emerge from long fissures. The basalts that built the Columbia Plateau completely buried preexisting topography over an area of 130,000 km^2. Other famous examples of such flood basalts occur in the Deccan Plateau of India and the Parana Plateau of Brazil and Paraguay.

Fissure eruptions tend to be more prevalent in areas where the earth's crust is subjected to tensional forces, that is, forces that tend to pull the crust apart. Crustal tension may result in vertical cracks, through which magma may force its way to the surface. Such fissures are particularly characteristic of the mid-oceanic ridge (and fissure) systems. Truly prodigious amounts of lavas have poured from these world-encircling features.

Central Eruptions and Volcanic Cones

Shield Volcanoes

We are accustomed to thinking of volcanic eruptions not in terms of fissure eruptions but as the discharge of lavas and gases from some sort of central vent or group of vents. The results of such eruptions are the familiar volcanic cones. Depending on the nature of the lava, the landforms produced from central eruptions differ in shape. Viscous lavas form cones with steep slopes. However, if lavas are relatively thin and free-flowing, they spread widely, producing a broadly convex or shield-shaped volcano such as Mauna Loa (Fig. 4–35). Hawaii (the "Big Island") is the upper part of five coalesced **shield volcanoes** that include Mauna Loa, Mauna Kea, Kilauea, Kohala, and Hualalai. Hawaii, which rises 9.6 km above its subsea base, has been built from the accumulation of over 50,000 km^3 of lava flows and pyroclastics. It is the largest pile of geologically young volcanic material on our planet.

Cinder Cones

Cinder cones (Figs. 4–36 and 4–37) are volcanic peaks built entirely of cinders and other pyroclastics that have been explosively ejected from a vent. Cinder cones rarely attain heights in excess of 500 meters. Their slopes vary between 30° and 35°, depending on the maximal angle at which the debris remains stable and does not slide downhill—that is, the *angle of repose.* Paricutin in Mexico (see Fig. 4–13) is a particularly well known cinder cone, whose growth was witnessed by dozens of geologists called quickly to the scene. Paricutin began as a crack that opened up in the corn field of a farmer named Dionisio Pulido on February 22, 1943. By the very next day, the eruption had built a cone 8 meters high. Within the span of a month, the ash, cinders, and lapilli had built a mountain of debris that towered 300 meters above the surrounding countryside.

Composite Cones

When both lava and pyroclastics are alternately emitted through a vent, a **composite cone** or **stratovolcano** is constructed of alternating pyroclastic layers and lava flows. These are the most common of the large continental volcanoes. They are represented by such familiar peaks as Mt. Rainier and Mount St. Helens in Washington, California's Mt. Shasta (Fig. 4–38), Mayon (on Luzon in the Philippines), Fujiyama (Japan), Vesuvius (Italy), Etna (Sicily), and Stromboli (on the Aeolian Islands north of Sicily).

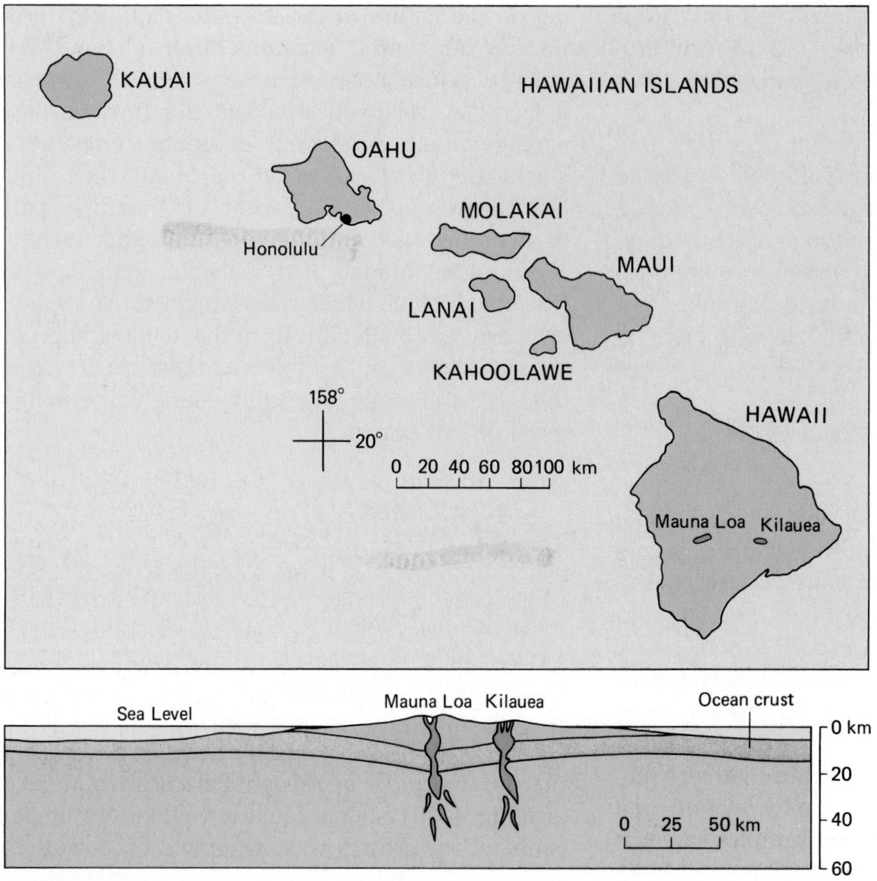

Figure 4–35 Map of the Hawaiian Islands and cross-section through Hawaii showing the broad, gently convex profile of Mauna Loa.

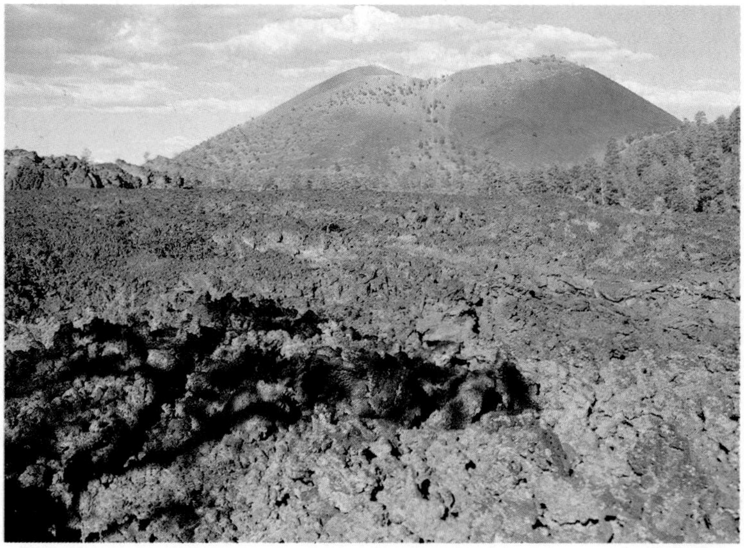

Figure 4–36 Sunset Crater, a cinder cone with crater, Sunset Crater National Monument. Note aa in foreground.

Figure 4–37 Vertical aerial photograph of large cinder cone, San Francisco volcanic field, northern Arizona. Lava flow issuing from the cone is 7 km long and more than 30 meters thick. (Courtesy of the U.S. Geological Survey.)

Volcanic Domes

Some volcanoes are nearly circular rounded domes that form from slow extrusions of highly siliceous, and hence more viscous, lavas. Such lava piles up primarily around the vent. **Volcanic domes** frequently develop within the crater walls produced by earlier more active eruptions (Fig. 4–39A). Pointed, monolithic projections of highly viscous lava sometimes squeeze upward through the surface of a dome. These so-called volcanic spines (Fig. 4–39B) may reach heights of hundreds of meters.

Craters and Calderas

Most people would have no difficulty defining a **crater** as a circular depression at the summit of a volcanic vent (Fig. 4–40). Craters also have steep inner sides and a total diameter less than three times their depth. They are rarely more than 1.6 km in diameter. Some craters form when eruption has ceased and lava that has not flowed down the slopes sinks slowly back into the vent. Craters may also form when the material at the top of the vent is blasted out by the force of volcanic explosions.

Crater Lake in Oregon is a familiar example of a caldera (Fig. 4–41). **Calderas** are roughly circular in outline, have relatively flat floors and steep surrounding walls, and are generally over 1.6 km in diameter. Calderas take their name from a Spanish word meaning kettle. They may be formed by collapse, by explosion, or possibly by a combination of these mechanisms. Geologists like the late Howel Williams of the University of California be-

Figure 4–38 Mt. Shasta, California, a prominent composite volcanic peak in the Cascade Range. The smaller peak in the background on the left side is called Shastina.

A

B

Figure 4–39 (A) Volcanic dome built up inside the crater of Mount St. Helens. (B) A 60-m-tall spine forced upward through the dome several days after the main eruption in 1980. (A, courtesy of the U.S. Geological Survey; B, courtesy of L.E. Davis.)

lieve that calderas like Crater Lake form when eruptions drain a large portion of the magma from the magma chamber beneath the volcano. Unsupported by underlying molten rock, the magma chamber roof might collapse to form a caldera (Fig. 4–42).

The caldera that formed on Bandai-san volcano in Japan is an example of caldera formation as a result of explosive activity. In 1888, a stupendous explosion blew off the summit and part of the northern slope of Bandai-san, leaving a caldera over 2.5 km wide. The inner walls of Bandai-san rise an impressive 350 meters above its floor.

The largest caldera known on earth is located about 65 km west of Yellowstone Lake in Yellowstone National Park. The basin is about 80 km in diameter and was formed about 2 million years ago by collapse following the extrusion of an enormous volume of silica-rich lava. The lava and extrusions from two subsequent eruptions

Figure 4–40 The crater of Masaya volcano in Nicaragua. This volcano had nearly continuous activity from 1529 to 1946. (Courtesy of R.E. Wilcox and the U.S. Geological Survey.)

Figure 4–41 Crater Lake in Oregon is the finest example of a caldera in the United States. The small cinder cone projecting above the lake surface was built during the last eruption and is called Wizard Island. (Courtesy of L.E. Davis.)

solidified but are still quite hot. They provide the heat responsible for the thermal springs and geysers for which Yellowstone National Park is famous.

Diatremes

Diatremes are vertical pipes or necks that have been filled with angular rock fragments. Because recent diatremes are subsurface features, geologists must necessarily study ancient ones that have been uncovered by erosion or examine parts of them exposed in deep diamond mines. They are thought to have been formed by gas-rich magmas, which, upon approaching the surface, explosively eject their gases. The torrents of gas tear chunks of lava and surrounding rock from the vent walls and bring up toward the surface fragments from the deeper zones of the crust and upper mantle. Thus, the rocks of diatremes provide a sample of materi-

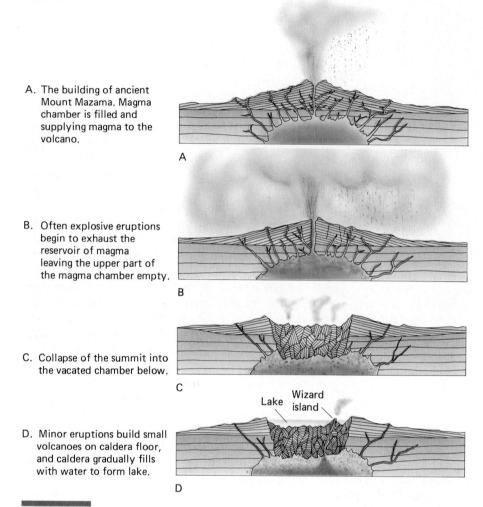

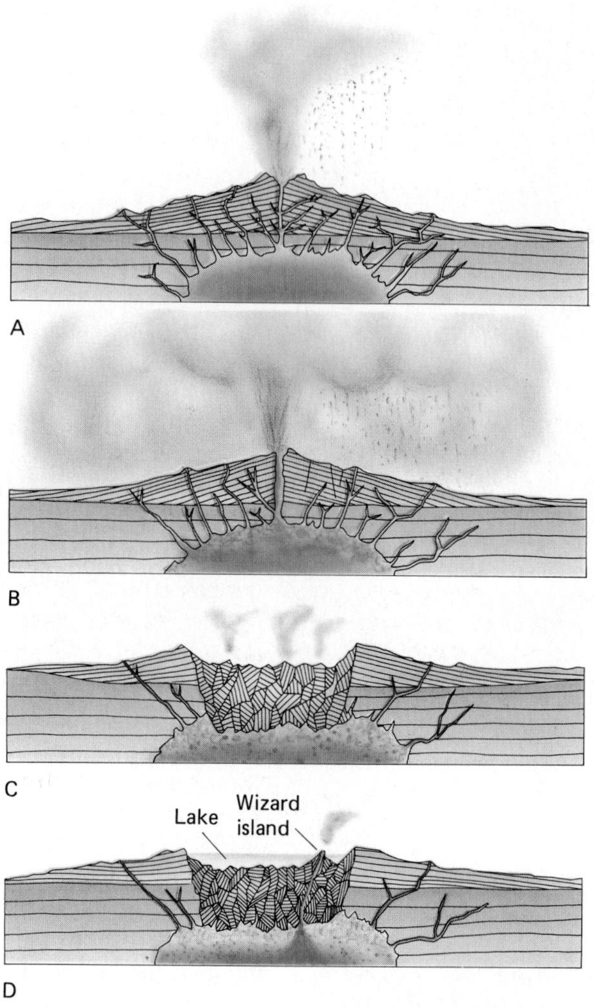

A. The building of ancient Mount Mazama. Magma chamber is filled and supplying magma to the volcano.

B. Often explosive eruptions begin to exhaust the reservoir of magma leaving the upper part of the magma chamber empty.

C. Collapse of the summit into the vacated chamber below.

D. Minor eruptions build small volcanoes on caldera floor, and caldera gradually fills with water to form lake.

Figure 4–42 The origin of the caldera at Crater Lake, Oregon. (After Williams, H., *Crater Lake: The Story of Its Origin.* Bulletin, Dept. of Geological Sciences, vol. 25. University of California Publications, Berkeley, 1941.)

als far below depths that can be reached by drilling and are therefore studied carefully by geologists seeking clues about the nature of the earth's interior.

Volcanic features include lava plateaus, shield volcanoes, cinder cones, composite cones, volcanic domes, and diatremes.

Extraterrestrial Volcanism

Volcanic activity is by no means a phenomenon restricted to the earth. Along with meteorite impact, volcanism has been an important process in the development of other planetary surfaces and atmospheres. About 60 per cent of Mars, for example, appears to be covered with volcanic rocks. At least 20 separate centers of volcanic activity have

Figure 4–43 High-resolution Viking Orbiter I spacecraft photograph of a large lava flow on Mars. Analysis of the height of the flows suggests the lavas were probably basaltic. Distance across the photograph is approximately 10 km. (Courtesy of R. Arvidson, NASA Space Imagery Center, Department of Earth and Planetary Sciences, Washington University.)

been recognized on its surface, and there are many isolated volcanoes as well.

All the volcanoes on Mars are shield volcanoes whose gentle slopes are composed of thin flows of what was once low-viscosity lava (Fig. 4–43). Mars also has the distinction of having the largest known volcano in the solar system. Olympus Mons (Fig. 4–44), as it is named, is 600 km in diameter at its base and rises 23 km above the surrounding plain. It is easily large enough to contain all of Manhattan Island, and its diameter measures five times that of Hawaii. Such an enormous accumulation of volcanic material in one place would probably not have occurred if Mars had a lithosphere composed of moving plates as on earth. On earth, volcanoes grow only as long as they remain over a fixed magma source in the mantle. Because the plates bearing the volcanoes move, the amount of magma is distributed in a chain of smaller volcanoes formed as the plate glides over the magma source.

Although often larger, Martian volcanoes have generally the same shape as those on earth. Tentatively, they can be interpreted as having been formed by similar processes. Mars also has huge lava flows (see Fig. 4–43), some of which extend over 1000 km from their apparent source vents. X-ray fluorescence analyses of the fine rock particles at the two Viking Lander locations indicate a primary lava of olivine basaltic composition. Such a lava would have been extremely fluid.

Venus is a planet rather similar to the earth in size and mean density and therefore may have an interior somewhat like that of the earth. Volcanoes also exist on Venus. One huge peak topped by a crater 40 km across has been located by radar, and clusters of smaller peaks, as well as what appear to be areas of extensive lava flows, have also been detected.

Hundreds of observations derived from the Apollo missions have furthered immensely our understanding of the moon's history. Volcanic activity is very much a part of that history (Fig. 4–45). The moon formed, as did the earth, about 4.6 billion years ago. From about 4.2 to 3.9 billion years ago meteorite bombardment produced the majority of the lunar craters. By about 3.8 billion years ago, the interior of the moon had become sufficiently hot as a result of radioactivity to initiate a 700-million-year episode of extensive volcanic activity. Lava flowed out onto the lunar surface, filling the large basins developed earlier by meteorite bombardment, and formed the basalt-floored maria. The era of volcanism appears to have ended by about 3.1 billion years ago. Our only satellite has been relatively quiet since then.

Jupiter's moon Io is another body in our solar system noteworthy for its volcanic activity. Altogether, eight active volcanoes have been detected on Io. Plumes of sulfurous gas and ash rise 300 km above some of these volcanoes.

Volcanoes are common features on many of the planets and their satellites in our solar system.

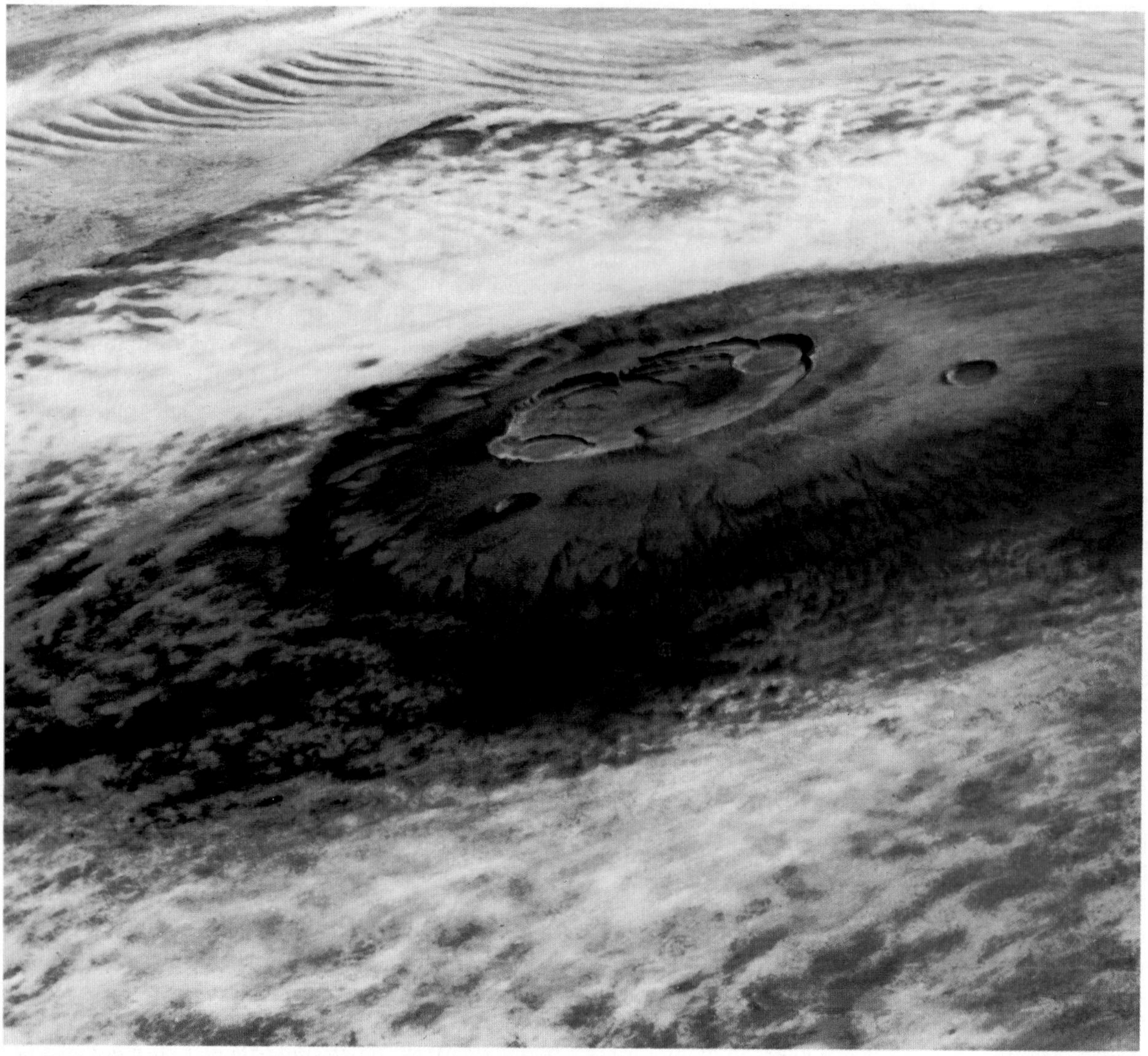

Figure 4–44 The gigantic Martian volcano Olympus Mons, photographed by Viking Orbiter in July 1976. The original black-and-white photograph has been colored in by the artist. (Courtesy of NASA/JPL; from Abell, G.O., *Realm of the Universe,* 3rd ed. Philadelphia, Saunders College Publishing, 1983.)

Predicting Volcanic Eruptions

It has been estimated that over a quarter of a million people have lost their lives as a result of recorded volcanic eruptions. Today, even more than in centuries past, heavily populated areas surround some volcanoes. What can be done to warn nearby populations of impending eruptions, and how may some of the effects of volcanism be ameliorated?

There is no reliable way of knowing whether long-inactive volcanoes are extinct or simply dor-

Figure 4–45 Volcanic cones, craters, and lava flows provide abundant evidence of past volcanic activity on the moon. The angular shadows result from the low angle of the sun and are normal shadows behind volcanoes. (Courtesy of R. Arvidson, Department of Earth and Planetary Sciences, Washington University.)

mant. However, methods have been developed that can be used to predict eruptions of volcanoes that are considered still active. Careful monitoring of volcanoes for such warning signals as preliminary minor eruptions, increase in ground temperatures, earthquakes, change in magnetic properties caused by heating of rocks at depth, tilting of the ground, and the appearance of new hot springs and steaming vents can be useful in foretelling the coming of a major eruption.

The primary instrument used to detect the multiple minor earthquakes that precede many eruptions is the seismograph (see Fig. 10–14). The U.S. Geological Survey has seismometer stations on 16 volcanoes in North and South America and Iceland. One or more *tiltmeters* supplement the seismometers at these stations. A tiltmeter is a highly sensitive bubble-type level with associated electronic devices. It records changes in slope of as little as 1 cm in 10 km. The upward movement of magma before an eruption often causes a slight swelling of the overlying ground surface, and this change is readily detected by the tiltmeter. Both seismometers and tiltmeters are automatic devices that report their activity by radio through the Earth Resources Technology Satellite (ERTS) to a designated monitoring center.

Even with such sophisticated devices as seismometers and earthquake counters, the problem of predicting the precise time of a volcanic eruption is not solved. For example, six days elapsed between the occurrence of earthquakes and the 1973 eruption of Fuego volcano in Guatemala, whereas earthquakes near Paricutin began two weeks before the eruptions.

Although there is no way to halt an impending volcanic eruption, there are certain measures that can be taken to guard against destruction caused by lava flows, mudflows, explosive debris, and tsunami. Simply knowing that lava and mudflows move down valleys should suggest that a topographic analysis would be useful in selecting sites for buildings and towns. For eruptions along coastlines, lava flows can sometimes be slowed by spraying with cold ocean water. This method proved useful during the 1973 eruption of Helgafell in Iceland.

In all volcano-prone regions, construction and building codes should be formulated and enforced in the location and design of structures. Where pyroclastic accumulations are likely to be heavy, roofs on buildings should be steeply pitched as is done in areas of heavy snows. If crater lakes exist, which, if breached, might trigger mudflows, then tunnels should be installed to siphon off the dangerous waters. Whenever possible, known active volcanoes should be set aside as national parks, so as to limit the growth of towns around their perimeters. It is evident that as populations continue to expand, volcanoes will become potentially even more hazardous. Most of us are fully aware that such hazards are an ongoing concern for people living in Hawaii or Alaska but overlook the danger to the increasing numbers of people living close to volcanoes like Mount St. Helens or Mt. Hood in the northwestern United States. Few geologists would be willing to state that presently inactive volcanoes of the Cascades are extinct. However, by monitoring and planning, anxiety and the hazards can be reduced. Such practices are particularly necessary for such United States cities as Tacoma and Seattle.

Hazards associated with volcanic eruptions can be greatly reduced by sensible planning relating to land use and construction.

Benefits from Volcanoes

Although volcanoes have dealt people some severe blows, they are not without their benefits. They have provided many areas with the raw materials for exceptionally fertile soils. Volcanoes provide some of the earth's most spectacular scenery and have built inhabitable land areas such as the Aleutians, Hawaiian Islands, and Iceland. The magmas that feed volcanic eruptions are capable of producing steam that can be used to drive electric turbines. There are, for example, several geothermal power plants now in operation that generate electricity from underground waters that have been heated by volcanic processes. These include "The Geysers" field in northern California (Fig. 4–46), the Larderello field in northern Italy, two Japanese sites, and an area near Wairaki in New

Figure 4–46
Geothermal energy facility at the Geysers in Sonoma County, California. (Courtesy of the U.S. Geological Survey.)

Figure 4–47 Tephra litters much of the ground at Craters of the Moon National Monument in Idaho. The small particles are mostly ash (to 4 mm in diameter) and lapilli (4 to 32 mm in diameter). The larger fragment is a volcanic bomb. (Courtesy of L.E. Davis.)

Zealand. The homes of more than 50,000 persons in Reykjavik, Iceland, are heated with this natural resource.

Volcanoes and Climate

Benjamin Franklin once stated that volcanoes are able to affect the weather. He was quite correct in this opinion, for volcanic dust and ash thrown into the air and circulated through the atmosphere can indeed lower temperatures by a small but measurable amount. The fine particles block the sun's rays and thereby reduce the amount of heat that reaches the ground. Usually, a single eruption is able to lower temperatures by no more than several tenths of a degree. Should there be a period of frequent eruptions around the globe, however, the combined effect could be greater. Although a veil of volcanic particles in the atmosphere is not considered capable of causing an ice age by itself, many geologists believe it could have a significant effect in lowering average global temperatures when combined with other factors.

Whereas volcanic dust may cause cooling, long-term volcanic emissions of carbon dioxide would have an opposite effect. For example, warm climatic conditions were widespread during the Early Paleozoic and Mesozoic, and may have been the result of a carbon dioxide build-up caused by increased global volcanism during periods of active sea floor spreading.

Summary

A volcano is an opening from which molten rock, solidified pieces of molten rock, and volatiles (gases and vapors) escape from the interior of the earth. Lava and pyroclastics may accumulate around the opening so as to form a volcanic mountain or may spread from fissures across extensive areas to form lava plateaus.

The distribution of volcanoes around the world is largely controlled by plate tectonics. Large numbers of volcanoes occur along convergent plate boundaries where subduction carries oceanic basalt and sediment to deep levels where partial melting and other magmatic processes can occur. Many other volcanoes are found along the divergent boundaries of spreading centers. Here, the melting temperatures of upper-mantle rocks are lowered because of reduced pressure associated with separating tectonic plates. Partial melting produces basaltic melts that find their way to the surface and erupt along mid-oceanic ridges. Occasionally these sea-floor volcanoes build themselves up above sea level to form volcanic islands. A fewer number of volcanoes do not develop along plate margins. These intraplate volcanoes form over rising plumes of hot mantle rock called hot spots.

The products of volcanism include volatiles (of which water is the most important), lavas, and fragments or particles of solidified lava termed pyroclastics or tephra (Fig. 4–47). Tephra may be consolidated to form pyroclastic rocks such as agglomerate, volcanic breccia, and welded tuffs (ignimbrites). Ash is tephra consisting of particles finer than 2 mm in diameter. When mixed with water, ash and other tephra may be mobilized in the form of devastating mudflows called lahars.

Among all volcanic rocks, basalt is the most abundant. It underlies the oceans and has formed extensive basalt lava plateaus on land. Basalt is a dark rock composed of silicate minerals that are relatively rich in iron, magnesium, and calcium. Rhyolite is the compositional opposite of basalt. It is light in color, contains very little iron and magnesium, and is rich in potassium and sodium feldspars and quartz. The composition of andesite, a volcanic rock second in abundance to basalt, is intermediate between basalt and rhyolite. In general, andesites contain approximately equal amounts of calcium and sodium feldspars. The volcanic rock obsidian is a glass formed from a melt that solidified too rapidly for crystals to form.

Landforms produced by volcanoes include cinder cones, composed of unconsolidated tephra; shield volcanoes, whose gentle slopes result from successive coverings of low-viscosity lava; composite cones, formed of alternating layers of tephra and lava flows; and lava plateaus, formed of low-viscosity lavas rising from fissure systems.

Terms to Understand

aa
agglomerate
andesite
basalt
caldera
cinder cone
columnar joint
composite cone
crater
diatreme
flood basalt (plateau basalt)
lahar
lava fountain
lava tunnel
nuée ardente
obsidian
pahoehoe
pumice
pyroclastic (tephra)
rhyolite
scoria
shield volcano
spatter cone
stratovolcano
tsunami
vesicle
volatile
volcanic breccia
volcanic dome
volcanic tuff
welded tuff (ignimbrite)

Questions for Review

1. Why is a high-silica lava more viscous than a low-silica lava? Why is explosive volcanic activity frequently associated with high-silica lava?

2. What are the most abundant gases emitted by volcanoes? What is the source of these gases?

3. Explain the role of gases in causing explosive volcanic activity.

4. What are the different types of pyroclastic materials ejected from a volcano during an explosive eruption? What is the name of a volcano composed almost entirely of cinders?

5. How do fissure flows differ from volcanoes? Have fissure flows produced an appreciable amount of igneous rocks during the geologic past? Cite an example.

6. In terms of their origin and structure, why do shield volcanoes, cinder cones, and stratovolcanoes differ in appearance?

7. By what two mechanisms may volcanic caldera originate?

8. What benefits, if any, has humanity derived from volcanic activity?

9. How is plate tectonics related to the following kinds of volcanic activity:

a. Volcanic activity in the Cascade Range of the United States.

b. Intraplate volcanism such as in the Hawaiian Islands.

c. Volcanism along mid-oceanic ridges.

10. Compare the minerals you would expect to find in a basalt with those in a rhyolite (refer to Figure 3–13).

11. How might the nature and products of volcanism differ in interplate volcanoes located on continents compared with those located on oceanic plates?

12. Most of the water released as steam during the eruption of Mt. St. Helens is considered *recycled* water. What is its source?

Supplemental Readings and References

Bullard, F.M., 1977. *Volcanoes of the World.* Austin, Tx., Univ. of Texas Press.

Decker, R.W., and Decker, B., 1981. *Volcanoes.* New York, W.H. Freeman & Co.

Harris, S.L., 1987. *Fire and Ice, the Cascade Volcanoes.* 3d ed., New York, Harcourt, Brace, and Jovanovich.

MacDonald, G.A., 1972. *Volcanoes.* Englewood Cliffs, N.J., Prentice-Hall, Inc.

Oakeshott, G.B., 1976. *Volcanoes and Earthquakes . . . Geologic Violence.* New York, McGraw-Hill.

Williams, H., and McBirney, A.R., 1979. *Volcanology.* San Francisco, W.H., Freeman & Co.

Weathering and Soils

5

The spectacular scenery of Monument Valley in southeastern Utah is dominated by majestic buttes such as the one shown here. Commonly, buttes are capped by a resistant layer of rock, in this case a thick unit called the Cedar Mesa Sandstone. Beneath this massive sandstone are clayey beds of the Halgaito Shale.

The earth, like the body of an animal, is wasted at the same time that it is replaced. It has a state of growth and augmentation; it has another state which is that of diminution and decay.

James Hutton Theory of the Earth, *1795*

Introduction

Over 3 billion years ago, water fell upon the earth much as it does today. Solid rocks were reduced to grains of sand by the action of water, ice, and chemically active solutions. We know this from clues left in ancient rocks whose great antiquity has been established by radioactive dating. The processes by which rocks are broken into smaller particles and chemically decomposed have continued unabated down through the eras of geologic time. We term these most enduring of geologic processes **weathering**. Weathering is of enormous importance to all of us. It causes solid rock to crumble and thereby facilitates erosion. Weathering provides sediment for the making of sedimentary rocks. Without weathering there would be no soil for the growth of plants that are vital to ourselves and all other forms of life.

Weathering is a phenomenon that combines two categories of processes. On the one hand are those processes that result in the fragmentation of once-solid rock without chemical change, as when water freezes and expands in crevices so as to cause the rock to break apart, or when rocks are wedged apart by the growth of roots from trees and shrubs. These processes of physical disintegration are designated **mechanical weathering**. In contrast, **chemical weathering** involves the chemical decomposition of rocks as their minerals react with water and air. Mechanical and chemical weathering work together in the destruction of rocks at the earth's surface, although one may predominate over the other. In areas of extreme cold or aridity, for example, mechanical weathering may predominate, but in warm wet regions, chemical decomposition is the predominant kind of weathering.

The term weathering refers to the chemical and mechanical changes that occur in rocks as a result of their exposure to the atmosphere.

Although mechanical breakup and chemical decay of a rock are partners in weathering, it is advantageous to discuss these two components of weathering separately, because disintegration involves largely mechanical processes, whereas decomposition entails the chemical interaction of water, gases, and minerals.

Mechanical Weathering

The chief agents of **mechanical weathering**, or **disintegration,** are frost action, temperature changes, pressure release due to unloading, crystal growth, and the wedging action of plant roots. Many different factors influence the efficiency of these agents in causing disintegration. The composition and texture of the rocks being weathered and the presence of joints, fractures, and voids clearly affect the rate at which solid rock can be reduced to rubble. Mechanical weathering is also affected by climate, topography, and the length of time over which weathering agents have been operating. By definition, mechanical weathering involves the physical disintegration of solid masses of rock into loose fragments. There is little chemical change in the rock itself. Thus, a chemical analysis of the disintegrated rock would be similar to that of the parent rock.

Frost Action

Water expands by about 9 per cent when it freezes. If the water freezes in a confined space, pressure caused by the expansion may cause rock masses to be pushed apart and ruptured. Such **frost action** has an important effect in mechanical weathering in that the freezing water is capable of exerting thousands of pounds of pressure per square inch. Many of us have become aware of the effects of frost action during severe winters when concrete roads and sidewalks are cracked by alternate freezing and thawing. The way freezing water disrupts solid rock is similar to the manner in which it breaks up streets and sidewalks. After water has filled a crack in the rock, the water at the lip of the crack may freeze (Fig. 5–1). Once this has occurred, the water deeper in the crack is sealed off, and as it begins to freeze and expand, it pushes against the walls of the fracture and thereby widens the space between those walls. In a subsequent thaw, fragments of rock may slip down into the crack and act as wedges to hold it open. These processes are most prevalent in areas where water is abundant and where temperatures drop to below freezing at night and then warm during the day. Mountainous regions in temperate zones have this kind of daily temperature change, and in such rugged regions frost action is an important weathering process.

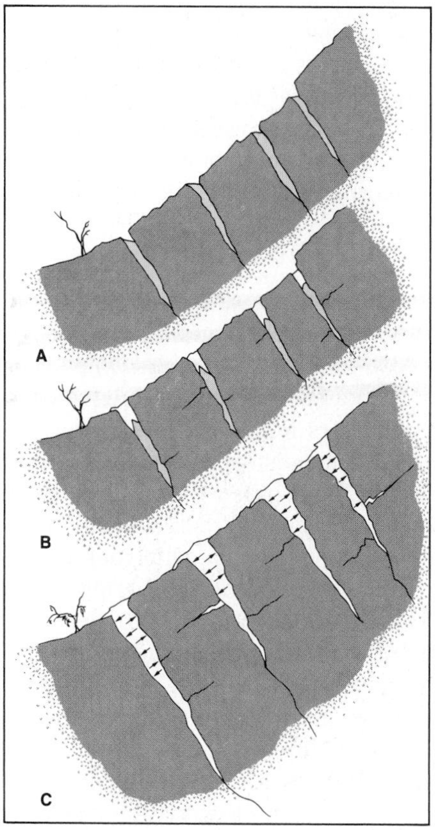

Figure 5–1 Three steps in the process of widening fractures and joints by the action of freezing water. (A) Water enters and fills rock fractures and joints. (B) As temperatures decline, ice forms at the surface and thereby forms a seal over underlying water. (C) Continued cooling eventually causes water at deeper levels to form ice. The transformation involves an increase in volume of 9 per cent, resulting in pressure on the walls of fractures. Joints and fractures are pried open and widened.

Unloading

Quarrymen and miners are well aware that when the heavy weight of rock is removed from a mine with consequent loss of support to the surrounding rock, the outside pressure may sometimes cause a veritable explosion of rock fragments into the excavations. In nature, erosion may similarly **unload** or remove large volumes of surface rock, thus lightening the load on deeper rock and allowing it to expand. Because the mass of rock is still confined on all sides, it can respond to the release of pressure only by expanding upward. As it expands it will rupture and form joints that are roughly parallel to the surface topography. This separation of rock into slabs or sheets by fractures that generally parallel the ground surface is called **sheeting** (Fig. 5–2). Sheeting is often seen in granite and other massive crystalline rocks that have been laid bare by glaciation or running water. It can also be readily observed along the upper walls of quarries, where it may even facilitate the removal of the stone. Sheeting provides planar passageways along which solution and frost action may proceed vigorously. Half Dome at Yosemite National park and Stone Mountain, Georgia, are examples of mountains whose shape has been controlled, at least partially, by sheeting.

Saline Crystal Growth

The growth of crystals, especially crystals of sodium chloride, calcium sulfate, or magnesium sulfate have been found to cause scaling in exposed rock surfaces and to dislodge grains and crystals from their parent rock. The process is particularly effective in porous rocks subjected to alternate wetting and drying. During the drier periods, crystals form and exert large expansive stresses as they grow. In addition, some salt minerals already crystallized in pore spaces undergo expansion when water is added, thereby promoting further destruction. Disintegration of building stone because of the growth of crystals in pore spaces has been a vexing problem for architects involved in the preservation of historically important buildings. An example is provided by London's House of Parliament where magnesium and calcium sulfates crystallize along zones of weakness in the dolomitic building stone and cause spalling and crumbling at an alarming rate. The sulfates form as a result of chemical reactions between the calcium and magnesium in the rock and atmospheric pollutants.

Figure 5–2 Sheet structure or sheeting in granite in the Anza desert east of San Diego, California.

Root Wedging

When one observes the growth of a plant from a seed, it is apparent that even a frail seedling is capable of exerting sufficient force to push aside relatively firm soil. As they grow into rock fractures and expand, the roots of large plants such as trees can exert correspondingly greater forces and are capable of widening cracks and accelerating the rate of disintegration by significant amounts (Fig. 5–3). When the plant dies and the roots decay, they leave openings in which freezing water may accumulate and further widen the void space. Plants also react with rocks chemically, as will be described in the section on chemical weathering.

Mechanical weathering is accomplished primarily by frost action, unloading, root wedging, and stresses induced by the growth of salt or gypsum crystals in pores and crevices.

Figure 5–3 The roots of this tree have cracked this large boulder apart. Root action is a powerful weathering agent. (Photograph by Grant Lashbrook.)

Chemical Weathering

Chemical weathering refers to the decomposition of rocks at or near the earth's surface and under relatively low temperatures (Fig. 5–4). Dur-

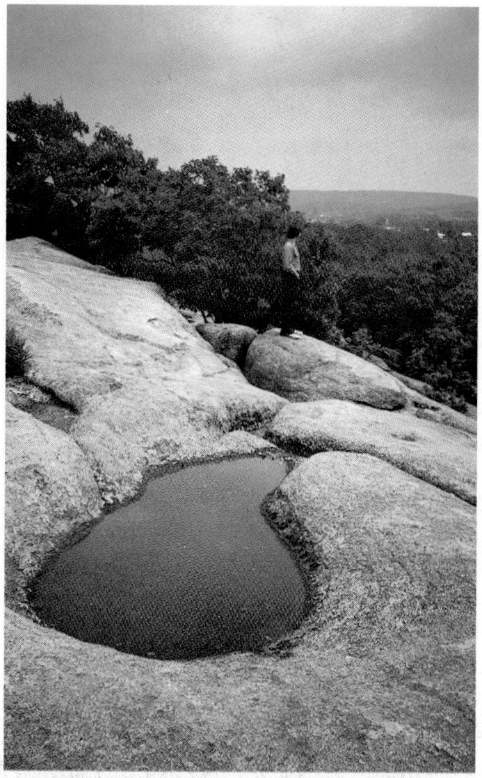

Figure 5–4 Weathering pits in granite. These pits are formed when water collects in depressions following rainfall. Carbon dioxide from the rainwater, as well as from lichens, gives the water a slight acidity, which accelerates chemical weathering.

ing chemical weathering, water and chemically active water solutions as well as oxygen and carbon dioxide from the atmosphere attack the minerals composing rocks. The susceptible parent minerals are frequently altered to softer secondary minerals, as when feldspars are converted to clay minerals and lose certain ions to solution. Chemical weathering is an exceedingly important process, for it leads to the formation of soils and in some parts of the world has resulted in the concentration of iron, aluminum, uranium, gold, and tin.

Relation of Disintegration to Decomposition

The processes of chemical and mechanical weathering are not truly independent of one another. Because of climatic factors, especially the amount and frequency of rainfall, one or the other process may be dominant in a particular location. For the most part, however, decomposition and disintegration are interacting processes. For example, disintegration greatly promotes decomposition by reducing large intact rock masses to accumulations of smaller fragments. This fragmentation provides more total surface area for chemical attack than was originally present in the larger mass of rock. One can see what happens readily by imagining a block of rock (Fig. 5–5) measuring 2 meters on each edge. Such a block would have a total of 24 square meters of surface area. If the block is split once along each of three perpendicular planes, each resulting block would have 6 square meters of surface area or a total of 48 square meters for all eight pieces. Chemically reactive solutions are able to decompose only surfaces they can reach, and so

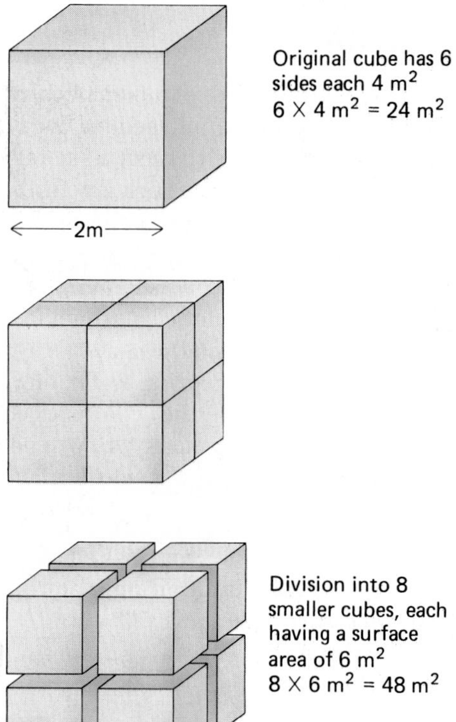
Original cube has 6 sides each 4 m^2
$6 \times 4\ m^2 = 24\ m^2$
2m
Division into 8 smaller cubes, each having a surface area of 6 m^2
$8 \times 6\ m^2 = 48\ m^2$

Figure 5–5 By splitting a cube of rock as shown here, the surface area exposed to chemical agents is doubled. It is for this reason that finer-grained materials decompose more rapidly than coarser-grained materials of identical composition.

the rate of chemical attack is enhanced by increasing surface area. One can demonstrate this fact by applying a few drops of dilute hydrochloric acid first to an unfragmented piece of calcite and then to a similar piece of calcite that has been pulverized. Because of the greater amount of total surface area, the pulverized rock will display the more vigorous effervescence.

Mechanical weathering increases the surface area on which chemical reactions may occur, and thereby enhances chemical weathering.

Processes of Chemical Weathering

People sometimes are surprised to learn that many rocks exposed at the earth's surface are really not in chemical equilibrium with their environment but rather are unstable and are slowly but continuously reacting with atmospheric components to dissolve or change to new substances that are more stable at the earth's surface. For most common silicate minerals, weathering proceeds in accordance with a generalization called the "rule of stability," which states that a mineral approaches stability most closely in an environment similar to that in which it formed. Intrusive igneous rocks, of course, are formed under conditions of high temperature, high pressure, and a deficiency of free oxygen and fresh water. When such rocks are laid bare on the continents, they find themselves in an environment of low temperature, low pressure, and abundant oxygen and water. Chemical change is inevitable, and the minerals most susceptible to change are those that formed under physical and chemical conditions most removed from conditions at the earth's surface. For example, among the common rock-forming silicate minerals, olivine forms at high temperatures and pressures early in the crystallization of a magma. Consequently, it rapidly weathers in the environments that exist at the earth's surface. Quartz forms much later under less extreme conditions of temperature and pressure and is less susceptible to weathering. In reference to the Bowen Reaction Series (discussed in Chapter 3), it is apparent that minerals nearer the top of the series generally weather more rapidly than those near the base (Fig. 5–6).

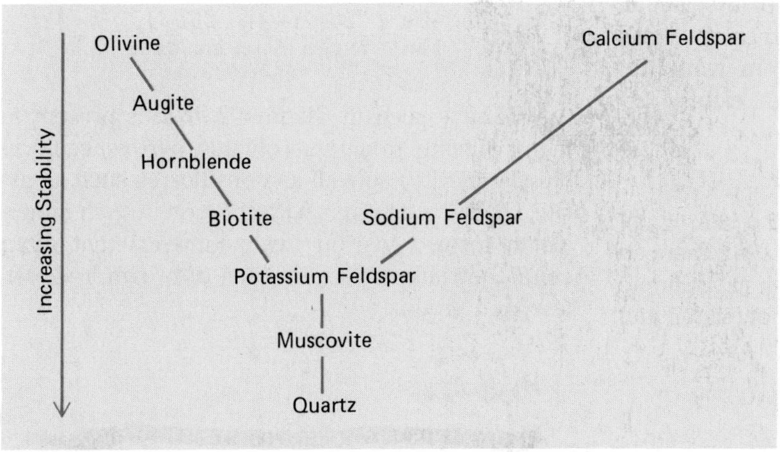

Figure 5–6 From a study of the weathering of various kinds of igneous rocks, the mineralogist S.S. Goldich in 1938 determined that the weathering sequence for the common rock-forming minerals coincides with Bowen's order of crystallization of minerals from a silicate melt. Thus quartz is the most resistant to decomposition of the common silicate minerals. (Reprinted from Goldich, S.S., *A Study of Rock Weathering, Journal of Geology, 46*:17–58, 1938, by permission of the University of Chicago Press.)

The order of stability of minerals found in igneous rocks is approximately the reverse of their order of crystallization from a cooling granitic melt. Thus, those minerals at the top of Bowen's reaction series weather more rapidly than those at the bottom.

In Chapter 2, we noted how silicon–oxygen tetrahedra are joined together by sharing some oxygen ions and that the number of metallic ions needed to neutralize the crystal is reduced by this sharing. This increased oxygen sharing by silicon results in a greater number of strong bonds between silicon and oxygen, and this in turn greatly increases the ability of the mineral to resist weathering. For example, the ratio of oxygen to silicon in olivine is 4, in pyroxene 3, in hornblende 2.7, in biotite 2.5, and in quartz 2. The diminishing ratios correlate nicely with greater resistance to weathering. All of the oxygen atoms are shared by silicon atoms in quartz, and this partially accounts for its great resistance to weathering.

Chemical weathering involves a variety of processes that are interrelated, that may occur simultaneously, and that utilize water, oxygen, and carbon dioxide. Solution, oxidation, hydrolysis, hydration, and chemical changes induced by plants are all processes involved in chemical weathering. These processes operate most vigorously in warm regions having abundant vegetation and rainfall.

Solution

Solution weathering involves the dissolving away of solid rock. It is a process that is enhanced by the presence of organic acids such as carbonic acid, which results from the union of water and carbon dioxide.

$$\underset{\text{water}}{H_2O} + \underset{\text{carbon dioxide}}{CO_2} \longrightarrow \underset{\text{carbonic acid}}{H_2CO_3}$$

Water for the above reaction may be supplied by rainfall, and the carbon dioxide is available from the atmosphere, from respiration by plant roots, and from the decay of organic matter. Although relatively weak, carbonic acid is an effective agent in dissolving rocks. As an example, consider its effect on the calcite of which limestone is formed.

$$\underset{\text{calcite}}{CaCO_3} + \underset{\text{carbonic acid}}{H_2CO_3} \longrightarrow \underset{\text{calcium ion}}{Ca^{2+}} + \underset{\text{two bicarbonate ion}}{2HCO_3^-}$$

Here, the carbonic acid solution reacts with calcite to form a calcium ion and two bicarbonate ions that are removed from the rock by percolating water. The process leaves cavities where there was once solid mineral matter. Chemical weathering by solution accounts for the caverns and other solution features that may characterize humid regions underlain by limestones.

Carbonic acid, which forms when carbon dioxide dissolves in water, is an important and effective agent for chemical weathering.

Oxidation

When oxygen combines with a substance, the process is **oxidation**. A familiar example is the combination of iron and oxygen to produce rust.

$$\underset{\text{iron}}{4Fe} + \underset{\text{oxygen}}{3O_2} \longrightarrow \underset{\text{iron oxide}}{2Fe_2O_3}$$

As we have seen in Chapter 2, iron is present in many silicate minerals (olivine, pyroxene, hornblende, biotite) as well as nonsilicates such as pyrite. Oxygen combines with the iron in such minerals to form new iron oxide minerals that often color the rock in various hues of brown, red, yellow, and green.

Hydrolysis

Hydrolysis is a chemical reaction between a mineral and water. The hydrogen ion contained in water is a potent wreaker of the atomic structure of silicate minerals. It is small and therefore able to enter the lattice easily. In addition it bears a concentrated electric charge capable of disrupting the charge balance in the mineral, thereby displacing ions such as potassium or calcium.

Hydrogen ions are made available for hydrolysis reactions when some small proportion of water molecules undergo dissociation:

$$\underset{\text{water}}{H_2O} \rightleftharpoons \underset{\text{hydrogen ion}}{H^+} + \underset{\text{hydroxyl ion}}{OH^-}$$

or when atmospheric carbon dioxide dissolves in water:

$$\underset{\text{water}}{H_2O} + \underset{\text{carbon dioxide}}{CO_2} \rightleftharpoons \underset{\text{hydrogen ion}}{H^+} + \underset{\text{bicarbonate ion}}{HCO_3^-}$$

The hydrogen ions are then free to attack silicate minerals such as orthoclase feldspar.

$$\underset{\text{orthoclase}}{2KAlSi_3O_8} + \underset{\text{two hydrogen ions}}{2H^+} + \underset{\text{water}}{9H_2O} \longrightarrow$$

$$\underset{\text{clay mineral (kaolinite)}}{Al_2Si_2O_5(OH)_4} + \underset{\text{silicic acid}}{4H_4SiO_4} + \underset{\text{two potassium ions}}{2K^+}$$

In the above reaction, the potassium ions displaced during hydrolysis may be carried away in solution, used by plants, or incorporated into clay minerals. Although a small portion of the silica is removed in solution, the greater amount remains in the clay-rich weathered residue.

Hydration

Hydration is a process whereby water is absorbed by a mineral and incorporated into the weathering product. For example, the mineral anhydrite ($CaSO_4$) may take in water to become gypsum ($CaSO_4 \cdot nH_2O$), or hematite (Fe_2O_3) may be converted to limonite ($Fe_2O_3 \cdot nH_2O$). Hydration is an important process in the development of clay and accounts for the presence of water within many clay minerals. An important aspect of hydration is that the hydrated mineral, because of the water it has taken up, is larger than the parent mineral. The increase in volume causes growing hydrated crystals to exert pressure on the walls of the spaces they occupy, and such pressure may contribute to rock disintegration.

Biochemical Weathering

Rooted plants also play a role in the weathering of materials at the surface of the earth. During photosynthesis, plants utilize the energy of the sun's radiation to combine water and carbon dioxide and produce tissue necessary for growth and maintenance. Part of the hydrogen from moisture taken in by the plant is released at the surface of rootlets (Fig. 5–7) where it is exchanged for ions of calcium, magnesium, and sodium from adjacent particles of clay and other minerals. These exchanged

Figure 5–7 Plant rootlets provide a continuing supply of hydrogen ions, which are exchanged for cations from adjacent mineral particles. These exchanged ions are required for plant nutrition. The process accelerates weathering of minerals in soils. (Modified from Keller, W.D., *Principles of Chemical Weathering,* Columbia, Missouri, Lucas Bros., Publishers, 1957.)

ions are required by the plant for nutrition. The hydrogen ions transferred to the clay particles render the clay slightly acidic, and this triggers weathering reactions with neighboring feldspars and other silicates. Eventually, some of these minerals also decompose to clay, thus perpetuating the weathering process.

Even tiny plants may contribute to weathering. An example is provided by the small encrusting organisms called lichens. Lichens are symbiotic mixtures of fungi and algae. They are often the first forms of life to colonize a newly exposed surface of rock. (The green encrustations visible in Fig. 5–4 are lichens). Lichens use organic acids and tiny, hairlike rootlets to dissolve iron and other elements directly from the rock. They are capable of dislodging grains by alternately swelling after rainfall and shrinking during dry periods.

Once a residue of weathered material begins to accumulate as a result of the various weathering processes, millions of bacteria, fungi, and algae colonize the clay-rich residue, and further contribute to weathering and soil development. These microbial organisms are able to accelerate the rate at which iron is dissolved. They also produce carbon dioxide, which combines with water to provide carbonic acid. Their remains and metabolic products add to the store of organic matter that combines with clay and other fine sediment to form soil.

Rates of Weathering

The Role of Climate

How rapidly a particular exposure of rock will weather is dependent on several different interacting factors in the environment. Climate is important because it controls temperature and rainfall. Warmer temperatures, of course, promote both chemical and biochemical reactions. Even an increase of as little as 10°C can double reaction rates. Similarly, high rainfall accelerates rates of weathering by increasing the possibilities for solution and reaction and by helping to wash away surface debris and expose fresh surfaces for chemical attack. Both higher temperatures and an abundance of moisture promote the growth of plants, and luxuriant plant growth assists in weathering by extracting ions from minerals and adding chemically reactive compounds. The importance of temperature and rainfall is at least partly evident in the thickness of loose material or **regolith** that develops over bedrock under various climatic conditions. In temperate regions having level topography, the layer of regolith is usually only about a meter deep, whereas in similar topography of the tropics, it may form a blanket tens of meters thick. Arctic areas are only thinly covered by regolith.

Chemical weathering proceeds most vigorously in climates characterized by relatively high temperatures and abundant rainfall. It proceeds more slowly in regions of cold or arid climate.

Parent Rock

Depending on their composition, texture, and such features as fractures and joints, rocks exposed at the earth's surface may exhibit a wide range of resistance to weathering. Limestones and dolostones, for example, are highly susceptible to solution (Fig. 5–8). In regions having high rainfall, these rocks weather more rapidly than in arid regions. This observation suggests that it is difficult to list rocks in the order of their resistance to weathering because a rock might be more durable than another under one set of conditions and the same rock less durable under different conditions. As a general rule, igneous and metamorphic rocks are more resistant to weathering than sedimentary rocks. A silica-cemented quartz sandstone, however, is among the most enduring of rock types.

Statues and other sculptured edifices that have actual dates carved into the stone permit one to assess rates of weathering. In cities, because of the higher levels of carbon and sulfur dioxides, chemical weathering rates are generally higher than in nonindustrialized areas. Figure 5–9 provides an example of damage to statues in cities. The bust of Beethoven was erected in a St. Louis park in 1884. In spite of many attempts to preserve the Italian Carrara marble from which the bust was carved, it

Figure 5–8 A dinosaur (Triceratops) carved in limestone peers from the ivy covering part of the geology building at Washington University in St. Louis. The original smooth surface of the limestone is now rough and pitted because of differential chemical weathering enhanced by carbon dioxide in rainwater. The particles of calcite that stand out in relief are relatively dense nonporous shell fragments of marine invertebrates. The carving and the building it decorates are about 70 years old. (Photograph by Herb Weitman.)

continued to decompose. The composer's gruff expression has been so altered that St. Louisans refer to the statue as "poor Beethoven."

Acid Rain

Rates of chemical weathering are, to be sure, affected by the acidity of the solutions that attack rocks and minerals. As described above, even unpolluted rainwater contains carbon dioxide and is slightly acidic. In recent years, however, this weak natural carbonic acid has been augmented by strong acids formed where rain has fallen through air polluted with oxides of nitrogen and sulfur. This acidic precipitation includes both nitric and sulfuric acid. It is called **acid rain**. Acid rain commonly exceeds the natural acidity of rainwater by more than 200 times, and in a few instances by over 1000 times. If not neutralized by alkaline rocks and soils, the acid accumulates in lakes and streams where it is often lethal to fish and other forms of aquatic life. Acid rain also damages the foliage of plants and reduces the quality of soil by **leaching** away nutrients. The acid solutions attack limestone, marble, and concrete building materials and damage valuable works of art. Presently, millions of dollars are being expended in acid rain research in the hope that a solution to the problem can be found. Aside from local chemical treatment of particular areas, however, a long-term solution must involve reduction in atmospheric pollutants that form acid rain.

Figure 5–9 The ravages of chemical weathering are clearly visible on this statue of the composer Ludwig van Beethoven erected in 1884 in Tower Grove Park, St. Louis, Missouri. The statue was carved in Italian Carrara marble. (Courtesy of the Washington University Center of Archaeometry.)

Rainfall containing sulfuric and nitric acids derived from air pollutants is termed acid rain. Acid rain is harmful to life, soils, monuments, and buildings.

Spheroidal Weathering

Spheroidal weathering is a term used to describe the spalling away of concentric surficial shells of the rounded surface of a boulder or rock mass. Such "onion-skin" weathering is believed to result primarily from the mechanical effects of chemical weathering. When feldspars decompose, the clay product has a greater volume than the parent feldspar. The increase in volume disrupts the interlocking texture of mineral grains in the rock and causes breakage and separation of the layer of partially weathered rock near the surface. Corners of roughly rectangular blocks broken loose along joints are decomposed along three surfaces simultaneously and are progressively rounded by spalling until the rock mass assumes the spheroidal form (Fig. 5–10).

Spheroidal weathering operates most effectively on relatively small rock masses such as those of boulder size. The kind of spheroidal weathering involving a larger-scale breaking off of concentric plates from bare rock surfaces is referred to as **exfoliation**. Rocks that have experienced exfoliation may resemble those having sheet structure (see Fig. 5–2), but while sheet structure is caused by release of pressure when overlying rock masses are removed by erosion, all forms of spheroidal weathering result from physical and chemical weathering. In mountainous regions composed of massive rocks such as granite, exfoliation may produce rounded summits called **exfoliation domes** (Fig. 5–11).

The Decay of Granite

The mineral alterations involved in chemical weathering are nicely illustrated by the decay of a granitic parent rock. As described in Chapter 3, granite is a common igneous rock composed of about two parts orthoclase, one part quartz, one part plagioclase, and small amounts of ferromagnesian minerals (see Fig. 6–2). Where masses of granite have been exposed for a long period of time, one may find places where the weathered products of the granite have not been washed away but have accumulated as a clayey granular residue called **grus** (Fig. 5–12). On close examination, this material is found to consist of quartz grains, partly decayed and clay-coated feldspars, rust-colored particles of partially decayed ferromagnesian minerals, and clay. The quartz grains in the weathered material are relatively unaltered because of the great resistance quartz has to chemical attack. As other

Figure 5–10 Spheroidal weathering in granite boulder (A) near Virginia Dale, Colorado; in massive sandstones (B) along highway 150 east of Bessemer, Alabama; in the Balolatchie Conglomerate (C) near Girven, Ayrshire, Great Britain; and in the basalt (D) of Ireland's Giant's Causeway. (Courtesy of J.C. Brice.)

minerals in the granite decompose around them, quartz grains are freed from the rock matrix, later to be transported and deposited as sand, and perhaps ultimately to become components of sandstone. Granite provides a good example of **differential weathering**, by which its various components weather at different rates.

We have noted that feldspars are aluminosilicates of potassium, sodium, and calcium (see Table 2–5). In the weathering process, potassium, sodium, and calcium are largely dissolved and carried away in solution. Subsequently they may be combined with other elements and incorporated into sedimentary rocks. At least some of the potassium is retained in newly forming clay minerals. The remaining aluminum and silicon in the feldspars become the chief ingredients of clay, and this accounts for the coating of clay frequently found around feldspar grains and the clay residue found adjacent to bodies of weathering granite. Later, this same clay may find its way into the making of sedimentary rocks like shale and claystone.

Figure 5–11 Exfoliation dome, along Tioga Pass, Yosemite Valley, California.

Figure 5–12 Grus, the clayey granular residue resulting from weathering of granite. The parent rock is a variety of granite called granodiorite. Sierra Nevada, California.

Quartz and feldspar are the two most abundant components of granite. On weathering of granite, feldspars decay to clay minerals. Quartz is resistant to decay, and accumulates as quartz sand.

During the decomposition of the ferromagnesian minerals present in the granite source rock, potassium, sodium, and calcium are dissolved in the same way as in the feldspars. Again, aluminum and silicon go into the construction of clay minerals. The ions that remain are iron and magnesium. The iron combines readily with oxygen to form iron oxide minerals such as hematite (Fe_2O_3) and hydrous iron oxide minerals like goethite, $HFeO_2$. These iron minerals color sediment and rocks in tints of yellow, orange, red, and brown. Finally, the magnesium that is derived from parent ferromagnesian minerals may find its way into limey sediments or become a component of certain clay minerals.

Soil

In terms of importance to human populations, the most significant aspect of weathering is in the development of soil. **Soil** can be defined as weathered material that will support the growth of rooted plants. A heap of quartz sand is not soil according to this definition, for it will not support vegetation. Nor is the so-called "lunar soil" of the moon a true soil. It is merely disintegrated mineral matter.

Soil is a vital natural material that would not develop were it not for the activities of a myriad of bacteria, fungi, worms, and insects. Actually, soil can be considered an intricate mixture of mineral solids, water, gases, dissolved substances, remains of dead organisms, and multitudes of thriving organisms.

Why is a certain amount of organic material necessary in order for weathered material to support the growth of rooted plants? One reason is that the organic material supplies nutrients required for vigorous plant growth. Plants need nitrogen in order to synthesize protein, but they are unable to take free nitrogen (N_2) from the air. Certain soil bacteria, however, are able to use this free nitrogen and convert it into usable fixed nitrogen compounds (nitrates). Also, the decaying vegetation and animal waste contain ammonia (NH_3), the nitrogen of which can be utilized by most plants.

Soils are mixtures of fragmented and weathered rocks and minerals, organic matter, air, and water that can support plant life.

The material that colors the upper parts of many soils gray or black is called **humus**. Humus is organic material that is so thoroughly decayed that one cannot discern the nature of the parent organisms. This essential organic mixture increases the porosity and water-holding capacity of the soil, provides a buffer against rapid changes in acidity, and assists in the retention of chemicals needed as plant nutrients.

Factors Governing Soil Development

Five factors are involved in soil formation. These are climate, parent material, topography, time, and biologic activity. Of these factors, climate is of greatest importance. Temperature and rainfall, of course, influence not only rates of weathering but also the nature of vegetation. Given enough time, soil-forming processes operating within a given climate will prevail over differences in parent material and provide a basically climate-controlled soil type. The reverse is also true. Identical rocks in temperate high-rainfall regions and in semitropical arid regions will produce quite different soils.

The influence of parent material on the formation of soils is of secondary importance but can be observed in soils that have recently developed from unaltered bedrock. Such soils still retain textural and mineral components that are derived from the parent rock but that may become obscured after the soil has evolved to a further stage. In some cases, various influences of parent material will persist in the developing soil. Soils developed above quartz sandstones may tend to be more acidic than soils formed on limestones, and these differences will be reflected by the natural vegetation living on each soil type. Not all soils are developed from underlying consolidated bedrock. Many fine agricultural soils are formed on loosely consolidated sediment laid down by streams, winds, glaciers, or the waters of lakes.

Topography also influences the ultimate nature of soils. On steep slopes, for example, erosion is usually more vigorous and the soil layer generally thinner. Also, water falling on slopes is partially lost to runoff, and therefore there is less water available for percolation into deeper layers of soil.

The factor of time in soil formation is often not fully appreciated by those who bulldoze away soil for construction or mining and expect nature to quickly restore this important resource. The development of soil is an exceedingly slow process. Depending on climatic conditions, several hundred to several thousands of years are required to produce a fertile soil. Agricultural soils formed during this lengthy natural process may be lost to erosion in only a few years, particularly in areas where soil conservation measures are not followed. The problem is not trivial, for an estimated 3 billion metric tons of good agricultural soil is lost each year from agricultural fields in the United States alone (Fig. 5–13).

Figure 5–13 Fertile topsoil being washed away in an untended field in eastern Missouri.

Soil develops by the weathering of bedrock, or of transported materials. Its characteristics are determined by climate, topography, vegetation, parent materials, and time.

Soil Horizons

Soils are not uniform in texture and composition from bedrock to the surface but rather consist of a number of fairly distinct layers or **soil horizons**, which differ in composition and physical characteristics. The stack of soil horizons is called a **soil profile** (Fig. 5–14). A simple profile, such as might develop in a humid region on granitic rock, has three distinct divisions. Soil scientists have named these layers the A, B, and C horizons. The A horizon, also known as "top soil," which lies immediately beneath the surface, is characterized by a high content of organic matter. It is also a zone in which soluble compounds are dissolved and, along with fine clay particles, carried downward to be deposited in the underlying B horizon. Because of these losses, the A horizon is referred to as leached or eluviated, whereas the B layer is sometimes called the washed-in or illuviated zone. Iron and clay minerals tend to accumulate in the B layer. In humid regions one can often recognize the B horizon by its more brownish color and clayey texture. The C horizon of the soil profile consists of partially altered parent material. In it one finds evidence of chemical weathering, but soil development has not progressed to the level at which the

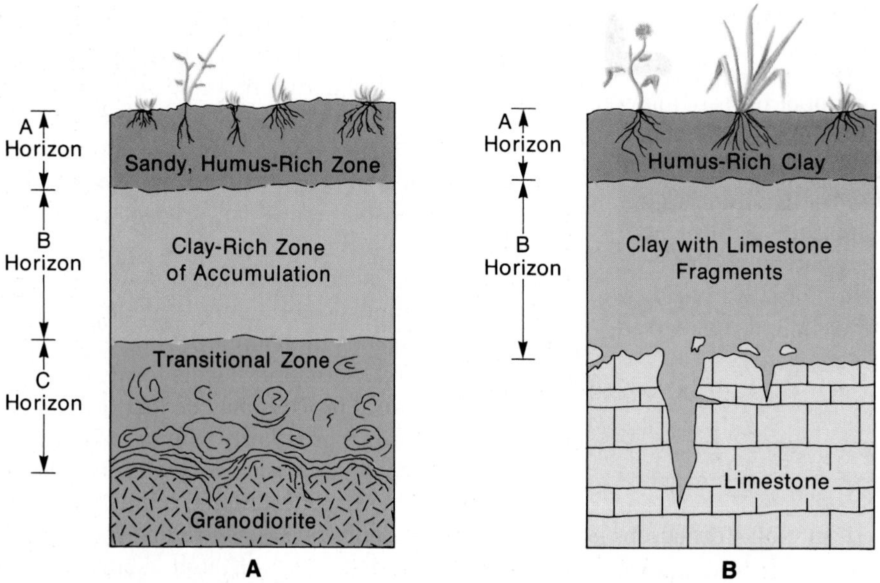

Figure 5–14 Soil profiles developed on (A) granodiorite and (B) limestone.

original characteristics of the bedrock are unrecognizable. Beneath the C horizon lies unaltered parent material.

A soil profile consists of the layers of altered material that occur in sequence from the surface to unaltered bedrock. Most soil profiles have three layers: an uppermost A horizon, an intermediate B horizon, and a lower C horizon.

Soils developed on limestones in humid regions often do not exhibit a clear division between the B and C horizons. In such soils, the A horizon consists of dark humus-rich material that changes downward into relatively light-colored clay and then stained and partly dissolved limestone (Fig. 5–14B). The clays in these soils are residues of clay impurities left when the limestone was dissolved.

How does the sequence of soil horizons develop in a humid region underlain by granitic rock? Imagine that glaciation has recently stripped away all loose sediment over a granite outcrop. Initially, disintegration and decomposition of the granite will produce a sandy layer composed mostly of grains of quartz and feldspar along with clay derived from the weathering of silicate minerals. As the layer of granular material thickens, plants establish themselves, and organic material from a variety of sources becomes incorporated into the surface layer. Meanwhile, with each rainfall, soluble ions and clay particles are flushed downward from the surface layer into a lower level. This lower level soon takes on the clayey character of the B horizon. While these events are taking place, solutions reaching still deeper continue the chemical attack on the parent material, thus perpetuating the C horizon.

Major Kinds of Soils

Soil scientists employ a rather complex nomenclature for the great variety of soil types present at one place or another around the globe. Fortunately, most of the names in these lengthy classifications can be avoided if one desires only a general understanding of a few major kinds of soils. For example, the soils that predominate in the eastern United States and southeastern Canada are called **pedalfers** (Fig. 5–15). The name was fabricated to emphasize the fact that in such soils aluminum *(al)* and iron *(fer)* have been leached from the A horizon and deposited in the underlying B horizon.

Figure 5–15 Pedalfer soil developed on siltstones and sandstones of Mississippian age in northern Alabama. The rustlike color of pedalfers reflects their rich content of iron.

Pedalfers are clayey soils that develop in regions having an abundance of rainfall.

In the more arid regions of the world, less humus develops in soils, and there is less opportunity for solution and leaching. In the absence of water, chemical weathering is slower, and less clay is produced. Usually, there is insufficient ground water to flush soluble materials such as calcite out of the B horizon, so it accumulates there as soil moisture is lost by evaporation. Because of the persistence of calcite in these soils, they are called **pedocals** (*pedon*, soil, and *cal* for calcium carbonate) (Fig. 5–16).

Frequently in dry areas, there is not only insufficient water to cause downward leaching, but there is an upward movement of water because of the high rate of evaporation at the surface. Mineral matter dissolved in the diminishing soil water is precipitated at the surface as a hardpan or **caliche** layer.

In the hot and humid regions of the tropics, the characteristic soils are **laterites**. The term "laterite" is derived from the Latin word *latere* or brick, and originally referred to the use of this material to make bricks in India and Cambodia. Laterites have a relatively thin organic layer covering a reddish leached layer, which is often underlain by a still darker red layer. In laterites, oxidation and hydrolysis have been so intense that feldspars and ferromagnesian minerals are completely decayed. Not only is calcium carbonate removed, but also silica. Only the most insoluble compounds, mainly aluminum and iron oxides, accumulate in these soils.

Figure 5–16 Pedocal soil exposed by gully erosion, eastern Nebraska. The light coloration in the B horizon results from accumulation of calcium carbonate.

Although laterites may support a lush growth of natural tropical vegetation, they are not good agricultural soils. When forested areas with lateritic soils are cleared and plowed, the thin organic cover tends to be rapidly oxidized in the prevailing warm climates. A thick accumulation of organic matter such as that found in rich black soils of more temperate regions cannot develop. After a few years of tilling and planting, the organic component of the soil is so depleted that fields must be abandoned.

In some laterites, the concentration of either iron or aluminum may reach levels that permit the deposits to be profitably mined. The iron-rich laterites originate from the weathering of parent materials that also contained iron, although not concentrated into ore bodies. Extensive lateritic ores of iron occur in Cuba, Columbia, Venezuela, and the Philippines. If there is little iron in the parent material and an abundance of aluminum, then laterites rich in hydrated aluminum oxides may form. Such materials are called **bauxites**. Ancient bauxite deposits are mined in Guyana, Ghana, northern Queensland, and Arkansas. Concentration of bauxite takes place in tropical areas of low relief where temperatures exceed 25°C most of the time and where there is an abundance of water for leaching and chemical reactions. At the present time, bauxite is the only ore from which it is economically practical to extract aluminum.

Soil Erosion

It takes nature thousands of years to produce a fertile soil. That same soil can be lost to erosion in only a few decades. Clearly, as we attempt to feed our expanding world population, care must be taken to prevent unnecessary erosional losses of this vital resource.

Soil erosion results primarily from uncontrolled runoff of surface water. In areas that have never been tilled, soil is protected by a cover of plants, and there is an approximate balance between the slow loss of soil by erosion and the development of new soil from underlying parent material. Whenever the protective shield of vegetation is broken, however, erosion will follow (Fig. 5–17).

Of course, some of the erosion resulting from farming is unavoidable. We must have food, and so we must break into the natural vegetative cover. There are, however, farming methods that reduce the loss of topsoil. For example, farmers now plow and plant their crops in rows that follow contours so that rainwater is retained rather than allowed to run off (Fig. 5–18). Efforts are made to reduce the amount of time between preparation of the soil for planting and planting itself so that the bare soil will not remain exposed to erosion for long periods. *Strip-cropping* is also practiced (Fig. 5–19). This technique involves planting alternate bands of erosion-resistant crops such as clover and alfalfa with open-spaced crops like corn. Soil losses in the erosion-susceptible corn strip are trapped in the adjacent strip of denser crops. Where fields are temporarily not utilized, soil-holding "cover" crops are planted, not only to retard erosion but to improve the soil by adding nitrogen. Because ordinary grass is an effective retardant for erosion, overgrazing by cattle and sheep must be prevented.

One of the most troublesome erosional problems faced by farmers is the control of gullies. Gullies start in either natural depressions, plow furrows, or vehicle tracks. With regard to the latter, off-road recreational vehicles compress the soil along the track, rendering it impermeable. As a re-

Figure 5–17 Gullying and consequent loss of soil on land that has been cleared of its protective covering of plants because of an urban development project. St. Louis County, Missouri.

Figure 5–18 A field in which crops have been planted in rows that parallel topographic contours and in which terraces have been built to help retain rainwater. (From U.S.D.A., SCS la-2707.)

Figure 5–19 Strip cropping. (From U.S.D.A., SCS ND-589.)

sult, rain water is prevented from percolating into the soil, and flows down the track where it promotes gullying (Fig. 5–20). Once the process has begun, the gullys extend themselves toward higher ground by a process called **headward erosion**. The gully head receives rainwash from a wide area. This water converges so as to flow rapidly into the steep depression at the upper end of the gully, eroding vigorously as it enters the channel. It is this rapid erosion that causes the gully to cut ever farther in the headward or upslope direction. To halt this destructive process, steps can be taken that prevent water from entering the gully. Normally this is done by excavating a shallow diver-

Figure 5–20 Severe gullying on a hillside near Colorado Springs, Colorado, caused by runoff within the tracks produced by off-road vehicles.

sion ditch around the gully head to trap its potential water supply and "starve" the gully. Because gullies grow by headward erosion, disposing of old refrigerators and automobiles in these ditches has no effect at all in retarding their growth.

Good management practices designed to inhibit erosional loss of fertile topsoil is essential to the continued productivity of soils. Food shortages and rural decay are the consequence of poor soil management.

Summary

Rocks exposed at the earth's surface are continuously subjected to the destructive forces of weathering. *Weathering* is a term used to describe the changes in surface materials that occur when they interact with water, air, and organisms. In general, two categories of processes operate during weathering. The first of these is called *disintegration* or *mechanical weathering* and acts primarily to break rock down into smaller sizes without directly causing a change in composition. Disintegration is accomplished in many ways. When water in rock crevices expands upon freezing, for example, it widens existing cracks and opens new ones. Erosion is constantly at work removing rock at the surface, and this unloading causes rocks to rupture and form joints as they expand under the lessened pressure. The growth of mineral crystals and moisture-seeking plant roots may also help to wedge rocks apart. Acting together, all these forms of disintegration reduce great blocks of rock to small particles and thereby greatly increase the total amount of surface area available for attack by chemically potent solutions.

This chemical activity constitutes the second kind of weathering known as *decomposition.* Unlike disintegration, decomposition does indeed alter a rock's composition. The essential substance for decomposition is water. Water is an effective solvent for some minerals and can chemically react with many more. For example, when water becomes acidic by dissolving carbon dioxide and sulfur dioxide from the air or soil, it is capable of decomposing a variety of silicate and nonsilicate minerals. Limestones are particularly susceptible and may experience extensive solution in the formation of caverns. Feldspars attacked by rainwater will yield clay, soluble carbonate, and residual silica. Iron oxides and clay minerals as well as soluble carbonates are the products of decomposition of ferromagnesian minerals. The main chemical processes operating during decomposition are solution, oxidation, hydrolysis, hydration, and chemical changes induced by plants.

Rates of weathering differ from place to place around the globe depending largely on climate and the nature of the rock being weathered. In general, chemical weathering is enhanced by abundant moisture with high temperatures. Minerals in rocks vary in their resistance to decomposition. Calcite yields readily to chemical attack, whereas quartz is extremely resistant. Among the common rock-forming silicate minerals, those near the top of the Bowen Reaction Series weather more rapidly than the minerals near the bottom.

The most beneficial effect for humanity of weathering is that it produces the nonorganic components of soil. *Soil* is weathered unconsolidated material capable of supporting the growth of plants. It develops as a result of a complex interaction of weathering processes and biologic activity. Climate, the parent rock materials, degree of slope, amount of time, and soil organisms are all factors in determining the characteristics of soils. In the United States and Canada, the two main types of soils are pedalfers and pedocals. *Pedalfer soils* are prevalent in regions having frequent rainfall. They are well-leached soils in which aluminum and iron as well as clay particles accumulate beneath the organic-rich surface layer. Because there is abundant moisture, soluble carbonates are flushed out of these soils. In the drier regions of North America, pedocal soils are common. *Pedocals* are rich in calcium compounds. There is insufficient water in arid regions to dissolve and wash away these salts, and so they accumulate in the soils.

In wet tropical climates, chemical leaching of parent materials is rapid. Not only are carbonates carried away, but even colloidal silica. The soils produced by these conditions are called laterites. *Laterites* are characterized by unusually high concentrations of iron and aluminum hydroxides. In a few places around the world, these metallic compounds are highly concentrated in laterites and are commercially exploited.

Terms to Understand

acid rain
bauxite
caliche
chemical weathering
decomposition
differential weathering
disintegration
exfoliation
exfoliation dome
frost action
grus
headward erosion
humus
hydration
hydrolysis
insolation
laterite
leaching
mechanical weathering
oxidation
pedalfer
pedocal
regolith
sheeting
soil
soil horizon
soil profile
solution weathering
spheroidal weathering
unloading
weathering

Questions for Review

1. In what way does an increase in the mechanical weathering of a rock mass increase its susceptibility to chemical weathering?

2. Describe what happens during the chemical weathering of plagioclase feldspar.

3. What are the differences between pedalfer and pedocal soils? Account for the origin of these differences.

4. What are laterites? How do they differ from pedalfers?

5. In what way is the susceptibility of a silicate mineral to weathering related to its position in the Bowen Reaction Series?

6. It has been observed that cemetery headstones made of the same marble and erected at the same time in country and in large city cemeteries weather more rapidly in the city locations. What is the likely explanation for this?

7. A plot of forested land in the eastern U.S. and one in the Amazon rain forest are both cleared for agriculture. Which plot will have the longest life as a productive soil and why?

8. Among the common sedimentary rocks, limestone, shale, and silica-cemented sandstone, which is likely to be most susceptible to chemical weathering in a humid temperate climate and why?

9. Among the common igneous rocks, granite, rhyolite, and basalt, which is likely to be most susceptible to chemical weathering in a humid temperate climate and why?

10. Given identical granitic rocks in Alaska and Georgia, and assuming the same amount of rainfall, in which area would decomposition of the rock be most rapid? Why?

11. A molecule of water is described as a dipolar molecule. What is the meaning of this expression? Which "ends" of the water molecule would attract a sodium ion in solution?

Supplemental Readings and References

Birkeland, P.W., 1984. *Soils and Geomorphology.* New York, Oxford University Press.

Carroll, D., 1970. *Rock Weathering.* New York, Plenum Press.

Hunt, C.B., 1972. *Geology of Soils.* San Francisco, W.H. Freeman & Co.

Keller, W.D., 1969. *Chemistry in Introductory Geology,* 4th ed. Columbia, Mo., Lucas Bros.

Ollier, C., 1969. *Weathering.* New York, American Elsevier.

Paton, T.R., 1978. *Formation of Soil Material.* Boston, Geo Allen & Unwin.

Winkler, E.M., 1973. *Stone: Properties and Durability in Man's Environment.* New York, Springer-Verlag.

Sedimentary Rocks

6

The carbonate rock dolostone forms a resistant ledge over a massive quartz sandstone. The cave was excavated for the extraction of the sandstone used in making glass. Pacific, Missouri.

In the course of time the level of the sea became lower, and as the salt water flowed away this mud became changed into stone.

Leonardo da Vinci (1452–1519). From his notebooks.

Introduction

As rocks are decomposed and disintegrated at the earth's surface, their altered products are carried to lakes and oceans as gravel, sand, mud, and soluble salts. The loose material is termed *sediment.* It may consist of rock and mineral particles, organic materials, or substances precipitated from sea water. When consolidated and cemented into hard masses, sediments become *sedimentary rocks.* Sediment is commonly deposited in layers, and as a result, the majority of sedimentary rocks have a layered or stratified appearance (Fig. 6–1). This stratification is often the result of slight differences in color, composition, or texture. It may also be caused by temporary cessation of deposition, or even minor erosion at the surface of deposition.

Sedimentary rocks constitute only about 12 per cent by volume of the earth's continental crust, yet these rocks cover 75 per cent of the total land area. They provide our best documentation of the earth's history and are the source of a great many materials necessary to modern civilization. Some of these materials, like clay and building stones, are commonplace. Others, like placer deposits of diamonds, chromite, and gold, occur only in limited amounts at a few localities. Aluminum, a metal derived from a residual mineral named bauxite, is another extensively exploited sedimentary resource. The fossil fuels such as coal and petroleum, however, are the most important of all sedimentary materials today. Indeed, oil, coal, and natural gas power our industrialized world.

From Sediment to Sedimentary Rock

The most significant factors involved in the origin of sedimentary rocks are *weathering,* which produces sediment; *transportation* and *deposition* of that sediment by water, ice, or wind; and the *lithification* necessary to convert loose particles of sediment into solid rock. Each of these factors may alter the composition or textural features of sediment and thereby provide a variety of different kinds of sedimentary rocks.

The solid particles carried by wind or water will be deposited whenever there is insufficient energy to carry them further. For example, if the velocity of dust-laden wind abates, there will be insufficient energy to carry particles of a given size, and those particles will be dropped. Similarly, if a stream's velocity is checked, as when entering a standing body of water, the stream also loses energy and is unable to continue to carry the material formerly carried at the higher velocity.

A reduction in a stream's velocity does not, of course, affect the dissolved materials as it does suspended solid particles. Material carried in solution is deposited by a process called **precipitation**, in which dissolved material is changed to a

Figure 6–1 Strata of the Proterozoic Rocknest Formation in the Northwest Territory of Canada consists primarily of cyclically repeated carbonate (dolostone) and shale layers. (Courtesy of Paul Hoffman).

solid and separated from the liquid in which it was formerly dissolved. For example, calcium carbonate, the principal mineral in the widespread sedimentary rock known as limestone, may be precipitated from water that contains calcium in solution as indicated below:

$$\underset{\text{(dissolved calcium ions)}}{Ca^{2+}} + \underset{\text{(dissolved bicarbonate ions)}}{2\,HCO_3^-} \rightleftharpoons \underset{\text{(calcium carbonate)}}{CaCO_3\downarrow} + \overbrace{\underset{\text{(water)}}{H_2O} + \underset{\text{(carbon dioxide)}}{CO_2}}^{H_2CO_3}$$

The bicarbonate ions that participate in the above reaction can be derived from the ionization of carbonic acid. As indicated by the arrows, the reaction is reversible. If carbon dioxide is removed from sea water, the reaction will proceed toward the right, and calcium carbonate will be precipitated. If, however, carbon dioxide is added to sea water, then the amount of carbonic acid in the water would build, and the reaction would proceed to the left. This would result in a chemical environment not conducive to calcium carbonate precipitation, and one in which existing calcium carbonate might begin to dissolve. As is evident here, the precipitation of calcium carbonate is a complex and delicate process in nature. It is influenced by organisms that utilize or liberate carbon dioxide, by processes that alter the acidity or alkalinity of the water, by the presence of organic compounds, and by the presence of sulfur, phosphorus, and magnesium.

Weathering and the Derivation of Sediment

In the preceding chapter we saw how the agents of chemical and mechanical weathering cause the decay and disintegration of solid bedrock. We noted how the weathering of a common igneous rock like granite can contribute grains of quartz and feldspar for making sandstone, clay for shale, and calcium carbonate for limestone (Fig. 6–2). Rocks other than granite, of course, yield different kinds and abundances of weathered products. Basalt, which contains no quartz, cannot yield quartz upon weathering. Similarly, one cannot expect the weathering of limestone to yield clay, except for the clay that was already present as an impurity. Another important control over the kind of sediment produced by weathering is climate. Warm humid climates enhance chemical weathering, causing more complete decay of feldspars and loss of ions to solution. More clay is produced, and the more resistant grains of quartz constitute the bulk of the unweathered residue.

Not all the materials comprising a sediment or sedimentary rock are mineral grains or clay particles. The dissolved ions of calcium (Ca^{2+}), sodium (Na^+), carbonate (CO_3^{2-}), and chloride (Cl^-) may be precipitated from solution to form sedimentary rocks or components of sedimentary rocks. Many animals build their shells of calcium carbonate, and accumulations of seashells on the ocean floor constitute sediment that can be consolidated into sedimentary rock. Similarly, the great masses of calcium carbonate secreted by corals in reefs can also be considered organic sedimentary rock.

Sedimentary rocks are derived from the weathering and erosion of parent materials, followed by their transportation, deposition, and lithification.

Sediment Transport and Deposition

Once weathering products have been formed from preexisting rocks, the next stage in the sequence of events leading to sedimentary rock is the removal and transport of those products. Many denudational agencies, including running water, moving ice, and wind, assist in this removal. The wind (Fig. 6–3) is an effective agent in picking up and blowing away the smaller and lighter particles. Glacial ice (Fig. 6–4) can move very large pieces of rock and carry an immense load of coarse sediment. Streams (Fig. 6–5) are also exceptionally effective in carrying not only solid particles of sediment but invisible dissolved salts as well. Ultimately, sediment-laden streams flow into lakes or the sea and their load of sediment is deposited. It may form sandy beaches, silty floodplains, and sometimes muddy boggy areas of estuaries and deltas.

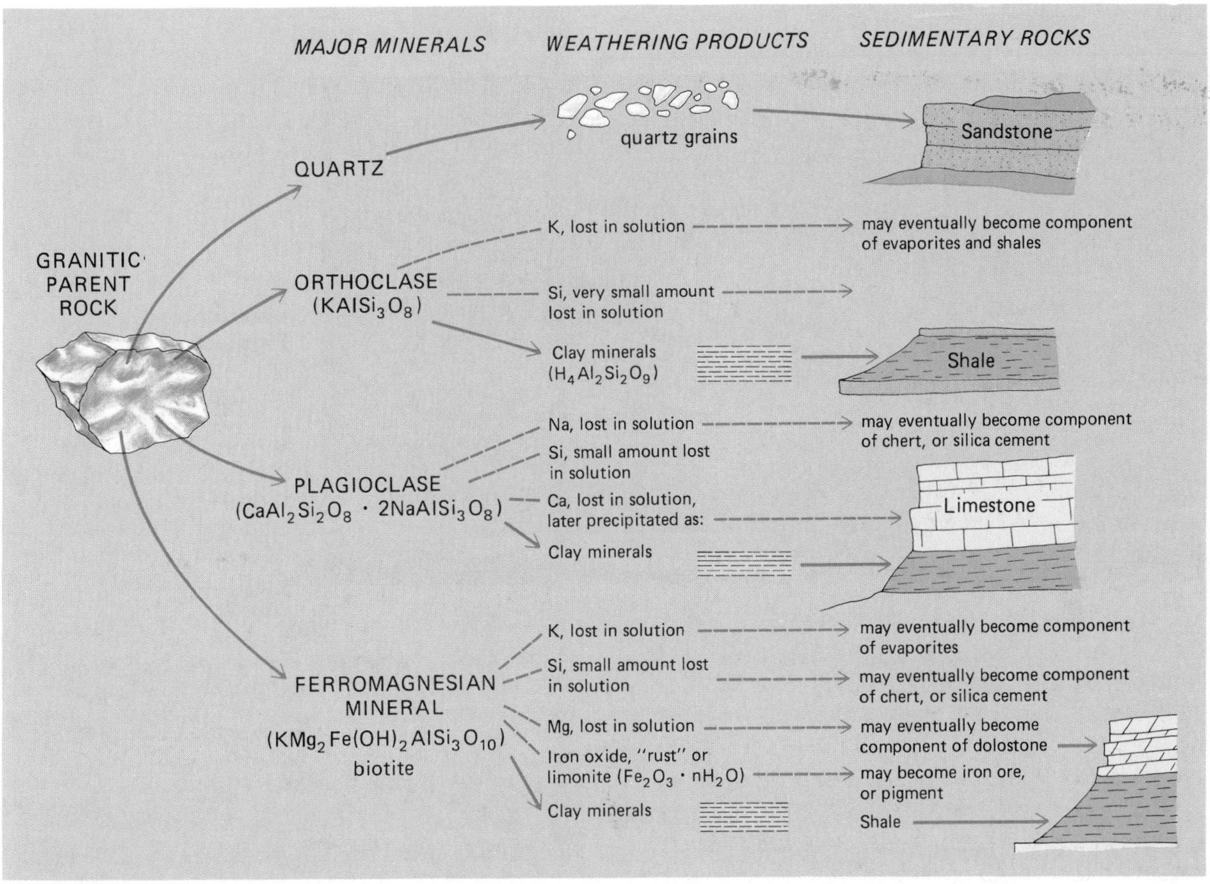

Figure 6–2 Conceptual diagram showing how the weathering of granitic rock yields quartz grains for quartz sandstone, clay for shale, and calcium for limestone. If weathering is not too severe, detrital grains of feldspar will also be included in sands and sandstones. For simplicity, minor mineral components are not included.

Figure 6–3 Fine silt and clay being transported by wind. (From the U.S.D.A.)

Figure 6–4 Taylor Glacier, Antarctica, with associated glacially transported deposits. Note large basaltic sill in the Paleozoic and Mesozoic rocks in the distant mountain. (Courtesy of P.B. Larson and the U.S. Geological Survey.)

Figure 6–5 Sediment-laden Tanana River, Alaska. The river has so great a load of sediment that it has deposited large amounts of silt, sand, and clay in its own bed, resulting in a braided stream pattern. (Courtesy of J.C. Brice.)

The work of moving water (streams, marine currents, and waves), wind, and glaciers is not at all confined to merely transporting sediment. In subsequent chapters, we will learn the importance of these geologic agents in extracting and modifying the particles they carry. They are the tools used by nature to sculpture our landscapes.

Lithification

Many changes take place in sediment after it has been deposited. Mineral grains may be dissolved away, some may grow by additions of new mineral matter, and the shapes of particles may be distorted by compaction. The result of some of these changes is to convert sediment into sedimentary rock. The conversion process is called **lithification.** *Compaction* and *cementation* are the principal means by which unconsolidated sediment is lithified.

The reduction in pore spaces in a rock as a result of the pressure of overlying rocks or pressures of earth movements is termed **compaction**. During compaction, individual grains are pressed tightly against one another, causing the expulsion of water and rearrangement of particles. The result is often the conversion of loose sediment into hard rock. Compaction is greatest and most important as a lithification process in finer-grained sediments like clay and mud.

Cementation involves the precipitation of minerals in the pore spaces between larger particles of sediment. The precipitated mineral, which most frequently is either calcium carbonate ($CaCO_3$), silica (SiO_2), or ferric iron oxide (Fe_2O_3), is called the **cement** (Fig. 6–6). Cement is added to a sediment *after* deposition. It differs from a rock's **matrix** (Fig. 6–7), which consists of clastic particles (often clay) that are deposited at the same time as the larger grains and help to hold the grains together.

The process of cementation may modify the grains in a sedimentary rock. For example, quartz may be precipitated onto rounded quartz grains to form a strong mosaic of quartz crystals containing the ghostly boundaries of the originally rounded grains within them. Calcium carbonate skeletal debris may be reorganized into a crystalline aggregate (Fig. 6–8). The migration of watery solutions through sediment favors these processes, as does deep burial and consequent increases in pressure and temperature.

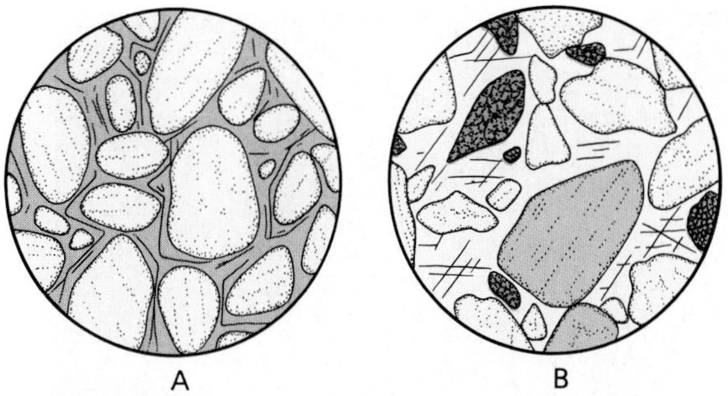

Figure 6–6 Two common types of cement in sandstones. (A) Quartz sandstone composed of well-sorted rounded quartz grains tightly cemented by quartz (SiO_2) outgrowths. (B) Sandstone composed of quartz, feldspar, and rock fragments cemented by coarse sparry calcite. Both are drawings of thin sections as viewed under the microscope. Diameter of each area is 1.0 mm.

Environments of Deposition

The environment of deposition is the location where sediment is being deposited. Each environment of deposition is characterized by geographic and climatic conditions that modify or determine the properties of sediment that is deposited within

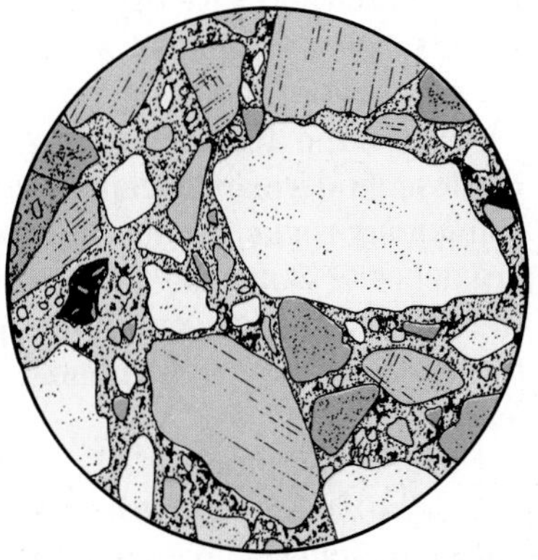

Figure 6–7 Thin section of sandstone composed of poorly sorted angular grains of quartz (gray), feldspars (green), and rock fragments (red). This sandstone lacks cement. Spaces between grains are filled with a matrix of clay and silt.

Figure 6–8 Coarsely crystalline limestone formed by recrystallization of detrital fragments of calcium carbonate.

it. Some sediments, such as chemical precipitates in water bodies, are solely the products of their environment of deposition. Their component minerals originated and were deposited at the same place. Other sedimentary accumulations come from a distant source and have experienced particular kinds of transport. Geologists are keenly interested in the sediment of today's depositional environments because the features they find in the deposits of these modern areas can also be seen in ancient sedimentary rocks. Comparing present-day sedimentary deposits to ancient sedimentary rocks permits one to reconstruct conditions in various parts of the earth as they were hundreds of millions of years ago.

The three major environments of deposition are the *marine*, *continental*, and *transitional* environments. Each contains a number of specific subenvironments. In a very general way, the marine environment may be divided into shallow marine (water depth less than 200 meters) and deep marine (greater than 200 meters). As illustrated in Figure 6–9, the shallow marine environment extends from the shore to the outer edge of the continental shelves. The kind of sediment deposited in this environment depends on many factors, including climate, parent bedrock, distance from shore, elevation of the adjacent land area, water depth, and the presence or absence of carbonate-secreting organisms. Because of wave turbulence and currents, sediment deposited in the shallow marine environment tends to be coarser than material laid down in the deeper realms of the ocean. Sands, silts, and clays are common. Where there are few continent-derived sediments and the seas are relatively warm, lime muds of biochemical origin may be the predominant sediment. Coral reefs are also characteristic of warm shallow seas.

In the deep marine environment far from the continents, only very fine clay, volcanic ash, and the calcareous or siliceous remains of microscopic organisms settle to the ocean floor. The exceptions are sporadic occurrences of coarser sediments that are carried down continental slopes into the deeper realms of the ocean by dense masses of mud- and silt-laden water called **turbidity currents** (see Fig. 6–36).

The shoreline of a continent is the transitional zone between marine and nonmarine environments. Here one finds the familiar shoreline accumulations of sand or gravel we call beaches. Mud-covered tidal flats that are alternately inundated and drained of water by tides are also found here, as are deltas. Deltas form when streams enter bodies of standing water, experience an abrupt loss of velocity, and drop their load of sediment. Provided shoreline currents and waves do not remove the deposits faster than they are supplied, the delta will grow seaward. In general, progressively finer sediment will be deposited in progressively deeper water as the current provided by the stream diminishes. At the same time, the stream channel is extended over former deposits, periodically chokes in its own debris, and breaks out to form new branches of the delta.

In addition to deltas, the transitional zone includes such features as barrier islands, which are built parallel to shorelines by wave and current action, and lagoons, which are often found between the mainland and a barrier island. Swamps are also frequent features of low-lying areas adjacent to the sea.

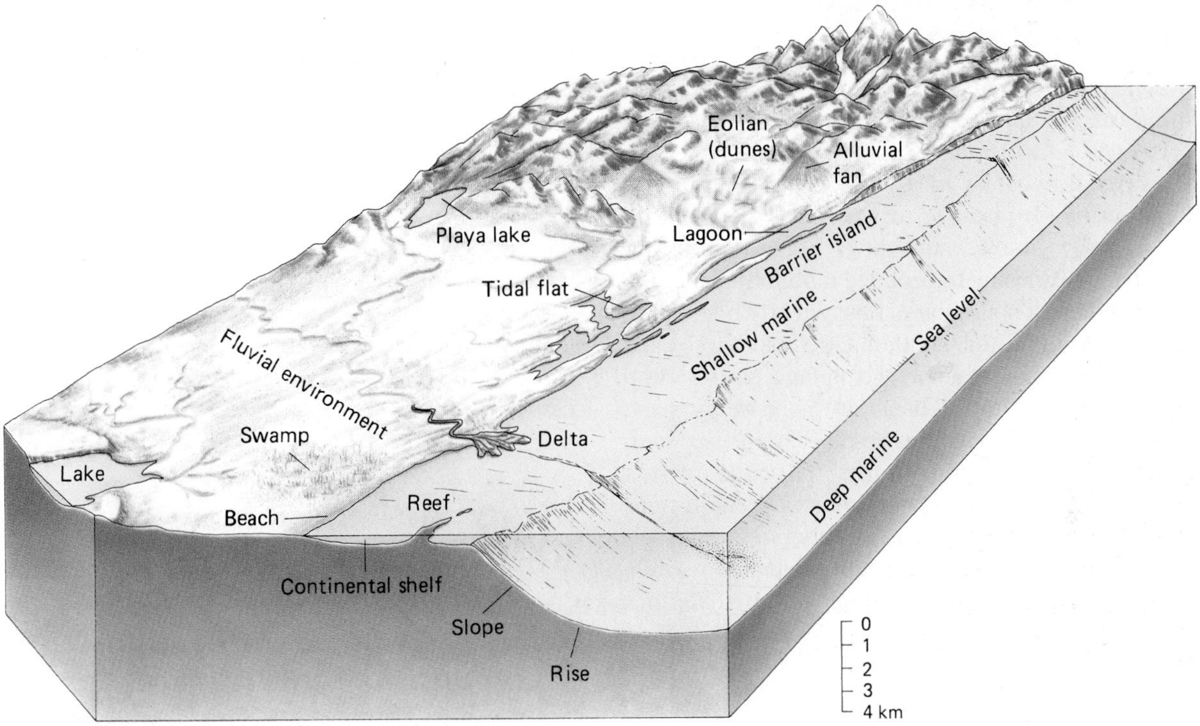

Figure 6–9 Diagram illustrating some marine and continental environments of deposition.

Continental environments of deposition include river floodplains, alluvial fans, lakes, glacial, and eolian environments. The silt, sand, and clay found along the banks, bars, and floodplains of streams are familiar to most of us. In general, stream deposits develop as elongate lenses that are oriented downstream and that grade abruptly from one particle size to another. In another setting, stream-transported materials may accumulate quickly when a rapidly flowing river emerges from a mountainous area onto a flat plain. The result of the abrupt deposition is an alluvial fan. Somewhat quieter deposition occurs in lakes, which are ideal traps for sediment. Silt and clay are common lake sediments, although a variety of sediment is possible, depending on water depth, climate, and the character of the surrounding land areas. The playa lakes of arid regions are shallow temporary lakes. Periodically they become dry as a result of evaporation. Perhaps the most dramatic environments of deposition are found in areas having active glaciers. Glaciers transport and deposit huge volumes and large fragments of rock detritus. Deposits are characteristically unsorted mixtures of boulders, gravel, sand, and clay. Where such materials have been reworked by glacial meltwater, however, they become less chaotic and resemble stream deposits.

Whereas glaciers can move materials of great size, wind is much more selective in the particle size it can transport. Environments where wind is an important agent of sediment transport and deposition are called eolian environments. They are characterized by an abundance of sand and silt, little plant cover, and strong winds. These characteristics typify some desert regions.

Environments of deposition may be continental, marine, or transitional. Each of these major environments of deposition is recognized by distinctive features of the sediment that has accumulated within it.

The Influence of Tectonics on Sedimentation

One final factor, called **tectonics**, also has an effect on the characteristics of sedimentary rock formed at a particular location. Tectonics refers to the crustal behavior of a part of the earth over a long period of time. For example, a region may be tectonically stable, subsiding, rising gently, or more actively rising to produce mountains and plateaus. Where a source area has recently been compressed and uplifted, an abundance of coarse clastics derived from the rugged upland source area will be supplied to the basin. In the geologic past, such a tectonic setting has resulted in the accumulation of great "clastic wedges" of sediment that thickened and became coarser toward the former mountainous source area. In other tectonic settings of the past, the source area has been stable and topographically more subdued, so that finer particles and dissolved solids became the most abundant components being carried by streams.

A tectonic setting that includes a rapidly rising source area will supply a large quantity of land derived sediment to the basin of sedimentation. A low-lying, stable land adjacent to a shallow sea is likely to deposit well sorted sands and carbonates in the adjacent basin.

The tectonic setting influences not only the size of clastic particles being carried to sites of deposition but also the thickness of the accumulating deposit. For example, if a former marine basin of deposition had been provided with an ample supply of sediment and was experiencing tectonic subsidence, enormous thicknesses of sediments might accumulate. Over a century ago, James Hall, an early New York geologist, recognized that the thick accumulations of shallow-water sedimentary rocks in the Appalachian region required that crustal subsidence accompany deposition. His reasoning was quite straightforward. It was easy to visualize filling a basin that was 12,000 meters deep with 12,000 meters of sediment. However, if the basin was only a few hundred meters deep, the only way to get tens of thousands of meters of sediment into it would be to have subsidence occurring simultaneously with sedimentation.

In a marine basin of deposition that is stable or subsiding very slowly, the surface on which sedimentation is occurring is likely to remain within the zone of wave activity for a long time. Wave action and currents will wear, sort, and distribute the sediment into broad blanket-like layers. This type of sedimentation will continue until there is either a change in sea level, or in the amount of sediment entering the depositional basin. For example, if the supply of sediment becomes too great for currents and waves to transport, deltas are likely to form.

Kinds of Sedimentary Rocks

The Basis for Identification: Composition and Texture

Sedimentary rocks are identified and named according to their composition and texture (Table 6–1). In regard to composition, the three most abundant mineral components of sedimentary rocks are clay minerals, quartz, and calcite. Evaporite minerals and dolomite, although less abundant, nevertheless form an appreciable portion of many sedimentary sequences. Sedimentary rocks nearly always also contain variable amounts of limonite and hematite. Also, most are mixtures of two or more components in which one mineral may predominate. Thus, sandstones composed mostly of quartz grains nearly always contain some clay or calcite, and limestones made mostly of calcite nearly always are contaminated with clay and quartz grains (Fig. 6–10).

Texture refers to the size and shape of individual mineral grains and to their arrangement in the rock. A rock that has a **clastic** texture (from the Greek *klastos*, "broken") is composed of particles of clay, silt, sand, and gravel, or fragments of parent rock or fossils that have been moved individually from their place of origin. In contrast to these clastic rocks, **nonclastic** rocks form by chemical or biochemical precipitation within a sedimentary basin.

Table 6–1 Classification of Sedimentary Rocks

Origin	*Rock Name*	*Texture*	*Composition*
Clastic	Conglomerate	Rounded particles greater than 2 mm	Mostly gravel-size pieces of quartz, chert, quartzite, or any rock type
	Breccia	Angular particles greater than 2 mm	Particles of quartz or any rock type
	Sandstone	Particles range in size from 0.0625 to 2 mm in diameter	Varies according to kind of sandstone. Quartz usually dominant. Arkose must contain 25% feldspar. Composition of graywacke includes considerable clay and rock fragments.
	Siltstone	Particles range in size from 0.0039 to 0.0625 mm in diameter	Generally composition similar to that of associated sandstones
	Shale	Particles less than 0.0039 mm in diameter	Quartz and clay minerals
Biogenic and/or chemical	Limestone	Varied textures, including micritic, oölitic, bioclastic, etc.	Calcite ($CaCO_3$)
	Dolostone	Varied textures, often porous, fine-grained	Dolomite ($CaMg(CO_3)_2$)
	Evaporites	Usually crystalline	Gypsum ($CaSO_4 \cdot 2H_2O$) Halite (NaCl) Sylvite (KCl)
	Coal	Compressed particles of partly decomposed plant matter	A complex of organic compounds

Texture and composition are the basis for the classification of sedimentary rocks. Texture refers to the size, shape, and arrangement of a rock's component grains.

Sedimentary rocks made of the remains of plants and animals are categorized as **biogenic**. Coal, for example, can be considered a sedimentary rock derived from the accumulation of plant remains. Limestones composed predominantly of the skeletal remains of invertebrate animals would also be considered biogenic.

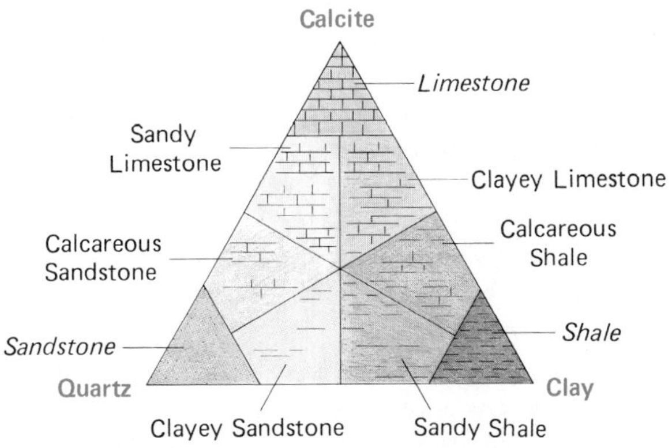

Figure 6–10 Triangular diagram suggesting the gradations that can exist between pure shales, sandstones, and limestones.

Clastic Sedimentary Rocks

Categories of Clastic Rocks

The component fragments of clastic sedimentary rocks range in size from microscopic particles to huge boulders (Table 6–2). The clastic group includes such familiar rocks as **conglomerates, sandstones, siltstones,** and **shales**. Texture, and particularly particle size, is the key to naming the clastic rocks. Thus, conglomerate is a rock composed of rounded particles larger than 2 mm in diameter (Figs. 6–11 and 6–12). **Breccias** (Figs. 6–13 and 6–14) are composed of particles that are angular but similar in size to conglomerate particles. The particles making up a conglomerate or breccia

Table 6–2 Size Range of Sedimentary Particles

Wentworth Scale (in millimeters)	*Fractional Equivalents (in millimeters)*	*Particle Name*
		Boulders
256		
128		Cobbles
64		
32		
16		Pebbles
8		
4		
		Granules
2		
		Very coarse sand
1.0		
		Coarse sand
0.5	1/2	
		Medium sand
0.25	1/4	
0.125	1/8	Fine sand
0.0625	1/16	Very fine sand
0.0313	1/32	
0.0156	1/64	
		Silt
0.0078	1/128	
0.0039	1/256	
		Clay

Figure 6–11 Pebble conglomerate.

can tell geologists something about the parent bedrock. For example, pebbles of granite in a conglomerate would indicate an igneous source area. In sandstones, grains range between 0.0625 and 2.0 mm in diameter. Siltstones, which tend to resemble sandstones in composition, are composed of still smaller particles. Shales are the finest of clastic rocks. They are composed of particles of silt and clay far too small to be seen without the aid of a microscope.

Figure 6–12 Conglomerate exposed near the southeast end of the Malone Mountains, Hudspeth County, Texas. (For scale, ruler in center is 10 cm long.) (Courtesy of C.C. Albritton, Jr., the U.S. Geological Survey.)

Grain Size and Sorting in Clastic Rocks

Geologists are interested in the sizes of component grains in clastic rocks, not only as a means of identifying them, but also because the size of particles can provide clues to the environment of deposition. For example, it is obvious that a stronger current of water (or wind) is required to move a big particle than a small one. Therefore, the size distribution of grains tells the geologist something about the turbulence and velocity of currents. It can also be an indicator of the kind and extent of transportation. If sand, silt, and clay are supplied by streams to a coastline, the turbulent inshore waters often winnow out the finer particles, so that gradation from sandy nearshore deposits to offshore silty and clayey deposits may result (Fig. 6–15). Sandstones formed from such inshore sands may retain considerable porosity and provide void space for petroleum accumulations. For this reason, one often finds petroleum geologists assiduously making maps showing grain size of deeply buried ancient formations to determine areas of coarser and possibly more permeable clastic rock.

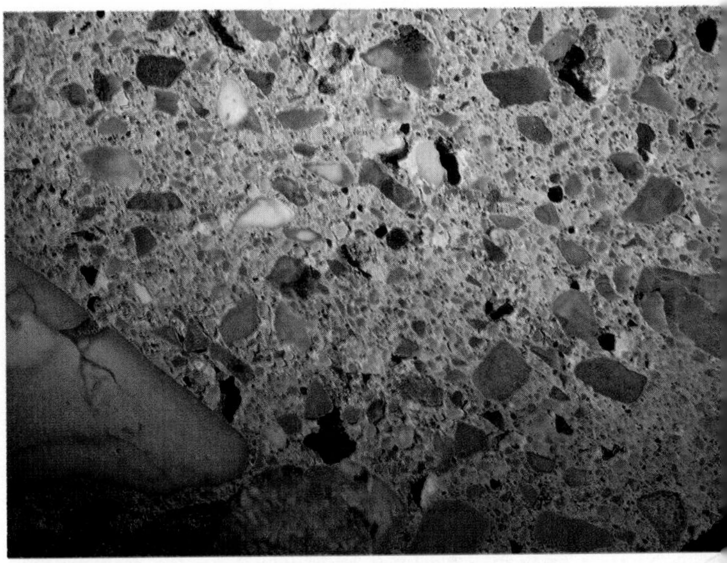

Figure 6–13 A sawed and polished slab of breccia.

Figure 6–14 Outcrop of very coarse breccia. This breccia is part of the Old Red Sandstone Formation and is exposed in Scotland. (Courtesy of the Institute of Geological Sciences, London.)

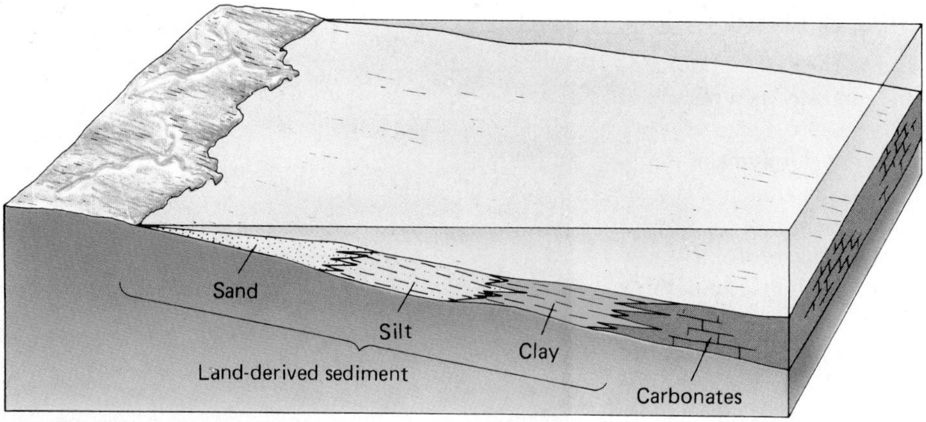

Figure 6–15 Idealized gradation of coarser nearshore sediments to finer offshore deposits.

Another aspect of a clastic rock's texture that involves grain size is sorting. **Sorting** refers to the uniformity of particle size in the sediment. Rocks composed of particles that are all about the same size are said to be "well sorted" (see Fig. 6–6A), and those that include grains with a wide range of sizes are termed "poorly sorted" (Fig. 6–16). Sorting often provides clues to conditions of transportation and deposition. Wind, for example, winnows the dust particles from sand, leaving grains that are all of about the same size. Wind also sorts the particles that it carries in suspension. Only rarely is the velocity of wind sufficient to carry grains larger than 0.2 mm. While carrying grains of sand size, winds sweep finer particles into the higher regions of the atmosphere. When the wind subsides, well-sorted silt-sized particles drop and accumulate. In general, windblown deposits are better sorted than deposits formed in an area of wave action, and wavewashed sediments are better sorted than stream deposits. Poor sorting occurs when sediment is rapidly deposited without being selectively separated by currents into sizes.

Figure 6–16 Thin section of a poorly sorted sandstone (graywacke) observed under the microscope with crossed polarizers. Note wide range of grain sizes. Width of field 9.0 mm.

The shape of particles in a clastic sedimentary rock can also be useful in determining its history, for it may reflect the amount of transportation or recycling a sediment has experienced. Shape can be described in terms of rounding of particle edges and sphericity (how closely the grain approaches the shape of a sphere) (Figs. 6–17 and 6–18). A particle becomes rounded by having sharp corners and edges removed by impact with other particles.

Often the transportational and depositional history of a clastic sedimentary rock can be inferred from its grain size, grain shape, and sorting characteristics.

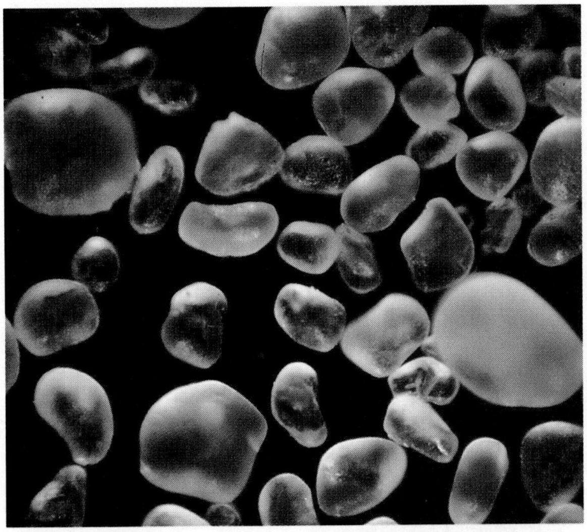

Figure 6–17 Well-rounded grains of quartz viewed under the microscope. From the St. Peter Formation near Pacific, Missouri.

Primary Sedimentary Structures in Clastic Sediments

A **primary sedimentary structure**, such as bedding or ripple marks, is one that forms during the deposition of sediment. In contrast, secondary sedimentary structures, such as folds and faults, develop after deposition. Primary structures are extremely useful to geologists interested in reconstructing ancient environments. For example, **mud cracks** indicate drying after deposition. These conditions are common on valley flats and in tidal zones. Mud cracks (Figs. 6–19 and 6–20) develop by shrinkage of mud or clay on drying and are most abundant in the subaerial environment. **Cross-bedding** is an arrangement of beds or laminations in which one set of layers is inclined relative to the others (Figs. 6–21 and 6–22). The cross-bedding

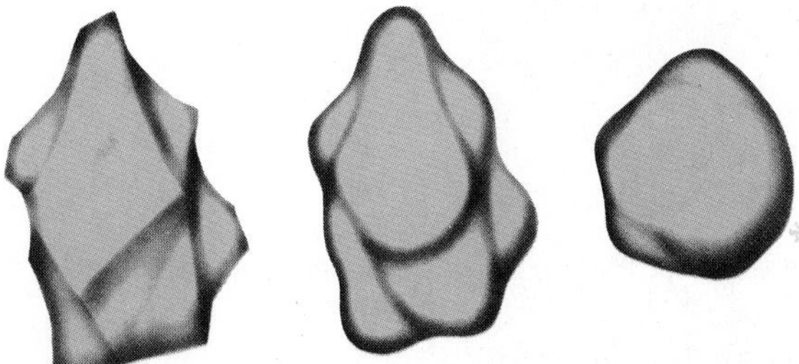

Figure 6–18 Shape of sediment particles. (A) An angular particle (all edges sharp). (B) A rounded grain that has little sphericity. (C) A well-rounded highly spheric grain. Roundness refers to the smoothing of edges and corners, whereas sphericity measures the degree of approach of a particle to a sphere.

Figure 6–19 Modern mudcracks formed in the soft clay around the margins of an evaporating pond. (Courtesy of L.E. Davis.)

Figure 6–20 Ancient mudcracks in the Jurassic Navajo Formation, Zion National Park, Utah. (Courtesy of L.E. Davis.)

Figure 6–21
Cross-bedding in the Nubian Sandstone of Egypt. (Courtesy of D. Bhattacharyya)

Figure 6–22
Cross-bedding accentuated by weathering in the Navajo Sandstone, north of Kanab, Utah. The cross-bedding at this exposure is of the tabular-planar type. (Courtesy of the U.S. Geological Survey.)

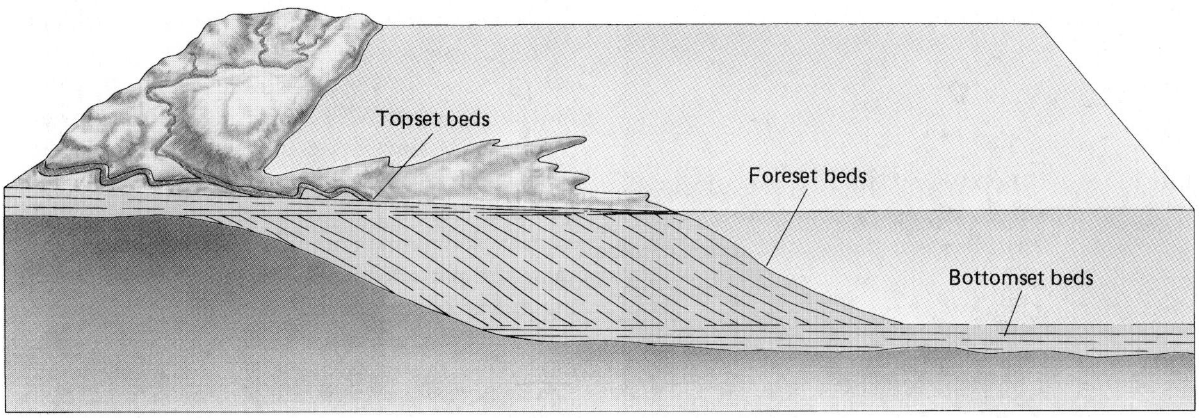

Figure 6–23 Cross-bedding in a delta. The succession of inclined foreset beds is deposited over bottomset beds that were laid down earlier. Topset beds are deposited by the stream above the foreset beds.

units can be formed by the advance of a delta (Fig. 6–23) or a dune. A depositional environment dominated by currents is inferred from cross-bedding. The currents may be wind or water. In either medium, the direction of the inclination of the sloping beds is a useful indicator of the direction taken by the current. By plotting these directions on maps, geologists have been able to determine the pattern of ocean currents and prevailing winds at various times in the geologic past.

Graded bedding consists of repeated beds, each of which has the coarsest grains at the base and successively finer grains nearer the top (Figs. 6–24 and 6–25). Although graded bedding may form simply as the result of faster settling of coarser, heavier grains in a sedimentary mix, it appears to be particularly characteristic of deposition by **turbidity currents**. As noted earlier, turbidity currents are masses of water containing large amounts of suspended material; the currents, having been made more dense than surrounding water by suspended muddy sediment, flow turbulently down slopes below relatively clearer water. Such currents are characteristically triggered by submarine earthquakes and landslides that may occur along steeply sloping regions of the sea floor. The forward part of the turbidity current contains coarser debris than does the tail. As a re-

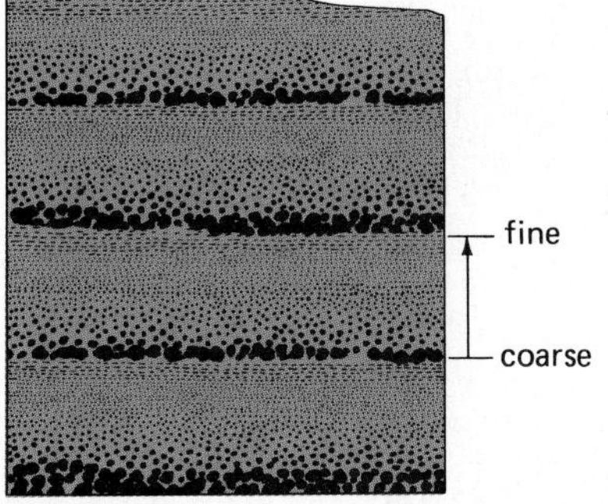

Figure 6–24 Graded bedding.

Figure 6–25 A close-up view of these Proterozoic sandstones (graywackes) reveals graded bedding and other evidence of deposition in turbidity currents. Great Slave Lake, Canada. (Courtesy of P. Hoffman.)

Figure 6–26 Ripple marks formed in sand on intertidal zone at Puerto Penasco, Sonora, Mexico. (Courtesy of Guillermo A. Salas, Universidad De Sonora.)

sult, the sediment deposited at a given place on the sea bottom grades from coarse to fine as the "head" and then the "tail" of the current passes over it.

Ripple marks are commonly seen sedimentary features that developed on modern sand deposits (Fig. 6–26), or along the surfaces of bedding planes (Fig. 6–27). *Symmetric ripple marks* (Fig. 6–28A) are formed by the oscillatory motion of water beneath waves. *Asymmetric ripple marks* are formed by air or water currents and are useful in indicating the direction of movement of currents. For example, ripple marks form at right angles to current directions; the steeper side of the asymmetric variety faces the direction in which the medium is flowing. Although there are instances of ripple marks developed at great depths on the sea floor, more frequently these features occur in shallow water areas.

Any sediment composed of small, discrete, sand-size particles may form ripple marks, including sands composed of shell fragments or other carbonate particles and the tiny spheres known as oöids (see Fig. 6–44).

Sedimentary features such as cross-bedding, mud cracks, and ripple marks can be studied in modern sediments, and used to interpret these features in ancient sedimentary rocks.

Sandstones

Of all the clastic sedimentary rocks, sandstones have been studied the most completely and provide the greatest amount of information about ancient environmental conditions. The mineral composition of the grains in sandstones provides important information about the source areas for the sediment as well as the history of the sediment

Figure 6–27 Ripple marks on bedding surface, Tensleep Formation of Pennsylvanian age, Wyoming. (Courtesy of L.E. Davis.)

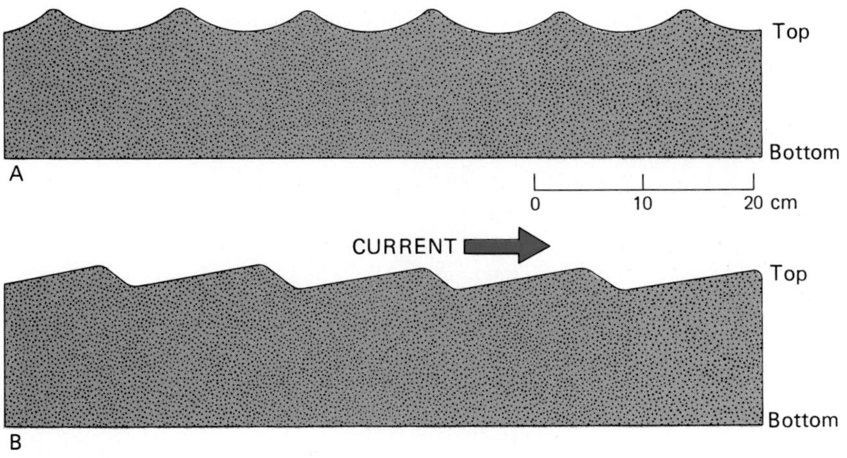

Figure 6–28 Profiles of ripple marks. (A) Oscillatory ripples. (B) Aqueous current ripples.

prior to deposition. Often, by closely studying the grains, one can ascertain whether the source material was metamorphic, igneous, or sedimentary. The mineral content also provides a rough estimate of the amount of transport and erosion experienced by the sand grains. Rigorous weathering and long transport tend to reduce the less stable feldspars and ferromagnesian minerals to clay and iron compounds and to cause rounding and sorting of the remaining quartz grains. Hence, one can assume that a sandstone rich in these less durable and angular components underwent relatively little transport and other forms of geologic duress. Such sediments are termed *immature* and are most frequently deposited close to their source areas. On the other hand, quartz can be used as an indicator of a sandstone's maturity; the higher the percentage of quartz, the greater the maturity.

In addition to providing an indication of a rock's maturity, composition is an important factor in the classification of sandstones into **quartz sandstones, arkoses, graywackes**, and **lithic sandstones** (sometimes termed subgraywackes) (Fig. 6–29).

Quartz sandstones (Fig. 6–30) are clastic rocks characterized by dominance of quartz with little or no feldspar, mica, or fine clastic matrix. The quartz grains are well sorted and well rounded. They are most commonly held together by such cements as calcite and silica. Chemical cements such as these tend to be more characteristic of "clean" sandstones like quartz sandstones and are not as prevalent in "dirtier" rocks containing clay. The presence of a dense clayey matrix seems to retard the formation of chemical cement, perhaps because fine material fills pore openings where crystallization might occur.

Calcite cement may develop between the grains as a uniform finely crystalline filling, or large crystals may form, and each may incorporate hundreds of quartz grains. Silica cement in quartz sandstone commonly develops as overgrowths on the original quartz grains (Fig. 6–31).

Quartz sandstones reflect deposition in stable, quiet, shallow water environments such as the shallow seas that inundated large parts of low-lying continental regions in the geologic past or some parts of our modern continental shelves (Fig. 6–32). These sandstones, as well as clastic limestones, exhibit sedimentary features, such as cross-bedding and ripple marks, which permit one to infer shallow-water deposition.

Quartz sandstone is composed predominantly of quartz grains that are well-sorted and well-rounded.

Sandstones containing 25 per cent or more of feldspar (derived from erosion of a granitic source area) are called **arkoses** (Fig. 6–33). Quartz is the most abundant mineral, and the angular-to-suban-

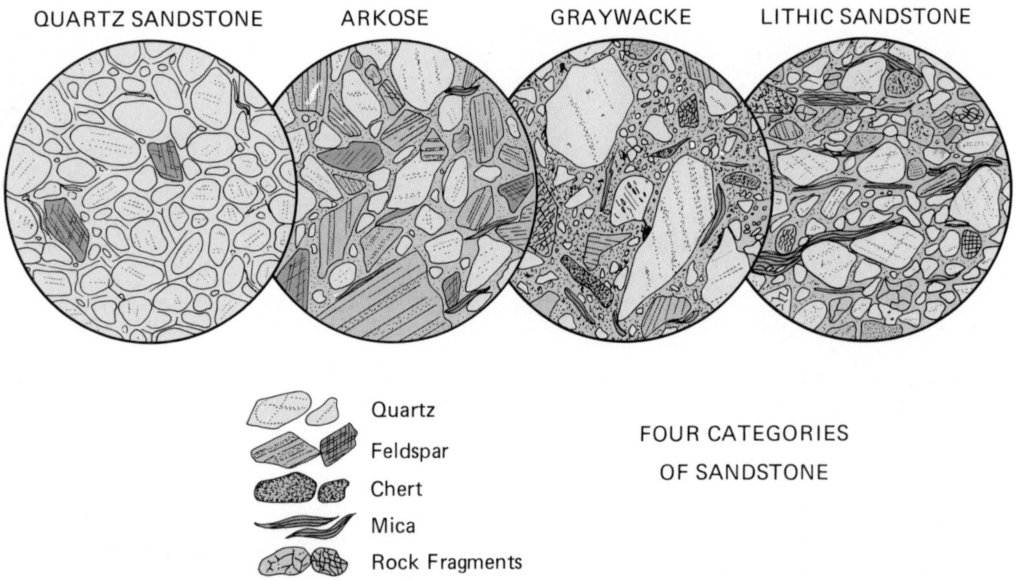

Figure 6–29 Four categories of sandstone as seen in thin section under the microscope. (Diameter of field is about 4 mm.)

Figure 6–30 Quartz sandstone. The color in this rock is caused by minor amounts of iron oxide.

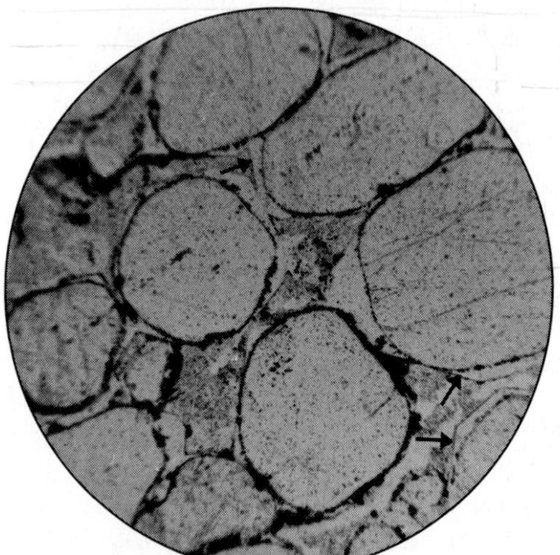

Figure 6–31 Thin section of a well-sorted quartz sandstone as seen under the microscope. Quartz grains have developed silica overgrowths, which are visible as ghostly crystal outlines. (Diameter of area is 1.0 mm.)

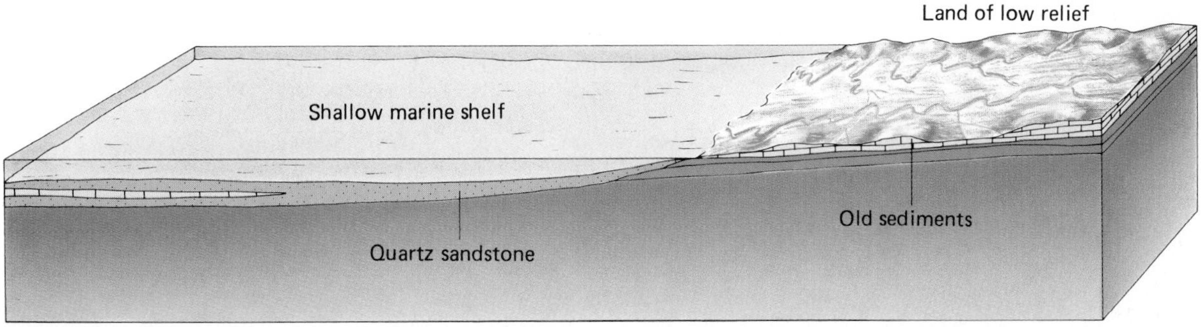

Figure 6–32 Idealized geologic conditions under which quartz sandstone may be deposited. There is little tectonic movement in this environment. Water depth is shallow, and the basin subsides only very slowly.

gular grains are bonded together by calcareous cement, clay minerals, or iron oxide. The presence of abundant feldspars and iron imparts a pinkish-gray coloration to many arkoses. In general, arkoses are coarse, moderately well-sorted sandstones. They may develop as basal sandstones derived from the erosion of a granitic coastal area experiencing an advance of the sea, or they may accumulate in fault troughs or low areas adjacent to granitic mountains (Fig. 6–34).

Figure 6–33 Thin section of an arkose viewed through a petrographic microscope. The clear grains are mostly quartz, whereas the grains that show stripes or a plaid pattern are feldspars. The matrix consists of kaolinite clay and fine particles of mica, quartz, and feldspar. (Courtesy of the U.S. Geological Survey; photo by J.D. Vine.)

Arkose contains at least 25 per cent feldspar. It tends to be less well-sorted, and its grains are more poorly rounded as compared to quartz sandstones.

Graywackes (from the German term *wacken*, meaning "waste or barren") are immature sandstones consisting of significant quantities of dark, very fine-grained material (Fig. 6–35). Normally, this fine matrix consists of clay, chlorite, micas, and silt. There is little or no cement, and the sand-sized grains are not in close contact because they are separated by the finer matrix particles. The matrix constitutes approximately 30 per cent of the rock, the remaining coarser grains consisting of quartz, feldspar, and rock particles. Graywacke has a dirty, "poured in" appearance. The poor sorting, angularity of grains (see Fig. 6–16), and heterogeneous composition of graywackes indicate an unstable source and depositional area in which debris resulting from accelerated erosion of highlands is carried rapidly to subsiding basins. Graded bedding (see Fig. 6–25), interspersed layers of volcanic rocks, and cherts (which may indirectly derive their silica from volcanic ash) further attest to dynamic conditions in the area of deposition. The inferred tectonic setting is unstable, with deposition occurring offshore from an actively rising mountainous region (Fig. 6–36). Graywackes and associated shales and cherts may contain fossils of deep-water organisms, indicating deposition at great depth. Such shallow-water sedimentary structures as cross-bedding and ripple marks are

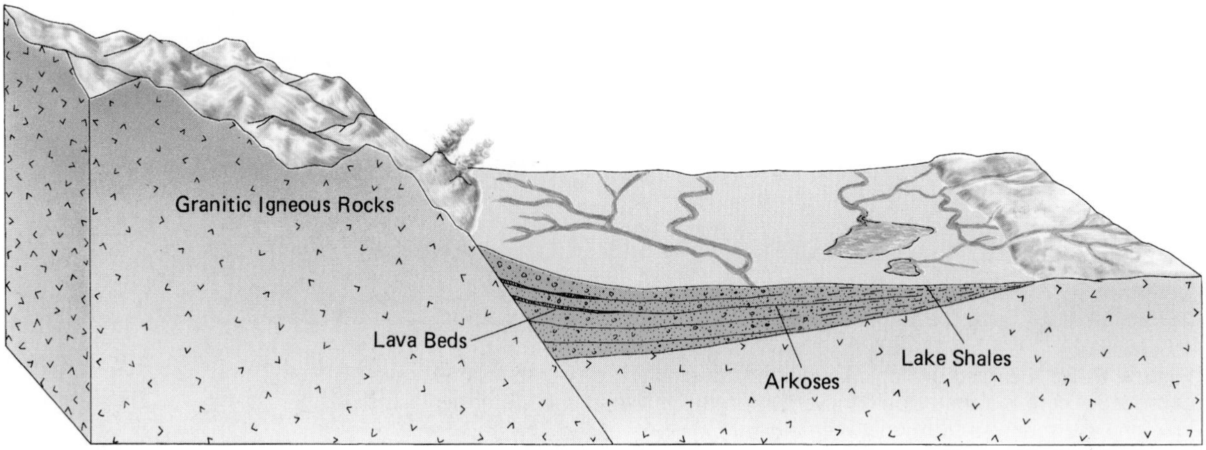

Figure 6–34 Geologic environment in which arkose may be deposited.

rarely found. To the experienced petrologist, graywackes clearly indicate dynamic, unstable conditions.

Graywacke is composed of grains of quartz, feldspar, and rock surrounded by a fine, clay-rich matrix.

Quartz sandstone, arkose, and graywacke are rather distinct kinds of sandstones. A rock that is more of a transitional type with regard to composition and texture is termed a **lithic sandstone** (subgraywacke). In lithic sandstones (see Fig. 6–29), feldspars are relatively scarce, whereas quartz, muscovite, chert, and rock fragments are abundant. There is a fine-grained detrital matrix that does not exceed 15 per cent, and the remaining voids are filled with mineral cement or clay. In lithic sandstone the quartz grains are better rounded and more abundant, the sorting is better, and the quantity of matrix is lower than in graywacke.

Figure 6–35 Graywacke.

The characteristic environment for lithic sandstones are deltaic coastal plains (Fig. 6–37), where they may be deposited in nearshore marine environments or swamps and marshes. Coal beds and micaceous shales are frequently associated with lithic sandstones.

Lithic sandstones have a high content of fine-grained rock fragments, mostly from preexisting shales, slates, cherts, or volcanic rocks. They are often associated with deltaic environments.

Shale, Mudstone, and Siltstone

Shale (Figs. 6–38 and 6–39) is a fine-grained laminated sedimentary rock composed of clay minerals and other clay and silt-size particles. Shales take their name from the French word *escale,*

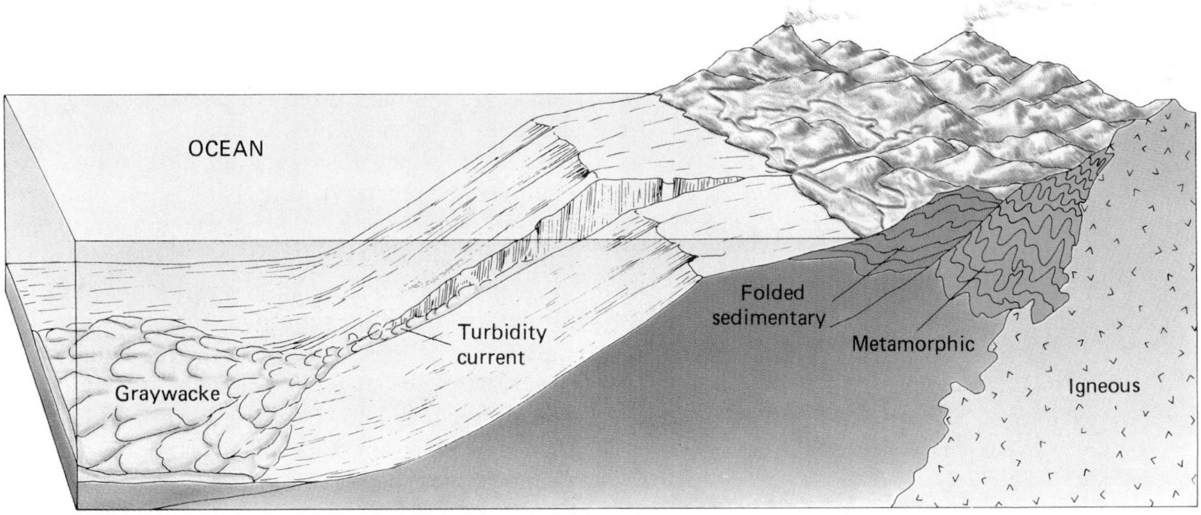

Figure 6–36 Tectonic setting in which graywacke is deposited. Frequently graywackes are transported by masses of water highly charged with suspended sediment. Because of the suspended matter, the mass is denser than surrounding water and moves along the sloping sea floor or down submarine canyons as a turbidity current. The graywacke sediment characteristically accumulates in deep-sea fans at the base of the continental slope.

meaning scale. The name refers to the tendency of shales to split readily into thin layers parallel to bedding planes. This property is termed **fissility** (Latin, *fissilis*, split) and probably results from the parallel alignment of tiny flat flakes of mica and clay minerals. Such particles are naturally deposited with their flat surfaces parallel to the floor of the depositional basin. Under compaction, even particles that were not originally in horizontal alignment may be brought into this position by the vertical pressure from overlying beds.

Some clay-rich rocks do not exhibit fissility and are often thick-bedded and massive. The general name for such nonfissile clayey rocks is **mudstone**. Another fine-grained sedimentary rock that generally lacks fissility is **siltstone**. Siltstones are composed largely of silt-sized particles of quartz and feldspar and contain less clay than shales or mudstones.

Carbonate Sedimentary Rocks

Limestones

Limestones are the most abundant of carbonate sedimentary rocks (Fig. 6–40). Although limestone

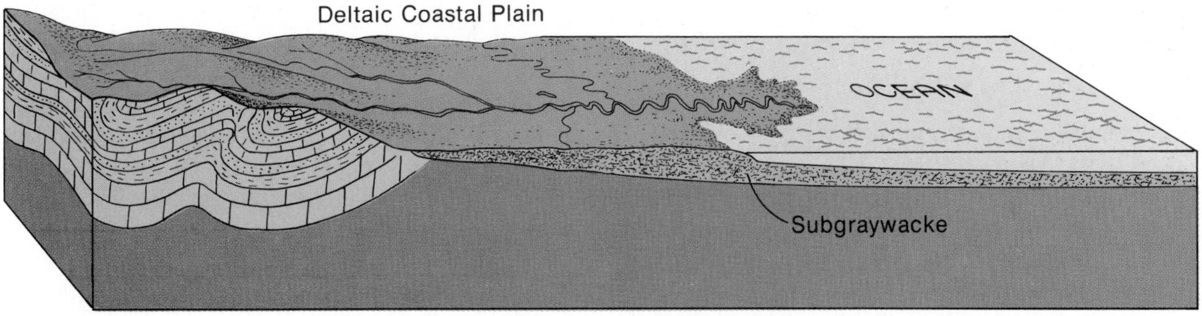

Figure 6–37 Deltaic environment in which subgraywackes may be deposited.

A

B

Figure 6–38 (A) A hand specimen of shale. This variety of shale contains waxy hydrocarbons and is called oil shale. (B) A thin section of shale cut perpendicular to bedding. The brown color results from the organic matter that is mixed with the clay and other fine particles.

lake deposits do occur, most limestones originated in the seas. *Nearly always, the formation of these marine limestones appears to have been either directly or indirectly associated with biologic processes.* In some limestones, the importance of biology is obvious, for the bulk of the rock is composed of readily visible shells of molluscs and skeletal remains of corals and other marine organisms (Fig. 6–41). In other limestones, skeletal remains are not present, but nevertheless the calcium carbonate that forms the bulk of the deposit was precipitated from sea water because of the life processes of organisms living in that water. For example, the relatively warm clear ocean waters of tropical regions are usually saturated with calcium carbonate. In this condition, only a slight increase in temperature, loss of dissolved carbon dioxide, or influx of supersaturated water containing $CaCO_3$ "seeds" can bring about the precipitation of tiny crystals of calcium carbonate. Organisms do not appreciably affect temperature, but through photosynthesis myriads of microscopic marine plants remove carbon dioxide from the water and thus may "trigger" the precipitation of calcium carbonate. The precipitate is thus an inorganic product of organic processes.

In general, limestones consist of one or more of a combination of such components as micrite, bioclasts, intraclasts, oöids, and carbonate spar. **Micrite** (Fig. 6–42A) is lithified mechanically deposited lime mud that may serve as a matrix between detrital pieces of limestone or form the entire substance of finely textured limestone. For limestones composed almost entirely of micrite, one can infer deposition in quiet water. **Bioclasts** (Fig. 6–43) are skeletal fragments of marine invertebrates, whereas **intraclasts** are fragments derived from the erosion of older limestones and other recycled and variously shaped pieces of calcium carbonate. **Oöids** (Fig. 6–44) are small spherical sand-sized grains formed by precipitation of carbonate around a nucleus. An oölite is a rock composed of oöids. The carbonate in oöids is added in concentric layers as the tiny spheres are rolled about on the sea floor by currents. Tidal areas of the coasts of Florida and the Bahamas have oöids forming today. Oölitic sediment frequently displays cross-bedding and ripple-marks just as do sandstones.

Carbonate spar is a clear crystalline carbonate that is normally deposited between the clasts as a cement or has developed by replacement of calcite. These components of limestone as seen through the microscope are shown in Figure 6–42. They permit classification of particular rocks as micritic limestone, intraclastic or bioclastic limestone, oölitic limestone, or sparry limestone.

Figure 6–39 Shales are generally less resistant to weathering and erosion than well-cemented sandstones. As a result, they may form shale slopes, such as these that occur beneath the massive sandstone strata exposed on either side of the Delores River valley, Colorado Plateau south of Grand Junction, Colorado.

Figure 6–40 Roadcuts exposing limestone strata are familiar to travelers on interstate highways throughout the Mississippi valley region.

Figure 6–41 The limestone known as coquina is composed entirely of the skeletal fragments (bioclasts) of marine invertebrate animals.

Although all limestones are composed of calcite (calcium carbonate), they differ in texture and the nature of component particles.

There are still other varieties of limestone. **Chalk** (Fig. 6–45) is a soft porous variety that is composed largely of tiny skeletal elements of unicellular, planktonic, calcareous algae. **Lithographic limestone** is a dense micritic variety once widely used in printing illustrations. **Travertine** (dripstone) is an inorganic precipitate with crystalline texture formed around springs and in caves by the loss of CO_2 in water containing calcium bicarbonate (Fig. 6–46).

Dolostone

Dolostone is nonclastic rock composed largely of the mineral dolomite, which is a calcium and magnesium carbonate. As found in exposures, dolostone is not easily distinguished from limestone. The usual field test for distinguishing dolostone from limestone is to apply cold dilute hydrochloric acid. Unlike limestone, which bubbles readily, dolostone will effervesce only slightly unless powdered.

The origin of dolostone is somewhat problematic. The mineral dolomite is not secreted by organisms in shell-building. Direct precipitation from sea water does not normally occur today, except in a few environments where the sediment is steeped in abnormally saline water. Because of these limitations, most geologists believe that the dolostones of the geologic past formed from limestones. The transformation involved replacement of some of the calcium in calcite by magnesium carried into buried limestone masses by percolating water.

$$\underset{\text{magnesium in solution}}{Mg^{++}} + \underset{\text{calcite in limestone}}{2\,CaCO_3} \rightarrow \underset{\text{dolomite}}{CaMg(CO_3)_2} + \underset{\text{calcium in solution}}{Ca^{++}}$$

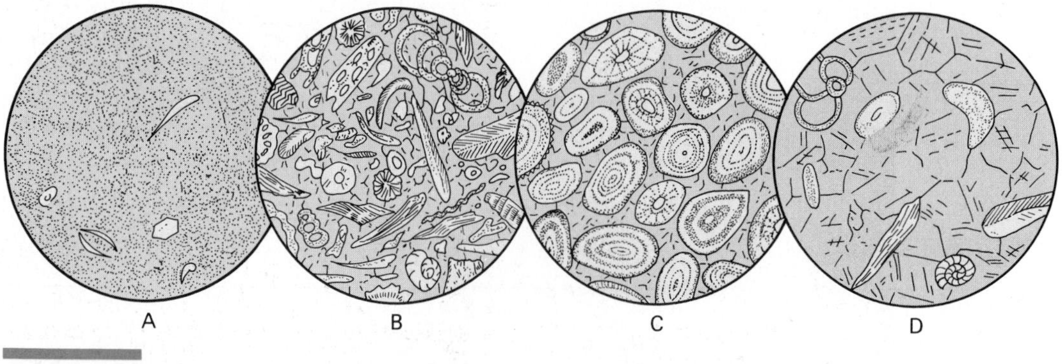

Figure 6–42 Textures of limestones as seen in thin section under a microscope. (A) Aphanitic limestone or micrite. (B) Bioclastic limestone with fine-grained sparry calcite as cement. (C) Oölitic limestone. (D) Sparry or crystalline limestone.

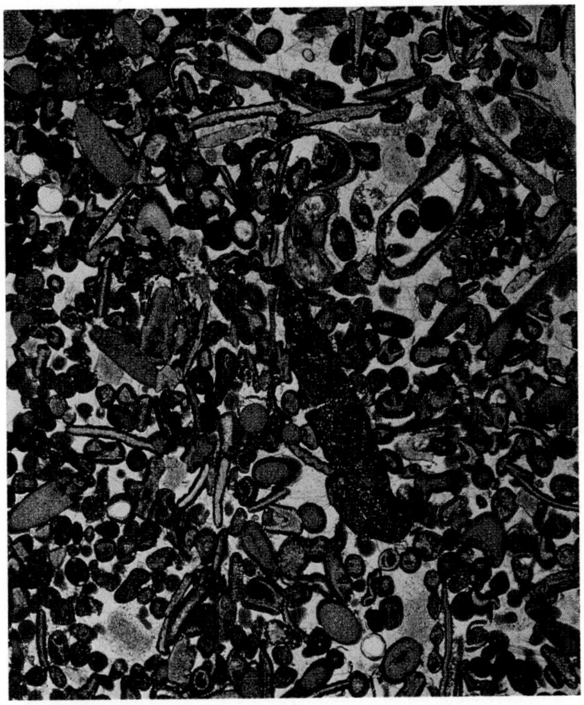

Figure 6–43 Bioclastic, oölitic limestone. The oöids and shell particles are cemented by carbonate spar (the clear material between grains). The specimen is from the Bedford Limestone of Mississippian age. The Bedford is widely used as a building stone.

Figure 6–44 Modern oöids from the Bahama Banks. Diameter of field is 5 mm. (Courtesy of J.C. Brice.)

Chert

Chert (Fig. 6–47) is a very finely crystalline (microcrystalline) form of quartz that occurs as extensive beds (Fig. 6–48) as well as nodules (Fig. 6–49) in limestones. Most cherts are at least 90 per cent silica, with the remaining 10 per cent consisting of clay, calcite, or hematite impurities. Water is usually present in amounts less than 1 per cent.

Figure 6–45 Chalk cliffs between Swanage and Ballard Point, Dorset, England.

Figure 6–46 Travertine (dripstone) is a type of limestone familiar to visitors to large caverns. Luray Caverns, Virginia.

Bedded cherts are thought to have formed from accumulations of siliceous remains of diatoms and radiolaria and from subsequent reorganization of the silica into a microcrystalline quartz. Additional silica for the formation of chert is also sometimes derived from the dissolution of volcanic ash. In fact, many sequences of bedded cherts are found in association with layers of ash and submarine lava flows.

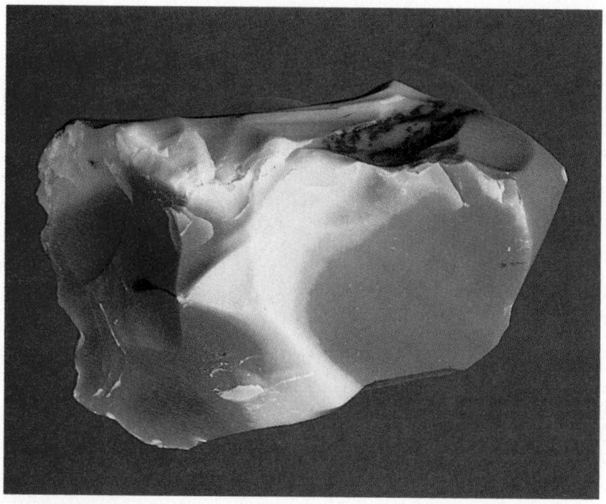

Figure 6–47 Chert. This "grainy" variety of chert is called novaculite. It is formed by very low grade metamorphism of common chert. Its texture makes it useful as a grinding stone.

Figure 6–48 Chert interbedded with red bands of hematite. Wadi Kareim, Egypt. (Courtesy of D. Battacharyya.)

Figure 6–49 White nodular chert in tan limestone. Fern Glen Formation west of St. Louis, Missouri.

The origin of nodular chert is a subject of debate among geologists. A favored theory is that the nodules form as replacements of carbonate sediment by silica.

Evaporites

In Chapter 2, we noted that **evaporites** are chemically precipitated rocks that are formed as a result of evaporation of saline water bodies. Only about 3 per cent of all sedimentary rocks consist of evaporites. Evaporite sequences of strata are composed chiefly of such minerals as gypsum ($CaSO_4 \cdot 2H_2O$), anhydrite ($CaSO_4$), halite (NaCl) and sylvite (KCl), and associated calcite and dolomite. Extensive ancient deposits of evaporites are currently being commercially worked in Michigan, Kansas, Texas, New Mexico, Germany, and Israel.

Evaporites are important raw materials for the chemical and construction industries. Gypsum, for example, is used in the manufacture of portland cement, plaster, and plaster products (Fig. 6–50A and B). Halite is mined in large quantities for use in the removal of snow and ice from streets, use in foods, and in the manufacture of the hydrochloric acid that is the mainstay of many industrial processes. Sylvite, called potash in the industry, is used along with phosphates and nitrates to manufacture

A B

Figure 6–50 (A) Selenite gypsum ($CaSO_4 \cdot 2H_2O$). (B) "Drywall" wallboard used in construction consists of plaster of Paris prepared from gypsum, sand, and water and sandwiched between two pieces of cardboard. (Courtesy of J.C. Brice.)

fertilizers. The name potash actually is derived from the ancient process of leaching ashes from plants in iron pots. The fact that one could obtain potash in this way led to its recognition as a substance important to plant growth. With the world's increasing food demands, it is not surprising that potash production has been increasing at a rate of about 10 per cent a year for the past decade.

The ideal conditions required for precipitation of evaporites are arid climates and a physiographic situation that would provide periodic additions of sea water to the evaporating marine basin. In the Gulf Coastal Region of the United States, as a result of the pressure of overlying rocks, deeply buried deposits of salt have flowed plastically upward to form underground salt domes (Fig. 6–51). In the process, the salt arched overlying strata, thereby producing structures into which petroleum sometimes migrated and became entrapped.

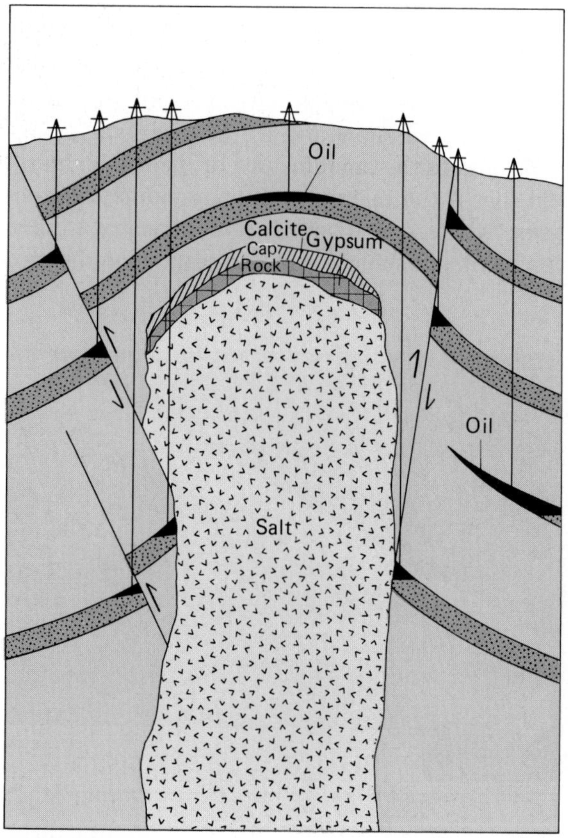

Figure 6–51 Salt dome, illustrating possibilities for oil entrapment in domelike structures (top center) by faults and by pinchout of oil-bearing strata.

If large quantities of lake or sea water are evaporated, halite, gypsum, and other so-called evaporites will be precipitated from solution.

Coal

Coal is a carbonaceous rock resulting from the accumulation of plant matter in a swampy environment combined with alteration of that plant tissue by both biochemical and physical processes until it is converted to a consolidated carbon-rich material. The biochemical and physical changes may produce a series of products ranging from peat and lignite to bituminous and anthracite coal. For coal to form, plant tissue must be accumulated underwater or be quickly buried because plant matter, if left exposed to air, is readily oxidized to water and carbon dioxide. With underwater accumulation or quick burial of plant material, a major part of the carbon can be retained.

As the world's supply of natural gas and petroleum diminishes, coal will become an increasingly more important source of energy. Because of its importance, coal will be discussed more fully in the chapter dealing with the earth's geologic resources (Chapter 19).

Color in Sedimentary Rocks

The color of a sedimentary rock is one of the first things noticed by an observer. Yet for sedimentary rocks, color is not very useful in identification. Color, however, may provide clues to a rock's composition and environment of deposition.

Hues of brown, red, and green are frequently formed in sedimentary rocks as a result of their content of iron oxides. Few, if any, sedimentary rocks are free of iron, and less than 0.1 per cent of this element can color a sediment a deep red. The iron pigments not only are ubiquitous in sediments but also are difficult to remove in most natural solutions.

Strata colored in shades of red, brown, or purple by iron are designated **red beds** by geologists. Oxidizing conditions required for the development of iron oxides are more typical of nonmarine than of marine environments. Thus, most red beds are floodplain, alluvial fan, or deltaic deposits. Some, however, are originally reddish-colored sediment carried to and deposited in the sea.

Black and dark gray coloration in sedimentary rocks—especially shales—usually results from the presence of organic carbon compounds or iron sulfides. The occurrence of an amount of organic carbon sufficient to result in black coloration implies an abundance of organisms in or near the former depositional areas as well as environmental circumstances that kept the remains of those organisms from being completely destroyed by oxidation or bacterial action.

The principal coloring agents in sedimentary rocks are carbon, which colors rocks gray and black, and oxides of iron, which color rocks in shades of yellow, brown, red, and green.

Fossils

Just as modern environments of deposition include animals and plants, so also did ancient depositional sites. The remains or traces of those organisms are frequently preserved as **fossils** (Fig. 6–52). Fossils may constitute only a minor feature of the rock, or, as is the case with many limestones, the fossils may be the dominant constituents (Fig. 6–53).

Former living organisms may become fossils as the result of several natural processes. The largest number of fossils are remains of marine creatures that died and were covered by the rain of sediment that continuously falls onto the sea floors. Less commonly, plant and animal remains are covered with windblown silt or volcanic ash, immersed in tar or quicksand, or engulfed by lava flows. For preservation, quick burial and the possession of a mineralized skeleton is usually required. Even then, only a very small percentage of animals are ever preserved. Once enclosed in sediment, the

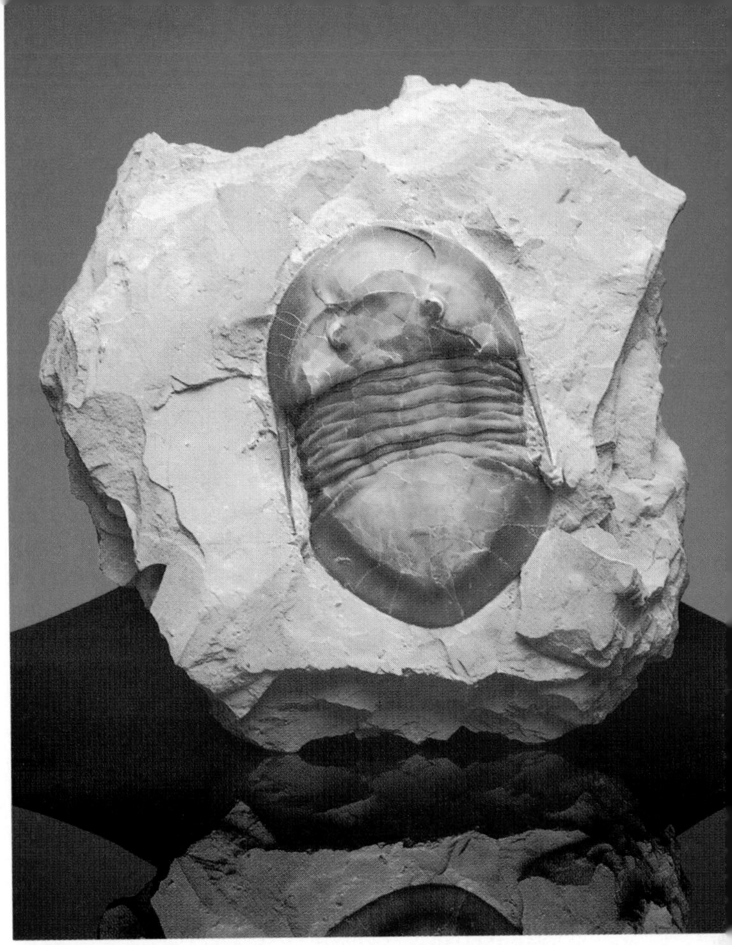

Figure 6–52 The trilobite *Isotelus brachycephalus* from strata of Ordovician age near Cincinnati, Ohio. The specimen is 23 cm in length. (Courtesy of Ward's Natural Science Establishment, Inc., Rochester, New York.)

Figure 6–53 Fossiliferous limestone. The fossil shells are brachiopods.

original hard parts may remain largely unaltered, or they may be replaced by mineral matter of a different composition. Pores may be filled with lime or silica brought in by underground water. Parts or organs may be dissolved completely, leaving behind empty molds. Molds may later be filled so as to form casts that resemble the original organic part. All of these processes for changing shells, bones, wood, and other organic matter into durable traces of former life are termed *petrifaction.*

For the student of sedimentary rocks, fossils are exceptionally useful as indicators of ancient environments of deposition. A limestone containing the remains of corals, for example, must have been deposited in shallow warm marine waters, whereas preserved impressions of oak leaves in a sandstone would lead one to suspect the enclosing rock was of continental origin and that the area of deposition had a climate similar to that where oak thrives today. Fossils may shed light on temperature, salinity, depth of water, and presence or absence of bottom currents in the depositional environment. In studies of broader scope, fossils provide information about the distribution of seas, shores, and mountains as they existed in the geologic past. They inform geologists about past changes in climate, the location of former land connections between continents, and changes in the positions of continents. Fossils document the long history of life and are indices to the age of the rock in which they are found. In the chapter titled "Time and Geology" we will discuss why fossils are so valuable in constructing a geologic time scale and how they may be used to determine the age and equivalency of strata around the world.

The two areas in which fossils have their greatest value, are in determining environments of deposition of ancient rocks, and in determining the age of those rocks.

The Sedimentary Record

Rock Units

Rock units are bodies of rock that can be subdivided and recognized wherever they occur on the basis of observable objective criteria. They can be identified in the field and mapped as units that are distinctive and different from neighboring units. The fundamental rock unit is the **formation**, which may be defined as a mappable lithologically distinct body of rock having recognizable contacts with other formations (Fig. 6–54).

Maps that record the distribution and nature of sedimentary as well as igneous and metamorphic rock units and the occurrence of geologic features like folds and faults are called geologic maps. To prepare a geologic map, formations must first be matched or correlated from place to place. In this way, the areal or lateral extent of each rock unit can be determined.

Formations are generally given two names: (1) a geographic name that refers to a locality where the formation is well exposed or where it was first described and (2) a rock name if the formation is primarily of one lithologic type. For example, the "St. Louis Limestone" is a carbonate formation named after exposures at St. Louis. Where formations have several lithologic types within them, the locality name may be followed by the word *formation.* Distinctive smaller units within formations may be split out as *members,* and formations may be combined into larger units called *groups* because of related lithologic characteristics (or by their position between distinct stratigraphic breaks). In mapping formations and other rock

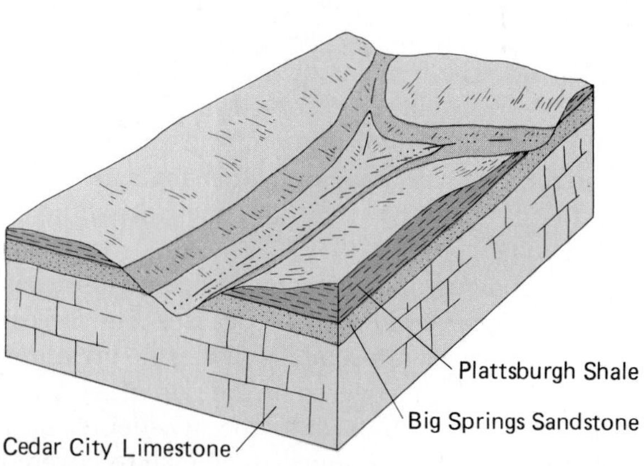

Figure 6–54 Formations. The diagram shows three formations. In practice, these formations would be formally named, often after a geographic location near which they are well exposed. For example, the three formations shown here might be designated the "Cedar City Limestone," "Big Springs Sandstone," and "Plattsburgh Shale."

units, geologists recognize that the unit may or may not be the same age everywhere it is encountered. The near-shore sands deposited by a sea slowly transgressing (advancing over) a coastal plain may deposit a single blanket of quartz sandstone. However, it will be older where the transgression began and younger where it ended (Fig. 6–55).

The basic unit of stratigraphy is a rock unit called a formation. Every formation has distinctive characteristics of color, composition, texture, or fossil content that permits one to map its areal extent.

Illustrating Geology

To aid in the synthesis and interpretation of field observations of sedimentary and other rocks, geologists prepare a variety of maps, graphs, and charts designed to show relations of rock bodies to one another, their thickness, the manner in which they have been disturbed, and their general composition. The most important graphic devices for communicating such information are columnar sections, cross-sections, and geologic maps.

Columnar sections show the vertical sequence of strata (layers of rock) at any given locality. After geologists have compiled columnar sections at several locations within a region, they may next prepare a *composite geologic column* (Fig. 6–56) that combines the data from all of the various individual sections. Individual columnar sections are also used in the construction of cross-sections. As implied by their name, *cross-sections* show the vertical dimension of a slice through the earth's crust. Some cross-sections, namely the stratigraphic type, emphasize the age or lithologic equivalence of strata. The vertical measurements for such stratigraphic cross-sections are made from a horizontal line representing the surface of a definite bed or stratum. That surface is called the *datum.* In nature, the surface used as datum may be inclined or folded, and so stratigraphic sections do not validly show the tilt or position of beds relative to sea level. Stratigraphic cross-sections are most effective in showing the way beds correlate and vary in thickness from exposure to exposure or well to well. In order to show the way beds are folded, faulted, or tilted, a *structural cross-section* can be prepared (Fig. 6–57). In drawing the structural cross-section, the datum is a level line parallel to sea level, and the tops and bottoms of rock units are plotted according to their true elevations. If the vertical and horizontal scales are similar, the attitude of the beds will be correctly depicted. Many times, however, it is useful to have a larger vertical then horizontal scale in order to emphasize geologic features.

Geologic maps (Fig. 6–58) show the distribution of rocks of different kinds and ages that lie directly beneath the regolith. Assume for a moment that the regolith were miraculously removed from a certain region, so that bedrock would be exposed everywhere. Next imagine that the surfaces of each rock unit were painted a different color and the entire region photographed vertically from an airplane. The resulting photograph would constitute a simple geologic map.

The actual preparation of a geologic map is basically similar to the construction of most other kinds of maps. The essential ingredients are a series of observations made at particular geographic locations. For geologic maps, the observations are

(Text continues on p. 164.)

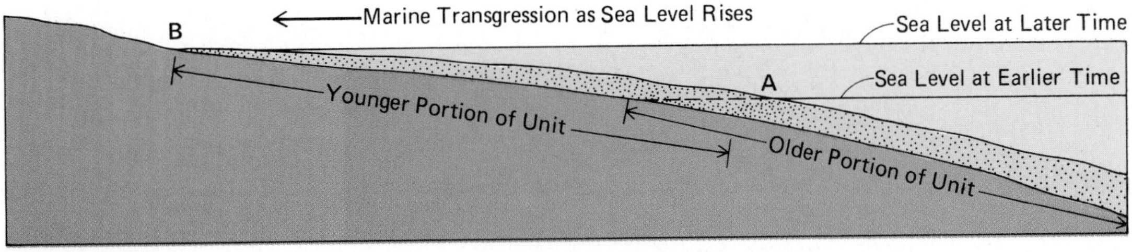

Figure 6–55 Diagram showing how the original deposits of a formation may vary in age from place to place.

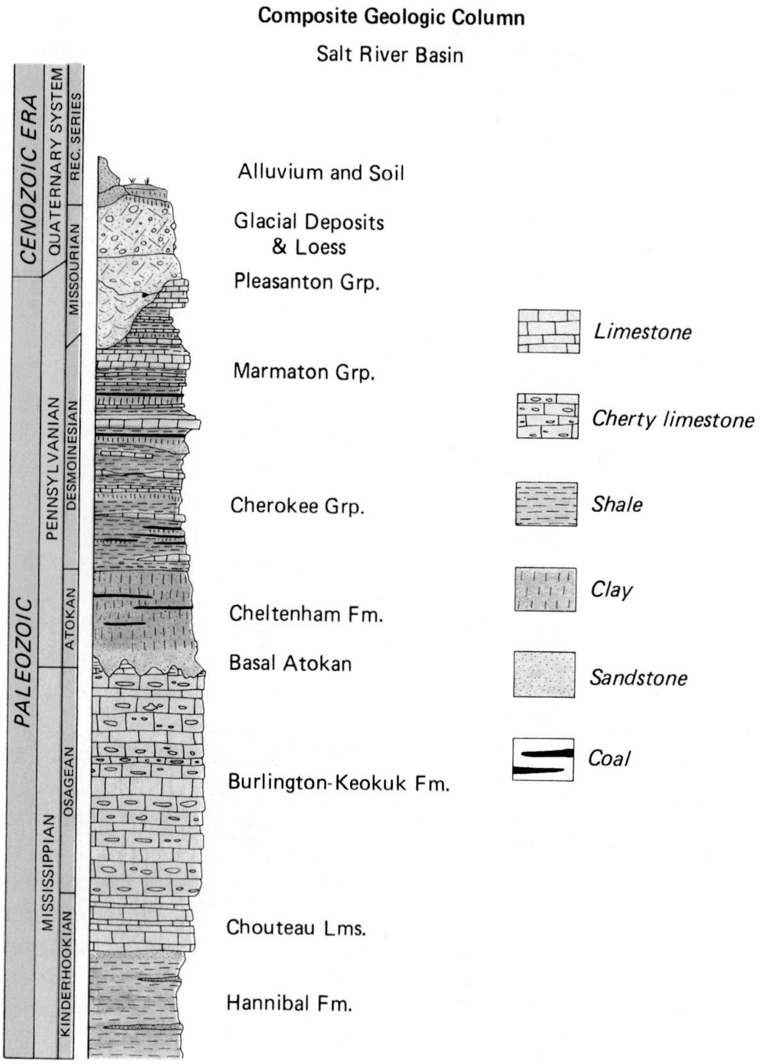

Figure 6–56 Example of a composite geologic column.

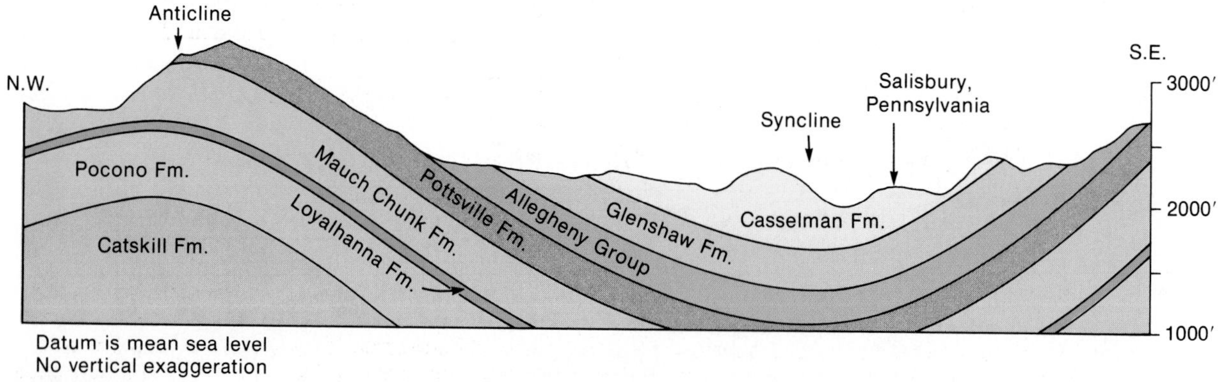

Figure 6–57 Geologic structural cross-section across Paleozoic rocks in the Appalachian Mountains, southeastern Pennsylvania, extending northwestward from the town of Salisbury.

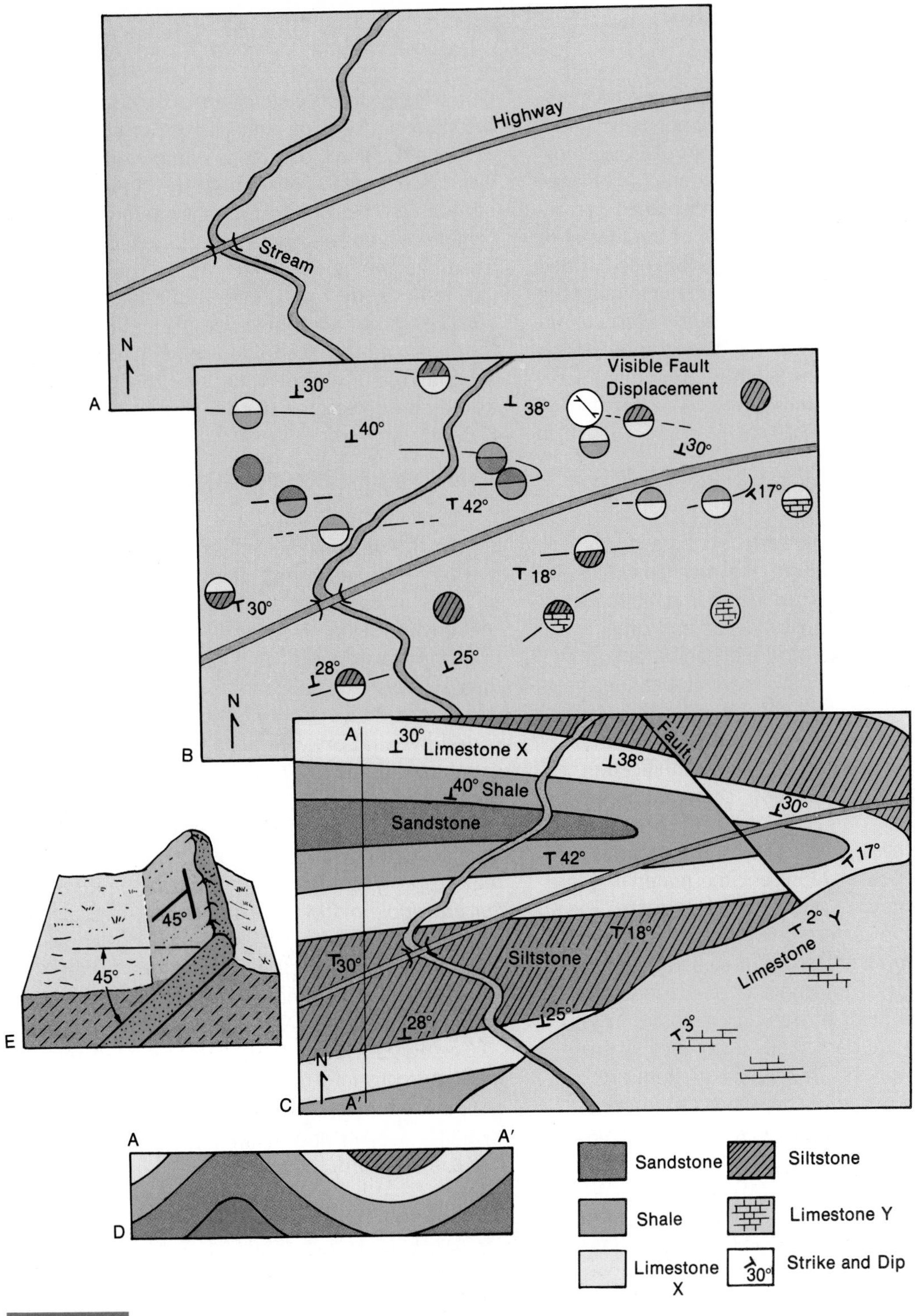

Figure 6–58 Steps in the preparation of a geologic map. (A) A suitable base map is selected. (B) The locations of rock exposures of the various formations are then plotted on the base map. Special attention is given to exposures that include contacts between formations; where they can be followed horizontally, they are traced onto the base map also. *Strike* (the compass direction of a line formed by the intersection of the surface of a bed and a horizontal plane) and *dip* are measured wherever possible and added to the data on the base map. After careful field study and synthesis of all the available information, formation boundaries are drawn to best fit the data. (C) On the completed map, color patterns are used to show the areal pattern of rocks beneath the cover of soil. (D) A cross-section is shown along line A—A′. (E) A block diagram illustrates strike and dip.

available in natural exposures, boreholes, quarries, and roadcuts. The geologist notes not only the locations of individual exposures but also the locations of formation boundaries ("contacts"). Places where rocks are disrupted and displaced are recorded, along with measurements of the angles of beds that are tilted and bent. Symbols are added for further clarification. In most geologic mapping today, the field observations are plotted directly on aerial photographs or on the topographic maps that were prepared from such photographs. A recent technique called radar imagery is capable of penetrating dense vegetation to accurately reveal the patterns of rock units at the ground surface.

Once the geologic map is completed, the geologist can tell a good deal about the geologic history of an area. The rock formations can be considered "chapters" in the sequence of geologic events. From the geologic map shown in Figure 6–58, one can deduce that the rock formations had been compressed into a folded pattern and then broken (faulted) and that the folded layers were of greater age than the level limestone layer that covers them in the southeast area of the map.

Summary

From the time when the earth acquired its atmosphere and hydrosphere, the mechanical and chemical breakdown of the original igneous rocks began. The weathering of igneous and other rocks through chemical and mechanical action produces gravel, sand, silt, clay, lime-bearing materials, and salts. These products, after they are deposited in the sea or in continental environments, are bound together by rock lithification processes such as compaction, cementation, and crystallization to form *conglomerates, breccias, sandstones, shales, limestones,* and *evaporites.*

Sedimentary rocks represent the material record of environments that once existed on earth. Their composition and texture provide clues to the nature of the source areas that provided the sediment, the kinds of transporting media, and the general environment of deposition. Fossils in sedimentary rocks are particularly instructive as environmental indicators. They tell us if strata are marine or nonmarine, if the water at the depositional site was deep or shallow, or if the climate was warm or cool. They also can be used to ascertain the geologic age of strata. Color in a sedimentary rock sometimes helps one to decide whether it was deposited in nonmarine or terrestrial environments. From the size and shape of grains in clastic rocks, geologists can sometimes infer whether the sediment was deposited in quiet water or turbulent currents, whether it had been sorted by wind or wave action, or whether it had experienced many or few cycles of erosion and deposition.

Similar kinds of information can sometimes be elucidated by the study of such sedimentary features as *graded bedding, cross-bedding,* and *current ripple marks.* Sandstones are particularly useful in investigations of ancient depositional environments. *Quartz sandstones, graywackes, arkoses,* and *lithic sandstones* each accumulate in particular paleogeographic and tectonic situations.

Geologists usually divide successions of sedimentary rocks into rock units that are sufficiently distinctive in color, texture, or composition to be recognized easily and mapped. Such rock units are called *formations.*

Terms to Understand

arkose
bioclast
biogenic
breccia
carbonate spar
cement
cementation
chalk
chert
clastic
coal
compaction
concretion
conglomerate
cross-bedding
crystallization
dolostone
evaporite
fissility
formation
fossil
graded bedding
graphic limestone
graywacke
intraclast
lithic sandstone
lithification
lithographic limestone
matrix
micrite
mud cracks
nonclastic

oöid
precipitation
primary sedimentary structure
quartz sandstone
red beds
ripple marks (symmetric and asymmetric)
rock unit
sandstone
shale
siltstone
sorting
stratum
tectonics
travertine
turbidity current

Questions for Review

1. Define sedimentary rocks. What is the derivation of the term *sedimentary* (consult your dictionary)?

2. What characteristics of sedimentary rocks are most useful in their classification and identification?

3. What are some of the changes that take place in a sediment after deposition that lead ultimately to lithification?

4. What is the distinction between the following:
   a. Clastic and nonclastic (chemical) sedimentary rocks
   b. Conglomerate and breccia
   c. Sandstone and clastic limestone
   d. Mature and immature sandstones
   e. Rock units and time-rock units

5. What interpretations about the history of a sedimentary rock can sometimes be obtained from an examination of:
   a. Degree of grain rounding
   b. Cross-bedding
   c. Grain size
   d. Graded bedding

6. How does matrix in a rock differ from cement? What kinds of cement can bind a sedimentary rock together? Which kind of cement is likely to be the most durable?

7. Why are sands and silts deposited in nonmarine desert environments rarely black or dark gray in color?

8. Under what circumstances might a reddish siltstone containing fossils of marine invertebrates have originated?

9. What sort of history might you infer for a sandstone composed of very angular grains, poorly sorted, and indurated with a 30 per cent matrix of mud?

10. Compare the probable tectonic environments in which a graywacke and a quartz sandstone might be deposited.

11. Why, in general, are fine-grained sediments often deposited farther from a shoreline than coarse-grained sediments?

12. A geologist studying a sequence of strata discovers that limestone is overlain by shale, which in turn is overlain by sandstone. What might this signify with regard to the *advance or retreat* of a shoreline with time?

13. What features or properties of sedimentary rocks indicate relatively shallow water deposition?

14. What are turbidity currents? How might turbidity current deposits be recognized?

15. Are carbonate sediments more likely to accumulate in warmer or colder parts of the ocean? Why?

16. What inference can be made about the environment of deposition of limestones composed almost entirely of micrite?

17. What is the most important difference between dolostone and limestone?

18. What is an arkose? What inferences can be made about the source materials for arkosic sands?

19. What single characteristic or property distinguishes a shale from a claystone?

Supplemental Readings and References

Boggs, S. Jr., 1987. *Principles of Sedimentology and Stratigraphy.* Columbus, Ohio, Merrill Publishing Co.

Brenner, R.L., and McHargue, T.R., 1988. *Integrative Stratigraphy.* Englewood Cliffs, N.J., Prentice-Hall, Inc.

Friedman, G.M., and Sanders, J.E., 1978. *Principles of Sedimentology.* New York, John Wiley & Sons.

Huxley, T.H., 1893. On a Piece of Chalk, in *Collected Essays of Thomas H. Huxley.* London, Macmillan & Co. (Reprinted as a separate volume in 1965 by Scribners, New York.)

Laporte, L.F., 1968. *Ancient Environments.* Englewood Cliffs, N.J., Prentice-Hall, Inc.

Metamorphism and Metamorphic Rocks

7

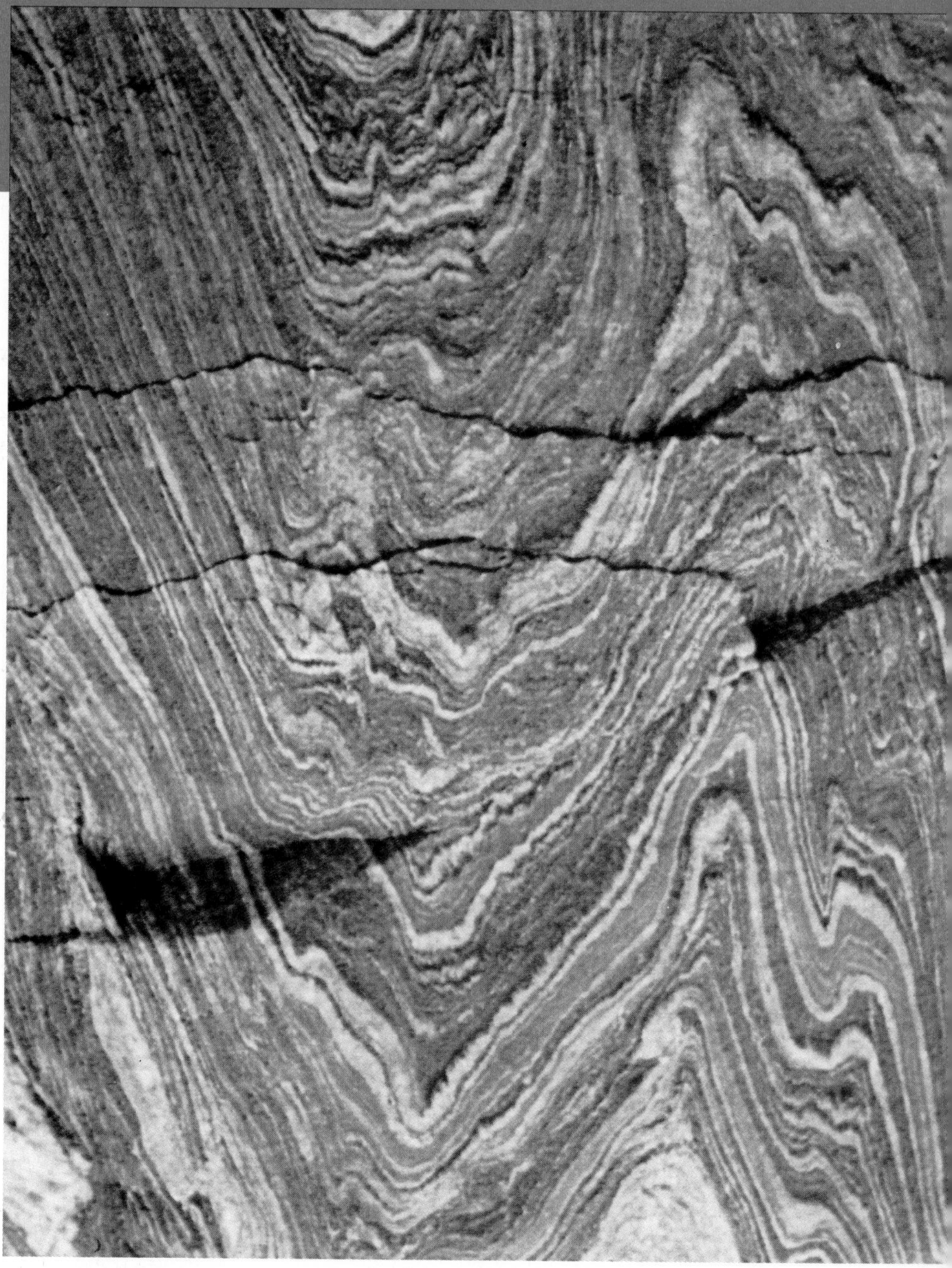

The 3.6-billion-year-old Amitsoq Gneiss, Lile Narssuaq, Greenland. (Courtesy of M.S. Smith.)

This rock . . . has experienced great metamorphoses, making the cement crystalline and schistose, and elongating and flattening the pebbles.

Edward Hitchcock. Metamorphism of the Newport Conglomerate, 1861.

Introduction

In the latter half of the 18th century, James Hutton of Edinburgh, regarded by many as the founder of modern geology, advanced the opinion that certain bodies of hard crystalline rock were once clastic sedimentary layers that had been altered to distinctly different kinds of rocks. Hutton's interpretation was accepted by the eminent Sir Charles Lyell. In the 1833 edition of his *Principles of Geology,* Lyell proposed using the term *metamorphic* for such rocks. He referred to the process by which metamorphic rocks are formed as **metamorphism**. In his work, Lyell correctly indicated that the changes that lead to the formation of metamorphic rocks do not include weathering, lithification, or melting. Rather, metamorphic rocks are transformed from previously existing parent rocks while in the solid state by heat, pressure, and the reactions of chemically potent aqueous solutions. This interpretation illustrates Lyell's keen insight into geologic processes; for, unlike sedimentary and volcanic rocks, metamorphic rocks cannot be seen as they form.

The alterations involved in metamorphism may be mostly mechanical, as when grains and crystals are deformed. In other cases, the deformation may be accompanied by a recrystallization of some or all of the original minerals. Such changes involve little or no loss or gain in bulk chemical composition.

In general, the characteristics of metamorphic rocks are dependent on the kinds and intensities of metamorphic agencies acting on the parent rock, as well as the nature of the parent rock itself. In the pages ahead, we will consider these agencies and discuss the origin and distinctive features of the main categories of metamorphic rocks.

Metamorphic rocks consist of previously formed rocks that have been changed in composition and texture by heat, pressure, and chemical action.

The Agents of Metamorphism

The three principle agents of metamorphism are heat, pressure, and chemically active fluids. Alone or in combination, these agents operate at various intensities to produce metamorphic rocks having distinctive textures and compositions.

Heat

Heat is a major cause of metamorphism. It can reduce the ability of a rock to withstand deformation, and it causes an increase in the rate of chemical reactions that facilitate the production of new mineral assemblages. The heat for metamorphism may be provided by nearby intrusions of magma, or it may be associated with compression of the crust or deep burial. The rate of increase in temperature at increasing depths in the earth is called the **geothermal gradient**. Measurements made in deep mines and wells indicate that the geothermal gradient for the crust is about 30°C per km of depth. This rate, however, varies from place to place, and is, not unexpectedly, greater near centers of active volcanism. The geothermal gradient is such that at depths of about 35 km temperatures are high enough to melt rock (Fig. 7–1).

Because particular mineral-forming chemical reactions occur only within a specific range of temperature, heat influences the ultimate mineral composition of metamorphic rocks. Also, as temperature rises, ions within the atomic lattice of

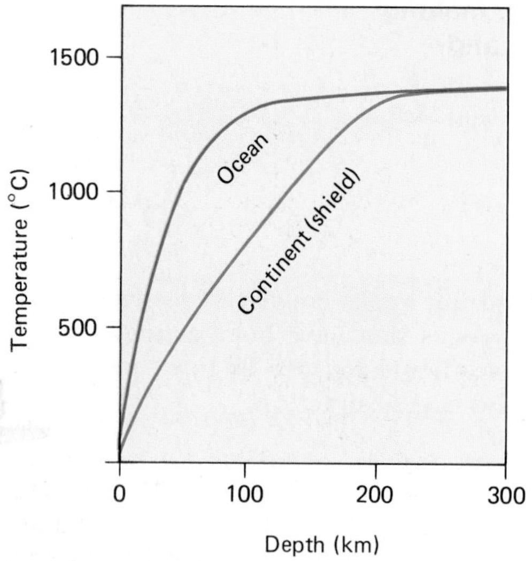

Figure 7–1 The increase in temperature with depth beneath a continental shield area (red) and ocean basin (green). (Courtesy of Geoffrey F. Davies.)

minerals become increasingly agitated. Eventually, they may move to new locations and arrange themselves into structural forms stable under the newer thermal conditions. As an example, consider what might happen to clay minerals in shale strata invaded by a granitic magma. Clay minerals are hydrous aluminum silicates. Heat applied to these minerals from the hot magma would cause a loss of the water contained in the clay and a conversion of clay into a common metamorphic mineral named andalusite. Andalusite is a nonhydrous aluminum silicate having an atomic structure that is stable under the new conditions. The released water in the above example, together with gases such as carbon dioxide, may also participate in the many chemical reactions associated with metamorphism.

As noted in Chapter 3, if temperatures in a rock mass rise high enough, the melting points of constituent minerals will be exceeded and magma will begin to form. If the melting becomes pervasive, igneous rather than metamorphic rocks will result.

Pressure

The tremendous pressures that exist several kilometers below the earth's surface (about 1000 kg/cm^2 at a depth of 4 km) or that are associated with collisions of tectonic plates can cause profound changes in deeply buried sedimentary rocks, preexisting metamorphic rocks, or igneous rocks. Mineral grains, for example, may recrystallize into new minerals with more tightly packed atomic structure and therefore greater density. Where mineral grains are in contact, the squeezing action may cause melting or solution at the points of contact and precipitation of material along the sides of grains that experience less pressure. It also facilitates the development of lineated and banded textures characteristic of many metamorphic rocks. (Fig. 7–2).

In addition to the pressure of overlying rock, which affects rocks uniformly, pressure may also be applied to a rock mass along certain preferred directions. Such directional pressure may cause rocks to shear during deformation. Shearing is common in near-surface rocks, which are cooler and more brittle. When pressure is applied to such rocks they do not respond plastically, but break into thin slabs that slide past one another like playing cards in a stack that is pushed over. The shearing that occurs in mountain building grinds preexisting minerals and thereby provides more surface area for chemical reactions. It also causes mixing of rock types that results in the production of new metamorphic minerals.

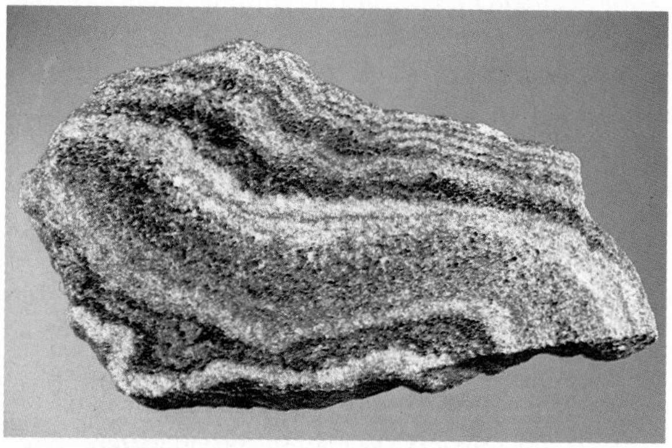

Figure 7–2 Gneiss with cut and polished surfaces exhibiting contorted bands of quartz and feldspar (light) and biotite (dark).

Chemically Active Solutions and Gases

In varying amounts, liquids and gases are always present in regions where metamorphism takes place. They play an important role in metamorphism by increasing the efficiency of recrystallization, serving as solvents, and accelerating the rate of chemical reactions. Laboratory experiments have shown that most minerals react so slowly to increases in temperature that metamorphic reactions would require lengthy spans of time to run their course. By adding only a minute amount of water to a laboratory capsule containing the experimental minerals, however, reaction rates are dramatically increased. One reason for this increase is that ions in a fluid medium can be brought into close proximity with each other and reach appropriate sites in the atomic structure of minerals more readily than in a dry environment. In some instances, water may enter into the composition of newly forming minerals such as mica, amphibole, and chlorite. An example is provided by the reaction below, in which the attractive green mineral serpentine is formed.

$$\underset{\text{(olivine)}}{5\,Mg_2SiO_4} + \underset{\text{(water)}}{4\,H_2O} \rightarrow \underset{\text{(serpentine)}}{2H_4Mg_3Si_2O_9} + \underset{\text{(removed in solution)}}{4\,MgO + SiO_2}$$

The water associated with metamorphic reactions may be derived from several possible sources. Some is water entrapped in parent sedimentary rocks at the time of their deposition. Another source is magma, from which may emanate large quantities of watery liquids and vapors. Smaller amounts of water may be given off by hydrous minerals as they begin to experience the effects of heat and pressure.

In addition to water, the gas carbon dioxide also promotes metamorphism. Carbon dioxide is readily liberated during the heating of limestone. This leaves the remaining oxide of calcium free to combine with silica or other impurities in the limestone to form calcium-bearing metamorphic minerals.

Water plays an important role in metamorphism. It facilitates chemical reactions, serves as a solvent, and transports ions.

Two Major Kinds of Metamorphism

The majority of metamorphic rocks around the world are products of either contact or regional metamorphism. All the agents of metamorphism operate in each category but with differing intensity and effect.

Contact Metamorphism

Contact metamorphism takes its name from the fact that it occurs at or near the contacts between hot igneous material and country rock. Heat is the most important agent in contact metamorphism, although pressure and chemical activity assist in producing alterations. Temperatures would, of course, be highest immediately adjacent to the igneous mass. Thus, as the distance increases between the contact of the igneous body and the country rock, the effect of the metamorphism decreases. Around many intrusions, roughly concentric zones of decreasing metamorphism called **metamorphic aureoles** can be recognized and mapped (Fig. 7–3).

The width of the metamorphic aureole and the amount of change that occurs during contact metamorphism is dependent on the composition, temperature, size, and shape of the hot igneous material, its viscosity and gas content, and the composition and permeability of the country rock (Fig. 7–4). In general, the larger and more acidic the igneous mass, the greater the effects of contact metamorphism.

Contact metamorphism has been responsible for some of the world's richest ore deposits. These include copper deposits in Arizona, Utah, and Korea; zinc ores in New Mexico and Ontario; iron ores (magnetite and hematite) in the Ural mountains; tin deposits in Finland; and graphite deposits in the Adirondacks.

A high temperature is the essential requirement for contact metamorphism. Directed pressure on the rock is not a requirement.

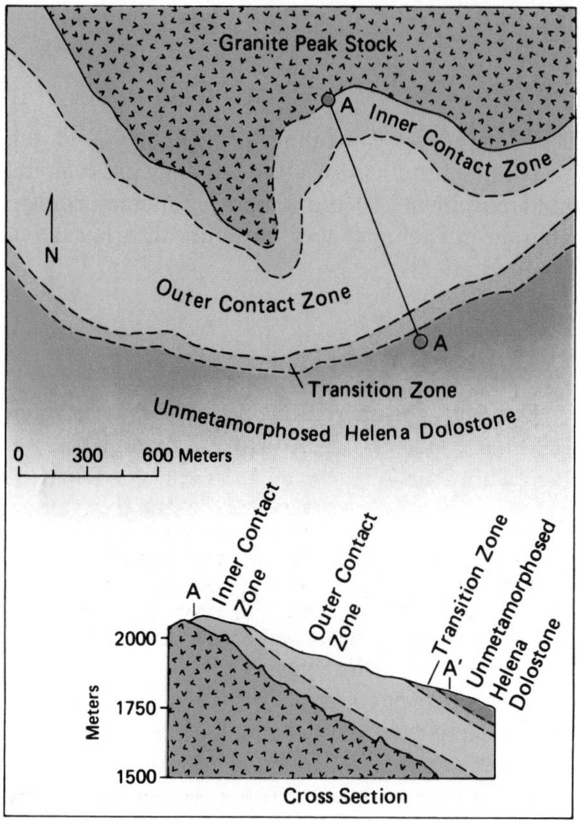

Figure 7–3 The Granite Peak aureole, Clark Co., Montana. This metamorphic aureole is developed around a granite stock that has intruded into dolostone country rock. Particular assemblages of metamorphic minerals occur within each contact metamorphic zone. (Simplified from Melson, W.G., *American Mineralogist*, 51:404, 1966, Fig. 1.)

Regional Metamorphism

Regional metamorphism may be defined as metamorphism that has developed over areas of many thousands of square kilometers as a result of intense compression associated with the convergence of tectonic plates. Most tracts of regionally metamorphosed rocks contain evidence of a complex history that began with thick accumulations of sedimentary and volcanic rocks at the margins of continents. As these materials accumulated, they were exposed to increasing temperatures and pressures that promoted extensive deformation and recrystallization. Ultimately, batholithic masses of magma were generated, adjacent to which contact metamorphism occurred. Clearly, all three of the agents of metamorphism participate in regional metamorphism. Today, tracts of regionally metamorphosed rocks attest to collisions of tectonic plates and delineate the locations of former mountain ranges. To account for the exposure of such rocks at the earth's surface, several kilometers of overlying rock must have been eroded. Nearly every continent has such metamorphic regions. Where they are exceptionally broad and composed of rocks metamorphosed during several episodes of deformation, they are termed **shields**. Shields are ancient terrains that may include former belts of deformation that have been reduced to low-lying regionally convex plains during repeated cycles of uplift and erosion. Every continent has one or more shields (Fig. 7–5). North America, for example, has the great Canadian shield, which covers over 3 million square miles of the North American continent.

Most metamorphic rocks are products of regional metamorphism, and are located within the deformed zones of present and former mountain ranges. They are exposed where the once deeply buried cores of mountains have been uncovered by erosion.

The tremendous forces associated with regional metamorphism frequently cause movements of crustal blocks against one another along great breaks or fractures called faults. The movements of one mass of rock relative to its neighboring masses along such faults cause fragmentation and pulverization as well as mechanical distortion of rock. The term **shear metamorphism** is often used in reference to this type of metamorphic alteration. Usually, the forces that cause this disruption are applied rapidly, and there is little rise in temperature except for that generated by friction. Even this amount of heat, however, can result in the growth of micaceous minerals in the rock. For these crushed and shattered rocks, the term **cataclastic** (from the Greek for "broken down") is quite suitable. In some cataclastic rocks, crushing has been so intense that component particles have been reduced to microscopic sizes. Subsequently,

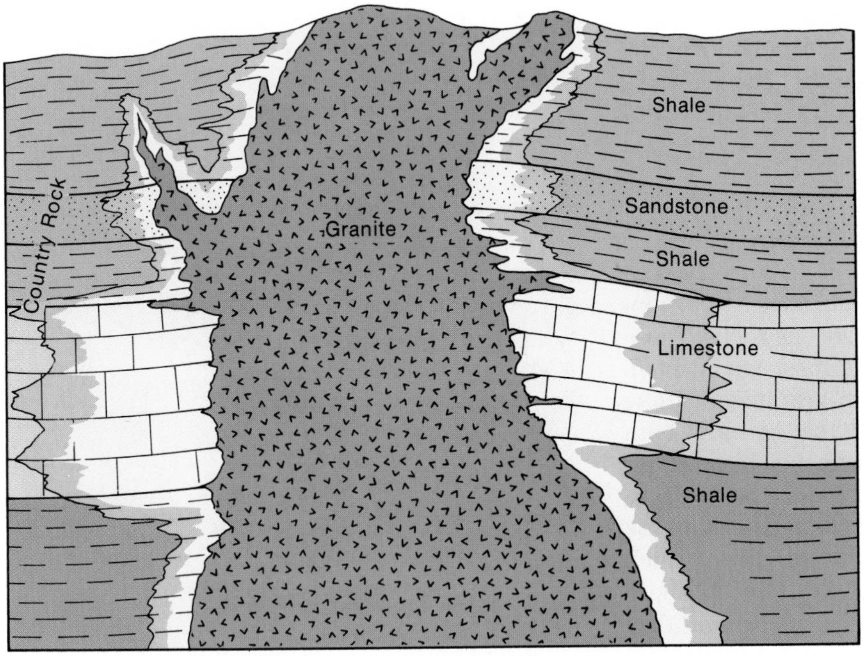

Figure 7–4 Metamorphic aureole developed around a granitic intrusion. The intensity of metamorphism diminished outward from the granite margin, and the width of aureole increased adjacent to more chemically reactive and permeable country rock.

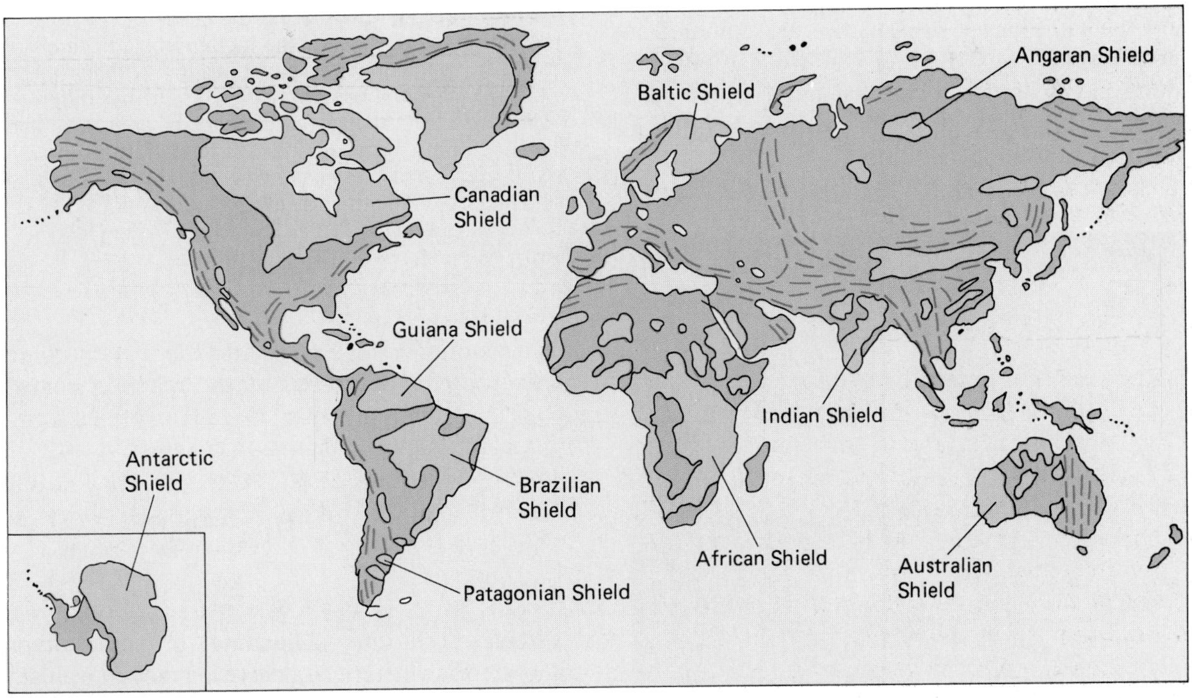

Figure 7–5 Shields of the world (shown in green). Mountain belts are indicated by broken red lines.

the tiny particles are partially recrystallized to form a durable hard rock known as **mylonite**. The name is derived from a Greek word meaning "mill" and is appropriate for these rocks composed of finely ground materials.

Metamorphic Minerals

Quartz, hornblende, feldspar, and mica are all minerals that compose ordinary igneous rocks. These same minerals are among the most abundant in metamorphic rocks. There are also minerals like chlorite (Fig. 7–6) and garnet that occur in both igneous and metamorphic rocks, but that are far more common in the latter. Some minerals ordinarily occur only in metamorphic rocks. Examples are andalusite, sillimanite, and kyanite, all of which have the composition Al_2SiO_5. Although these three minerals have the same composition, they vary in the shape of the crystals they form, and each develops within its own characteristic pressure and temperature range (Fig. 7–7). The hydrous iron aluminum silicate known as staurolite (Fig. 7–8) as well as almandite garnet (Fig. 7–9) and the pink garnet grossularite are common in metamorphic rocks. Spinel and wollastonite are found in calcareous rocks that have experienced metamorphism. Parent rocks (**protoliths**) rich in

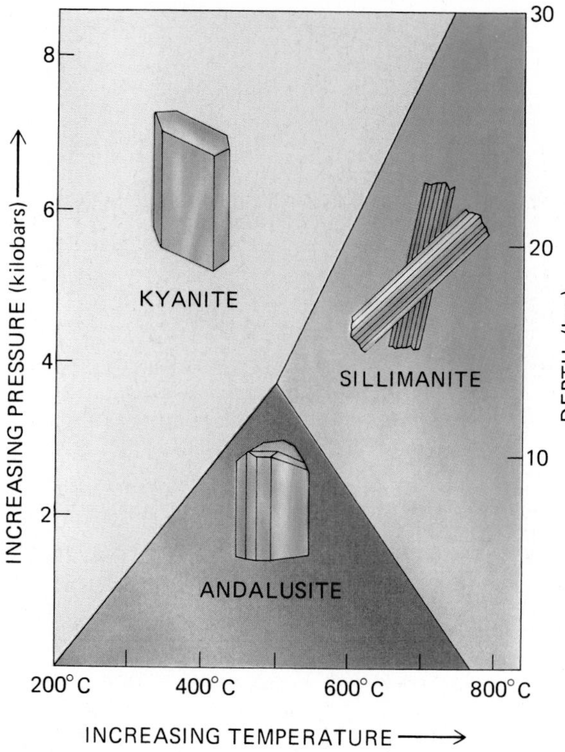

Figure 7–7 Graph illustrating the range of temperature and pressure in which the metamorphic minerals andalusite, kyanite, and sillimanite form. All three minerals have the same composition ($A1_2SiO_5$) but differ in crystal form. Note that the occurrence of andalusite and sillimanite in a rock would indicate temperatures in excess of 500°C and pressures below 4 kb. Kyanite and sillimanite occur together at temperatures above 500°C also but at pressures greater than 4 kb. (After Holdaway, M.J., *Am. J. Sci.*, 271:97–131.)

Figure 7–6 Chlorite schist. The metamorphic mineral chlorite gives this rock its name.

magnesium and silica may yield the metamorphic mineral talc, as indicated by the reaction below.

$$\underset{\text{dolomite}}{3\,CaMg(CO_3)_2} + \underset{\text{quartz}}{4\,SiO_2} + \underset{\text{water}}{H_2O} \longrightarrow$$

$$\underset{\text{talc}}{Mg_3Si_4O_{10}(OH)_2} + \underset{\text{calcite}}{3\,CaCO_3} + \underset{\text{carbon dioxide}}{3\,CO_2}$$

Whether chlorite, sillimanite, or some other mineral forms in a rock undergoing metamorphism depends not only on the composition of the parent rock but also on the intensity of heat and pressure. Particular minerals develop under specific condi-

Figure 7–8 The metamorphic mineral staurolite. This specimen is composed of two crystals that have grown to penetrate each other at right angles. Two or more crystals united in a definite way such as this are known as twin crystals. (Courtesy of W.T. Schaller and the U.S. Geological Survey.)

tions. The amount of heat and pressure required for the formation of some of the simpler metamorphic minerals can be determined in the laboratory. In such experiments, silicate powders are mixed with water vapors and gases and raised to high temperatures and pressures within specially designed containers called "hydrothermal bombs." The result is the formation of metamorphic minerals at specific levels of temperature and pressure. Experiments of this kind have helped geologists to designate certain minerals as indicative of a particular degree of metamorphism or **metamorphic grade**. These minerals are dubbed **index minerals**. The presence of chlorite in a metamorphic rock, for example, would suggest that relatively low temperatures and pressures occurred during regional metamorphism. With somewhat higher temperatures, staurolite would signal the attainment of intermediate metamorphic grade, and sillimanite would develop under the still higher temperatures of high metamorphic grade (Fig. 7–10).

Metamorphic grade refers to the extent to which a body of metamorphic rock differs from its parent rock. For example, if shale is converted to slate, metamorphism is low grade. If changed to a garnet or staurolite schist, middle grade metamorphism is indicated. If converted to a sillmanite-rich schist, the rock would be considered high grade.

Metamorphic Rocks

Metamorphic rocks are at least as varied as are igneous and sedimentary rocks. Their variety is related to the diversity of parent materials as well as to the kind and intensity of metamorphic agencies. As is true with igneous and sedimentary rocks, the basis for the classification of metamorphic rocks is composition and texture (Table 7–1). As far as possible, compositional and textural characters are selected that permit placement of metamorphic rocks into categories that reflect similar parentage, history, and kind of metamorphism.

Figure 7–9 Layer of almandite garnets (red minerals) in a garnet gneiss, western coast of Greenland. The garnets are 1.0 to 3.0 cm in diameter. (Courtesy of M.S. Smith.)

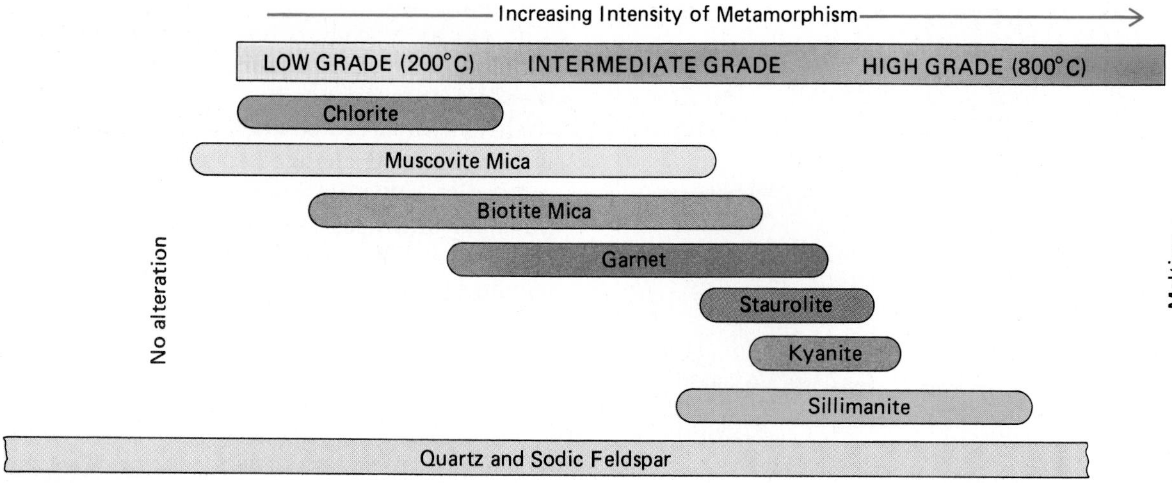

Figure 7–10 Generalized diagram depicting the metamorphic minerals that will develop during the progressive metamorphism of shale. As the parent rock (protolith) is subjected to low-grade metamorphism, tiny chlorite and muscovite minerals will make their appearance. The shale will be metamorphosed to slate. With a higher grade of metamorphism and higher temperatures, muscovite will be joined by biotite, garnet, and possibly staurolite. As temperatures approach the level of high-grade metamorphism, kyanite and sillimanite appear. Beyond 800°C, the rock may be melted.

Texture of Metamorphic Rocks

Most metamorphic rocks can be distinguished from igneous and clastic sedimentary rocks on the basis of texture. Because mineral grains in metamorphic rocks have not developed in sequence from a melt, they do not exhibit the same kind of disoriented or chaotic interlocking fabric characteristic of igneous rocks. Also, because they are not composed of detrital grains, they do not resemble clastic sedimentary rocks. Metamorphic minerals grow in the solid state and often are aligned parallel to one another so as to fit together like a wall constructed of well-fitted irregular elongated stones. Sometimes certain mineral grains will grow considerably larger than the average grain size. Such larger grains are called **porphyroblasts** (Fig. 7–11), and the texture is called porphyroblastic (Fig. 7–12).

An important textural attribute of most but not all metamorphic rocks is **foliation** (Fig. 7–13). Foliation (Latin, *foliatus*, "leaved") can be described as a parallel arrangement or distribution of minerals in a metamorphic rock. This alignment of mineral grains may cause the rock to split into more or less flat pieces (Fig. 7–14). Whether foliation is fine or coarse depends on the size and shape of the component grains. Conspicuously foliated rocks usually contain abundant flat minerals such as mica and chlorite. Coarse or imperfect foliation is usually the result of the growth and segregation of rather blocky minerals such as feldspar and quartz into roughly parallel bands and lenses (Fig. 7–15). Some metamorphic rocks are composed of

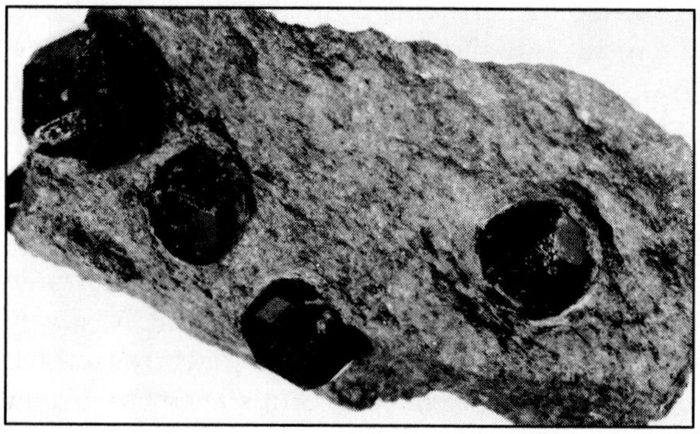

Figure 7–11 Porphyroblasts of almandite garnet in a mica schist. (Courtesy of the Institute of Geological Sciences, London.)

Table 7–1 Origin and Characteristics of Metamorphic Rocks

Category	Metamorphic Rock	Texture and Appearance	Typical Mineral Composition	Type of Metamorphism	Parent Rock
Foliated	Slate	Aphanitic, smooth, dull foliation planes; gray, black, green, or purple	Clay minerals, mica, chlorite	Regional	Shale
	Phyllite	Aphanitic, although some grains may be visible; foliation planes commonly wrinkled and more lustrous than slate	Mica, chlorite, quartz	Regional	Shale
	Schist	Phaneritic, distinctly foliated, platy (mica, chlorite) or needlelike (hornblende) minerals commonly segregated into layers	Mica, chlorite, talc, graphite, hornblende, garnet, staurolite	Regional	Shale, basalt or gabbro, impure sandstones, impure limestones
	Gneiss	Phaneritic irregular foliation composed of relatively robust minerals; foliation less distinct than in schists	Quartz, feldspars, garnet, mica, hornblende, staurolite	Regional	High-silica igneous rocks and sandstones
Non-foliated or weakly foliated	Marble	Phaneritic, fine to coarsely crystalline; often variegated (marbled)	Calcite, dolomite	Regional or contact	Limestones or dolostones
	Quartzite	Phaneritic, sugary-textured	Quartz	Regional or contact	Quartz sandstones
	Greenstone	Aphanitic, scattered dark visible crystals; dark green	Chlorite, epidole, amphiboles	Regional	Low-silica volcanic rocks
	Amphibolite	Phaneritic, similar to amphibole schist but foliation less apparent	Hornblende and plagioclase	Regional	Low-silica igneous rocks
	Hornfels	Aphanitic to fine phaneritic, grains, equidimensional	Andalusite cordierite, mica, quartz	Contact	Usually shales or mudstones

prismatic or needlelike crystals such as hornblende that are parallel to one another. The term used to describe this type of texture is **lineation**.

A few metamorphic rocks do not exhibit foliation. For example, the metamorphic rocks marble and quartzite, which will be described in the next section, characteristically have no readily visible directional arrangement of mineral constituents.

Kinds of Metamorphic Rock

There are hundreds of different kinds of metamorphic rocks, but those that occur extensively at the earth's surface comprise a relatively short list. It is convenient to divide these common varieties into two groups, namely, those that exhibit foliation and those that do not.

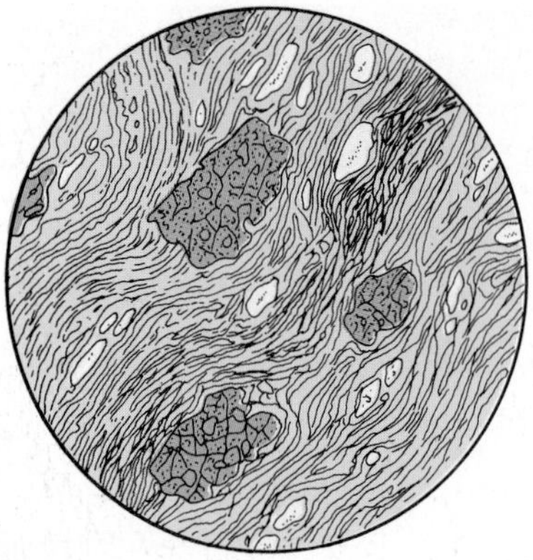

Figure 7–12 Porphyroblastic texture as viewed in thin section with the aid of a microscope. The large, dark, ragged porphyroblasts are staurolite grains. Clear grains are quartz and the flaky materials consist of biotite and muscovite. The rock is a staurolite-mica schist. Area of field is 5 mm.

Foliated Metamorphic Rocks

Slate. **Slate** (Fig. 7–16) is a product of regional metamorphism of clayey rocks, particularly shale. Its origin involves the transformation of clay minerals into microscopic crystals of sheet-structure silicates such as muscovite. The small size of these flakey minerals, and the persistence of relic planes of bedding, attest to the relatively low grade of metamorphism under which slate forms.

As a result of the parallel growth of the minute flakes of mica in slate, the rock characteristically will split into rather smooth thin slabs. This trait is called **slaty cleavage**. Slaty cleavage is caused by the growth of the sheet structure silicates at right angles to the directions from which pressures were applied (Fig. 7–17). Measurements of the orientation of slaty cleavage in outcrops can be used in reconstructing the deformational history of a metamorphic region. Planes of slaty cleavage need have no relationship to bedding planes in the shales from which slates are usually derived.

Because of its uniform cleavage and impermeable nature, slate is an ideal material for classroom blackboards, pool table tops, and roofing tiles. Carbon and iron sulfide are the main coloring agents in black slates, whereas red and green varieties derive their color from iron oxide impurities.

Phyllite. **Phyllite** is a metamorphic rock that is similar to slate in composition. It forms as the result of the metamorphism of shales and other clay-rich rocks. If one looks at phyllites closely, it is possible to see tiny glistening flakes of chlorite and muscovite that have formed from clay minerals. These micaceous minerals give the parting surfaces of phyllites a characteristic wrinkled surface

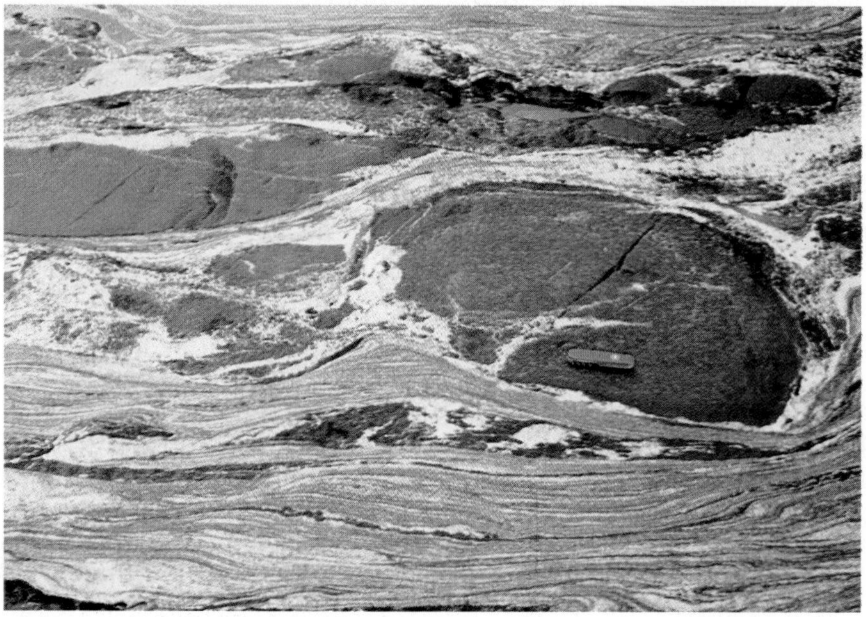

Figure 7–13 Well-developed foliation in a gneiss (the Amitsoq Gneiss) exposed at Lile Narssuaq on the west coast of Greenland. The dark masses are remnants of a basaltic layer that has been stretched, thinned, and broken at irregular intervals by the tectonic forces that caused the metamorphism. (Courtesy of M.S. Smith.)

Figure 7–14 Steeply inclined mica schists splitting apart along foliation planes. Great Smoky Mountain National Park, North Carolina-Tennessee.

Figure 7–15 Coarse (gneissic) foliation developed in a quartz–feldspar–biotite gneiss.

Figure 7–16 Slate.

Figure 7–17 These layers of slate from the Kodiak Peninsula of Alaska exhibit well-developed slaty cleavage. Note that the cleavage planes are vertical and are not parallel to bedding, as in the fissility of shale. (Courtesy of the U.S. Geological Survey.)

and luster that is not seen in slates. Phyllite represents an intermediate degree of metamorphism between slate and schist.

Schist. The flakey or needlelike minerals in **schist** are sufficiently large to be readily visible to the unaided eye (Fig. 7–18). They impart a kind of foliation that is described as schistose. **Schistose foliation** is normally associated with medium-grade metamorphism. The foliation in schists is distinctly less uniform than in either slate or phyllite, and traces of relic bedding are rarely present. Thin bands of flakey minerals such as mica (Fig. 7–19), chlorite, kyanite, or talc occur in parallel alignment and cause the schistosity. In some schists, lathlike, needlelike, or fibrous minerals may also cause foliation. Schists may develop during the regional metamorphism of igneous, sedimentary, or lower-grade metamorphic rocks. The majority of schists, however, appear to have been derived from rocks rich in clay. Schists are named according to the most conspicuous mineral in the rock. Thus, there are mica schists, chlorite schists, amphibole schists, and many other types of schist.

Gneiss. **Gneisses** are coarse-grained coarsely banded metamorphic rocks (see Fig. 7–14). Their foliation results from the segregation of minerals into bands rich in quartz, feldspar, biotite, and amphibolites. Usually the ferromagnesian minerals

A

B

Figure 7–18 Two mica schists. (A) A muscovite mica schist containing large porphyroblasts of almandite garnet. The muscovite imparts a silvery luster to the rock. (B) A biotite schist. The black minerals are biotite, and lighter components are feldspars, quartz, and muscovite.

Figure 7–19 Thin section of biotite schist cut perpendicular to foliation and viewed with polarized light. Note parallel alignment of the brightly colored mica flakes. Width of field 0.9 mm.

are concentrated into darker bands, leaving feldspars and quartz as the principal components of the lighter-colored layers. Mica is usually present but is commonly not as conspicuous as in schists. There is also far less tendency for the rock to split along foliation planes as in schists. Gneiss may be formed from the severe or high-grade metamorphism of a variety of preexisting igneous, metamorphic, or sedimentary rocks, although high-silica igneous rocks and sandstones are the common parent materials.

In areas of regional metamorphism, it is not uncommon to find rock exposures consisting of thin bands of granite alternating with irregular masses and bands of gneiss (Fig. 7–20). Such zones of mixed rocks are most commonly found in the transitional areas between outcrops of granite and purely metamorphic rocks. These gneissic yet granitic-looking rocks are called **migmatites** (from the Greek *migma,* "mixture"). They are thought to represent a stage in a progressive series that begins with the metamorphism of mostly sedimentary rocks to produce schists and gneisses and then leads to the development of migmatites and ultimately to granite.

In the order of increasingly coarser foliation, the most common foliated metamorphic rocks are slate, phyllite, schist, and gneiss.

Nonfoliated or Weakly Foliated Metamorphic Rocks

As is implied by their name, the nonfoliated metamorphic rocks show little, if any, preferred orientation of grains. Their occurrence is the result of the absence of directed pressures, as in the aureoles of igneous intrusions, and/or the prevalence of

Figure 7–20 This migmatite of Precambrian age is part of the Cherry Creek metamorphic complex of Montana. (Courtesy of Eva Moldovanyi.)

minerals such as quartz and calcite, which are rather equidimensional in form and thus unlikely to contribute to foliation.

Marble. Marble (Fig. 7–21) is a fine to coarsely crystalline rock composed of calcite or dolomite. It is a relatively soft rock and can be scratched easily with a steel nail. Limestones and dolostones are the parent rocks for marbles. The fossils often present in such carbonate sedimentary rocks are nearly always destroyed during the metamorphism that produces marble. In its purest form, marble may be snowy white, but many varieties are beautifully variegated (marbled) with colors resulting from impurities derived from the parent rock. Small amounts of iron oxide or hydroxide produce red, brown, and yellow hues, whereas green colors are imparted by amphiboles, talc, and serpentine. Organic matter may tint marbles gray or black. It is the lack of foliation, the uniformly crystalline character, the beauty, and the relative softness of marbles that have caused them to be used by sculptors and architects down through the centuries.

Quartzite. Quartzite is a firm, compact rock composed of intergrown quartz grains (Fig. 7–22) that are so firmly united that fracture usually occurs across grains rather than around them. Quartzites (Fig. 7–23) are very hard, and a hefty blow of the rock hammer is needed to break them. They are derived from quartz sands and sandstones, and can develop under either regional or contact metamorphic conditions.

Greenstone. Greenstones are dark green rocks having a generally aphanitic texture interrupted by scattered dark porphyroblasts. The minerals largely responsible for the green color in these rocks are chlorite, epidote, and actinolite. Greenstones are derived from low-silica igneous rocks such as basalts that have been subjected to relatively low-grade metamorphism. Great elongate outcrops of these rocks occur in ancient igneous terrains and have come to be known as *greenstone belts.* Geologists speculate that these greenstone belts may be the metamorphosed relics of a primordial basaltic crust that was later partially engulfed by intrusions of granitic magma.

Amphibolite. Amphibolites are dark-colored medium- to coarse-grained metamorphic rocks. The two essential and most abundant minerals in amphibolites are hornblende and plagioclase. In some amphibolites, the longer axes of these minerals are in parallel arrangement so as to impart a lineation texture. More commonly, little or no parallel orientation is visible in hand specimens.

Rocks rich in ferromagnesian minerals such as basalt are the usual parent rocks for amphibolites. Less commonly, amphibolites are derived from calcareous rocks that contain silica, magnesium, and iron as impurities. In general, amphibolites differ from typical greenstones in lacking or having only poorly developed foliation and in having a coarser texture.

Hornfels. Hornfels are fine-grained, nonfoliated, usually dark-colored, dense metamorphic rocks composed of a mosaic of rather equidimensional grains. They are formed near the contacts of intrusions, and their mineral composition depends

Figure 7–21 Marble. On close examination one can discern the lustrous cleavage surfaces of the calcite crystals that compose this pink marble from Georgia. (The 1.0-cm square is for scale.)

Figure 7–22 Photomicrograph of a thin section of quartzite from the Valmy Fm., Humboldt County, Nevada. Width of field, 65 mm. (Courtesy of R.J. Roberts and the U.S. Geological Survey.)

Figure 7–23 Hand specimen of an attractive green quartzite.

on the nature of the country rock near the magma. The majority of hornfels appear to have been derived from clay-rich rocks.

Metamorphic Zonation

Regional metamorphism has affected the rocks of all the ancient mountain-building belts of the earth. Depending on their age, some of these belts still exist as mountainous tracts or partially denuded uplands. In the shield regions of continents, often all that remain are the eroded stumps of ancient ranges. Nevertheless, it is possible to discern the dimensions and locate the central and marginal parts of these ancient mobile belts by recognizing progressive changes in associations of metamorphic minerals. One of the earliest attempts to zone ancient metamorphic terrain was undertaken by the British geologist George Barrow in 1893. Barrow showed that metamorphic rocks adjacent to plutonic bodies in the Grampian Highlands of Scotland were coarsely foliated feldspar-rich gneisses containing the high-grade mineral sillimanite. These rocks were in turn surrounded first by schists containing kyanite and then by schistose rocks bearing crystals of staurolite. Still farther outward, the rocks were rich in biotite and chlorite and exhibited barely preserved relic stratification. It was evident to Barrow that the different mineral zones recorded a gradually decreasing intensity of metamorphism outward from the centers of maximal compression.

In the years following the publication of Barrow's findings, maps depicting metamorphic zones have been prepared for scores of areas around the world. The geologists preparing these maps frequently followed the suggestion of another British geologist named C. E. Tilley, who in 1925 suggested a "rule" for defining the boundary of metamorphic zones. Tilley reasoned that the first appearance of a particular metamorphic index mineral recorded the place where a definite temperature, required for the origin and stability of that specific mineral, had been reached. Thus, a line connecting these locations could be interpreted as a line of equal metamorphic intensity. Such a line has been named as **isograd**.

Characteristically, metamorphic zones bounded by isograds appear on maps as broad bands representing sequentially more severe conditions of metamorphism. On Figure 7–24, for example, one notes a western band of relatively weakly metamorphosed rocks containing the index mineral chlorite (see Fig. 7–10). Moving eastward one crosses parallel bands of schists containing first appearances of biotite and then garnet. To the east of these zones, isograds marked by such high-grade metamorphic minerals as kyanite, andalusite, and sillimanite occur. They clearly imply a higher intensity of metamorphism just west of Vermont's eastern boundary.

A method for classifying areas of metamorphic rock according to the conditions under which they formed involves the mapping of metamorphic facies. A **metamorphic facies** is an association of rocks characterized by a definite set of minerals formed under specific metamorphic conditions (Fig. 7–25). The facies is named after its most characteristic rock type or mineral. For example, under relatively low temperatures and pressures, the green minerals chlorite and epidote (a hydrous calcium, iron, and aluminum silicate) may form from a parent rock such as basalt. The abundant presence of these minerals produces greenish schists that are called **greenschists**. An assemblage of rocks in which greenschists predominate is designated a **greenschist facies**. The presence of the greenschist facies implies a low grade of regional metamorphism. At somewhat higher temperatures but far greater pressures the blueschist facies develops. The next higher grade of metamorphism is represented by the **amphibolite facies**, in which hornblende, garnet, and plagioclase are abundant. At still higher pressures and temperatures, sillimanite, garnet, and pyroxene mark the occurrence of the **granulite facies**. At extremely high pressures, pyroxene and garnet-rich rocks of the **eclogite facies** may form. Eclogites are also thought to occur in parts of the earth's mantle.

Metamorphic facies are recognized by their association of metamorphic index minerals formed under particular temperature and pressure conditions.

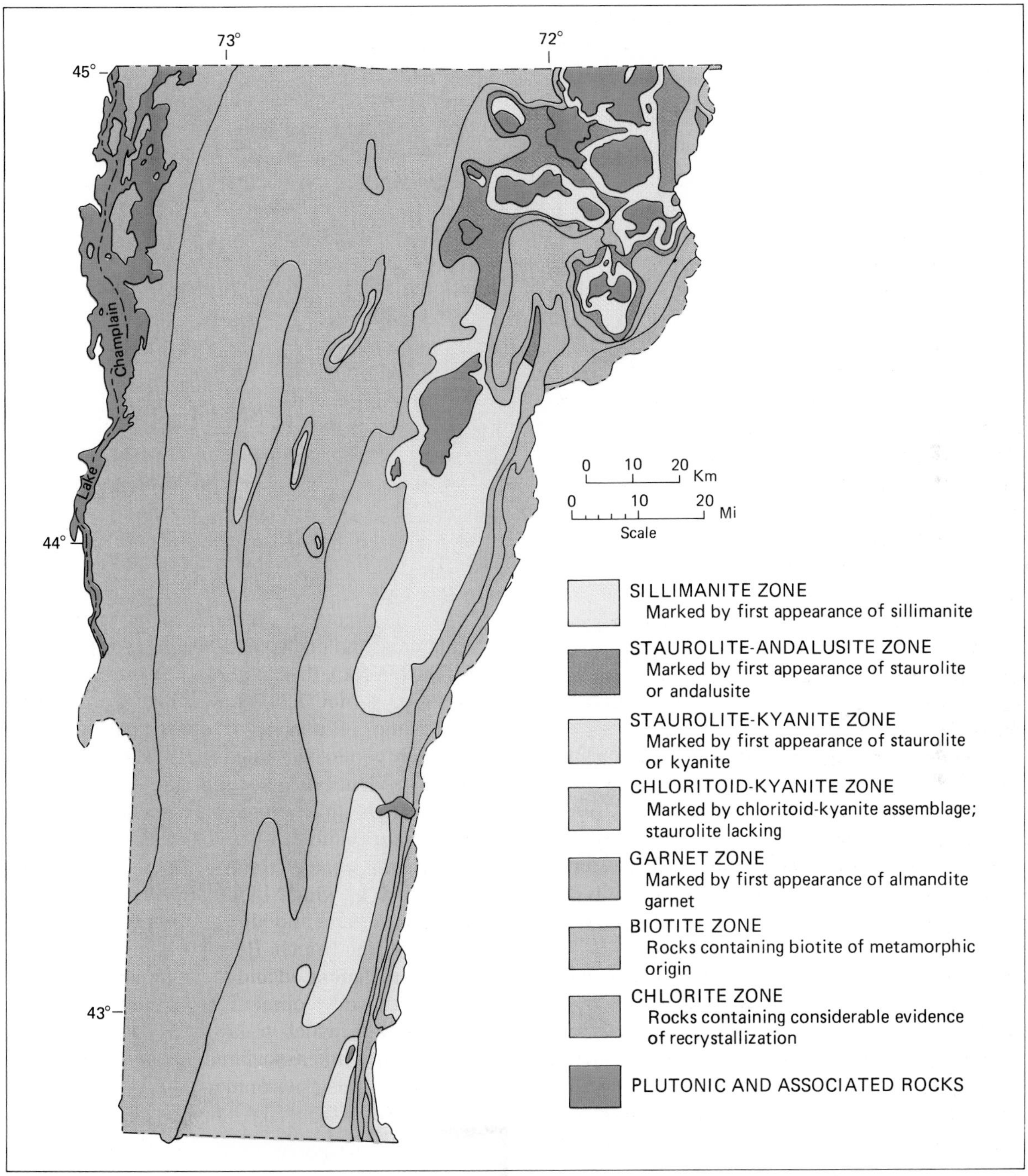

Figure 7–24 Metamorphic map of Vermont depicting zones of regional metamorphism. (Simplified from "Centennial Geologic Map of Vermont." Copyright © 1961, State of Vermont.)

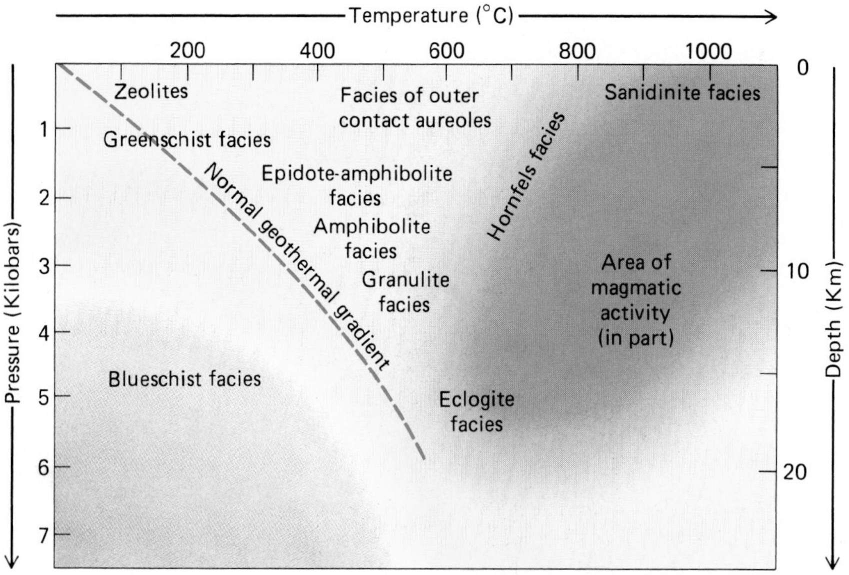

Figure 7–25 Relation of mineral facies to temperature and pressure. (After T.F.W. Barth.)

Metamorphism at the Margins of Tectonic Plates

In Chapter 1 we described the earth's lithosphere as consisting of a number of slowly moving plates. In those parts of the globe where the plates are moving apart, magma rises from below and is added to the separating plate margins as new crust. At other margins, plates either move laterally past one another or they converge. Where convergence occurs, compressional forces squeeze rocks and sediments on the opposing plates and provide the temperature and pressure conditions necessary for many different kinds of metamorphic facies.

As an example, consider the events that occur when an oceanic plate encounters one bearing a continent (Fig. 7–26). In such an encounter, the cool oceanic crust with its cover of sediment begins a relatively rapid descent along the subduction zone. The descent is so rapid that the oceanic crust and sediment reach depths characterized by high pressure before there has been enough time to gain appreciable heat. Thus, these materials find themselves in an environment of high pressure but relatively low temperature.

This environment is ideal for the formation of blueschist, a rock that takes its name from its rich content of a blue amphibole mineral known as glaucophane. Blueschist zones are carefully mapped by geologists, who recognize that they can be used to identify the location and orientation of ancient subduction zones. One such area exists today as the California Coast range.

Unlike the zone of high pressure and modest temperature in which blueschist forms, zones of both high pressure and high temperature are found inland from the trench. Here great masses of molten rock are generated and move upward to supply lavas for the development of volcanic arcs. The Sierra Nevada, which today is composed of granitic intrusives and associated metamorphic rocks, developed in this environment.

The temperature, pressure, and chemical conditions required to produce features of most metamorphic rocks are accounted for by the theory of plate tectonics.

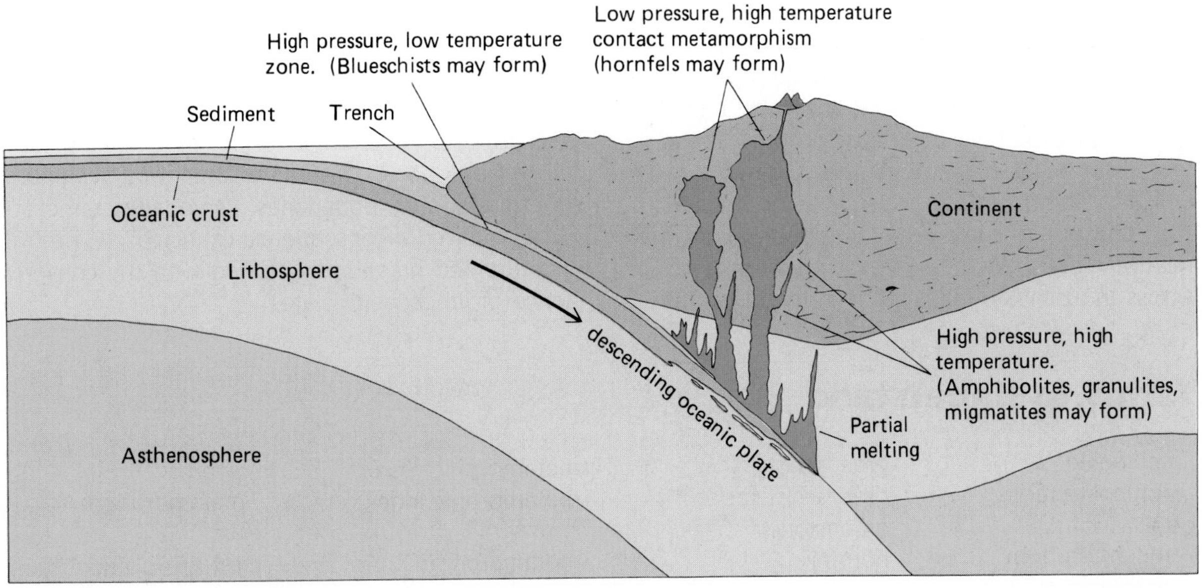

Figure 7–26 Metamorphism associated with tectonic-plate convergence.

Summary

Metamorphism may be defined as the sum of the processes by which previously existing rocks of any kind undergo mineralogic and textural changes in the solid state in response to changing chemical and physical conditions.

The term is not intended to include changes associated with weathering or the ordinary lithification of sedimentary rocks. Rather, the metamorphism of rocks involves *recrystallization,* chemical change, and *mechanical deformation.* In high-pressure metamorphic environments, atoms in older minerals tend to regroup into denser more compact minerals that are stable under the new conditions. Particular minerals will form only at very high pressures and temperatures, while others form at moderate or low pressures or temperatures. Thus metamorphic minerals can be used to infer the conditions responsible for producing metamorphic rocks. Those minerals that are particularly useful in this way are called *metamorphic index minerals.*

The agents of metamorphism include *heat, pressure,* and *chemically active solutions.* Certain chemical reactions involved in the recrystallization of minerals only occur at specific temperatures. For this reason, and also because it accelerates chemical reactions, heat plays an essential role in metamorphism.

The enormous pressures associated with mountain building or deep burial can result in plastic deformation of rocks, pulverize mineral components, and promote the growth of denser minerals. The texture developed by a metamorphic rock may result from the manner in which pressure has been applied to the parent rock. If it is applied uniformly to all sides, the pressure is *lithostatic* and is not likely to cause a preferred alignment of growing crystals. Directed pressure promotes the growth of minerals in the direction of least stress and results in a banding or layering of minerals that is called *foliation.*

Chemically potent solutions are great promoters of metamorphism. Even a small amount of water facilitates the movement of ions to appropriate sites in the atomic structures of metamorphic minerals.

The major kinds of metamorphism are contact and regional. As implied by its name, *contact metamorphism* occurs around the margins of magmatic bodies. Beginning at the contact, progressively less intense zones of metamorphism may develop, resulting in a metamorphic aureole. It is not just the heat that causes contact metamorphism but also chemically active gases expelled by the magma.

In contrast to contact metamorphism, which is relatively local, *regional metamorphism* may occur across thousands of square kilometers. Metamorphic rocks in these regional belts are well foliated and divisible into distinct *metamorphic zones or facies* characterized by minerals that developed in response to particular conditions of pressure and temperature. These broad bands lie roughly parallel to the trends of ancient mountain ranges and lithospheric plate boundaries. Regional metamorphism is often a consequence of increases in temperature and pressure associated with the convergence of lithospheric plates.

Terms to Understand

amphibolite
amphibolite facies
cataclastic metamorphism
contact metamorphism
eclogite facies
foliation
geothermal gradient
gneiss
granulite facies
greenschists
greenschist facies
greenstone
hornfels
index minerals
isograd
lineation
marble
metamorphic aureole
metamorphic facies
metamorphic grade
metamorphic index mineral
metamorphism
migmatite
mylonite
phyllite
porphyroblast
protolith
quartzite
regional metamorphism
schist
schistose foliation
shear metamorphism
shield
slate
slaty cleavage

Questions for Review

1. What are the major causes or agents of metamorphism? Do these agents operate independently or are they interrelated?
2. Why does the presence of even small amounts of water generally increase the rate and amount of metamorphism in rocks already subjected to high temperature and pressure?
3. What is the difference between shear stress and lithostatic pressure? Which is more likely to cause the development of foliation?
4. What is slaty cleavage? How does the origin of slaty cleavage differ from the origin of fissility in shale?
5. The metamorphic minerals sillimanite, andalusite, and kyanite are all found in a metamorphic rock. What are the temperature and pressure conditions under which the rock formed?
6. What factors influence the amount of alteration resulting from contact metamorphism?
7. How does the development or origin of porphyroblasts differ from the origin of phenocrysts in porphyritic rocks?
8. What two criteria are of greatest importance in the classification of metamorphic rocks?
9. Name a probable parent rock for each of the metamorphic rocks listed below:
   a. Marble
   b. Slate
   c. Gneiss
   d. Quartzite
   e. Amphibolite
   f. Chlorite schist
10. What are migmatities? What may be their relationships to granite?
11. To a large extent, metamorphic rocks develop as closed chemical systems. What is the meaning of this statement?
12. What are isograds? How are isograds defined on a map depicting outcrops of metamorphic rocks?

Supplemental Readings and References

Barth, T.F.W., 1952. *Theoretical Petrology.* New York, John Wiley & Sons.

Best, M.G., 1982. *Igneous and Metamorphic Petrology.* San Francisco, W.H. Freeman.

Ernst, W.G., 1969. *Earth Materials.* Englewood Cliffs, N.J., Prentice-Hall, Inc.

Turner, F.G., 1968. *Metamorphic Petrology.* New York, McGraw-Hill Book Co.

Winkler, H.G.F., 1979. *Petrogenesis of Metamorphic Rocks.* 5th ed. New York, Springer-Verlag.

Time and Geology

8

When erosion has removed part of a rock sequence and younger rocks are deposited over the erosional surface, the temporal break in the geologic record is called an *unconformity*. This is a view of the "great unconformity" as it was named by John Wesley Powell who explored the Grand Canyon of the Colorado River by boat in 1869. The view is from Kaibab Trail. The beds extending horizontally across the middle of the photograph were deposited about 500 million years ago. The underlying rocks are approximately 2000 million years old. Thus the unconformity represents a 1500-million-year gap or hiatus in the geologic record. (Courtesy of D. Bhattacharyya.)

Each grain of sand, each minute crystal in the rocks about us is a tiny clock, ticking off the years since it was formed. It is not always easy to read them, and we need complex instruments to do it, but they are true clocks or chronometers. The story they tell, numbers the pages of earth history.

Patrick M. Hurley How Old is the Earth? *Doubleday Anchor Books, 1959*

Introduction

We humans have a fascination with the concept of time. Geology instructors are particularly aware of this fascination, for they are often asked the ages of various rock and mineral specimens brought to the university by students and returning vacationers. If informed that the samples are tens or even hundreds of millions of years old, the collectors are often pleased, but they are also perplexed. "How can this fellow know the age of this specimen by just looking at it?" they think. If they insist on knowing the answer to that question, they may next receive a short discourse on the subject of geologic time. It is explained that the rock exposures from which the specimens were obtained have long ago been organized into a standard chronologic sequence based largely on the way strata occur one on top of another, the stage of evolution as indicated by fossils, and actual ages in years obtained from the study of radioactive elements in the rock. Thus, the geologists must have a vast fund of background knowledge and experience in order to recognize particular rocks as being of a certain age. The science that permits geologists to accomplish this feat is called **geochronology**. With an understanding of geochronology, many formerly isolated bits of information about the earth's past can be placed in proper sequence, and a history can be written.

Relative Geologic Time

There are two ways of thinking about geologic time. On the one hand, we may wish to know the actual number of years ago that a particular geologic event occurred (such as an intrusion of granite or deposition of limestone). This kind of specific time is called *quantitative geologic time* (also called *absolute* geologic time). *Relative geologic time*, on the other hand, requires not that one know a precise age, but rather whether a given rock body is older or younger than another. Before the discovery of radioactivity, relative dating provided the only method available to early geologists as they began to piece together episodes in the history of the earth. Even with the advent of quantitative methods, the principles of relative dating continue to be widely employed in the day-to-day geologic study of the earth and other planets in our solar system.

Relative time places an event in earth history as occurring either before or after another event. Quantitative time places an event in earth history as having occurred a specific number of years ago.

Superposition

The concept of relative geologic dating of rocks was introduced about three centuries ago by a Danish physician named Nils Stensen. Following the custom of his day, Stensen latinized his name to Nicolaus Steno. He settled in Florence, Italy, where he became physician to the Grand Duke of Tuscany. Because the Duke did not require Steno's constant attention, the physician had ample time to explore the countryside, visit quarries, and examine strata. Around the year 1669, his geologic studies led him to formulate a basic principle of geochronology. It is called the **principle of superposition**.

The principle of superposition states that in any undisturbed sequence of sedimentary rocks (or other materials deposited on a surface, such as volcanic ash or lava), the oldest layer will be at the bottom, and successively higher layers will be successively younger (Figure 8–1). It is a rather obvious axiom, which probably had been understood

Figure 8–1 Example of a superpositional sequence of undisturbed Triassic and Jurassic strata, Capital Reef National Monument, Utah. How would you express the relative age of the green shale? (Courtesy of W.B. Hamilton.)

by many naturalists even before Steno. Yet Steno was the first to explain the concept formally. The fact that his principle was self-evident does not diminish its geologic importance. Furthermore, the order of succession of strata is not always apparent in regions where layers have been steeply tilted or overturned (Fig. 8–2). In such instances, the geologist must examine the strata for clues useful in recognizing their uppermost layer. The way fossils lie in the rock, as well as evidence provided by mud cracks and ripple marks, are particularly useful clues when one is trying to determine "which way was up."

Superposition refers to the natural sequence in which sediments (or lava flows) accumulate in layers one above the other, with the oldest at the bottom.

Figure 8–2 Steeply dipping strata grandly exposed in the Himalayan Mountains. It is often difficult to recognize the original tops of beds in strongly deformed sequences such as this. (Courtesy D. Bhattacharyya.)

James Hutton on Time and Process

The factor of time enters into every geologic process. It requires time to weather rocks, to form crystals, to raise mountains, and to erode canyons. The length of time available for the action of geologic processes such as erosion influences the configuration of landforms resulting from those processes (Fig. 8–3). When the science of geology was in its infancy, however, all rocks were thought to be of about the same age, and all geologic features were believed to have been formed at the same time. Furthermore, the earth was generally believed to be only several thousand years old. It was James Hutton who first recognized the immensity of geologic time and its importance in geologic processes. In his *Theory of the Earth,* which he presented to the Royal Edinburgh Society in 1785, Hutton showed irrefutably that hills and valleys were not everlasting but that the surface of the earth undergoes constant change as a result of mountain-building forces, erosion, and sedimentation. The earth, said Hutton, "is thus destroyed in one part, but it is renewed in another." Hutton watched as sediment was transported to sites of deposition and noted the deposition of clay and sand. It became apparent to him that ancient sedimentary rocks had formed in the same way and that prodigious amounts of time were required to deposit the great thicknesses of sediment represented in the rock record. His understanding of the vastness of geologic time stands as a milestone in the history of geology.

By observing the slowness of the processes which form and destroy rocks, James Hutton came to recognize the great age of the earth.

Hutton built his geologic theories around a belief that "the past history of our globe must be explained by what can be seen to be happening now." This simple but powerful idea was eventually named **uniformitarianism** by a geologist named William Whewell. Charles Lyell (1797–1875) became the principal advocate and interpreter of uniformitarianism.

Figure 8–3 A view of Arches National Park, Utah. At one time the surface of the ground in this area was at the level of the summit of the sandstone walls (called "fins") and prominent cliffs. Erosion over a long period of time has steadily removed the rock, leaving only resistant remnants. The sandstone forming the fins is the Entrada Formation of Jurassic age. (Courtesy of J.R. Gill and the U.S. Geological Survey.)

Uniformitarianism

Perhaps because it is so general a concept, **the principle of uniformitarianism** has been reinterpreted in a variety of ways by scientists and theologians from Hutton's generation down to our own. Some of the ideas about what uniformitarianism now implies would seem strange to Hutton himself. If the term uniformitarianism is to be used in geology (or any science), one must clearly understand what is uniform. The answer is that the physical and chemical laws that govern nature are uniform. Hence, the history of the earth may be deciphered in terms of present observations on the assumption that natural laws are invariant with time. These so-called natural laws are merely the accumulation of all our observational and experimental knowledge. They permit us to predict the conditions under which water becomes ice, the behavior of a volcanic gas when it is expelled at the earth's surface, or the effect of gravity on a grain of sand settling to the ocean floor. Uniform natural laws govern such geologic processes as weathering, erosion, transport of sediment by streams, movement of glaciers, and the movement of water into wells.

Hutton's use of what later was termed uniformitarianism was simple and logical. By observing geologic processes in operation around him, he was able to infer the origin of particular features he discovered in rocks. When he witnessed ripple marks being produced by wave action along a coast, he was able to state that an ancient rock bearing similar markings was once a sandy deposit of some ancient shore. And if that rock now lay far inland from a coast, he recognized the existence of a sea where Scottish sheep now grazed.

Uniformitarianism permits geologists to explain past events by making comparisons to what can be seen at present.

Hutton's method of interpreting rock exposures by observing present-day processes was given the catchy phrase "the present is the key to the past" by Sir Archibald Geikie (1835–1924). The methodology implied in the phrase works very well for solving many geologic problems, but it must be remembered that the geologic past was sometimes quite unlike the present. For example, before the earth had evolved an atmosphere like that existing today, different chemical reactions would have been involved in the weathering of rocks. Life originated in the time of that primordial atmosphere under conditions that have no present-day counterpart. As a process in altering the earth's surface, meteorite bombardment was once far more important than it has been for the last three billion years or so. Many times in the geologic past continents have stood higher above the oceans, and this higher elevation resulted in higher rates of erosion and harsher climatic conditions compared with intervening periods when the lands were low and partially covered with inland seas. Similarly, at one time or another in the geologic past volcanism was more frequent than at present.

Nevertheless, ancient volcanoes disgorged gases and deposited lava and ash just as present-day volcanoes do. Modern glaciers are more limited in area than those of the recent geologic past, yet they form erosional and depositional features that resemble those of their more ancient counterparts. All of this evidence suggests that present events do indeed give us clues to the past, but we must be constantly aware that in the past, the rates of change and intensity of processes often varied from those to which we are accustomed today, and that some events of long ago simply do not have a modern analogue.

The eighteenth-century concept of uniformitarianism was not the only contribution James Hutton made to geology. In his *Theory of the Earth*, published in 1785, he brought together many of the formerly separate thoughts of the naturalists who preceded him. He showed that rocks recorded many episodes of upheaval separated by quieter times of erosion and sedimentation. In his own words, there had been a "succession of former worlds."

Unconformities: Gaps in the Time Record

On the Isle of Arran (an island of Scotland) and also at Jedburgh, Scotland, Hutton came across exposures of rock where steeply inclined older strata had been beveled by erosion and covered by gently inclined younger layers (Figs. 8–4 and 8–5). It was clear to Hutton that the older sequence was not only tilted but also partly removed by erosion before the younger rocks were deposited. The erosional surface meant that there was an interval of time not represented by strata. It represented a chronologic gap or hiatus. Hutton did not suggest a name for this feature, which was later termed an **unconformity**. More specifically, the exposure studied by Hutton was an **angular unconformity** because the lower beds were tilted at an angle to the upper. For Hutton, the rocks at Siccar Point provided ample evidence for the kind of geologic change he described in *Theory of the Earth.* The lower strata had been deposited, uplifted, and tilted, and then subjected to erosion. Much later, sedimentation was renewed on the old erosional surface. There was indeed "a succession of worlds," that could be revealed by the careful interpretation of unconformities.

The principal kinds of unconformities are depicted in Figure 8–6. Of the types shown, the angular unconformity (Fig. 8–7) provides readily appar-

(Text continues on p. 196.)

Figure 8–4 Angular unconformity at Siccar Point, southeastern Scotland. It was here that the historical significance of an unconformity was first realized by James Hutton in 1788. (Courtesy of E.A. Hay.)

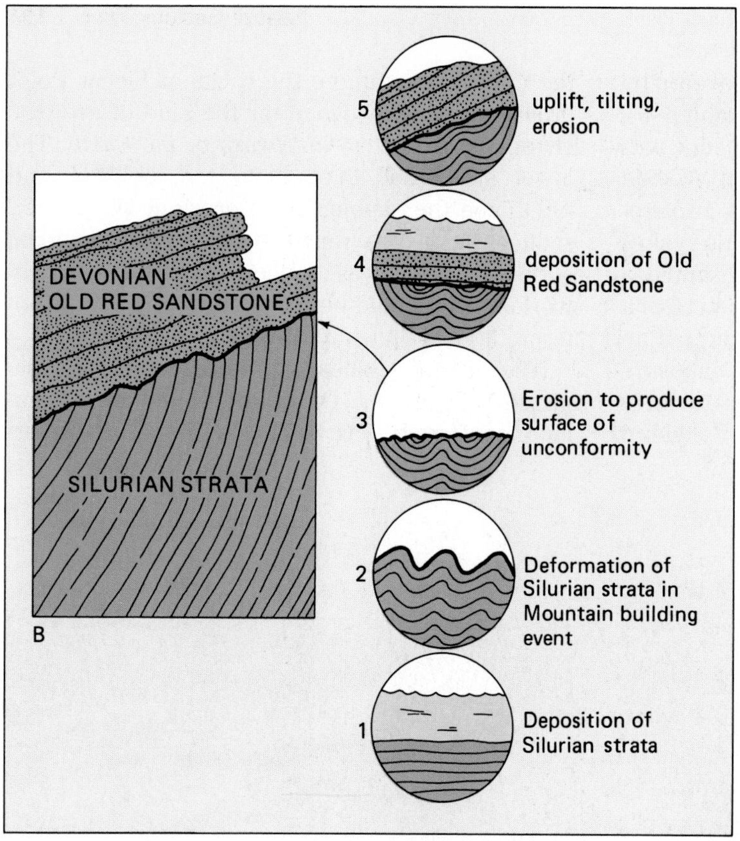

Figure 8–5 The angular unconformity at Siccar Point, Scotland. The drawing at the bottom indicates the historical sequence of events that occurred in this area during the Silurian and Devonian.

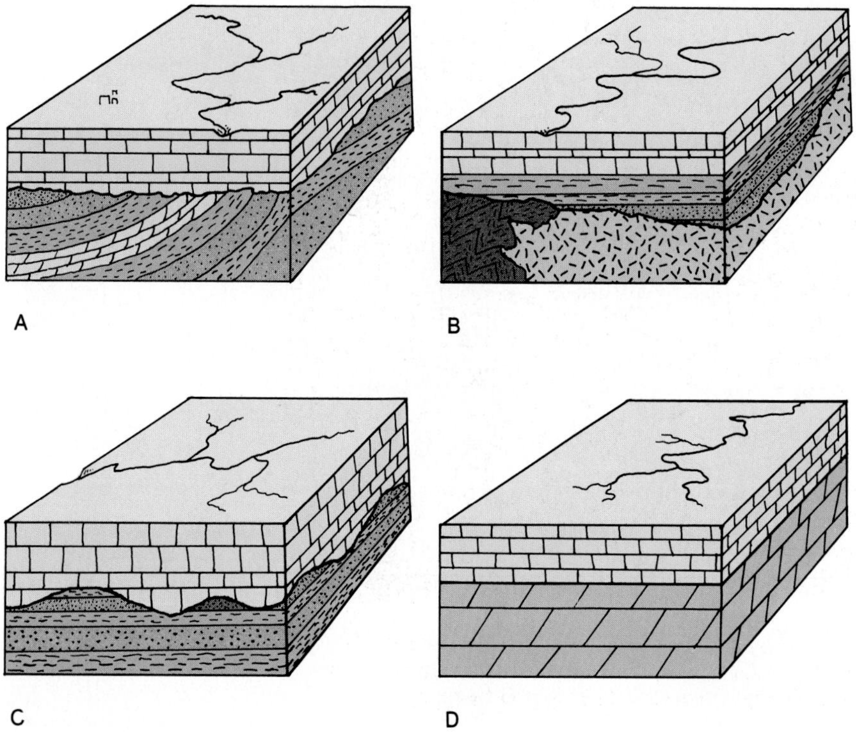

Figure 8–6 Types of unconformity. (A) Angular unconformity. (B) Nonconformity. (C) Disconformity. (D) Paraconformity.

Figure 8–7 An angular unconformity separates beds of the Precambrian (Proterozoic) Uncompahgre Sandstone from overlying, nearly horizontal Devonian strata of the Elbert Formation at Box Canyon Falls, Ouray, Colorado. (Photograph by Thomas E. Williams, with permission.)

ent evidence for crustal deformation. Such deformation is not evident in the **disconformity**. Here the erosional surface lies between parallel strata and may indicate only a simple withdrawal and subsequent return of the sea. **Nonconformities** are erosional surfaces separating igneous or metamorphic rocks from overlying sedimentary beds (Fig. 8–8). In many instances, the igneous and metamorphic rocks were emplaced or metamorphosed deep within the roots of ancient mountain ranges that subsequently experienced repeated episodes of erosion and uplift. Eventually, the igneous and metamorphic core of the mountains lay exposed and provided the surface upon which the younger strata were deposited. The final type of unconformity is termed a **paraconformity** and consists only of a bedding plane between parallel strata. Paraconformities record a pause in deposition and therefore, like other unconformities, represent a gap in the geologic record. Paraconformities can be recognized only when fossils or minerals that can be dated radiogenically indicate that there is an unrecorded segment of geologic time.

An unconformity is a surface between a group of beds and the rocks beneath, representing an interval of time when erosion occurred, or when deposition had ceased for a long period of time.

Biotic Succession

Until about 1800, geologists did not fully grasp how to use their knowledge of local geology to develop a wider view of the relative sequence of events for an entire region or even a continent. They were able to understand the sequence of older to younger rocks at local exposures of strata but could not relate their findings to lithologically different strata that were exposed at distant locations. This problem was to be resolved by a self-taught geologist named William Smith (1769–1839).

William Smith was an English surveyor and engineer who devoted 24 years to the task of tracing out the strata of England and representing them on a map. Small wonder that he was called "Strata

Figure 8–8 The erosional surface of this nonconformity is inclined at about 45° and separates Precambrian rhyolitic rock from overlying Upper Cambrian Bonneterre Dolostone. Taum Sauk Mountain, southeastern Missouri. (Courtesy of D. Bhattacharyya.)

Smith." He was employed to locate the routes of canals, to design drainage for marshes, and to restore springs. In the course of his work, he independently came to understand the principles of stratigraphy, for they were of immediate use to him. By knowing that different types of stratified rocks occur in a definite sequence and that they can be identified by their composition and texture (lithology), the soils they form, and the fossils they contain, he was able to predict the kinds and thicknesses of rock that would have to be excavated in future engineering projects. His use of fossils (Fig. 8–9) was particularly significant. Prior to Smith's time, collectors rarely noted the precise beds from which fossils were taken. Smith, on the other hand, carefully recorded the occurrence of fossils and quickly became aware that certain rock units could be identified by the particular assemblages of fossils they contained. He used this knowledge first to trace strata over relatively short distances and then to extend his "correlations" over greater distances to strata of different lithology that were inferred to be the same age because they contained the same fossils. Ultimately, this knowledge led to the **principle of biotic succession**, which stipulates that the life in the earth's long history was unique for any given period and that the fossil remains of life permit geologists to recognize contemporaneous deposits around the world.

William Smith did not know why organisms had changed through time, but neither was such understanding necessary for the success of his work. He did his mapping 60 years before the publication of Darwin's *Origin of Species.* Today, we recognize that different kinds of animals and plants succeed one another in time because life has evolved continuously. Because of this continuous change, or

A B

Figure 8–9 Among the fossils used by William Smith in correlating Jurassic strata in England were ammonoid cephalopods. These were marine animals that lived in a coiled, chambered conch. Although ammonoid cephalopods are extinct, their nautiloid cephalopod relatives survive in the form of the graceful chambered nautilus of modern seas. The two ammonoids shown here have unusual preservation. The specimen on the left has retained its original mother-of-pearl, whereas the specimen on the right has had its calcium carbonate conch replaced by pyrite (fool's gold).

evolution, only rocks formed during the same age could contain similar assemblages of fossils. Beginning with the work of Smith, geologists determined the succession of fossil animals and plants in many areas around the world where the strata are undeformed and their superposition is clearly apparent. The faunal and floral succession determined for these areas was then verified at many additional localities. Ultimately, this knowledge of the correct succession was used to ascertain the relative ages of rocks even in areas where the original superpositon of strata was uncertain because of deformation.

Figure 8–10 A dike of light-colored granitic rock cuts across dark metamorphic rocks in the Beartooth Mountains of Wyoming. Cross-cutting relationships indicate that the granitic rocks are younger than the metamorphic rocks into which they intrude.

According to the principle of biotic succession, fossils of each successive body of strata will be recognizably different from those in strata above and below.

Inclusions and Cross-Cutting Relations

In 1830, a geologist wrote a book that presented under one title the most important geologic concepts of the day. His name was Charles Lyell, and his book was the classic *Principles of Geology.* It grew to three volumes and became immensely important in the Great Britain of Queen Victoria. In this work can be found additional criteria useful in establishing relative ages of rock units. For example, Lyell discussed the general principle that a geologic feature that cuts across or penetrates another body of rock must be younger than the rock mass penetrated. In other words, the feature that is cut is older than the feature that crosses it (Fig. 8–10). This observation is now termed the **principle of cross-cutting relationships**. The generalization applies not only to rock units but also to geologic structures like faults (a break in rock along which movement has occurred) and unconformities. Thus, in Figure 8–11, the break in rocks represented by fault *b* was formed after the deposition of strata *d*. Because the intrusion of magma *c* cuts across *b* and *d*, it is younger (later) than both of these features. By superposition, deposition of beds *e* was the last event.

Cross-cutting relationships indicate that a rock is younger than any rock it cuts.

Another generalization to be found in Lyell's *Principles* relates to **inclusions**. Lyell, like Hutton before him, logically discerned that fragments (such as xenoliths) within larger rock masses are older than the rock masses in which they are enclosed. Thus, whenever two rock masses are in contact, the one containing pieces of the other will be the younger of the two. In Figure 8–12, the beds have been tilted by mountain-building forces into a vertical orientation. The cobbles of granite within the sandstone tell us that the granite is older and

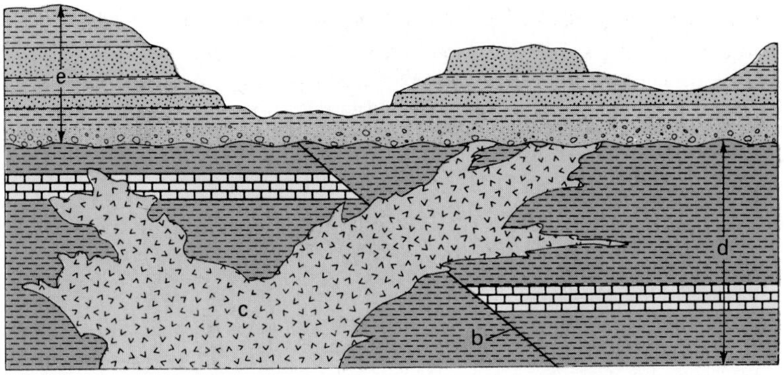

Figure 8–11 Time sequence of events from spatial relations of strata, faults, and intrusions.

that the eroded granite fragments were incorporated into the sandstone. In Figure 8–13, the granite was intruded as a melt into the sandstone. Because there are inclusions of country rock in the granite, the granite must be the younger of the two units.

Correlation of Sedimentary Rocks

In the preceding section, we alluded to William Smith's skills in matching the strata at one end of the country to those of the other. Without such skills we would know only the history of individual localities, and we would not be able to relate that local history to other areas. A history of the entire earth could not be written, nor could a standard geologic time scale be assembled without an ability to match strata from locality to locality.

The process of matching strata at one place to those of another is called **stratigraphic correlation**. Formations of rock may be correlated on the basis of their physical features, the fossils they contain, or their position in the stratigraphic sequence.

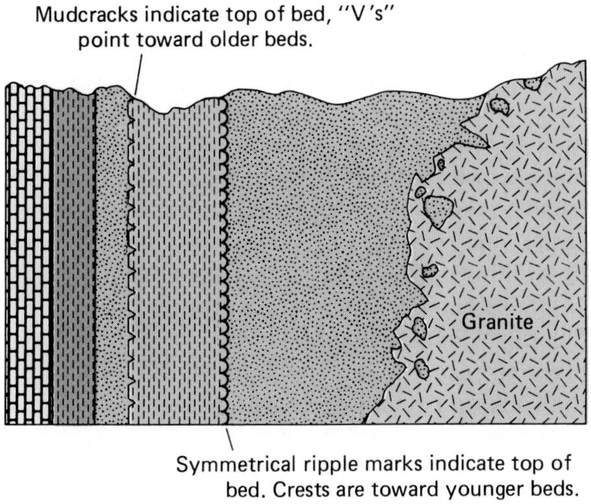

Figure 8–12 Inclusions of sandstone in granite indicate that sandstone is the older unit. (After being deposited, strata were tilted to their present vertical orientation.)

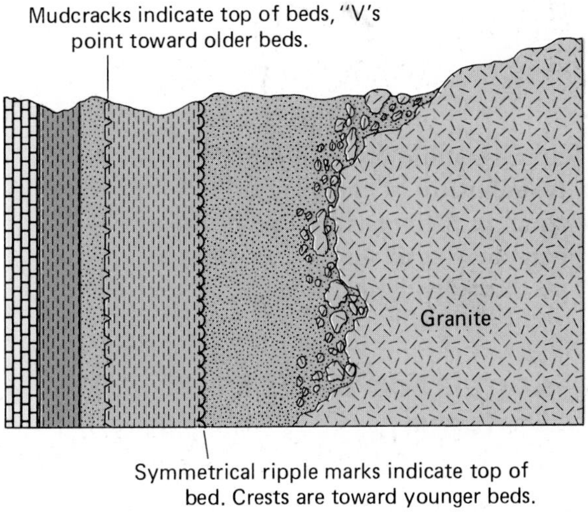

Figure 8–13 Granite inclusions (country rock) in sandstone indicate that granite is the older unit. (Originally horizontal strata have been tilted to a vertical orientation.)

Figure 8–14 Mapping and correlating sedimentary strata are easier in arid regions because of thinner soil cover and less vegetation. Here one can view the Upper Triassic cliff-forming Wingate Sandstone, underlain by the red Chinle Formation, which overlies black Precambrian metamorphic rocks in Colorado National Monument, near Grand Junction, Colorado. During episodes of uplift and erosion, Paleozoic rocks were eroded away, so that the Mesozoic beds lie directly on Precambrian rocks. (Courtesy of Robert J. Weimer.)

Stratigraphic correlation involves the determination of the equivalence of formations in different locations through a comparison of distinctive lithologic or fossil characteristics.

Correlation Based on Physical Features

In the most obvious kind of correlation, formations from widely separated outcrops are matched on the basis of composition, texture, color, and weathered appearance. Of course, such lithologic features must be used with caution as they may change from place to place within the same formation. In many arid regions the problem of lateral change of a lithologic feature causes less concern, because there is little soil and vegetation covering strata, and one can physically trace the formation by walking continuously along the outcrop (Fig. 8–14). Sometimes the lithology of the rock is not sufficiently distinctive to permit correlation from place to place. In such situations, the position of the rock unit relative to strata immediately above or below may suffice to make the correlation (Fig. 8–15).

A simple illustration of how correlations are used to build a composite picture of the rock record is provided in Figure 8–16. A geologist working along the sea cliffs at locality 1 recognizes a dense oölitic limestone (formation F) at the lip of the

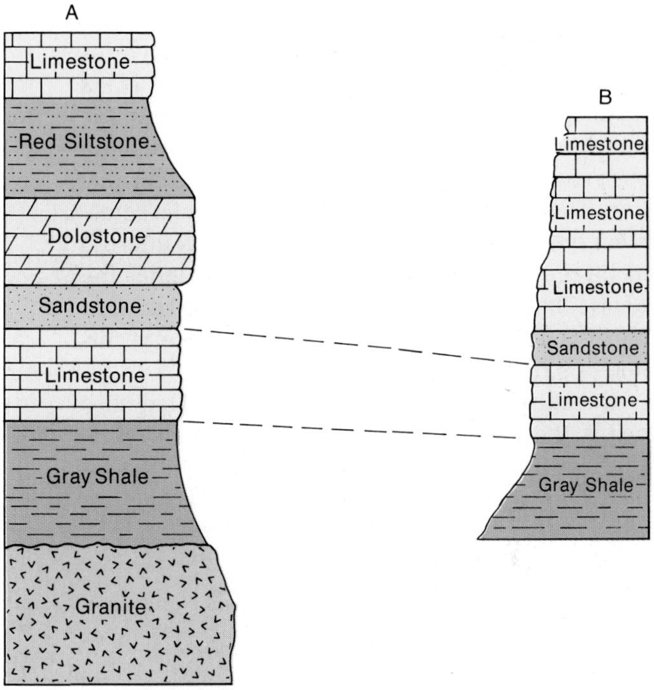

Figure 8–15 If the lithology of a rock is not sufficiently distinctive to permit its correlation from one locality to another, it position in relation to distinctive rock units above and below may aid in correlation. In the example shown here, the limestone unit at locality A can be correlated to the lowest of the four limestone units at locality B because of its position between the gray shale and the sandstone units.

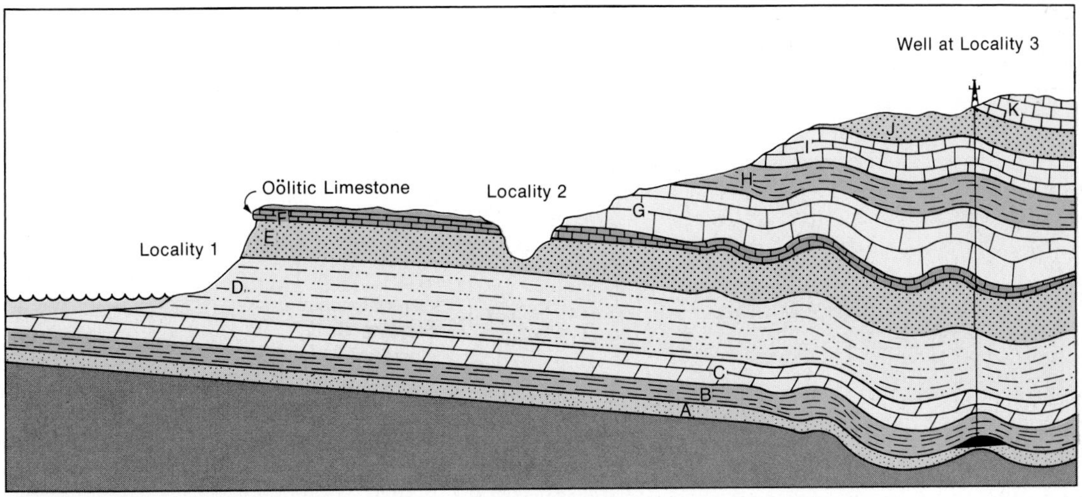

Figure 8–16 An understanding of the sequence of formation in an area usually begins with examination of surface rocks and correlation between isolated exposures. Study of samples from deep wells permits the geologist to expand the known sequence of formations and verify the areal extent and thickness of both surface and subsurface formations.

cliff. The limestone is underlain by formations E and D. Months later, the geologist continues his survey in the canyon at locality 2. Because of its distinctive character, he recognizes the oölitic limestone in the canyon as the same formation seen earlier along the coast and makes this correlation. The formation below F in the canyon is somewhat more clayey than at locality 1 but is inferred to be the same because it occurs right under the oölitic limestone. Working his way upward toward locality 3, the geologist maps the sequence of formations from G to K.

Questions still remain, however. What lies below the lowest formation thus far found? Perhaps years later an oil well, such as that at locality 3, might provide the answer. Drilling reveals that formations C, B, and A lie beneath D. Petroleum geologists monitoring the drilling of the well would add to the correlations by matching all the formations penetrated by the drill to those found earlier in outcrop. In this way, piece by piece, a network of correlations is built up across an entire region.

Correlation Based on Fossils

One of the limitations inherent in correlating rocks by their lithologic attributes is that rocks of similar lithology have been deposited repeatedly over the long span of geologic time. Thus, there is the danger of correlating two formations that look alike but were deposited at quite different times. Fossils are important in preventing such mismatching (Fig. 8–17).

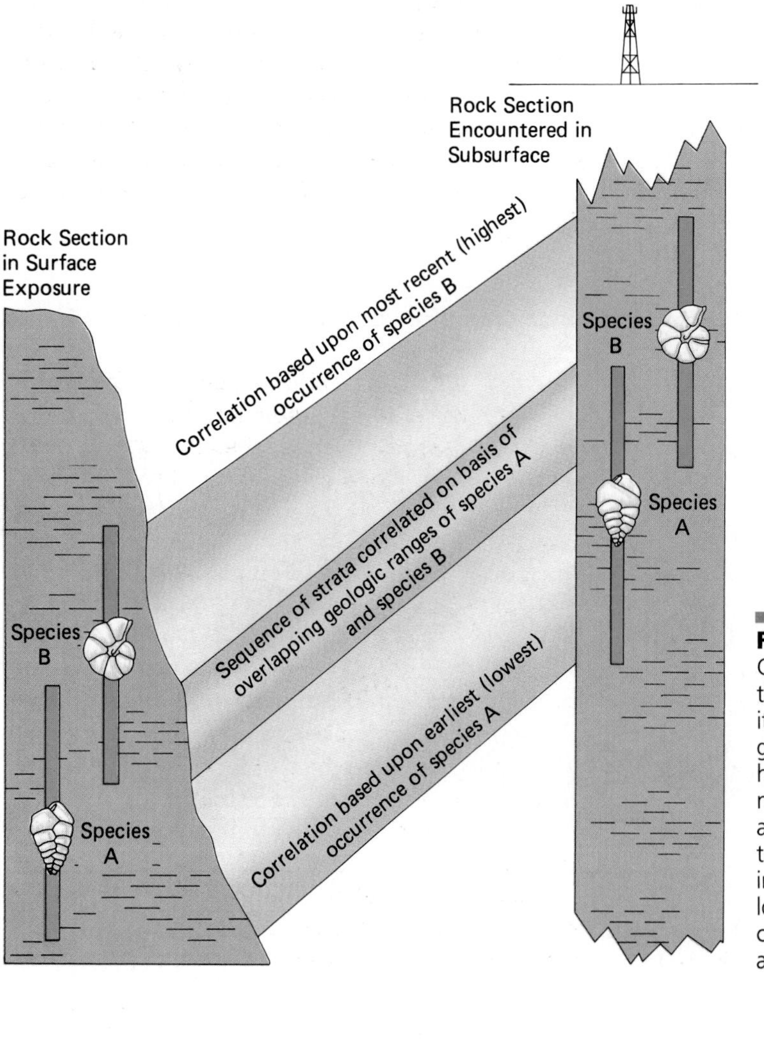

Figure 8–17
Correlation by fossils. In this simplified illustration, it is assumed that the geologic ranges of fossils have been verified at numerous other localities and do not represent fortuitous faunal migrations into or out of the two localities. (The fossil species shown are protozoans called foraminifers.)

Not all fossils are useful for making correlations. Some are rare, restricted to a few localities, or derived from organisms that populated the earth over a great length of geologic time. These kinds of fossils have only limited value in correlation. There are, fortunately, other fossils that are abundant, are widely dispersed around the globe, and are the remains of animals or plants that lived during a relatively short span of geologic time. Such short-lived but widespread fossils, called **guide** or **index fossils** (Fig. 8–18), are especially useful in correlating strata.

The Standard Geologic Time Scale

The early geologists had no way of knowing how many time units would be represented in the completed geologic time scale. Consequently, the time scale grew piecemeal, in an unsystematic manner. Units were named as they were discovered and studied. Sometimes the name for a unit was borrowed from local geography, from a mountain range in which rocks of a particular age were well exposed, or from an ancient tribe of Welshmen. Sometimes the name was suggested by the kinds of rocks that predominated. When a relatively well exposed and complete sequence of previously unnamed strata was discovered, it was carefully defined and designated the **type section**. This terminology continues in use today. The type section is a standard that may be referred to when identifying a stratigraphic unit at a different locality (Fig. 8–19).

As noted in Chapter 1 and in Figure 8–20, the history of the earth is divisible into large segments termed **eons**. Eons in turn are divided into **eras**, eras into **periods**, and periods into **epochs**. Eras, periods, epochs, and divisions of epochs called **ages** all represent intangible increments of time. They are geologic **time units**. The rocks formed during a specified interval of time are called **time–rock units** (Table 8–1). For example, strata laid down during a given *period* compose a time–rock unit called a **system**. Thus, one may properly speak of climatic changes during the Cambrian *period* as indicated by fossils found in rocks of the Cambrian *system*. Each of the geologic systems is recognized largely by its distinctive assemblage of fossil organisms that differ from other fossils in both older and younger systems. **Series** is the time–rock term used for rocks deposited during an epoch, while **stage** represents the tangible rock record of an age.

The steps leading to the recognition of time–rock units began with the use of superposition in establishing age relationships. Local sections of strata were used by early geologists to recognize beds of successively different age and, thereby, to record successive evolutionary changes in fauna and flora. (The order and nature of these evolutionary changes could be determined because higher layers are successively younger.) Once the faunal and floral succession was deciphered, fossils provided an additional tool for establishing the order of events. They could also be used for correlation, so that strata at one locality could be related to the strata of various other localities. No single place on earth contains a complete sequence of strata from all geologic ages. Hence, correlation to the type sections of many widely distributed local sections was necessary in constructing the geologic time scale (Fig. 8–21).

Quantitative Geologic Time

After developing a geologic time scale based on relative age, it is understandable that scientists would seek a way to assign absolute ages in years to the various periods and epochs. From the time of Hutton, geologists were convinced that the earth was indeed very old. But precisely how old? A way to answer that question awaited the discovery of radioactivity by Henri Becquerel in 1896.

Radioactivity

The radioactivity discovered by Henri Becquerel was derived from elements like uranium and thorium, which are unstable and break down or decay to form other elements or other isotopes of the same element. Any individual uranium atom, for example, will eventually decay to lead if given a sufficient length of time. To understand what is meant by "decay," let us consider what happens to

A

B

C

D

Figure 8–18 Examples of guide fossils. (A) An Ordovician horn coral (*Streptelasma*). (B) A Devonian brachiopod (*Mucrospirifer*). (C) A Cretaceous pelecypod (*Trigonia*). (D) A Mississippian crinoid (*Taxocrinus*).

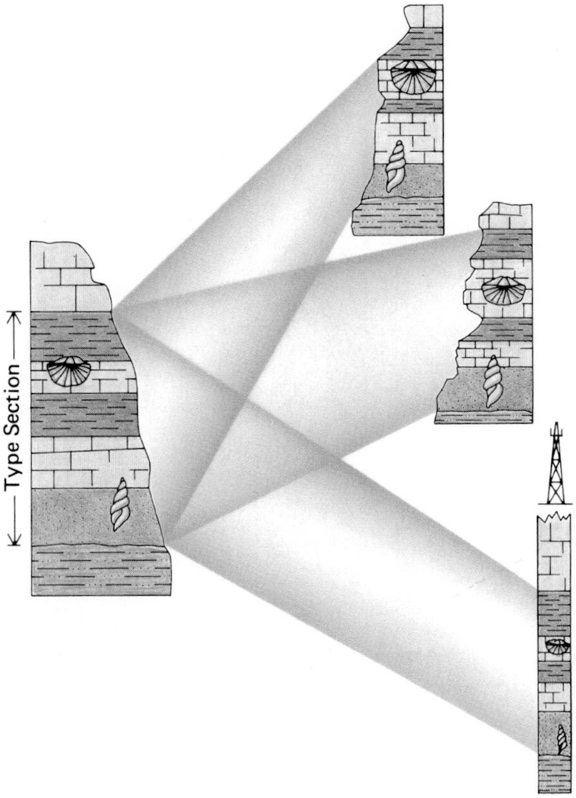

Figure 8–19 The *type section* is a carefully defined and described standard with which strata at other locations can be compared.

a radioactive element like uranium 238 (^{238}U). Uranium 238 has an atomic mass of 238. The "238" represents the sum of the weights of the atom's protons and neutrons (each proton and neutron having a mass of 1). Uranium has an atomic number (number of protons) of 92. Such atoms with specific atomic number and weight are sometimes termed **nuclides**. Sooner or later (and entirely spontaneously) the uranium 238 atom will fire off a particle from the nucleus called an **alpha particle**. Alpha particles are positively charged ions of helium. They have an atomic weight of 4 and an atomic number of 2. Thus, when the alpha particle is emitted, the new atom will now have an atomic weight of 234 and an atomic number of 90. The new atom, which is formed from another by radioactive decay, is called a **daughter element**. From the decay of the parent nuclide, uranium 238, the daughter nuclide, thorium 234, is obtained (Fig. 8–22). A shorthand equation for this change is written:

$$^{238}_{92}U \longrightarrow {}^{234}_{90}Th + {}^{4}_{2}He$$

This change is not, however, the end of the process, for the nucleus of thorium 234 (^{234}Th) is not stable. It eventually emits a **beta particle** (an electron discharged from the nucleus when a neutron splits into a proton and an electron). There is now an extra proton in the nucleus but no loss of atomic weight because electrons are essentially weightless. Thus, from $^{234}_{90}Th$ the daughter element $^{234}_{91}Pa$ (protactinium) is formed. In this case, the atomic number has been increased by 1. In other instances, the beta particle may be captured by the nucleus, where it combines with a proton to form a neutron. The loss of the proton would decrease the atomic number by 1.

A third kind of emission in the radioactive decay process is called **gamma radiation**. It consists of a form of invisible electromagnetic waves having even shorter wavelengths than X-rays.

The decay of uranium 238 involves the emission of alpha and beta particles, as well as gamma radiation.

The rate of decay of radioactive isotopes is uniform and is not affected by changes in pressure, temperature, or the chemical environment. Therefore, once a quantity of radioactive nuclides has been incorporated into a growing mineral crystal, that quantity will begin to decay at a steady rate with a definite percentage of the radiogenic atoms undergoing decay in each increment of time. Each radioactive element has a particular mode of decay and a unique decay rate. As time passes, the quantity of the original or parent nuclide diminishes, and the number of the newly formed or daughter atoms increases, thereby indicating how much time has elapsed since the clock began its timekeeping. The "beginning," or "time zero," for any mineral containing radioactive nuclides would be the moment when the radioactive parent atoms became part of a mineral from which daughter elements could not escape. The retention of daughter elements is essential, for they must be counted to

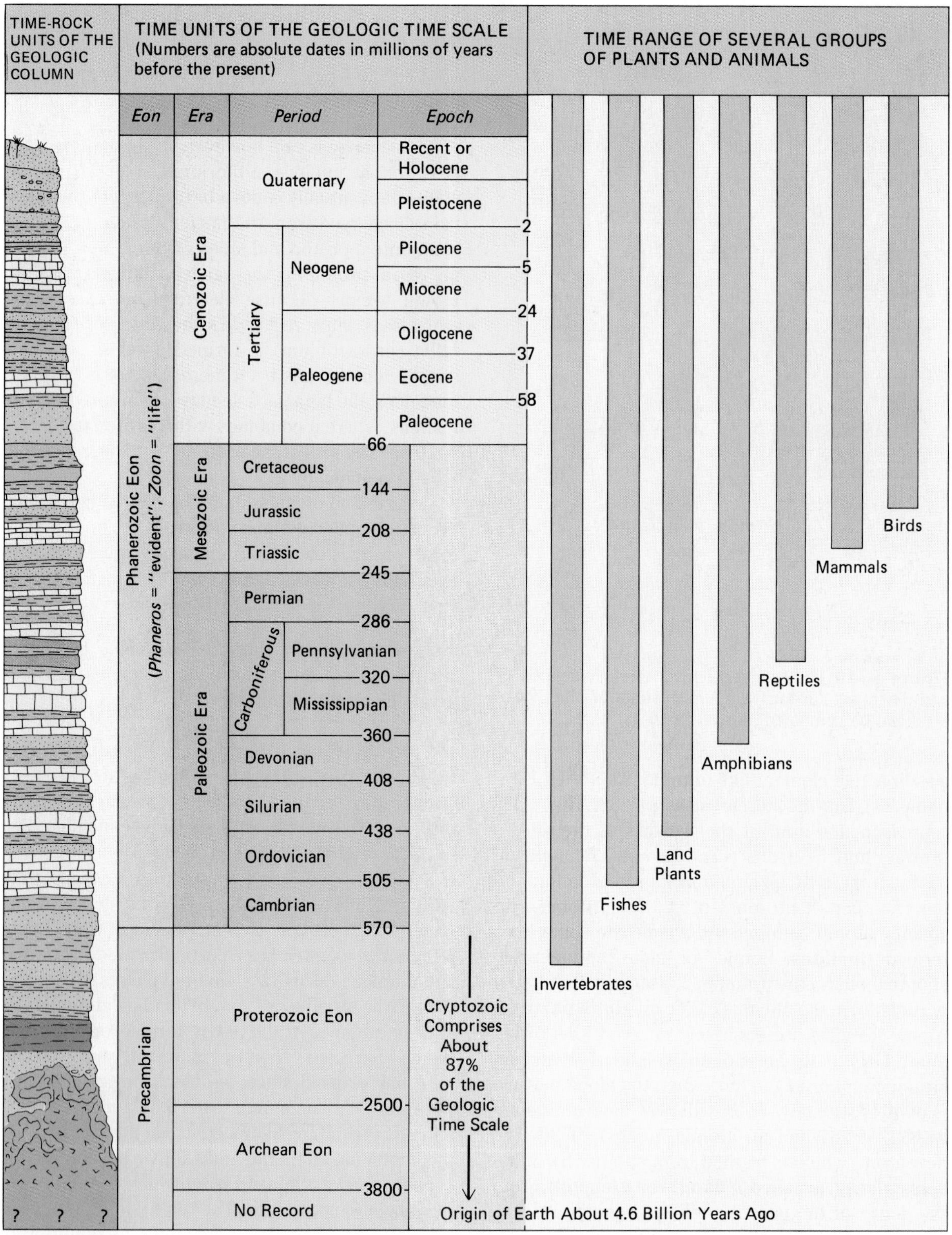

Figure 8–20 Geologic time scale. (From A.R. Palmer, The decade of North American geology, geologic time scale, *Geology* 11:503–504, 1983.)

Table 8–1 Terms Commonly Used for the Time and Time–Rock Divisions

Time Divisions	*Time–Rock Divisions*
ERA	ERATHEM
PERIOD	SYSTEM
EPOCH	SERIES
AGE	STAGE
	ZONE

determine the original quantity of the parent nuclide (Fig. 8–23).

If a sample being analyzed for its radiometric age has lost daughter products, it will yield a radiometric age younger than its true age.

The determination of the ratio of parent to daughter nuclides is usually accomplished with the use of a **mass spectrometer**, an analytic instrument capable of separating and measuring the pro-

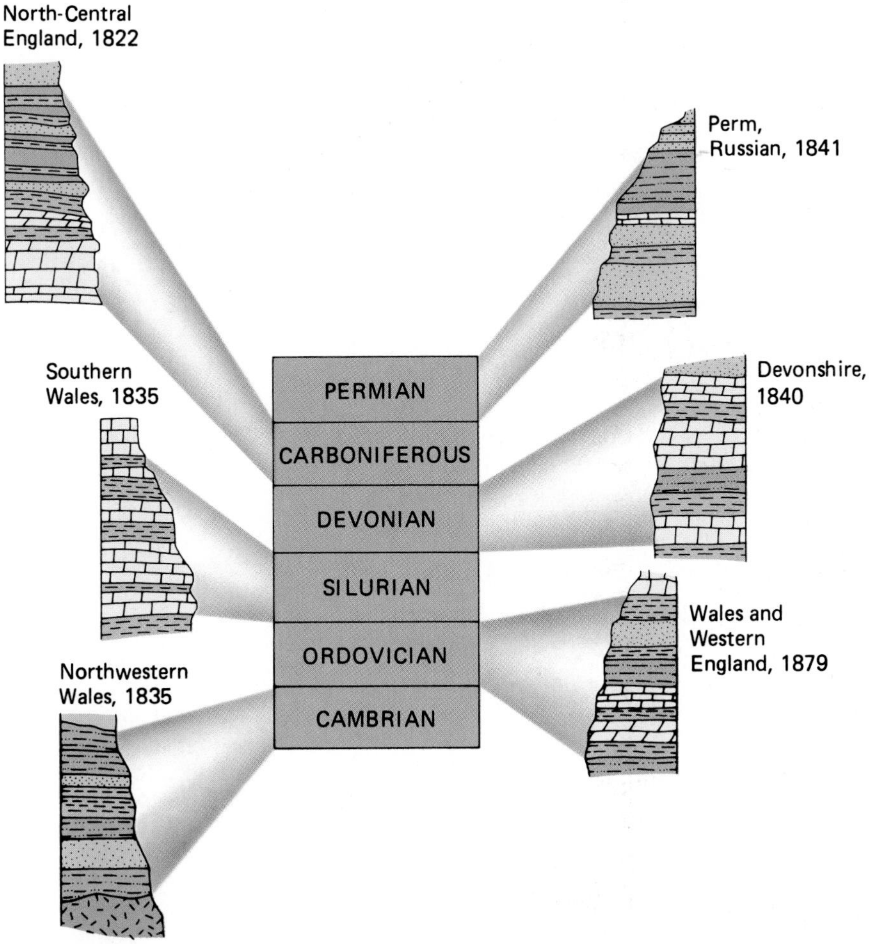

Figure 8–21 The standard geologic time scale for the Paleozoic and other eras developed without benefit of a grand plan. Instead it developed by the compilation of "type sections" for each of the systems.

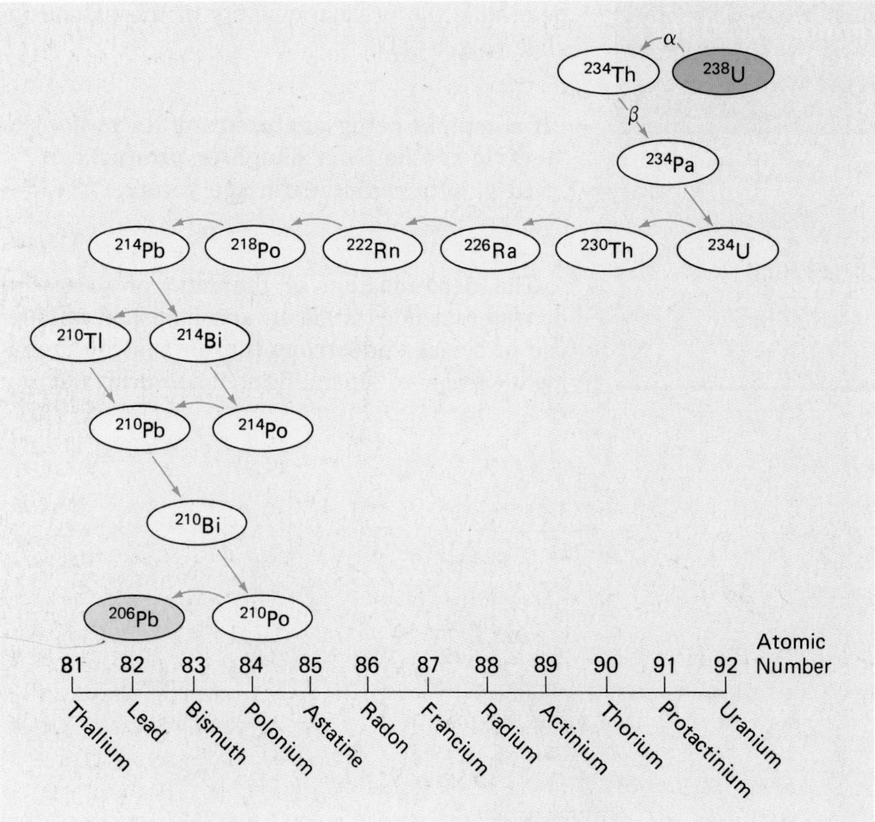

Figure 8–22 Radioactive decay series of uranium 238 (^{238}U) to lead 206 (^{206}Pb).

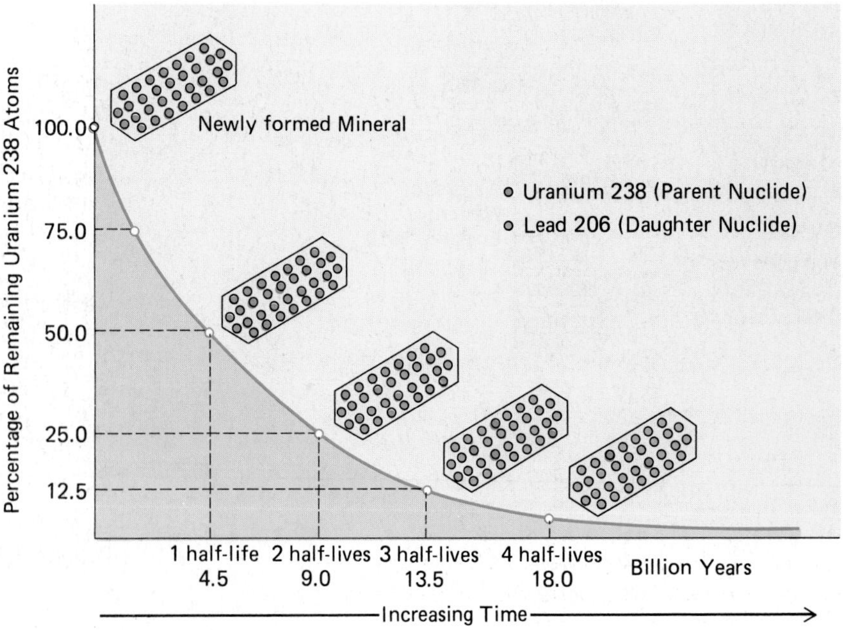

Figure 8–23 Rate of radioactive decay of uranium 238 to lead 206. During each half-life, one half of the remaining amount of the radioactive element decays to its daughter element. In this simplified diagram, only the parent and daughter nuclides are shown, and the assumption is made that there was no contamination by daughter nuclides at the time the mineral formed.

portions of minute particles according to their mass differences. In the mass spectrometer, samples of elements are vaporized in an evacuated chamber where they are bombarded by a stream of electrons. This bombardment knocks electrons off the atoms, leaving them positively charged. A stream of these positively charged ions is deflected as it passes between plates that bear opposite charges of electricity. The degree of deflection depends on the charge-to-mass ratio. In general, the heavier the ion, the less it will be deflected (Fig. 8–24).

Of the three major families of rocks, the igneous clan is by far the best for radiometric dating. Fresh samples of igneous rocks are less likely to have experienced loss of daughter products, which must be accounted for in the age determination. Igneous rocks can provide a valid date for the time that a silicate melt containing radioactive elements solidified.

In contrast to igneous rocks, the minerals of sediments can be weathered and leached of radioactive components, and age determinations are far more prone to error. In addition, the age of a detrital grain in a sedimentary rock does not give an age of the sedimentary rock but only of the parent rock that was eroded much earlier.

Dates obtained from metamorphic rocks may also require special care in interpretation. The age of a particular mineral may record the time the rock first formed or any one of a number of subsequent metamorphic recrystallizations.

Once an age has been determined for a particular rock unit, it is sometimes possible to use that data to approximate the age of adjacent rock masses. A shale lying below a lava flow that is 110 million years old and above another flow dated at 180 million years old must be between 110 and 180 million years of age (Fig. 8–25). Discovery of fossils within the shale might permit one to assign it to a particular geologic system or series, and by correlation, to extend the age data around the world (Fig. 8–26).

Half-Life

One cannot predict with certainty the moment of disintegration for any individual radioactive atom in a mineral. We do know that it would take an infinitely long time for all of the atoms in a quantity of radioactive elements to be entirely transformed to stable daughter products. Experimenters have also shown that there are more disintegrations per increment of time in the early stages than in later stages (see Fig. 8–23), and one can statistically forecast what percentage of a large population of atoms will decay in a certain amount of time.

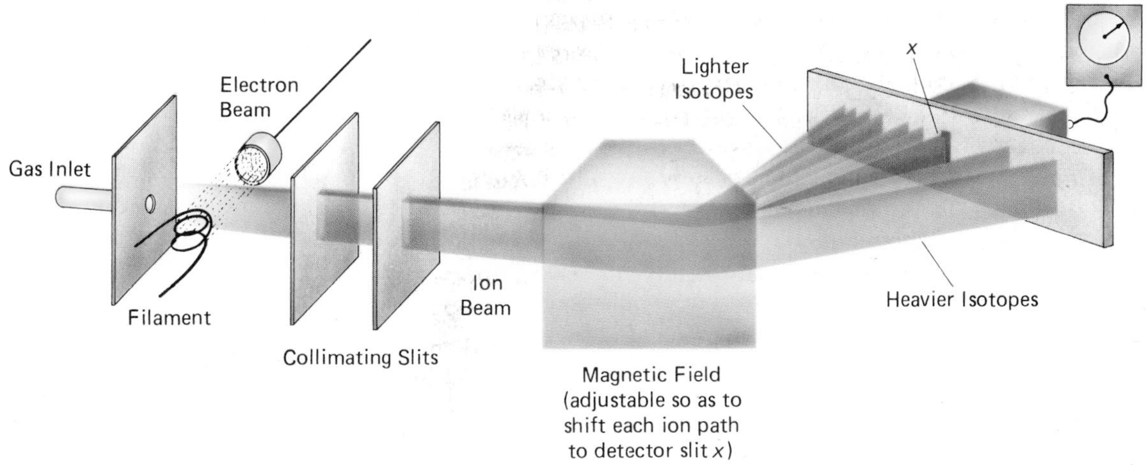

Figure 8–24 Schematic drawing of a mass spectrograph. In this type of spectrograph the intensity of each ion beam is measured electrically (rather than recorded photographically) to permit determination of the isotopic abundances required for radiometric dating.

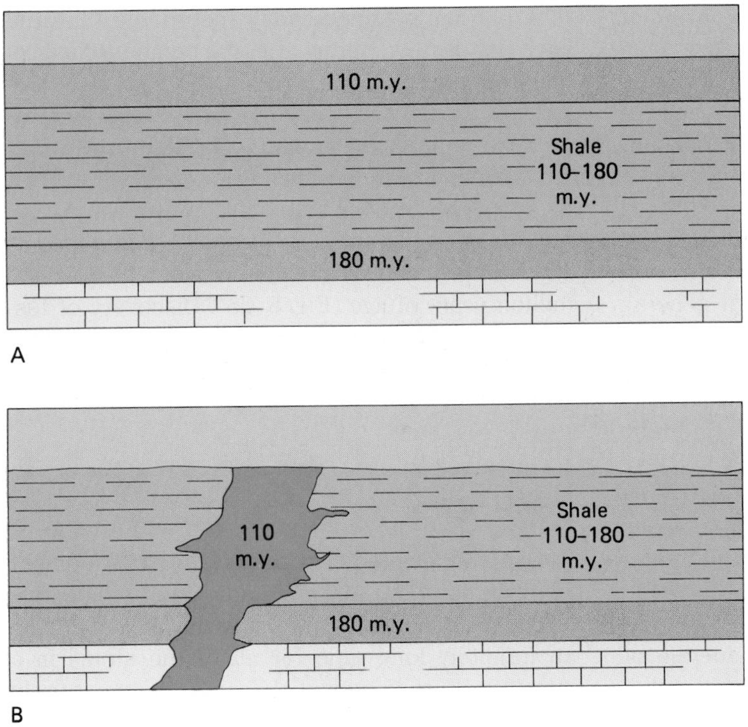

Figure 8–25 Igneous rocks that have provided absolute radiogenic ages can often be used to date sedimentary layers. (A) The shale is bracketed by two lava flows. (B) The shale lies above the older flow and is intruded by a younger igneous body. (Note: m.y. = million years.)

Because of these features of radioactivity, it is convenient to consider the time needed for half of the original quantity of atoms to decay. This span of time is termed the **half-life**. Thus, at the end of the time constituting one half-life, ½ of the original quantity of radioactive element still has not undergone decay. After another half-life, ½ of what was left remains or ¼ of the original quantity. After a third half-life, only ⅛ would remain, and so on.

Every radioactive nuclide has its own unique half-life. Uranium 235, for example, has a half-life of 704 million years. Thus, if a sample contains 50 per cent of the original amount of uranium 235 and 50 per cent of its daughter product, lead 207, then that sample is 704 million years old. If the analyses indicate 25 per cent of uranium 235 and 75 per cent of lead 207, two half-lives would have elapsed, and the sample would be 1408 million years old (Fig. 8–27.)

The rate of decay of a radioactive element is expressed as its half-life, or the time required for half of the atoms in a sample of that element to decay.

The Principal Geologic Timekeepers

At one time, there were many more radioactive nuclides present on earth than there are now. Many of these had short half-lives and have long since decayed to undetectable quantities. Fortunately for those interested in dating the earth's most ancient rocks, there remain a few long-lived radioactive nuclides. The most useful of these are uranium 238, uranium 235, rubidium 87, and potassium 40 (Table 8–2). There are also a few short-lived radioactive elements that are used for dating more recent events. Carbon 14 is an example of such a short-lived isotope. There are also short-lived nuclides that represent segments of a uranium or thorium decay series.

Timekeepers that Produce Lead

Dating methods involving lead require the presence of radioactive nuclides of uranium or thorium that were incorporated into rocks when they originated. To determine the age of a sample of mineral or rock, one must know the original number of par-

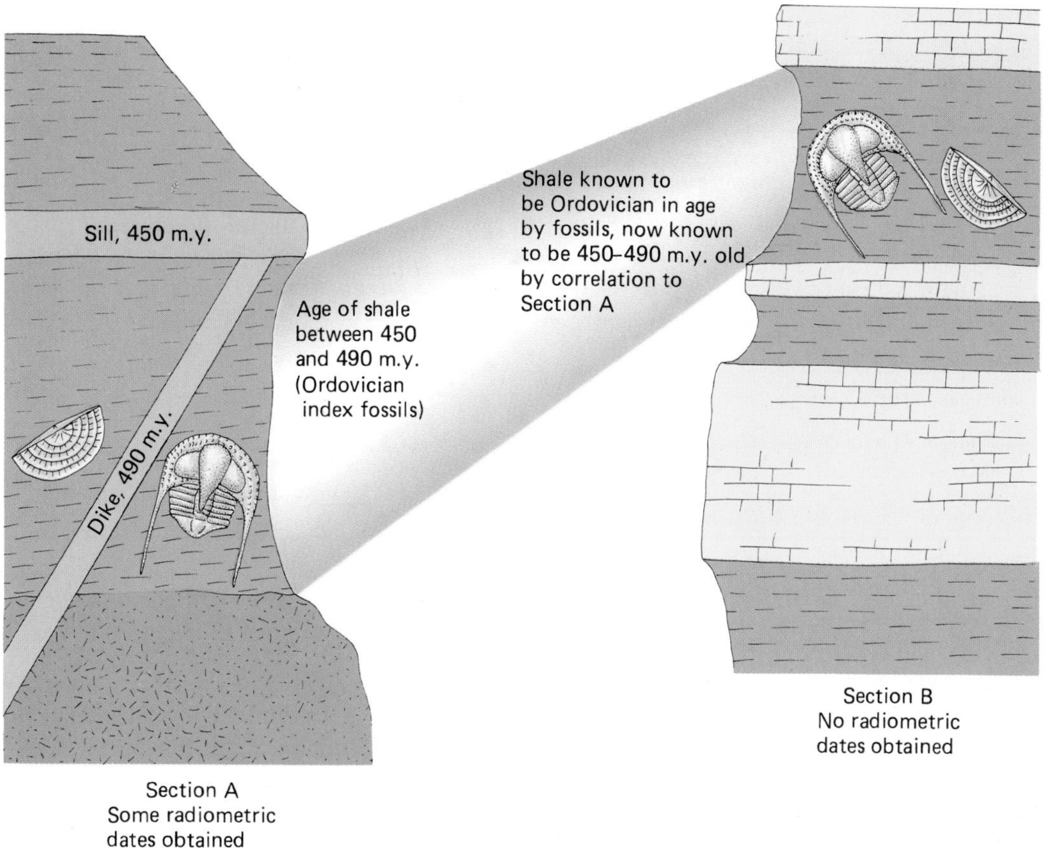

Figure 8–26 The actual age of rocks that cannot be dated radiometrically can sometimes be ascertained by correlation.

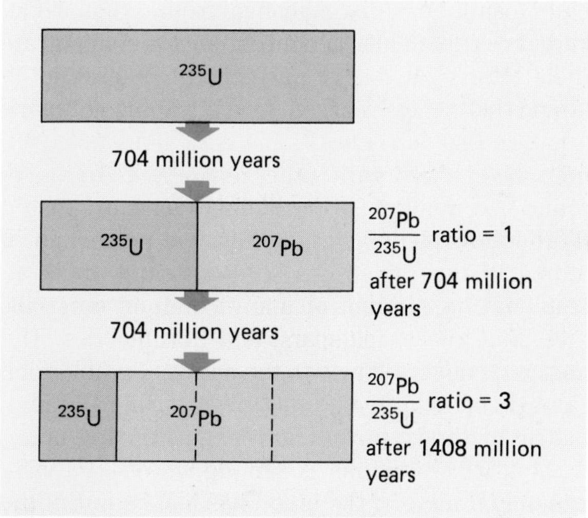

Figure 8–27 Radioactive decay of uranium 235 to lead 207.

ent nuclides as well as the number remaining at the present time. The original number of parent atoms should be equal to the sum of the present number of parent atoms and daughter atoms. The assumptions are made that the system has remained closed, so that neither parent nor daughter atoms have ever been added or removed from the sample except by decay, and that no daughter atoms were present in the system when it formed. The presence, for example, of original lead in the mineral would cause the radiometric age to exceed the true age. Fortunately, geochemists are able to recognize original lead and make the needed corrections.

As we have seen, different isotopes decay at different rates. Geochronologists take advantage of this fact by simultaneously analyzing two or three isotope pairs as a means to cross-check ages and detect errors. For example, if the $^{235}U/^{207}Pb$ radiometric ages and the $^{238}U/^{206}Pb$ ages from the same

Table 8–2 Some of the More Useful Nuclides for Radioactive Dating

Parent Nuclide*	Half-Life (years)†	Daughter Nuclide	Materials Dated
Carbon 14	5730	Nitrogen 14	Organic materials
Uranium 235	704 million (7.04×10^8)	Lead 207 (and helium)	Zircon, uraninite, pitchblende
Potassium 40	1251 million (1.25×10^9)	Argon 40 (and calcium 40)‡	Muscovite, biotite, hornblende, volcanic rock, glauconite, K-feldspar
Uranium 238	4468 million (4.47×10^9)	Lead 206 (and helium)	Zircon, uraninite, pitchblende
Rubidium 87	48,800 million (4.88×10^{10})	Strontium 87	K-micas, K-feldspars, biotite, metamorphic rock, glauconite

*A *nuclide* is a convenient term for any particular atom (recognized by its particular combination of neutrons and protons).
†Half-life data from Steiger, R.H., and Jäger, E., 1977. Subcommission on geochronology: convention on the use of decay constants in geo- and cosmochronology. *Earth and Planetary Science Letters* 36:359–362.
‡Although potassium 40 decays to argon 40 and calcium 40, only argon is used in the dating method because most minerals contain considerable calcium 40, even before decay has begun.

sample agree, then one can confidently assume the age determination is valid.

Radiometric ages that depend on uranium–lead ratios may also be checked against ages derived from lead 207 to lead 206. Because the half-life of uranium 235 is much less than the half-life of uranium 238, the ratio of lead 207 (produced by the decay of uranium 235) to lead 206 will change regularly with age and can be used as a radioactive timekeeper (Fig. 8–28). This is called a lead–lead age, as opposed to a uranium–lead age.

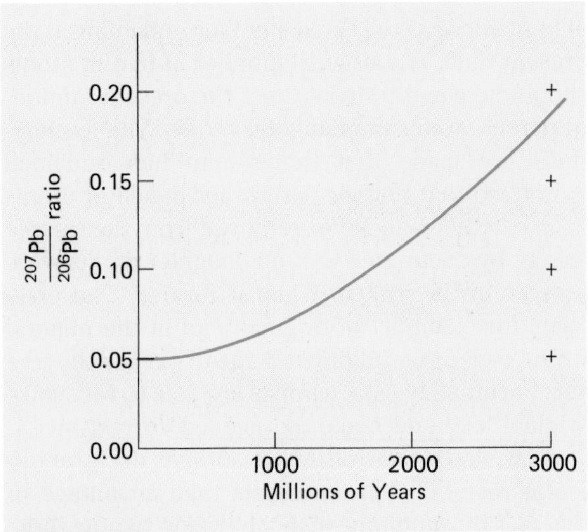

Figure 8–28 Graph showing how the ratio of lead 207 to lead 206 can be used as a measure of age.

The Potassium–Argon Method

Potassium and argon are another radioactive pair widely used for dating rocks. By means of electron capture (causing a proton to be transformed into a neutron), about 11 per cent of the potassium 40 in a mineral decays to argon 40, which may then be retained within the parent mineral. The remaining potassium 40 decays to calcium 40 (by emission of a beta particle). The decay of potassium 40 to calcium 40 is not useful for obtaining radiometric ages because radiogenic calcium cannot be distinguished from original calcium in a rock. Thus, geochronologists concentrate their efforts on the 11 per cent of potassium 40 atoms that decay to argon. One advantage of using argon is that it is inert—that is, it does not combine chemically with other elements. Argon 40 found in a mineral is very likely to have originated there following the decay of adjacent potassium 40 atoms in the mineral. Also, potassium 40 is an abundant constituent of many common minerals, including micas, feldspars, and hornblendes. The method lends itself also to the dating of whole rock samples of solidified lavas. However, like all radiometric methods, **potassium–argon dating** is not without its limitations. A sample will yield a valid age only if none of the argon has leaked out of the mineral being analyzed. Leakage may indeed occur if the rock has experienced temperatures above about 125°C. In specific localities, the ages of rocks dated by this method reflect the last episode of heating rather than the time of origin of the rock itself.

The half-life of potassium 40 is 1251 million years. As illustrated in Figure 8–29, if the ratio of potassium 40 to daughter products is found to be 1 to 1, then the age of the sample is 1251 million years. If the ratio is 1 to 3, then yet another half-life has elapsed, and the rock would have a radiogenic age of two half-lives, or 2502 million years.

The Carbon 14 Method

Techniques for age determination based on content of radiocarbon were first devised by W. F. Libby and his associates at the University of Chicago in 1947. The method is an indispensable aid to archaeologic research and is useful in deciphering the very recent events in geologic history. Because of the short half-life of carbon 14—a mere 5730 years—organic substances older than about 50,000 years contain very little carbon 14. New techniques, however, allow geologists to extend the method's usefulness back to almost 100,000 years.

Unlike uranium 238, carbon 14 is created continuously in the earth's upper atmosphere. The story of its origin begins with **cosmic rays**, which are extremely high energy particles (mostly protons) that bombard the earth continuously. Such particles strike atoms in the upper atmosphere and split their nuclei into smaller particles, among which are neutrons. Carbon 14 is formed when a neutron strikes an atom of nitrogen 14. As a result of the collision, the nitrogen atom emits a proton, captures a neutron, and becomes carbon 14 (Fig. 8–30). Radioactive carbon is being created by this process at the rate of about two atoms per second for every cm^2 of the earth's surface. The newly created carbon 14 combines quickly with oxygen to form CO_2, which is then distributed by wind and water currents around the globe. It soon finds its way into photosynthetic plants because they utilize carbon dioxide from the atmosphere to build tissues. Plants containing carbon 14 are ingested by animals, and the isotope becomes a part of their tissue as well.

Eventually, carbon 14 decays back to nitrogen 14 by the emission of a beta particle. A plant removing CO_2 from the atmosphere should receive a share of carbon 14 proportional to that in the atmosphere. A state of equilibrium is reached in which the gain in newly produced carbon 14 is balanced by the decay loss. The rate of production of

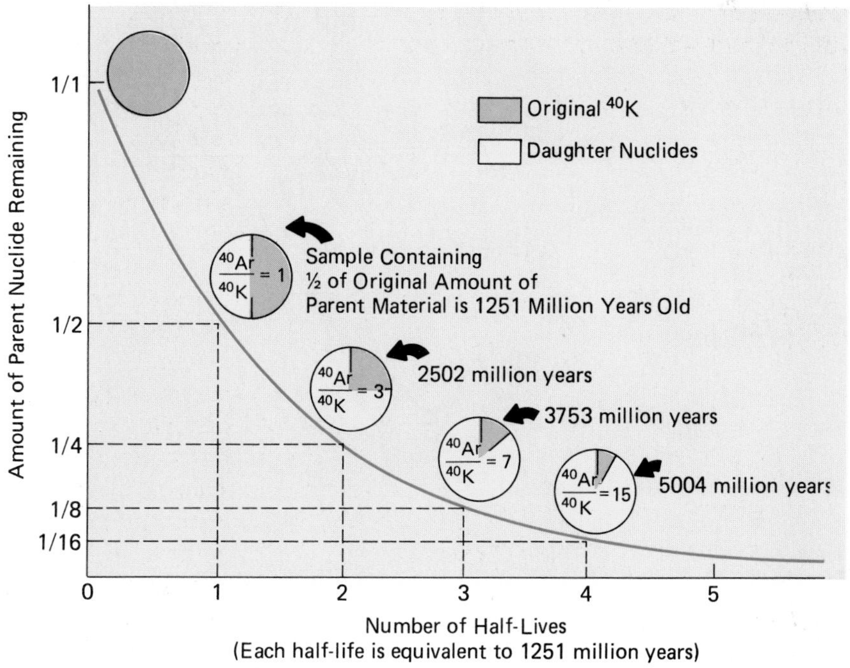

Figure 8–29 Decay curve for potassium 40.

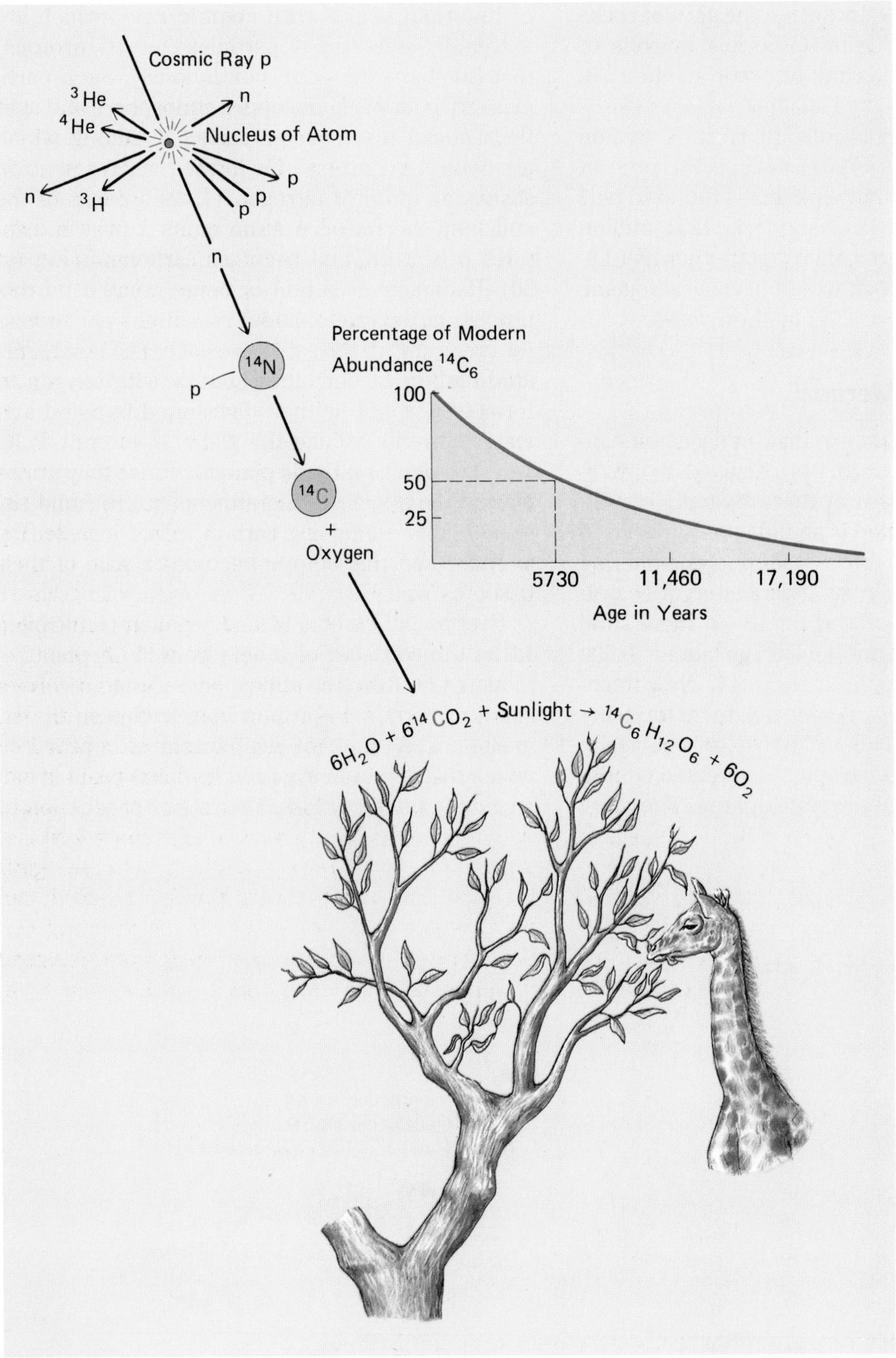

Figure 8–30 Carbon 14 is formed from nitrogen in the atmosphere. It combines with oxygen to form radioactive carbon dioxide and is then incorporated into all living things.

carbon 14 has varied somewhat over the past several thousand years (Fig. 8–31). As a result, corrections in age calculations must be made. Such corrections are derived from analyses of standards such as wood samples, whose exact age is known from tree ring counts.

The age of some ancient bit of organic material is not determined from the ratio of parent to daughter nuclides, as is done with previously discussed dating schemes. Rather, the age is estimated from the ratio of carbon 14 to all other carbon in the sample. After an animal or plant dies, there can be no further replacement of carbon from atmospheric CO_2, and the amount of carbon 14 already present in the once living organism begins to diminish in accordance with the rate of carbon 14 decay. Thus, if the carbon 14 fraction of the total carbon in a piece of pine tree buried in volcanic ash were found to be about 25 percent of the quantity in living pines, then the age of the wood (and the volcanic activity) would be two half-lives of 5730 years each or 11,460 years.

The carbon 14 technique had considerable value to geologists studying the most recent events of the Pleistocene Ice Age. Prior to the development of the method, the age of sediments deposited by the last advance of continental glaciers was surmised to be about 25,000 years. Radiocarbon dates of a layer of peat beneath the glacial sediments provided an age of only 11,400 years. The method has also been found useful in studies of groundwater migration and in dating the geologically recent uppermost layer of sediment on the sea floors.

The Age of the Earth

Anyone interested in the total age of the earth must decide what event constitutes its "birth." Most geologists assume "year 1" commenced as soon as the earth had collected most of its present mass and had developed a solid crust. Unfortunately, rocks that date from those earliest years have not been found on earth. They have long since been altered and converted to other rocks by various geologic processes. The oldest earth materials known are grains of the mineral zircon taken from a sandstone formation in western Australia. The zircon grains are 4.1 to 4.2 billion years old. They were probably eroded from granitic crust and deposited, along with other detrital grains, in river sands. Other very old rocks on earth include 3.4-billion-year-old granites from the Barberton Mountains of South Africa, 3.7-billion-year-old granites of southwestern Greenland, and metamorphic rocks of about the same age from Minnesota.

Meteorites, which many consider to be remnants of a shattered planet that originally formed at about the same time as the earth, have provided uranium–lead ages of about 4.6 billion years. From such data and from estimates of how long it would take to produce the quantities of various lead isotopes now found on the earth, geochronologists feel that the 4.6-billion-year age for the earth can be accepted with confidence. Substantiating evidence for this conclusion comes from returned moon rocks. The ages of these rocks range from

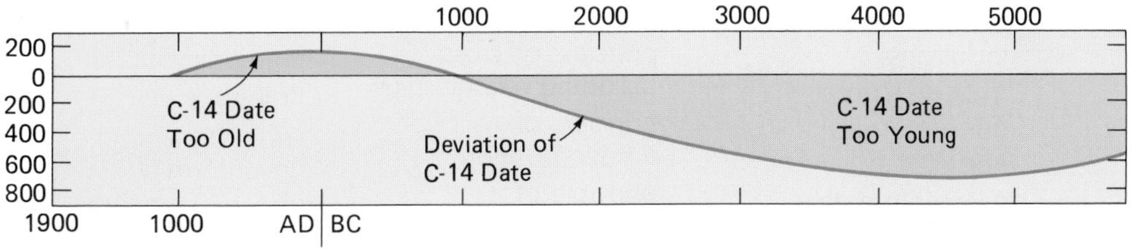

Figure 8–31 Deviation of carbon 14 ages to true ages from the present back to about 5000 B.C. Data are obtained from analysis of bristle cone pines from the western United States. Calculations of carbon 14 deviations are based on half-life of 5730 years. (Adapted from Ralph, E.K., Michael, H.N., and Han, M.C., Radiocarbon dates and reality, *MASCA Newsletter* 9:1, 1973.)

3.3 to about 4.7 billion years. The older age determinations are derived from rocks collected on the lunar highland, which may represent the original lunar crust. The moons and planets of our solar system probably originated as a result of the same cosmic processes and at about the same time.

Summary

In no other science does time play as significant a part as in geology. Time provides the frame of reference necessary to the interpretation of events and processes on the earth. In the earlier stages of its development, geology was dependent on relative dating of events. James Hutton helped scientists visualize the enormous periods needed to accomplish the events indicated in sequences of strata, and the geologists who followed him pieced together the many local stratigraphic sections, using fossils and superposition. A scale of *relative geologic time* gradually emerged.

A method for determining quantitative geologic time was achieved only after the discovery of radioactivity at about the turn of the 20th century. Scientists found that the rate of decay by radioactivity of certain elements is constant and can be measured and that the proportion of *parent* and *daughter elements* can be used to reveal how long they have been present in a rock. Over the years, continuing efforts by investigators as well as improvements in instrumentation (particularly of the mass spectrometer) have provided many thousands of age determinations.

The radiometric transformations widely used in determining absolute ages are *uranium 238 to lead 206, uranium 235 to lead 207, potassium 40 to argon 40,* and *carbon 14 to nitrogen 14.* Methods involving uranium–lead ratios are of importance in dating the earth's oldest rocks. The short-lived carbon 14 isotope that is created by cosmic ray bombardment of the atmosphere provides a means to date the most recent events in earth history. For rocks of intermediate age, schemes involving potassium–argon ratios, are useful. A figure of 4.7 billion years for the earth's total age is now supported by ages based on meteorites and on lead ratios from terrestrial samples.

Terms to Understand

age
alpha particle
angular unconformity
beta particle
cosmic rays
daughter element
disconformity
eon
epoch
era
gamma radiation
geochronology
guide fossils (index fossils)
half-life
inclusions
mass spectrometer
nonconformity
nuclide
paraconformity
period
potassium–argon dating
principle of biotic succession
principle of cross-cutting relationships
principle of superposition
principle of uniformitarianism
radioactivity
series
stage
stratigraphic correlation
system
time unit
time-rock unit
type section
unconformity
uniformitarianism

Questions for Review

1. By what methods did geologists attempt to determine the age of the earth before the discovery of radioactivity? Why were these methods inadequate?

2. What types of radiation accompany the decay of radioactive isotopes?

3. In making an age determination based on the uranium–lead method, why should an investigator select an unweathered sample for analysis?

4. How do the isotopes carbon 12 and carbon 14 differ from one another in regard to the following: (a) number of protons, (b) number of electrons, and (c) number of neutrons? What is the origin of carbon 14?

5. Define half-life. Why is this term used instead of an expression like "whole-life" in radiometric dating? What are the half-lives of uranium 238, potassium 40, and carbon 14?

6. Pebbles of basalt within a conglomerate yield a radiometric age of 300 million years. What can be said about the age of the conglomerate? Several miles away, the same conglomerate stratum is bisected by a 200-million-year-old dike. What now can be said about the age of the conglomerate in this location?

7. Has the amount of uranium in the earth increased, decreased, or remained about the same over the past 4.6 billion years? What can be stated about changes in the amount of lead?

8. State the estimated age of a sample of mummified skin from a prehistoric human that contained 12.5 per cent of an original quantity of carbon 14.

9. How are dating methods involving decay of radioactive elements unlike methods for determining elapsed time by the funneling of sand through an hour glass?

10. State the effect on the radiometric age of a zircon crystal being dated by the potassium–argon method if a small amount of argon 40 escaped from the crystal.

11. If an intrusion of granite cuts into or across several strata, which is older, the granite or the strata?

12. Minerals suitable for radiometric age determinations are usually components of igneous rocks. How, then, can quantitative ages be obtained for sedimentary formations?

13. If an erosional surface (unconformity) cuts across or truncates folded strata, did the erosion occur before or after the strata were folded?

14. What is the essential difference between a time–rock unit and a time-unit? Give an example of each.

Supplemental Readings and References

Berry, W.B.N., 1968. *Growth of a Prehistoric Time Scale.* San Francisco, W.H. Freeman & Co.

Eicher, D.L., 1976. *Geologic Time.* 2nd ed. Englewood Cliffs, N.J., Prentice-Hall, Inc.

Faul, H., 1966. *Ages of Rocks, Planets, and Stars.* New York, McGraw-Hill Book Co.

Hamilton, E.I., 1965. *Applied Geochronology.* New York, Academic Press, Inc.

Harbaugh, J.W., 1968. *Stratigraphy and Geologic Time.* Dubuque, Iowa, William C. Brown Co., Publishers.

Ojakangas, R.W., and Darby, D.G., 1976. *The Earth, Past and Present.* New York, McGraw-Hill Book Co.

Toulmin, S., and Goodfield, J., 1965. *The Discovery of Time.* New York, McGraw-Hill Book Co.

Folds, Faults, and Mountains

9

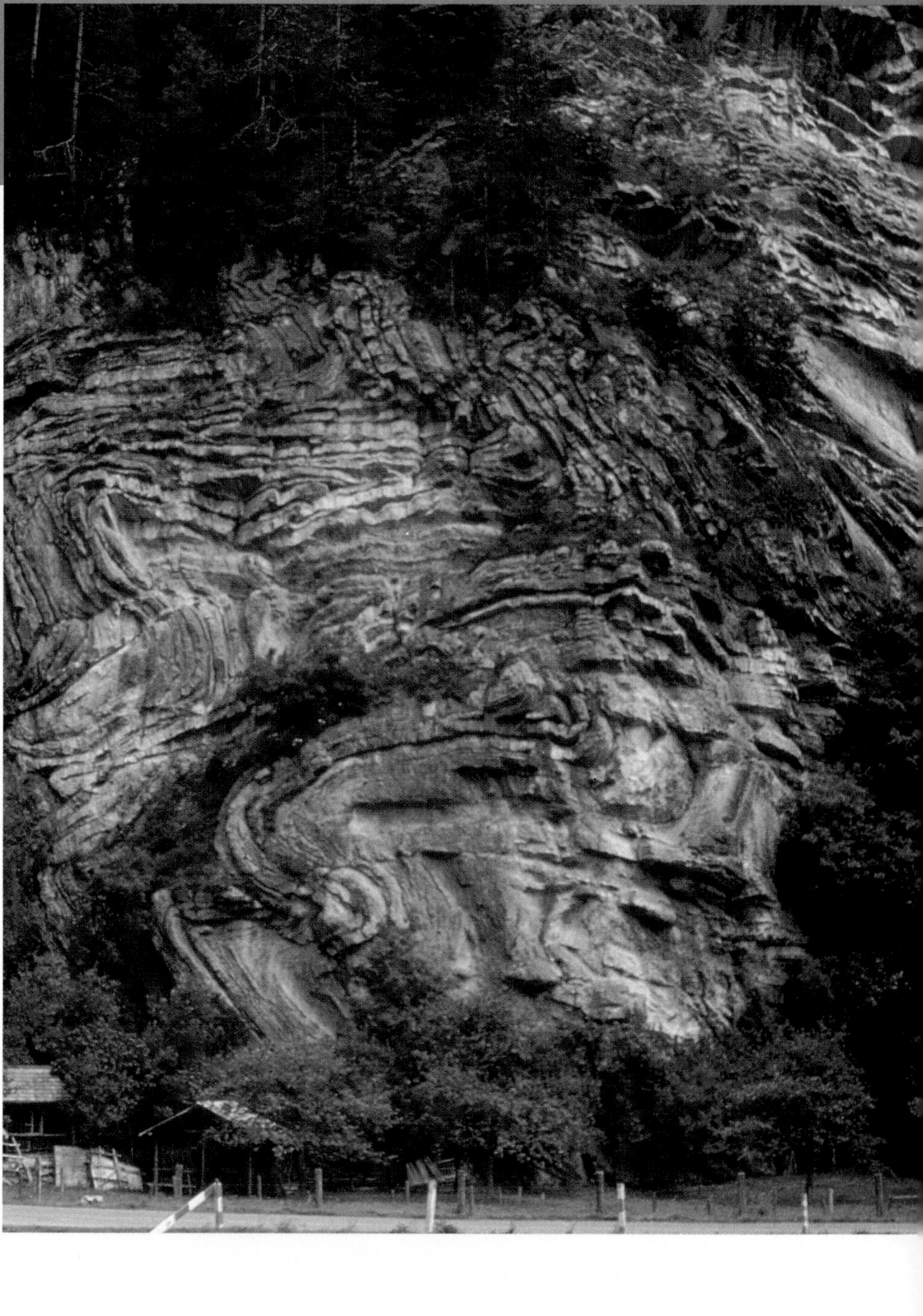

Severely folded strata exposed on north side of Rhone River valley west of Sierre, Switzerland.

It is not uncommon to find the mass of rock on one side of a fissure thrown up above or down below the mass with which it is in contact on the other side.

Sir Charles Lyell, 1882

Introduction

The expression "solid as a rock" reflects a belief that rocks are sturdy and unyielding. In certain ways, rocks are indeed strong. Yet like most other solids, rocks possess limits of strength that, if exceeded, will cause them to fail. Compressional, tensional, or shearing forces acting on rocks may bend or break them, providing the rich variety of folds, faults, fractures, and joints that give each geologic region on earth its distinctive character. Frequently, these so-called geologic structures also serve as traps for petroleum and natural gas, or provide avenues for solutions that deposit important ore minerals. On a grander scale, the careful study of folds and faults reveals the forces that have affected particular expanses of the earth's crust. Such studies improve the geologist's understanding of the behavior of tectonic plates and provide valuable insight into the birth and development of mountains.

Geologic Structures

Almost from the moment they solidify, rocks begin to respond to a variety of deformational forces. They are compressed by a heavy burden of overlying rocks, squeezed or stretched along the margins of moving tectonic plates, and subside or rise as they strive to reach equilibrium with denser rocks that lie far beneath the earth's surface. All of these events contribute to the making of **geologic structures**.

Deformation of Rocks

In order to better understand how rocks are deformed, it will be useful to fortify ourselves with a few important concepts. The first of these is the concept of stress. **Stress** is a force applied over a given area of the rock mass. There are three different kinds of stress, namely, compressional, tensional, and shear. As is apparent from its name, compressive stress tends to compress or squeeze the rock body on which it is acting. Tensional stress, on the other hand, exists when forces tend to pull a rock apart. Shear stress results from forces acting parallel, but in opposite directions (Fig. 9–1).

The strain exhibited by a rock is the tangible evidence for stress. **Strain** is the change in shape or size of a rock as it responds to stress. The change may be temporary or permanent depending upon the amount of stress and the strength of the rock. A rock's strength is the stress needed to cause it either to break or to permanently deform.

In rocks, as in other solids, stress can result in three kinds of strain or deformation. As stress builds, a rock may initially experience **elastic strain**. Like the strain seen in a stretched rubber band, a rock that is elastically strained is potentially capable of returning to its original size and shape. If, however, increasing amounts of stress are applied to a rock, the elastic limit (Fig. 9–2) for the rock will be reached, beyond which it will experience plastic deformation. Plastic deformation results in permanent changes in size and shape (Fig. 9–3). Furthermore, as indicated by the leveling off of the stress versus strain curve, small increases in stress will now cause relatively greater deformation. Finally, the rock will no longer be able to adapt to stress, and it will break.

Every rock has a limit beyond which it cannot continue to respond to stress by bending (strain), and will therefore fracture.

Not all rocks respond to stress by plastic deformation. Some rocks are **brittle**, so that when

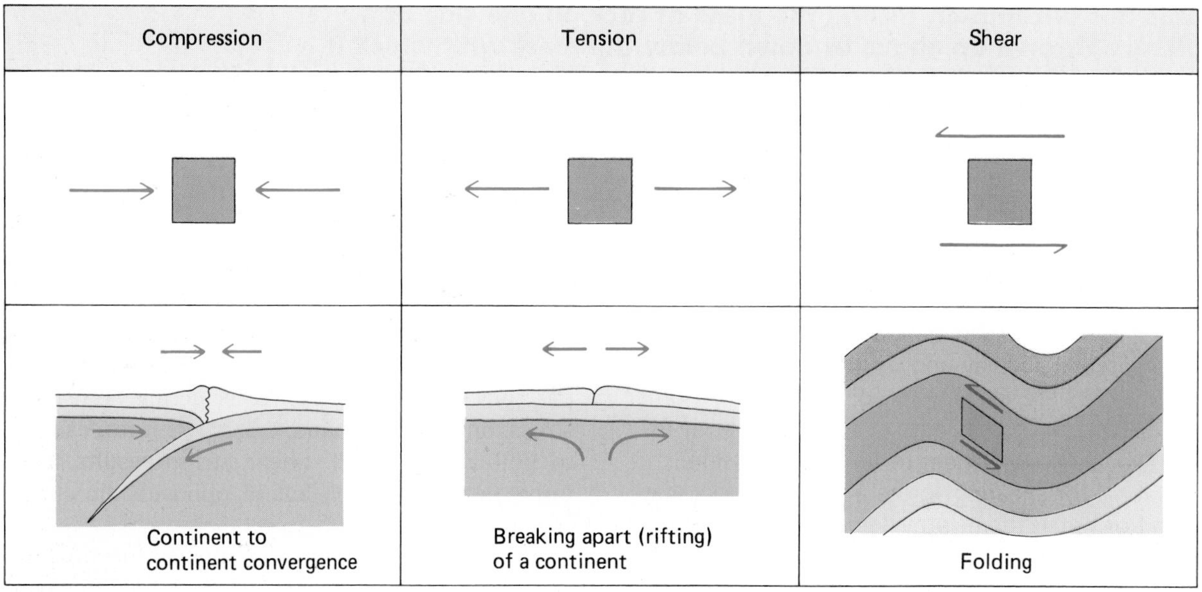

Figure 9–1 Arrows in upper figures represent compression, tension, and shear. Geologic situations where each is likely to occur are depicted beneath each.

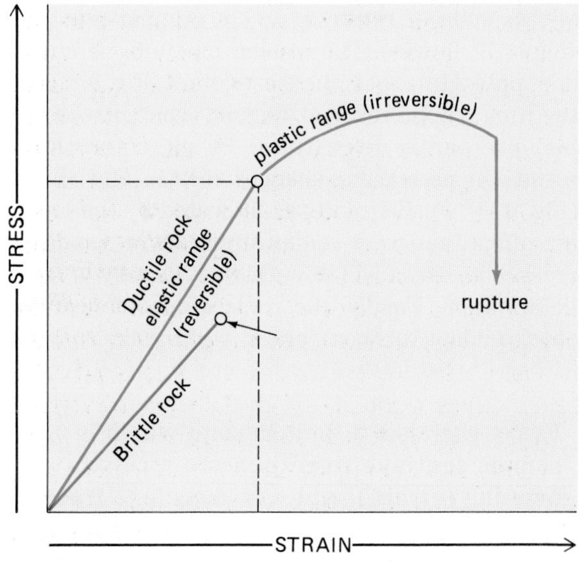

Figure 9–2 Graph of stress versus strain for hypothetical materials.

stress is applied they simply break rather than deform plastically. Other rocks are **ductile** and will undergo considerable smooth deformation before they rupture. It is also possible for a particular kind of rock to behave in a brittle manner when near the surface of the earth (where confining pressures and temperatures are relatively low), and in a plastic manner at depth (where confining pressures and temperatures are high). Thus, temperature and pressure are additional factors governing the way rocks deform. Other factors include the type of rock itself, and the amount of time over which stress is applied. A rock may not respond plastically if stress is rapidly applied, but may undergo extensive plastic deformation in response to low, but long-sustained stress.

Structural Terms

Geologists employ the word **attitude** whenever they wish to refer to the way in which a stratum, fault surface, or other geologic feature is oriented in space (Fig. 9–4). Are the rocks horizontal or tilted? If they are tilted, how much and in what compass direction? The two terms *strike* and *dip* are used to convey answers to these questions.

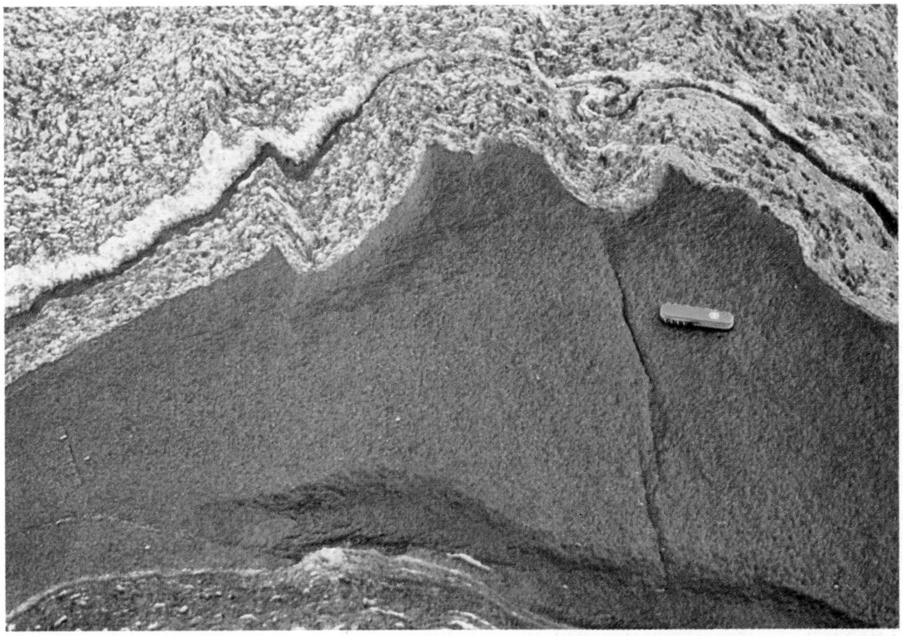

Figure 9–3 Plastic deformation of a once-planar dark basalt dike. The dike intrudes a feldspar-rich gneiss, the Amitsoq Gneiss, near Lile Narssuaq, Greenland. (Courtesy of M.S. Smith.)

Figure 9–4 A geologist measures the amount of tilt in strata exposed along a road cut in Jefferson County, Colorado. (Courtesy of the U.S. Geological Survey.)

Strike is the compass direction of a line formed by the intersection of the surface of an inclined stratum with an imaginary horizontal plane (Fig. 9–5). A moment of reflection on this definition tells us that if one faces toward one end of the line, one will obtain a different compass reading than when facing the other direction. If one end points to the northwest, the opposite end must point to the southeast. In order to ensure uniformity in notation, geologists record only the end of the strike line that makes an acute angle with true north. Thus, the strike notation N.35°W. indicates a stratum having a strike of 35° west of true north.

Dip refers to the maximal angle of slope of a tilted stratum measured directly downward from the horizontal plane. In addition to providing a measure of the steepness at which a bed is tilted, dip also refers to the direction toward which the bed is tilted. The *direction of dip* is perpendicular to the strike. A ball placed on the exposed surface of a dipping bed would roll directly down the bed's direction of dip. In recording the attitude of an inclined stratum on a map, a symbol such as that shown in Figure 9–5 is used to express the strike as well as the angle and direction of dip. In a geologic report, the geologist might abbreviate the attitude of a stratum with the notation N.15°E., 38°S.E.

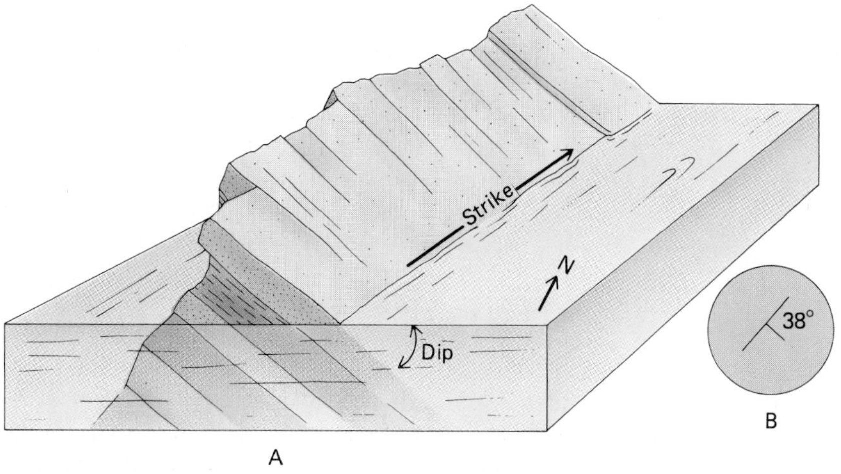

Figure 9–5 Strike and dip of tilted strata. (A) The strike is represented by the line of intersection of a horizontal plane and the inclined (dipping) bed. In this illustration the horizontal plane is represented by the water surface of a lake. (B) The dip direction is perpendicular to the strike. Dip angle is 38°. The symbol for strike and dip is shown in B.

Figure 9–6 Folded dolostone strata of the Rocknest Formation of the Proterozoic age, Northwest Territories, Coronation Gulf, Canada. (Courtesy of Paul Hoffman.)

Readers would then know that the stratum in question had a strike of 15° east of north and was sloping (dipping) at an angle of 38° in a direction toward the southeast.

Strike and direction of dip are measured perpendicular to each other. Thus, a bed that dips either east or west has a north-south strike.

Folds

A **fold** is a bend or flexure in layered rocks (Fig. 9–6). Although folds may occur in igneous and metamorphic rocks, they are more common in beds of sedimentary rock that were originally laid down in relatively horizontal layers. Long ago geologists recognized this simple fact and correctly inferred that if strata were found in other than a horizontal attitude, then crustal movements had affected them. Folding is the most common kind of deformation in layered rocks, and folds are well developed in all the great mountain systems of the globe. They are found in all sizes, from great sweeping flexures measuring many tens of kilometers across to tiny accordian-like crenulations. Inclined parts of folds, exposed here and there as isolated fold remnants (Fig. 9–7), are carefully located on geologic maps. Data collected in this way are used to reconstruct the original appearance and nature of deformation of various parts of the crust. Although folds may have a distinctive outcrop pattern at the earth's surface, they are actually subsurface structures, and do not necessarily correspond to topographic features. A level plain may exist above highly folded strata, a valley above an uparching of beds (Fig. 9–8), or a hill above a downfold (Fig. 9–9).

The principal categories of folds are anticlines, synclines, domes, basins, and monoclines (Fig. 9–10). **Anticlines** are uparched rocks in which the oldest rocks are in the center and the youngest rocks are on the flanks (Fig. 9–11). **Synclines** are downwardly folded rocks that have youngest beds in the center and oldest rocks on the flanks (Fig. 9–12). Folds tend to occur together, rather like a series of petrified wave crests and troughs. The erosion of anticlines and synclines characteristically produces a topographic pattern of ridges and valleys (Fig. 9–13). The ridges develop where resistant rocks project at the surface, whereas valleys develop along zones underlain by more easily eroded rocks (Fig. 9–14).

A special kind of fold in which there is a single steplike bend in otherwise horizontal beds is termed a **monocline** (see Fig. 9–10C). Monoclines may be simple bends in pliable strata caused by vertical displacement of underlying rocks.

The **axis** of a fold is a line drawn along the points of maximum curvature of each bed. In most anticlines, the axis would correspond to the ridge line of a roof. The axis may be horizontal, or it may be inclined, in which case it is said to **plunge**.

(Text continues on p. 227.)

Figure 9–7 Steeply dipping shale and graywacke strata, Ouachita mountains of Oklahoma.

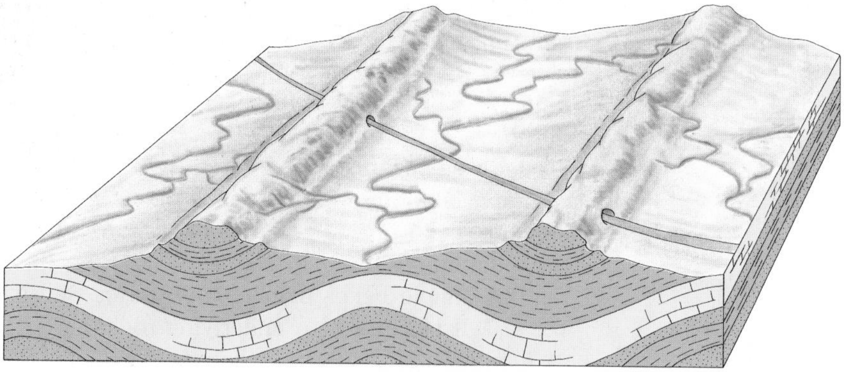

Figure 9–8 Anticlines (upfolded strata) do not necessarily correspond to ridges on the earth's surface. Here the ridges are formed by resistant layers of sandstone that are part of synclinal (downfolded) strata. Such features are common in the folded Appalachians.

Figure 9–9 Geologic structures often have no relationship to surface topography. Here a downfold in beds (syncline) lies beneath a high ridge. Sideling Hill road cut, 6 miles west of Hancock, Maryland, along U.S. 48, Ridge and Valley Province. (Courtesy of K.A. Schwartz.)

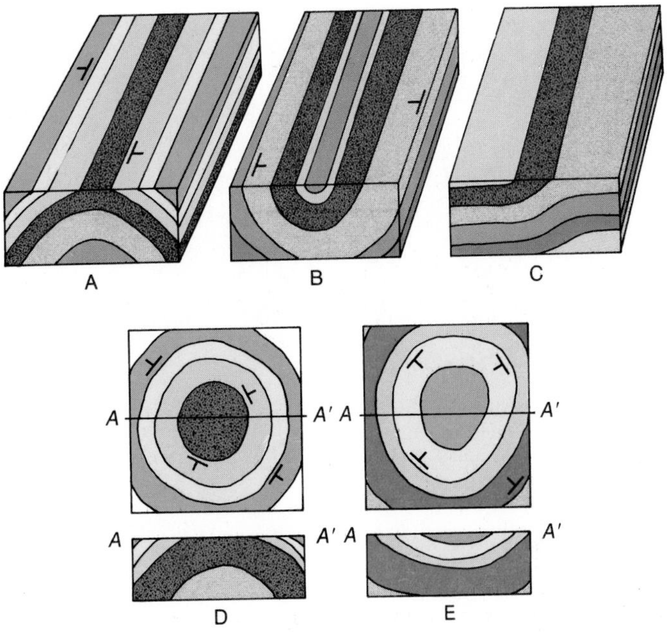

Figure 9–10 Types of folds. (A) Anticline. (B) Syncline. (C) Monocline. (D) Map view and cross-section of a dome. Notice older strata are in center of outcrop pattern. (E) Map view and cross-section of a basin. Younger rocks occur in central area of outcrop pattern.

Figure 9–11 Anticline exposed in the north bank of the Sun River, Teton County, Montana. (Courtesy of R.R. Mudge and the U.S. Geological Survey.)

Figure 9–12 Small syncline in the Ordovician Knox Limestone, Sevier County, Tennessee. (Courtesy of W.B. Hamilton.)

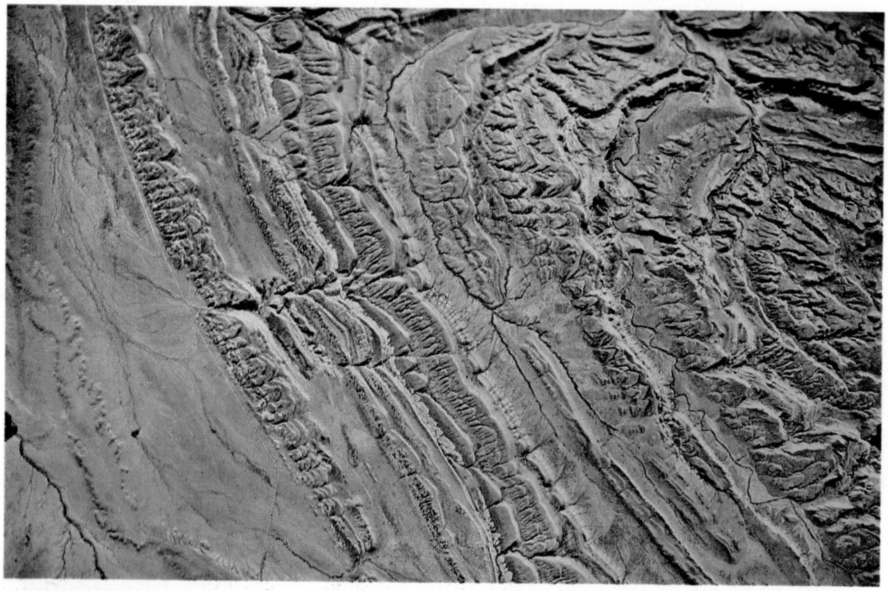

Figure 9–13 An anticlinal fold in Wyoming. Because of the sparseness of vegetation and soil cover, rock layers are clearly exposed and differences in resistance to erosion cause layers to be etched into sharp relief. (Courtesy of the U.S. Geological Survey.)

Figure 9–14 Ouachita mountains of Arkansas. The landscape here has been developed on folded rocks. Because rainfall here is considerably greater than in the area shown in Figure 9–13, minor differences in rock resistance are obscured by soil and vegetation. (Courtesy of the U.S. Geological Survey.)

Some folds plunge in one direction at one end and in a different direction at the other. Such folds are said to be doubly plunging (Fig. 9–15). In topography underlain by non-plunging folds, ridges and valleys tend to be parallel. Zig-zag patterns of ridges and valleys are more characteristic of plunging folds. The **axial plane** (Fig. 9–16) of a fold is an imaginary surface connecting axial lines in successive beds. It may be flat or curved and have any orientation from horizontal to vertical.

In deciphering the configuration and location of a fold, geologists carefully note strike and dip on a base map at as many localities as is feasible (see Fig. 6–58). It is then possible to reconstruct the appearance of the folds prior to erosion. Where rock exposures are sparse, an understanding of where to expect relatively older or younger layers of an erosionally beveled anticline or syncline can be very useful. As illustrated in Figure 9–10, sedimentary beds are necessarily deposited so that the oldest beds are at the bottom of a sequence, and sequentially younger beds occur above the initial layer. If the layers were then arched into an anticline and the top "sliced off" by erosion, the geologist mapping the anticline would encounter successively younger beds when walking from the axis toward the outside of the fold. In the case of a syncline, younger beds would be found along the axis and older beds toward the outside.

Anticlines and synclines are commonly produced by compressional stress.

Folds may vary in symmetry as well as size. The curve of the arch may be broad and rounded or sharp and angular. A **symmetric fold** is one in which the axial plane is a plane of symmetry. Thus the two limbs are mirror images of each other. As implied by their name, *asymmetrical* folds are not symmetrical, and the axial plane is not a plane of symmetry. Compressional forces may push a fold over and onto its side so that the axial plane is inclined and both limbs dip in the same direction but at different angles. Such structures are called **overturned folds** (Fig. 9–17). If the overturning is so complete that the axial plane is horizontal, then the fold is called **recumbent**. During the formation of the Alps, immense recumbent folds were developed. Large and complex recumbent folds of the Alpine type are also called **nappes**.

Domes and **basins** are similar to anticlines and synclines, except that they have elliptical to roughly circular outcrop patterns. Many resemble short, doubly plunging anticlines. A dome (Figs. 9–18 and 9–19) consists of uparched rocks in which the beds dip in all directions away from the center point. In contrast, a basin is a downwarp in which beds dip from all sides toward the center or bot-

(Text continues on p. 230.)

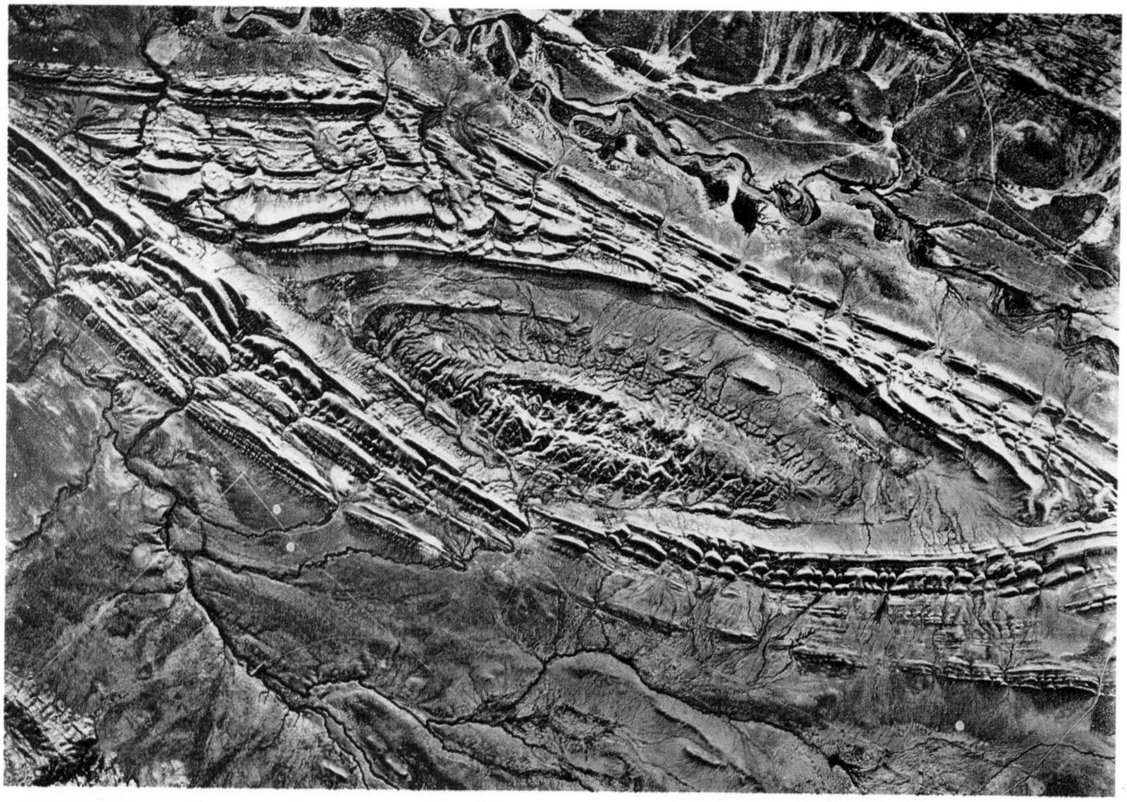

Figure 9–15 Vertical air photo of a doubly plunging anticline. Note how the beds dip away from the axis on the limbs. Big Horn basin, northern Wyoming. (Courtesy of the U.S. Geological Survey.)

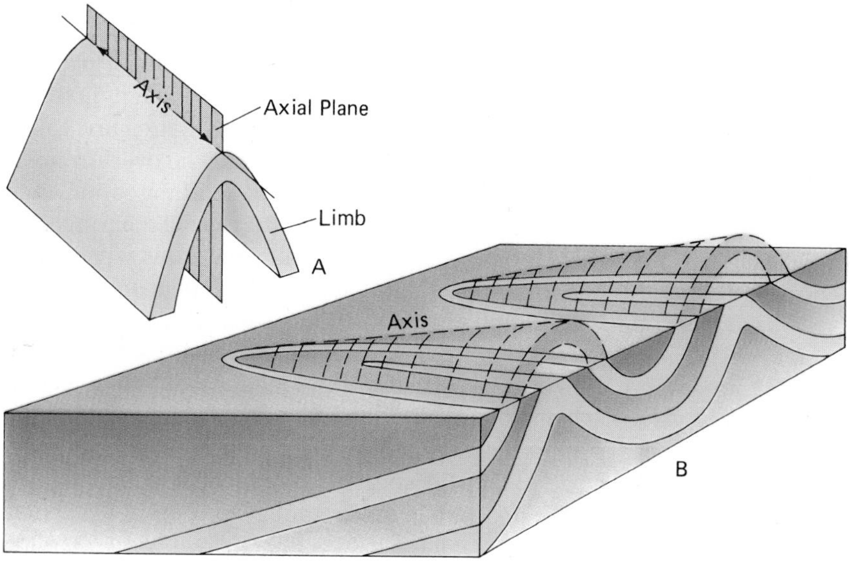

Figure 9–16 (A) Diagram of an anticline illustrating nomenclature of *axis, axial plane,* and *limb.* (B) Map pattern of plunging folds idealized on a level surface. These folds have an inclined axis.

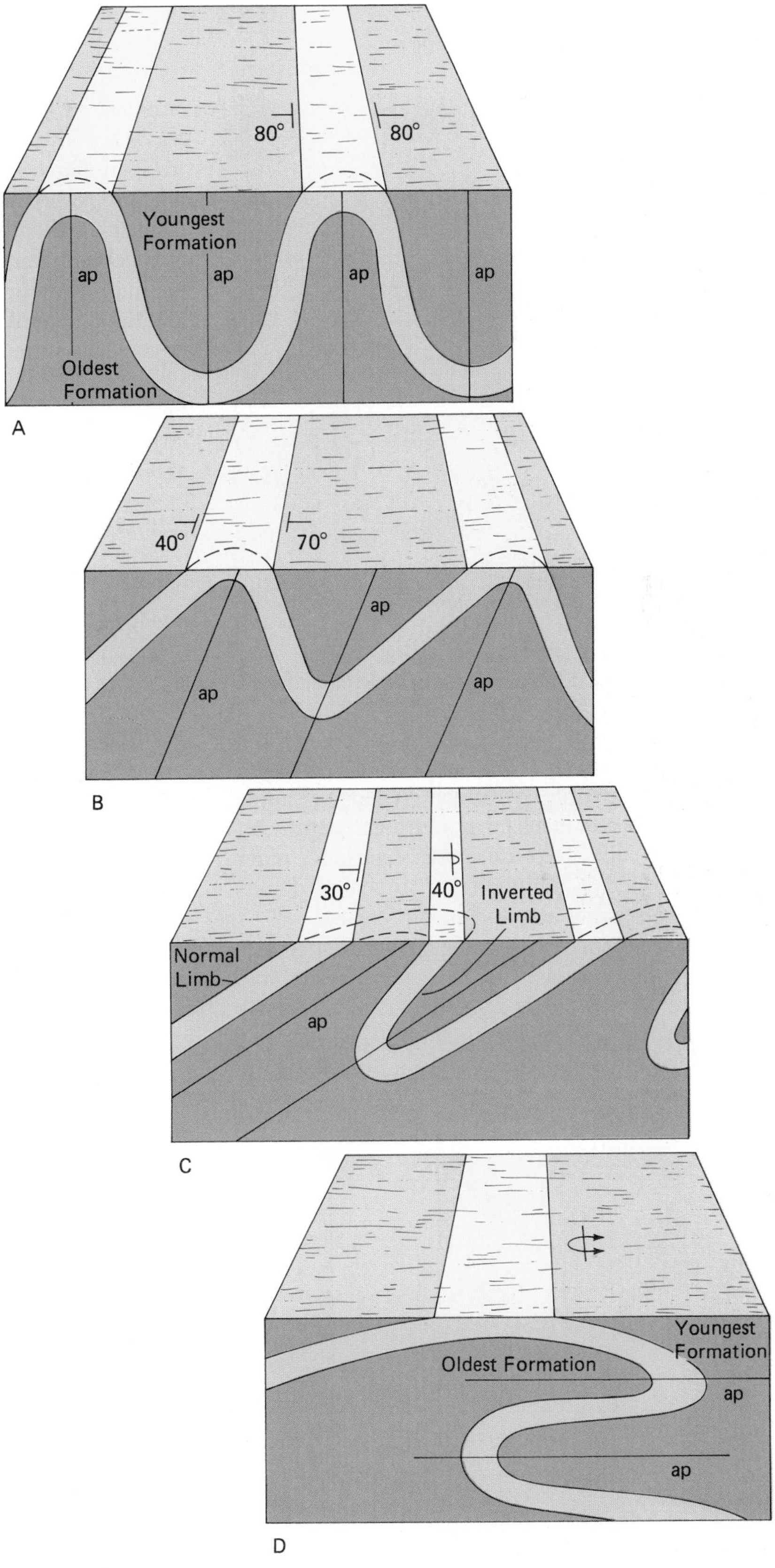

Figure 9–17 Four varieties of folds. (A) Symmetric folds (B) Asymmetric folds. (C) Overturned folds. (D) Recumbent folds. (ap = axial plane.)

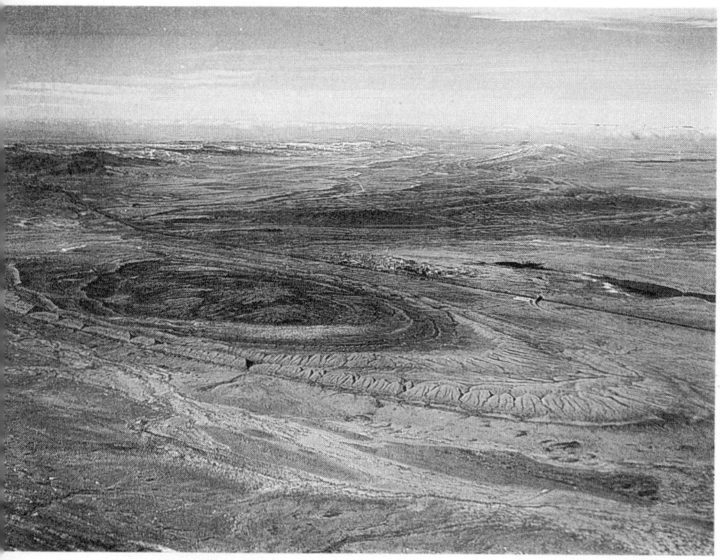

Figure 9–18 Structural dome at Sinclair, Carbon County, Wyoming. (Courtesy of the U.S. Geological Survey and J.R. Balsey.)

tom of the structure. The truncated beds of simple domes and basins may form a circular pattern of ridges and valleys. At the earth's surface, older beds are found in the center of truncated domes and younger strata at the center of basins.

Folds are mapped by geologists in order to better understand the past geologic events of a region. There are, however, more practical reasons for determining the precise locations and geometry of folds. The attitude of strata has an influence on the stability of the bedrock on which buildings and roads are constructed. Further, folded rocks may trap supplies of petroleum and provide underground reservoirs of artesian water. Geologists tracing the locations of coal seams and other mineral deposits must understand the pattern of folding in order to predict the presence of resources at locations where erosion, younger beds, or soil cover has obscured their presence (Fig. 9–20). Cross-sections and structure contour maps are indispensable for this work. In the preparation of a structural contour map, the elevation of a key bed is obtained from surface exposures or the records of wells. These elevations are plotted at their correct locations on a base map, and structure contour lines are drawn through points of equal elevation (Fig. 9–21). The result is a map that shows the position of the key bed relative to sea level across the map area.

Faults

Recognition of Faults

Faults are fractures along which there has been movement (Fig. 9–22). That is, adjacent rock masses have slipped past one another in response to tension, compression, or shearing stress. The various categories of faults are named according to the relative directions in which slippage occurs. Geologists rely upon offsets in strata, veins, or other geologic structures in order to determine the relative directions of movement. For those faults

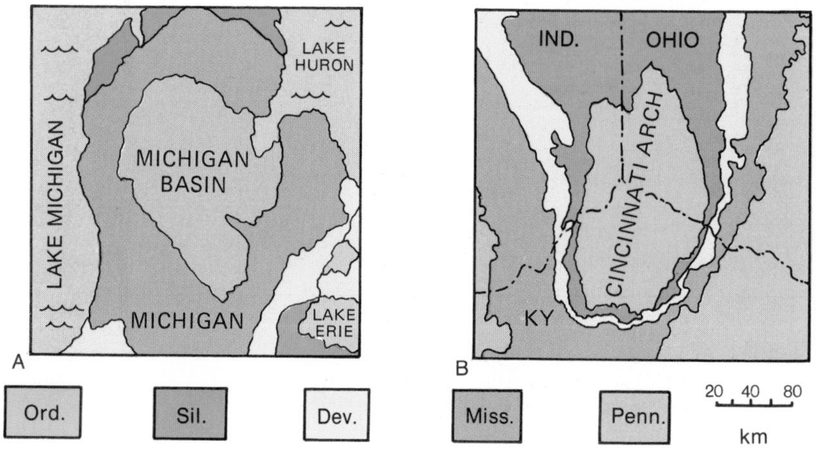

Figure 9–19 In an erosionally truncated basin like the Michigan Basin (A), youngest beds are centrally located. In a dome-like structure such as the Cincinnati Arch (B), oldest beds are located in the center. (Compare with Figure 9–10 D and E.)

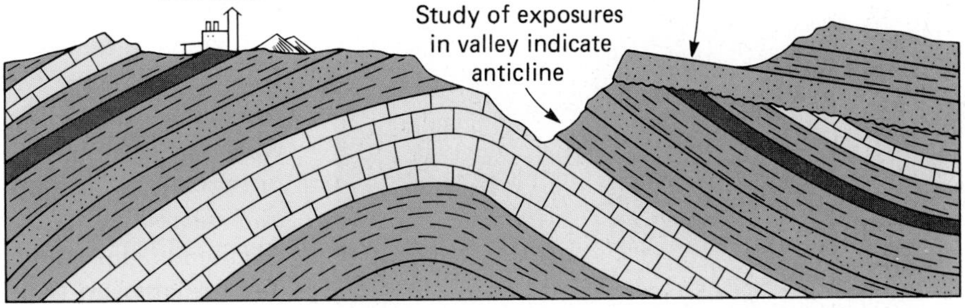

Figure 9–20 Knowledge of the pattern of folding in an area is useful in the exploration for coal beds and layered ore deposits. The geologist, noting angle and direction of dip and superpositional sequence of strata, can predict the location of coal seams or mineral deposits where they are not exposed at the surface.

that have experienced recent movements, small escarpments (fault scarps) may show where one segment of the surface moved upward relative to the other (Fig. 9–23). There may also be displacements of physiographic features such as ridges and stream channels (Fig. 9–24).

The plane of dislocation along which movements occur during faulting is called the **fault plane**. As rock masses grind past one another along the fault plane, they produce smooth surfaces with parallel scratches called **slickensides** and pulverized rock material known as **fault gouge**.

Categories of Faults

Fault planes may sometimes provide permeable zones along which hot mineralized water solutions

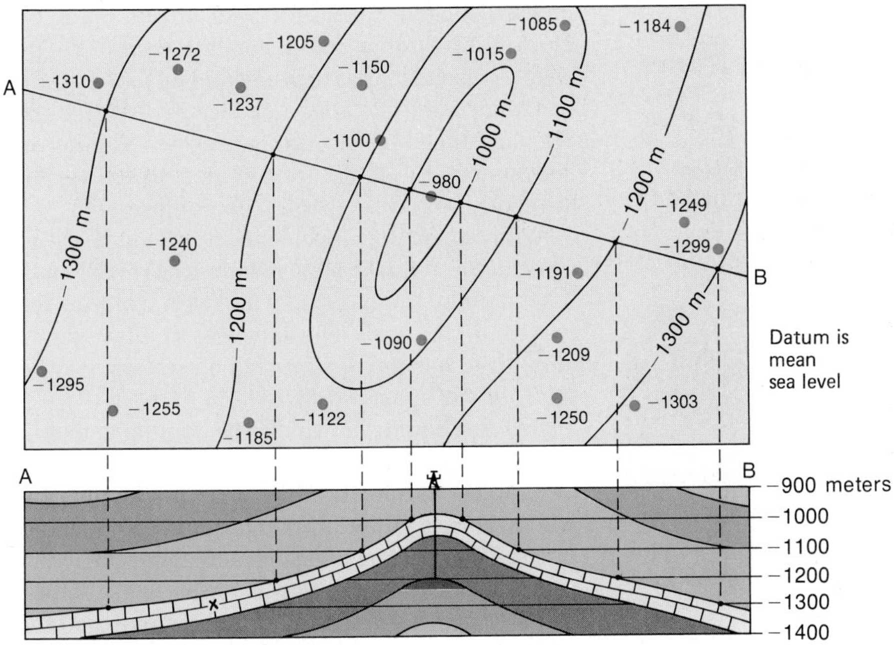

Figure 9–21 Structure contour map *(above)* of deeply buried layer of limestone (x). Both the map and the A-B cross-section *(below)* indicate the strata are folded into an anticline (up-arch) and that the crest of the anticline would be encountered about 980 meters below sea level.

Figure 9–22 Closeup view of the edge of two fault planes that cut through beds of sandstone and siltstone near Dry Gulch, Idaho. Beds on the right have slipped downward relative to beds on the left. Note the small splinter fault just to the left of the more prominent fault. The offset in this normal faulting is indicated by the bed marked X, which is offset to X'. (Courtesy of the U.S. Geological Survey.)

Figure 9–23 When movement along a fault having vertical displacement is relatively rapid, erosion may sculpture the scarp into triangular facets, such as the prominent one seen here north of Bitterwater, California. (Courtesy of J.C. Brice.)

can penetrate and may, in relatively rare instances, deposit valuable ores. Long ago, miners tunneling along sloping fault planes helped to provide a nomenclature that even today serves to identify certain kinds of faults. The side of the fault on which the miners stood was appropriately dubbed the **footwall**. Above their heads was the **hanging wall**. These terms are put to work in the definition of two major categories of faults, namely, *normal and reverse faults.* In a **normal fault** (see Fig. 9–22) the hanging wall appears to have moved downward relative to the footwall (Fig. 9–25). Normal faulting occurs most frequently in rocks that have been subjected to horizontal tensional forces—that is, forces that tend to stretch the earth's crust. A system of normal faults with parallel fault planes may have topographic expression as linear uplands separated by troughlike valleys (Fig. 9–26).

The topographically depressed down-faulted segments that are bounded by normal faults are termed **grabens** after the German word for grave, and the high-standing blocks that generally stand above the grabens are called **horsts**.

In **reverse faults**, the hanging wall has moved upward relative to the footwall. If one holds two blocks of wood cut like those in Figure 9–25, it is easy to visualize that they may be pushed together or compressed in order to cause reverse faulting. Similarly, regions of earth's crust containing numerous reverse faults (and folded strata as well) are likely to have experienced compression.

With regard to movement along faults, we are referring to *relative movements.* In a reverse fault, for example, the hanging wall has moved up relative to the footwall. The same result, however, occurs if the hanging wall remained stationary while the footwall moved down.

Reverse faults in which the fault plane is inclined 45° or less from the horizontal plane are termed **thrust faults** (see Fig. 9–25). In other words, thrust faults are simply low-angle reverse faults. They may have displacements of many kilometers. Such thrusting may carry immense slabs of older crustal rocks great distances, so that they come to rest on younger rocks. Erosion may remove most of the overthrust sheet, leaving behind isolated erosional remnants called **klippen** (singular *klippe;* Fig. 9–27). In some thrusts, the upper plate is thin and composed of relatively flat rocks.

(Text continues on p. 235.)

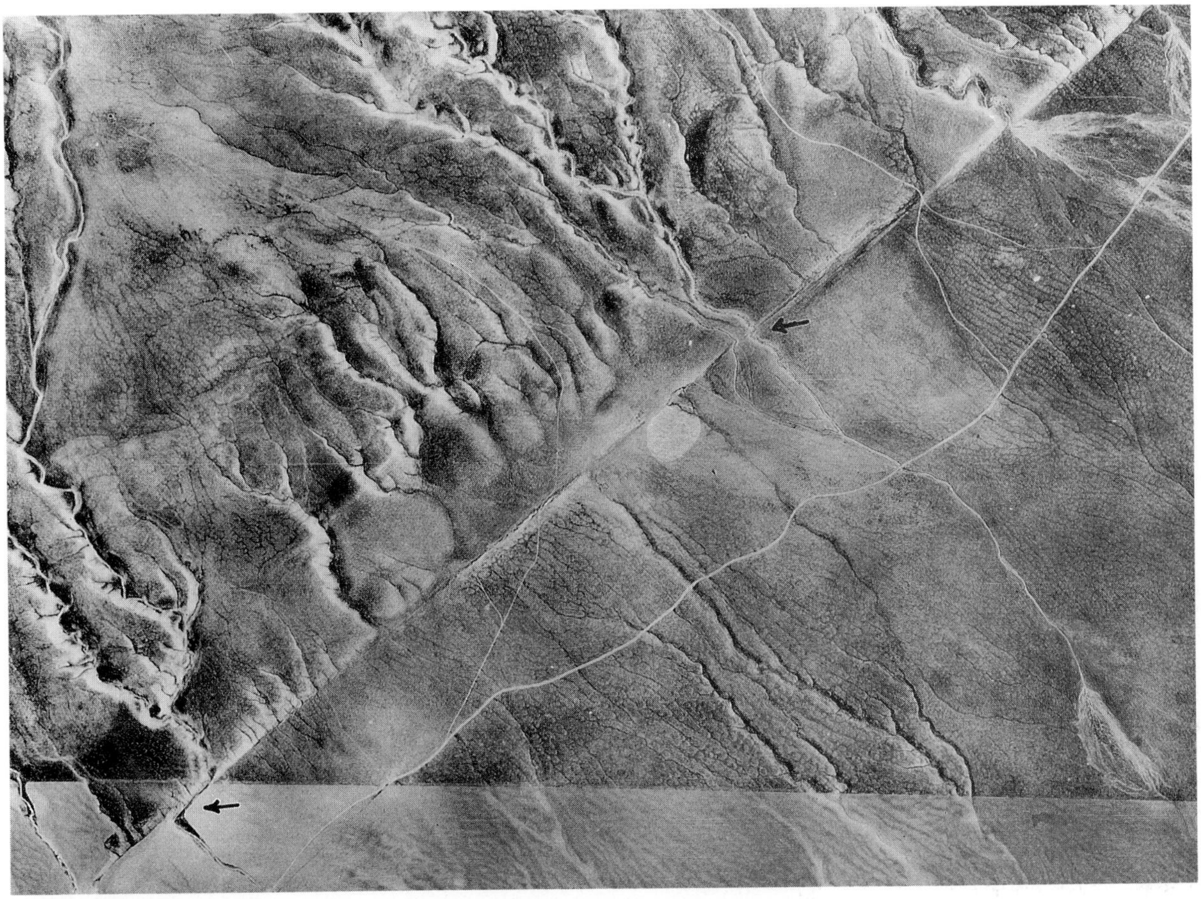

Figure 9–24 Part of an air photo mosaic showing offset streams (arrows), scarps, and trenches along the trace of the San Andreas fault, San Luis Obispo County, California. (Courtesy of the U.S. Geological Survey.)

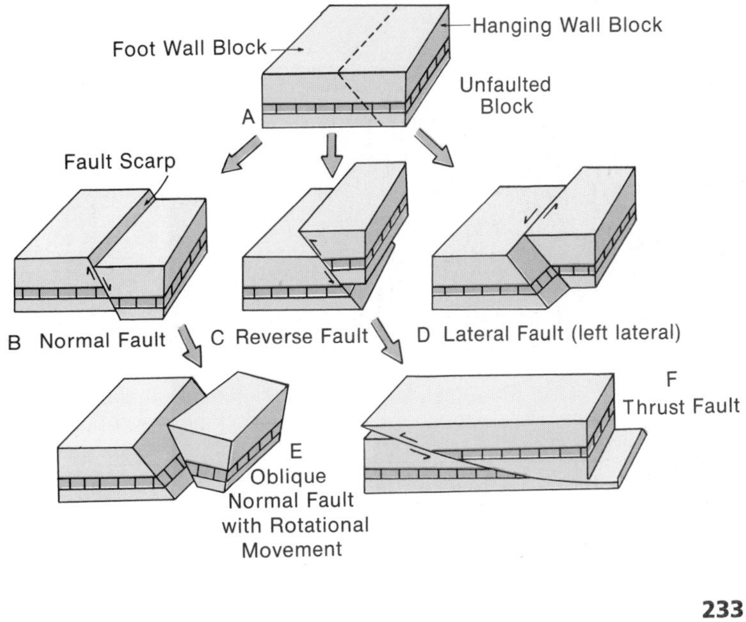

Figure 9–25 Types of faults. (A) The unfaulted block with the position of the potential fault shown by dashed line. In nature, movements along faults may vary in direction, as shown in (E). A thrust fault (F) is a type of reverse fault that is inclined at a low angle from the horizontal.

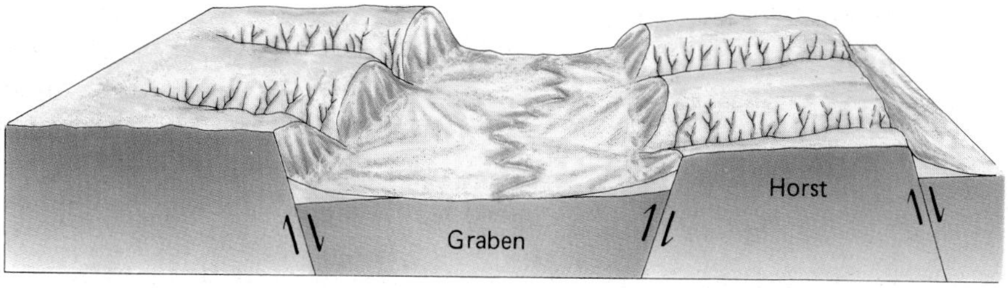

Figure 9–26 Graben and horst topography. Note the underfit stream meandering across the broad, fault-controlled valley.

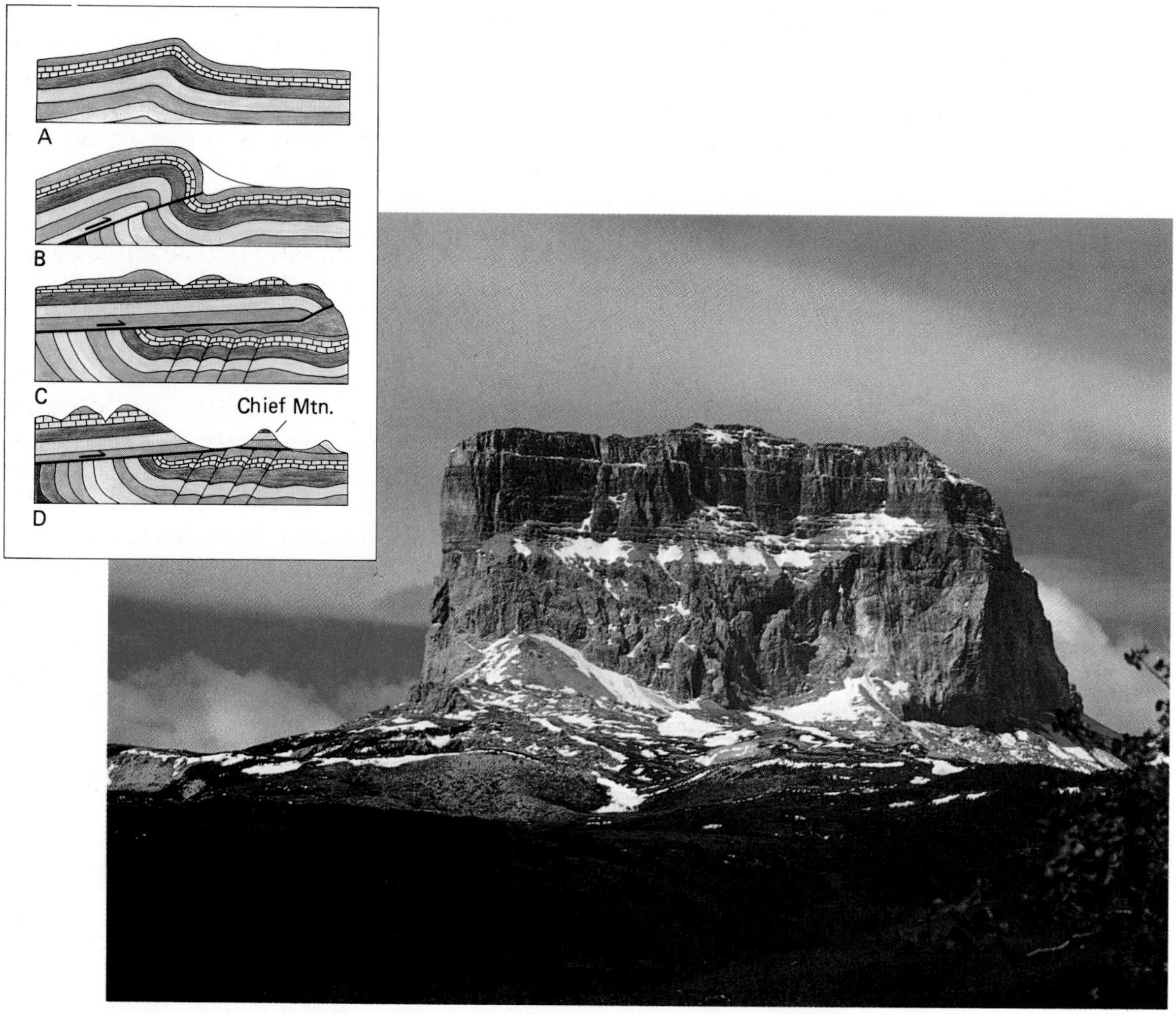

Figure 9–27 Chief Mountain, Glacial National Park, Montana, is a well-known example of a *klippe.* Here a mass of Precambrian rock has been pushed along a thrust fault over a sequence of Cretaceous strata. The fault, called the Lewis overthrust, has a horizontal displacement of 24 km. The inset illustrates the sequence of events. (A) Compression begins. (B) Anticline is overturned and thrust eastward. (C) Thrust has moved its maximal distance. (D) Erosion removes most of overthrust slab, leaving Chief Mountain as a klippe. (Courtesy of the U.S. National Park Service.)

Other thrusts are formed during extreme folding, in the course of which the lower limb of an overturned fold breaks and the fold is pushed out as a thrust fault. In the Alps and Himalayas such overthrust folds occur on a spectacular scale.

Another type of faulting involves primarily horizontal movements along the strike of the fault plane. Such features are called **strike-slip faults**, although the informal term *lateral fault* is also sometimes applied. The fault plane in a strike-slip fault is usually steep, and the displacement along the fault planes may be tens or even hundreds of kilometers. The San Andreas fault is a strike-slip fault. It is over 1000 km long, and the total lateral movement along the fault has been estimated by geologists to be over 500 km. Displacement on strike-slip faults is designated either right lateral or left lateral. In order to decide which kind is being observed, one must simply look across the fault zone to see if the opposite block moved to the right or left. The San Andreas in California is a right lateral fault. Of course, there is nothing in nature that restricts fault movements to either a purely vertical or horizontal direction, and so there are many faults that are formed as a result of oblique and rotational movements. For example, an oblique-slip fault exhibits displacement obliquely across the fault plane. Rotational faults occur where one of the fault blocks turns or rotates in a scissorlike motion against the other.

Reverse and thrust faults usually result from compressional stress, normal faults from tensional stress, and strike-slip faults from either.

Joints

Rocks at or near the earth's surface are nearly always broken by cracks or fractures. Such fracturing tends to break a rock mass into a network of blocks (Fig. 9–28). If there has been little or no slipping of one block relative to its neighbor, then the crack between the blocks is termed a **joint**. Joints occur most frequently as a consequence of either crustal tension or shearing forces. Tensional joints may develop in rocks whenever they experience a shrinkage in volume as a result of the loss of moisture or heat. Usually, the contraction takes place around many centers within the rock mass, and fractures develop at right angles to the directions of tension. Everyone has seen such tensional joints developed where layers of mud have dried out. The joint pattern is polygonal (many-sided). Earlier we noted that polygonal jointing is also characteristic of basaltic lava flows, sills, and dikes. More coarsely crystalline rocks such as granite also develop jointing. Although in many granite bodies this jointing appears to be related to cooling, it is evident from the increase in the number of joints near the surface of granite outcrops that jointing may result from the relief of pressure that occurs when a weighty burden of overlying rock has been removed by erosion.

Joints differ from faults in having little or no movement of rocks parallel to the fracture.

Quarrymen use joints in rock to facilitate the removal of quarry stone. Joints also affect the movement of underground water and must be considered in the design of dams. In the geologic past, these natural breaks in rocks have controlled the movement of ore-bearing hydrothermal solutions.

Figure 9–28 Vertical joints exposed on the surface of a limestone bed. The two sets of joints developed here intersect at nearly right angles. (Courtesy of B. Stinchcomb.)

Mountains and Their Origin

Geologic structures are found in all kinds of geographic settings, but they are particularly common in the earth's great mountain belts (Fig. 9–29). This is not unexpected, for most mountains are the magnificent end products of the same deformational forces that break rocks and bend them into anticlines and synclines.

Mountains are land masses of irregular relief that stand conspicuously above their surroundings. There are many kinds of mountains. The simplest are *volcanic mountains*, which result from the accumulation of lava and pyroclastics around a volcanic vent. If an area happens to be underlain by rocks that are particularly resistant to erosion, that area may develop into residual mountains (Fig. 9–30), while surrounding areas underlain by softer rocks are eroded away more rapidly. Some mountains are the result of high-angle fault displacements (Fig. 9–31). The mountains of the Basin and Range Province of the western United States are **fault-block mountains** of this kind (Fig. 9–32). Some of these north-south trending ranges are over 1500 meters high (Fig. 9–33).

Mountains in which the predominant structures are folds are called **fold mountains** (Fig. 9–34). These constitute the major mountain belts of the world. They are not, however, composed exclusively of folds. Faulting of all kinds, volcanoes, and igneous and metamorphic core complexes may all occur within fold mountain ranges.

Folds are the predominant geologic structures in the world's major mountain systems.

The Plate Tectonic Model of Mountain-Building

In order to understand how mountain belts develop, it will be useful to imagine a time about 200 million years ago when North America was part of an immense supercontinent. In response to the forces that drive tectonic plates, a great rift developed in the supercontinent, and North America

(Text continues on p. 239.)

Figure 9–29 Multiple anticlines visible near summit area of Northern Caribou mountains, British Columbia. Rocks are mostly metamorphosed graywacke. The anticlines are part of a larger uparching, termed an anticlinorium (Courtesy of G. Ross.)

Figure 9–30 The St. Francois mountains of Missouri are residual mountains composed of granitic and rhyolitic rocks that are more resistant to erosion than the sedimentary rocks that surround and once covered them. A major component of the elevation of these mountains results from uplift.

Figure 9–31 The magnificent Teton mountains have been sculptured by erosion from a crustal block recently uplifted along a near-vertical fault. Several of the peaks would challenge even the most experienced mountaineers.

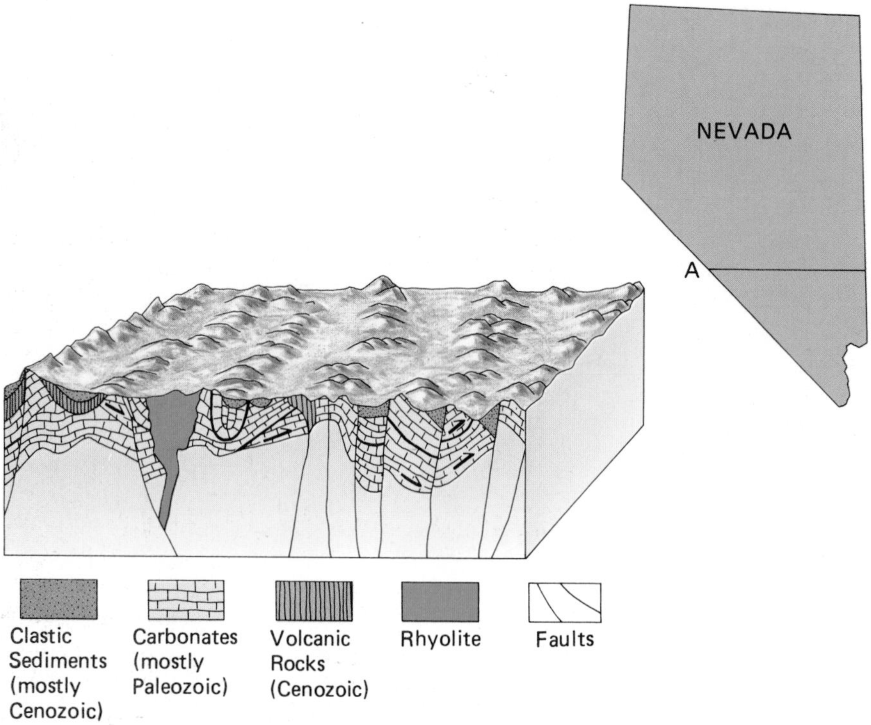

Figure 9–32 Geologic section across the Basin and Range Province along line A——A′ in southern Nevada.

Figure 9–33 The Nopah Range, visible in the distance, is one of many ranges that alternate with basins (foreground) in the Basin and Range Province.

began to drift slowly westward. As it moved away, the ocean filled the widening space left behind. Along the trailing margin of North America, sediment accumulated on the continental shelf and rise (Fig. 9–35). Relatively shallow water sandstones, shales, and limestones were laid down on the shelf, whereas the deeper continental rise area experienced deposition of siliceous sediments and turbidites (graywackes). Today we find such rocks among the loftiest peaks of mountain ranges, yet they originated below sea level. The events leading to their elevation and their deformation involved a change in the pattern of moving plates. As has frequently occurred in the geologic past, a passive continental margin with its load of sediment may be confronted by another plate that converges with it (Fig. 9–36). The accumulation of sediment deposited on the passive margin would then be compressed as if in a huge vise. This compression, as well as subduction of the opposing plates, would result in melting at depth, the development of volcanic arcs, and extensive metamorphism.

A plate tectonics model such as the one described above appears to explain most of the structural and sedimentological features of ranges like the Appalachians. Figure 9–37 shows that the sequence of events begins with the accumulation of thick bodies of sediment on the rear, or trailing, edges of diverging plates. The modern analogy is the continental shelf of eastern North America. The ocean basin widens as the continents drift apart. At a later period, the movement of crustal plates is reversed, the ocean narrows, and is ultimately squeezed out of existence as the plates collide and crumple the accumulated sediment. As suggested by the geologic history of eastern North America, the plates may separate again, reestablishing ocean tracts.

The events described above can be deciphered from the rocks of most of the earth's great mountain ranges. The sequence has been termed a **Wilson cycle**, as a tribute to J. Tuzo Wilson, one of the pioneers in developing plate tectonics theory. A Wilson cycle begins with the opening of an ocean basin by continental rifting (and rafting), followed by closing of that same ocean basin accompanied by subduction of oceanic lithosphere along continental margins. It may also culminate in continent-to-continent collision.

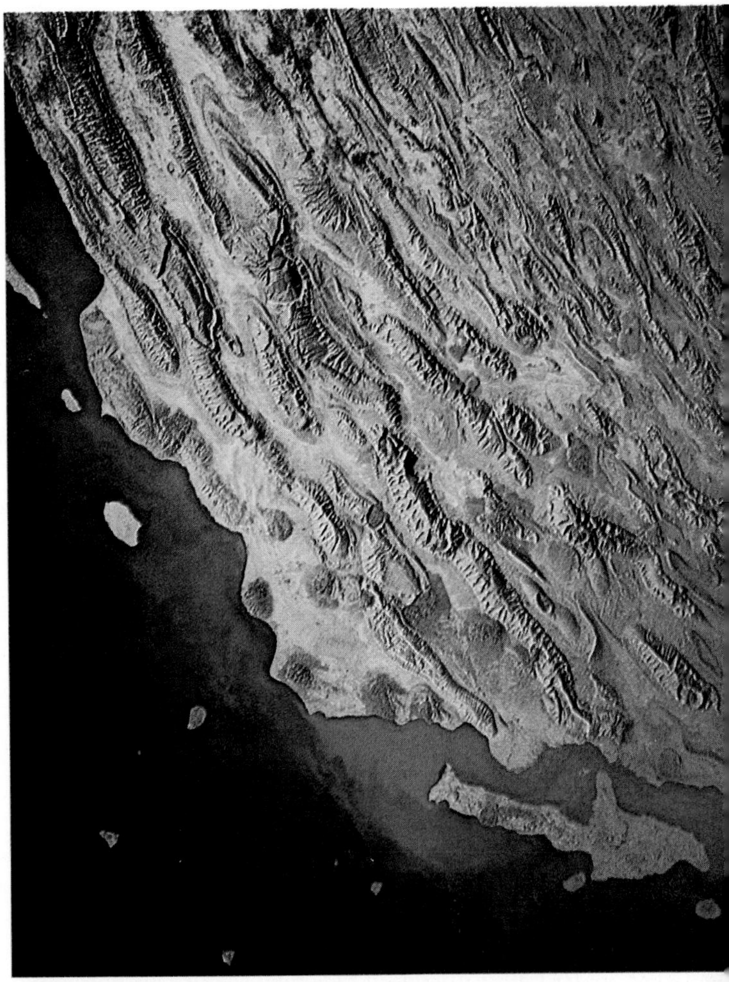

Figure 9–34 Gemini-3 photograph of the Zagros fold belt east of the Persian Gulf near the Straits of Hormuz, southern Iran. These mountains formed during the collision of Arabia and Iran only about 10 million years ago. The anticlines (and synclines that are now masked by alluvial sediments) are doubly plunging. The dark circular areas are salt domes formed from Precambrian and early Paleozoic salt layers that pierced overlying strata. (Courtesy of NASA.)

Isostatic Adjustments

From the moment strata are raised above sea level by mountain-building forces, and continuing for hundreds of millions of years thereafter, the relentless processes of weathering and erosion work to reduce the elevation of mountains. In time great volumes of soil and rock are removed, and the underlying mantle is relieved of its mountain-

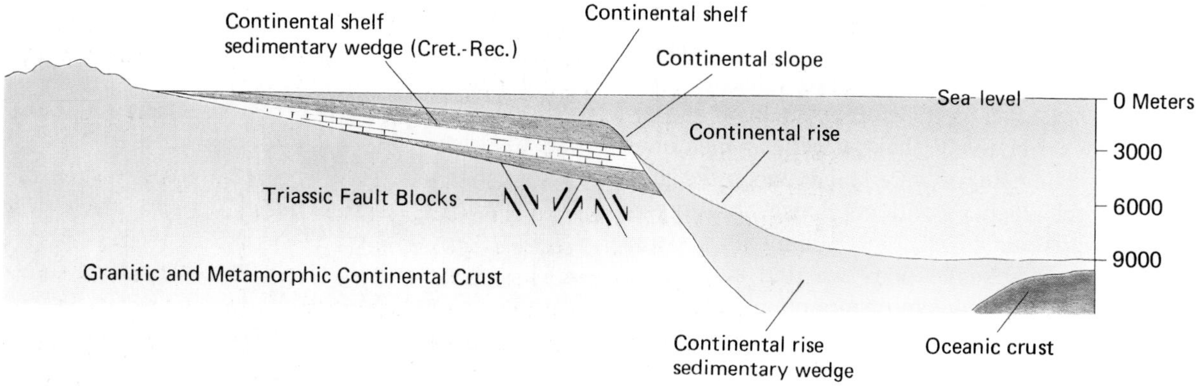

Figure 9–35 Sediments accumulating on the passive eastern coast of the United States (vertical scale exaggerated).

ous burden of rock. In response, the crust gradually rises, rather like a boat that rises in the water as its cargo is unloaded. In this way, mountains may continue to experience uplift and assume new positions of gravitational balance long after their orogenic climax has occurred.

The periodic uplift and denudation make it possible for geologists to view the once deeply buried interior of an ancient mountain range. Eventually, there is insufficient lightweight "root" remaining for further upward adjustments, and the old mobile tract becomes a stable part of the continental platform.

Of course, the passage of each stage in the history of a mountain belt may require an incredible length of time. The passive margin that became the Appalachian Mountains received sediment for over 300 million years. The orogenic phase, however, is relatively short, comprising only a few tens of millions of years. Longest of all is the erosional finale, which may continue over a span of several hundred million years.

The example of a boat rising in the water as it is unloaded provides a rough analogy to a geologic condition known as isostasy. **Isostasy** refers to the condition of balance that exists among segments of the earth's crust as they come into flotational equilibrium with denser mantle material. Those parts of the earth's crust that are thicker and lighter will stand at higher elevations, and those that are thinner and heavier will lie at lower levels. Seismic data and gravity calculations indicate that the masses of continents are thicker and less dense than oceanic crustal materials. Hence, the continents stand higher than the ocean basins. A little "bathtub geology" will help to illustrate the concept. Assume you have a large cube of a low-density wood such as pine (representing a continent) and a smaller cube of heavy oak (representing the oceanic crust). When placed in a tub of water, the pine, of course, will float higher. Next, if one places an identical second block of pine precisely on top of the first, the pine blocks will sink somewhat, but the vertical difference between the upper surfaces of the pine and oak cubes will be greater than before. Thus, both the density of the block of material and its size (thickness) are important in determining the level of flotational equilibrium. Imagine further that the upper pine block represents a mountainous region of the continent. Lift the upper pine block off the lower, and notice that the lower block will rise to a new level. By crude comparison, we might predict a rise in a mountain range as erosional processes remove the "upper block" and upset the gravitational balance.

The geology of mountain systems indicates that their history began with a depositional phase, followed by an orogenic phase, and culminated in a long period of erosion and isostatic adjustment.

Of course, mountainous regions would not experience repeated uplifts were it not for the fact that a large part of their mass extends downward

(Text continues on p. 243.)

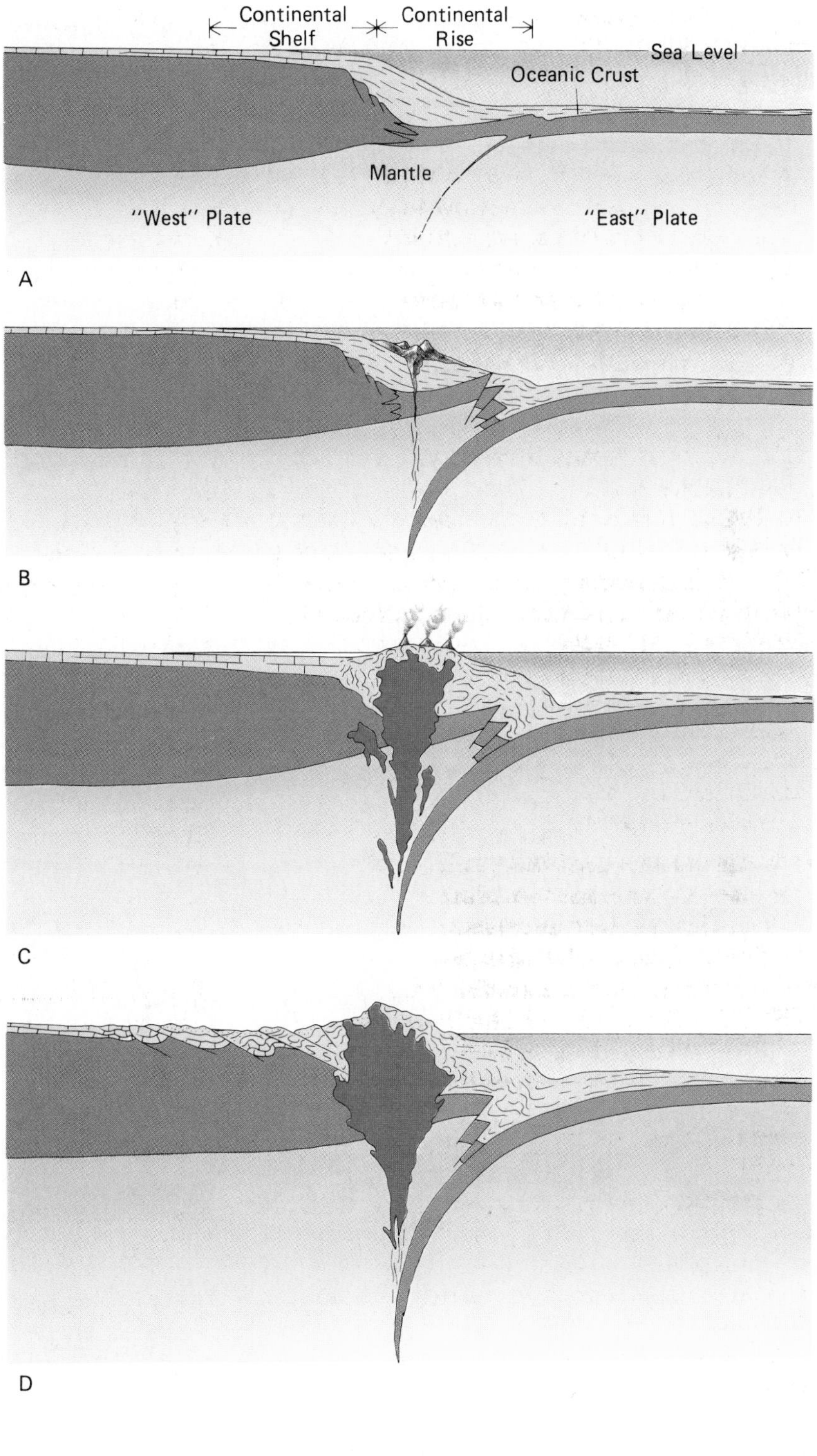

Figure 9–36
Plate-tectonics model for the origin of a mountain belt. (A) Trailing margin of a continent sags and receives thick accumulation of sediments as it drifts to the left. Carbonates on continental shelf grade seaward to graywackes and shales and finally to deep-sea sediments. (B) Lithospheric plate bearing oceanic crust collides and is subducted beneath plate-bearing continent. When subducted plate has descended to about 100 km, basaltic-type lavas, generated by melting of the lithosphere, rise and are extruded as submarine volcanics. A trench begins to form in the zone of subduction. (C) As subduction continues, granitic-type magmas, formed from molten subducted materials, rise. Their buoyancy and heat cause doming at the surface, and this doming is accompanied by metamorphism and deformation. Graywackes accumulate adjacent to the rising mountain chain. (D) Subduction and orogeny continue. Sheets of folded and metamorphosed rocks are thrust faulted toward the craton, and new granites are welded into the older continental crust. (Modified from Dewey, J.F., and Bird, J.M. Mountain belts and the new global tectonics. *J. Geophys. Res.,* 75: 2625–2647, 1970. Copyright, The American Geophysical Union.)

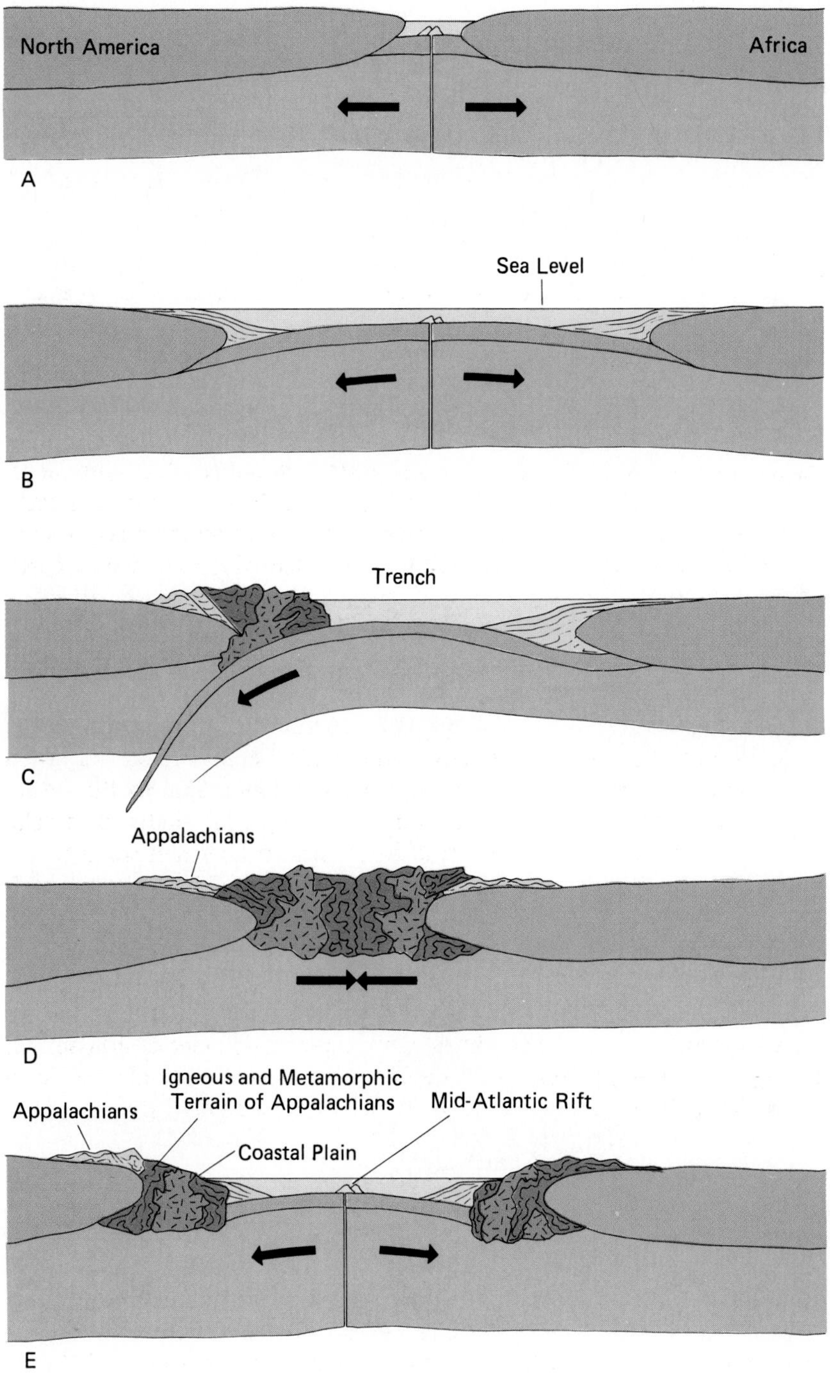

Figure 9–37
Plate-tectonics model for events leading to the formation of the Appalachian mountains and present Atlantic Ocean, as proposed by Robert S. Dietz. (A-B) Continents separate and ancestral Atlantic Ocean is inserted; sediment accumulates on margins of continents. (C-D) Plates converge, subduction zone is formed, and the geocline is collapsed. Continents are sutured together. (E) Rifting occurs again, and new deposits form along continental margins. (After R. Dietz, 1972.)

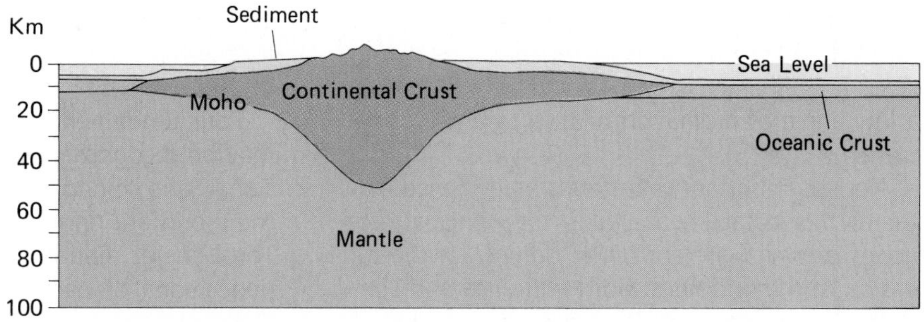

Figure 9–38 According to the theory of isostasy, the light-weight continental crust (mean density about 2.9 g/cc) is in flotational equilibrium with the denser (3.3 g/cc) rather plastic mantle below. Seismology indicates that the continental crust varies in thickness, being thicker under great mountain ranges than under lowlands.

below the average level of the crust (Fig. 9–38). Mountains have "roots," somewhat like icebergs that have most of their mass below sea level. Several independent lines of evidence confirm the existence of mountain roots. One historically important clue came from the unexpected behavior of a plumb bob during an 1854 triangulation survey of India. A plumb bob is a pointed weight on a string. When freely suspended, the bob points approximately to the center of the earth. If, however, prominent hills or mountains are nearby, the plumb bob will be pulled by gravity toward these highland areas. An important discovery of the survey of India was that the deviation from vertical of the plumb bob suspended over the plain at the foot of the Himalayan Mountains was appreciably less than it should be if the mountain mass was simply a load heaped on the earth's surface. Indeed, the deflection of the plumb bob could be adequately explained only if mountains had deep roots of density similar to that of their exposed rocks and projecting downward into a subcrustal material of higher density. This interpretation was confirmed by measurements of the value of gravity over mountains and adjacent lowlands and by experiments on the strength of rocks, which clearly indicated that a crust of uniform thickness could not possibly support the weight of a great mountain system.

Just as isostasy operates *within* a continent, it also operates within the ocean basins. As noted earlier, new oceanic crust is added to the ocean floor at spreading centers. This material is relatively hot and hence less dense than when it has cooled. As a result, near spreading centers (mid-oceanic ridges), the surface of the lithosphere stands at a higher elevation than it would if cold. As it moves away in the course of sea floor spreading, it slowly cools, and thereby becomes more dense. By the time a given segment of lithosphere arrives at a subduction zone, it may already have experienced considerable subsidence.

Summary

The crust of the earth is not the stable *terra firma* it sometimes seems to be. We are reminded of this fact by earthquakes as well as the many occurrences of rocks that have been severely deformed and broken by compressional and tensional forces. These forces bend rocks if they are "bendable" or ductile, and break them if they are brittle. The result is geologic structures such as faults and folds.

Faults are breaks in crustal rocks along which there has been a displacement of one side relative to the other. According to the directions of that movement, faults are classified as *normal, reverse (thrust),* or *lateral.* Intermediate types include *rotational* and *oblique faults.* Normal faults result from tensional forces and are recognized by the apparent downward movement of the hanging wall relative to the footwall. In reverse faults, the hanging wall appears to move up relative to the footwall. A thrust fault is merely a variety of reverse fault characterized by a fault plane that has a low angle of inclination relative to a horizontal plane.

No less important than faults as evidence of the earth's instability are folds, the principal categories of which are *anticlines, domes, synclines, basins,* and *monoclines.* Both anticlines and domes have uparched strata but differ in that domes are roughly symmetric, with beds dipping more or less equally away from some central point. Synclines and basins are composed of down-folded strata, with the basin being more or less circular and characterized by beds that dip inward toward the center. In monoclines, strata dip for an indefinite length in one direction and then return to their former, usually horizontal, attitude.

Four fundamental types of individual mountains may be recognized: volcanic, block-faulted, erosional, and folded. The great mountain systems of the world are primarily linear folded mountains. These major mountain systems have experienced a depositional stage involving the accumulation of thick sequences of sediment along the passive margins of continents. This depositional phase is followed by mountain building that is usually the result of an encounter with another tectonic plate. Following the development of a mountain belt, it undergoes lengthy periods of erosion and successive isostatic uplifts.

Terms to Understand

anticline
attitude (structural)
axial plane (of fold)
axis (of fold)
basin (structural)
brittle
dip (direction & angle)
dome
ductile
elastic strain
fault
fault-block mountains
fault gouge
fault plane
fold
fold mountains
footwall
geologic structure
graben
hanging wall
horst
isostasy
joint
klippe
monocline
nappe
normal fault
overturned fold
plunge
recumbent fold
reverse fault
slickensides
strain
stress
strike
strike-slip fault
symmetric fold
syncline
thrust fault
Wilson cycle

Questions for Review

1. Name, describe, and sketch three kinds of faults. Which of these tend to be most abundant in regions of crustal compression?

2. What are the principal kinds of folds? Are tracts of closely folded rocks the result of compressional or tensional forces?

3. In Figure 9–4, what is the approximate angle of dip of the beds being examined by the geologist? If the direction of dip was found to be N.30°W., what then would be the strike?

4. Which side of the more prominent fault (the one to the right) in Figure 9–22 is the hanging wall? What two lines of evidence indicate the direction of movement of the left side?

5. How might you be able to differentiate between a dome and a basin on a geologic map by using the age relationships of the outcrops?

6. What is the meaning of the term isostasy? At what time in the history of a mountain belt are isostatic adjustments likely to be most frequent?

7. What type of faults would you expect to find along a divergent zone between lithospheric plates? Along a shear zone?

8. Distinguish between the following:

a. stress and strain
b. fault and joint
c. brittle and ductile behavior in rock masses

9. What factors may influence whether a rock behaves in a brittle or ductile manner?

10. What evidence might you seek in the field if you were trying to determine if a fault was currently or recently active?

Supplemental Readings and References

Billings, M.P., 1972. *Structural Geology,* 3rd ed. Englewood Cliffs, N.J., Prentice-Hall, Inc.

Davis, G.H., 1984. *Structural Geology of Rocks and Regions.* New York, John Wiley & Sons.

Dietz, R.S., 1972. *Geosynclines, mountains, and continent building. Sci. Am.* 228:24–33.

Francheteau, J., 1983. The oceanic crust. *In The Dynamic Earth.* San Francisco, W.H. Freeman & Co., pp. 39–54. (Originally appeared as an article in *Sci. Am.,* Sept. 1983.)

Park, R.G., 1982. *Foundations of Structural Geology.* Glasgow, Blackie Publications.

Suppe, J., 1985. *Principles of Structural Geology.* Englewood Cliffs, N.J., Prentice-Hall, Inc.

Earthquakes and Earth Structure

10

Ruins of apartment buildings in the city of Leninakan following the December 1988 earthquake in Soviet Armenia near the Soviet–Turkish border. The earthquake, registering 6.9 on the Richter scale, caused widespread devastation and tens of thousands of fatalities in the villages and cities of this Caucasus mountain region. (Wide World Photo, with permission.)

The deep interior of the earth is inaccessible, and no rays of light penetrate to let us see what is below the surface. But rays of another kind penetrate and carry with them their messages from the interior.

I. Lehman (1959)

Introduction

In this age of artificial satellites and spacecraft, our attention is often directed toward the remarkable discoveries resulting from the exploration of outer space. We tend to forget that there is still much to learn about the "inner space" of our own planet. This is particularly true of that part of the earth that lies beneath the crust. Our deepest wells penetrate only about 10 of the 6300 km that separate us from the center of the planet. Basaltic lavas sometimes provide a glimpse of materials that originated 50 or 60 km below the surface, and some diamond-bearing kimberlites may have risen from depths of 200 km. For the most part, however, our knowledge of the earth's insides has been indirectly inferred from the study of earthquake waves (seismic waves). In this chapter, we shall examine the cause and effects of earthquakes and review what they tell us about the hidden interior of our planet.

When the Earth Shakes

Ours is a living, dynamic planet. Each year this fact is emphasized for us by over a million earthquakes. At least 50 of these cause loss of life and severe damage to property (Figs. 10–1 and 10–2). A lesser number are major catastrophes. A horrifying recent example is the December 1988 earthquake in Soviet Armenia. In less than two minutes, that quake transformed the bustling city of Leninaken (population 290,000) into a dust-shrouded wasteland of collapsed building, the dying, and the dead.

What actually happens during an earthquake? What does it feel like to be in one? Perhaps the best way to answer these questions is to describe the events of a few noteworthy earthquakes.

The Good Friday Earthquake, Alaska, 1964

It was about 5:36 in the afternoon on March 27 in Anchorage, Alaska. Except for a few snowflakes that floated down from partly cloudy skies, the weather had been pleasant, and there was a general mood of geniality as Alaskans looked forward to their Easter holiday and the coming of spring. A housewife was preparing dinner while her two children played in the neighborhood park across the street. Robert B. Atwood, editor of the Anchorage *Daily Times*, was relaxing at home by improving his skill at playing the trumpet. In a real estate office nearby, a new resident of the city was preparing to purchase a home in a tract called Turnagain Heights (Fig. 10–3). Moments later the newcomer and the other unsuspecting citizens of Anchorage were jarred by the strongest earthquake recorded in North America during historic time. In the home of the young housewife, dishes rattled and walls swayed at awkward angles. Terrified, she rushed out of the house to retrieve her two children, but fell headlong into a crack that had unexpectedly opened in the loose sliding soil. She was certain she would lose her life in the next moment, but the floor of the trench heaved upward so that her children and a neighbor were able to bring her to safety.

Across town, Robert Atwood ran outside his shaking house and watched in panic as his driveway and yard broke into large dizzily sliding blocks. He tried to leap from block to block, but a fissure opened beneath him, and he tumbled downward. Sand and clay rained down upon him. He tried to free himself, but found his right arm anchored in the sand. Suddenly he realized that his hand still clutched the trumpet he had been playing moments before. He released his hold on the instrument and scrambled out of the ditch. Not far away, the customer for a new home in Turnagain

Figure 10–1 Damage caused by the 1985 Mexico City earthquake. This steel frame building with brick walls was shaken off its foundation (Courtesy of the U.S. Geological Survey.)

Figure 10–2 Graben formed on Fourth Avenue during the 1964 Anchorage earthquake. Before the earthquake, stores on the right were level with stores at the far left. (Courtesy of the U.S. Geological Survey.)

Figure 10–3 Damage from earthquakes may be indirect, as when they trigger landslides. The Turnagain Heights landslide in Anchorage, Alaska, was triggered by the 1964 Anchorage earthquake. Failure occurred along slip surfaces in the Bootlegger Cove clay. Some of Anchorage's finest homes were destroyed.

Heights stared in disbelief as he watched the new house he was about to purchase slide down the hillside and splinter into a mass of ruin.

The Good Friday earthquake was not centered in Anchorage but rather about 130 km southeast of the city beneath the frigid waters of Prince William Sound (Fig. 10–4). Other cities, such as Valdez and Seward, were also damaged, although not as severely. The greater damage at Anchorage can be attributed to the fact that the city is built on a thick blanket of loose glacial sediment—gravel, sand, and clay. These unstable materials are poorly bonded by frozen water. Beneath the glacial deposits, one finds a thick bed of weak plastic clay known to geologists as the Bootlegger Cove clay. When the tremendous shaking began, the Bootlegger Cove clay began to flow downward toward Cook Inlet, causing great cracks and landslides in the overlying glacial material. Indeed, most of the damage to Anchorage was either directly or indirectly due to slumps, slides, and settling of the ground.

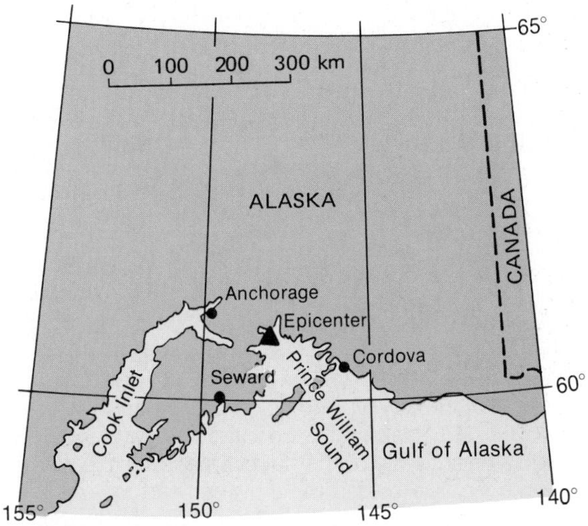

Figure 10–4 Approximate location of epicenter of the Alaskan earthquake of 1964.

The Earthquakes of 1811–12, New Madrid, Missouri

We are accustomed to hearing about earthquakes in places like California, Alaska, Japan, Peru, and Iran. An earthquake seems an improbable event for persons living in the low-lying stable heartlands of continents. No region, however, is immune to earthquakes. In the 53 days between December 16, 1811, and February 7, 1812, this fact was powerfully demonstrated to Indians and settlers living in the area that surrounds the village of New Madrid, near the southeastern tip of Missouri (Fig. 10–5). At that time, nearly 2000 earthquakes were counted, many of which were probably greater than the quakes that destroyed San Francisco in 1906. For the relatively few inhabitants of New Madrid county, the events were an extended nightmare. As the earth heaved and swayed, cabins collapsed on their occupants, stout trees snapped in two, and great chasms opened and belched forth fountains of sand, water, and sulphurous fumes. Upheaval of broad domelike mounds temporarily dammed the Mississippi, and the backed-up water produced the effect of the river seeming to flow backwards for a time. The surging floods carried logs, furniture, and boats upstream and deposited much of the debris in the tree tops.

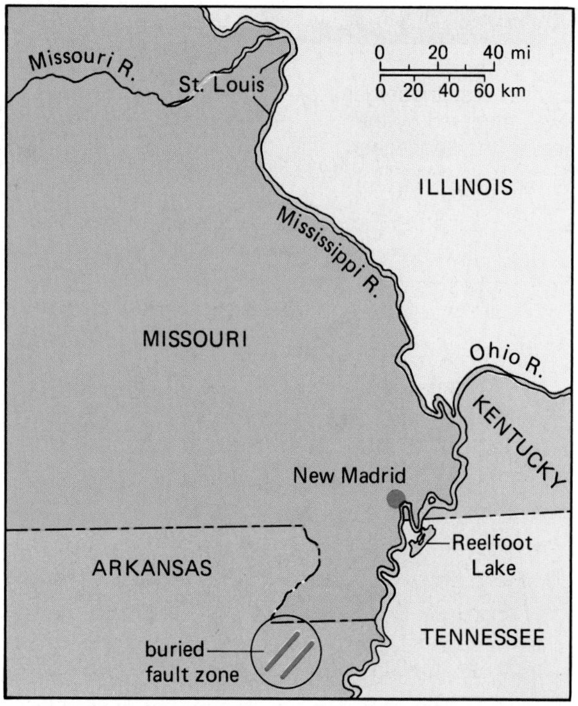

Figure 10–5 New Madrid, Missouri, region in which the 1811–1812 earthquakes occurred. Movement within a recently discovered fault zone in northeast Arkansas may have caused these historic earthquakes. The fault zone had not been discovered earlier because it is deeply buried by sediments of the Mississippi Valley. U.S. Geological Survey geologists discovered the fault zone by studying artificially induced seismic waves.

Ground subsidence accompanying the earthquake converted fertile forests into deep bogs and formed swamps in a zone from Cape Girardeau, Missouri, 190 miles southward into Arkansas. Until subsequently reclaimed by the construction of drainage canals and levees, the area was appropriately dubbed "Swampeast Missouri." Other areas also experienced the sinking of narrow strips of land, which then filled with water. In Tennessee, one of these troughs became the 20-mile-long and seven-mile-wide Reelfoot Lake.

The shocks from the New Madrid earthquake were felt from New Orleans to Chicago and from the east to west coasts of the United States. Indian legend and geologic structures in the area indicate that other earthquakes had occurred prior to 1811. Even within the present decade, citizens of Cairo, Illinois, Memphis, Tennessee, and St. Louis are occasionally jolted into remembering that earthquakes can occur in the so-called stable interior of the U.S.A. They are, however, much less frequent in the midcontinental region than along the continent's western margin, which is a lithospheric plate boundary. Nevertheless, there is no cause for complacency, for a major earthquake could occur in this region at any time. Because of the increased population, the destruction and loss of lives would be far greater than in 1811.

The 1906 San Francisco Earthquake

The most infamous of American earthquakes struck San Francisco at about 12 minutes past 5 on the morning of April 18, 1906.

Citizens awoke in stark terror amid the awful roar of collapsing buildings and clanging of jostled church bells. For many, crushed by collapsing roofs and masonry that crashed through ceilings, conscious moments were brief. Those who managed to make their way out into the open and away from falling objects recalled that the earthquake, which had begun with relatively small movement, increased to a jarring crescendo at the end of about 40 seconds. The shocks abruptly stopped for a few moments and then struck again even more violently for about a half minute. This main shock was then followed by smaller earthquakes, termed aftershocks. Although the destruction to buildings by shaking was extensive, the main damage came from fire, as gas from broken gas mains was ignited. Water lines were snapped by the earth movements, and firemen could not effectively battle the conflagration that raged through the city. For three days, however, firemen and volunteers worked with wet mops, dynamite, and shovels until they had finally put out the fires. Slowly, the thousands who had fled the city returned. It was then apparent that the parts of the city that had experienced the greatest destruction from the earthquake itself had been built on loose soils used to fill ravines and swampy areas (Fig. 10–6). Even today, parts of San Francisco (and other cities) are built on loose materials dumped along the coastline to extend the land into the sea. Such areas are in maximal danger from earthquakes.

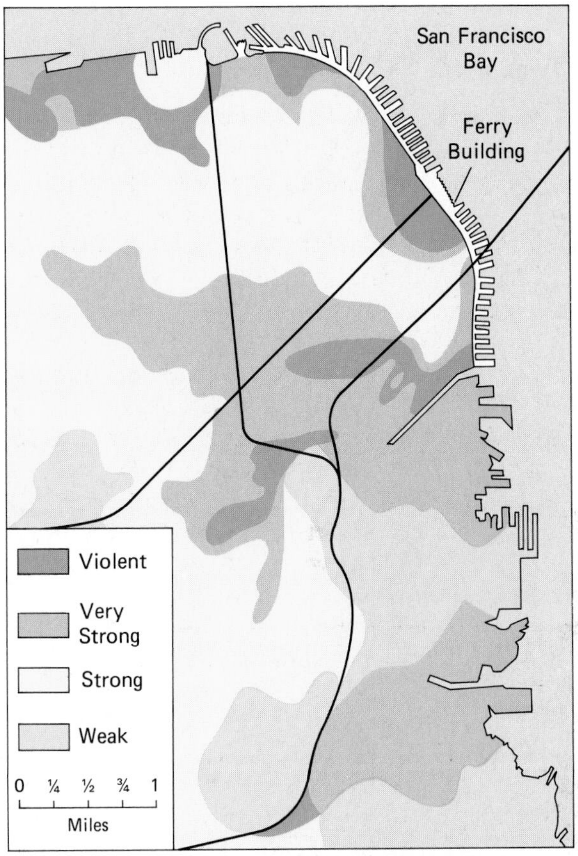

Figure 10–6 Distribution of the damage due to the 1906 San Francisco earthquake. Land recovered from the bay, reclaimed and filled swampland, river bottoms, and other areas of thick unconsolidated soil experienced the most damage. Buildings on bedrock experienced relatively little damage. (From Watkins, J.S., Bottino, M.L., and Morisawa M., *Our Geological Environment,* Philadelphia, W.B. Saunders Company, 1975.)

The cause of the San Francisco earthquake was sudden movement along the notorious San Andreas fault (Figs. 10–7 and 10–8). This great scar in the earth's crust can be traced nearly 1000 km from Cape Mendocino to the Gulf of California. The San Andreas fault, along with adjoining networks of faults, is responsible for most of the earthquakes felt in California. The damage to San Francisco (as well as San Jose, Santa Rosa, and Stanford University) during the 1906 quake was primarily the result of horizontal displacement along the San Andreas fault. The amount of horizontal movement varied from place to place up to a maximum of 6.5 meters, most of which occurred as the oceanward side of the fault was jolted northwestward.

In Chapter 1 we described the slowly moving tectonic plates that compose the earth's outer shell. Where such plates slip past one another, faulting can be expected. The San Andreas fault is one such plate margin (Fig. 10–9). It marks the boundary between the North American and Pacific plates. The Pacific plate carries a splinter of North America with it as it creeps northwestward at a rate of about 7.5 cm a year (relative to the main body of the continent). Thus, if movement continues for the next 20 million years, Los Angeles will have traveled past San Francisco and have continued to a point far to the north of that city. Californians are well aware that movement along the San Andreas is neither continuous nor smooth. The masses of rock on either side of the San Andreas and other faults tend to lock and then slip suddenly (Fig. 10–10) releasing the stress that has accumulated and sending severe jolts throughout the land.

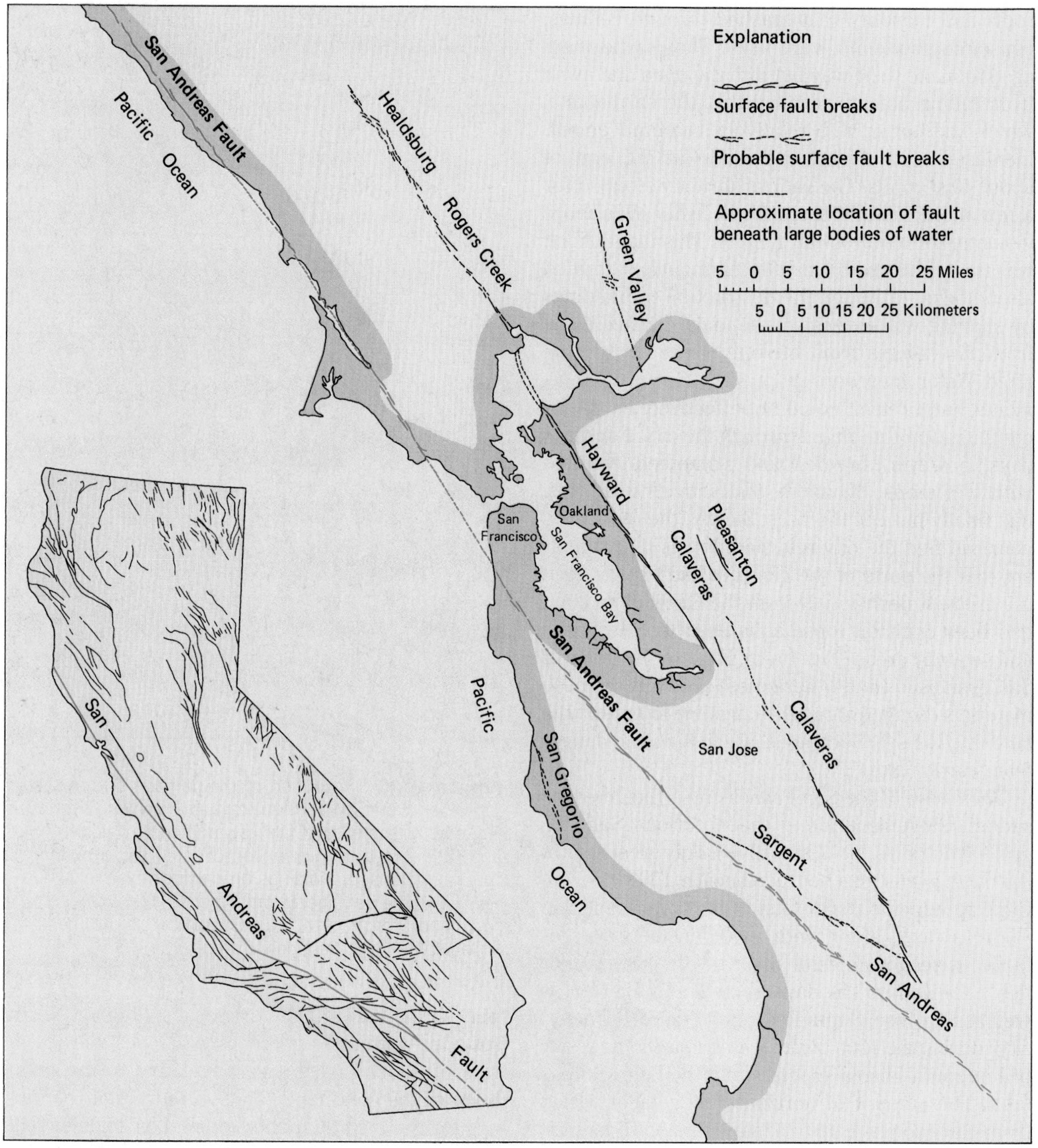

Figure 10–7 (A) San Andreas and nearby historically active faults in the San Francisco Bay Region. (B) Continuation of the San Andreas to the northwest and southeast (From Watkins, J.S., Bottino, M.L., and Morisawa M., *Our Geological Environment,* Philadelphia, W.B. Saunders Company, 1975.)

Earthquake Distribution and Tectonics

If one examines a map showing the location of earthquake epicenters (Fig. 10–11), it is readily apparent that earthquakes do not occur randomly around the globe. Rather, earthquake epicenters tend to be concentrated along relatively narrow zones. In fact, 80 per cent of the total earthquake energy for the earth is released in the zones that border the Pacific Ocean. A second set of zones extends from southern Europe through Turkey, Iran, northern India and on toward Burma, and accounts for nearly another 15 per cent of earthquake energy. Clearly, there is some sort of disruptive activity along these great earthquake trends. According to the concepts of plate tectonics, which will be examined in detail in the next chapter, the zones of high seismicity are believed to delineate the margins of great moving plates of lithosphere. The plates collide, pull apart, or slide past neighboring plates, and these interactions along plate margins generate the stress in crustal rocks that causes earthquakes.

A lesser number of earthquakes occur within the central regions of lithospheric plates. These earthquakes tend to be of lower magnitude, although occasionally an event as large as the New Madrid earthquake may occur. The cause of intraplate earthquakes is not well understood. Many seem to be the result of the erosional removal of vast quantities of surface materials. Without the weighty overburden, segments of continental crust may rise bouyantly and generate stresses in the surrounding rocks. Old faults may be reactivated by this motion, and formerly stable rock masses ruptured. Geologists also speculate that some intraplate earthquakes may represent the early warnings of the breaking apart of continents.

The majority of earthquake foci are located at the edges of tectonic plates, with a fewer number between plate margins.

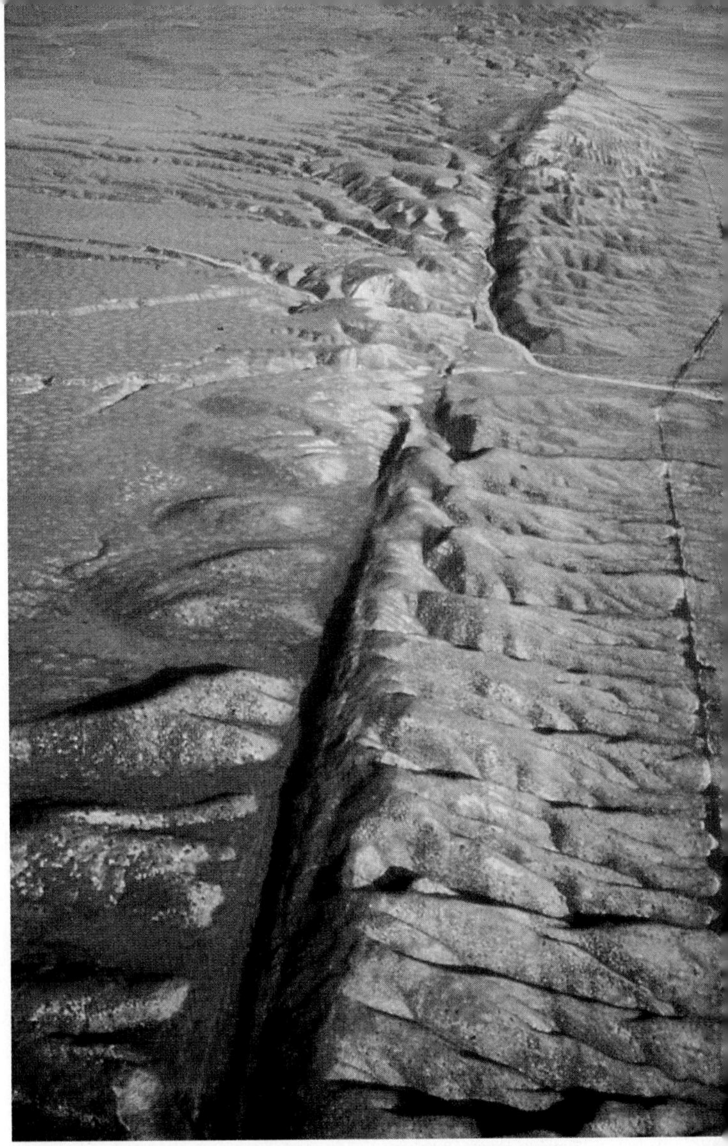

Figure 10–8 Oblique aerial view of the San Andreas fault zone along the Carrizo Plain, San Luis Obispo County, California. A linear valley has been eroded along the main trace of the fault zone. The dark line at the right is a fence along which tumbleweed has accumulated. (Courtesy of R.E. Wallace and the U.S. Geological Survey.)

The Cause of Earthquakes

Children of the Wanyamwasi tribe of Africa learn from their elders that the earth is a great disc supported on one side by a lofty mountain and on the other by an amorous giant. The giant's lovely wife stands beside him holding up the sky. Occasionally, the giant gets the notion to hug his wife,

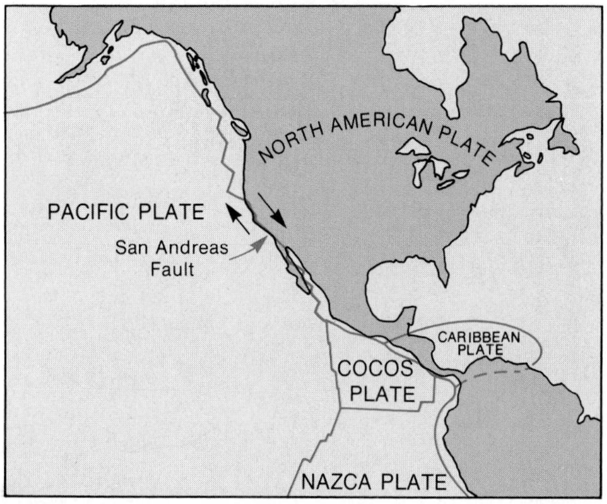

Figure 10–9 The San Andreas fault in California is a transform fault along with the Pacific plate slides northwestward relative to the North American plate at a rate of about 7½ cm per year.

thereby causing the momentarily neglected earth to totter. Thus a romantic impulse, say the Wanyamwasi, can cause an earthquake.

The Wanyamwasi legend is only one expression of the centuries-old search by peoples of every culture for the cause of earthquakes. Today, of course, most people would say earthquakes result when rocks break or there are sudden movements along "locked segments of faults."

In the previous chapter, we learned that **stress** is a force applied over a specific area of a rock so as to compress it, pull it apart, or cause one side to be bent relative to the other. We further noted that the effect of stress was to cause **strain**, which is a measure of the amount a body is deformed by stress (see Fig. 9–2). During long periods of strain, energy sufficient to cause earthquakes may slowly accumulate in rocks. When the stress that accompanies strain accumulates beyond the competence of the rock, there is a rupture. Blocks of crustal rocks snap back toward equilibrium, and this movement produces an earthquake. The "snap back" movement is sometimes referred to as **elastic rebound**.

Elastic Rebound

The mechanism involved in elastic rebound was originally described by the Californian geologist H. F. Reid, who based his model on surveys obtained for land adjacent to the San Andreas fault both before and after the 1906 earthquake. In some ways elastic rebound resembles what happens when one slowly bends a wooden yardstick. The energy applied to bending the stick is stored within it as the wood responds to the stress by bending elastically. When the elastic limit of the wood is finally exceeded, it breaks and the splintered ends whip violently back and forth. (In some instances,

Figure 10–10 Rows of lettuce offset suddenly by about 11 inches during the magnitude-6.5 Imperial Valley earthquake of October 15, 1979, in California. (Courtesy of the U.S. Geological Survey).

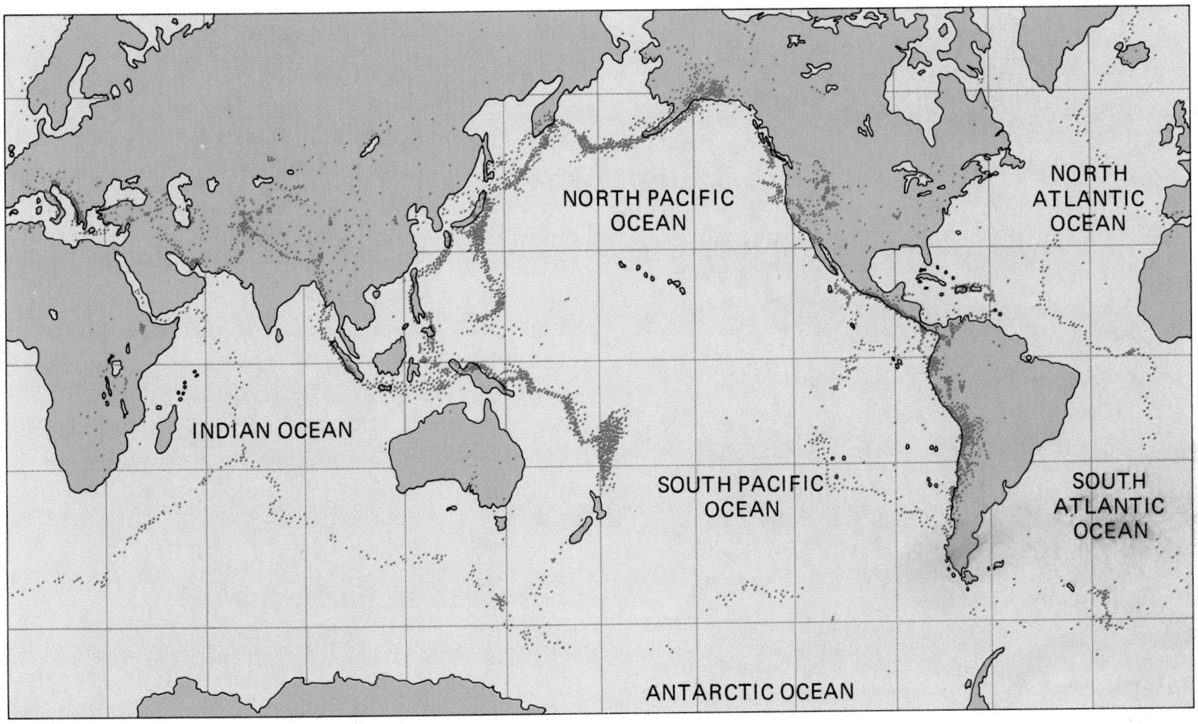

Figure 10–11 World distribution of earthquakes occurring at depths of 0 to 700 km. Notice the major earthquake belt that encircles the Pacific and another that extends eastward from the Mediterranean toward the Himalayas and East Indies. The mid-oceanic ridges are also the site of many earthquakes. (From Turk, J., and Turk, A., *Physical Science with Environmental and Other Practical Applications,* Philadelphia, W.B. Saunders Company. 1977.)

of course, the wood may not break but experience permanent deformation, as rocks sometimes do.)

Figure 10–12 is a simplified depiction of elastic rebound. Block A shows an area having a **fault** along which there has been no movement for a long period of time. The two lines perpendicular to the fault can be thought of as fence lines. Gradually, stress builds along the fault, and the rock begins to show strain (Block B). For a time, because of friction or cementation, the opposing sides of the fault remain locked together. Eventually, however, so much strain accumulates that the frictional bond along the fault fails at its weakest point (Block C). The strain is suddenly released as elastic strain.

The sudden jolt and snap-back that occur send a series of vibrations or earthquake waves that travel outward in all directions. The release of stress at one point may cause ruptures at other places along a large fault system, so that earthquake activity may continue for days or weeks following the main shock. A preexisting fault is not always a requirement for an earthquake. Stresses may build up in unfaulted rocks, and earthquakes may be generated by initial faulting.

Deep earthquakes may result from sporadic slippage in rocks that are simultaneously undergoing plastic deformation. Beyond a depth of about 700 km, however, failure occurs by creep, and earthquakes do not occur.

Earthquakes most commonly occur when rocks break and move abruptly along a fault so as to relieve the strain that had accumulated in the rock.

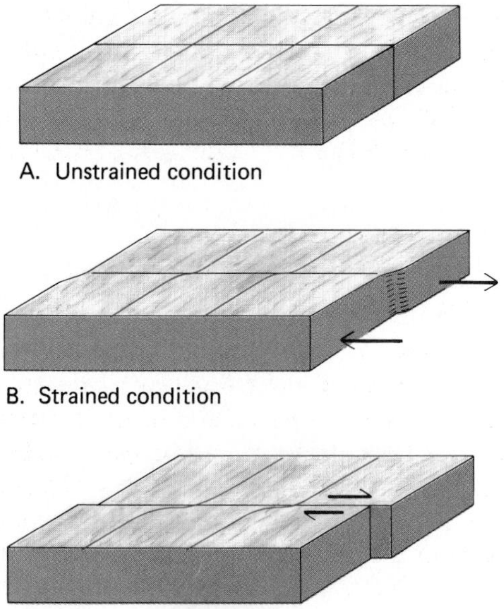

Figure 10–12 Model of the elastic-rebound mechanism. The red lines can be imagined as fence lines.

Seismographs and Earthquake Waves

Anyone who has gone through the harrowing experience of a major earthquake will recall sensing the confusion of side-to-side and up-and-down blows and vibrations. However, it is difficult to differentiate particular kinds of movements within the jumble of jerks and jolts. Actually, the various movements are manifestations of different kinds of elastic waves that travel through and around the earth from the point where rock breakage or slippage occurred. With the help of **seismographs** (Fig. 10–13), it is possible to isolate the major kinds of elastic wave motion involved in earthquakes.

Recording Earthquakes

It is easy to record the passage of an automobile or train as we stand in one place and observe that movement. We have objects on the ground to use as a frame of reference in relation to which the motion is readily seen. However, during an earthquake, we move in unison with the earth and there is no stationary frame of reference. To solve this

Figure 10–13 Interior of a modern seismograph station. The pier on the left holds three short-period seismometers (rear corner) driving galvanometers, which reflect onto the drum in the foreground. Long-period seismometers and galvanometers are on the pier to the right. Both piers are isolated from the floor of the vault. (Photograph courtesy of Earth Physics Branch, Department of Energy, Mines, and Resources, Canada.)

problem, instruments have been developed that make use of *inertia*, or the tendency of a heavy body to "stay put" as the earth moves around it. To accomplish this feat, the heavy body is suspended from springs or flexible wires, which tend to "absorb" earth movements. The body is able to lag behind, maintaining its position as the earth moves around it.

Seismographs are constructed with a mass or weight that rests as independently of earth motion as possible. This has been achieved in several ways. For recording vertical movements, the seismograph contains a weight suspended by springs from an overhead support (Fig. 10–14A). Horizontal movements are recorded by seismographs containing a horizontal pendulum having a weight on a rigid arm supported by a wire. The pendulum is freely attached to a socket joint from the supporting column (Fig. 10–14B). In either type of seismograph, the weight tends to remain at rest while the support vibrates with the earth. The difference in motion can then be recorded and measured in many different ways. One simple device employs a delicate pen attached to the weight. The pen traces a continuous line on a revolving drum that is itself fixed to the supporting framework of the instrument. The drum turns as a clock so that the arrival times of a vibration can be determined. Of course, the actual seismographs placed in earthquake-monitoring stations are far more complex than the simple instruments illustrated in Figure 10–14. The ray of light used to record vibrations on the photographic paper is reflected off a series of mirrors to amplify even very small motions of the ground. Other devices serve to dampen the natural swinging motion of the pendulum and filter out unrelated background motions. To analyze all movements of earthquakes fully, at least three seismographs must be in simultaneous operation.

Earthquake Intensity and Magnitude

Soon after an earthquake has been reported, people want to know something about its intensity. Therefore, attempts have been made to formulate standard scales of intensity and magnitude. **Magnitude** is a term that is used to describe the amount of energy released by an earthquake, whereas **intensity** measures the effect of the earthquake on structures and people at a particular place. Intensity scales are qualitative. They record the observable effects of earthquakes as ranging from slight tremors detected by only a few people (ranked I on a scale of I to XII) to total destruction (XII on the intensity scale). (A widely used standard for judging earthquake intensity is the modified **Mercalli Scale** provided in Table 10–1.) Intensity values can be plotted on maps (Fig. 10–15) by drawing lines called **isoseismal lines**, which connect points of equal earthquake intensity. Such maps are useful in delineating areas in which the effects of earthquakes were similar. Intensity scales, however, have rather obvious limitations. They cannot be used in uninhabited areas because there would be no structures subjected to earthquake damage. Also, the intensity assigned to an area is likely to

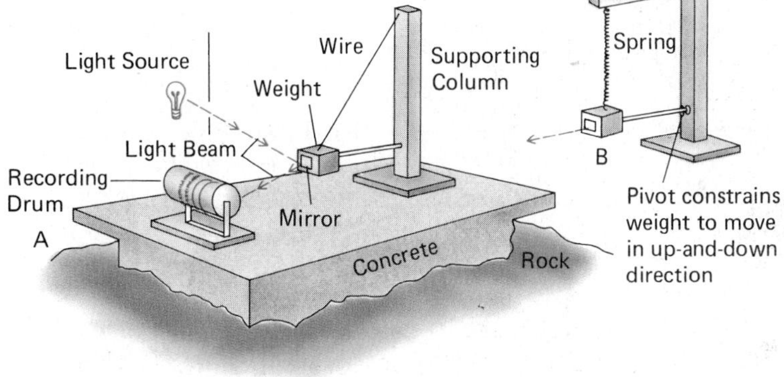

Figure 10–14 (A) A sketch of a seismograph that is recording horizontal earth motion. A light spot on the boom moves across photographic paper on the recording drum as the boom oscillates. (Springs are omitted for clarity.) By reorienting the drum and modifying apparatus as shown in B, vertical ground motion may be recorded.

Table 10–1 Modified Mercalli Scale of Earthquake Intensity

I. Not felt except by a very few under especially favorable circumstances. (About 2–3 on Richter scale.)

II. Felt only by a few persons at rest, especially on upper floors of buildings. Delicately suspended objects may swing.

III. Felt quite noticeably indoors, especially on upper floors, but many people do not recognize it as an earthquake. Standing motor cars may rock slightly. Vibration like that from passing truck.

IV. During the day felt indoors by many, outdoors by few. At night some awakened. Dishes, windows, doors disturbed; walls make creaking sound. Sensation like heavy truck striking building. Standing motor cars rocked noticeably.

V. Felt by nearly everyone; many awakened. Some dishes, windows, etc., broken; a few instances of cracked plaster; unstable objects overturned. Disturbances of trees, poles, and other tall objects sometimes noticed. Pendulum clocks may stop.

VI. Felt by all, many frightened and run outdoors. Some heavy furniture moved; a few instances of fallen plaster or damaged chimneys. Damage slight. (About 5 to 6 on Richter scale.)

VII. Everybody runs outdoors. Damage negligible in buildings of good design and construction; slight to moderate in well-built ordinary structures; considerable in poorly built or badly designed structures; some chimneys broken. Noticed by persons driving cars.

VIII. Damage slight in specially designed structures; considerable in ordinary substantial buildings, with partial collapse; great in poorly built structures. Panel walls thrown out of frame structures. Fall of chimneys, factory stacks, columns, monuments, walls. Heavy furniture overturned. Sand and mud ejected in small amounts. Changes in well-water levels. Disturbs persons driving motor cars.

IX. Damage considerable in specially designed structures; well-designed frame structures thrown out of plumb; great in substantial buildings, with partial collapse. Buildings shifted off foundations. Ground cracked conspicuously. Underground pipes broken.

X. Some well-built wooden structures destroyed; most masonry and frame structures destroyed with foundations; ground badly cracked. Rails bent. Landslides considerable from river banks and steep slopes. Shifted sand and mud. Water splashed over banks.

XI. Few, if any, masonry structures remain standing. Bridges destroyed. Broad fissures in ground. Underground pipelines completely out of service. Earth slumps and land slips in soft ground. Rails bent greatly.

XII. Damage total. Waves seen on ground surfaces. Lines of sight and level distorted. Objects thrown upward into the air. (About 8 on Richter scale.)

Modified from Richter, C. F., *Elementary Seismology*, San Francisco and London, W. H. Freeman & Co., 1958.

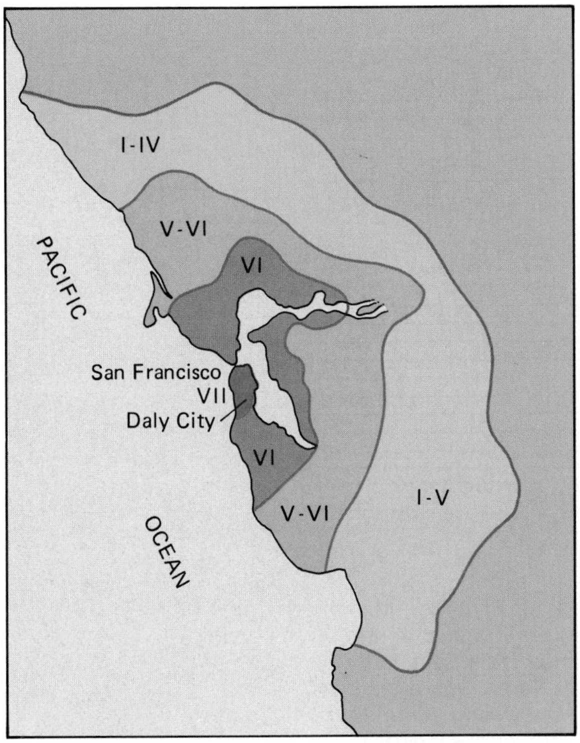

Figure 10–15 Isoseismal map prepared by the U.S. Coast and Geodetic Survey following the 1957 Daly City, California, earthquake. A maximum intensity of VII was observed. (Courtesy of the U.S. Coast and Geodetic Survey.)

be influenced by the psychology of the inhabitants, the quality of construction, and local soil stability.

In an attempt to remedy the problems associated with qualitative scales, C. F. Richter devised a system that measures the magnitude of an earthquake in terms of the motion recorded by a seismograph of certain specifications at a standard distance (100 km) from the earthquake source. Calculations permit one to determine the equivalent responses of other seismographs at any distance. In this quantitative scale, one measures the seismic wave amplitude released by the shock rather than the intensity or degree of destructiveness. The Richter scale has a logarithmic basis, so that an increase in one whole number corresponds to an earthquake 10 times stronger than one indicated by the next lower number. This translates into an approximate 30-fold increase in the amount of energy released. Thus, magnitude 7 represents ground motion about 10 times that of magnitude 6, and 31.7 times as much energy released. A magnitude-7 earthquake would represent 100 times the ground motion and 900 times the energy released of the magnitude-5 earthquake.

The 8.5 magnitude of the 1964 Alaskan earthquake was the greatest ever recorded in North America. An earthquake rated 2.5 on the Richter scale would hardly be noticed, but one of magnitude 4.5 may cause local damage. An earthquake with magnitude greater than 8.9 has never been recorded, suggesting that there is a limit to which rocks can accumulate strain energy before they break or slip.

The Richter scale is a numerical, logarithmic scale of earthquake magnitudes in which the difference between two consecutive whole numbers reflects a tenfold increase in the amount of ground motion.

Seismic Waves

From the study of **seismograms** (the records of seismographs), geologists have recognized that earthquakes move through the earth as waves. Although there are several different kinds of waves, the three that are of most importance are *primary*, *secondary*, and *surface* waves. They are defined by describing the motion of a "particle" of rock that lies in the path of the wave.

Primary Waves

Primary waves take their name from the fact that they are the speediest of the three kinds of earthquake waves and therefore the first to arrive at a seismograph station after there has been an earthquake (Fig. 10–16). They travel through the upper crust of the earth at speeds of 4 to 5 km per second. Near the base of the crust they speed along at 6 or 7 km per second. In these primary waves (also called P-waves for brevity), pulses of energy are transmitted in such a way that the movement of rock particles is parallel to the direction of propagation of the wave itself. Thus, a given particle of rock set in motion during an earthquake is driven into its neighbor and bounces back. The neighbor

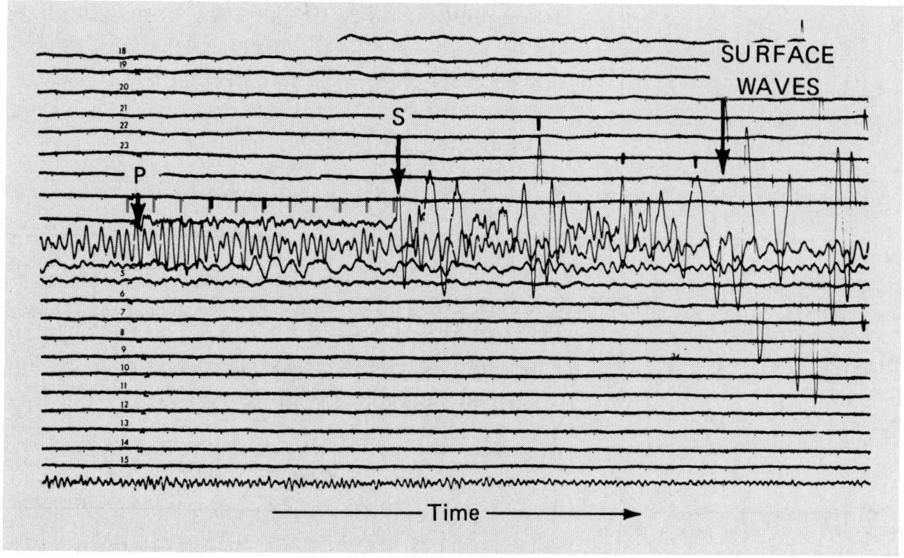

Figure 10–16 Record of a magnitude-6 earthquake that occurred in Turkey, March 28, 1964, and was recorded by the vertical component seismograph located in northwestern Canada. Time increases from left to right. Small tick marks represent 1-minute intervals. P means primary waves, S indicates secondary waves. Notice that the first S-waves arrive about 10 minutes after the arrival of the first P-waves. (Courtesy of the Earth Physics Branch, Department of Energy, Mines, and Resources, Canada.)

strikes the next particle and rebounds, and subsequent particles continue the motion. The result is a series of alternate compressions and expansions that speed away from the source of shock. Thus P-waves are similar to sound waves in that they are *longitudinal* and travel by compression and rarefaction (Fig. 10–17). It is an accordian-like "push-pull" movement that can be transmitted through solids, liquids, and gases. Of course, the speed of P-wave transmission will differ in materials of different density and elastic properties. They tend to die out with increasing distance from the earthquake source and will echo or reflect off rock masses of differing physical properties.

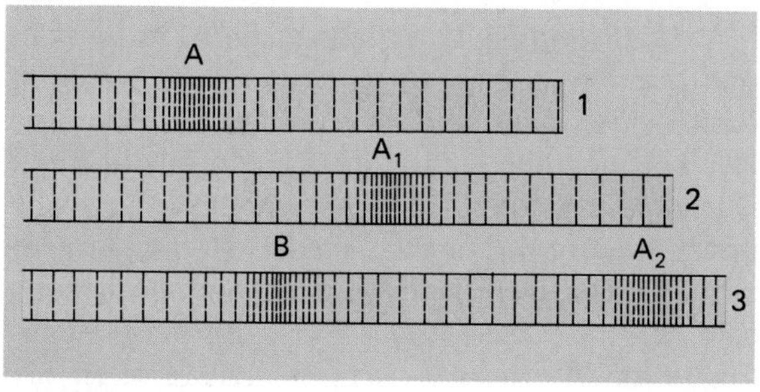

Figure 10–17 Movement of primary wave. In 1, the compression has moved the particles closer together at A. In 2, the zone of compression has moved to A_1. In 3, the zone of compression has moved to A_2 and a second compressional zone (B) has moved in from the left.

Secondary Waves

Another part of the energy released at an earthquake source is carried away by slower-moving waves called **secondary waves** (S-waves). These waves are also called **shear** or **transverse**. They travel 1 or 2 km per second slower than do P-waves. The movement of rock particles in secondary waves is at right angles to the direction of propagation of the energy (Fig. 10–18). A demonstration of this type of wave is easily managed by tying a length of rope to a hook and then shaking the free end. A series of undulations will develop in the rope and move toward the hook—that is, in the direction of propagation. Any given particle or

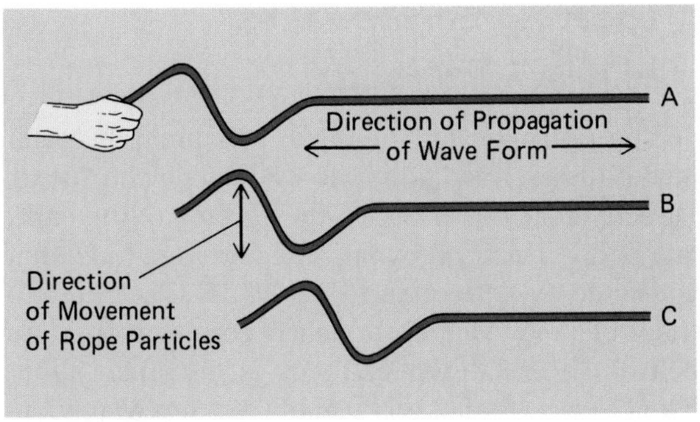

Figure 10–18 Analogy of propagation of S-waves by displacement of a rope. In rocks, as in this rope, particle movement is at right angles to the direction of propagation of the wave. A, B, and C show the displacement of the crest from left to right at successive increments of time.

point along the rope, however, will move up and down in a direction perpendicular to the direction of propagation. It is because of their more complex motion that S-waves travel more slowly than P-waves. They are the second group of oscillations to appear on the seismogram (see Figure 10–14). Unlike P-waves, secondary waves will not pass through liquids or gases.

Both P- and S-waves are sometimes also termed **body waves** because they are able to penetrate deep into the interior or body of our planet. Body waves travel faster in rocks of greater elasticity, and their speeds therefore increase steadily as they move downward into more elastic zones of the earth's interior and then decrease as they begin to make their ascent toward the earth's surface. The change in velocity that occurs as body waves invade rocks of different elasticity results in a bending or refraction of the wave. The many small refractions cause the body waves to assume a curved travel path through the earth (Fig. 10–19).

Not only are body waves subjected to refraction, but they may also be partially reflected off the surface of a dense rock layer in much the same way as light is reflected off a polished surface. Many factors influence the behavior of body waves. An increase in the temperature of rocks through which body waves are traveling will cause a decrease in velocity, whereas an increase in confining pressure will cause a corresponding increase in wave velocity. As mentioned earlier, in a fluid where no rigidity exists, S-waves cannot propagate, and P-waves are markedly slowed.

Surface Waves

Surface waves are large-motion waves that travel through the outer crust of the earth. Their pattern of movement resembles that of waves caused when a pebble is tossed into the center of a pond. They develop whenever P- or S-waves disturb the surface of the earth as they emerge from the interior. There are actually several different types of motion in surface waves. *Rayleigh surface waves*, for example, have an elliptical motion that is opposite in direction to that of propagation, whereas *Love surface waves* vibrate horizontally and perpendicular to wave propagation (Fig. 10–20). Surface waves are the last to arrive at a seismograph station. They are usually the primary cause of the destruction that can result from earth-

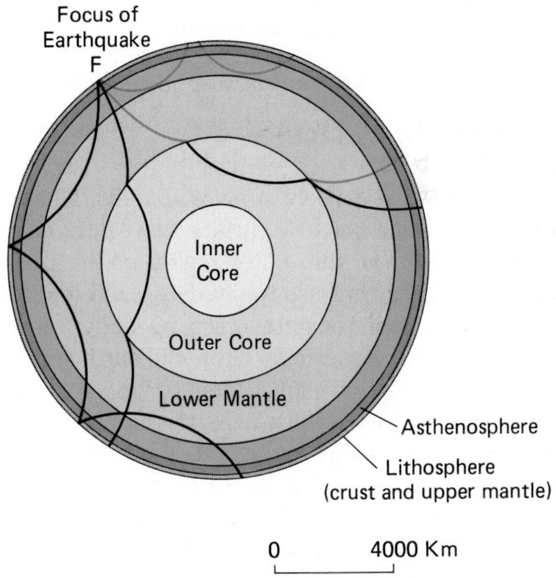

Figure 10–19 Cross-section of the earth showing paths of some earthquake waves. P-waves (black) that penetrate to the core are sharply refracted as shown in the path from F to 1. S-waves (red) end at the core, although they may be converted to P-waves, traverse the core, and emerge in the mantle again as P- and S-waves. Both P- and S-waves may also be reflected back into the earth again at the surface. (From U.S. Geological Survey publication, *The Interior of the Earth.*)

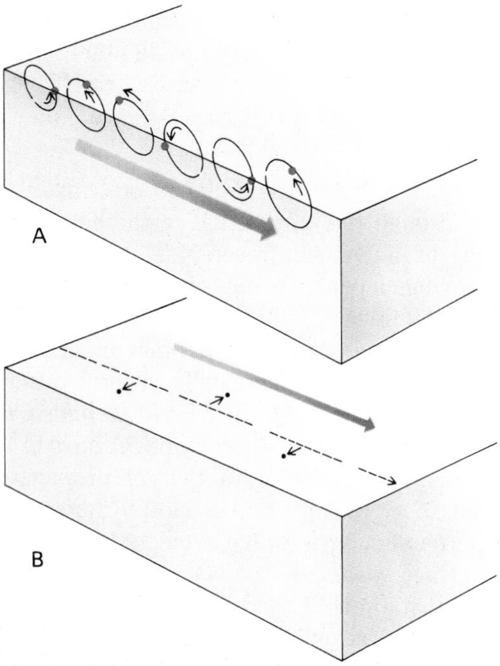

Figure 10–20 (A) Elliptical particle motion in Rayleigh surface waves (ground surface shown without displacement). (B) Particle motion in Love wave is in horizontal plane and perpendicular to direction of wave propagation.

quakes affecting densely populated areas. This destruction results because surface waves are channeled through the thin outer region of the earth, and their energy is less rapidly dissipated into the large volumes of rock traversed by body waves. Indeed, surface waves may circle the earth several times before friction causes them to fade. As is the case with ordinary water waves, the motion of surface waves diminishes with depth. It has been demonstrated that the total depth to which surface wave motion can exist is about equal to the distance between two surface wave crests. Thus, surface waves that are 2000 meters from crest to crest will shake a section of crust about 2000 meters thick.

The most common types of seismic waves emanating from the focus of an earthquake are P-waves, S-waves, and surface waves.

Locating Earthquakes

Epicenter and Focus

The point within the earth at which an earthquake disturbance is initiated is called the **focus**. Directly over the focus on the surface of the earth is a point, specified by latitude and longitude, that is termed the **epicenter** (Fig. 10–21). The depth to the focus may vary from nearly zero to more than 600 km. If the depth of focus is less than about 8 km, the earthquake is usually not felt at any appreciable distances from the epicenter. Deep-focus earthquakes, however, are detected at great distances from their epicenters. Seismologists have developed a classification of earthquakes according to their depth of focus. A *shallow-focus* earthquake has its focus at a depth between 0 and 70 km. Foci that lie between 71 and 300 km define *intermediate-focus* earthquakes, while *deep-focus* earthquakes have a focal depth greater than 300 km. Earthquakes originating at depths greater than 700 km rarely occur because the rocks at such depths are relatively plastic and do not accumulate strain.

Distance to the Epicenter

As a first step in determining the location of the source of an earthquake, one must find the distance between the recording station and the epicenter. The study of seismograms provides the key to determining this distance. Typically, the record of an earthquake can be divided into three major parts. The record begins with relatively simple oscillations produced by the arrival of primary waves. These smaller-scale tracings are followed abruptly by waves of somewhat greater amplitude representing incoming secondary waves. The surface waves come along last and produce the largest and most complex patterns of all (see Fig. 10–16). How might the pattern of seismic traces be used to find the distance to the epicenter? The answer to that question was provided in the early 1900s by John Milne, the founder of modern seismology. Milne discovered that the time separation on a seismogram between the starting points of the P- and S-waves is greater for a distant earthquake than for one located nearby. For example, the time that elapses between the arrival of the P- and the

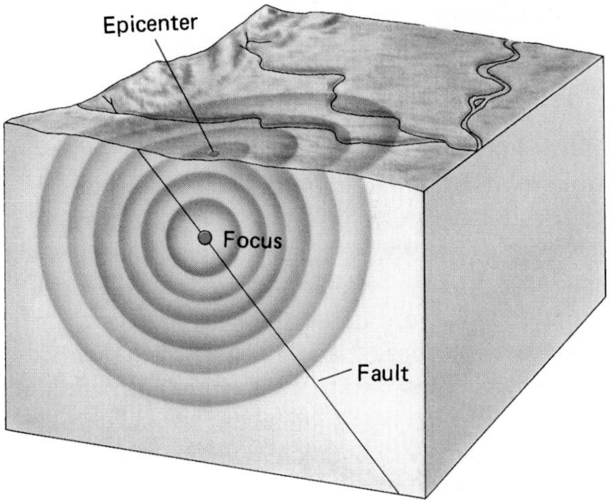

Figure 10–21 The relation between the *focus* and *epicenter* of an earthquake. The focus is the point of initial movement and the epicenter is the point on the surface directly above the focus. The location of an earthquake is usually described by the geographic position of its epicenter and by its focal depth.

arrival of an S-wave (sometimes termed the "S minus P interval") for a focus 500 km away is about 30 seconds. For a focus located 1000 km away, 100 seconds would separate the arrival times of P- and S-waves. Thus, the time interval between the arrival of P- and S-waves is different for every distance between the earthquake center and the seismograph. The situation is comparable to a phenomenon we have all experienced when we see lightning in the sky but hear the clap of thunder a moment or two later. The time that elapses between seeing the lightning and hearing the thunder would be greater at greater distances from the location of the lightning bolt.

Geophysicists have determined how rapidly seismic waves travel by observation of earthquakes whose source locations and times of occurrence are well known. The information obtained from these observations is used to construct **time–distance graphs** that can then be employed in locating earthquake epicenters. Figure 10–22 is such a graph. Assume that a seismograph records the arrival of a P-wave at 8:00 A.M. and the arrival of the S-wave 2 minutes later. On the edge of a piece of paper, the P minus S interval of 2 minutes can be marked off by using the time-travel scale on the left side of the time-distance chart. This interval can then be vertically fitted into just the right size space between the P- and S-wave curve. Then, by reading down onto the horizontal scale, one can ascertain that the earthquake was 1600 km away from the seismograph station. A line extended horizontally from the point where the 2-minute mark touches the P-wave curve to the left edge of the graph indicates that the P-wave required 3 minutes and 30 seconds to reach the seismograph. Thus, the time of the actual shock at the focus occurred 3 minutes and 30 seconds before 8:00 A.M. Newspapers might now announce the time of occurrence of the earthquake and its distance, but precisely *where* did it occur?

Direction to Epicenter

To find the precise location of the epicenter, distance determinations from three or more seismograph stations are needed. On a map, the location of each seismograph station is plotted. A circle is

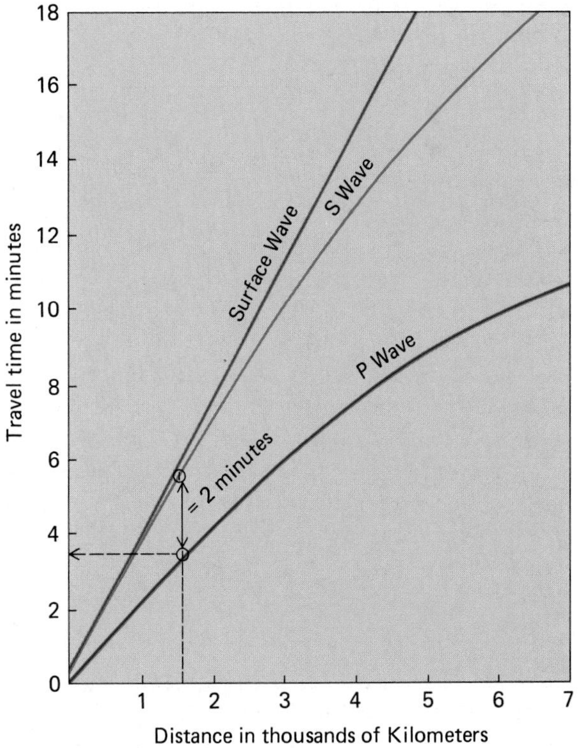

Figure 10–22 Time–distance graph.

drawn around the location that has the radius of the seismograph-to-epicenter distance (Fig. 10–23). The earthquake must have occurred somewhere on that circle. Communications from two or more other stations are used as the basis for plotting similar circles, and the point where three circles intersect locates the epicenter. Large earthquakes are recorded at hundreds of different earthquake stations. Therefore, there are an abundance of measurements to use in checking the validity of the epicenter's location.

Seismic Waves and the Earth's Interior

The Divisions of Inner Space

Most of what we know about the earth's deep interior is derived from the interpretation of countless seismograms documenting recent and past earthquakes around the globe. From such studies, geophysicists have found evidence of the gradual change in rock properties with depth as well as the relatively abrupt boundaries between major internal zones. Boundaries where seismic waves experience an abrupt change in velocity or direction are called **discontinuities**. Two widely known breaks of this kind are named, after their discoverers, the Mohorovičić and Gutenberg discontinuities.

The discontinuity discovered by **Mohorovičić** (pronounced Mō-hō-rō-vĭtch′-ĭtz) was based on his observation that seismograph stations located about 150 km from an earthquake received earthquake waves sooner than those nearer to the focus. Mohorovičić reasoned that below a depth of about 30 km there must be a zone having physical properties that permit earthquake waves to travel faster. That layer is the upper mantle.

In Figure 10–24, the shallow direct waves (A), although traveling at only about 6 km per second, are the first to arrive at the closer seismograph station. At the farther station, however, the deeper but faster-traveling wave (C) has caught up and moved ahead of the shallow direct wave, and is therefore the first to arrive at seismograph station 2. The situation is analogous to what many of us do when we take a longer freeway route to our destination

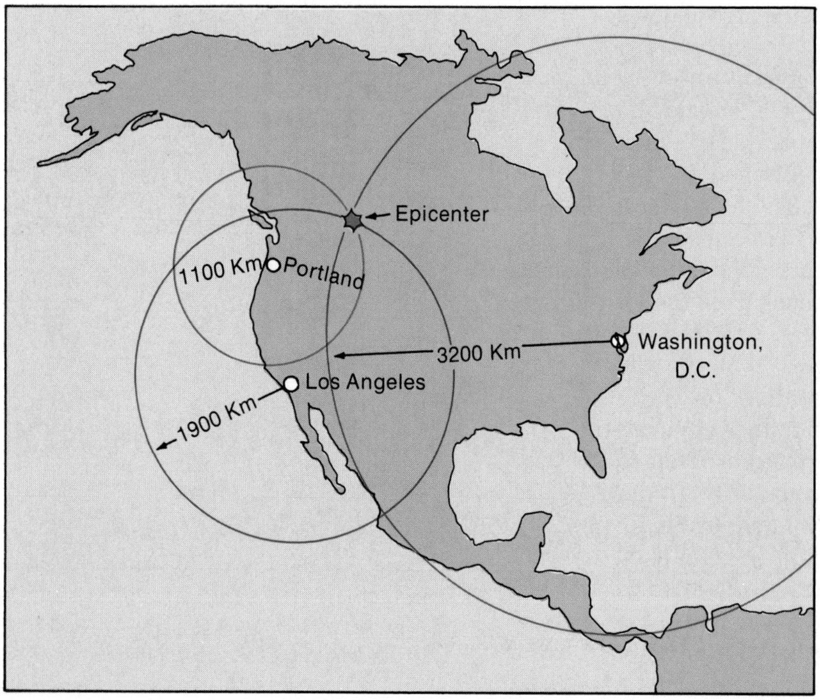

Figure 10–23 Method used for locating the epicenter of an earthquake. The distance to the epicenter from seismograph stations in Washington D.C., Los Angeles, and Portland was first determined from the P minus S time interval. This distance is used as the radius to draw circles around each station in order to find the intersection of all three. The epicenter was determined to be near Saskatoon, Saskatchewan, Canada.

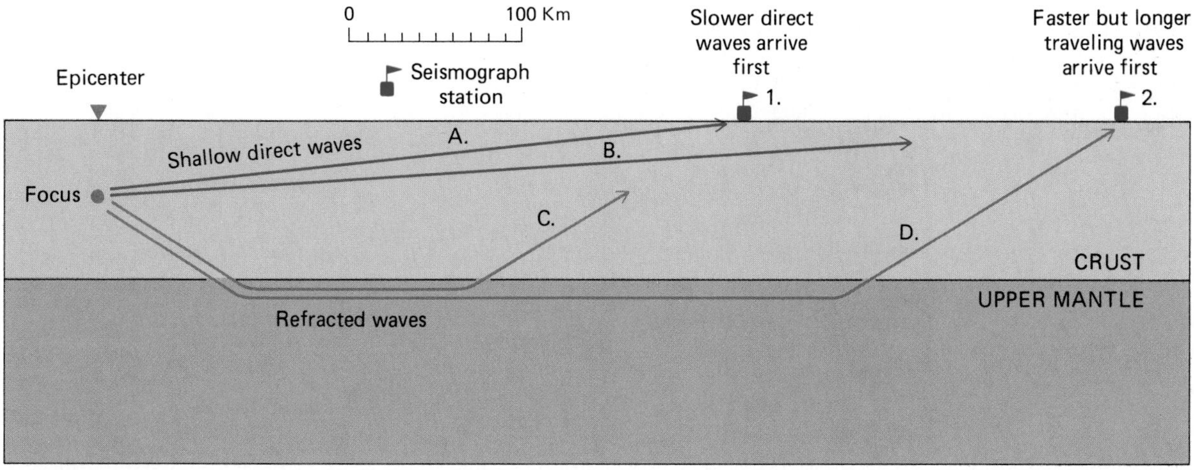

Figure 10–24 Mohorovičić's conclusion about the location of the base of the earth's crust was based on this interpretation of the travel paths of early and late-arriving seismic body waves.

than a more direct route through city streets, because the greater speed on the highway allows us to arrive at the destination sooner.

The Mohorovičić discontinuity lies at about 30 or 40 km below the surface of the continents, and at lesser depths beneath the ocean floors (Fig. 10–25). It is composed of rocks rich in iron and magnesium silicates.

The **Gutenberg discontinuity** is located nearly halfway to the center of the earth at a depth of 2900 km. Its location is marked by an abrupt decrease in P-wave velocities and the disappearance of S-waves. The Gutenberg discontinuity marks the outer boundary of the earth's core (Fig. 10–26).

The study of seismic waves is the most effective means of interpreting the nature of the earth's interior.

The Core

Inferences from Body Waves

For many years before the development of modern seismology, geologists correctly inferred that the earth had a very dense central **core**. They had determined that the overall specific gravity of the earth was 5.5. The average specific gravity for sur-

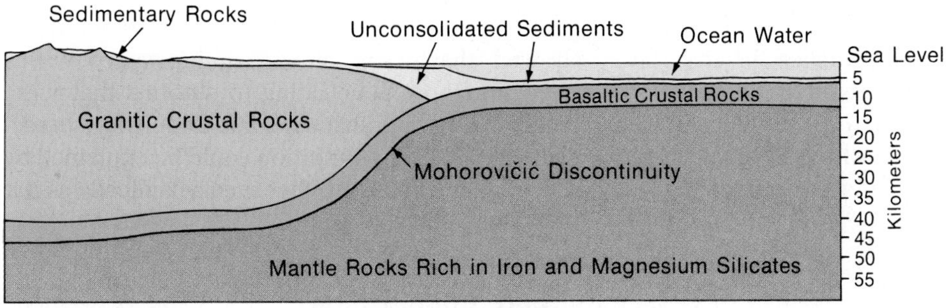

Figure 10–25 Generalized cross-section of a segment of the earth's crust showing location of the Mohorovičić discontinuity.

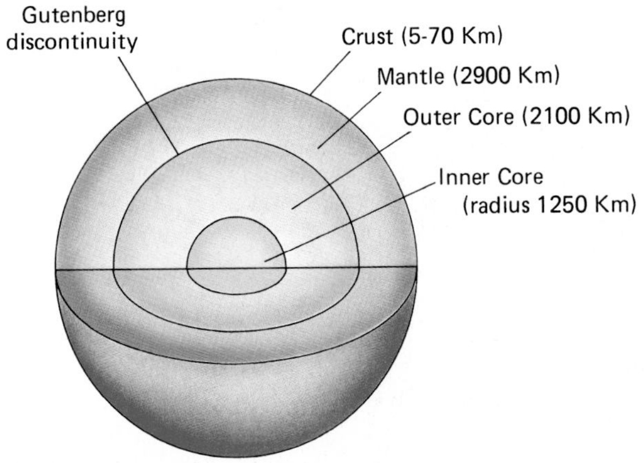

Figure 10–26 The interior of the earth. The crust appears as a line at this scale.

face rocks, however, was only about 2.7. Thus, they reasoned, the inside of the earth must be composed of much denser material than the outer surface on which we live.

As noted above, the precise boundary of the core was determined by the study of earthquake waves. Seismology has also provided a means for discerning subdivisions of the core and deciphering some of its physical properties. For example, at a depth of 2900 km (the core boundary) the S-waves meet an impenetrable barrier, while at the same time P-wave velocity is drastically reduced from about 13.6 km/second to 8.1 km/second. Earlier, we noted that S-waves are unable to travel through fluids. (Fluids cannot sustain shear.) Thus, if they were to enter a fluidlike region of the earth's interior, they would be absorbed there and would not be able to continue. Geophysicists believe this is what happens to S-waves as they enter the outer part of the core. As a result, the secondary waves generated on one side of the earth fail to appear at seismograph stations on the opposite side, and this observation is the principal evidence for an outer core that behaves as a fluid. The outer core barrier to S-waves results in an *S-wave shadow zone* on the side of the earth opposite an earthquake. Within the **shadow zone**, which begins 105 degrees from the earthquake focus, S-waves do not appear.

Unlike S-waves, primary waves are able to pass through liquids. They are, however, abruptly slowed and sharply refracted as they enter a fluid medium. Therefore, as primary seismic waves encounter the molten outer core of the earth, their velocity is checked and they are refracted downward. The result is a *P-wave shadow zone* that extends from about 105° to 143° from the focus. Beyond 143°, P-waves are so tardy in their arrival as to further validate the inference that they have passed through a liquid medium. At the upper boundary of the core, P-waves are also reflected back toward the earth's surface (Fig. 10–27). Such P-wave echos are clearly observed on seismograms.

The radius of the core is about 3500 km. The inner core is solid and has a radius of about 1220 km, which makes this inner core slightly larger than the moon. A transition zone approximately 500 km thick surrounds the inner core. Most geologists believe that the inner core has the same composition as the outer core and that it can exist only as a solid because of the enormous pressure at the center of the earth.

Evidence for the existence of a solid inner core is derived from the study of hundreds of seismograms produced over several years. These studies showed that weak late-arriving primary waves were somehow penetrating to stations that were within the P-wave shadow zone. Geophysicists recognized that this penetration could be explained by assuming the inner core behaved seismically as if it were solid.

Core Composition

The earth has an overall density of 5.5 g/cm^3, yet the average density of rocks at the surface is less than 3.0 g/cm^3. This difference indicates that mate-

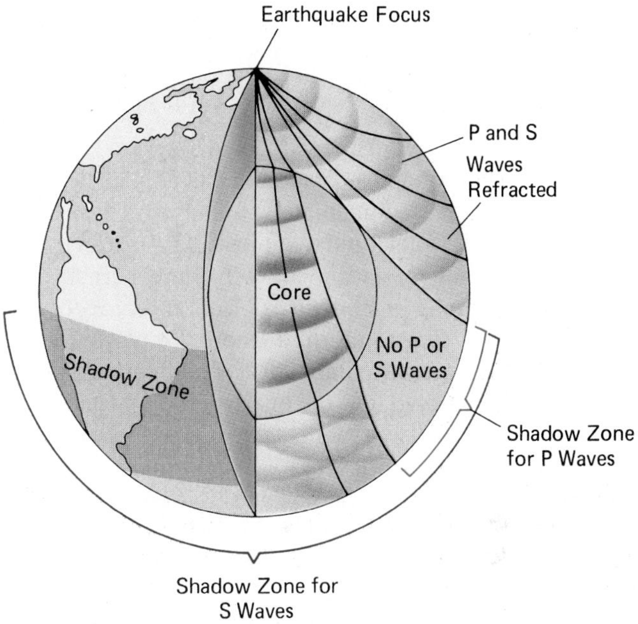

Figure 10–27
Refraction of seismic body waves. The S-wave shadow zone is believed to be the result of absorption of S-waves in the liquid outer core. When the P-wave enters the liquid outer core, it slows and is bent downward, giving rise to the P-wave shadow zone within which neither P- nor S-waves are received.

rials of high density must exist in the deep interior of the planet in order to achieve the 5.5 g/cm^3 overall density. Calculations indicate that the rocks of the mantle have a density of about 4.5 g/cm^3 and that the average density of the core is about 10.7 g/cm^3. Under the extreme pressure conditions that exist in the region of the core, iron mixed with nickel would very likely have the required high density. In fact, laboratory experiments suggest that a highly pressurized iron-nickel alloy might be too dense and that minor amounts of such elements such as silicon, sulfur, or oxygen may also be present to "lighten" the core material.

Support for the theory that the core is composed of iron (85 per cent) with lesser amounts of nickel has come from the study of meteorites. A large number of these samples of solar system materials are iron meteorites that consist of metallic iron alloyed with a small percentage of nickel. Some geologists believe that iron meteorites may very well be fragments from the core of a shattered planet. Their abundance in our solar system suggests that the existence of an iron-nickel core for the earth is plausible.

There is another kind of evidence for the earth's having a metallic core. Anyone who understands the functioning of an ordinary compass is aware that the earth has a magnetic field. The planet itself behaves as if there were a great bar magnet embedded within it at a small angle to its rotational axis. Geophysicists believe that the earth's magnetic field may, in some way, be associated with electric currents. This interpretation is favored by the discovery, over 160 years ago, that a magnetic field is produced by an electric current flowing through a wire. The silicate rocks of the lithosphere and mantle are not good conductors, however, and therefore unlikely materials for the development of electromagnetism. In contrast, iron is an excellent conductor. If scientists are correct in the inference made from density considerations and study of meteorites that the core is mostly iron, then the magnetic field lends credence to that inference.

An iron and nickel composition for the earth's core is indicated by the earth's density, the composition of meteorites, and the earth's magnetic properties.

The Origin of the Core

How does the existence of a dense metallic core relate to theories on the origin of the earth? A currently favored theory would have the earth and other planets created from the atoms, molecules, and particles circulating within a turbulent cosmic dust cloud. According to this concept, the earth would have acquired most of its materials while relatively cool, and its various elements would have been mixed and dispersed from its surface to its center. How, then, did layering develop? The answer that comes immediately to mind is that after the earth had accumulated most of its matter, it became partially or entirely molten. As in a great blast furnace, much of the excess iron and nickel percolated downward to form the core. Remaining iron and other metals combined with silicon and oxygen and separated into the overlying less dense mantle. Still lighter components may have separated out of the mantle to form an uppermost crustal layer. Geologists do not insist on complete melting for this differentiation of materials into layers. Under conditions of high pressure and temperature, elements might have migrated to their appropriate levels by the slow diffusion of ions through solid materials. Indeed, it would be difficult to account for the present abundance of volatile elements on earth if it had melted completely, for under such circumstances these materials would be driven off. Even for partial melting and solid diffusion to have occurred, however, heat would have been required. That heat may have been supplied in various ways. Certainly, the decay of radioactive elements must have been an important thermal source. In addition, heat was probably supplied by the kinetic energy of debris still showering the proto-planet, from gravitational compression, from solar radiation, and even from tidal friction generated by the nearby moon.

The Mantle

Materials of the Mantle

As was the case with the earth's core, our understanding of the composition and structure of the **mantle** is based on indirect evidence. As inferred from seismic data, the mantle's average density is about 4.5 g/cm^3, and it is believed to have a stony, rather than metallic, composition. Oxygen and silicon probably predominate and are accompanied by iron and magnesium as the most abundant metallic ions. The iron- and magnesium-rich rock **peridotite** approximates fairly well the kind of material inferred for the mantle. A peridotitic rock not only would be appropriate for the mantle's density but also is similar in composition to stony meteorites as well as rocks that are thought to have reached the earth's surface from the upper part of the mantle itself. Such suspected mantle rocks are indeed rare. They are rich in olivine and pyroxenes and contain small amounts of certain minerals, including diamonds, that can form only under pressures greater than those characteristic of the crust.

Layers of the Mantle

The mantle is not merely a thick homogeneous layer surrounding the core but is itself composed of several concentric layers that can be detected by studying earthquake data. In Figure 10–28, note that within the mantle there are three zones of increase in wave velocity. These sudden increases

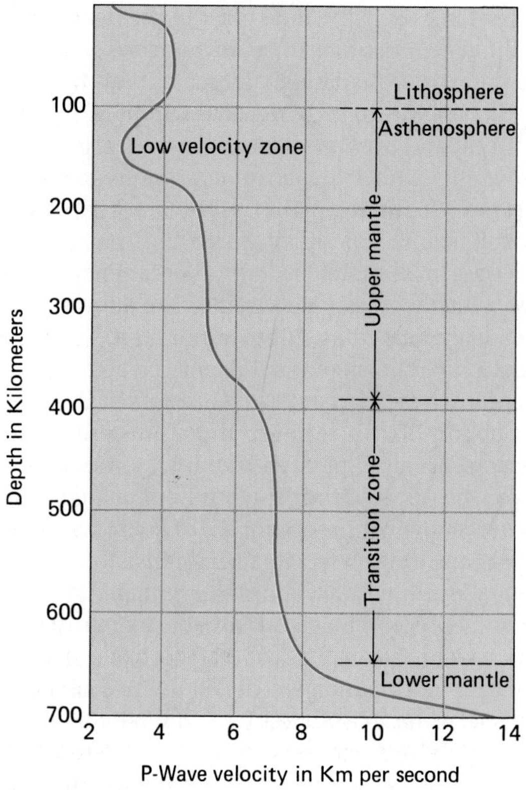

Figure 10–28 Generalized graph of average P-wave velocities vs. depth in the outer 70 km of the earth, showing the location of the low-velocity zone.

cannot be explained as simply the result of pressure increases with depth. Such increases would be too gradual to cause seismic wave velocity to increase so abruptly. There must, therefore, be some change in the physical nature of the material. One of the zones is encountered at about 400 km and is taken to mark the base of the **upper mantle**. Beneath the upper mantle is the **transition zone**, which extends downward to about 650 km. The **lower mantle** lies beneath the transition zone and above the core.

The composition of the lower mantle is difficult to infer. A plausible guess that is supported by seismic evidence is that it consists mostly of silicates and oxides of magnesium and iron. In the high-temperature and high-pressure environment of the lower mantle, iron, magnesium, silicon, and oxygen atoms are rearranged into denser and more compact crystals. For example, near the earth's surface, olivine and pyroxene are relatively stable minerals. However, in the high-temperature and high-pressure environment below a depth of about 400 km, olivine is likely to be converted to minerals whose atoms are more closely packed (Fig. 10–29). Similar dense crystals have been formed from olivine in high-pressure laboratory experiments. Rocks composed of such dense minerals would be capable of causing the increase in seismic velocities that characterize the mantle.

The upper mantle is of particular importance because its evolution and internal movements affect the geology of the crust. The most remarkable feature of the upper mantle is the **low-velocity zone**. As suggested by its name, this is a region in which there is a decrease in the speed of S- and P-waves (see Fig. 10–28). The low-velocity zone occupies an upper region of the larger **asthenosphere**. The rock here is near or at its melting point, and magmas are generated.

Geophysicists believe the seismic waves are slowed in the low velocity zone, not because of a decrease in density, but rather because they enter a region that is in the state of a crystalline-liquid mixture. In such a "hot slush," perhaps 1 to 10 per cent of the material would consist of pockets and droplets of molten silicates. Such material would be capable of considerable motion and flow, and would serve as a slippery mobile layer on which overlying lithosphere plates could move.

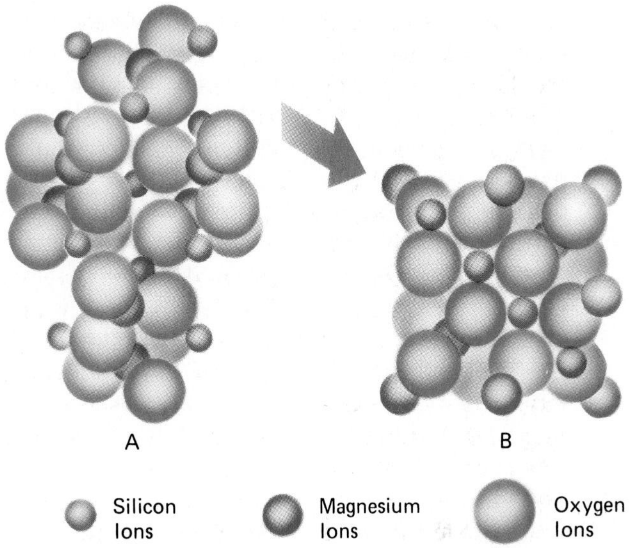

Figure 10–29 High and low pressure forms of the mineral olivine. At shallow depths in the mantle, olivine is stable in the form shown in (A) (this is the form of olivine found at the earth's surface). When the pressure reaches a critical value corresponding to a depth of about 400 km, the molecule collapses into a more dense form (B) in which oxygen ions are more closely packed.

The Crust of the Earth

The **crust of the earth** is seismically defined as all of the solid earth above the Mohorovičić discontinuity. It is the thin rocky veneer that constitutes the continents and the floors of the oceans. The crust is not a homogeneous shell in which low places were filled with water to make oceans and higher places make continents. Rather, there are two distinct kinds of crust, which, because of their distinctive compositions and physical properties, determine the very existence of separate continents and ocean basins.

The Oceanic Crust

Beneath the varied topography of the ocean floors lies an **oceanic crust** that is approximately 5 to 12 km thick and has an average density of about 3.0 g/cm^3. Three layers of this oceanic crust can be recognized. On the upper surface is a thin layer of unconsolidated sediment that rests on the irregular surface of the igneous basement layer. The second layer consists of basalts that had been extruded underwater. The nature of the deepest layer of oceanic crust is not clear. Many suspect that it is metamorphosed basaltic mantle material that has become somewhat less dense by chemically combining with sea water. In rocks of the oceanic crust, we find a concentration of such common elements as iron, magnesium, and calcium. They are included in the plagioclase feldspars, amphiboles, and pyroxenes of basalts. The upper mantle is the ultimate source for the lavas that formed the oceanic crust.

The Continental Crust

Properties of the Continental Crust

At the boundaries of the ocean basins, the Mohorovičić discontinuity plunges sharply beneath the thicker **continental crust**. Depth to the Moho beneath the continents averages about 35 km, although it may be considerably deeper or shallower in particular regions. The continental crust is not only thicker than its oceanic counterpart but also less dense, averaging about 2.7 g/cm^3. As a result, continents "float" higher on the denser mantle than the adjacent oceanic crustal segments. Somewhat like great stony icebergs, the roots of continents extend downward into the mantle.

The concept of light crustal rocks "floating" on denser mantle rocks was given the name *isostasy* in the previous chapter. Were it not for isostacy, mountain ranges would gradually subside, for there are no rocks having sufficient strength to bear the heavy load of mountain ranges. Thus, mountains are not supported by the strength of the crust but rather are in a state of flotational equilibrium with denser underlying rocks.

Although the continental crust is referred to as being "granitic," it is really composed of a variety of rocks that approximate granite in composition. Igneous continental rocks are richer in silicon and potassium and poorer in iron, magnesium, and calcium than igneous oceanic rocks. Also, extensive regions of the continents are blanketed by sedimentary rocks.

Origin of the Continental Crust

One of the major questions relating to the history of the earth is how the lighter granitic masses that constitute the continents developed. Many geologists believe the development of the crust involved a series of events that began about 4 billion years ago with upwellings of lava derived from the partially molten upper mantle. This initial crust of lava was then subjected to repeated episodes of remelting, during which time lighter components were separated out and distributed near the earth's surface. Wherever uplands existed, as along volcanic island arcs, the solidified lavas were subjected to erosion and the ordinary processes of oxidation, carbonation, and hydration that accompany chemical weathering. The products of this weathering and erosion were the earth's earliest sediments, which were then altered by rising hot gases and silica-rich solutions from below. Recycling and melting of this well-cooked and now lighter mix of earth materials led ultimately to the rocks of granitic character that formed the nuclei of continents. The new continents might then have provided a source for additional sediment that would have collected along continental margins. Subsequently, these sediments also might have been metamorphosed and melted during orogenic events, and then welded or "accreted" onto the initial continental nuclei.

The earth's early continental crust was the product of weathering and erosion of preexisting basaltic crust, deposition of the sediment derived from that crust, and conversion of the sediment to rocks of granitic character during episodes of metamorphism and remelting.

Earthquake-Related Hazards

Hardly anyone would dispute the notion that earthquakes are potentially hazardous phenomena. The vibrations that accompany earthquakes trigger landslides, cause the collapse of buildings, roads, and bridges, generate tsunami (seismic sea waves), and contribute to outbreaks of fires. There are also hazards of disease, panic, and the breakdown of order caused by the interruption of vital services. The severity of any of the hazards depends not only on the magnitude of the earthquake but on the general geology of the areas affected and the time and place of occurrence.

Tsunami

For the people of Lisbon, November 1, 1755 was a day of unprecedented terror. On that day, with a thunderous terrifying roar, the city was convulsed by an earthquake that was to claim 60,000 lives and demolish five out of every six buildings. Panic-stricken citizens rushed down to the open wharfs in order to avoid falling masonry and fire but within minutes were hurled into oblivion as a 15-meter wall of water smashed into the harbor, destroying everything in its path. That wall of water was the advancing front of a seismic sea wave or **tsunami**.

Tsunami frequently accompany earthquakes that occur near coastlines (Fig. 10–30). For example, the tsunami associated with the Alaskan earthquake was triggered by the sudden movement of the sea floor in Prince William Sound. Seventy miles away from the quake center at the Port of Valdez, onrushing tsunami lifted the steamship Chena 9 meters skyward and then plunged it to the floor of the bay. Miraculously the steamer was then able to right itself. Only three crewmen were killed (one suffered a heart attack and two were crushed by lurching cargo). People on shore were less fortunate, as the initial and then second wave swept them away without a trace and reduced the waterfront facilities to a mass of splintered wood and twisted metal.

Although tsunami may be caused by landslides along a coastline and by volcanic eruptions, the majority are generated by earthquakes. Often they develop when segments of the sea floor subside abruptly or are suddenly thrust upward along faults. Waves are generated in the water mass in much the same way as they would be when a water-filled metal tub is struck by a sharp blow. In the open ocean, tsunami are known to travel at speeds in excess of 400 km per hour. They are scarcely noticed on the open ocean, however, be-

Figure 10–30 Tsunami damage of the port facilities of Seward, resulting from the tsunami generated by the 1964 Alaskan earthquake. (Courtesy of the U.S. Geological Survey.)

cause the wavelength of the tsunami is increased to as much as 100 km, and the wave height over so broad a distance is not visibly evident. On approaching the shallow zone near a coastline, however, the bottom of the tsunami is retarded by friction. As the wave is slowed in the shallower zone, the enormous volume of water tends to pile up. The result is an increase in wave height to 30 or 40 meters.

Tsunami travel rapidly but, because of the great distances involved, may require several hours to cross an ocean. Therefore, it is feasible to set up a warning system to alert coastal areas of the possibility of property damage and loss of life from an approaching tsunami. Such a warning system for the Pacific, called the SSWWS (Seismic Sea Wave Warning System), has been developed by the U.S. Coast and Geodetic Survey with the cooperation of several federal agencies and the Armed Forces. A network of seismograph stations alert scientists in Honolulu of the location and magnitude of earthquakes. If the interpretation of these data indicates the possibility of a tsunami, its estimated arrival times at different locations are calculated with the use of tsunami travel-time charts, and warnings are issued. Of course, for locations close to the earthquake, there may be insufficient time to provide a warning. It is also difficult to predict the force and destructiveness of a tsunami when it arrives at a particular locality, as these factors are related to the geographic and bathymetric configuration of the coastline, the direction of the approach of the tsunami, and the state of the tides.

Inhabitants of coastal areas that have just experienced an earthquake would be well advised to evacuate to high ground. They should also be informed that the arrival of a tsunami may sometimes be signaled by a temporary withdrawal of the sea along a coastline. Those who foolishly venture out on the vacated sea floor to gather stranded fish may receive a fatal surprise.

Seiches

A **seiche** (pronounced sāsh) is a long wave set up in an enclosed body of water like a lake or reservoir or in a partially enclosed bay. Many seiches are caused by tides, winds, or currents, but earthquakes can also cause the pendulum-like oscillation of water that is characteristic of seiches. During the Alaskan earthquake of 1964, a seiche was set up in Kenai Lake that resulted in a maximal rise of lake levels of 9 meters. The bulge of water flooded inland, knocking down large trees and stripping the soil down to bedrock. Most of the earthquake-induced seiches for which there are adequate records did not result in great loss of life or property. Nevertheless, this type of geologic hazard should be evaluated in the planning of housing developments and recreational areas around reservoirs.

Ground Displacements

One of the most pervasive misconceptions about the effects of earthquakes is the terrifying notion that bottomless canyons suddenly open in solid rock, swallow up screaming victims, and then close up to crush them. Actually, there have been very few authenticated deaths caused by people falling into fissures. Fissures do sometimes open up in the unconsolidated materials overlying bedrock but not in solid rock. Usually these cracks appear as a result of settling, slumping, or sliding triggered by the much deeper ruptures in solid rock that caused the earthquake. As noted previously, the extreme fracturing at Anchorage in 1964 and San Francisco in 1906 was largely in the surface layer of unconsolidated materials.

Landslides

Earthquake-induced landslides are more dangerous to humans than are the development of cracks in the ground. As an example of the magnitude of this hazard, one need only recall the 1970 catastrophe in Peru that was responsible for burying tens of thousands of villagers in Yungay and Ranrahirca. Another well-known landslide caused by an earthquake occurred near midnight in Madison Canyon, Montana, on August 17, 1959. An earthquake dislodged a great chunk of the mountain above the Madison River, and a chaotic jumble of fragmented rock roared across the valley, blocking the river and burying 19 vacationers at the Rock Creek Campsite under more than 60 meters of debris. The landslide effectively dammed the Madison River, and a lake, appropriately dubbed

Earthquake Lake, formed behind the slide. To the east of the slide area stands Hebgen Dam, which holds back the water of Hebgen Lake. As a result of the earthquake, the reservoir was thrown into seiche movement. The first four oscillations sloshed water completely over the top of the dam, and oscillatory motion was still discernible 11 hours later.

Liquefaction

Often during earthquakes, fine-grained water-saturated sediments may lose their former strength and form into a thick mobile mudlike material. The process is called **liquefaction**. The liquefied sediment not only moves about beneath the surface but may also rise through fissures and "erupt" as mud boils and mud "volcanoes." As described earlier, most of the destruction to buildings at Turnagain Heights during the 1964 Alaskan earthquake could be related to the flowage of liquefied soils. Following a 1964 earthquake near Niigata, Japan, liquefied sediment flowed out from beneath tall buildings, causing many of them to tilt slowly over onto their sides. People trapped on the roof of one building were able to walk down one side of the structure to safety.

Fire

Injury from falling objects, burial in landslides or collapsed buildings, and fire pose the greatest dangers to life during earthquakes. We have already noted the ravages of fire during the San Francisco earthquake. For many years that disaster was referred to as the "Great San Francisco Fire," as if the earthquake itself was of lesser importance. Yet another example of this hazard is provided by the earthquake that devastated the cities of Yokahama and Tokyo in 1923. The quake struck during the noon hour, when midday meals were being prepared. Hundreds of fires broke out almost instantaneously. The panic-stricken citizens of Tokyo crowded into small open areas only to die of the heat and suffocation as the flames from surrounding buildings consumed most of the available oxygen. When the fires were finally extinguished, there were more than 100,000 dead, 40,000 seriously injured, and nearly a half million houses demolished. The Japanese had learned a costly lesson. The rebuilt cities have broad streets to accommodate fire trucks, fire-fighting systems designed to function even during earthquakes, auxiliary water systems, and buildings constructed to resist damage.

Defense Against Earthquakes

If there is an earthquake near your home this year, will you know what to do to lesson the danger to your family and yourself (Table 10–2)? Is it possible to reduce the hazards of earthquakes? Can anything be done in earthquake-prone areas to lesson the intensity of future earthquakes? Geophysicists and engineers are aggressively examining these questions, and their research holds great promise.

Planning for Earthquakes

Every major earthquake stimulates interest in revising building codes, initiating more rigorous zoning rules (Fig. 10–31), and providing for emergency water and power. The newer buildings in earthquake-prone areas of the U.S., Russia, China, and Japan are designed to resist damage from shaking. The effectiveness of these efforts can be judged by comparing the damage from earthquakes that occur in underdeveloped countries with the lesser damage caused by recent earthquakes of the same magnitude that have occurred in Japan and the United States. Even in the world's most modern cities, however, there is no reason for complacency. Many of our most cherished structures were built long before the advent of seismology, and far too many more recently constructed buildings lack the necessary safeguards, either because of costs, public apathy, or ignorance. Zoning plans frequently fail to include considerations of the nature, water content, and strength of the materials upon which construction is planned. In general, solid rock is a much safer foundation material than granular soil or sediment because such loose material tends to magnify seismic wave amplitudes. On some occasions construction is planned directly over active faults.

Table 10–2 Earthquake Safety Tips

The following checklist of action to take in the event of an earthquake may be clipped and posted for handy reference.

Before

1. Store emergency supplies: food, water, first aid kit, flashlight, and battery-powered radio.
2. Take a practical first-aid course.
3. Locate main switches and valves that control the flow of water, gas, and electricity into your house. Know how to operate them.
4. Support community programs that inform the public and emergency personnel about earthquake preparedness.
5. Take action to strengthen or eliminate structures that are not earthquake-resistant.
6. Support "parapet ordinances" that would remove dangerous unreinforced overhangs and cornices from buildings.
7. Support building codes that require earthquake-resistant construction and careful foundation preparation and grading.
8. Support land-use policies that recognize and allow for the potential dangers of active fault zones.
9. Heavy furniture above the fifth floor in tall buildings should be bolted to the floor.
10. Require guard rails across the inside of plate-glass windows that extend to the floor.
11. Support basic research into the cause and mechanism of earthquakes and fault movement.

During

1. Don't panic even if your are frightened.
2. If you are indoors, stay there. Get under a desk, table, or doorway.
3. Do not rush outside. Falling debris has caused many deaths.
4. Watch for falling plaster, bricks, and other objects.
5. If you are outside, move away from buildings and power lines; stay in the open.
6. If you are in a moving car, stop as soon as it is safe. Remain in the car.

After

1. Check your family or the people near you for injuries.
2. Inspect your utilities for damage to water, gas, or electrical conduits. If they are damaged, turn them off.
3. Extinguish open flames.
4. Do not use the telephone except to report an emergency.
5. Turn on your battery-powered radio for emergency information.
6. Don't go sightseeing.
7. Stay away from damaged structures; aftershocks can cause the collapse of weakened structures.
8. Stay away from beaches and waterfront areas subject to seismic sea waves (commonly called "tidal waves").

From *California Geology* 24(11):216, 1971; published by The California Division of Mines and Geology.

Clearly, we cannot start anew in designing earthquake-proof cities. We can, however, continuously improve cities and make intelligent decisions regarding new construction. Buildings being erected in earthquake-prone areas should include special supports and braces to provide good horizontal strength so that they can resist the "whiplash" effect caused by ground shifting laterally beneath them and the side-to-side shaking that occurs during earthquakes. At all levels, the structure should be anchored and bonded so that the parts of the building move as a unit, although with limited flexibility.

Figure 10–31 Housing developments obliterate the surface expression of the San Andreas fault near San Francisco. The solid line traces the axis of the fault. (Courtesy of the U.S. Geological Survey.)

The Prediction of Earthquakes

If it were possible to predict the precise time, size, and place of earthquakes, thousands of lives could be saved by evacuation. Although the prediction of earthquakes within months or days has been reported in China and Japan, such predictions are still rarely possible for most areas. Seismic risk maps (Fig. 10–32) developed from compilations of the distribution and intensity of earthquakes cannot be used for prediction but are at least helpful in evaluating seismic risks for purposes of building design and insurance.

Long-term earthquake prediction of a rather vague nature can be made, and such forecasts help to increase the public's awareness of the need for precautions. For example, studies conducted along the San Andreas fault in California suggest a break will occur within the next two decades. Forces have been building up along the fault since the 1906 earthquake. Calculations suggest there is now enough of this stored energy to propel one side of the fault at least 5 meters—an amount approaching the 6.5 meters of slippage that occurred in 1906.

In order to predict a major earthquake more precisely, scientists must learn to detect and evaluate the often subtle changes that occur in the physical characteristics of the rocks that are experiencing earthquake-induced stress. Sometimes, the rocks respond to the build-up of stress by a series of staccato-like slippages that generate a cluster of preliminary small earthquakes called *foreshocks*. It has been shown that foreshocks have indeed preceded some intense earthquakes. There are, however, other ways that the crust may signal the buildup of stress. Many of these clues are measurable. For example, if parts of the crust are either compressed or stretched, rocks will experience changes in density, water content, and magnetism. They may be tilted, raised, or lowered by amounts that can be measured with the help of sensitive instruments. In the investigations of areas having a high seismic risk, instrument stations are arranged in a network so as to record automatically these crustal changes. The stations in an array contain not only seismographs but tiltmeters, strain gauges, magnetometers, gravity meters, and electrical resistivity devices (for measuring changes in water content). The data are carefully monitored and their importance in predicting earthquakes assessed.

Careful observation of changes in rock properties, tilt of the ground surface, changes in water content, and changes in the magnetic and electrical properties of rocks have the potential for prediction of earthquakes.

Reducing Earthquake Intensity

Because of the enormity of the forces that cause earthquakes, the very thought of exercising control over them seems absurd. Nevertheless, geologists whose sanity has never been questioned are seriously seeking ways to reduce the intensity of future earthquakes. One method for possibly "defusing" earthquakes came to light in the early 1960s. At that time, water containing chemical waste from the U.S. Army Rocky Mountain Arsenal was being discarded by pumping it down a 3800-meter drill hole into the cracks and joints of an ancient body of deeply buried granite. To the surprise and embarrassment of the Army, it was discovered that

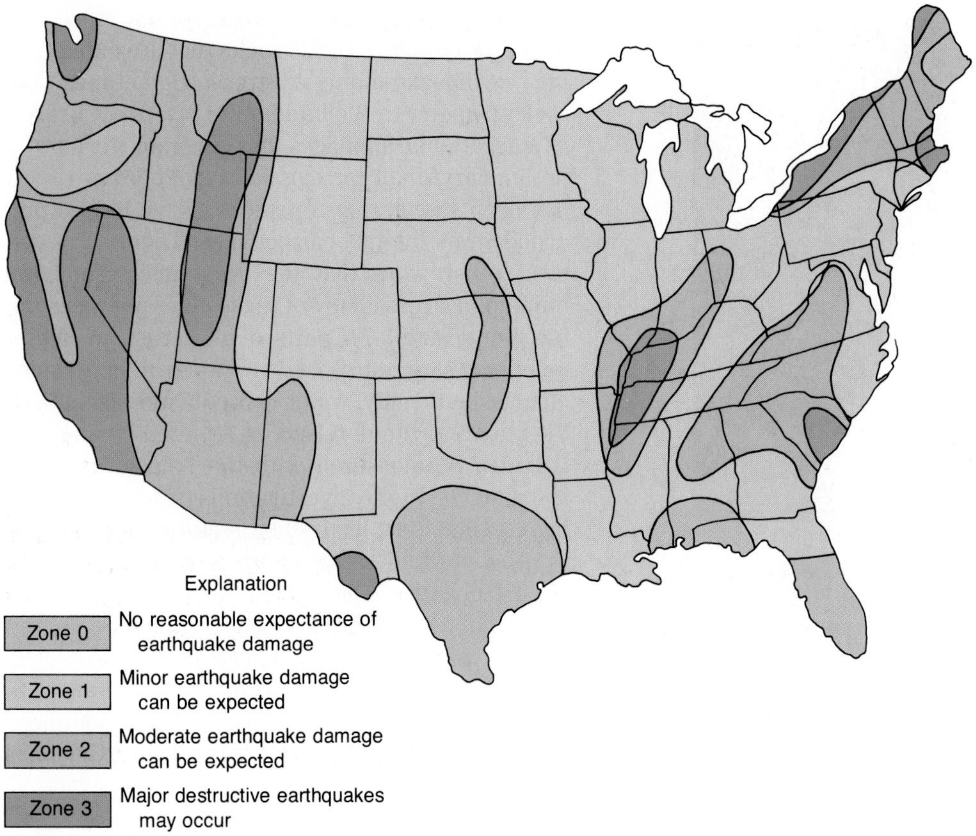

Figure 10–32 Earthquake risk map of the United States. (Courtesy of the U.S. Geological Survey.)

the project was causing earthquakes. Furthermore, the frequency of earthquakes correlated closely with the rates at which the fluid was pumped into the granite (Fig. 10–33). Study of the earthquake records indicated the tremors originated from vertical slippages along faults in the granite. The interpretation of the data seemed self-evident. Before the pumping of the fluid into the granite, the faults had been locked by irregularities in their surfaces and by mineral cement. The injected fluids widened joints and lubricated zones of slippage. The result was to allow movement to relieve minor stress buildup.

The Rocky Mountain Arsenal episode suggested to geologists that stress accumulating over a long period of time along a major fault might be relieved gradually (in a series of small shocks) by pumping water into the fault zone. In this way, a major disaster might be averted. It might also be feasible to strengthen the locked segments of a fault by removing water and thereby temporarily prevent an earthquake. These intriguing possibilities were soon to be tested in a controlled experiment at the Rangely oil field in northwestern Colorado. Prior to the experiment, water had been injected into oil-bearing sandstones in the field in an effort to recover residual oil. The practice had caused earthquakes that originated in a fault that traversed the field at depth. In order to test the relation between water injection and earthquakes fully, a network of seismographs was set up in the field and continuously monitored from 1969 to 1973. During this period, water was alternately pumped into the ground and then withdrawn. The scientists were rewarded with a dramatic correlation. Earthquakes occurred when water was pumped into the subsur-

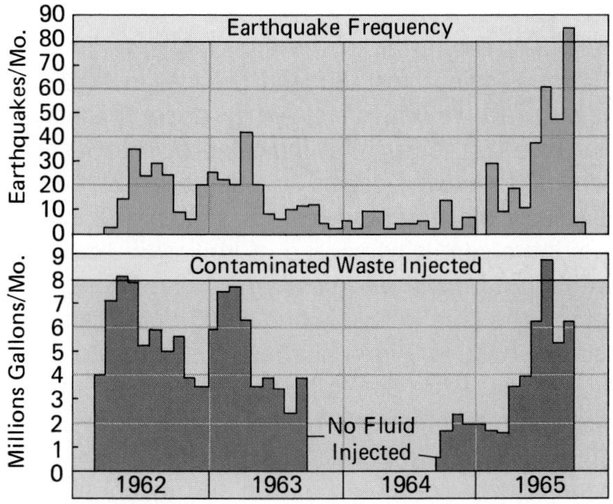

Figure 10–33 In 1965, geologist David Evans showed this correlation between the injection of fluid into a waste disposal well at the Rocky Mountain Arsenal and earthquake frequency. (From Watkins, J.S., Bottino, M.L., and Morisawa, M., *Our Geological Environment*, Philadelphia, W.B. Saunders Company, 1975.)

face, and earthquakes ceased when the water was pumped out.

The information derived from the Rocky Mountain Arsenal and Rangely studies suggested a scheme for relieving stress along active faults. In theory, deep wells might be drilled several km apart up and down the length of great faults. Next, water might be pumped into the wells in one segment of the fault and withdrawn in the next adjacent segment. The lubricated segment would slip a short distance, but the slippage would be halted in the "dry" section. The process could then be repeated farther along the fault by injecting water in the dry segment and withdrawing water from the next segment farther up the fault line. In this way, stress along the fault could be relieved segment by segment. There might be many small and harmless slippages rather than a disastrous break. The method, however, has not been adequately tested, and it carries with it the risk of generating larger-than-expected earthquakes.

Summary

Earthquakes are vibrations of the earth's surface caused by sudden rupture of rock masses or displacements along faults within and below the crust. They are frequent phenomena. It has been estimated that at least a million earthquakes occur each year, and at least 150,000 of these are felt by persons nearby.

Seismology is the study of earthquakes, and the principal tool of the seismologist is the seismograph. These instruments are constructed with a heavy weight suspended so that it will be disturbed as little as possible by earth movements. Earth movements are recorded on a time-recording rotating drum so that the arrival of earthquake waves can be determined to a fraction of a second.

The three most important kinds of seismic waves propagated during an earthquake are *primary waves*, *secondary waves*, and *surface waves*. A seismic wave is defined by describing the motion of a particle in its path. In primary waves, for example, the direction of vibration is in the same direction as the path of propagation of the wave. They have a "push-pull" or compressional progression through the rock. In secondary waves, the direction of vibration is perpendicular to the direction of propagation. Thus, secondary waves have a sinuous shearing or transverse motion. Secondary waves can travel in solids only, since there is no shear resistance in liquids or gases. The third type of wave, called a surface wave, spreads out along the surface of the earth only. They are slower-moving than either primary or secondary waves and have very complex motion.

Primary waves travel faster than secondary waves and thus arrive first at a seismograph station. On the seismogram, the time that has elapsed between the arrival of the primary and secondary waves is readily determined. This interval is different for each distance between the seismograph and earthquake center and therefore can be used to calculate the distance from the seismograph station to the source of the waves. Using the calculated distance to the source of waves from three or more different seismograph stations, the location of the earthquake can be determined. A circle with radius equal to the calculated distances is drawn around each of the three stations, and the intersection of the three circles locates the earthquake center. The point within the earth at which the waves originated is called the *focus*, whereas the location directly above the focus is the earthquake's *epicenter*.

There are two commonly used methods for measuring the strength of an earthquake. Intensity scales are based on the noticeable effects of an earthquake on buildings and people. Magnitude scales attempt to measure the amount of energy actually released by the earthquake and thus are not influenced by the character of buildings, the nature of surficial geologic materials, or the descriptions of frightened people. The widely used *Richter scale* defines the magnitude of an earthquake in terms of the motion documented by seismographs.

Geophysicists continue to investigate the cause of earthquakes. One theory suggests that forces build up in the crust gradually and that rocks will resist those forces for a time by bending. However, soon a critical value is reached and breakage occurs, releasing the pent-up energy. The concept has been named the *elastic rebound theory*. Presently scientists are experimenting with methods of causing the strained rocks to release their stored energy in a series of small movements rather than one large and potentially disastrous one.

Much of what we infer about the earth's deep interior is derived from the study of earthquake waves. For example, abrupt changes in the velocities of seismic waves at different depths provide the basis for a three-fold division of the earth into a central core, a thick overlying mantle, and a thin enveloping crust. Sudden changes in earthquake wave velocities are termed discontinuities. The *core* of the earth, as indicated by the *Gutenberg discontinuity*, begins at a depth of 2900 km. It is likely that the outer core is molten and that the entire core is composed of iron with small amounts of nickel and possibly either silicon or sulfur. The core probably originated during an episode of heating, when heavier constituents of protoplanet were drawn by gravity to the center, and lighter components rose to the surface. Constituting about 80 per cent of the earth's volume, the *mantle* is composed of iron and magnesium silicates such as olivine and pyroxene. A warm, rather plastic zone in the mantle is recognized by lower seismic wave velocities. The presence of this low-velocity layer, is an important element in the modern theory of plate tectonics. The low velocity layer may function as a weak plastic layer on which horizontal motions in more rigid surface layers can occur.

The seismic boundary that separates the mantle from the overlying crust is the *Mohorovičić discontinuity*. It lies far deeper under the continents than under the ocean basins. Thus, the continental crust is thicker than the oceanic crust. There are compositional and density differences as well. The continental crust has an overall granitic composition and is less dense than the oceanic crust, which is composed of basaltic rocks.

Terms to Understand

asthenosphere
body waves
continental crust
core
discontinuities
earth's crust
elastic rebound
epicenter
fault
focus
Gutenberg discontinuity
intensity
isoseismal
isoseismal lines
liquefaction
low-velocity zone
lower mantle
magnitude
mantle
Mercalli Scale
Mohorovičić discontinuity
oceanic crust
peridotite
primary (compressional) seismic wave
secondary (transverse or shear) seismic wave
seiche
seismogram
seismograph
shadow zone
strain
stress
surface seismic wave
time–distance graph
transition zone
tsunami
upper mantle

Questions for Review

1. What is the cause of earthquakes? Include in your answer a description of the elastic rebound theory.

2. What is the difference between the terms *stress* and *strain* as used in the studies of the strength of rocks? What is meant by the terms *elasticity* and *elastic limit*?

3. Geologists describe earthquakes in terms of their *intensity* and *magnitude*. What is the difference in meaning between these two terms?

4. Why is it unlikely that earthquakes having magnitudes higher than 9.0 will occur?

5. What are the three major kinds of earthquake waves? Which are also termed "body waves?" Describe the motion of the two kinds of body waves.

6. A seismogram records an interval of 3 minutes between the arrival of P-waves and S-waves. With the use of the time-travel curves (Fig. 10–22), determine the distance between the seismograph station and the epicenter.

7. What major internal zones of the earth would be penetrated if one were able to drill a well from the North Pole to the center of the earth? What major seismic discontinuities would be crossed?

8. What is a tsunami? Why are they rarely detected by ships at sea?

9. What is the evidence for the fluidity of the earth's outer core and for the solidity of the inner core?

10. What is the inferred composition of the core and of the mantle? What is the basis for inferring these compositions?

11. How does the continental crust differ from the oceanic crust in composition, thickness, and density?

12. What is the low velocity layer, and how can it be detected?

Supplemental Readings and References

Bolt, B.A., 1982. *Inside the Earth*. San Francisco, W.H. Freeman & Co.

Bolt, B.A., 1988. *Earthquakes*, 2d ed. San Francisco. W.H. Freeman & Co.

Bolt, B.A., Horn, W.L., and Macdonald, G.A., 1977. *Geological Hazards*, 2nd ed. New York, Springer-Verlag.

Hodgson, John H., 1964. *Earthquakes and Earth Structure*. Englewood Cliffs, N.J., Prentice-Hall, Inc.

Oakeschott, G.B., 1975. *Volcanoes and Earthquakes: Geologic Violence*. New York, McGraw-Hill Book Co.

Preis, F., 1975. Earthquake prediction. *Sci. Am.* 232:14–23.

Raleigh, C.B., Healy, J.H., and Bredehoeft, J.D., 1976. An experiment in earthquake control at Rangely, Colorado. *Science* 191:1230–1237.

Richter, C.F., 1958. *Elementary Seismology*. San Francisco, W.H. Freeman & Co.

Verney, P., 1979. *The Earthquake Handbook*. London, The Paddington Press.

Vitaliano, D.B., 1973. *Legends of the Earth*. Bloomington, Indiana, Indiana University Press.

Crustal Plates in Motion: Plate Tectonics

11

The central igneous and metamorphic core of the Himalayan Mountains. The Himalayas are the highest and youngest of the earth's major mountain belts. They were formed by the collision of the Indian tectonic plate with the Eurasian plate. (Courtesy of D. Bhattacharyya.)

What was solid earth has become the sea, and solid ground has issued from the bosom of the waters.

Ovid (43 B.C.–18 A.D.) Metamorphoses

Introduction

As implied in the epigram from the work of the poet Ovid 2000 years ago, Roman intellectuals recognized that parts of the sea floor had risen from the sea and that dry land had sunk beneath the waves. Much later, Leonardo da Vinci (1452 to 1519), Robert Hooke (1635 to 1703), and Nicholaus Steno (1638 to 1687) also wrote of such geologic changes. Steno, a Danish scientist working in Italy, taught that because marine sediments are laid down in horizontal layers, wherever they are found as folded strata, crustal movements must have taken place. In the three centuries since Steno's death, geologists have mapped extensive tracts of our continents and intensely studied the structure and rocks of mountain systems. They soon recognized that large-scale deformations of the earth's crust were not isolated local events, but that widely separated mountain ranges exhibited similarities of structure and history that were difficult to explain. Recently, however, a theory emerged that gave geologists a vibrant coherent explanation for the similarities. It is called **plate tectonics**, and it has taken on the dimensions of a scientific revolution. Plate tectonics provides a framework for associating a multitude of observations about rocks, earthquakes, and landforms into an integrated view of the whole earth.

The idea of plate tectonics was late in coming. Perhaps the most important reason for its tardy arrival was simply that, until recently, there were large regions of the globe that could not be adequately studied. This was particularly true of the vast realms beneath the oceans. For two centuries, the continents, which constitute only about 30 per cent of the earth's surface, received most of the attention. Technologic problems and formidable costs so limited exploration of the ocean floors that oceanographic research was possible only for relatively few earth scientists.

The lack of adequate information about the ocean floors began to be corrected in the years following World War II. Research related to naval operations produced submarine detection devices that also proved useful in measuring magnetic properties of rocks. In the mid-1940s to late 1950s, the need to monitor atomic explosions resulted in the establishment of a worldwide network of seismometers. This network provided precise information about earthquake locations. The magnetic field over large portions of the sea floor was also soon to be charted by the use of newly developed and delicate magnetometers. Other technologic advances ultimately permitted scientists to examine rock that had been dated by radiometric methods and then to determine the nature of the earth's magnetic field at the time those rocks had formed. Geologically recent reversals of the magnetic field were soon detected, correlated, and accurately dated. A massive federally funded program to map the bottom of the oceans was launched, and depth information from improved echo depth sounding devices poured into data collecting rooms to be translated into maps and charts. A new picture of the ocean floor began to emerge (see Fig. 1–12). It was at once awesome, alien, and majestic. Great chasms, flat-topped submerged mountains, broad abyssal plains, and imposing volcanic ranges appeared on the new maps and begged an explanation. How did the volcanic mid-oceanic ridges and deep-sea trenches originate? Why were both so prone to earthquake activity? Why was the Atlantic Ridge so nicely centered and parallel to the coastlines of the continents on either side? As the topographic, magnetic, and geochronologic data accumulated, the relationship of these questions became apparent. An old theory called **continental drift** was re-examined, and the new, encompassing theory of plate tectonics was formulated.

Continental Drift

It requires only a brief examination of the world map to notice the remarkable parallelism of the continental shorelines on either side of the Atlantic Ocean. If the continents were pieces of a jigsaw puzzle, it would seem easy to fit the great "nose" of Brazil into the re-entrant of the African coastline. Similarly, Greenland might be inserted between North America and northwestern Europe. It is not surprising, therefore, that earlier generations of map gazers also noticed the fit and formulated theories involving the breakup of an ancient supercontinent.

In 1858 there appeared a work titled *La Création et ses Mystères Dévoilés.* Its author, A. Snider, postulated that before the time of Noah and the biblical flood, there existed a great region of dry land. This antique land developed great cracks encrusted with volcanoes, and during the great Deluge, a portion separated at a north–south trending crack and drifted westward. Thus, North America came into existence. Near the close of the 19th century, the Austrian scientist Eduard Suess became particularly intrigued by the many geologic similarities shared by India, Africa, and South America. He formulated a more complete theory of a supercontinent that drifted apart following fragmentation. Suess called that great land mass **Gondwanaland** after Gondwana, a geologic province in east-central India.

The next serious effort to convince the scientific community of the validity of these ideas was made in the early decades of the 20th century by the energetic German meteorologist Alfred Wegener. His book, *Die Entstehung der Kontinente und Ozeane* ("The Origin of the Continents and Oceans"), is considered a milestone in the historical development of the concept of continental drift.

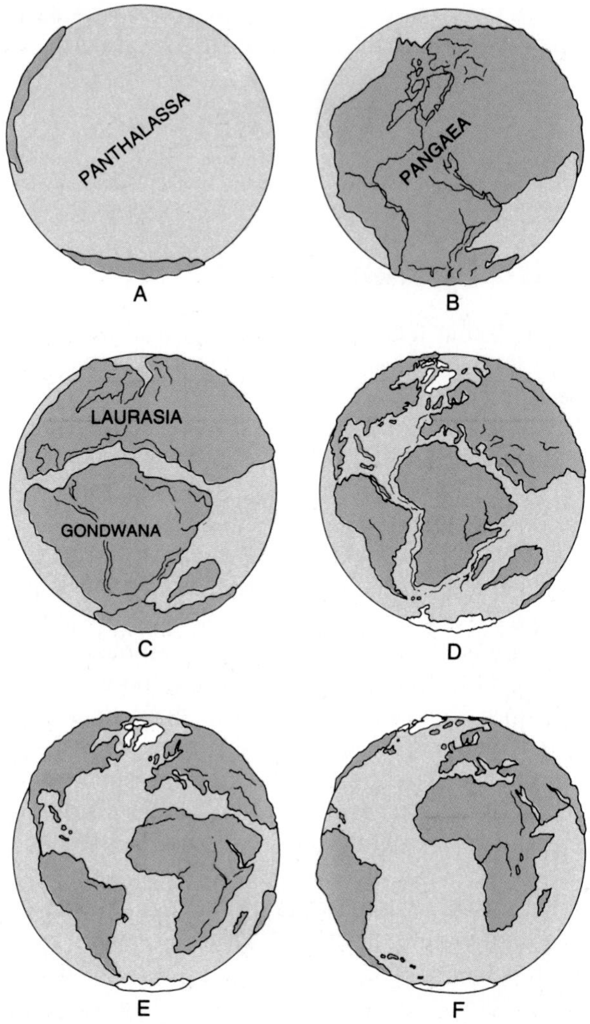

Figure 11–1 (A and B) Alfred Wegener's model of the earth as it was during the Permian. At this time there was one ocean (Panthalassa) and one continent (Pangaea). (C) By Triassic time the continent had begun to split into a northern Laurasia and southern Gondwana. (D) Dismemberment continued, and by about the beginning of the Mesozoic Era, the earth resembled drawing E. (F) Further widening of the Atlantic and northward migration of India brings the earth to its present state.

Wegener's hypothesis was quite straightforward. Building on the earlier notions of Eduard Suess, he argued again for the existence in the past of a super-continent that he dubbed **Pangaea**. That portion of Pangaea that was to separate and form North America and Eurasia came to be known as **Laurasia**, whereas the southern portion retained the earlier designation of Gondwanaland. According to Wegener's perception, Pangaea was surrounded by a universal ocean named **Panthalassa**, which opened to receive the shifting continents when they began to split apart some 200 million years ago (Fig. 11–1). The fragments of Pangaea drifted along like great stony rafts on the denser material below. In Wegener's view, the bulldozing forward edge of the slab might be expected to crumple and produce mountain ranges like the Andes.

The assiduous investigations of Wegener were not to go unchallenged. Criticism was leveled chiefly against his notion that the continents slid along through denser oceanic crust in the manner of giant granitic "icebergs." The eminent and scientifically formidable physicist Sir Harold Jeffries calculated that the ocean floor was far too rigid to allow for the passage of continents, no matter what the imagined driving mechanism. It is now known that Jeffries was correct in asserting that continents cannot—and do not—plow through oceanic crust. Shortly, it will be shown that continents do move, but they move only as passive passengers on large rafts of lithosphere that glide over a comparatively soft and plastic upper layer of the earth's mantle. Nevertheless, much of the evidence Wegener and others had assembled can be used to substantiate both the old and the newer concepts. We should have a look at this evidence.

The most convincing evidence for continental drift remains the geographic fit of the continents. Indeed, the correspondence is far too good to be fortuitous, even when one considers the expected modifications of shorelines resulting from erosion, deformation, or intrusions following the breakup of Pangaea some 200 million years ago. A still closer match results when one fits the continents together to include the continental shelves, which are really only submerged portions of the continents. Such a computerized and error-tested fitting of continents was carried out by Sir Edward Bullard, J. E. Everett, and A. G. Smith of the University of Cambridge (Fig. 11–2). Remarkably, this work showed that over most of the boundary the average mismatch was no more than 1°—a snug fit indeed.

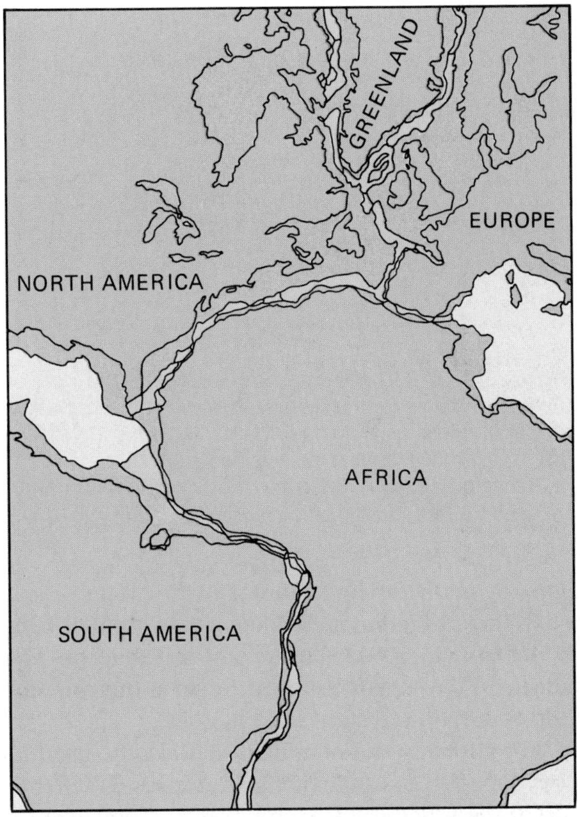

Figure 11–2 Fit of the continents as determined by Sir Edward Bullard, J.E. Everett, and A.G. Smith. The fit was made along the continental slope (green color) at the 500-fathom contour line. Overlaps and gaps (shown in pink) are probably the result of deformations and sedimentation after rifting. (Adapted from E.C. Bullard, et al. *Philos. Trans. R. Soc. Lond.* 258:41, 1965.)

Another line of evidence favoring the drift theory involves sedimentologic criteria indicating similarity of climatic conditions for widely separated parts of the world that were once closely adjacent to one another. For example, in such widely separated places as South America, southern Africa, India, Antarctica, Australia, and Tasmania, one finds glacially grooved rock surfaces and deposits of glacial rubble developed in the course of late Paleozoic continental glaciation. The deposits of poorly sorted clay, sand, cobbles, and boulders are called **tillites**. They, along with the grooves and scratches on rock surfaces that were apparently beneath the moving ice, attest to a great ice age that affected Gondwanaland at a time when it was as yet unfragmented and lying at or near the south polar region (Fig. 11–3). Furthermore, if the directions of the grooves on the bedrock are plotted on a map, they indicate the center of ice accumulation and the directions in which it moved. Unless the southern continents are reassembled into Gondwanaland, this center would be located in the ocean, and great ice sheets do not develop centers of accumulation in the ocean. Hence, the existence of Gondwanaland seems plausible. In a few in-

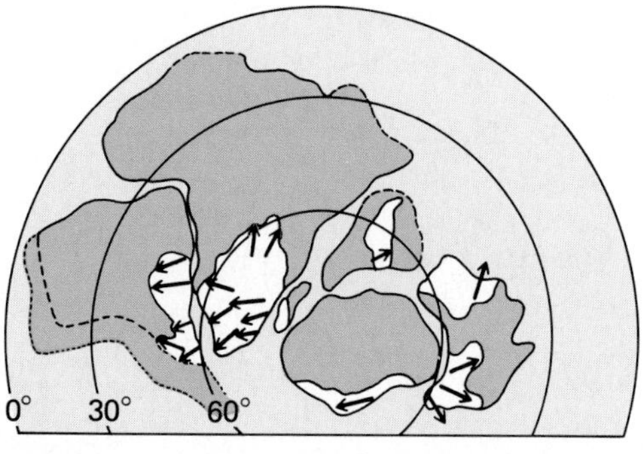

Figure 11–3 Reconstruction of Gondwanaland near the beginning of Permian time, showing the distribution of glacial deposits (shown in white). Arrows show direction of ice movement as determined from glacial scratches on bedrock. (Modified from W. Hamilton and D. Kinsley. *Geol. Soc. Am. Bull.* 78:783–799, 1967.)

stances, oddly foreign boulders in the tillites of one continent are found in the deposits of another continent now located thousands of miles across the oceans. Petrologists are able to trace their parent outcrops to the distant land masses.

Fossil plants and coal beds can also be used to test the concept of moving land masses. Trees that have grown in tropical regions of the globe characteristically lack the annual rings resulting from seasonal variations in growth. Exceptionally thick coal seams containing fossil logs from such trees imply a tropical paleoclimate. The locations of such coal deposits should approximate an equatorial zone relative to the ancient pole position for that age. Also, it is evident that such coal seams now being exploited in northern latitudes must have been moved to those locations from the equatorial zones along which the source vegetation accumulated.

Because of the decrease in solubility of calcium carbonate with rising temperatures, thick deposits of marine limestone also imply relatively warm climatic conditions. Arid conditions, such as those existing today on either side of the equatorial rain belt, can be recognized in ancient rocks by the presence of desert sandstones and evaporites. (Evaporites are chemical precipitates such as salt and gypsum that characteristically form when a body of water containing dissolved solids is evaporated.) Today such deposits are ordinarily formed in warm arid regions located about 30° north or south of the equator. If one believes continents have always been where they are now, then it is difficult to explain the great Permian salt deposits now found in northern Europe, the Urals, and the southwestern United States (Fig. 11–4). The evaporites had probably been precipitated in warmer latitudes before Laurasia migrated northward.

A somewhat similar kind of evidence can be obtained by examining the locations of Permian reef deposits. Modern reef corals are restricted to a band around the earth that is within 30° of the equator. Ancient reef deposits are now found far to the north of the latitudes at which they had originated.

At least some of the paleontologic support for continental drift was well know to Suess and Wegener. Usually, in the Gondwana strata overlying the tillites or glaciated surfaces, there can be found nonmarine sedimentary rocks and coal beds containing a distinctive assemblage of fossil plants. Named after a prominent member of the assemblage, the plants are referred to as the ***Glossopteris*** flora (Fig. 11–5). Paleobotanists who have supported the idea that the continents have shifted have argued that it would be virtually impossible for this complex temperate flora to have developed in identical ways on the southern continents as they are separated today. The seeds of *Glossopteris* were far too heavy to have been blown over such great distances of ocean by the wind.

Another element of paleontologic corroboration for the concept of moving land masses is provided by the distribution in the Southern Hemisphere of fossils of a small aquatic reptile named ***Mesosaurus*** (Fig. 11–6). An interpretation, based on its skeletal remains, of this animal's habits, as well as the nature of the sediment in which its fossils are found, strongly suggests that it once inhabited lakes and estuaries and was not an inhabitant of the open ocean. The discovery of fossil remains of *Mesosaurus* in both Africa and South America lends credence to the notion that these continents were once attached. Perhaps for a time, as they were just beginning to separate, the location of the present coastlines became dotted with lakes and protected bodies of water that harbored *Mesosaurus*. This seems a more plausible explana-

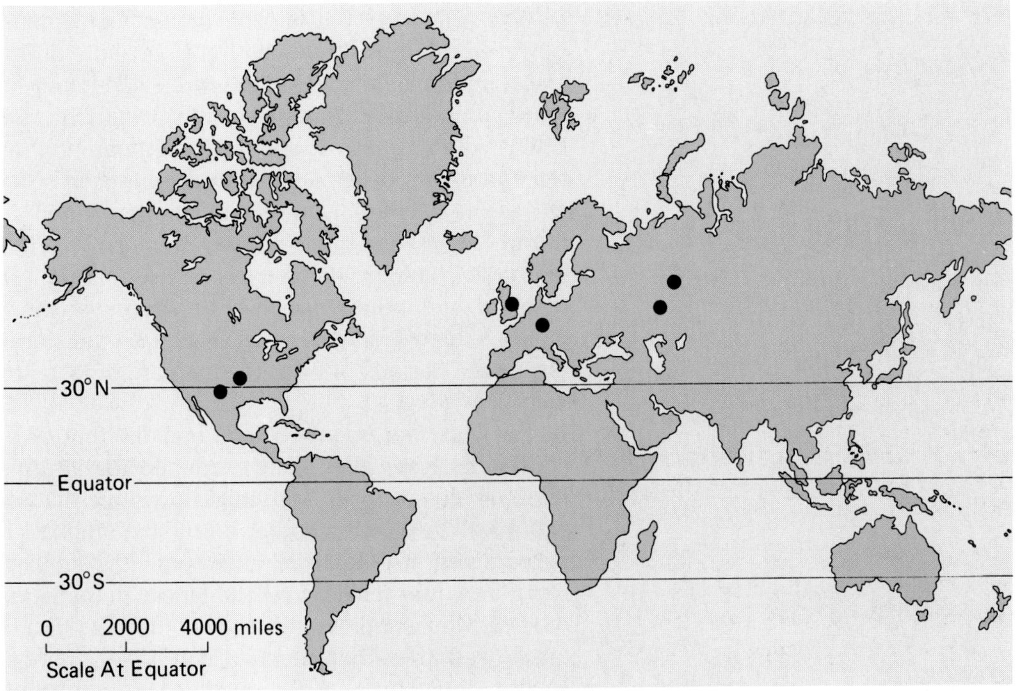

Figure 11–4 Location of prominent Permian evaporite deposits.

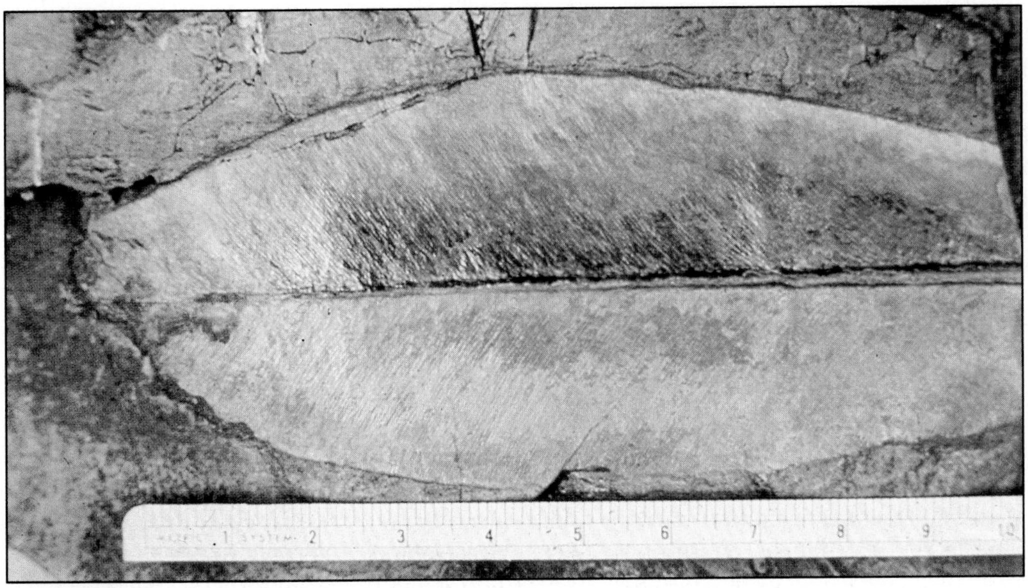

Figure 11–5 Fossil *Glossopteris* leaf associated with coal deposits and derived from glossopterid forests of Permian age. This fossil was found on Polarstar Peak, Ellsworth Land, Antartica. (Courtesy of the U.S. Geological Survey, J.M. Schopf, and C.J. Craddock.)

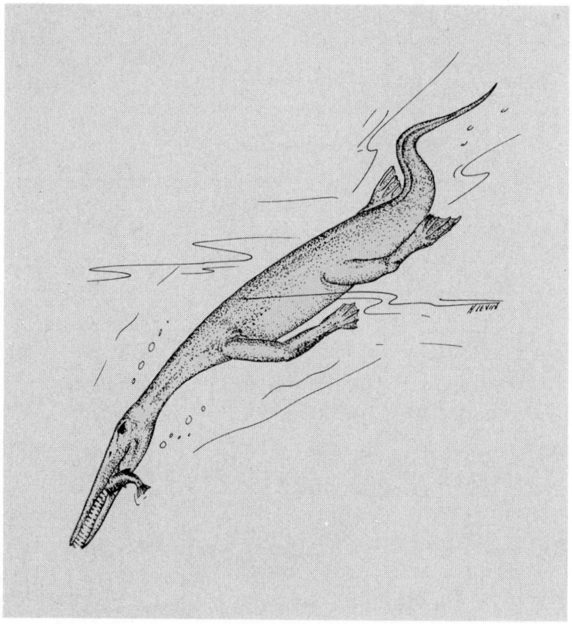

Figure 11–6 *Mesosaurus.*

tion than one requiring the creature to navigate the South Atlantic or to wend its way across formidable latitudinal climatic barriers by following the shorelines northward and around the perimeter of the Atlantic.

Before fragmentation of a supercontinent such as Laurasia, one would expect to find numerous similar plants and animals living at corresponding latitudes on either side of the line of future separation. Such similarities do occur in the fossil record of continents now separated by extensive oceanic tracts. Faunas of Silurian and Devonian fishes, for example, are comparable in such now-distant locations as Great Britain, Germany, Spitzbergen, eastern North America, and Quebec. These similarities, not only in fishes but also in amphibians, persist into the Carboniferous Period. In the Permian and periods of the Mesozoic Era, one again finds striking similarities in the reptilian faunas of Europe and North America. The fossil evidence implies not only the former existence of continuous land connections but also a uniformity of environmental conditions between the elements of Laurasia that once were at approximately similar latitudinal zones.

When one turns to Cenozoic mammalian faunas, however, one finds the situation to be quite different. Distinctive faunal elements are evident on separate continents and are especially evident in Australia, South America, and Africa. Apparently, as the continents became separated from each other by ocean barriers, genetic isolation resulted in morphologic divergence. The modern world's enormous biologic diversity is at least partially a result of evolutionary processes operating on more or less isolated continents. The faunas and floras during the periods prior to the breakup of Pangaea were less diverse.

The character, sequence, age, and distribution of rock units have also been examined for insights into concepts of drift. One might presume that locations close to one another on the hypothetic supercontinent would have environmental similarities that would result in resemblances in the kinds of rocks deposited. As indicated by the correlation chart of southern continents (Table 11–1), there is such a similarity in the geologic sections of now widely separated land masses. The sections begin with glacial deposits such as the Dwyka tillite of Africa and the Itararé tillites of South America. These cobbly layers are overlain by nonmarine sandstones and shales containing the *Glossopteris* flora and layers of coal. This sequence may indicate a warming trend from boreal glacial to more temperate climates. However, one must be careful with such interpretations, for our most recent continental glaciations occurred under a cool-temperate climatic regime. The next higher group of strata shows evidence of deposition under terrestrial conditions with an abundance of alluvial, eolian, and stream deposits that contain fossils of an interesting group of mammal-like reptiles. The uppermost beds of the sequences often include basalts and other volcanic rocks that are of similar age and composition in India and Africa but are somewhat younger in South America. The absence of the reptile-bearing Triassic strata and the Jurassic volcanics in Australia is believed to be the result of Australia's separating from Africa at an early date—probably early in the Permian—and long before the separation of Africa and South America. Australia must have retained just enough of a connection with Africa to permit the entry of dinosaurs and marsupials.

Table 11–1 Gondwana Correlations

SYSTEM	SOUTHERN BRAZIL	SOUTH AFRICA	PENINSULAR INDIA
Cretaceous	Basalt	Marine Sediments	Volcanics Marine Sediments
Jurassic	(Jurassic Rocks Not Present)	Basalt	Sandstone and Shale Volcanics Sandstone and Shale
Triassic	Sandstone and Shale with Reptiles	Sandstone and Shale with Reptiles	Sandstone and Shale with Reptiles
Permian	Shale and Sandstones with *Glossopteris* Flora Shale with *Mesosaurus*	Shale and Sandstones with Coal and *Glossopteris* Shale with *Mesosaurus*	Shale and Sandstones with Coal and *Glossopteris* Shale
Carboniferous	Sandstone Shale and Coal with *Glossopteris* Tillite	Sandstone Shale and Coal with *Glossopteris* Tillite	Tillite

Yet another way to test the notion of a supercontinent is to see whether or not geologic structures, such as the trends of folds and faults, match up when now-distant continents are hypothetically juxtaposed. Such correlative trends do exist. Folds and faults are often difficult to date, although if successful one can establish the contemporaneity of a fault lineage or fold axis now on separated continents. One may also examine the outcrop patterns of Precambrian basement rocks and discern correlative boundaries between now widely separated continents. A distinctive folded sequence of Precambrian strata in central Gabon, for example, can be traced into the Bahia Province of Brazil. Also, isotopically dated Precambrian rocks of West Africa can be correlated with rocks of similar age in northeastern Brazil.

Evidence in support of Wegner's theory of continental drift included the fit of continental coastlines, stratigraphic and structural similarities between continents, paleoclimatological reconstructions based on the global distribution of glacial and evaporite deposits, and the distribution of similar fossil groups.

The Testimony of Paleomagnetism

One can sympathize with Alfred Wegener in his losing battle with the imposing Sir Harold Jeffries. As previously noted, Wegener's ideas came along a half century too early. The scientific discoveries of the past two decades would have provided him with evidence that Jeffries would have had difficulty in refuting. The new information came from the study of magnetism that had been imparted to ancient minerals and rocks and preserved down to the present time. To understand this **paleomagnetism**, as it is called, it is necessary to digress for a moment and consider the general nature of the earth's present magnetic field.

The Earth's Present Magnetism

It is common knowledge that the earth has a magnetic field. It is this field that causes the alignment of a compass needle. The origin of the magnetic field is still a question that has not been fully resolved, but many geophysicists believe it is generated as the rotation of the earth causes slow movements in the liquid outer core. The magnetic lines of force resemble those that would be formed if there were an imaginary bar magnet extending through the earth's interior. The long axis of the magnet would be the conceptual equivalent of the earth's magnetic axis, and the ends would correspond to the north and south geomagnetic poles (Fig. 11–7). Although today the geomagnetic poles are located about 11° of latitude from the rotational axis, they slowly shift position. When averaged over several thousand years, the geomagnetic poles and the geographic poles do coincide. If we assume that this relationship has always held true, then by calculating ancient magnetic pole positions from paleomagnetism in rocks, we have coincidentally located the earth's geographic poles. It should be kept in mind, however, that such interpretations are based on the supposition that the rotational and magnetic poles have always been relatively close together. This seems a reasonable assumption based on the modern condition as well as on paleontologic studies that have shown inferred ancient climatic zones in plausible locations relative to ancient pole positions. Another assumption is that the earth has always been dipolar. Paleomagnetic studies from around the world thus far support this supposition.

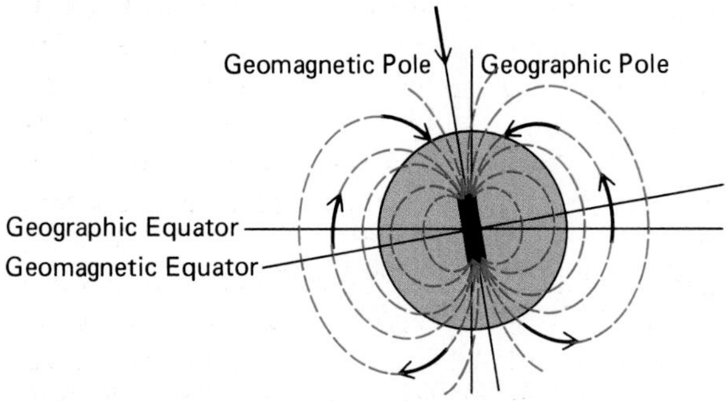

Figure 11–7 Magnetic lines of force for a simple dipole model of the earth's magnetic field.

Remanent Magnetism

The magnetic information frozen into rocks may originate in several ways. Imagine for a moment the outpouring of lava from a volcano. As the lava begins to cool, magnetic iron oxide minerals form and align their polarity with the earth's magnetic field. That alignment is then retained in the rock as its crystallization is completed. In a simple analogy, the magnetic orientations of the minerals responded as if they were tiny compass needles immersed in a viscous liquid. Because they are aligned parallel to the magnetic lines of force surrounding the earth, they not only point the way toward the poles (magnetic declination) but also become increasingly more inclined from the horizontal as the poles are approached (Fig. 11–8). This inclination, when detected in paleomagnetic analysis, can be used to determine the latitude at which an igneous body containing magnetic minerals cooled and solidified.

Another name for the magnetism frozen into ancient rocks is **remanent magnetism (RM)**. The type described previously, in which igneous rocks cool past the Curie temperature (also called the "Curie point") of its magnetic minerals, is further classified as *thermoremanent magnetism* The **Curie temperature** is simply that temperature above which a substance is no longer magnetic. A

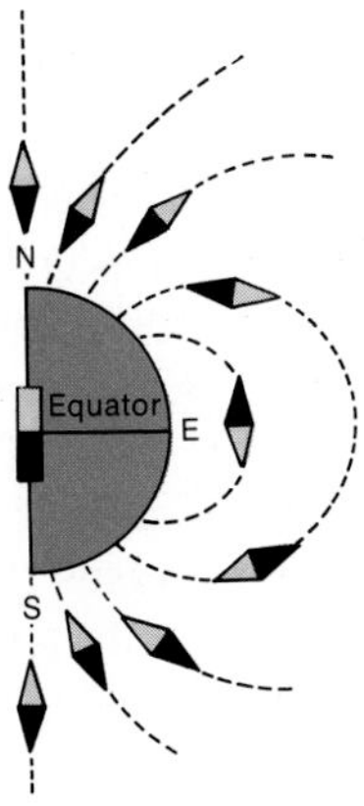

Figure 11–8 A freely suspended compass needle aligns itself in the direction of the earth's magnetic field. The inclination of the needle will vary from horizontal at the equator to vertical at the poles.

few minerals, the most important of which is magnetite (Fe_3O_4), have the property of acquiring RM. Magnetite is widespread in varying amounts in virtually all rocks. The manner in which RM is acquired is complicated but can be explained in a general way. Remanent magnetism in a mineral is ultimately due to the fact that some atoms and ions (charged atoms) have so-called magnetic moments; this means that they behave like tiny magnets. The magnetic moment of a single atom or ion is produced by the spin of its electrons. When magnetite takes on its RM, the iron ions align themselves within the crystal lattice so that their magnetic moments are parallel. Now most igneous rocks crystallize at temperatures in excess of 900°C. At these extreme temperatures, the atoms in the minerals have a large amount of energy and are being violently shaken. The vibrating shaking atoms are unable to line up until more cooling occurs. Finally, when the Curie temperature is reached (578°C for magnetite), enough energy has been lost to allow the atoms to come into alignment. They will stay aligned at still lower temperatures because the earth's field is not strong enough to alter the alignment already frozen into the minerals.

Thermoremanent magnetism develops in igneous rocks as they cool below the Curie temperature.

Igneous rocks are not the only kinds of earth materials that can acquire remanent magnetism. In lakes and seas that receive sediments from the erosion of nearby land areas, detrital grains of magnetite settle slowly through the water and rotate so that their directions of magnetization parallel the earth's magnetic field. They may continue to move into alignment while the sediment is still wet and uncompacted, but once the sediment is cemented or compacted, the depositional remanent magnetism is locked in.

Over the past two decades, geophysicists have been measuring and accumulating paleomagnetic data for all the major divisions of geologic time. Their results are partly responsible for the recent revival of interest in drift hypotheses. For example, when ancient pole positions were located on maps, it appeared that they were in different positions relative to a particular continent at different periods of time in the geologic past. Either the poles had moved relative to stationary continents, or the poles had remained in fixed positions while the continents shifted about. If the poles were wandering and the continents "stayed put," then a geophysicist working on the paleomagnetism of Ordovician rocks in France should arrive at the same location for the Ordovician poles as a geophysicist doing similar work on Ordovician rocks from the United States. In short, the paleomagnetically determined pole positions for a particular age should be the same for all continents. On the other hand, if the continents had moved and the poles were fixed, then we should find that pole positions for a particular geologic time would be different for different continents. The data suggest that this latter situation is the more valid.

Another way to view the data of paleomagnetism is to examine what are called **apparent polar wandering curves** (the name implies that the poles are moving, but as we have just noted, this is not likely). These curves are merely lines on a map connecting pole positions relative to a specific continent for various times during the geologic past. As shown in Figure 11–9, the curves for North America and Europe meet in recent time at the present North Pole. This means that the paleomagnetic data from recently formed rocks from both continents indicate the same pole position. A plot of the more ancient poles, however, results in two

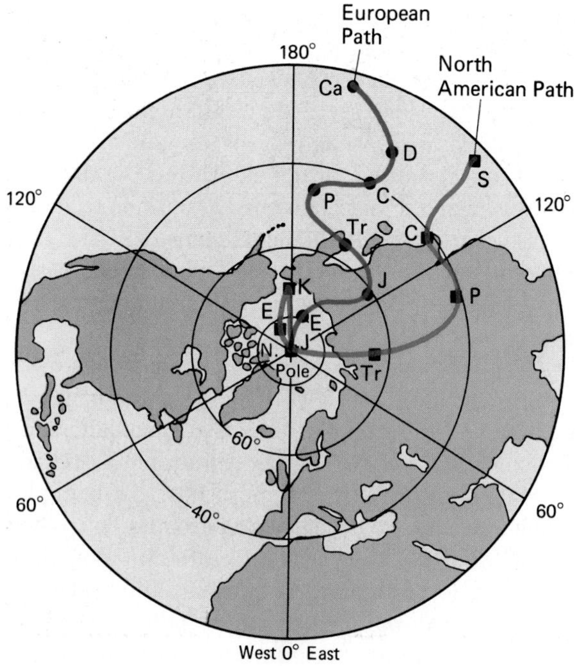

Figure 11–9 Apparent paths of polar wandering for Europe and North America. A scatter of points have been averaged to a single point for each geologic period: Ca, Cambrian; S, Silurian; D, Devonian; C, Carboniferous; P, Permian; Tr, Triassic; J, Jurassic; K, Cretaceous; E, Eocene. (After M.H.P. Bott. *The Interior of the Earth: Structure and Processes.* New York, St. Martin's Press, 1971.)

similarly shaped but increasingly divergent curves. If this divergence resulted from a drifting apart of Europe and North America, then one should be able to reverse the movements mentally and see if the curves do not come together. Indeed, the Paleozoic portions of the polar wandering curves can be brought into close accord if North America and its curve were to be slid eastward about 30° toward Europe. This sort of information gave new life to the old notion of continental drift, and contributed to the development of the theory of plate tectonics.

Apparent polar wandering curves for Europe and North America are offset by an amount that increases with increasing age. This offset disappears if the continents are moved back together.

Plate Tectonics

When considered at the general level, plate tectonics is a remarkably simple concept. The lithosphere, or outer shell of the earth (Fig. 11–10), is constructed of 7 huge slabs and about 20 smaller plates that are between them. The larger plates (Fig. 11–11) are approximately 75 to 125 km thick. Movement of the plates causes them to converge, diverge, or slide past one another, resulting in frequent earthquakes along plate margins. When the locations of earthquakes are plotted on the world map, they clearly define the boundaries of tectonic plates (see Fig. 10–12).

A tectonic plate containing a continent would have the configuration shown in Figure 11–12. Plates "float" upon a weak region of the upper mantle called the **asthenosphere** (from the Greek *asthenos*, "weak"). Geophysicists view the asthenosphere as a region of rock plasticity and flowage. Its presence was first detected by Beno Gutenberg on the basis of changes in seismic wave velocities. It should be noted that the boundary between the lithosphere and the asthenosphere does not coin-

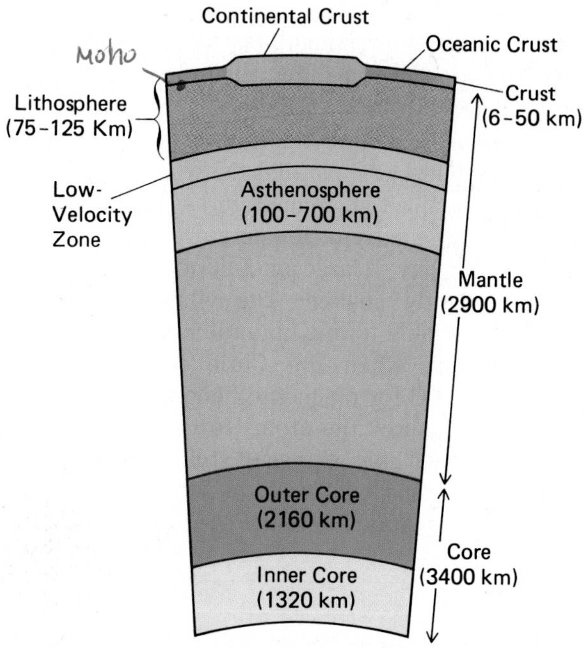

Figure 11–10 Divisions of the earth's interior. (For clarity, the divisions have not been drawn to scale.)

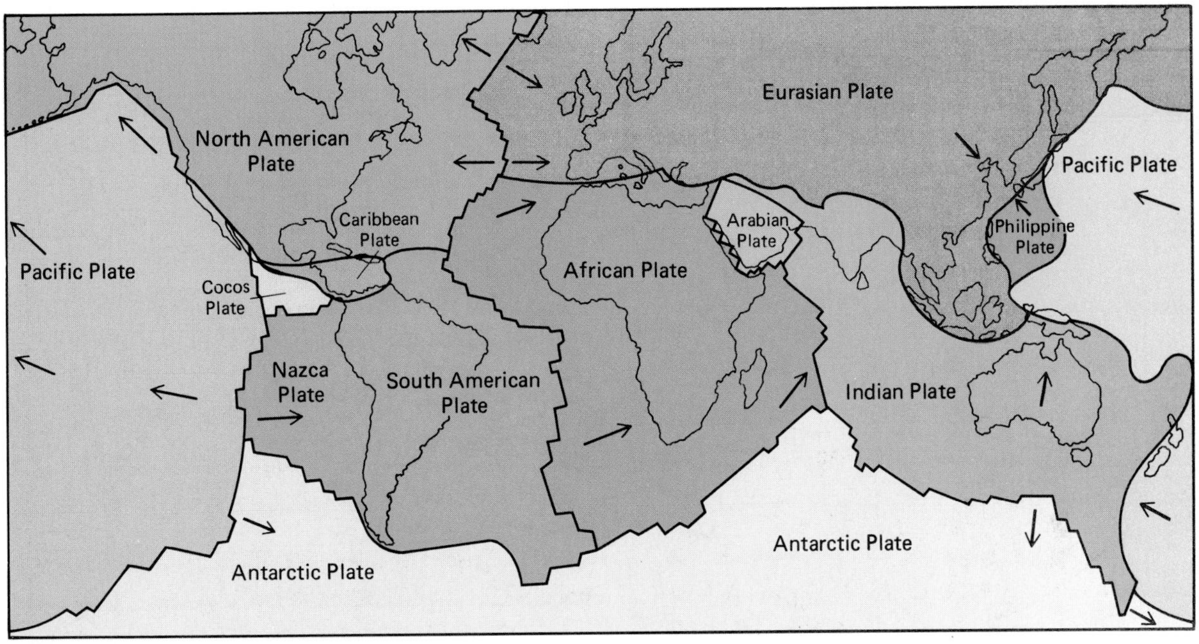

Figure 11–11 The earth's major tectonic plates. Arrows indicate general direction of movement of plates.

cide with the crust-mantle boundary but rather includes the upper mantle.

Plate Boundaries and Sea Floor Spreading

Central to the idea of plate tectonics is the differential movement of lithospheric plates. For example, plates tend to move apart at **divergent plate boundaries**, which may manifest themselves as mid-oceanic ridges complete with tensional ("pull-apart") geologic structures. Indeed, the mid-Atlantic ridge approximates the line of separation between the Eurasian and African plate on the one hand and the American plate on the other (Fig. 11–13). As is to be expected, such a rending of the crust is accompanied by earthquakes and enormous outpourings of volcanic materials that are piled high to produce the ridge itself. The void between the separating plates is also filled with this molten rock, which rises from below the lithosphere and solidifies in the fissure. Thus, new crust (i.e., new sea floor) is added to the **trailing edge** of each separating plate as it moves

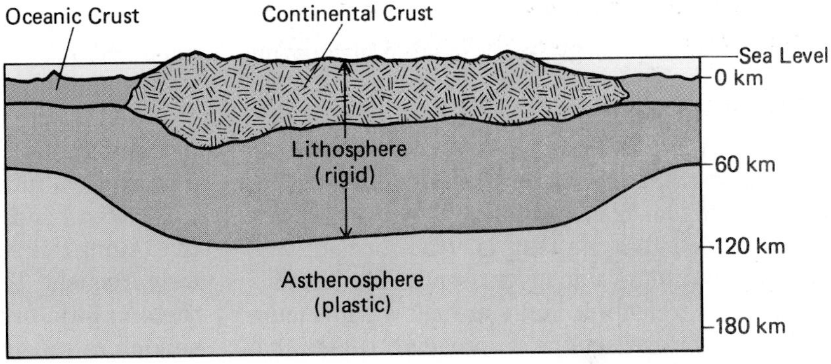

Figure 11–12 The outermost part of the earth consists of a strong relatively rigid lithosphere, which overlies a weak plastic asthenosphere. The lithosphere is capped by a thin crust beneath the oceans and a thicker continental crust elsewhere.

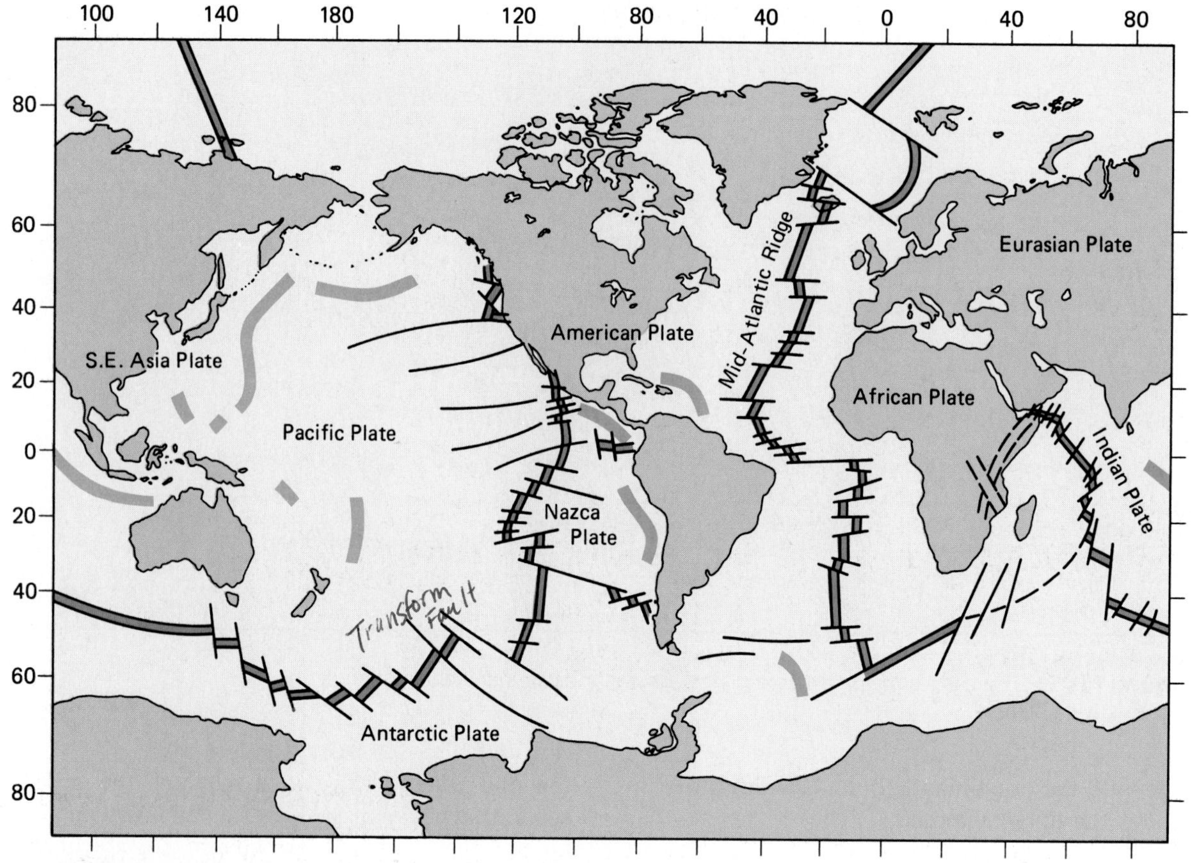

Figure 11–13 Locations of mid-oceanic ridges (red) and trenches (blue). Note fracture zones offsetting the ridges. Dashed lines in Africa represent the East African Rift Zone. (After B. Isacks, et al. *J. Geophys. Res.* 73:5855–5899, 1968. Copyright, The American Geophysical Union.)

slowly away from the mid-oceanic ridge (Fig. 11–14). The process has been appropriately named **sea floor spreading**. Zones of divergence may originate beneath continents, rupturing the overlying land mass and producing rift features like the Red Sea and Gulf of Aden.

The axis of spreading is not a smoothly curving line. Rather, it is abruptly offset by numerous faults. These features are termed **transform faults** and are an expected consequence of horizontal spreading of the sea floor along the earth's curved surface. Transform faults take their name from the fact that the fault is "transformed" into something different at its two ends. The relative motions of transform faults are shown in Figure 11–15. The ridge acts as a spreading center that exists to both the north and the south of the fault. The rate of relative movement on the opposite sides along fault segment X to X′ depends on the rate of extrusion of new crust at the ridge. Because of the spreading of the sea floor outward from the ridge, the relative movement is opposite to that expected by ordinary fault movement. Thus, at first glance, the ridge-to-ridge transform fault (Fig. 11–15A) appears to be a left lateral fault, but the actual movement along segment X–X′ is really right lateral. Notice, further, that only the X–X′ is really right lateral and that only the X–X′ segment is active. Along this segment, seismic activity is particularly frequent. To the west of X and the east of X′ there is little or no relative movement except for sinking as crustal materials cool.

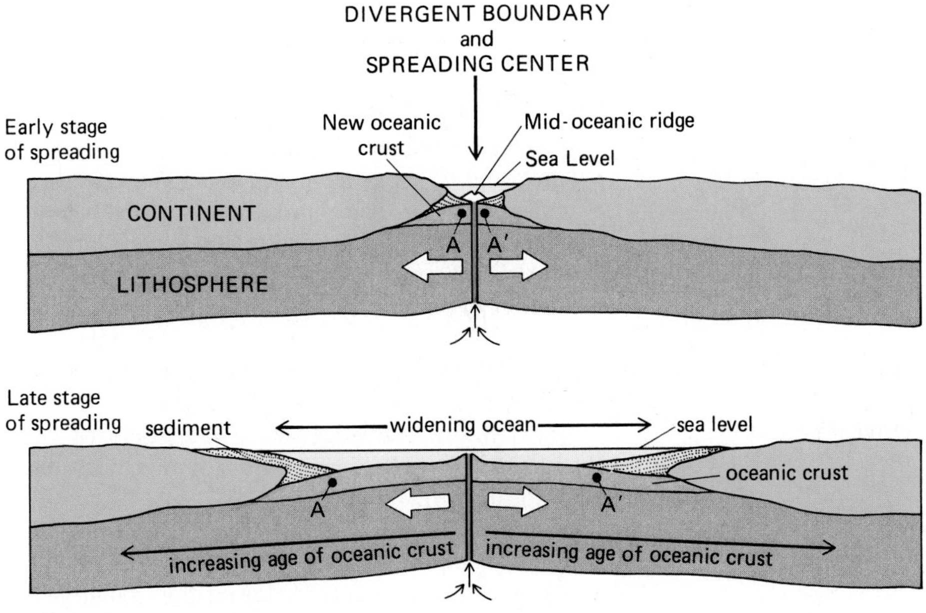

Figure 11–14 Sea floor spreading. As the rift widens, basaltic lavas rise to fill the space and solidify to become a new part of the trailing edge of a tectonic plate. A and A′ are reference points.

If plates are receding from one another at one boundary, they may be expected to collide or slide past other plates at other boundaries. Thus, in addition to the divergent plate boundaries that occur along mid-oceanic ridges, there are **convergent** and **shear boundaries**. Convergent plate boundaries develop when two plates move toward one another and collide. As one might guess, these convergent junctions are characterized by a high frequency of earthquakes. In addition, they are thought to be the zones along which folded mountain ranges or deep-sea trenches may develop.

The structural configuration of the convergent boundary is likely to vary according to the rate of spreading and whether the leading edges of the plates are composed of oceanic or continental crust. Geophysicists recognize that when the plates collide, one slab may slip and plunge below the other, producing what is called a **subduction zone**. The sediments and other rocks of this plunging plate are pulled downward (subducted), melted at depth, and, much later, rise to become incorporated into the materials of the upper mantle and crust. In some instances, the silicate melts provide the lavas for chains of volcanoes.

At subduction zones, ocean floor is forced downward along an inclined plane, creating trenches on the ocean floor, and generating earthquakes, volcanoes, and bodies of magma.

An example of a shear-plate boundary is the well-known San Andreas Fault in California. Along this great fault, the Pacific Plate moves laterally against the American Plate (Fig. 11–16). Shear-plate boundaries are decidedly earthquake-prone but are less likely to develop intense igneous activity. They are the active segments of transform faults along which no new surface is formed or old surface consumed.

Crustal Behavior at Plate Boundaries

We have noted that there are three kinds of plate boundaries: divergent, convergent, and shear. With regard to convergent boundaries, there are basically three events that may occur: 1) A plate bearing a continent at its leading edge may collide with

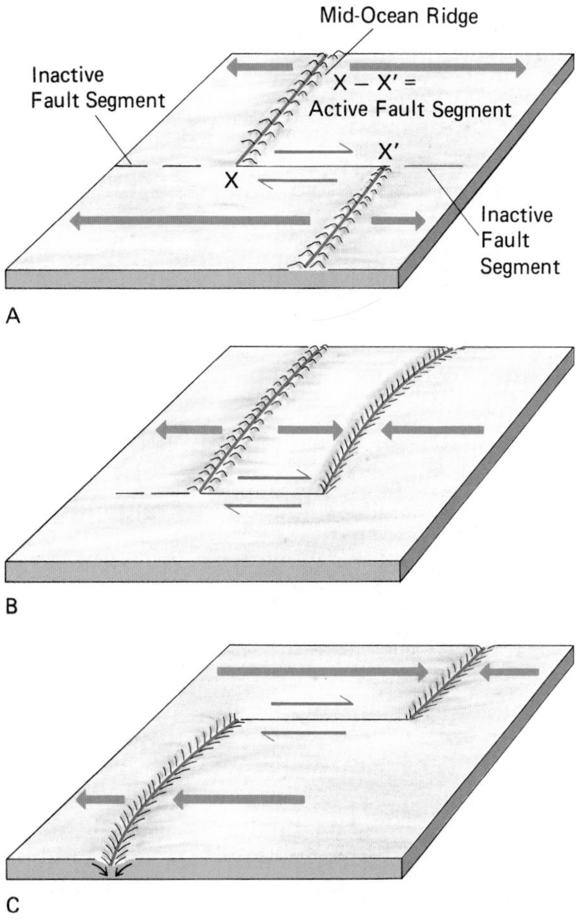

Figure 11–15 Three types of transform faults. (A) Ridge–ridge transform. (B) Ridge–trench. (C) Trench–trench.

another plate also bearing a continent at its leading edge; 2) a plate bearing oceanic crust may collide with a similar plate; or 3) a plate bearing oceanic crust may converge upon one bearing a continent.

Continent-to-Continent Collision

If the leading edge of a plate is composed of continental crust, and it collides with an opposing plate of similar composition, the result would be a fold and thrust mountain system containing stratovolcanoes, granitic plutons, and metamorphic belts. Such mountain systems may have notable symmetry, for relatively similar conditions would exist on both of the opposing plates. Because continental plate margins are too light and buoyant to be carried down into the asthenosphere, subduction does not occur in this type of collision (Fig. 11–17). Instead, the crust at the plate margins is deformed and may detach itself from deeper zones, and slabs of crust from one plate may be thrust under the other. The zone of convergence between the two plates, recognized by the severity of folding, faulting, and intrusive activity, is termed the **suture zone**.

Continental lithosphere is too thick and too light to be subducted. Hence, when continents converge, their margins are buckled, thickened, and deformed without subduction.

A dramatic example of an encounter between two continents occurred during the convergence of India and Eurasia (Fig. 11–18). The mighty Himalayas are the spectacular result of this smashup. The study of structures produced during this convergence has provided some interesting insights into the history of continent-to-continent encounters. It was at one time assumed that once a convergence had occurred, motion between opposing plates would abruptly cease. However, analysis of movement along faults associated with convergence and evidence provided by photographs taken by satellites, indicate that this is not true. The plates may continue to converge, albeit at a slower rate, long after making contact with one another. Some of the continued movement is the result of crustal shortening that occurs whenever a flat span of crust is crumpled into a smaller area. The amount of crustal shortening resulting from the India–Eurasia convergence has been estimated at about 1500 km. Geologists can account for 500 to 1000 km of that amount in the folds and thrust faults of the suture zone. The remaining 500 to 1000 km were apparently dissipated along major east-west trending lateral faults in China and Mongolia (Fig. 11–19). Thus, plate convergence may continue long after initial closure by lateral release of plate movement along strike-slip faults (lateral faults) in the region peripheral to the suture zone. Geologists find similar strike-slip faulting in older continent-to-continent convergences such as those that produced the Appalachians and Urals.

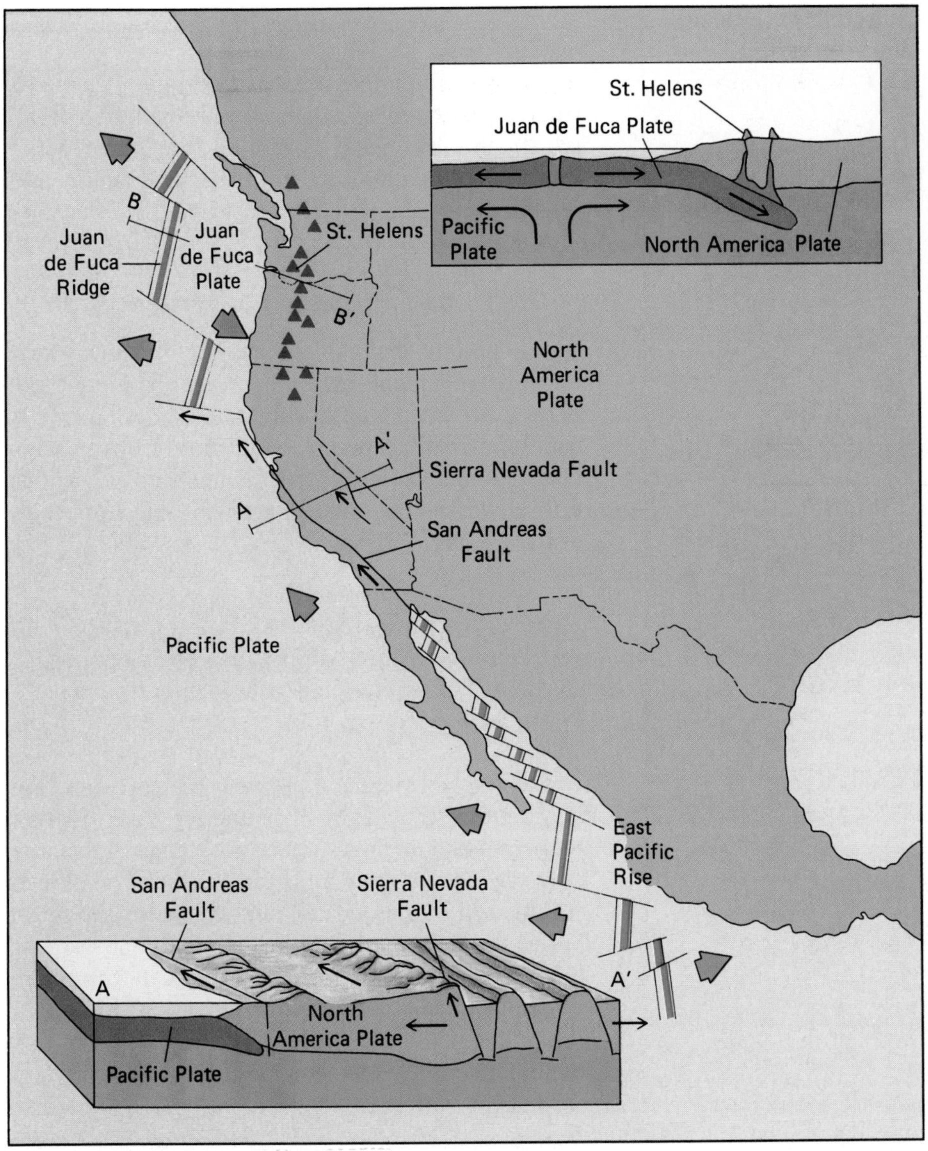

Figure 11–16 The juncture of the North American and the Pacific tectonic plates. The double lines are spreading centers. Note the trace of the San Andreas Fault. To the north in Oregon and Washington, the small Juan de Fuca Plate plunges beneath the North American continent to form the Cascades. (Courtesy of the U.S. Geological Survey.)

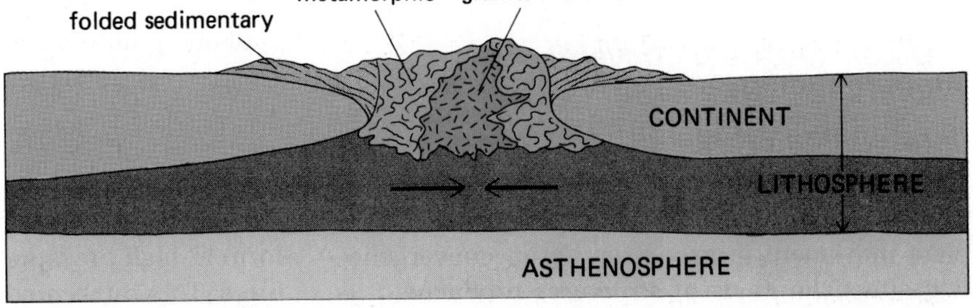

Figure 11–17 Collision of two tectonic plates, each bearing a continent.

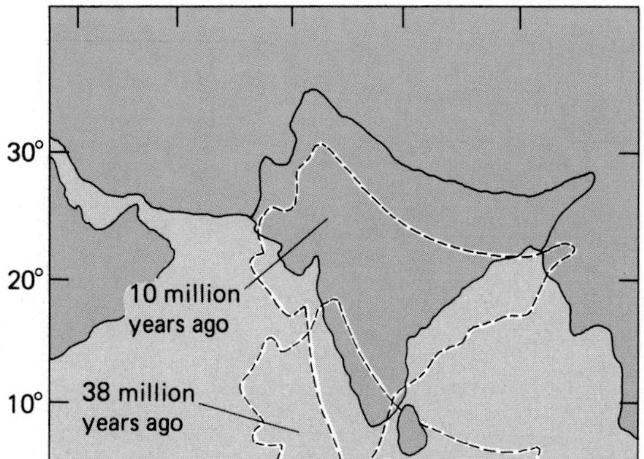

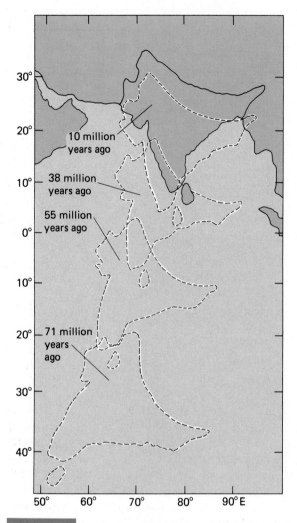

Figure 11–18 The northward migration of India relative to Eurasia, showing the position of the continent 71 million years ago (Cretaceous), 55 million years ago (Early Eocene), 33 million years ago (Oligocene), and 10 million years ago (Late Miocene). The configuration of the northern boundary of India is conjectural. (From P. Molnar and P. Tapponnier. Cenozoic tectonics of Asia: Effects of a continental collision. *Science* 189 (4201):419–426, Copyright 1975, AAAS.)

Ocean-to-Ocean Convergence

The second kind of convergence situation at plate boundaries involves the meeting of two plates that both have oceanic crust at their converging margins (Fig. 11–20). Although the rate of plate movement in an ocean–ocean convergence will affect the kinds of structures produced, it is likely that such locations will develop deep-sea trenches with volcanic island arcs developed parallel to the trenches. Examples are the Tonga-Kermadec Trench of the southwestern Pacific, and the Aleutian Trench west of Alaska.

Ocean-to-Continent Convergence

When a tectonic plate bearing oceanic crust encounters a plate margin bearing a continent, the denser oceanic crust is subducted (Fig. 11–21). A sea-floor trench is produced offshore from an associated range of mountains. Numerous volcanoes normally develop, and the lavas issuing from their eruptions are andesites.

The name andesite is taken from the Andes Mountains where this rock is the product of partial melting associated with subduction zones.

Regions of ocean–continent convergence are characterized by rather distinctive rock assemblages and geologic structure. As we have seen, the convergence of two lithospheric slabs results in subduction of the oceanic-plate, whereas the more buoyant continental plate maintains its position at the surface but experiences intense deformation, metamorphism, and melting. A great mountain range begins to take form as a result of all of this dynamic activity. At the same time, sediments and submarine volcanic rocks along the subduction zone are squeezed, sheared, and shoved into a gigantic, chaotic medley of complexly disturbed rocks called a **mélange**. Associated with the mélange one finds a distinctive assemblage of deep-sea sediments containing microfossils, sub-marine lavas, serpentinized peridotite, and gabbro. Altogether, these rocks constitute an **ophiolite suite** (Fig. 11–22). The ophiolite suite rock masses are actually splinters of the oceanic plate that were scraped off the upper part of the descending plate and inserted into the crushed forward edge of the continent. Ophiolites indicate the locations of ancient subduction zones, as do the presence of blueschists. As noted in Chapter 7, blueschists form at high pressures but relatively low temperatures. This rather unexpected combination of con-

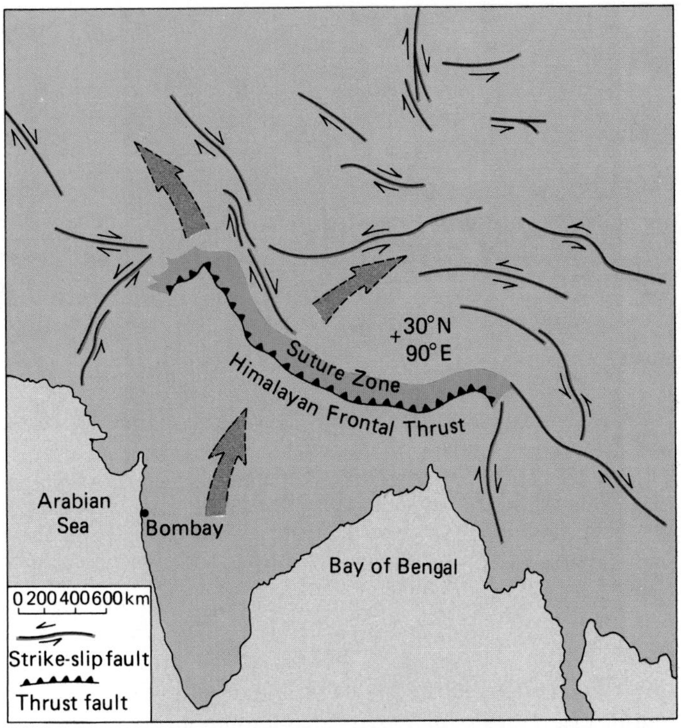

Figure 11–19 Major lateral (strike-slip) faults above and adjacent to the Indian suture zone. These faults may represent lateral transfer of some of the movement involved in the India–Eurasia convergence. Geologically recent movement along the faults implies that the plate movement involved in the convergence is still in progress. (Fault locations from P. Molnar and P. Tapponnier. Cenozoic tectonics of Asia: Effects of a continental collision. *Science* 189 (4201):419–426, Copyright 1975, AAAS.)

ditions is characteristic of subduction zones where the relatively cool oceanic crust and its cover of marine sediment plunges rapidly attaining high pressure before there has been enough time for heating.

What Drives it All?

Once we accept the fact that plates of lithosphere do move across the surface of the globe, then the next question to ask concerns how they are moved. The propelling mechanism is thought by some scientists to consist of large thermal convection cells that flow like currents of thick liquid and are provided with heat from the decay of radioactive minerals (Fig. 11–23A). The convection cells are believed to be located in the asthenosphere or, less likely, in the deeper parts of the mantle. Currents are thought to rise in response to heating from below and then to diverge and spread laterally. As they move laterally, the currents carry along with them the overlying slab of lithosphere and its surficial layers of sediment. Mantle material upwelling along the line of separation would join

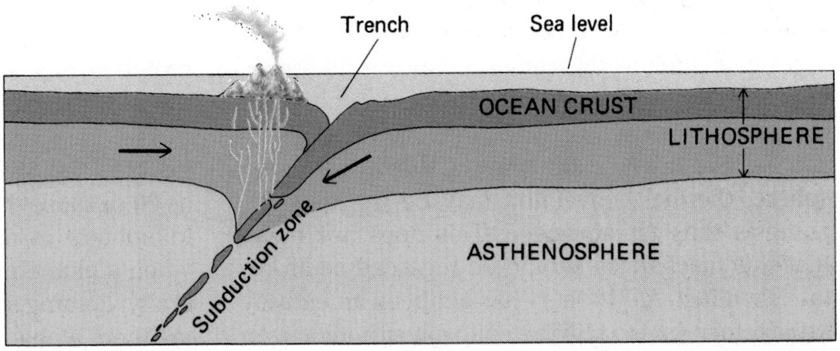

Figure 11–20 Convergence of two plates, both of which bear oceanic crust on their leading margins.

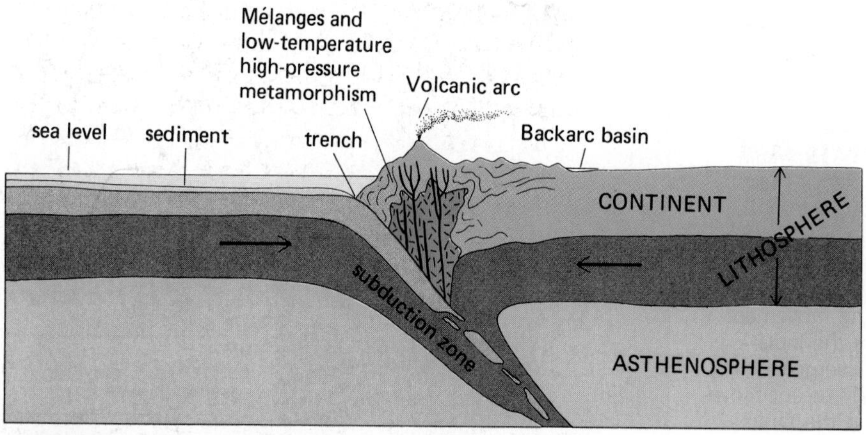

Figure 11–21 Convergence of the continental part of a plate with the oceanic part of another plate. The leading edge of the continental plate is crumpled, whereas the oceanic plate buckles downward, creating a trench offshore. The situation generalized here is similar to that off the west coast of South America where the Nazca Plate plunges beneath the South American Plate.

the trailing edges of the plates on either side. Ultimately, the convecting current would encounter a similar current coming from the opposite direction, and both viscous streams would descend into the deeper parts of the mantle to be reheated and moved toward the direction of an upwelling. Above the descending flow, one might expect to find subduction zones and deep-sea trenches, whereas mid-oceanic ridges would mark the locations of ascending flows.

It is possible only to speculate about the size and number of convection cells; it is very likely that they might have changed through the long course of earth history. Initially, there may have been only one large convecting system that swept together scumlike slabs of continental crust into a single land mass. Later systems may have been characterized by larger numbers of convection cells located closer to the base of the lithosphere. A concept such as this may account for an apparent increase in continental fragmentations in the Mesozoic compared with the Paleozoic.

As a mechanism for moving plates of lithosphere, thermal convection may be the best hypothesis thus far advanced. This does not mean that this mechanism should be regarded as an established fact. As yet there has not been an entirely satisfactory way to test the concept, although such currents may be physically plausible, and there is some tenuous evidence that they exist. For example, heat flows from the earth's interior at a greater rate along mid-oceanic ridges than from adjacent abyssal plains. Yet analyses of hundreds of measurements by geophysicists at Columbia University suggest that although heat flow along the Atlantic Ridge is 20 per cent greater than on the adjacent floors of the ocean basins, it is still not great enough to move the sea floor at the rates suggested by paleomagnetic studies. Another problem relates to the layering in the upper mantle that has been detected in seismologic studies. The mixing that accompanies convectional overturn would seem to preclude such layering.

The search for a mechanism to drive lithospheric plates has produced another convection-related model. In this so-called **thermal plume model** (see Fig. 11–23C) mantle material does not circulate in great rolls but rather rises from near the core–mantle boundary in a manner suggestive of the shape and motion of a thundercloud. Proponents of this theory suggest there may be as many as 20 of these thermal plumes, each a few hundred kilometers in diameter. According to the model, when a plume nears the lithosphere, it spreads laterally, doming surficial zones of the earth and moving them along in the directions of radial flow. The

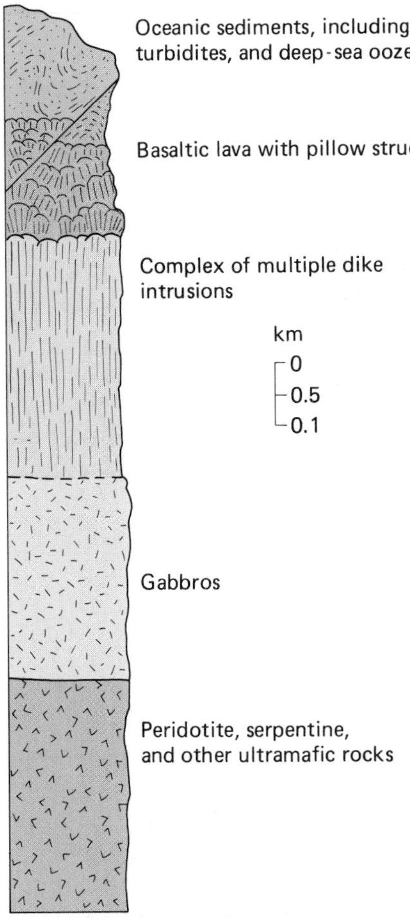

Figure 11–22 Idealized section of an ophiolite suite. Ophiolites are thought to be splinters of the ocean floor squeezed into the continental margin during plate convergence.

center of the Afar Triangle in Ethiopia (Fig. 11–24) has been suggested as the site of a plume that flowed upward and outward, carrying the Arabian, African, and Somali plates away from the center of a triple juction of rifts that meet at angles of about 120°. Two of the rifts, now occupied by the Gulf of Aden and the Red Sea, are still active and are presently widening. The third, the African Rift, is apparently now inactive, and may never admit the sea. Triple rifts in which two rifts are active and a third fails are characteristic of the break-up of continents. The two active rifts mark the locations of new continental margins.

Faced with the uncertainties in all existing theories for plate-moving forces, one should keep an open mind about other, perhaps less popular, hypotheses. Among other mechanisms proposed are "slab-pull" and "ridge-push." Slab-pull is thought to operate at the subduction zone, where the subducting oceanic plate, being colder and denser than the surrounding mantle, sinks actively through the less dense mantle, *pulling* the rest of the slab along as it does so (see Fig. 11–23B). The ridge-push mechanism, which may operate independently of, or along with, slab-pull, results from the fact that the spreading centers stand high on the ocean floor and have low-density roots. This configuration causes them to spread out on either side of the mid-oceanic ridge and transmit this "push" to the tectonic plate.

Tests of the Theory of Plate Tectonics

We have examined the various lines of evidence supporting Wegener's notion of continental drift. Nearly all of these clues also support the newer concepts embodied in the theory of plate tectonics. The earlier clues were based mostly on evidence found on land. The new theory, with its keystone concept of sea floor spreading, was developed from evidence gleaned from the sea floor.

Slightly before the time when geophysicists like Harry Hess of Princeton University began formulating ideas of sea floor spreading, other scientists were puzzling over findings related to paleomagnetism. The new data were obtained from sensitive magnetometers that were being carried back and forth across the oceans by research vessels. These instruments were able to detect not only the earth's main geomagnetic field but also local magnetic disturbances or magnetic anomalies frozen into the rocks along the sea floor. Maps were produced that exhibited linear anomaly bands of high and low field-magnetic intensities parallel to the west coast of North America (Fig. 11–25). Surveys of magnetic field strength on traverses across the mid-Atlantic ridge revealed a similar pattern of symmetrically distributed belts of anomalies (Fig. 11–26). Directly over the ridge, the earth's magnetic field was 1 per cent stronger than expected. Adjacent to this zone the field was somewhat weaker than would have been predicted, and the

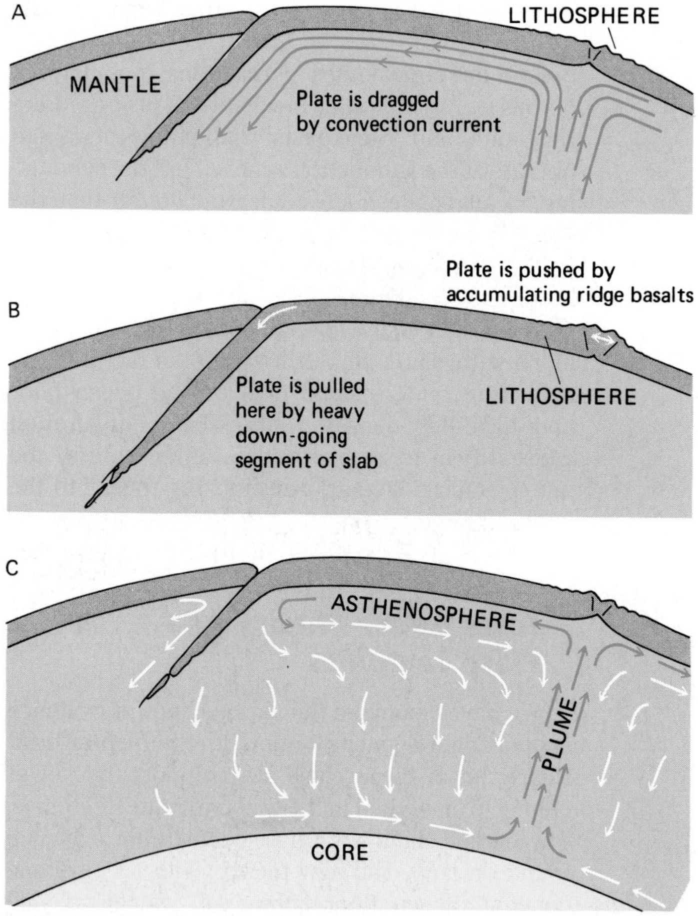

Figure 11–23 Three models that have been suggested as possible driving mechanisms for plate movements. (A) Plate dragged by roll-like convection movements in the mantle. (B) The "push–pull" model, in which plates are pushed laterally at spreading centers and pulled by the plunging cool and dense leading segment of the plate. (C) Thermal plume model, in which upward movement is confined to thermal plumes that spread laterally and drag the lithosphere along.

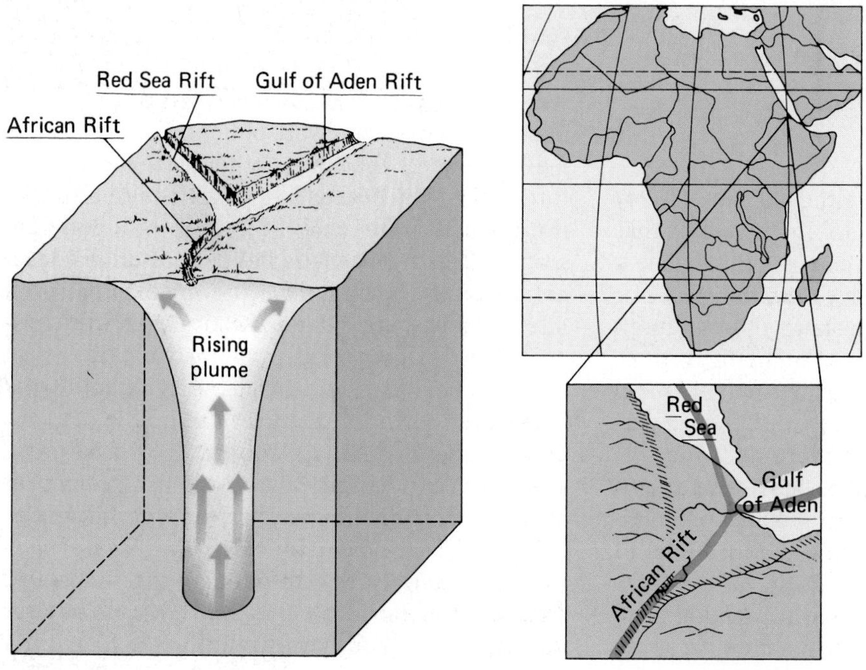

Figure 11–24 Rising plumes of hot mantle may cause severe rifts, often forming 120° angles with one another. An example is the Afar Triangle, shown at the south end of the Red Sea on the small map of Africa.

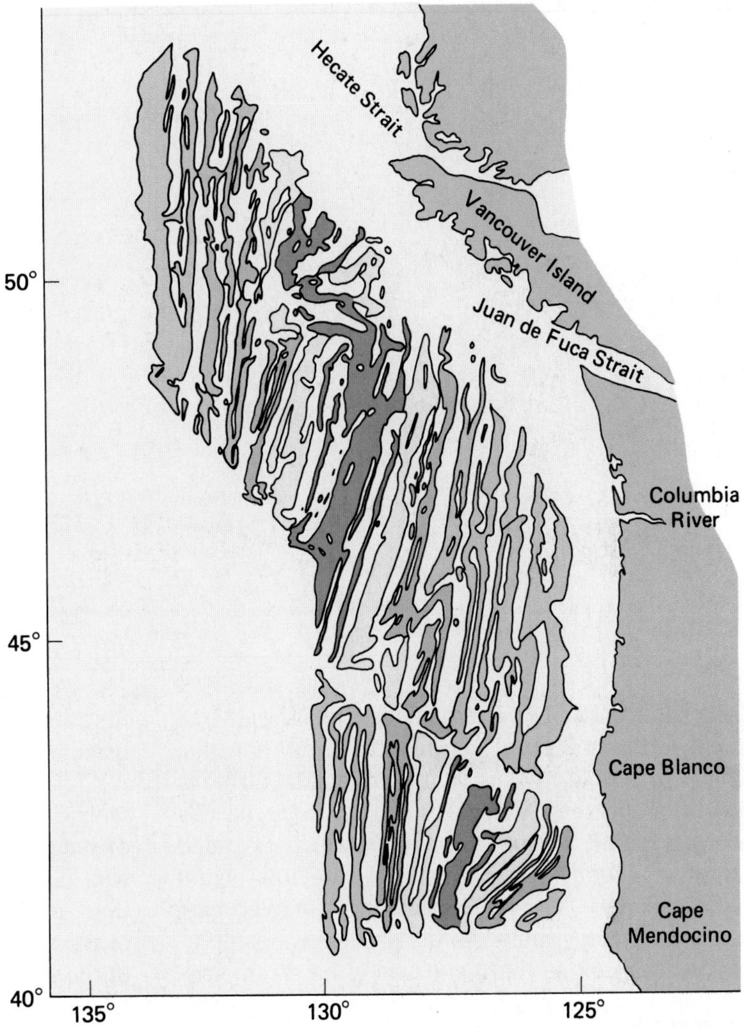

Figure 11–25 Magnetic field produced by the rocks on the floor of the northeast Pacific. Note the symmetry with respect to the ridges (the red areas). The ages of the rocks increase away from the ridges. The larger gray-green areas are about 8 to 10 million years old. (After Raff, A.D., and Mason, R.G., Magnetic survey of the West Coast of North America, 40° N-52° N. latitude. *Bull. Geol. Soc. Am.*, 72:1267–1270, 1961.)

changes from weaker to stronger and back again often occurred in distances of only a few score miles.

In the early 1960s, geophysicists had reached the conclusion that these variations in intensity were caused by reversals in the polarity of the earth's magnetic field. The magnetometers towed behind the research vessels provided measurements that were the sum of the earth's present magnetic field strength and the paleomagnetism frozen in the crustal rocks of the ocean floor. If the paleomagnetic polarity was opposite in sign to that of the earth's present magnetic field, the sum would be less than the present magnetic field strength. This would indicate that the crust over which the ship was passing had reversed paleomagnetic polarity. Conversely, where the paleomagnetic polarity of the sea floor basalts was the same as that of the present magnetic field, the sum would be greater, and a normal polarity would be indicated.

Since 1963, geophysicists have learned that these irregularly occurring reversals have not been at all infrequent. During the past 70 million years, the earth's magnetic field has seen many episodes when the polarity was opposite to that of today (Fig. 11–27). These changes in polarity were incorporated into the remanent magnetism of the lavas extruded at the mid-oceanic ridges. The lava would acquire the magnetic polarity present at the time of

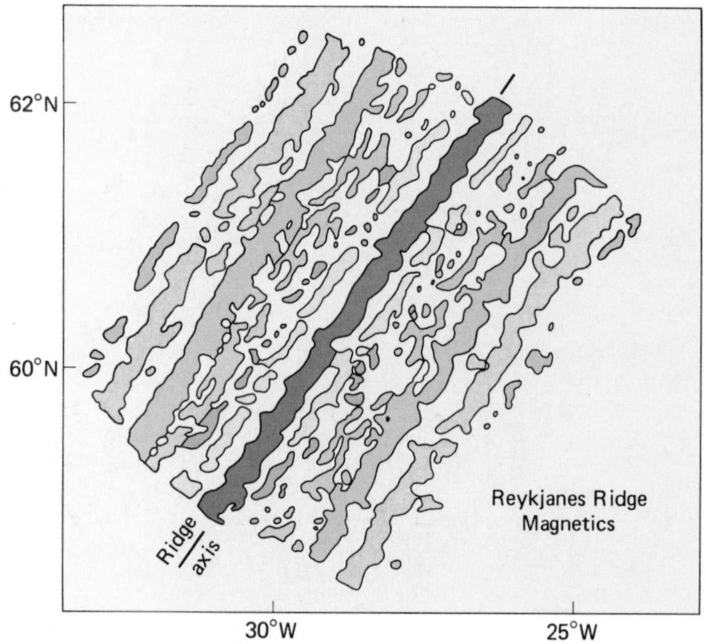

Figure 11–26 The magnetic field produced by the ocean-floor rocks over the Reykjanes Ridge near Iceland. The outlined areas represent predominantly normal magnetization. The ages of the rocks increase away from the ridge. (After Heirtzler, J.R., et al., Magnetic anomalies over the Reykjanes ridge, *Deep Sea Res.*, 13:427–443, 1966; and Vine, F.J., Magnetic anomalies associated with mid-ocean ridges, *in* Phinney, R.A. (ed.), *The History of the Earth's Crust, A Symposium*, Princeton, Princeton University Press. 1968.)

extrusion and then would move out laterally, as has been previously described. If during that time the earth's polarity was as it is today, the stripe is said to represent "normal" polarity. If the earth's normal polarity reversed, the band of extruded lavas that followed behind the previous band would acquire "reverse" polarity as it cooled past the Curie temperature. As the process repeated itself through time, the result would be symmetric, "mirror image" patterns of normal and reverse bands on either side of spreading centers like the mid-oceanic ridges (Fig. 11–28).

The discovery of normal and reverse magnetic stripes on the ocean floor was strong, substantiating evidence for sea floor spreading. What was needed next was a method for determining the age of large areas of the ocean floor so that the rate of ocean floor movement could be ascertained. Cores of basalt from the oceanic crust were not suitable for radiometric dating because basalt is altered, when hot, by contact with sea water. By lucky coincidence, a method for determining the age of the stripes was provided through the work of A. Cox, R. R. Doell, and G. B. Dalrymple. These geophysicists were investigating the magnetic properties of lava flows on the continents. They were able to accurately identify periodic reversals of the earth's magnetic field imprinted in superimposed layers of basalt (Fig. 11–29). Many of the layers in these superpositional sequences could be dated by radiometric methods. With such data, the time sequence of magnetic reversals for the past 5 million years was determined. The final step was to correlate these dates obtained from land basalts to the rocks of the sea floor. Confirmation for the ages of some of the oceanic basalts was obtained from the study of fossils in overlying sediments.

With knowledge of the age of particular normal or reverse magnetic stripes, it was now possible to calculate rates of sea floor spreading and sometimes to reconstruct the former positions of continents. Figure 11–30 illustrates how this can be done. For example, if one wishes to know the distance between the eastern coast of the United States and the northeastern coast of Africa about 81 million years ago, one brings together the traces of the two 81-million-year-old magnetic stripes, being careful to move the "sea floor" parallel to the transform faults, which indicate the direction of movement.

The velocity of plate movement is not uniform around the world. Plates that include large continents tend to move slowly. Their velocity relative to the underlying mantle rarely exceeds 2 cm per

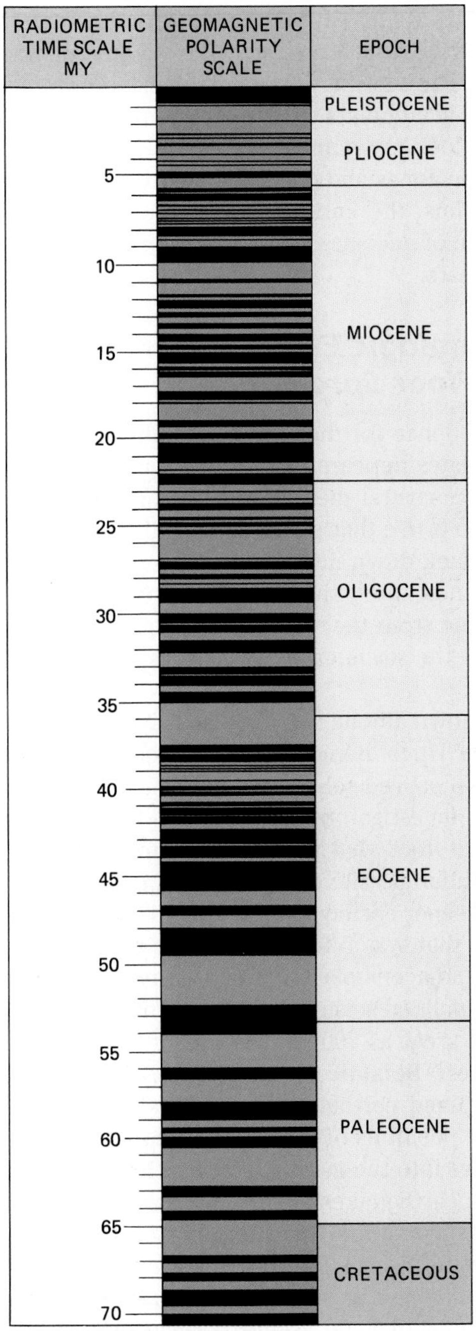

Figure 11–27 Reversals of the earth's magnetic field during the past 70 million years. The field was "normal," as today, during intervals shown in black, (Modified from J.R. Heirtzler et al. *J. Geophys. Res.* 73:2119–2136, 1968. Copyright, The American Geophysical Union.)

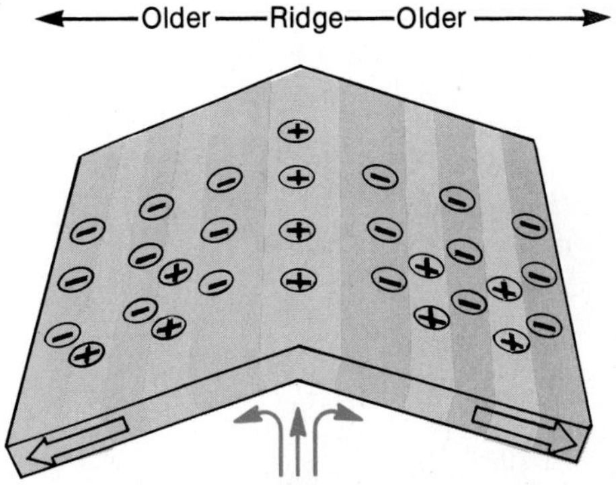

Figure 11–28 The normal (+) and reversed (−) magnetizations of the sea floor. Note the symmetry of the magnetizations with respect to the ridge. (From J.M. McCormick and J.V. Thiruvathukal. *Elements of Oceanography,* 2nd ed. Philadelphia, Saunders College Publishing, 1981.)

year. Plates that are largely devoid of large continents have average velocities between 6 and 9 cm per year. Figure 11–31 provides a summary of the rates and directions of sea floor spreading as determined by analyses of displacements along transform faults, magnetic anomalies, and other geophysical data.

The age of the ocean floor and rates of sea floor spreading can be determined from the study of magnetic reversal bands on the ocean floor and their correlation to the stratigraphy of magnetic reversals on land.

Evidence of Sea Floor Spreading from Oceanic Sediment

If newly formed crust joined the ocean floor at spreading centers and then moved outward to either side, then the sediments, dated by the fossil planktonic organisms they contain, could be no older than the surface on which they came to rest. Near a mid-oceanic ridge, the sediments directly over basalt should be relatively young. Samples of the first sedimentary layer above the basalt but progressively farther away from the spreading center should be progressively older (Fig. 11–32).

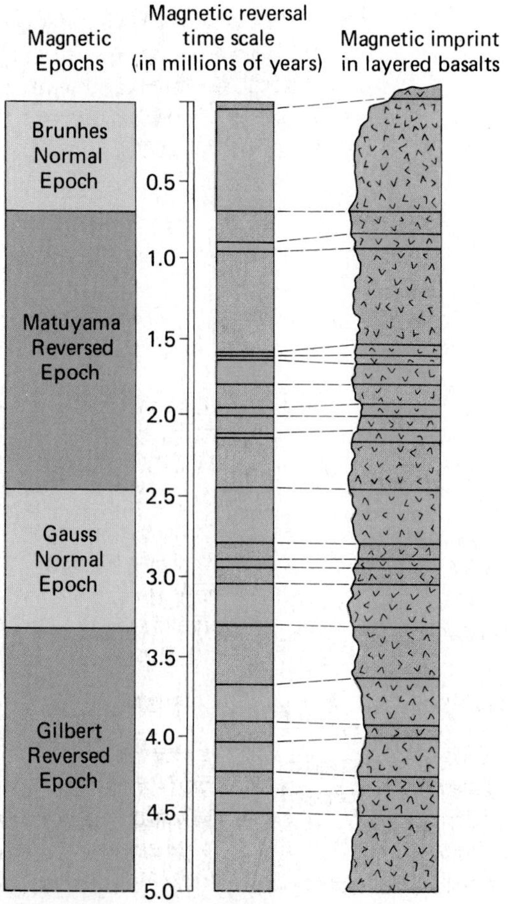

Figure 11–29 The imprint of normal and reversed magnetic polarity in a composite section of layered basalts from many continental localities. The magnetic reversal time scale is shown in the center column, and at the left are the *magnetic epochs* named after famous investigators of magnetic effects. The magnetic epochs reflect a predominance of one kind of polarity. Within the larger magnetic epochs are *magnetic events* of shorter duration. (Modified from Cox, A., Geomagnetic reversals, *Science,* 163:237–245, 1969.)

In a succession of cruises that began in 1969, the American drilling ship *Glomar Challenger* had collected ample evidence to prove that the previously stated conclusions on the age distribution of sediments are correct. The studies carried out by scientists associated with the drilling project confirmed earlier assumptions that there were no sediments older than about 200 million years on the sea floor, that the sediments on the sea floor were relatively thin, and that they became thinnest closer to the mid-oceanic ridges. Spreading provided a logical reason for these observations. A given area of oceanic plate surface was simply not in existence long enough to accumulate a thick section of sediments. The sea floor clays and oozes were conveyed to subduction zones and dragged back down into the mantle. Thus, the entire world ocean is virtually swept free of deep-sea sediments every 200 or 300 million years.

Seismologic Evidence for Sea Floor Spreading

The evidence for the actual movement of lithospheric plates appears to be fairly direct and susceptible to several methods of testing. It is more difficult to prove that plates of the lithosphere are dragged back down into the mantle and resorbed. The best indications that subduction actually does occur come from the study of earthquakes generated along the seismically active colliding edges of plates.

Well known among these studies are the investigations of Hugo Benioff, a seismologist and designer of improved seismographs. Benioff and subsequent investigators compiled records of earthquake foci that occurred along presumed plate boundaries. The charting of the data showed that the deeper earthquake foci occur along a narrow zone that was tilted at an angle of 30° to 60° under the adjacent island arc or continent (Fig. 11–33). This inclined seismic shear zone was traced to depths as great as 700 km (435 mi). It was named the **Benioff Seismic Zone** after its discoverer. These inclined earthquake zones are believed to define the positions of the subducted plates where they plunge into the mantle beneath the overriding plate. The earthquakes are the result of fracturing and subsequent rupture as the cool, brittle lithospheric plate descends into the hot mantle.

In addition to the deep and intermediate focus earthquakes along the Benioff Zone, there are intense and often destructive earthquakes at shallow depths where the edges of two rigid plates press against one another. The 1964 Alaskan earthquake originated at this shallow level. Such quakes are thought to occur along the shear plane between the subducting oceanic lithosphere and the continental lithosphere.

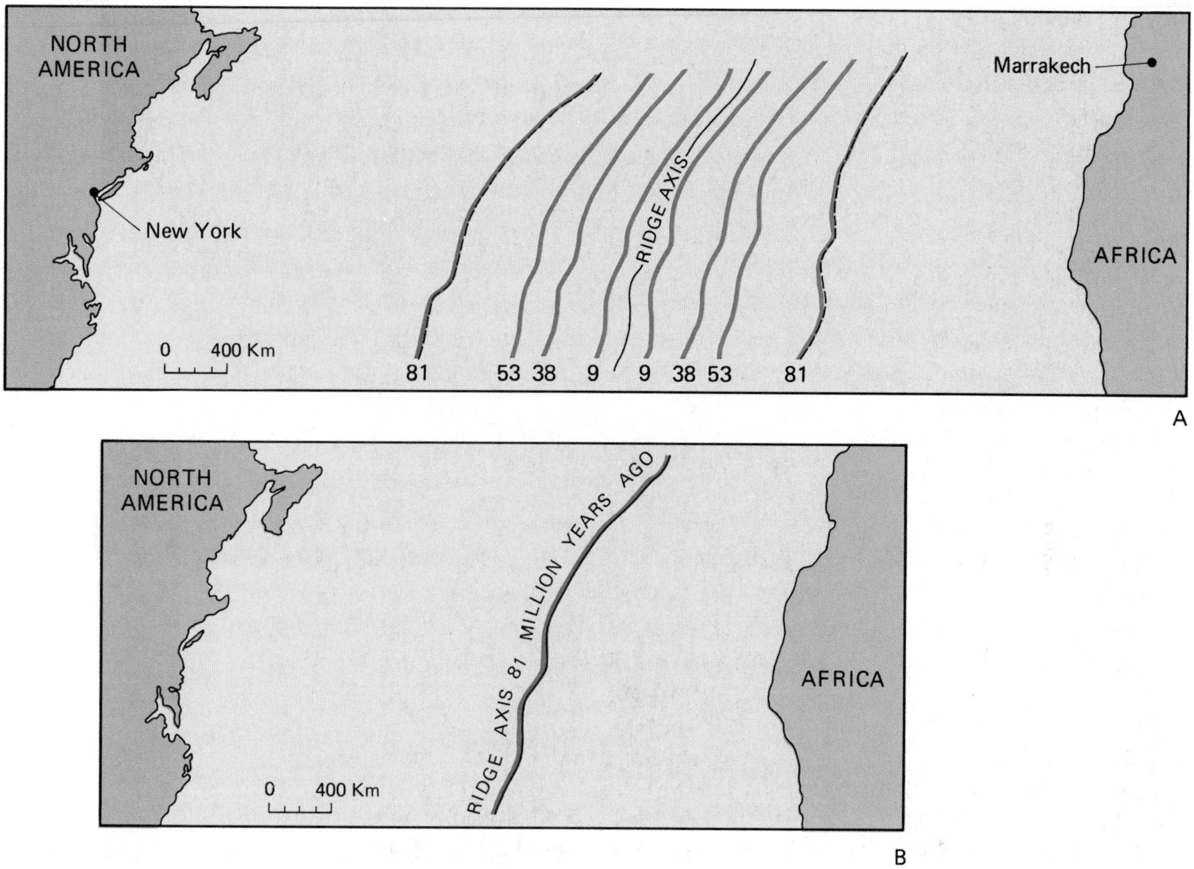

Figure 11–30 A method for determining the paleogeographic relations of continents. (A) The present North Atlantic with ages plotted for some of the magnetic stripes. To see the location of North America relative to Africa, say, 81 million years ago, the 81-million-year-old bands are brought together. (B) The result is the view of the much narrower North Atlantic of 81 million years ago. (Modified from Pitman III, W.C., and Talwani, M., *Geol. Soc. Am. Bull.* 83:619–644, 1972.)

In addition to trenches and andesitic volcanism, subduction zones are recognized by the presence of Benioff Zones along which numerous shallow, intermediate, and deep earthquake foci are located.

The Role of Gravity

Gravity measured over deep-sea trenches is characteristically lower than that measured in areas adjacent to the trenches. Geophysicists refer to such phenomena as negative gravity anomalies. A **gravity anomaly** is the difference between the observed value of gravity at any point on the earth and the computed theoretic value. Negative gravity anomalies occur where there is an excess of low-density rock beneath the surface. The very strong negative gravity anomalies extending along the margins of the deep-sea trenches can mean only that such belts are underlain by rocks much lighter than those at depth on either side. Because the zone of lower gravity values is narrow in most places, it is believed to mark trends where lighter rocks dip steeply into the denser mantle. Geophysicists assume that the less dense rocks of the negative zone must be held down by some force to prevent them from floating upward to a level

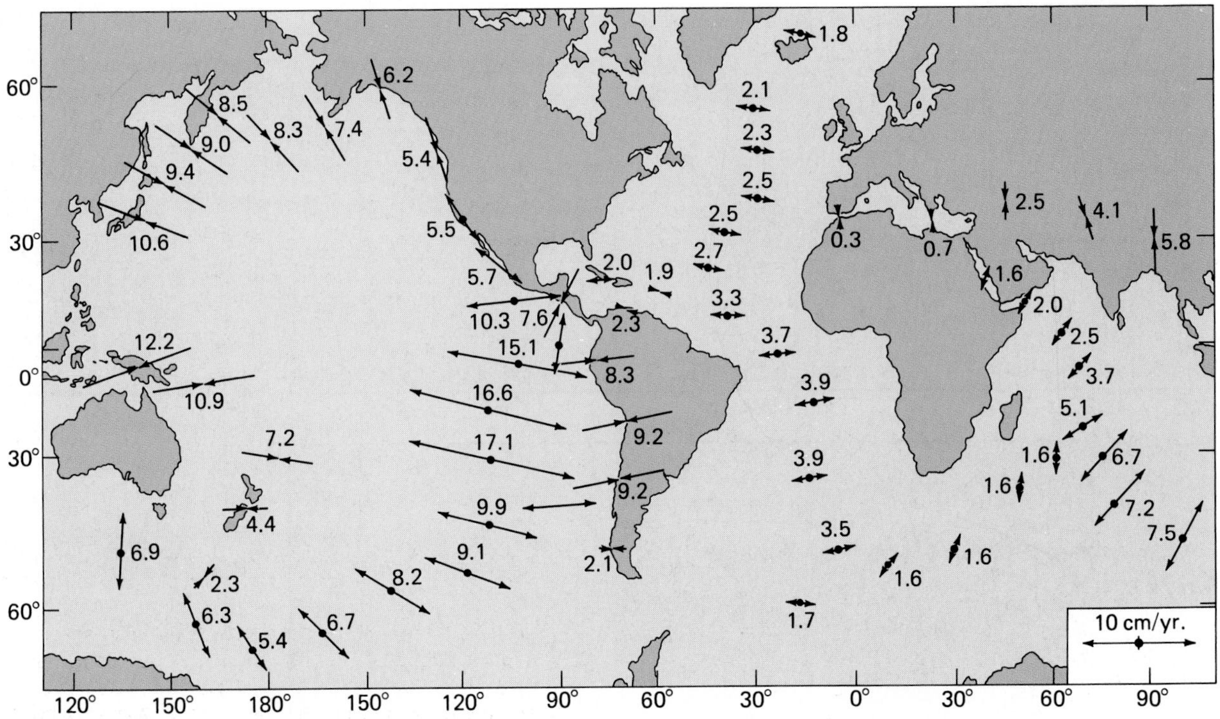

Figure 11–31 Rates and directions of sea floor spreading. (Results from J.B. Minster and T.H. Jordon. *J. Geophys. Res.* 83:5331–5354, 1978.)

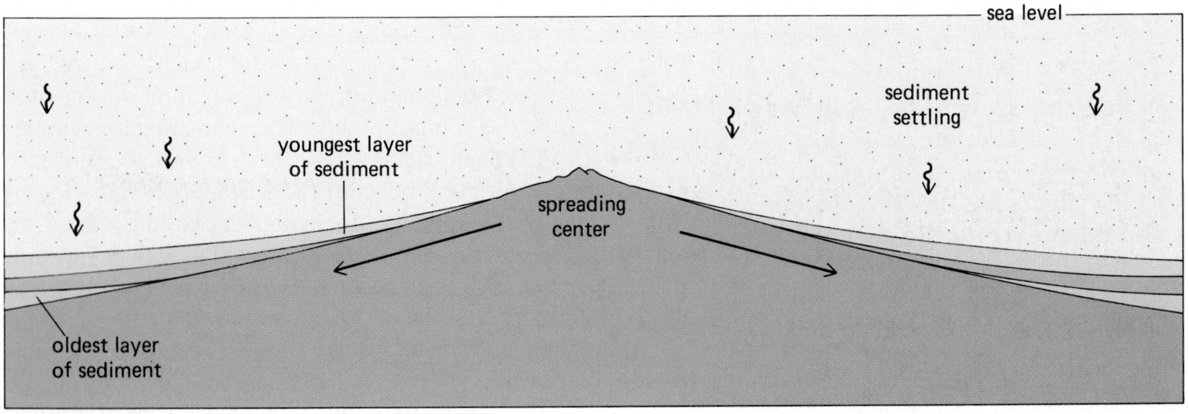

Figure 11–32 As a result of sea floor spreading, sediments near the mid-oceanic ridges are youngest, and sediments *directly above the basaltic crust* but progressively farther away from the ridge are sequentially older.

appropriate to their density. That force might be provided by a descending convection current.

Hot Spots

Anyone examining the superb topographic maps (see Figure 1–12) of the sea floor that have been produced over the past decade cannot help but notice the chains of volcanic islands and seamounts (submerged volcanoes) that rise in alignment on the sea floor. Because these volcanoes occur at great distances from plate margins, they must have originated differently from the volcanoes associated with mid-oceanic ridges or deep-sea trenches. Their striking alignment has been explained as being a consequence of sea floor spreading. According to this notion, intraoceanic volcanoes develop over a "hot spot" in the **asthenosphere**. The hot spot is fixed in location and may be a manifestation of one of the deep "plumes" of upwelling mantle rock described earlier. Lava from the plume may work its way to the surface to erupt as a volcano on the sea floor. As the sea floor moves (at rates as high as 10 cm per year in the Pacific), volcanoes form over the hot spot and expire as they are conveyed away; and new volcanoes form at the original location. The process may be repeated indefinitely, resulting in a great linear succession of volcanoes that may extend for thousands of kilometers in the direction that the sea floor has moved. The Hawaiian volcanic chain is believed to have formed in this manner from a single source of lava over which the Pacific Plate has passed on a northwesterly course. In support of this concept are radioactive dates obtained from rocks of volcanoes that clearly indicate that those farthest from the source are the oldest.

At the western end of the Hawaiian Islands is a string of submerged peaks called the Emperor Seamounts (Fig. 11–34). These submerged volcanoes trend in a more northerly direction than the Hawaiian Islands. However, both the seamounts and the islands are part of a single chain that has been bent as a result of a change in the direction of plate movement. Such an interpretation is supported by age determinations. The oldest of the Hawaiian Islands (near the bend) was formed about 40 million years ago. The seamounts continue the age sequence toward the end of the chain, where the peaks are about 80 million years old. Thus, 40 million years ago the Pacific Plate changed course and put a kink in the Hawaiian chain.

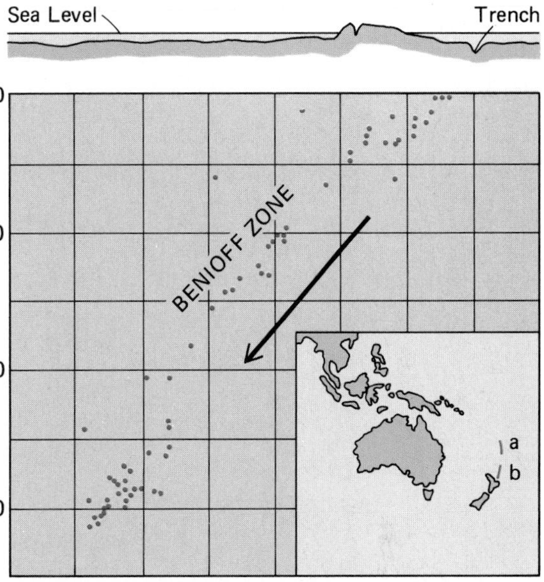

Figure 11–33 Vertical cross-section showing spatial distribution of earthquakes along a line perpendicular to the Tonga Trench and volcanic island arc. (Inset) The Tonga Trench (a) lies just to the north of the Kermodec Trench (b). (Simplified from B. Isacks, J. Oliver, and R. Sykes. *J. Geophys. Res.* 73:5855, 1968.)

The ocean around Hawaii is not the only part of the globe having hot spots. As indicated in Figure 11–35, hot spots are widely dispersed and occur beneath both continental and oceanic crust. Yellowstone National Park (Fig. 11–36) is a hot spot, which, unlike the Hawaiian Islands, is in the interior of a continent.

A hot spot is the surface expression of a mantle plume.

Lost Continents and Alien Terranes

We are accustomed to thinking of continental crust in terms of large land masses like North American or Eurasia. There are, however, many relatively small patches of continental crust scat-

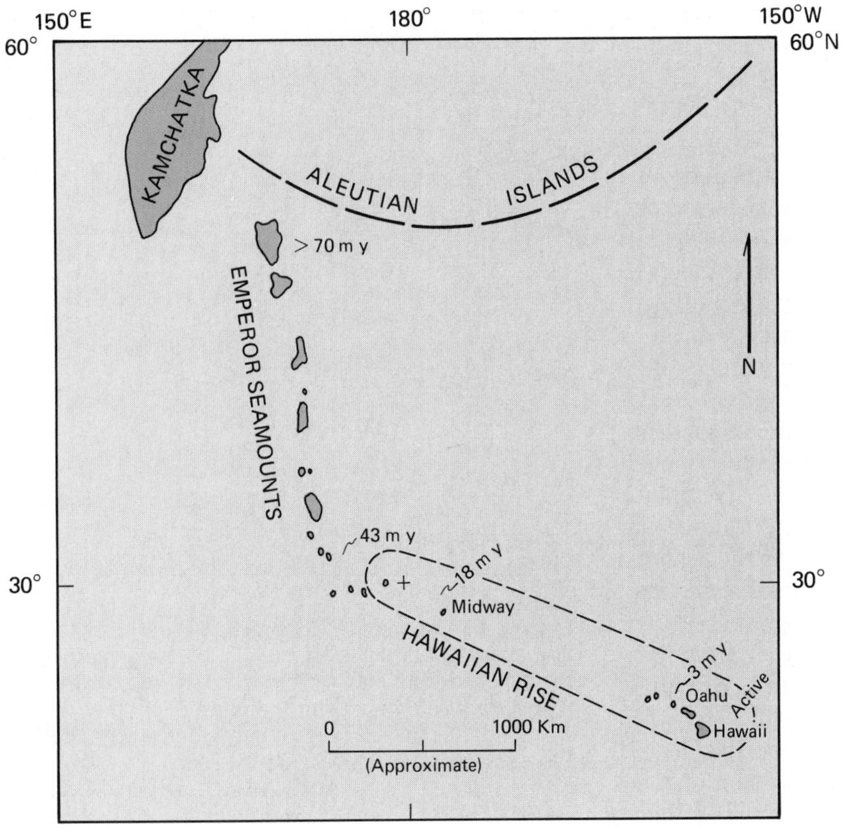

Figure 11–34 Bend in trend of the Hawaiian Island Emperior Seamount Chain was probably caused by change in direction of movement of the Pacific Tectonic Plate. (From J.S. Watkins, M.L. Bottino, and M. Morisawa. *Our Geological Environment.* Philadelphia, W.B. Saunders Company, 1975.)

tered about on the lithosphere. As long ago as 1915, Alfred Wegener described the Seychelles Bank (Fig. 11–37) in the Indian Ocean as a small continental fragment that had broken away from Africa. The higher parts of the Seychelles Bank project above sea level as islands, but many other such small patches of continental crust are totally submerged. Geologists use the term **microcontinents** for these bits of continental crust that are surrounded by oceanic crust. They are recognized by their granitic composition, by the velocity with which compressional seismic waves traverse them (6.0 to 6.4 km/second), by their general elevation above the oceanic crust, and by their comparatively quiet seismic nature.

It is apparent that microcontinents are merely small pieces of larger continents that have experienced fragmentation. As these smaller pieces of continental crust are moved along by sea floor spreading, they may ultimately converge upon the subduction zone at the margin of a large continent. Because they are composed of relatively low density rock and hence are buoyant, they are a difficult bite for the subduction zone to swallow. Their buoyancy prevents their being carried down into the mantle and assimilated. Indeed the small patch of crust may become incorporated into the crumpled margin of the larger continent as an exotic or alien block.

It is interesting to note that geologists found evidence of microcontinents long before the present theory for their origin was formulated. While mapping in Precambrian Shields, they came across areas that were incongruous in structure, age, fossil content, lithology, and paleomagnetic orientation when compared to the surrounding geology. It was as if these areas were small, self-contained, isolated geologic provinces. Often their boundaries were marked by major faults. Geologists designated these areas as **allochthonous terranes** to indicate that they had not originated in the places where they now rested. Allochthonous terranes have been recognized on every major land mass, with well-studied examples in the northeastern

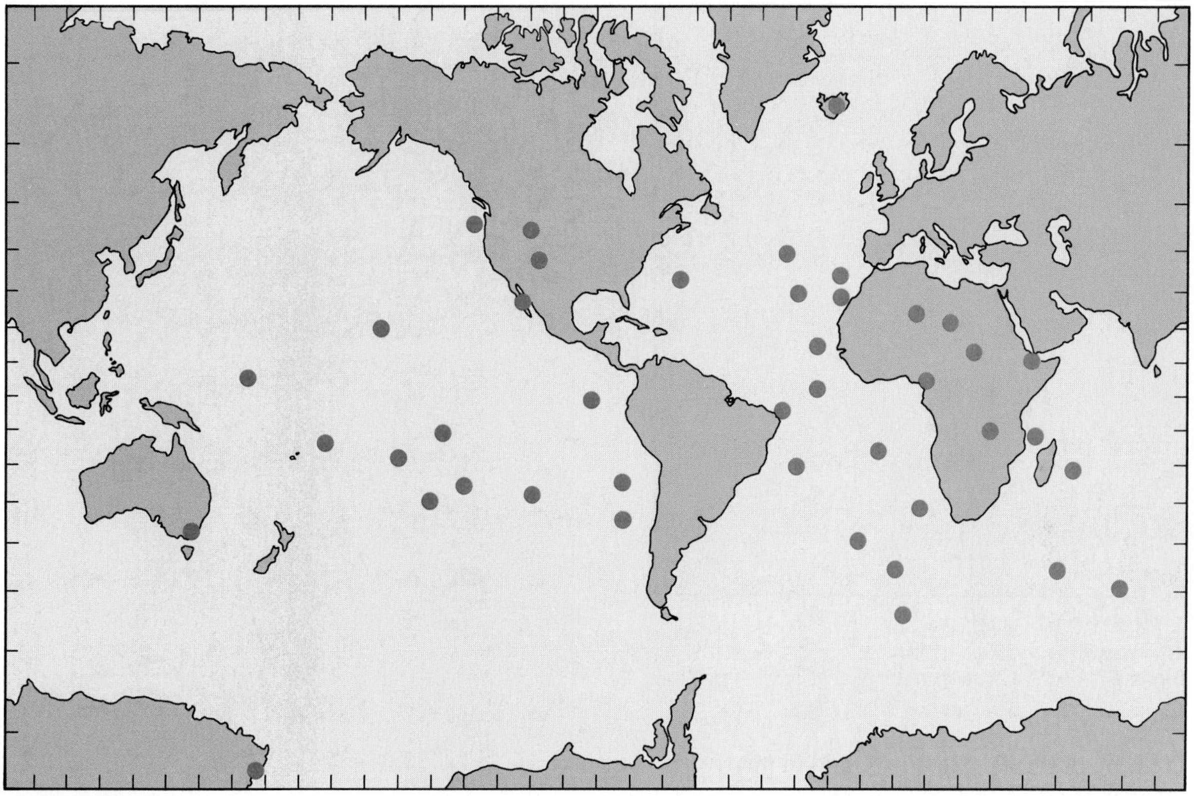

Figure 11–35 Locations of some of the major hot spots around the earth. (Courtesy of Tom Crough, Department of Geologic and Geophysical Sciences, Princeton University.)

Figure 11–36 Mammoth Geyser, Yellowstone Park, Wyoming. Yellowstone is a hot spot. Surface waters percolating through a system of deep fractures reach the hot rocks below and erupt in columns of hot water and steam.

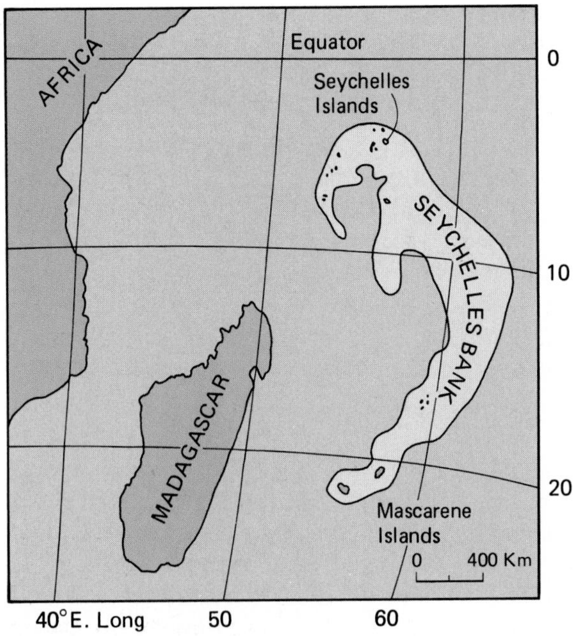

Figure 11–37 Location map of the Seychelles Bank.

U.S.S.R., in the Appalachians, and in many parts of western North America (Fig. 11–38). In particular, Alaska appears to be largely constructed of allochthonous terranes.

If splinters of continents can be transported on the spreading sea floor, so can pieces of oceanic crust. Particularly in the Cordilleran Mountain belt of North America, one finds allochthonous terranes that were apparently microplates of ocean crust containing volcanoes, seamounts, segments of island arcs, and other features of the ocean floor. All of these features were carried to the western margin of North America like passengers on a huge conveyor belt. As the plate that bore

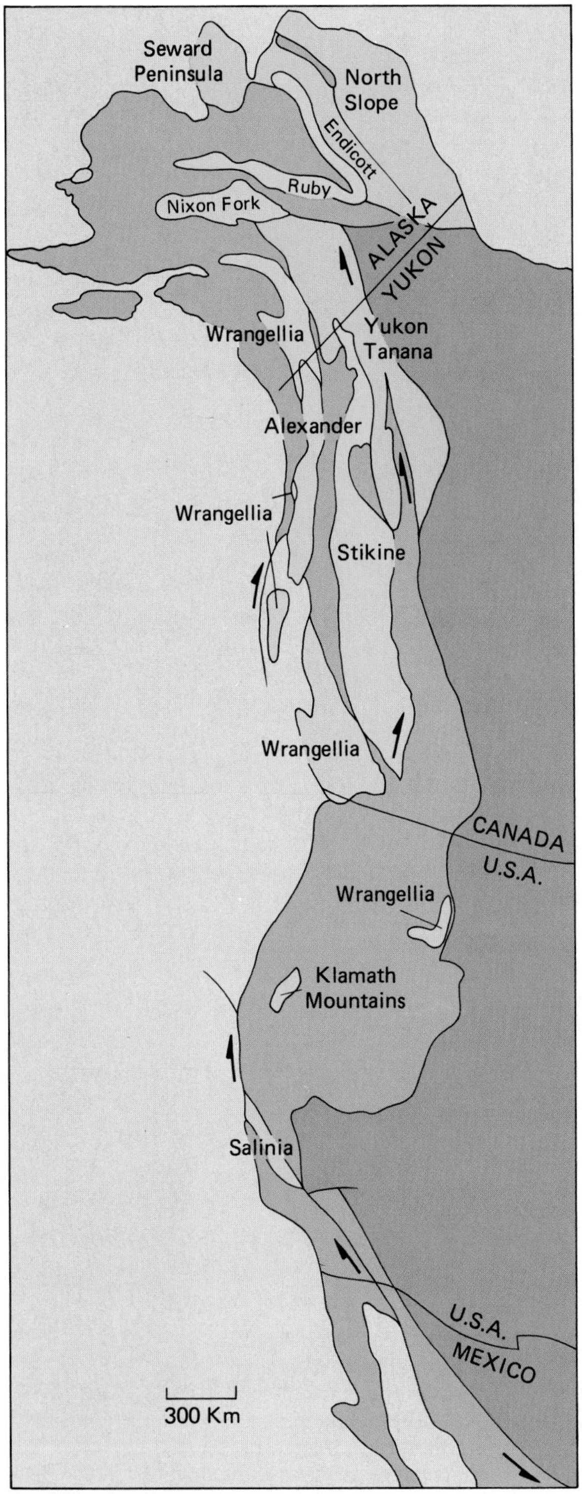

Figure 11–38 The larger allochthonous continental terranes in western North America that contain Paleozoic or older rocks. At some time prior to reaching their present locations, these terranes were continental fragments embedded in oceanic crust or microcontinents. Those colored yellow probably originated as parts of continents other than North America, whereas the darker-colored gray-green blocks are possibly displaced parts of the North American continent. (From Z. Ben-Avraham, *Am. Sci.* 69:298, 1981.)

them plunged downward at the subduction zone located along the continental margin, volcanoes, seamounts, and the other features were scraped off the descending plate and plastered onto the continental margin. The result was a vast collage of accreted oceanic microplates interspersed with similarly transported microcontinents. It appears that about 50 such allochthonous or exotic terranes exist in the Cordillera, all lying west of the edge of the continent as it existed about 200 million years ago. Clearly this process of successive additions of oceanic and continental microplates can significantly increase the rate at which a continent is able to grow. It is also apparent that active continental margins such as those bordering the Pacific Ocean are likely to grow faster than passive margins because of microplate accretion.

Allochthonous terranes are blocks of lithosphere that have been conveyed to major land masses, and incorporated within them by tectonic processes.

Unraveling the orogenic history of mountain ranges containing multiple allochthonous terranes is an enormously complex task requiring the cooperation of geologists, geophysicists, and paleontologists. Such a collaboration can often yield dramatic results. Recently, paleontologists and geologists working in the Wallowa Mountains of Oregon discovered a massive coral reef of Triassic age. The reef, however, clearly did not belong where it was found. It rested on volcanic rocks whose paleomagnetic orientation indicated that they had solidified from lavas at a latitude considerably closer to the equator.

Further investigation indicated that the fossil organisms in the reef were identical to those found in Triassic strata of the Austrian and German Alps. These strata were deposited prior to Alpine mountain building in an ancient sea that extended from the present east coast of Japan into the Mediterranean region. The similarity between the fossils of Oregon and those of the Alps strongly suggests that the reefs now in Oregon actually grew around the margins of volcanic islands in that sea and, in the subsequent 220 million years or so, were transported thousands of kilometers eastward until they collided with the North American continent. By that time, of course, the reef organisms had died, but their fossilized skeletons remained as the reef was accreted onto the western margin of North America.

Summary

We have seen how the many notions of drifting continents and sea floor spreading have been combined into the modern theory of plate tectonics. As with any new theory, there is much that remains to be tested and explained. Nevertheless, the central idea is convincing and has great appeal because of the way it serves to unify and relate geologic phenomena.

According to the views embodied in plate tectonics, the crust of the earth and part of the upper mantle compose a brittle shell called the lithosphere. The lithosphere is broken into a number of plates that override a soft plastic layer in the mantle called the asthenosphere. Heat from within the earth may create convection currents in the mantle, and such currents may provide a propelling force to move the plates. However, sliding of plates in response to gravity may also prove to be a mechanism for their movement.

Materials for the lithospheric plates originate along fracture zones typified by mid-oceanic ridges. Along these tensional features, basaltic lavas rise and become incorporated into the trailing edges of the plates, simultaneously taking on the magnetism of the earth's field as they crystallize. In recent geologic time, this process has provided a record of geomagnetic reversals that in turn permit geophysicists to determine the rates of sea floor spreading.

When plates collide, one of three kinds of tectonic behavior may occur. If it is an oceanic plate colliding with a plate bearing a continent, the oceanic plate will slide beneath the continental plate as a subduction zone, and be melted at depth. The zone of collision between the plates will be marked by volcanoes, an offshore trench, and an asymmetric, single-fold belt mountain range. Where a continent-to-continent collision occurs, mountains having a symmetrical double-fold belt develop. No subduction occurs in this type of collision because of the relatively lower density of continental crust. This lower density also explains why the continents are older than the oceanic crust, and why when continents collide they produce mountain ranges rather than trenches. The collision of an oceanic plate with another oceanic plate produces a volcanic island arc and deep sea trench.

Where continents lie above thermal plumes or divergent zones in the asthenosphere, the land mass may break apart and produce rift features like the Red Sea and Gulf of Aden. Often three rifts develop from a central point. Two of these may widen to admit the sea, and the third may fail and remain as a continental rift zone like the African Rift.

Aside from the suturing of one large land mass onto another, a continent may grow in either of two ways, both involving mountain building. One process involves orogenic compression and metamorphism of sediments that had accumulated in marginal geoclines. The second is by accretion of microcontinents and oceanic microplates bearing volcanoes, sea floor plateaus, and seamounts. These bits and pieces of crust are carried by sea floor spreading to marginal subduction zones, where they are accreted onto the edge of the continent as allochthonous terranes. The Cordillera of North America appear to be largely composed of a complex collage of such terranes, each having its own distinctive geology that can often be traced to distant source regions.

Plate tectonics has provided an interrelated network of answers for such questions as how continents grow, why basins of sedimentation experience upheaval to form mountains, what causes the majority of the world's earthquakes and volcanoes, what determines the locations of major mineral deposits, and why the history of life on earth is interrupted by episodes of crisis or of variation in biologic diversity.

Terms to Understand

allochthonous terranes
asthenosphere
Benioff Seismic Zone
continental drift
convergent boundary
Curie temperature
divergent plate boundary
Glossopteris
Gondwanaland
gravity anomaly
Laurasia
Mesosaurus
microcontinents
ophiolite suite
paleomagnetism
Pangaea
Panthalassa
plate tectonics
polar wandering curves
remanent magnetism
sea floor spreading
shear boundary
subduction zone
suture zone
thermal plume
tillite
trailing edge
transform fault

Questions for Review

1. What large-scale features of the sea floor seem to have developed in response to crustal tension?

2. What sea floor features seem related to crustal compression?

3. What is remanent magnetism? What is its origin, and how is it used in finding ancient pole positions? How do the data of paleomagnetism support the concept of moving continents?

4. How has the detection of reversals in the earth's geomagnetic field been used as support for the concept of sea floor spreading?

5. How does the age distribution of chains of intraoceanic volcanoes enhance the concept of sea floor spreading?

6. What relationship might thermal convection cells in the mantle have to regions of crustal tension and compression?

7. Why are the most ancient rocks on earth found on the continents, whereas only relatively young rocks are found on the ocean floors?

8. Compile a list of items that you might use to convince a skeptic that the present continents are drifted fragments of a larger supercontinent that broke apart.

9. Do mid-oceanic volcanoes have lavas of granitic or basaltic composition? What is the composition of lavas in continental mountain ranges adjacent to deep-sea trenches? How might one account for these compositions?

10. What is a gravity anomaly? Where in the ocean basins have negative gravity anomalies been detected? What is the meaning of these anomalies with regard to deep-sea trenches?

11. According to plate tectonic theory, how might the Himalayan Mountains have originated? The San Andreas Fault? The Dead and Red Seas?

12. According to the theory of plate tectonics, where is new material added to the sea floor, and where is older material consumed?

13. A moving tectonic plate must logically have sides, a leading edge, and a trailing edge. Where are the leading and trailing edges of the American Plate located?

14. Why are the paths of polar wandering different for North America compared with Europe? Why is "polar wandering" a rather deceiving expression?

Supplemental Readings and References

Ben-Avraham, Z., 1981. The movement of continents. *Am. Sci.* 69:291–299.

Cox, A., and Hart, R.B., 1986. Plate tectonics; How it works. Palo Alto, Calif., Blackwell.

Dietz, R.S., and Holden, J., 1970. Reconstruction of Pangaea; break-up and dispersion of continents, Permian to Recent. *J. Geophys. Res.* 75:4939–4956.

Hallam, A., 1973. *A Revolution in the Earth Sciences.* New York, Oxford Press.

Miller, R., and Eds., 1983. *Continents in Collision.* Alexandria, Va., Time-Life Books.

Pitman, W.C. III, and Talwani, M., 1972. Sea-floor spreading in the North Atlantic: *Geol. Soc. Am. Bull.* 83:619–646.

Sullivan, W., 1974. *Continents in Motion.* New York, McGraw-Hill Book Co.

Valentine, J.W., and Moores, E.M., 1974. Plate tectonics and the history of life in the oceans. *Sci. Am.* 230:80–89.

Wegener, A., 1929 (1966 translation). *The Origin of the Continents and Oceans.* New York, Dover Publications.

Weyman, D., 1981. *Tectonic Processes.* London, Geo. Allen & Unwin.

Wilson, J.T. (ed.), 1970. *Continents Adrift: Readings in Scientific American.* San Francisco, W.H. Freeman & Co.

Vink, G.E., Morgan, W.J., and Vogt, P.R., 1985. The earth's hot spots. *Sci. Am.* 252:50–58.

Mass Movements of Earth Materials

12

Rockfall in Sogne Fjord, Scandinavian Mountains, north coast of Norway.

"Immense masses of rock, many of which must have weighed several tons, were apparently dancing along, light as corks, on the surface, being transported by the earthly mass beneath."

Eye-witness account by J.S. Douglas of a mudflow in the San Joaquin Valley in 1905

Introduction

On the night of October 9, 1963, a resident of Casso, Italy, was awakened by a loud roar resembling the noise produced by an avalanche. The man's house was built on a mountain-side 250 meters above the reservoir behind the famous Vaiont Dam. He had heard these noises many times. Usually they subsided after a few seconds. On this night, however, the roar continued and increased in volume until at 10:35 P.M. a savage blast of wind rocked the house so fiercely that the roof was lifted from the walls. Rock and water poured through the windows. The man raced for the door as the ceiling collapsed on the bed he had occupied just moments before. By morning, the shaken Italian learned that he was one of a few survivors of a catastrophe caused by a landslide *on the opposite side* of the Vaiont Reservoir. In less than 1 minute the slide had carried 240 million cubic meters of rock into the reservoir and beyond to slopes over 200 meters up the other side of the valley (Fig. 12–1). In the reservoir, giant waves resulting from the slide spilled over the top of the dam and surged into the valley below. Surprisingly, the dam did not fail, but downstream over 2600 people lost their lives as the torrent swept over them.

The landslide that caused the Vaiont Dam catastrophe is an example of only one kind of gravitational transfer of earth materials down slopes. Geologists refer to all downslope movements, whether spectacularly fast or imperceptibly slow, as **mass wasting**. The term combines the notion of the development of rock waste by weathering with that of movement of the products of weathering downhill primarily under the influence of gravity (Fig. 12–2). Along with running water and

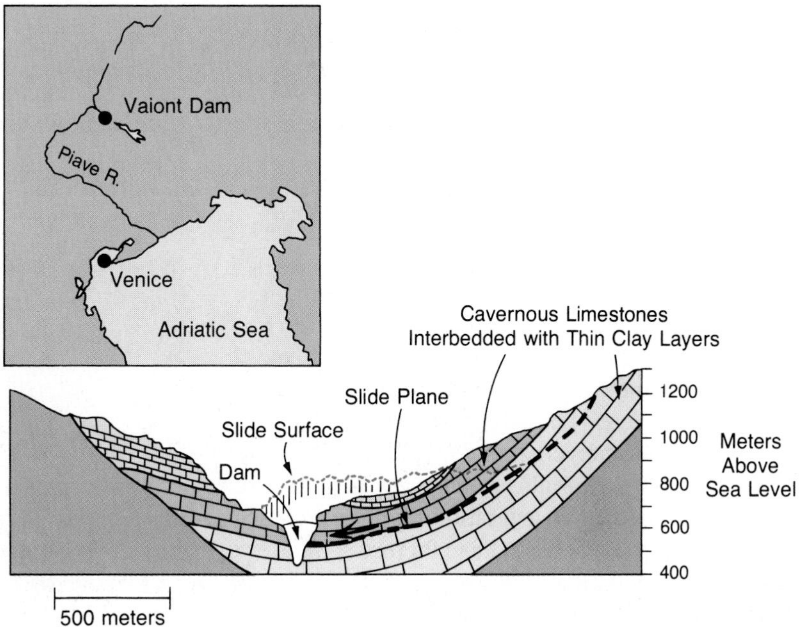

Figure 12–1 Location map and simplified cross-section of the Vaiont Reservoir. (Cross-section generalized from Kiersch, Vaiont Reservoir disaster, *Civil Engineering,* 34:32–39, 1964.)

Figure 12–2 Landslide above Emerald Bay, Lake Tahoe, Nevada. This 1955 landslide obliterated several hundred feet of highway, which has since been cleared.

moving ice, mass wasting operates continuously to give us the ever-changing landscapes of the earth.

Factors Governing Mass Wasting

Although the downhill movement of earth material is ultimately induced by gravity, many other factors influence the susceptibility of areas to mass movements, and the speed and frequency of such movements. Clearly, the steepness of slopes has a bearing on mass movements, as well as the composition of earth materials, water content, bedrock attitude, and nature of vegetation. For convenience of study, we will examine these factors separately, although nearly always two or more are combined in any single episode of mass movement.

Gravity and Slope

Gravity is an essential factor in many geologic processes, including those related to the behavior of streams, or of glaciers. However, its work in mass wasting is accomplished without the assistance of flowing water or moving ice. When masses of earth materials are loosened because of weathering, erosion, or changes in their physical properties, it is gravity that moves these materials to lower elevations (Fig. 12–3) where equilibrium may be, at least temporarily, restored.

Bodies of rock on a slope are attracted directly to the center of the earth by gravity. In addition, there are components of gravity that act on that same body of rock in both the downslope direction, and in a direction perpendicular to the slope.

Whether or not a given block of rock will be able to overcome friction and move downslope will be influenced most directly by the angle of the slope itself. Compare, for example, slopes X and Y in Figure 12–4. On both the X and the Y, the gravitational force drawing the boulder directly to the center of the earth is the same. On the steeper slope, however, the component of gravity directed along the slope has increased, whereas the component of gravity that assists in holding the rock against the slope has decreased. This shifting in the magnitude of forces is the reason for an observation all of us have made, namely, that objects slide down steeper slopes more readily than gentle ones. On slopes steeper than 45°, most unconsolidated materials will not remain in position but will begin to slide or flow downward. Surfaces covered by such loose materials tend to have maximal slope angles of between 25° and 40°.

The pull of gravity is the essential factor in the movement of bedrock, rock debris, or soil on a sloping surface.

Earth Materials

Some kinds of rocks, and even certain unconsolidated earth materials, are inherently better able to stay in position on a slope than others. Massive uniform-textured rocks like granite, basalt, and quartzite have interlocking grains that give them

Figure 12–3 Landslide scars in hillsides above Roosevelt Lake, Washington. The slide created a huge wave, which traveled across the lake and onto the opposite shore to a height of 20 meters above lake level. (Courtesy of the U.S. Geological Survey.)

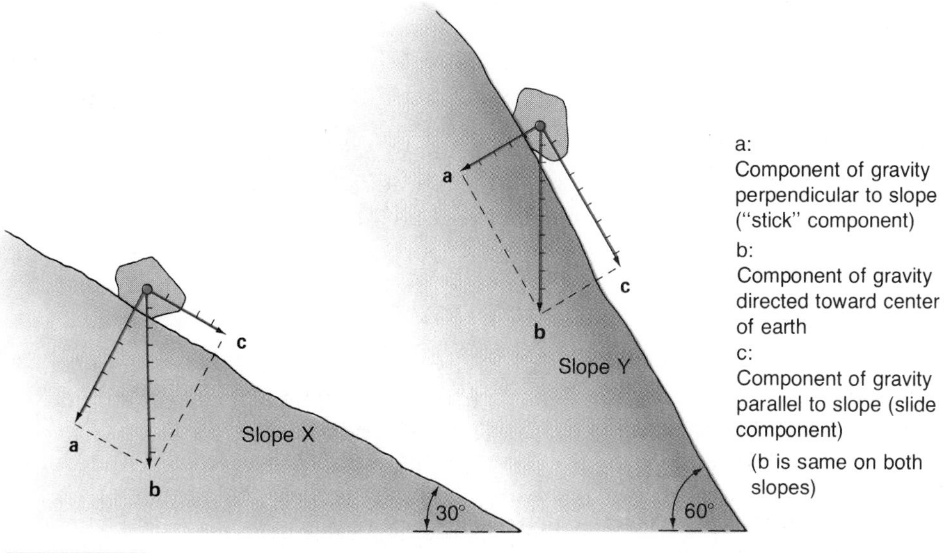

Figure 12–4 Illustration indicating how gravity acts on a boulder on a slope. The boulders are identical, and therefore the *b* component of gravitational force is the same, as indicated by similar length of the arrows directed toward the earth's center. Note that in slope Y, the *a* component (which helps keep the boulder in place) is shorter than on slope X, and the *c* component (which promotes sliding downslope) is longer than on slope X.

nearly equal strength in all directions. Such rocks hold their position against gravity until separated from the unaltered rock mass by weathering and the development of joints. Once loosened, of course, the rock tumbles or slides downslope.

Not all kinds of crystalline rocks have equal strength in all directions. Planes of schistosity and slaty cleavage in metamorphic rocks, as well as clayey seams separating the strata of massive limestones, serve as slip zones along which the overlying rock may move gravitationally on the lower material. If these planes of potential slippage lie parallel to the surface of a slope, great tabular masses of rock may lose their hold and slide down the valley side. Thus, in addition to the kind of rock involved in a mass movement, the composition and structural attitude of underlying material must also be considered in assessing the potential for mass wasting at any given location.

Unconsolidated earth materials such as soil, clay, sand, and gravel are also susceptible to mass wasting. The fact that clay becomes plastic and weak when it absorbs water is well known. Wet clays not only act as lubricants for mass movements but tend to flow out laterally from beneath burdens, thus causing subsidence and slumping. Some clays may resist mass movements for centuries, either because of weak cements or a disorderly arrangement of mineral particles. Such clays, however, may suddenly become unstable when disturbed by leaching, vibrations from earthquakes, or heavy traffic. The degree of uniformity of grain size (sorting) and the presence of root networks and mineral cements also affect the ability of coarse materials to keep their position on a slope.

Water

Water increases the possibility for mass wasting in a variety of ways. It adds weight, contributes to buoyancy, and decreases the cohesion of clay minerals. Wet clay may become a slippery layer over which rock masses may slide down slopes. Water also widens joints and bedding planes through solution, thereby releasing large and small bodies of rock to the force of gravity. Even the hydrostatic pressures exerted by underground water on overlying materials may disrupt equilibrium on a slope and trigger mass movements. Indeed, hydrostatic pressures were a significant factor in the Vaiont Dam disaster.

Water is of critical importance in causing mass movements. Its added weight and lubricating effect facilitate downslope movements of earth materials.

In considering the role of water in causing mass movements in unconsolidated materials such as soils, it is necessary to recognize that the amount of water is more important than its mere presence. For example, if dry sand is poured onto a level surface, it will form a cone having slopes of about 30° with the surface. For dry sand, 30° is the maximum slope possible, and is referred to as the **angle of repose** for dry sand. Next, if water is carefully added to the cone so as to wet the grains but not flood them, surface tension in the water film enveloping the sand grains will hold them together. It will now be possible cut vertical walls in the cone. (A child playing on the beach quickly recognizes the importance of using wet sand to build a sand castle.) However, if additional water is poured on the sand until all of the pore spaces are filled, the cone will collapse.

Because surface tension tends to hold sand grains in place, moist sand will resist downslope movement better than sand which is saturated with water.

Attitude of Bedrock, Joints, Faults, and Fractures

As described in Chapter 8, the term "attitude" in a geologic sense refers to the way in which geologic features like strata or fault planes are positioned in the earth's crust. The attitude of a sandstone stratum, for example, may be horizontal, vertical, or inclined at any angle. As is apparent from Figure 12–5, the inclination of layered rocks relative to a slope can have great influence on the susceptibility of the slope to mass wasting. If strata dip parallel to the slope surface (Figure 12–6), the chances of failure are considerably greater than if

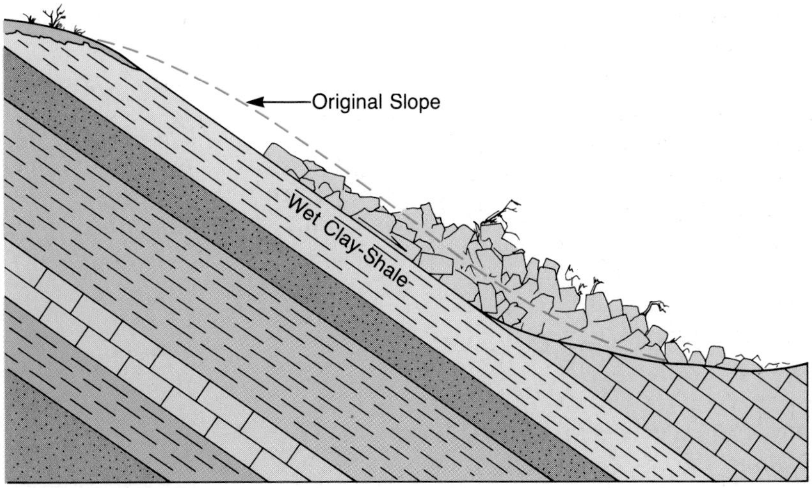

Figure 12–5 The possibility of downslope movement of stratified rocks is increased if the strata dip parallel to slopes and are underlain by weak materials such as wet shales.

beds dip in a direction opposite to the direction of slope.

Vegetation

Plants are able to protect soil and regolith against erosion and contribute to the stability of slopes by binding soil together within root systems. In areas where plant cover is sparse, mass wasting may be enhanced, particularly if slopes are steep and water is readily available. A rather unusual illustration of the role played by vegetation in influencing mass wasting occurred near Menton, France, several decades ago. Farmers in that area removed olive trees from slopes in order to plant carnations, which, at the time, brought a greater

Figure 12–6 Large-scale landslide that developed during the construction of Interstate 80 in Echo Canyon, Utah. The slide occurred along the bedding of plane of strata dipping generally toward the highway. (Courtesy of the Utah Geological Survey.)

profit. The carnation plants, however, lacked the deep roots of olive trees. The weakened slope failed, and the resulting landslide caused the death of 11 people.

Important controls over mass wasting are gravity, the degree of slope inclination, the nature of earth materials, water content, attitude of bedrock, and plant cover.

Initiating Causes of Mass Movements

Conditions favoring mass wasting such as those described in the preceding sections may exist for a long time without any movement actually occurring. Like the proverbial straw that broke the camel's back, some additional ingredient is needed to initiate movement. Among the many situations or events that trigger mass movements are removal of support for material on a slope, overloading, reduction of friction, and vibrations from earthquakes.

Removal of Support

Slopes that have remained stable for centuries can become susceptible to mass wasting simply by the removal of supporting materials near the base of the slope. Engineers are aware of this fact and attempt to avoid oversteepening slopes or excavating into the base of a hillside composed of weak materials. The base of a slope, after all, acts as a buttress supporting material farther uphill. The removal of that base during road construction is the cause of the failures nearly everyone has seen along the sides of highways (Fig. 12–7). Oversteepening of man-made slopes can also promote slope failure. Nevertheless, engineers sometimes oversteepen slopes to save land, and consequently, money. The practice may occur in road cuts, terraced housing developments (Fig. 12–8), and highway embankments (Fig. 12–9).

Although slope failures resulting from poor construction practices are common, geologists are especially interested in slope failures resulting from such natural causes as the continuous erosion and undermining of hillsides by streams (Fig. 12–10), waves, and glaciers.

Figure 12–7 Mass wasting (rotational slump) resulting from removal of lower part of slope during road construction, Marin County, California. (Courtesy of the U.S. Geological Survey and J. Schlocker.)

Overloading

The costly and frustrating problems of earth movements during the construction of the Panama Canal are almost legendary. Slumping of material along the sides of the excavation were partly the result of cutting the sides of the canal too steeply, as well as having to excavate through the highly plastic clays of the Cucaracha Formation. The unstable conditions were further aggravated, however, when materials excavated from the canal were foolishly dumped at the upper edge of the cut. The added weight was more than the already weak underlying materials could support, and **slumping** occurred with nerve-racking frequency. Overloading may also occur naturally, as when material is transferred onto a slope by landslides, avalanches, and snowfall. Rainwater seeping into soils adds its weight to that of the soil and thus may contribute to overloading the underlying materials.

Figure 12–8 Failure of an oversteepened subdivision slope in the backyard of a home in St. Louis County, Missouri. As trouble developed concrete piers were poured to support the back part of the house, but these also moved downslope with the soil mass.

Reduction of Friction

Reducing the frictional forces that hold a particle, a rock mass, or a mass of unconsolidated material on a slope, will, of course, trigger mass movements. We have already examined the effect of water in reducing friction. Frequently mass movements can be directly related to increases in water from precipitation or the melting of snow and ice. Leakage from reservoirs, ponds, irrigation canals, and even septic tanks may provide the water that reduces the cohesiveness of materials in slopes and causes them to fail (Fig. 12–11).

Earthquakes

Landslides and other forms of mass wasting are frequently initiated by the jolt or vibrations of earthquakes. As described in Chapter 9, earthquakes preceded the Madison Canyon Landslide of

Figure 12–9 Slump features along access road to Highway 64, St. Louis County, Missouri.

1959 (Fig. 12–12) as well as the destructive landslides accompanying the 1964 Alaskan earthquake. In addition to the landslide-triggering effects of earthquakes, mass movements may be initiated by vibrations from blasting in mines, quarries, and road cuts (see Fig. 12–6); by gun fire; and even by the passage of heavy vehicles.

Figure 12–10 Mass wasting (slump) in shale, caused by undercutting by the Bad River, Stanley County, South Dakota. (Courtesy of the U.S. Geological Survey and D.R. Crandell.)

Events that may trigger mass movements include removal of supporting materials at the base of slopes, overloading, reduction of friction by addition of water, and earthquakes.

Types of Mass Wasting

Earth materials on slopes may fail and move in many ways, and this provides a basis for classifying different types of mass movement. As indicated in Table 12–1, three factors are involved in defining a particular kind of mass movement; namely, the rate of the movement, whether the mass moves as a coherent body or is disturbed during movement, and the amount of water in the mass. In **flow**, the material near the surface of the mass moves more rapidly than lower layers, and there is considerable internal deformation. Movement along a clearly defined plane of slippage is called **slide**. The popular term **landslide** is used in a rather loose way for all forms of rapid downslope movement.

Creep

The slowest of all forms of mass wasting is **creep**. Creep is measured in mere centimeters per year. It can be defined as the slow downhill movement of regolith that results from continuous rearrangement of its constituent particles. The moving material may be more or less dry or contain only minor amounts of water. Creep is a form of flow, and therefore, the rate of movement is greatest

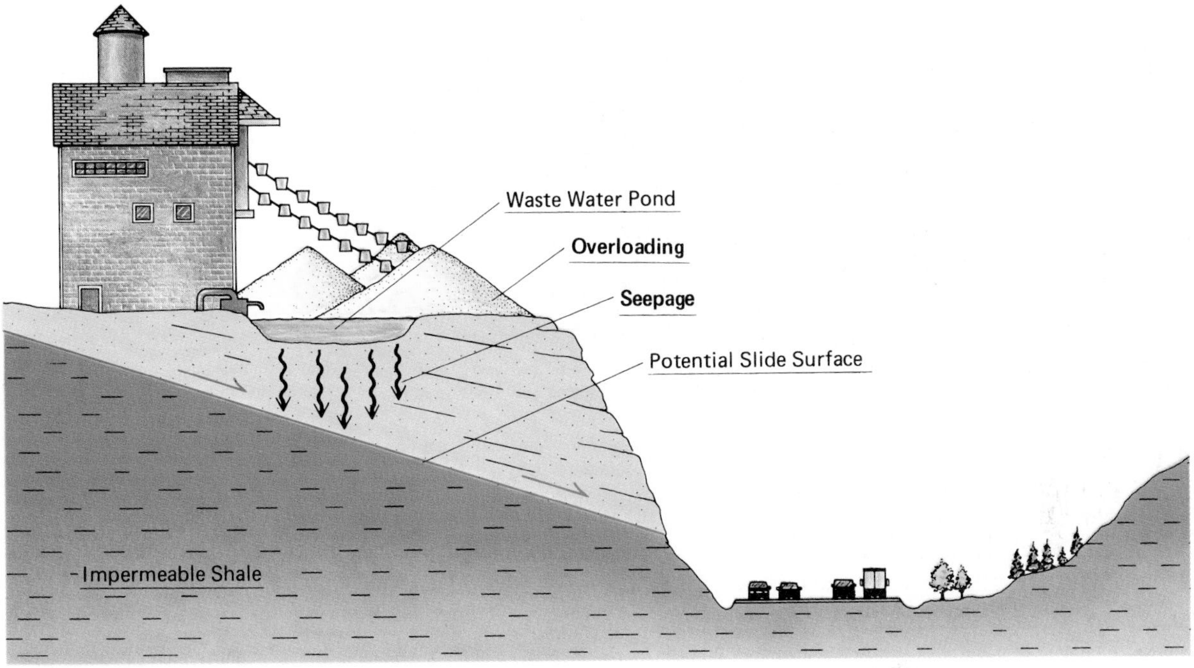

Figure 12–11 Overloading near the edge of a slope or roadcut, as well as seepage of water, provides the potential for slope failure.

Figure 12–12 A 1959 earthquake triggered the Madison Canyon landslide. The canyon is located in Montana, north of Yellowstone National Park. The slide debris dammed the Madison River, and water backed up behind the dam to form Quake Lake, seen in the foreground. One can see the scar of the landslide on the far side of the lake. (Courtesy of L.E. Davis.)

Table 12–1 Types of Mass Movement

	Rate of Movement	Amount of Water Present
Flow (Movement Distributed Throughout Material)		
Creep	Slow	Water not necessary
Rock glacier, rock, stream	Slow	Water not necessary
Solifluction	Fast	Water-saturated
Mudflow, debris flow	Slow or fast	Much water
Earthflow	Slow or fast	Much water
Slide (Movement as One Mass on a Slip Surface)		
Debris avalanche	Fast	Wet or dry
Slump	Fast	Wet or dry
Landslide, rockslide	Fast	Wet or dry
Fall (Free Fall of Rock or Soil)		

From Watkins, J.S., Bottino, M.L., and Morisawa, M., *Our Geological Environment*, Philadelphia, Saunders College Publishing, 1975.

near the surface and decreases with depth. Depending on the material, one can differentiate soil creep and rock creep.

One cannot stand in a field and observe creep in action. It is much too slow a process. One can, however, observe the cumulative effects of creep. These effects include pavements and stone walls that have been displaced, bent trees, fences forced out of alignment, and tilted fence posts and telephone poles (Fig. 12–13). If a slope is underlain by steeply dipping strata, the upper edges of the beds may be broken and bent as a result of creep (Fig. 12–14).

Creep may be regarded as the result of three mechanisms. The first of these is called **heaving** and involves repeated expansion of surface layers in a direction perpendicular to the slope. The expansions may result from the freezing of interstitial

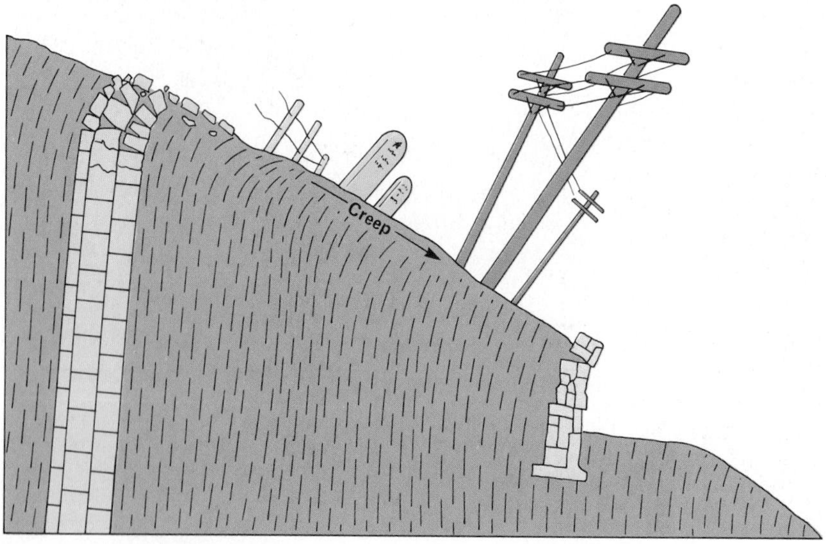

Figure 12–13 Indications of creep include downslope bending and dragging of rock layers as well as tilted and displaced posts, poles, and graveyard markers.

water or simply periodic wetting, which may cause certain clays to swell. When the creep material thaws or dries, the material drawn out perpendicular to the slope during the expansion is lowered by a component of gravity into a position slightly downslope from the initial position (Fig. 12–15). The second mechanism involved in creep is the plastic flowage of moist clays as they respond to the pull of gravity and weight of overburden. Finally, biologic agencies contribute to creep in several ways. The burrowing activities of soil animals and trampling of the surface by cattle and sheep may disturb the regolith sufficiently to offset its stability.

Solifluction

The term **solifluction** (Latin *solum*, "soil"; *fluere*, "to flow") literally means "soil flow." It applies to the flowage of soil and rock debris that are saturated with water and that may be subjected to alternate freezing and thawing. Solifluction may occur along slopes as gentle as only 5°. It is a process particularly prevalent in the saturated soils of colder regions of the earth. During the relatively short summer thaw of subpolar regions, meltwater may thoroughly saturate the upper mobile layer of regolith. This mucky upper layer may then slowly flow down even very gentle slopes on top of the still-frozen underlying zone. The movement is similar to that seen in aa lava, and may produce arcuate ridges and troughs at the lower part of forward-moving tongues of soil and rock.

The permanently frozen earth material on which solifluction may occur is called **permafrost**. Permafrost is not at all an unusual material, for it is widespread on lands within the Arctic Circle. The frozen layer of permafrost may reach a thickness of over 300 meters (Fig. 12–16). Those of us who live in temperate regions are unaccustomed to thinking about the many problems caused by permafrost conditions. Things we take for granted, like a piped-in supply of drinking water and underground sewage systems, are not always feasible in areas having permafrost. Water pipes must be encased in heated housings to prevent freezing. Ways to get rid of sewage before it freezes also require special planning. Even septic tanks must be warmed to increase the rate of digestion of waste.

Figure 12–14 Soil creep as indicated by the bent near-surface parts of vertical beds of chert, Monterey Formation, Berkeley Hills, California.

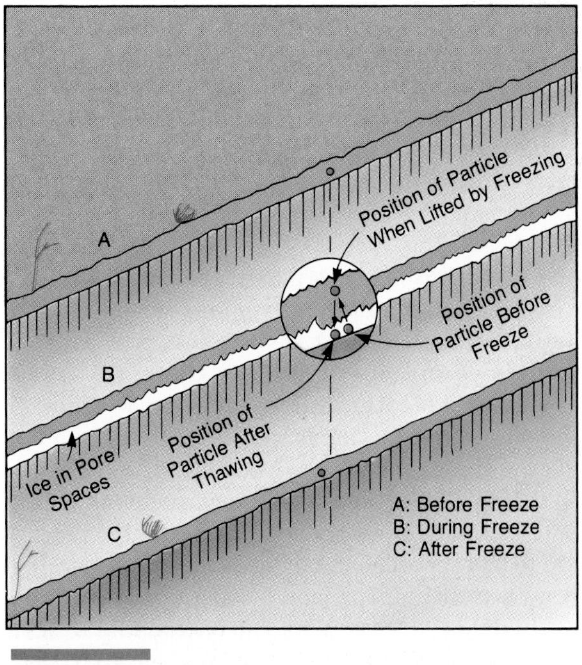

Figure 12–15 The effect of frost heaving on creep. Expansion of the water in porous material during freezing raises the material perpendicular to the slope. On thawing, the material is lowered to a position slightly downslope of the original position.

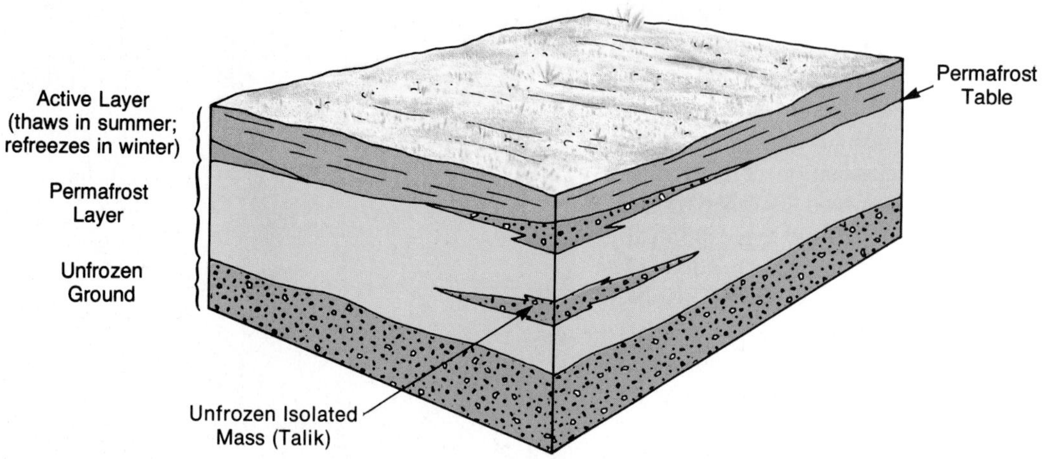

Figure 12–16 Typical section of permafrost terrain. (Modified from the U.S. Geological Survey circular *Permafrost*).

In the construction of the famed trans-Alaska pipeline, the supports for the elevated pipes were linked to refrigeration units to prevent the permafrost beneath the support from thawing and causing the support to sink.

Permafrost is an important reason why much of the Arctic tundra is so water-soaked. The subsurface layer of solid frozen ground is essentially impermeable, and so water in the mobile surface zone cannot escape underground. It remains at the surface where it saturates the regolith and facilitates movements such as solifluction.

Solifluction is particularly characteristic of arctic and subarctic regions where summer meltwater cannot penetrate into the impermeable permafrost zone, thus causing the saturated upper zone to flow even on gentle slopes.

In frigid polar and alpine environments, alternate thaw-and-freeze conditions not only contribute to mass movement but also cause considerable frost heaving, mixing, and sorting of the regolith. Such mixing is termed **cryoturbation** and results in the "stone rings" (Fig. 12–17) and polygonal networks of cobbles and boulders found across the surfaces of polar regions. Many such patterned areas (Fig. 12–18) are believed to develop when, during colder seasons, frost heaving pushes coarser sediment outward from rather regularly spaced centers. The finer material lags behind and may be further concentrated during thaw periods when meltwater at the surface flushes additional fine material into the central areas of the stone rings.

Rock Glaciers

Near or above the timberline in high mountains one can find tongues of coarse rock debris having an overall form suggestive of a valley glacier. These glacier-like masses of angular, unsorted rock waste are called **rock glaciers** (Fig. 12–19). Unlike true glaciers, which are composed of ice containing rock fragments, rock glaciers are primarily composed of rock debris with ice and meltwater filling the spaces between rock fragments. The rock glacier moves by means of internal deformation of the rock and ice mixture, as well as the mobility resulting from seasonal freezing and thawing. Measurements of rock glaciers in Alaska indicate that they are able to move at a rate of about 1 meter per year.

Earthflows

An **earthflow** is a sluggish, rather erratic flow of clayey or silty regolith down relatively gentle slopes. Water is present but usually not to the point of saturation. Earthflows can sometimes be recog-

Figure 12–17 Stone circles, such as these on Leidy Peak in the Uinta Mountains of Utah, have been a geologic enigma for many years. Recent year-long observations in Spitzbergen suggest that such circles are formed by a crude convection in which soil moves upward near the center of the circle, then laterally (carrying stones along with it), and then plunges downward, leaving the rocks on the surface to form the ring.

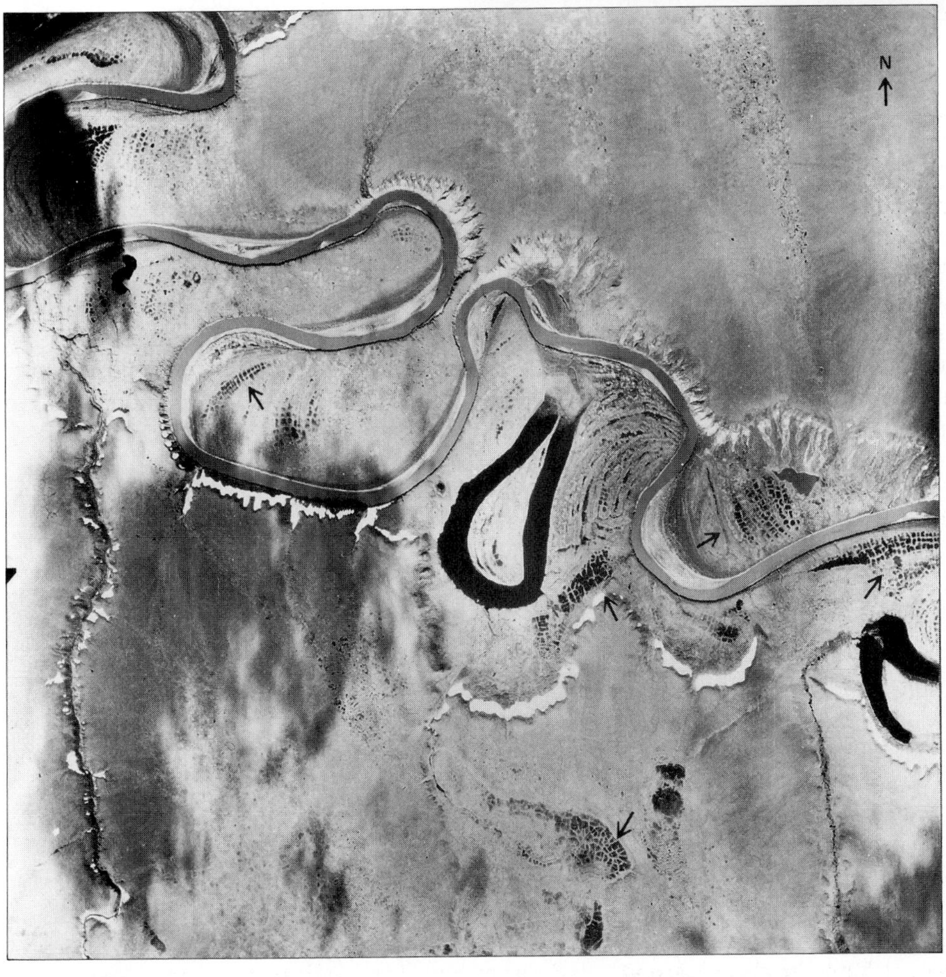

Figure 12–18 Polygonal ground on the flood plain of the Kogosukruk River, Alaska. Scale of air photograph 1:20,000. (Courtesy of the U.S. Geological Survey.)

Figure 12–19 Rock glaciers in the San Juan Mountains of Colorado. (Courtesy of Peter B. Larson.)

nized by a curved scarp that develops at the breakaway line on the slope and by the crescent-shaped bulges at the convex "toe" of the flow (Fig. 12–20). As the toe of the earthflow is pushed into narrow valleys, it may dam streams and produce lakes.

Earthflows are rather frequent occurrences along the valley of the St. Lawrence River and in parts of Scandinavia. In these areas, one finds relatively thick deposits of fine-grained sediments deposited in former marine embayments that existed shortly after the last advance of continental glaciers.

Mudflows

In general, mudflows are even more mobile than earthflows. They can be defined as the relatively fast flow of regolith that has the consistency of mud and behaves rather like a viscous fluid. Mudflows are normally supersaturated with water, so that they have a "soupy" consistency. They tend to travel in stream-like masses within the confines of valleys and canyons. On reaching the mouth of the canyon, they characteristically spread out in a broad apron (Fig. 12–21), covering fields and houses. In addition to a sudden abundance of water (as is available following a heavy thunderstorm), mudflow development is favored by a lack of vegetation and an abundance of loose regolith. Often these conditions are found in arid regions that are subjected to periodic torrential rains. One such region is southern California, where periodic brush fires leave the ground without protection against the next heavy rainfall. **Debris flows** are similar to mudflows except that they contain many large fragments of rocks and trees, all carried in

Figure 12–20 Small earthflow accompanied by slumping on a slope adjacent to a city street. Note that the toe of the flow has overridden the curb and will soon block part of the street.

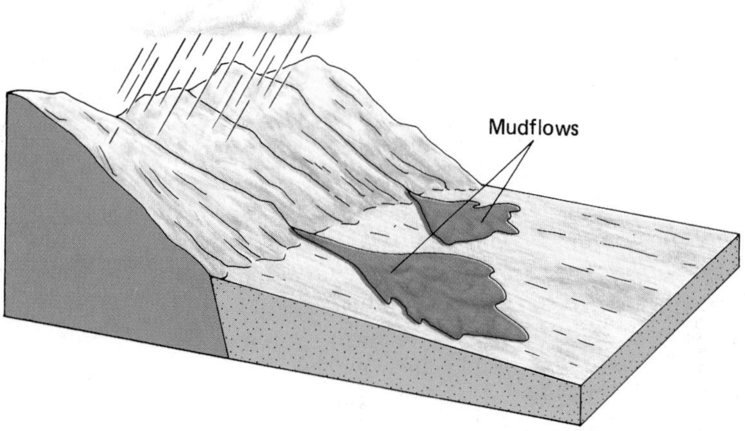

Figure 12–21 Setting for typical mudflows at the front of a mountain range or hilly terrain in an arid climate.

the rapidly moving mixture of sand, silt, clay, and water (Fig. 12–22). The mudflows described in Chapter 4 as **lahars** occur where heavy rains saturate slopes composed of volcanic ash.

Avalanches

The more familiar notion of the meaning of the term **avalanche** is that of a large mass of snow and ice in swift motion down a mountainside. An avalanche, however, need not consist only of snow and ice, but can incorporate unspecified amounts of regolith. Ice and snow may be entirely lacking in so-called *debris avalanches.* The intense mobility of debris avalanches results not only from a steep slope, but also from the included air made dense with dust and fine sediment. As the debris rushes downslope, it is preceded by a blast of outrushing

Figure 12–22 Effects of a large debris flow that occurred in the Hindu Kush Mountains of Pakistan in 1975. The left photograph shows the actual flow as it roared by geologist John Talent at a velocity of over 3 meters per second. The flow transported boulders over 2 meters in diameter. The right photograph shows some of these large boulders left behind by the debris flow. (With permission of R.J. Wasson; photographs by John Talent.)

air and dust that is capable of destroying buildings and trees in its path.

Avalanches are the most savage, unpredictable, and fast-moving of all forms of mass wasting. Certainly the fear that mountain dwellers have for avalanches is well justified. A dreadful example of their destructive power occurred in 1970, when a mass of ice weighing about 3 million tons broke away from a glacier located on the sides of Mount Huascaran in Peru. The great mass of ice plummeted down the mountainside, ricocheted off valley walls, and gathered within its surging mass additional millions of tons of rock, soil, and trees. Within only 7 minutes, the avalanche had traveled 12 miles. Preceded by blast waves that toppled trees and blew apart houses, it roared into the densely populated farming area at the base of the mountain, killing at least 3500 people instantly.

Slump

Slump is a form of slide motion in which a mass of rock or regolith slips along a concave upward-curved surface while at the same time rotating backward upon a horizontal axis (Fig. 12–23). As a result of this movement, the upper surface of the slump block may be tilted toward the slope. **Earthflows** may develop at the toe of the slump. Slumping is a common type of slide movement, and slumped regolith can frequently be seen in highway cuts, cliffs (Fig. 12–24), and along the banks of streams. Slumping provides sediment to streams, and thereby contributes to the formation of valleys.

Most slumps consist of backwardly rotated wedges that leave crescent-shaped scarps at their upper margins.

Rockslides and Rockfalls

The free sliding of recently detached segments of bedrock on a sloping joint or bedding plane constitutes a rockslide. **Rockslides** are likely to take place along the steep sides of mountains and may cause great damage. A particularly interesting example of a rockslide occurred in 1925 in Wyoming's Gros Ventre River Valley. The stratified rocks of the slopes on one side of the valley consisted of wet clay shales interbedded with limestones and sandstones. These strata dip toward the valley at an angle of approximately 20°. Thus, conditions were ideal for downdip and downslope slippage of overlying beds on the weak clay layers. Indeed, relatively slower movements such as earthflows had plagued the area for a century or more. The Gros Ventre slide occurred in the spring as meltwater from snow and ice seeped into the earthflow scars and saturated the underlying clays. Thus weakened, the side of the mountain broke loose, releasing nearly 48,000,000 cubic meters of rock waste, which swept all the way across the valley floor and up the opposite slope. When the dust had settled, the rock debris had created a 70-meter-high natural dam across the Gros Ventre River (Fig. 12–25). That dam failed two years later, causing a flood downstream, which destroyed most of the town of Kelly, Wyoming.

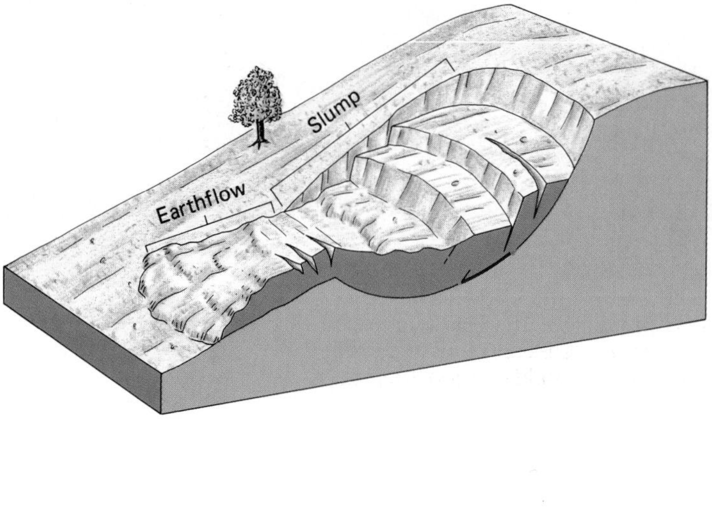

Figure 12–23 Idealized diagram showing characteristics of slump feature with earthflow at the toe.

Figure 12–24 Large slump at Pacific Palisades California, 1960, covered U.S. Highway 101. (Courtesy of California Division of Highways.)

When earth materials experience nearly vertical free fall, as from cliffs, the descriptive term used is **rockfall**. The more or less continuous rain of loosened rock fragments from cliff faces results in an accumulation of shattered rock at the base of cliffs. These piles of debris or **talus** form the talus slopes frequently seen at the base of rocky bluffs (Fig. 12–26). The angle of repose for talus slopes is remarkably constant at between 34° and 37°.

Valleys, Mass Wasting, and Sheetwash

Relation of Mass Wasting to Valley Development

Streams are the most important agents in the development of valleys. The erosive powers of running water, aided by the multitudinous impacts and abrasion of pebbles and other transported rock debris within the flowing stream, effectively excavate the channel. Yet if a stream were the only mechanism involved in forming valleys, they would have vertical rather than sloping valley sides (Fig. 12–27). A stream, after all, can erode only the surfaces with which it is in direct contact. Therefore other agencies, especially mass wasting, are constantly at work widening and shaping valleys. The stream acts as a conveyor belt, carrying away its own load as well as the debris supplied to it by mass wasting. Streams even contribute to the effectiveness of mass wasting by removing material at the base of the valley so as to undermine the slopes and cause slumping (Fig. 12–28) and debris slides. Weathering also assists mass wasting by reducing coherent rock to loose regolith, which is then susceptible to such processes as creep, solifluction, and earthflow.

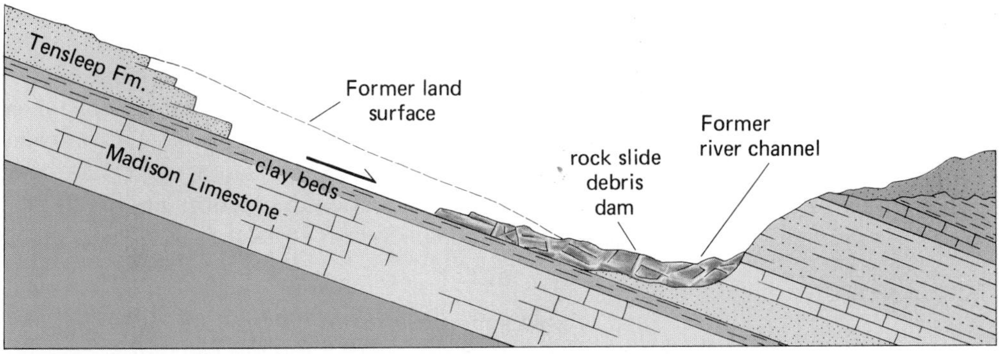

Figure 12–25 North-south section illustrating the Gros Ventre rockslide and resulting damming of the Gros Ventre River in Wyoming, on June 23, 1925. Heavy rains and melting snow had saturated the clay beds beneath the Tensleep Sandstone, causing an enormous mass of rock to become loosened and slide suddenly down the valley. (After W.C. Auden, U.S. Geological Survey.)

A

B

Figure 12–26 (A) Talus slopes on the inner rim of Crater Lake, Oregon. (B) Talus slopes along the canyon of the Salmon River in Idaho. (Courtesy of James C. Brice.)

Mass wasting provides a continuous supply of material to agents of long distance transport, such as streams and glaciers.

Erosion by Sheetwash

Another process that contributes significantly to valley widening is sheetwash. **Sheetwash** (also called sheet flow) is a thin sheet of flowing water that forms on slopes when rain falls faster than the soil can absorb it, or when the ground is already saturated. Such a sheet of water flowing down a sloping surface picks up and transports small particles in its path, much as rainfall removes dust from a sloping roof. The effectiveness of sheetwash erosion depends on several factors, among which are the kind and thickness of plant cover; the porosity, permeability, and cohesiveness of the soil; the duration and intensity of the rainfall; and dimensions and steepness of the slope over which the sheet of water is moving. With a steeper slope, the velocity of sheet flow increases and therefore also the erosive powers of the water. Length of slope is also important, because the sheet has to flow over some critical distance before it begins to erode. Slopes on which the vegetation cover has been removed by construction practices are highly susceptible to sheetwash erosion.

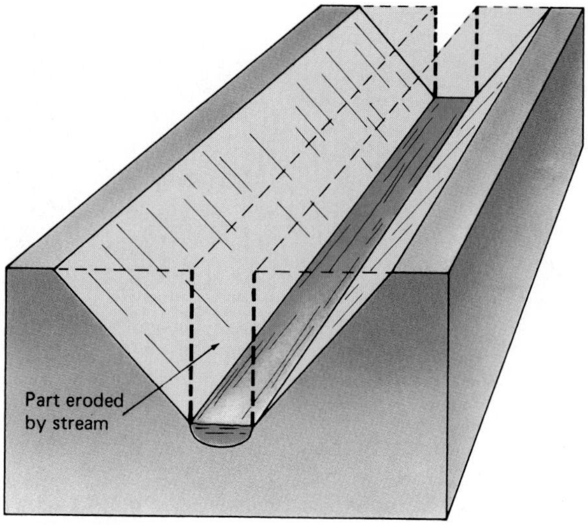

Figure 12–27 Streams do not act alone in the formation of valleys. They erode their own channel but also serve as conveyor belts in carrying away material derived from the valley sides by mass wasting and sheet wash. The volume of material contributed by mass wasting is represented by the space enclosed within the dotted lines.

Sheetwash is the unchanneled flow of water down slopes during rainfall.

Sheetwash will not flow indefinitely as a planar unit of uniform thickness. Friction and irregularities on the ground surface cause water to be concentrated into threads of current, which in turn

Figure 12–28 Streams serve to carry away earth materials brought to the channel by mass movement. In this example, the mass movement is slump along the Gardiner River just outside the north entrance to Yellowstone National Park. (Courtesy of J. Brice.)

erode narrow channels called **rills**. Rills in turn widen and deepen as water moves into them. Eventually, they develop into gullies and are integrated into a nearby stream system.

Erosion by Rainsplash

The thin film of water moving down a slope as sheetwash has limited velocity and mass and therefore is not of itself always able to appreciably erode and transport soil particles. Help is provided by irregularities in the ground surface and by clots of vegetation that temporarily dam small areas and subsequently result in rapid surges of water as the obstacle is surmounted or circumvented. Another potent erosional aid to sheetwash is the often overlooked but significant power of raindrops. When a raindrop strikes the ground, it splashes not only water into the air but suspended particles of sediment as well (Fig. 12–29). On a slope, the sediment derived from one point on the ground and thrown into the air will return to the ground at a point downslope from its point of origin. Thus, there is a net downhill transport of material. Also important is the fact that rainsplash loosens particles and makes them more readily available for transport by sheetwash.

The effectiveness of raindrop erosion depends on such factors as the size of the drops and the duration and intensity of the rain. Because this process operates best on bare ground, vegetation can prevent or inhibit raindrop erosion. The density and texture of the soil also have an influence, as suggested by the fact that fine sands are especially susceptible to raindrop erosion.

Averting Damage from Mass Movements

The yearly cost of landslides and other mass movements to governments (and the taxpayers who support governments) is truly astronomic. As a result,

Figure 12–29 The splash of falling raindrops on unconsolidated sediment effectively loosens it and detaches particles. (Courtesy of the U.S. Department of Agriculture.)

agencies involved in highway construction, railbed maintenance, and the location of utilities and buildings are increasingly concerned with the recognition of potential problem areas and with the development of measures for the prevention and control of mass movements. What are some of the things that can be done to reduce the costs and hazards of mass movements?

Terrain Analysis

One rather obvious way to avoid problems with landslides or earthflow is to avoid building on slopes that are likely to experience such mass movements. In the United States and Canada, geologic surveys are providing "environmental geology maps" that outline areas of unstable or potentially unstable slopes in urban areas. Such maps can be used by governmental agencies to establish zoning laws or to prohibit building in areas likely to experience damaging mass movements.

Landslide Prevention

In areas where construction necessitates cutting into the base of a potentially unstable slope, certain engineering techniques have been useful in preventing slides. If a roadbed is being excavated through an area underlain by steeply dipping strata, the removal of the rock above a potential slippage plane may be all that is needed to prevent a mass movement (Fig. 12–30). If the cut is being made through unconsolidated materials of low strength, the stability of the slope may be further improved by flattening the cut-slope angle. The cutting of benches (Fig. 12–31) can also help to stabilize a slope by lightening the load bearing on the toe of the slope. Benching of the slope into small step-like segments distributes the load, and in the event one of the smaller slope segments fails, the debris is trapped on the bench immediately below.

In many locations, slopes may be made more stable by planting vegetation, the roots of which help hold the regolith together and also take up excessive moisture. By the use of drainage channels, pipes, and diversion walls, the amount of water percolating into the slope (and thus weakening it) can be reduced. Finally, structural devices such as concrete buttresses (Fig. 12–32), rock bolts (Fig. 12–33), metal-bin cribbing (Fig. 12–34),

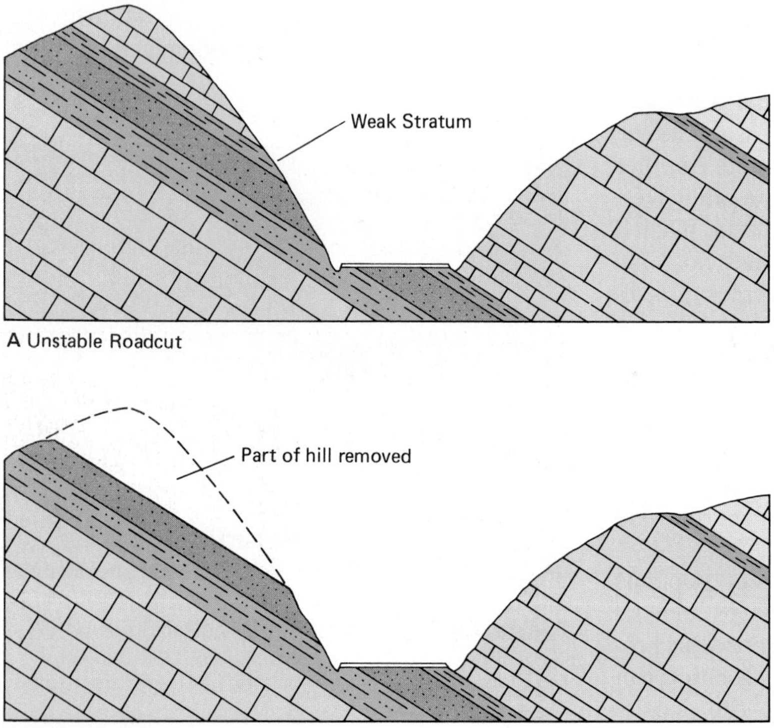

Figure 12–30 Removal of rock mass above plane of potential failure so as to protect highway from landslide.

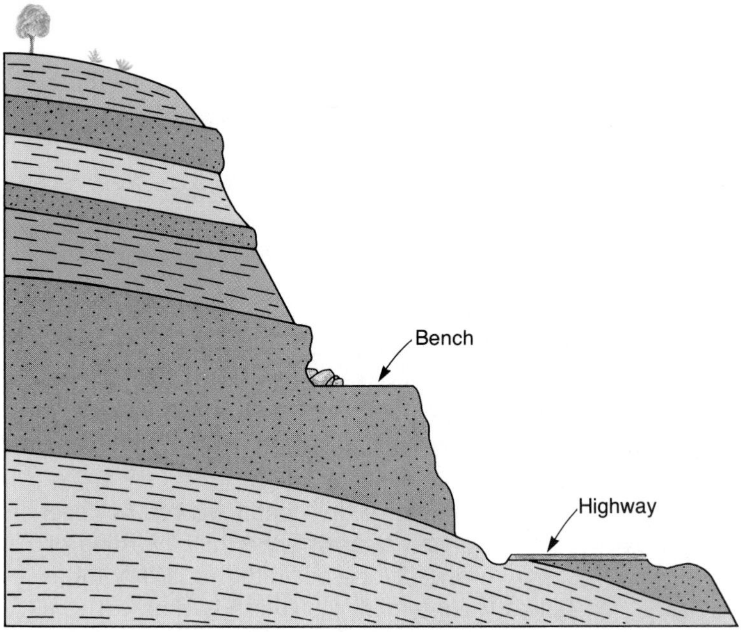

Figure 12–31 Bench cut into highway excavation will serve to lighten load on toe of cut and also serve to catch debris from the upper slope.

and walls of rock-filled wire crates called **gabions** can be used to hold back debris from landslides and other forms of mass wasting. Of course, some of these cures are expensive, and it is thus better to avoid building near potentially unstable slopes wherever possible.

Slopes prone to landslides, earthflows, slumping, or avalanching should not be used as building sites.

Figure 12–33 Rock bolts can be effective in preventing the slippage of bedrock along inclined planes of weakness. The system involves lowering a heavy cable into a drill hole and cementing it at the base within stable bedrock. A steel plate and nut are tightened onto the cable, thus anchoring the unstable beds above to the stable beds below.

Figure 13–32 Concrete wall and fence protect railway and highway from rockfall and bounding rocks. (Courtesy of Pennsylvania Department of Transportation.)

Figure 12–34 Metal-bin cribbing used to help stabilize a slope. (Courtesy of the U.S. Department of Commerce, Bureau of Public Roads.)

Summary

The mechanically fragmented and chemically decomposed rock and soil formed by weathering are ultimately moved to lower elevations under the impetus of gravity. This gravitational transfer of earth materials constitutes *mass wasting.* Although often imperceptible, mass wasting is of enormous importance in the degradation and leveling of the lands. Loose material is delivered by mass wasting processes to streams, glaciers, and ocean waves and then transported and redistributed by those agencies.

The effectiveness of gravity in mass wasting is influenced by such factors as degree of slope, nature of the materials on slopes, the amount of water in the material being moved, the kind and amount of vegetation, and the attitude of bedrock. If slopes happen to be in a relatively stable condition, mass movements may nevertheless be triggered by earthquakes or by the placing of extra loads of overburden on slopes, by the removal of material that helped support the slope, or by changing the physical characteristics of earth materials so that they move or slide more readily.

Mass movements may be either rapid or slow, depending on the nature of the conditioning factors noted above. Slow movements include *creep, earthflow, mudflow,* and *solifluction.* Creep is the slowest but most widespread of these movements. It can sometimes be detected by tilted or displaced fence posts and bent or broken upper edges of steeply dipping strata. *Frost heaving* is considered an important contributor to creep. If the amount of water in the overburden increases, creep may give way to an earthflow, and if water content in the moving material reaches the point of saturation, mudflows are a likely result. *Solifluction* occurs when such frozen ground melts from the top downward. The water-saturated surficial regolith then moves along on top of the still-frozen underlying layer.

Because of their potential for destruction and often spectacular power, rapid forms of mass wasting such as *slides, slumps,* and *avalanching* are more familiar to most of us. Such rapid movements are, of course, more prevalent in mountainous regions. *Rockfalls* are also frequent in rugged terrains, and the accumulated debris from this form of mass-wasting forms the talus slopes nearly always present at the base of cliffs and steep hills.

The sculpture of our continents has not been caused by the action of either mass wasting or stream erosion alone but rather is largely the result of the interaction of these two processes. Erosion by ice, wind, and waves plays an important but secondary role. As an illustration of the interaction of mass wasting and stream erosion, one may view the Grand Canyon of the Colorado River in Arizona. Only a fraction of the rock once occupying that enormous cleft was eroded by the stream itself. The bulk of the filling was first weathered and then passed downward to the stream as mass movements. On reaching the valley floor, the river picked up the material supplied to it, and moved it toward an ultimate resting place in the sea.

Mass movements, whether natural or induced by human activities, can be exceptionally costly. For this reason, informed governmental agencies have conducted studies that indicate areas particularly prone to mass wasting. Engineers should be deeply involved in the investigation of potentially unstable slopes at construction sites and along the intended route of roadbeds. Wherever the potential for destructive mass movements in an area is recognized, and those areas cannot simply be avoided, steps can be taken to prevent damage. Potentially dangerous drainage can be diverted, retaining walls constructed, or slopes flattened by excavation.

Terms to Understand

angle of repose
avalanche
creep
debris flow
earthflow
flow
gabions
heaving
lahar
landslide
mass wasting
mudflow
permafrost
rill
rock glaciers
rockfall

rockslide
sheetwash
slide
slump
solifluction
talus

Questions for Review

1. What is creep? Why is creep more rapid on a steep slope than a gentle slope? What is the difference between creep and earthflow?

2. How does falling rain accomplish erosion? Under what circumstances is falling rain most effective in causing erosion?

3. Under what geologic conditions would there be a high probability of landslides?

4. In what ways may massive downslope movements be initiated or "triggered?"

5. Suppose you were considering a hill slope as a possible site for a house. How might you determine whether or not mass movements had previously occurred on the slope?

6. Building sites on a hillside with a fine view are restricted to two locations. Both are underlain by shale strata, but at one site the strata dip steeply into the hill, and at the other the strata dip roughly parallel to the slope of the hill. Which homesite would you select and why?

7. What is meant by "angle of repose?" Which has a higher angle of repose, dry or moist (but not saturated) sand?

8. With respect to the mass of material that has moved, how does movement by flow differ from movement by slip?

9. How does the mass movement called mudflow differ from slump?

10. Why do landslides occur more frequently during the spring in mountain ranges?

11. Why do mudflows occur more often in arid and semiarid regions than in regions having abundant rainfall?

12. Explain how streams and mass wasting work together to erode the land surface.

13. What measures might be taken to help prevent mudflows in a region that is susceptible to this type of mass movement?

14. Mass movements and erosion have been prevalent on earth for billions of years. Why then do we still have regions of high mountains and irregular topography?

15. In what way are solifluction and permafrost related? What special problems associated with permafrost affect people and communities located in polar or near-polar regions?

16. Prepare a list of human activities that potentially may diminish the stability of slopes and cause slumps or landslides.

17. Of the common types of metamorphic rocks described in Chapter 7, which might be a potentially hazardous foundation for the pier of a large bridge or the abutment of a dam?

18. What is the origin of the peculiar stone rings and polygonal networks of cobbles that are found on the surface of the ground in polar regions?

19. As a precaution against potential damage caused by mass movements, what conditions would you seek to avoid in selecting a site for the construction of your dream home?

20. How might you distinguish between the deposits of an ancient landslide and those of an ancient cobble beach?

Supplemental Readings and References

Buol, S.W., Hole, F.D., and McCracken, R.J., 1973. *Soil Genesis and Classification.* Ames, Iowa, The Iowa State University Press.

Bloom, A.L., 1969. *The Surface of the Earth.* Englewood Cliffs, N.J., Prentice-Hall, Inc.

Bloom, A.L., 1978. *Geomorphology.* Englewood Cliffs, N.J., Prentice-Hall, Inc.

Eckel, E.B. (ed.), 1958. *Landslides and Engineering Practice.* Nat. Acad. Sci.-Natl. Res. Council Publication 544.

Keller, E.A., 1982. *Environmental Geology* 3d. ed. Columbus, Ohio, Charles E. Merrill Publ. Co.

McDowell, B., and Fletcher, J.E., 1962. Avalanche! 3,500 Peruvians perish in seven minutes. *Natl Geogr.* 121:855–880.

Ritter, D.F., 1978. *Process Geomorphology.* Dubuque, Iowa, Wm. C. Brown Co.

Schumm, S.A., 1966. The development and evolution of hillslopes. *J. Geol. Educ.*

Sharpe, C.F.S., 1938. *Landslides and Related Phenomena.* New York, Columbia University Press.

Varnes, D.J., 1978. *Slope Movement Types and Processes in Landslides: Analysis and Control.* Transportation Research Board, Natl. Acad. Sci.–Natl. Res. Council Spec. Rpt. 176.

Streams and Valleys

13

The meander pattern and point-bar deposits are well developed in this scenic view of the Rausch River in British Columbia. A glacier formerly filled the valley, and the stream is developed in glacial sediment. (Courtesy of G. Ross.)

Running water labors continually to reduce the whole of the land to the level of the sea.

Charles Lyell, 1830

Introduction

In southeastern Australia, there is a river that flows for a distance of about 2000 km from the highlands of Victoria to Encounter Bay south of Adelaide. The aborigines who live nearby tell an interesting tale about the origin of this stream and its valley. They describe how a narrow crevice was opened by an earthquake, and how subsequently, wherever it rained, the crevice would develop a thin trickle of water. The tale does not end at that point, however, for it is necessary to account for the considerable width and graceful bends of the stream and its valley. A second earthquake along with the best "fish story" of all time, is invoked to completely account for the characteristics of the stream. During the second quake, an immense fish squeezed upward into the cleft from somewhere deep within the earth. The trickle of water in the rift was quite inadequate for the great fish, and so with mighty strokes of its powerful tail it wriggled its way to the ocean, leaving behind the wide sinuous valley of the Murray River.

Actually, of course, the valley described above was eroded over a long period of time by the river that flows within it. The concept that, given sufficient time, even puny-looking streams have such erosive powers would have been difficult for the aborigines to grasp. It was also a concept that was not understood by intellectuals in 16th century Europe. Most naturalists of that time believed valleys formed as a consequence of unknown catastrophic events. An exception was a Saxon professor of mineralogy named Agricola (1494–1555). Agricola taught his students that streams cut the valleys in which they flow and even were responsible for the development of fertile floodplains. Agricola's view was echoed much later in the writings of James Hutton, whose masterful reasoning is evident in this statement published in 1802 by his friend and advocate James Playfair:

> *Every river appears to consist of a main trunk fed from a variety of branches, each running in a valley proportioned to its size, and all of them together forming a system of valleys communicating with one another, and having such a nice adjustment of their declivities, that none of them join the principal valley, either on too high or too low a level, a circumstance which would be infinitely improbable if each of these valleys were not the work of the stream which flows in it.*
>
> Illustrations of the Huttonian Theory

Today, there is no disagreement with the concept that, except for relatively few fault troughs, most valleys are formed by streams. The process of valley formation begins when water flowing down a slope as sheetwash is separated by surface irregularities into small channels that then develop into gullies. With further additions of water from rain or snow, gullies are deepened by the running water and widened by mass wasting until they begin to take on the appearance of small valleys. Headward erosion (Fig. 13–1) results in the lengthening of the older channels, while new ones are eroded on the slopes of the developing valleys. In their early stages of development, the streams are marked by many irregularities such as waterfalls and rapids. In time, however, the slopes of channels become more gentle, and the streams begin to widen their valleys by cutting laterally. Little by little, an areal pattern of broader primary streams and secondary tributaries begins to emerge.

The most abundant landforms on continents are valleys, and streams are essential to the development of nearly all valleys.

Figure 13–1 Aerial view of a small drainage basin. Streams are being extended toward the top of the photograph by **headward erosion.** San Bernardino Mountains, California. (Courtesy of U.S. Geological Survey.)

If we were to inform the tribe of aborigines mentioned earlier that the real factor behind the development of stream valleys was the sun, they would probably be as skeptical of our "sun story" as we are of their "fish story." Nevertheless, the sun is essential for the development of streams. The sun evaporates and lifts water that then falls from the atmosphere as rain or snow. Some of this water is returned to the atmosphere by evaporation or transpiration from plants. Another portion flows down slopes and in channels as runoff (Fig. 13–2). A third portion percolates into the ground and emerges later as springs at lower altitudes. This water also ultimately enters the stream system.

Thus streams are fed by water from runoff and springs (Fig. 13–3). The rate at which water is supplied to the stream determines its discharge or rate of flow. Discharge in turn affects a stream's velocity, width, depth, and at least partially, the amount of sediment it can transport. In this chapter, we will define and examine the relations among these interacting variables and attempt to understand the many facets of stream behavior. Streams, after all, are important to us humans. They provide us with drinking water, electric power, water for crops, fertile floodplains, and convenient routes for highways. Properly managed, they help us dispose of the waste products of our cities. Most important, they have sculptured for us esthetically pleasing landscapes to which we can escape whenever our view of the world becomes disagreeable because of too much concrete and too few trees.

Gravity: The Force Behind the Flow

Everyone knows that water flows from high places to lower ones. Gravity, of course, is the cause of the movement. Every molecule of water in a stream is attracted directly downward toward the earth's center by gravity, but the molecules are unable to move rapidly downward because they are supported by surrounding water molecules and the stream bed. Water in a stream is therefore forced to move along parallel to the slope of the stream bed under the influence of a component of gravity. As was the case in our consideration of mass wasting, the down slope component of gravity acting on a water molecule depends on the steepness of the slope (see Fig. 12–3). If the slope is vertical (as in a waterfall), the water will be pulled directly downward by the full force of gravity. As the angle of inclination of the slope from horizontal is reduced, the value of the component of gravity will also be lessened.

If the component of gravity acting on water molecules were unopposed in a stream, the water would quickly attain incredibly high rates of movement. Friction of the flowing water against the floor and walls of the channel, as well as internal friction among and between water molecules and suspended sediment, all serve to oppose the force of gravity.

The slope of the stream channel between any two specified points (and along which the component of gravity operates) is called the **gradient**. Stream gradients are usually expressed as a ratio as follows:

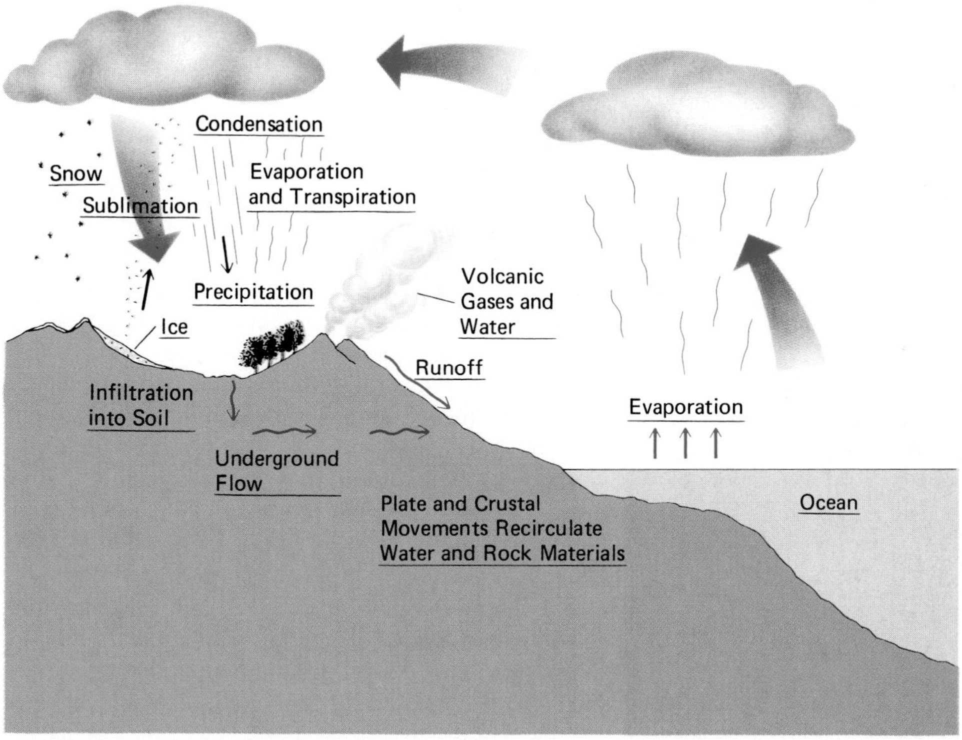

Figure 13–2 The hydrologic cycle.

$$\text{Gradient} = \frac{\text{Vertical Distance (in meters)}}{\text{Horizontal distance (in meters)}}$$

One can determine the gradient of part of a stream by using data on a topographic map. To do this, one first locates two adjacent contour lines connecting points of equal elevation. For example, Figure 13–4 depicts 180-meter and 190-meter contour lines crossing a stream at two points that are 2000 meters apart. If we assume the channel floor is parallel to the water surface, then the stream has fallen 10 meters vertically over the 2000-meter horizontal distance. The gradient of this segment of the stream would be 10/2000 or 0.005 (this means the stream falls 0.005 of a meter per meter of horizontal distance). The gradient of the stream in meters per kilometer can be obtained by multiplying the gradient by the number of meters in a km. Thus, 1000×0.005 equals 5 meters per km. The upper parts of streams that flow out of the Rocky Mountains have gradients of over 50 m/km.

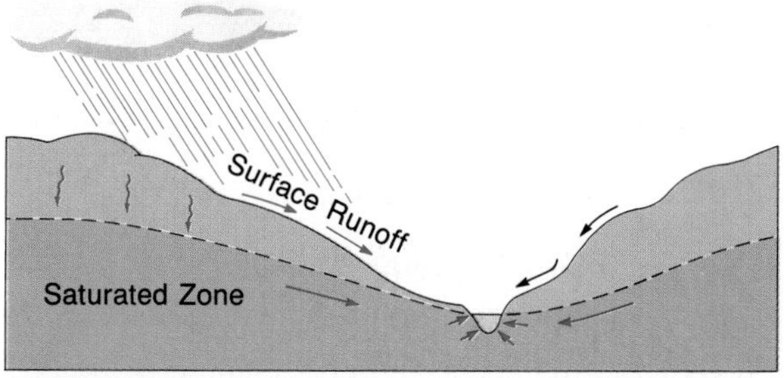

Figure 13–3 Streams are fed by water from runoff as well as from underground water that emerges above or within the channel.

The Dynamics of Streams

Streams are highly complex systems affected by many interacting variables. As is often the case with natural systems, if any one of the variables is altered, changes are also produced in other dependent variables. The most important variables are discharge, velocity, gradient, sediment load, bed roughness, and cross-sectional area of the stream. As an example of the interdependence of variables, if the discharge of a stream were to be suddenly increased after a rainfall, then the stream's velocity will increase, water may rise in the channel so as to provide a larger cross-sectional area, and the load of solid particles carried by the stream may increase. A stream's characteristics will also change along the length of its course. In most streams, the quantity of water, and transported sediment, the velocity, and cross-sectional area increase downstream, whereas the gradient decreases.

The important variables controlling stream behavior are discharge, velocity, gradient, channel size and shape, and load.

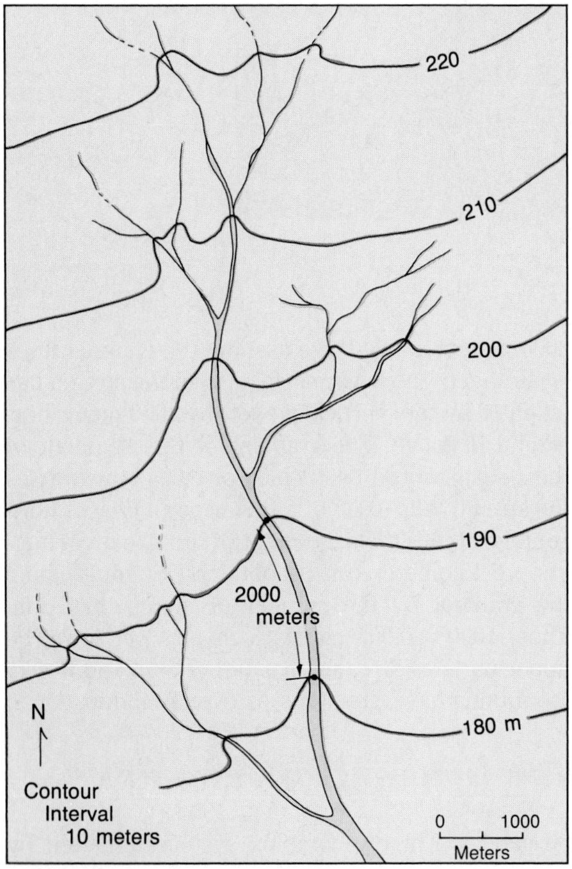

Figure 13–4 The gradient for a stream that descends 10 meters over a distance of 2000 meters would be 0.005.

Discharge

The amount of water passing through a given cross-section of a stream in a given time is termed **discharge**. Discharge is usually expressed in terms of cubic meters of water per second. Thus, the quantity of discharge, by convention designated Q, is obtained by multiplying a stream's cross-sectional area in square meters (A) by the average velocity (V) in meters per second. The area of a stream's cross-section is derived from measurements of average width and depth. Velocity of a stream is measured with a special device called a *current meter*, and the average velocity is calculated from many current meter readings. After the cross-sectional area and average velocity of a stream are calculated, it is a simple matter to determine discharge by use of the equation $Q = AV$. A stream with an average velocity of 0.5 meter per second and a cross-sectional area of 20 square meters would have a discharge of 10 cubic meters per second. The discharge equation indicates that a river with constant water volume would have to increase its velocity if its channel were restricted by either geologic features or artificial constructions.

A stream has no inherent ability to increase or decrease its discharge. Discharge is determined for the stream by the amount of water that enters the channel from runoff, springs, and tributaries. These contributions are in turn influenced by climate (which controls rates of evaporation and precipitation), topography, infiltration capacity of soils, and the susceptibility of the river bed to seepage. An example of the magnitude of discharge in a large stream is provided by the Mississippi River. At Vicksburg, the Mississippi River discharges its waters at rates of about 57,000 cubic meters per second when in flood. Small streams vary in discharge from about 0.3 to 280 cubic meters per second.

The succession of tributaries that enter a major stream are responsible for the downstream increase in discharge.

Velocity

The **velocity** of a stream is the speed at which water moves down the channel in a direction perpendicular to the channel cross-section. Velocity is controlled by several factors, including the gradient of the stream, its cross-section, the viscosity of the water (which is dependent on the amount of suspended matter and temperature), channel shape, and channel roughness. One can calculate the average velocity of a stream if its discharge and cross-sectional area are known from the relation:

$$V_{\text{av}} = \frac{Q}{A}.$$

The average velocity may differ considerably from the actual velocity measured at a particular point in the channel. A stream is likely to be flowing fastest about a third of the distance down from its surface. Water would be moving a bit slower at the air–water interface because of friction with the overlying air, and slowest of all along the bottom because of friction with the channel. A longitudinal segment of a stream between any two points is called a **reach**. If the section across which velocity is measured is a straight reach, the greatest velocity will be toward the middle of the stream. Around curves, however, maximum velocity shifts toward the outside of the bend. Figure 13-5 indicates the distribution of velocity along straight and curved reaches of a stream.

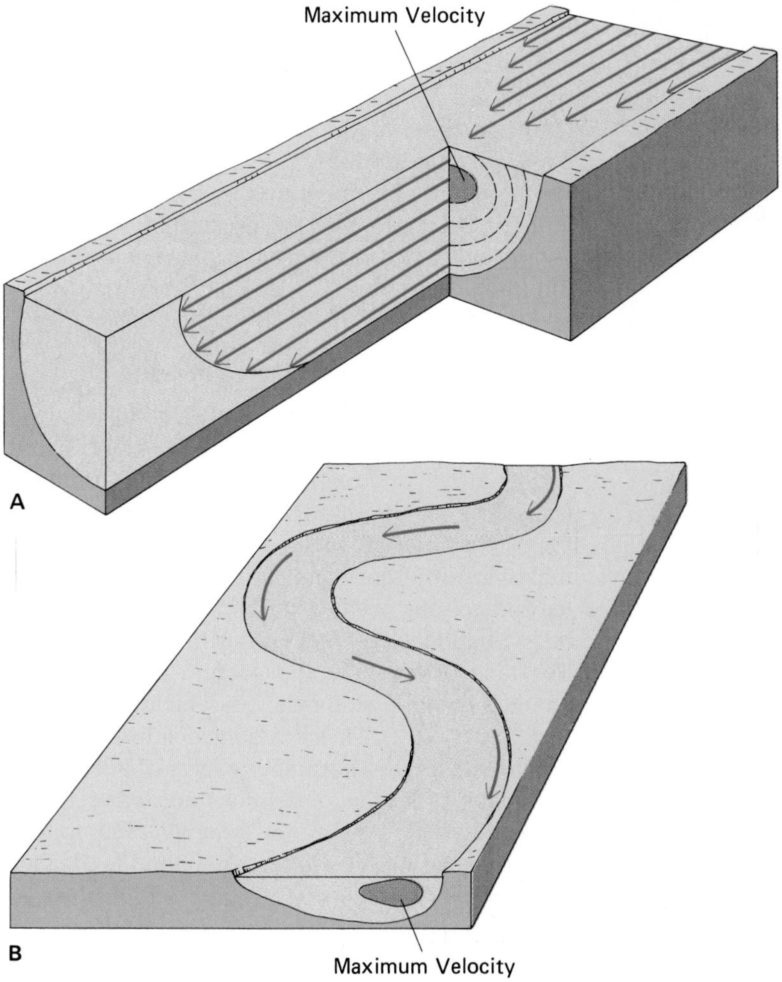

Figure 13–5 Velocity distribution along a straight section of an idealized river. (A) Upper right part of block indicates relative surface velocities, and sides of block indicate minimal velocity near the bottom of the stream where velocity is decreased because of friction. Velocity increases upward except for a decrease near the surface where velocity is again reduced because of frictional drag. (B) Maximal surface velocity is generally at the center in straight sections of a stream and shifts toward the outer bank along curves in the river.

Stream velocity is primarily controlled by two factors: gradient and volume of water.

Flow Pattern

When water is moving slowly and smoothly, the paths taken by individual water molecules may be either straight or gently curved and are parallel to the paths of neighboring molecules. This type of flow is called laminar. It was first demonstrated in 1883 by Osborne Reynolds, who introduced a few drops of dye to clear water flowing gently through a glass tube (Fig. 13–6). The dye moved with the water in smooth parallel lines. One can also produce this effect in a metal trough, which more closely resembles a straight reach of a river. If the velocity of flow in the trough is increased (by increasing either the gradient or discharge), the flow lines of dye will become wavy and chaotic. Water immediately above the bottom of the channel is slowed by friction with the bottom and therefore moves more slowly than adjacent water masses, and the variance throws the water into a chaotic movement called **turbulent flow**. Such swirling eddying turbulence is the dominant kind of flow in all natural streams. Turbulence is the result of high velocity and irregularities in the bed and bank of a stream. It is an important attribute of stream flow, for it enhances the ability of the stream to erode its bed and to keep particles suspended in the moving water mass.

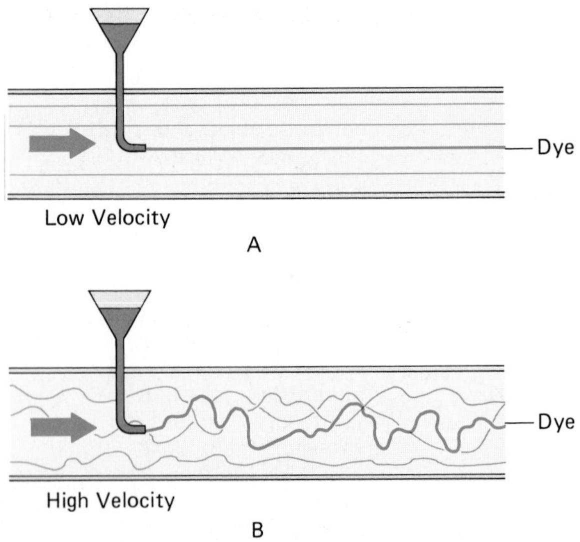

Figure 13–6 (A) In laminar flow, molecules move in parallel paths, and flow lines do not cross. (B) Turbulence is characterized by many irregular eddies superimposed on the general direction of flow.

Turbulent flow, in which water molecules travel in curved, erratic paths, is the characteristic kind of flow in streams.

Transportation and Erosion

The Load

The dissolved and solid rock waste carried by a stream constitutes its **load**. A stream's load is derived from several possible sources. A part of the load is carried to the channel by sheetwash and mass wasting. Additional load is provided by the stream itself as it undercuts its banks and erodes its bed. At locations where erosion by wind or ice is prevalent, these agencies may also contribute to the load carried by the stream.

Dissolved Load

It is easy to overlook the dissolved load carried by a stream. It is invisible and does not significantly affect a stream's behavior. Nevertheless, transportation of dissolved mineral matter is an important function of rivers. It has been estimated that nearly 4 billion tons of dissolved materials are carried to the ocean each year by streams. The most abundant components of this load are positive ions of calcium and sodium and negatively charged sulfate, chloride, and bicarbonate ions (see Table 17–2). These dissolved substances may precipitate as the thin mineral crust that forms on the inside of kettles. Soap does not lather well in mineral-rich or "hard water" and will react to form calcium stearate. The latter compound is the bane of fastidious housekeepers, for it is the substance of which bathtub rings are composed. As one might guess, streams that flow across relatively soluble limestone terrains are particularly rich in dissolved mineral matter.

Solid Matter

The solid grains in a stream are transported by suspension, saltation, and traction. Particles in **suspension** are carried along within the body of flowing water. They are kept from being deposited by upward-moving turbulent eddies. Only relatively small grains, such as those in the silt and clay sizes, are able to resist the pull of gravity and remain in the **suspended load**. Except in extremely fast-flowing steams, sand particles do not remain suspended but collect on the surface of the stream bed where they move along in a series of leaps (Fig. 13–7). The motion is termed **saltation**. Still larger sand grains, as well as granules, pebbles, and cobbles, are moved along by sliding and rolling motions. The process of transporting coarse sediment along the bottom of a stream, either by rolling or sliding, is called **traction**. All of the material carried by traction constitutes the stream's **bed load**. Because individual fragments in the bed load move only intermittently, the bed load moves down the channel more slowly than the water itself.

Bed load movement in a stream occur primarily during times of flooding.

Precisely which range of particle sizes is transported by suspension, saltation, or traction is, within limits, dependent on velocity. Should the velocity of a rapidly flowing stream be decreased, the larger particles in the suspended load might become part of the saltation load, and particles formerly moving by saltation would continue their downstream progress by sliding and rolling.

Competence and Capacity

The term **competence** is used to express the largest particle that a stream can transport under a given set of conditions. The competence of a river may change considerably from its headwaters to its mouth, and even at any single point along its course. Streams can transport much larger particles during times of flood when velocity of flow is at a maximum. In general, the diameter of a rock fragment that can be moved by the stream is proportional to the square of the velocity of flow. Thus, if the velocity is doubled, the stream can move particles four times as large as before. The explanation for this relationship is that each molecule of water not only strikes a grain of sediment twice as hard, but also twice as many molecules strike the grain during a given interval of time.

A second term used to describe sediment tranport of a stream is capacity. **Capacity** refers to the maximal amount of material the stream can carry under a given set of conditions. Because there are limits to the amount of load that can be included within a given volume of water, it is evident that a stream's capacity will be markedly influenced by discharge.

A stream's capacity is largely determined by its discharge. Competence is largely controlled by velocity.

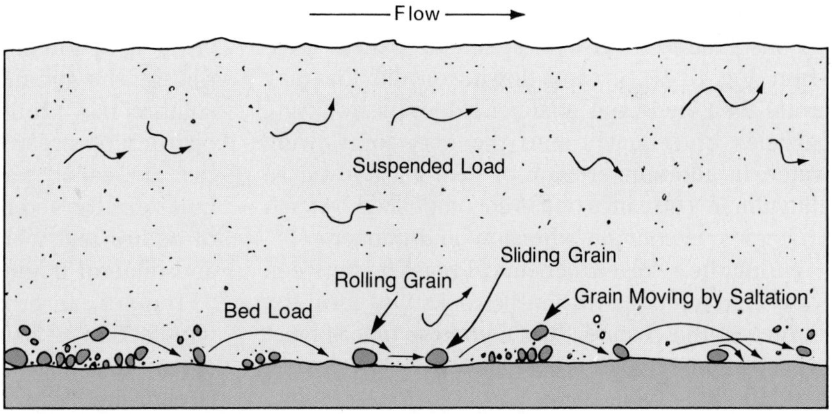

Figure 13–7
Transportation of sediment by a stream. The particles being moved by saltation, rolling, and sliding constitute the *bed load.* Particles in the *suspended load* are carried above the stream bed within the body of flowing water.

Erosion by Streams

Relation of Velocity to Erosion

Velocity is the most important factor in determining the rate of channel erosion by a stream. Erosion occurs whenever particles of sediment are lifted from or moved from the stream bed. Particles may be pushed forward by the force of the moving water against their upstream sides. They may also be lifted because of the reduced pressure on their upper surfaces as water flows over them. The forces that attempt to lift grains at rest and bring them into the transported load are opposed by gravity as well as the natural cohesive force that tends to hold together small particles at rest. Of these two forces, gravity is the more important for large particles, although cohesion plays an important role for the finer clay-size particles. Once deposited on the stream bed, the flat clay particles tend to cling to the channel floor. Water, moving against the thin edges of the clay particles, has little surface to push against. In general, if the channel surface is composed of clay, stream velocities must reach about 100 cm/sec before erosion can begin, whereas sand on the stream bed is readily lifted into the flow at velocities of only 10 cm/sec. This relationship is shown in Figure 13–8. Notice that the velocities needed to initiate movement are far greater for clay than for pebbles, because of the cohesiveness of clay.

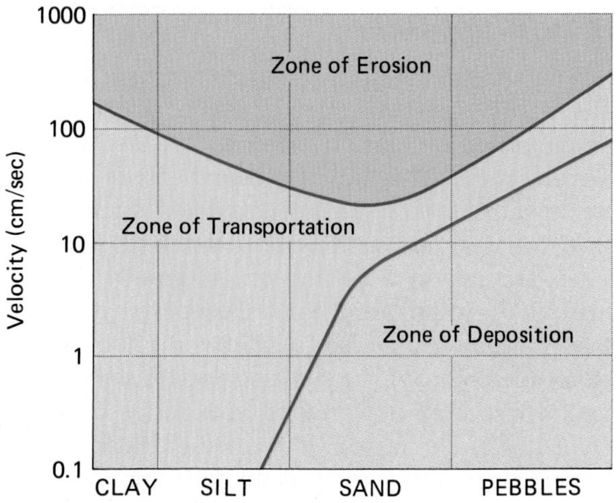

Figure 13–8 Graph showing the relationship of stream velocity to erosion, transportation, and deposition of various grain sizes. The upper curve indicates the minimum velocity required to erode particles, whereas the lower curve shows the velocity below which particles will be deposited. In the area between the two curves, grains will be carried by the stream in its bed or suspension load. Note that a higher velocity is needed to erode grains than to transport them. (After F. Hjulström, 1935. *Bull. Geol. Instit. Uppsala* 25:295.)

Erosional Processes

Stream erosion is the removal of rock and mineral matter from the bed of a river. The material in which the bed is cut may consist of bedrock, or, as is more common, the stream bed may consist of unconsolidated river-laid sediment called **alluvium** (Fig. 13–9). Streams flowing on alluvium may erode their beds and channel sidewalls by lifting particles of sediment into the mass of flowing water. In addition, erosion of bedrock as well as alluvium in a stream's bed is accomplished by such processes as solution, abrasion, and impact.

Solution involves a chemical reaction between water and minerals within the rocks that form the surface of the channel. It is a process that is most prevalent in regions where streams flow across particularly soluble rocks like dolostone or limestone. The dissolution of rock forming the bed of the stream contributes to the dissolved load. That contribution from the bed, however, is relatively minor when compared with the much larger amount of dissolved material derived from chemical weathering across the entire drainage area and carried to the streams by underground water or sheetwash.

In contrast to the largely chemical affects of solution, **abrasion** refers to the mechanical wearing away of the channel as a result of the collision and grinding action of the sand grains and pebbles carried by the stream. Acting together, abrasion and solution may produce basin-like depressions in the bedrock of streams. These smooth, rounded cavities are called **potholes** (Fig. 13–10). Most potholes are formed by the relentless circular motion of eddies and whirlpools that contain loose rock and mineral fragments as tools (Fig. 13–11).

Impact, an erosional process associated with wind action, is also important in stream erosion. It occurs whenever large particles carried by the stream are thrown forcibly against the sides and

Figure 13–9 Geologist at work making latex peels of alluvial sediment in order to better understand the depositional process and interpret ancient stream deposits. (Photograph by E.D. McKee, courtesy of the U.S. Geological Survey.)

bottom of the bed. If the stream bed is cut in bedrock, the many continuous blows from these propelled particles knock off fragments of rock and thereby contribute to the solid particles comprising the stream's load. In addition, fragments of rock in transport are themselves eroded by impact and abrasion.The process contributes to the rounding of pebbles and cobbles.

Base Level

Stream erosion, in combination with mass wasting, has done more to develop the landscapes of our continents than any other group of geologic processes. Everywhere we see the ravines, valleys, canyons, ridges, and floodplains that result in part from the activity of rivers. There are times when that activity is inspirational, as when streams topple over resistant ledges to form waterfalls. At other times, the activity of streams may be catastrophic, as when floods sweep away buildings and undermine bridges. But most of the work of streams goes on largely unnoticed, as the flowing water sculpts valleys and transports the material that once filled them to the ocean.

Figure 13–10 Pothole erosion in Pikes Peak Granite, Colorado Springs, Colorado.

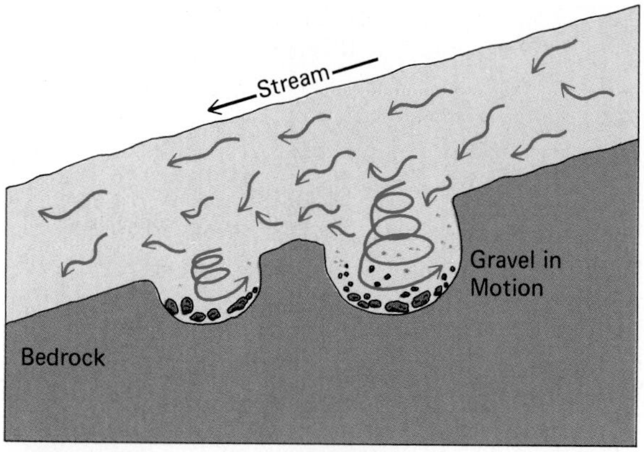

Figure 13–11 Pebbles and cobbles swirled about by turbulent water are effective tools in producing potholes in solid rock.

The degradational work of streams is truly immense. Each year, rivers carry an estimated 10 million metric tons of continental material to the oceans. This work, however, cannot continue indefinitely in a drainage basin that remains tectonically stable. Eventually a stream would cut its valley to sea level and would be unable to accomplish further erosion. This lower limit of erosion is called **base level**.

The significance of base level was recognized in 1869 by the one-armed Civil War veteran and geologist J. W. Powell. While leading a perilous boat expedition down the unexplored gorges of the Colorado River, Powell was understandably impressed with the formidable erosive abilities of that great stream. "How long," he wondered, "could the river continue to actively cut downward?" The answer seemed readily apparent. Eventually, the stream would have eroded its bed nearly to sea level, and with little or no slope, would cease its geologic work. "We may consider the level of the sea," wrote Powell, "to be a grand base level, below which the dry lands cannot be eroded." Powell's "grand base level" is now termed **ultimate base level**.

Powell also made studies of the mouths of tributaries entering the Colorado River and perceived that these confluences served as **temporary** or **local base levels** for each tributary. The tributary could cut its valley no deeper than the major stream into which it empties. Lakes, dams, and resistant rock formations may also form local base levels, which temporarily provide a downward limit to the erosive powers of streams.

The importance of base level as a control over stream behavior is more readily understood if one considers what occurs to a stream when its base level is either raised or lowered. It is common for base level to be artificially raised when a dam is constructed. The reservoir formed behind the dam becomes the stream's new local base level (Fig. 13–12B). Because of the higher base level, the gradient of the stream is reduced. Consequently, it cannot carry its load and deposits sediment in its channel. The channel is built higher by this action, and eventually again attains sufficient gradient to do its work.

Should base level be lowered, as might occur if the land surface were uplifted, or if a lake that served as local base level were drained, the stream would not have a greater gradient. Its ability to erode and transport would be improved, and it would erode its channel to conform to the new lower base level. The stream segment nearest the new base level would be the first to be affected, and erosion would progress upstream from the

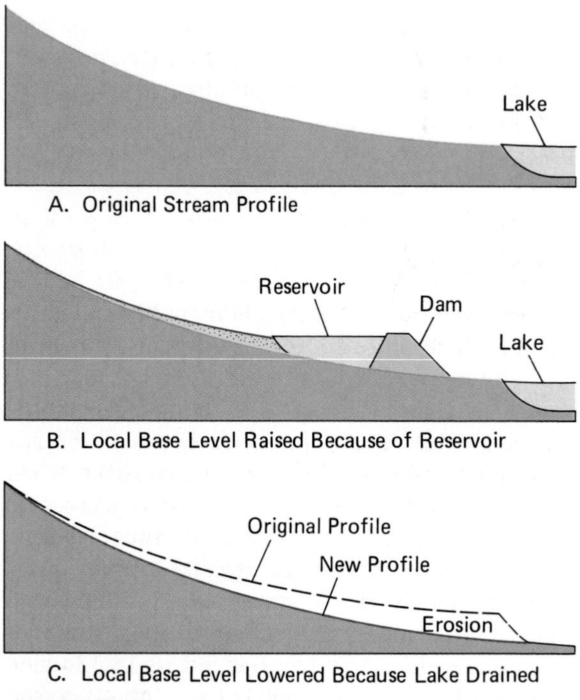

Figure 13–12 Adjustment of a stream to (B) raising of local base level, and (C) lowering of base level.

point at which base level had been lowered (see Fig. 13–12C).

Base level is the lowest point to which a stream flows. If base level is lowered, erosion will occur. If it is raised, deposition will occur.

Streams and Tectonics

If we recall the statement made earlier that rivers carry approximately 10 billion tons of solid and dissolved material to the oceans each year, and further remember that this transfer had been in progress for over 3.6 billion years, then one might wonder why entire continents have not been reduced to what Powell described as "the grand base level." The reason, of course, is that the lands have not been static. Mountain building associated with plate tectonic activity has produced a continuous supply of upland areas to balance those worn low by the earth's degradational agencies.

Graded Streams

If a stream of water begins to flow down a sloping surface, the water will cut a channel sufficiently large to contain the average or usual amount of water. In the early stages of development, the **gradient** of the stream is likely to be relatively steep, and the valley will develop a V-shaped cross-profile. The stream will work actively to erode its channel and eliminate the irregularities. As time passes, the gradient will be lessened while the load is steadily increased because of contributions from an expanding network of tributaries. Eventually the stream attains a condition in which the supply of sediment fed into the stream approximately equals the stream's capacity to transport that load. Indeed, most of the river's available energy will be consumed in this transport, and rapid downcutting of the channel ceases. The stream is now said to have reached the graded condition. A graded stream is ideally a stream in equilibrium. Any change in the supply of sediment, velocity, or discharge will result in a displacement of equilibrium in a manner that will tend to absorb the effects of the change. For example, a **graded stream** that becomes underloaded (as when sediment-trapping dams are constructed in its tributaries or headwaters) will have energy to spare and will compensate by erosionally deepening its bed so as to flatten the gradient and decrease velocity. As an alternative the stream might increase the number of bends (i.e., increase its sinuosity) and thereby lessen its gradient. Should a stream become overloaded, it may build up its bed initially by depositing the surplus load and thereby steepen its gradient (Fig. 13–13), or it might increase overall gradient by reducing its sinuosity. In either case, the steepened gradient would result in increased velocity and capacity.

The overall result of these "cut and fill" events is that the stream develops a concave-upward profile of equilibrium. The profile, technically termed the **longitudinal profile**, can be constructed by connecting points of elevation along the length of the stream (Fig. 13–14). The concave slope is a fundamental characteristic of graded streams. Near the headwaters, the gradient of the stream is steep and the river can do its work with the relatively small discharge available. Nearer the mouth, the gradient is lower yet the stream continues to accomplish its work because it has greater discharge and smaller particles to transport and less friction-inducing channel surface. In regard to this last factor, the reduced total friction from water contact with the channel in lower reaches means the stream can maintain a given velocity on a lesser gradient than can an upstream segment having a steeper gradient (Fig. 13–15).

A graded stream is one that is adjusted to the discharge and load inherent to its drainage system. Graded streams have concave-upward long profiles.

Yet another characteristic of graded streams is the progressive downstream decrease in the size of the particles transported. This decrease in particle sizes toward the stream's mouth is largely a consequence of sorting during transportation. Smaller particles are able to remain in motion more continuously than larger ones and so are moved downstream more rapidly. Also, very coarse sand and pebbles introduced into the stream at its headwaters tend to be mechanically reduced during transportation.

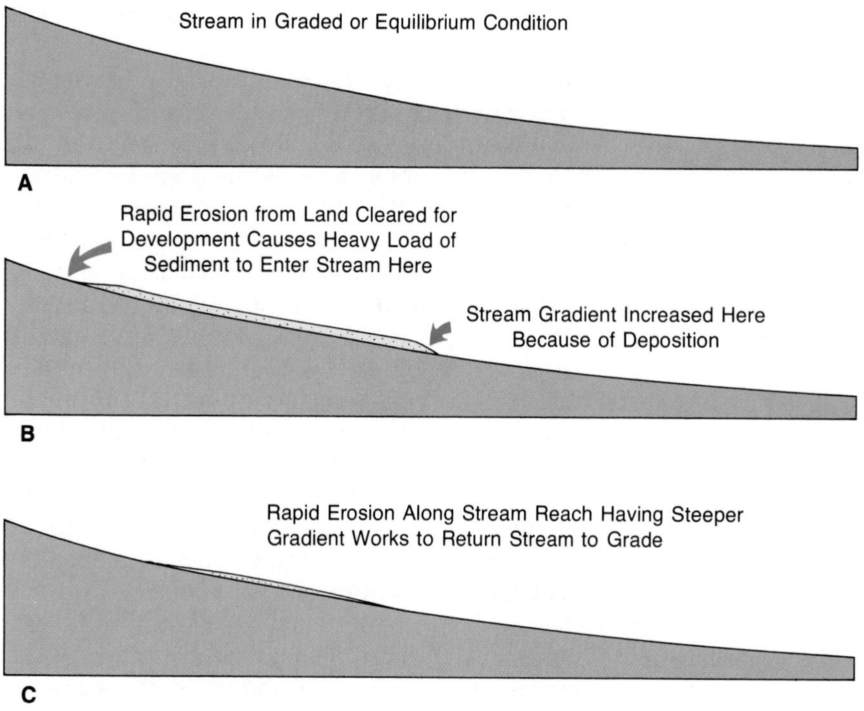

Figure 13–13 A graded stream that becomes overloaded will initially deposit the portion of the load it cannot carry. The result is a bed build-up. At the downstream edge of the build-up, the steeper gradient will favor rapid erosion, and the stream will gradually remove the sediment and return to the graded condition.

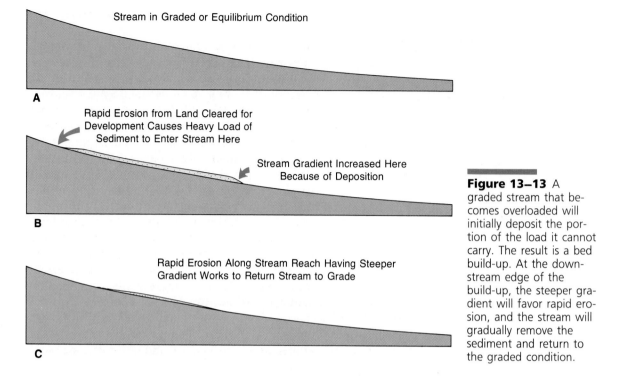

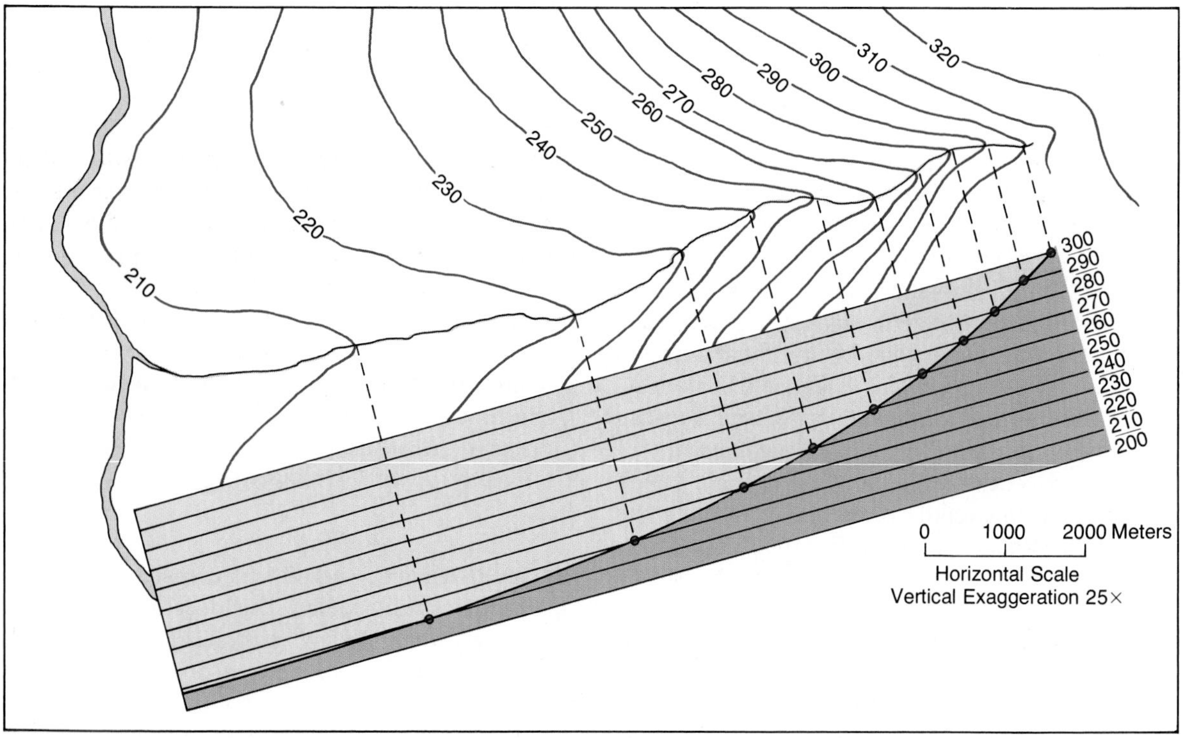

Figure 13–14 The longitudinal profile of a stream is constructed by locating points where topographic contour lines cross a stream channel and projecting those points onto a piece of ruled paper. Graded streams typically develop the concave-up longitudinal profile shown here.

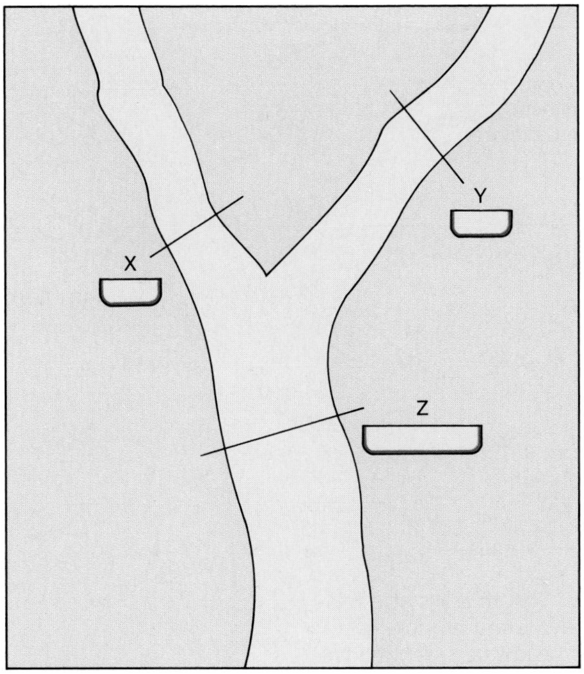

Figure 13–15 The cross-sectional area of tributary X plus Y equals that of the major stream Z. Yet the part of the channel surface in contact with flowing water (the wetted perimeter) of Z is less than the sum of X plus Y. Thus water flowing in Z experiences less frictional drag.

Consequent and Subsequent Streams

The first streams to develop on an undissected land surface will flow down the preexisting slope of the land according to the most favorable and unobstructed route. Such streams are termed **consequent streams** because their courses are a consequence of the original slope of the land surface (Fig. 13–16). The course of other streams may not be determined primarily by an original surface but rather by differences in the structure and resistance to erosion of the bedrock. Such streams are called **subsequent**. Subsequent streams lengthen themselves by headward erosion along belts of easily eroded rock. As the process of headward erosion proceeds, the head of the subsequent stream may ultimately extend into the drainage territory of rival streams that flow at slightly higher levels. When such connections are completed, the flow of the higher-level stream will be diverted into the channel of the subsequent stream, thereby increasing its discharge and erosive powers. This "capture" of one stream by another is termed **stream piracy** and is a commonplace occurrence, especially in the early evolution of drainage systems (Fig. 13–17).

Features Formed by Streams

Newly formed streams do not possess the valley flats and smooth gradients of older streams. As the river continues its work, however, gradients are reduced. The river will cease to cut downward as actively and begin to cut horizontally into the outside of bends. In the lower-velocity area that characterizes the inside of bends, deposition occurs. The process of cutting and filling is called **lateral**

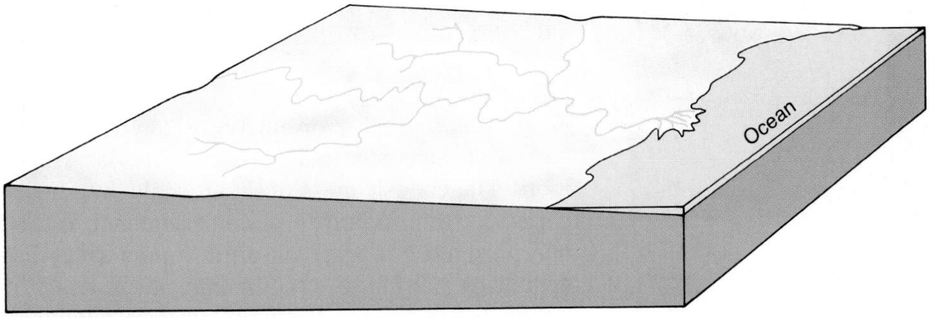

Figure 13–16 The uplifted portion of the continental shelf provides a sloping surface for the development of *consequent* streams. The courses of such streams are a *consequence* of the original land surface.

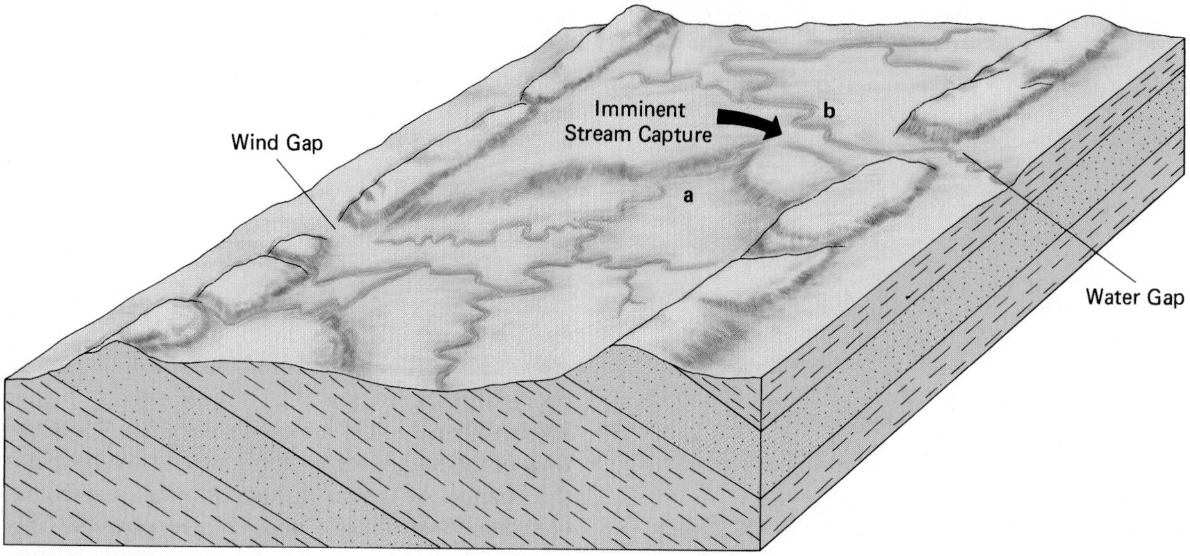

Figure 13–17 Both streams *a* and *b* are subsequent streams. Stream *a* lies at a relatively lower level than stream *b* and is extending its valley by headward erosion. Eventually, it will intersect the channel of stream *b* and capture its headwaters. A windgap is developed in the ridge on the left side of the block, and a watergap is present on the adjacent ridge.

Figure 13–18 The Ohio River near Cincinnati has developed a broad floodplain by processes of lateral planation involving erosion of banks on the outside of meanders and deposition on the inner bends where velocity is reduced. (Courtesy of J.C. Brice.)

planation (Fig. 13–18) and is responsible for the relatively smooth areas bordering a stream. These areas are called **floodplains** because they are inundated during times when the river overflows its banks.

Lateral cutting by a stream predominates along the outside of each meander bend.

As the stream continues to widen its valley by lateral planation, the pattern of alternate left and right bends begins to assume some regularity. The bends are now termed **meanders**, from a Greek world meaning "bends." In general, the meandering pattern (Figs. 13–19 and 13–20) is developed by streams that flow on unconsolidated but relatively cohesive silt and clay and on flat valley floors. The cohesiveness of the fine alluvium helps to restrict the stream to a single channel, rather than permitting it to break up into numerous distributaries. In flume experiments, meanders are readily formed from originally straight channels. Although not essential, the sinuous pattern may be initiated when a part of the bank collapses and thereby diverts the flow of water to the opposite bank. The natural spiral motion of the water mass (Fig. 13–21) tends to perpetuate the development of meanders downstream from the initiating obstacle.

Most meandering streams traverse broad valley flats. Some, however, have steep valley walls and lack floodplains. The meander pattern in such streams was developed by the stream long ago when it was flowing on a valley flat. Subsequently, through regional uplift or some other geologic change, the river was caused to cut down into its former floodplain but at the same time maintained its meander pattern. The result is meanders bounded by steep valley walls. Such meanders are said to be **incised** (Fig. 13–22). When accompanied by other substantiating lines of evidence, incised meanders may be considered evidence for uplift of the earth's crust.

Not all streams or stream segments (reaches) develop a regular meandering pattern. Some divert their flow into numerous shallow dividing and reuniting subchannels separated by transient islands and bars. Such streams are said to have a **braided pattern** (Fig. 13–23). The braided pattern appears to develop in streams having highly variable discharge, rapid rate of sediment supply, and banks composed of loose, noncohesive materials such as sand and gravel. (In contract, meanders develop best where banks are composed of cohesive silt and clay). Because of the susceptibility of sands and gravels to erosion, the braided stream becomes choked with sediment, and divides into many small channels called **distributaries**. Sporadic decrease in discharge among the distributaries may cause local deposition in channels, resulting in blockage and further separation into subchannels.

Figure 13–19 Symmetric meander pattern developed along a reach of the Wood River near Fairbanks, Alaska. The three center meanders are complicated by chutes where water has broken through a swale or low area in the deposits of the inside of the bend. (Courtesy of J.C. Brice.)

Figure 13–20 Meanders, abandoned meanders, point-bar deposits, and oxbow lakes are numerous along the floodplain of the Chatanika River, Alaska. The view is downstream from the place where the Alaskan Pipeline crosses the Chatanika Floodplain. (Courtesy of J. Brice.)

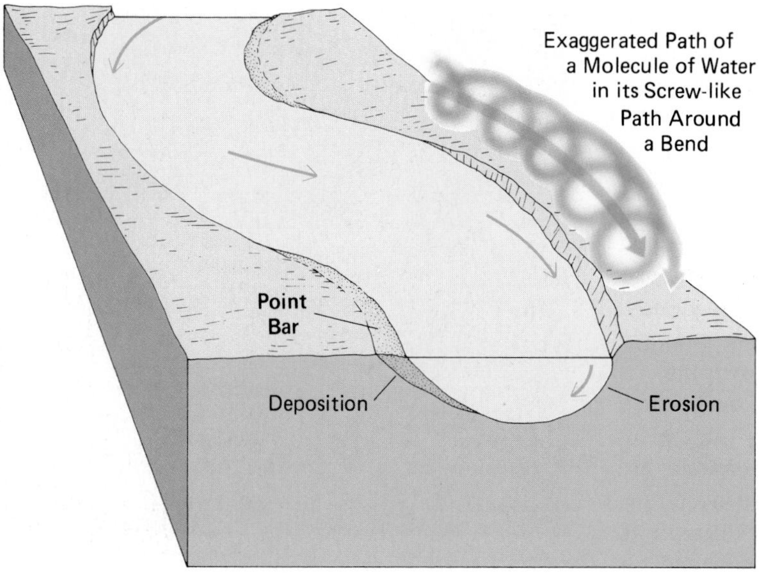

Figure 13–21 Water molecules in a sinuous channel not only move downstream but also follow a spiral-like path. (Modified from Leopold, L.S., and Langbein, W.B., 1960. *A Primer on Water,* U.S. Geological Survey, U.S. Gov't Printing Office 0-398-800.)

Figure 13–22 Incised meanders of the Colorado River south of the Henry Mountains in Utah. (Photograph by W.B. Hamilton, courtesy of the U.S. Geological Survey.)

Features of Floodplains

One may find a large number of secondary features formed along the valleys of meandering streams. These include point-bar deposits, cutoffs, oxbow lakes, and natural levees. **Point-bar** deposits are accumulations of sand and gravel along the inside of meander bends (Fig. 13–24). These deposits develop as a consequence of the velocity being less around the inner margin of a bend, so that the stream deposits part of its load. Point-bars provide the frequent sandy havens for weary canoeists and vacationing bathers. On the outside bends of the meanders, the force of inertia tending to cause the water to flow in a straight line causes the water to impinge upon its bank, to "pile up," and to actively erode.

The cutting along the outside bends of meanders with accompanying filling along inside bends results in slow migration of the meanders in a downstream direction (Fig. 13–25). The migration, if speeded up, might resemble the pattern produced where a snake had crossed an area of sand. Frequently, during the migration of meanders, two bends will come together to form a meander **cutoff** (Fig. 13–26). The abandoned loop of the channel is left as a curved remnant on the floodplain. Because of its resemblance to the yoke placed over the shoulders of oxen to enable them to pull wagons, these abandoned meanders have been dubbed **oxbow lakes** (Fig. 12–27). With time, the oxbow lake may become filled with sediment and disappear.

As meanders shift back and forth across floodplains, they produce oxbow lakes and cut-offs.

River floodplains are not completely level. Most floodplains have a raised crest or ridge adjacent to the channel. These ridges along the side of streams are called **natural levees** (Fig. 13–28). They are formed during flood stages when the water overflows its banks, is suddenly slowed, and drops part of its load. Along parts of the lower Mississippi, the natural levee is the highest tract of land for miles around. Natural levees may prevent tributary streams entering the floodplain from emptying into the channel of the main stream. As a result, the tributary must follow a course parallel to, and behind, the natural levee until it comes upon a place where it can enter the main channel (see Fig. 13–28). Such streams are called **Yazoo streams** after the river of that name in Mississippi.

Figure 13–23 Braided stream pattern in Tanana River downstream from Fairbanks, Alaska. (Courtesy of J.C. Brice.)

Figure 13–24 Point-bar developed on the inside of a bend in the Salt River, northeastern Missouri. (Courtesy of J.C. Brice.)

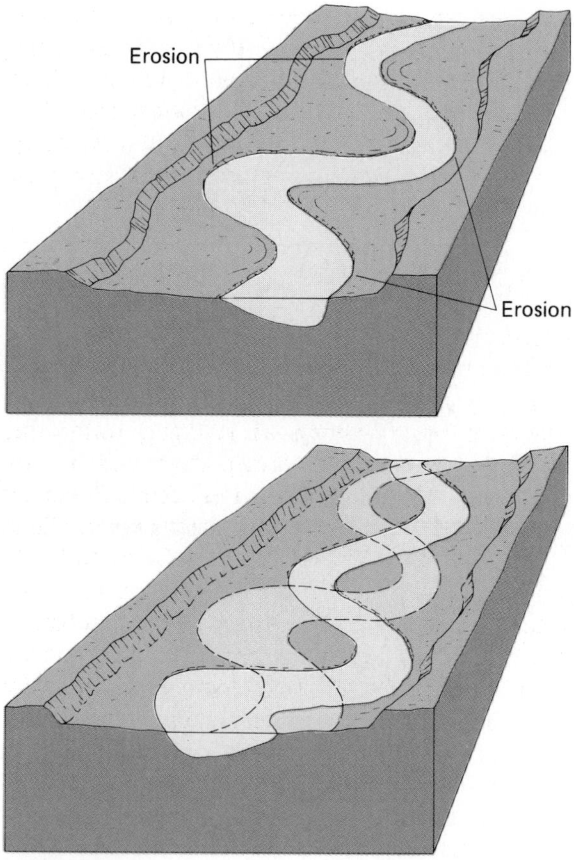

Figure 13–25 Migration of meanders in a downstream direction. Dotted line in lower block marks the earlier channel location.

Alluvial Fans

Alluvial fans are cone-shaped deposits composed of river-transported silt, sand, and gravel (Fig. 13–29). They are formed where streams emerge from a mountainous area and flow onto relatively level terrain. Stream velocity is reduced rather abruptly at such locations, largely because of widening of the channel and loss of water to the porous sediment beneath the stream bed. The stream is unable to carry its former load under these new conditions. Sediment is deposited within channels, causing water to spill over banks and form new channels, many of which disappear before reaching the foot of the fan because of water loss of rapid seepage. These events are repeated endlessly back and forth across the depositional area, eventually building the alluvial fans.

Stream Terraces

A stream system is dynamic and constantly changing. In any given area at a given time, it may be expending most of its energy in transporting its load and have little energy remaining for erosion of its bed. At other times, a reduction in load may result in energy available for active downcutting. Elevation of the channel because of tectonic movements or isostatic uplift, increases in discharge, or lowering of sea level are other factors that might cause a stream to begin eroding its bed. During the downcutting, areas of previously deposited alluvium of the floodplain are removed, leaving the undissected flat remnants temporarily undisturbed. These level remnants located along the margins of the valley are called **stream terraces** (Figs. 13–30 and 13–31). They may persist until eroded by the stream as it meanders endlessly back and forth across the valley. Occasionally old

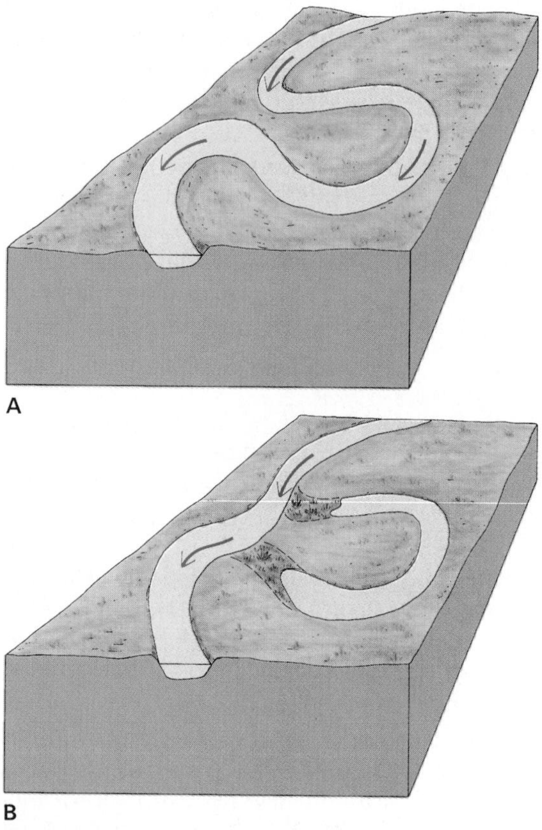

Figure 13–26 Development of a meander cutoff and an oxbow lake.

Figure 13–27 Oxbow lakes in the valley of the White River (a bend of which can be seen at the extreme left), Prairie County, Arkansas. (Courtesy of the U.S. Geological Survey.)

terraces are buried under more recent alluvium when conditions cause the river to cease active downcutting and begin a depositional phase. Such buried terraces are of particular interest to archaeologists seeking the level floodplain surfaces on which ancient peoples may have built their villages.

Deltas

Accumulations of sediment deposited by streams where they enter a large body of water like a lake or sea are called **deltas** (Fig. 13–32). On encountering the body of standing water, the stream's velocity and transportive powers are quickly reduced, and deposition occurs. Sediment accumulating in the channel reduces the gradient, blocks the channel, and causes the flow to be diverted and divided into new channels. The branches leading outward from the main channel are called **distributaries**. Several environmental conditions will favor the development of deltas. An abundant supply of sediment, as one might guess, is the most important condition. In addition, delta formation is enhanced by shallow depth of water, tectonic stability, and lack of strong waves or currents in the area of delta growth.

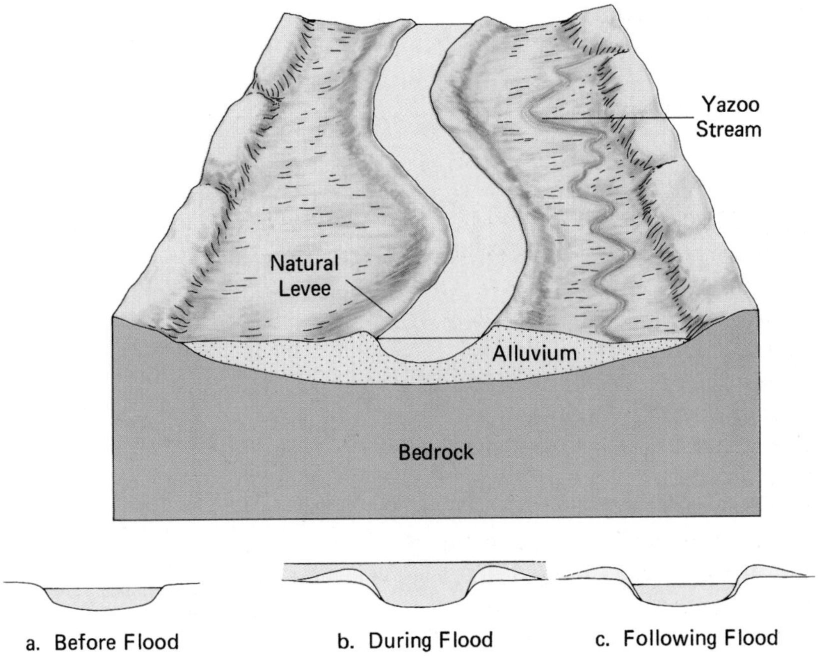

Figure 13–28 Floodplain of a stream having well-developed natural levees. The formation of the levees is indicated in the sequence of figures beneath the block diagram. Levees are developed when the river overflows its banks. Once the river is outside the channel, velocity is reduced and deposition builds a submerged ridge. After the flood, this ridge is left standing higher than the surrounding floodplain.

Figure 13–29 Alluvial fans at the foot of the Panamint Range, Inyo County, California. (Courtesy of the U.S. Geological Survey.)

Delta development is facilitated by abundant sediment supply from the stream, and limited dispersal of that sediment by shoreline waves or currents.

One of the most distinctive sedimentologic features of deltas is cross-stratification (Fig. 13–33). This feature begins to form as soon as the river enters the sea or lake and drops the coarser components of its load nearest the landward part of the delta. Usually sediment is swept over the initial pile-up, coming to rest as seaward sloping layers. These layers constitute the **foreset beds**. They make up the major portion of simple deltas. Finer sediments carried to the deltaic area by the stream remain in suspension longer and are swept out farther into the basin. These nearly horizontal (or only gently dipping) layers constitute the **bottomset beds**. They are covered by foreset beds as the delta progrades seaward. Above the foreset beds, **topset beds** may develop as deposition occurs on top of the advancing foreset beds. These same features, when found in sedimentary rocks, allow ge-

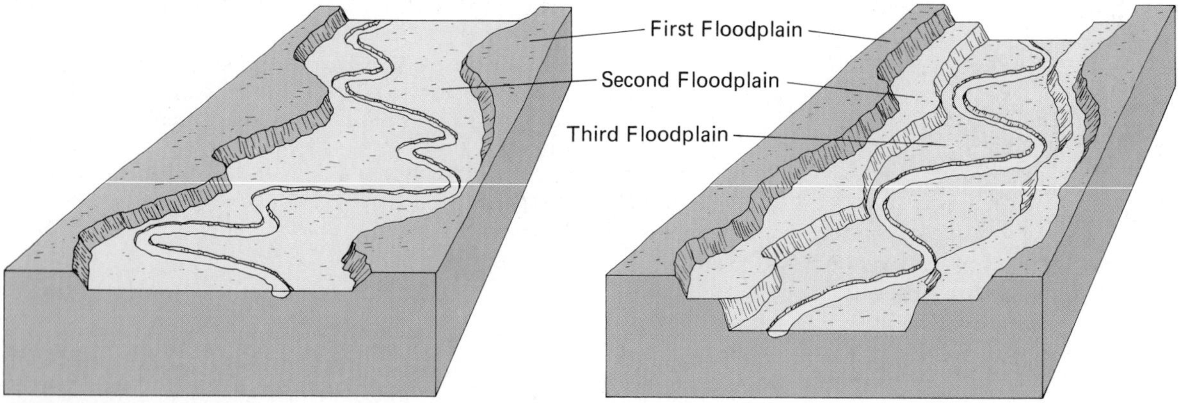

Figure 13–30 Development of terraces by periodic uplift. (A) Uplift has caused the stream to cut into its former floodplain and gradually develop a new floodplain. (B) With another pulse of uplift, the process is repeated, and yet a lower floodplain is produced. Remnants of the former floodplains form the terraces that border the valley.

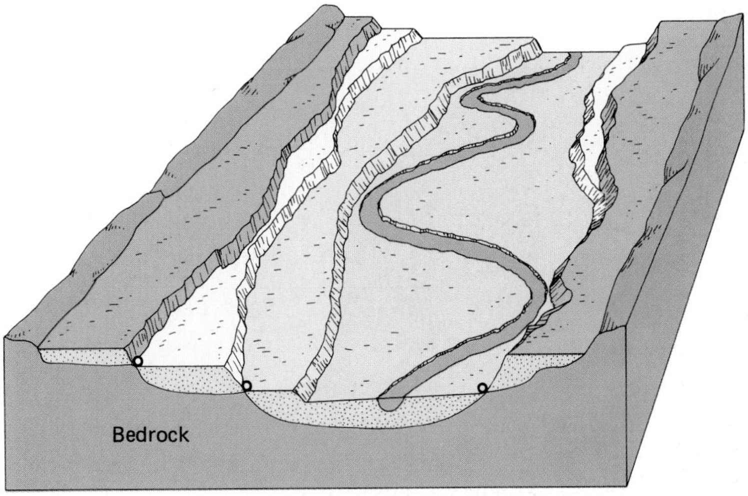

Figure 13–31 Unpaired terraces may be formed by the stream's continuous downcutting as it swings back and forth across its floodplain. In migrating across the floodplain, the stream gradually cuts to a lower level, and in returning, it cuts to a still lower level. In this way unpaired terraces are developed. Terraces represent the surfaces of earlier floodplains. They may be protected by partly buried spurs of resistant bedrock, which deflect the stream away from older terraces. (Small circles indicate location of such bedrock bulwarks.)

ologists to recognize ancient deltas. Of course, the depositional sequence is frequently interrupted by shifting of distributary channels, and is very complex in large deltas like that at the mouth of the Mississippi River.

Because of the fertility of their soils and the convenience of their coastal locations, modern deltas have always been centers of commerce and agriculture. Most of the agricultural lands of Egypt, for example, are located on the delta and floodplains of the Nile. The fertile fields of the Mekong Delta are a vital food source for the people of Cambodia and Vietnam.

Ancient deltas are nearly as important as their present-day counterparts but for different reasons. Deltas of bygone geologic epochs are found on the continents at many locations around the world, and a few lie beneath parts of the continental shelves. The intertonguing of coarse fluvial sediment with impermeable clay in these ancient deltas frequently produces ideal geologic situations for the entrapment of petroleum and natural gas. Coal beds also occur in association with the sediments of ancient deltas (see Chapter 6).

Waterfalls

Among the most aesthetically pleasing and spectacular features of streams are waterfalls (Fig. 13–34). A **waterfall** is a place in the stream where water descends vertically to a lower level. Waterfalls are usually a consequence of either bedrock characteristics or glaciation. Most commonly, waterfalls develop in places where the stream flows across an outcrop of a particularly resistant stra-

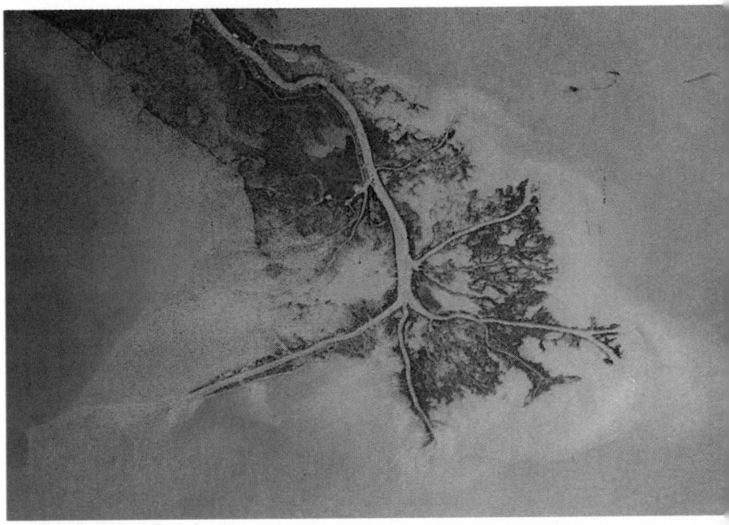

Figure 13–32 This photograph of the Mississippi Delta was taken in 1985 from the space shuttle. It shows why this delta is called a "birdfoot" delta. Long stringers of sediment deposited by the distributaries resemble a bird's toes reaching out into the Gulf of Mexico. Also visible because of its color is the main stream of the Mississippi as it disperses into its distributaries. (Courtesy of NASA.)

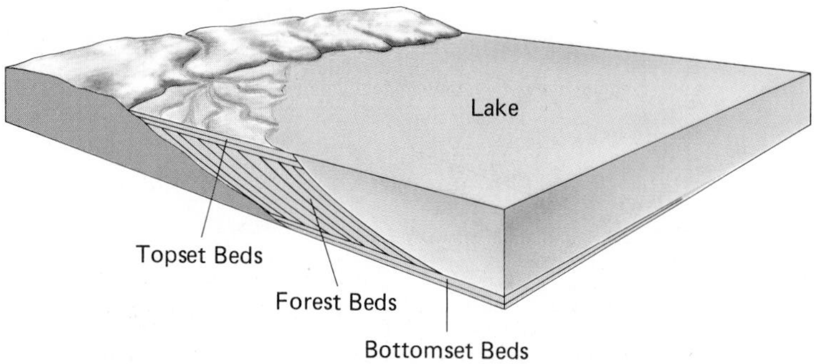

Figure 13–33 The ideal arrangement of beds in a small delta that is being built out into a lake. Relatively coarse material is deposited initially to form the inclined layers or foreset beds. Somewhat finer sediment is swept farther toward the center of the lake and deposited as bottomset beds. Foreset beds prograde over these bottomset beds. The stream deposits topset beds on top of the foreset beds. In large deltas that are built into the ocean, foreset beds are less steeply inclined, and there is a complex intertonguing of coarser channel deposits and interchannel deposits.

tum that is underlain by one or more layers of more easily erodable rock. Downstream from the outcrop of hard rock, the less resistant layer is more rapidly eroded, resulting in an abrupt drop in the elevation of the stream bed (Fig. 13–35).

Although the upper edge of a waterfall consists of resistant rock, it is nevertheless subject to erosion. It will also be undercut as water rushes over the rock surface and excavates the fragmented debris at the base of the fall. As a result, the waterfall gradually recedes upstream. Eventually, most waterfalls degenerate into rapids and disappear altogether.

Niagara Falls (Figs. 13–36 and 13–37), located between Lake Erie and Lake Ontario, developed when the retreating ice of the last glacial age uncovered an escarpment formed by southwardly dipping resistant dolostone. Weak shales beneath the dolostone strata are continuously being excavated in the pool beneath the falls. As a result, the falls at Niagara are receding at an average rate of about one meter per year.

Waterfalls in the eastern United States are abundant where eastward-flowing streams pass from highlands composed of igneous and metamorphic rocks onto coastal plain sediments, which are easily eroded. Even our colonial ancestors noted the general alignment of waterfalls at the contact between the two different kinds of rock and named

Figure 13–34 Comet Falls on the south side of Mt. Ranier, Mt. Ranier National Park, Washington. (Courtesy of D.R. Crandell and the U.S. Geological Survey.)

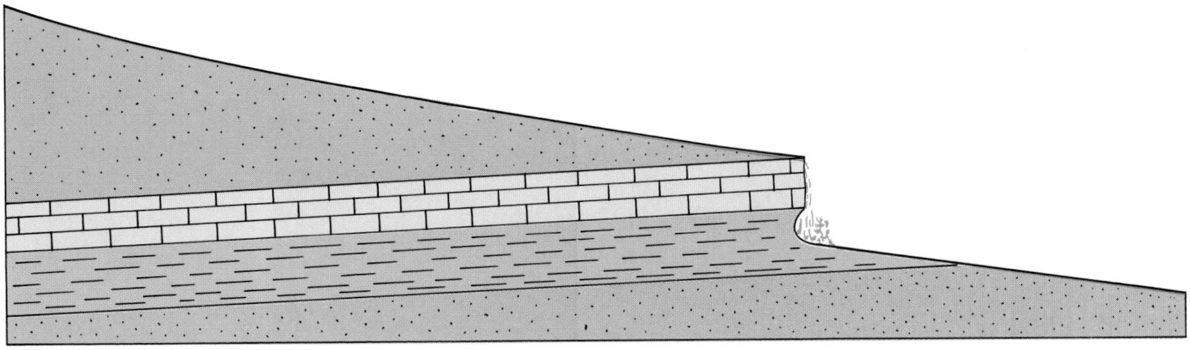

Figure 13–35 Where an outcrop of relatively resistant rock is underlain by less resistant rock, the weaker rock may be rapidly eroded by the stream and the more resistant layer undercut. If the resistant rock collapses and leaves a vertical face, water will plunge over the crest as a waterfall.

Figure 13–36 Niagara Falls, formed where the Niagara River flows from Lake Erie into Lake Ontario. The classic section of the American Silurian System is exposed along the walls of the gorge below the falls. (Courtesy of the Ontario Ministry of Industry and Tourism.)

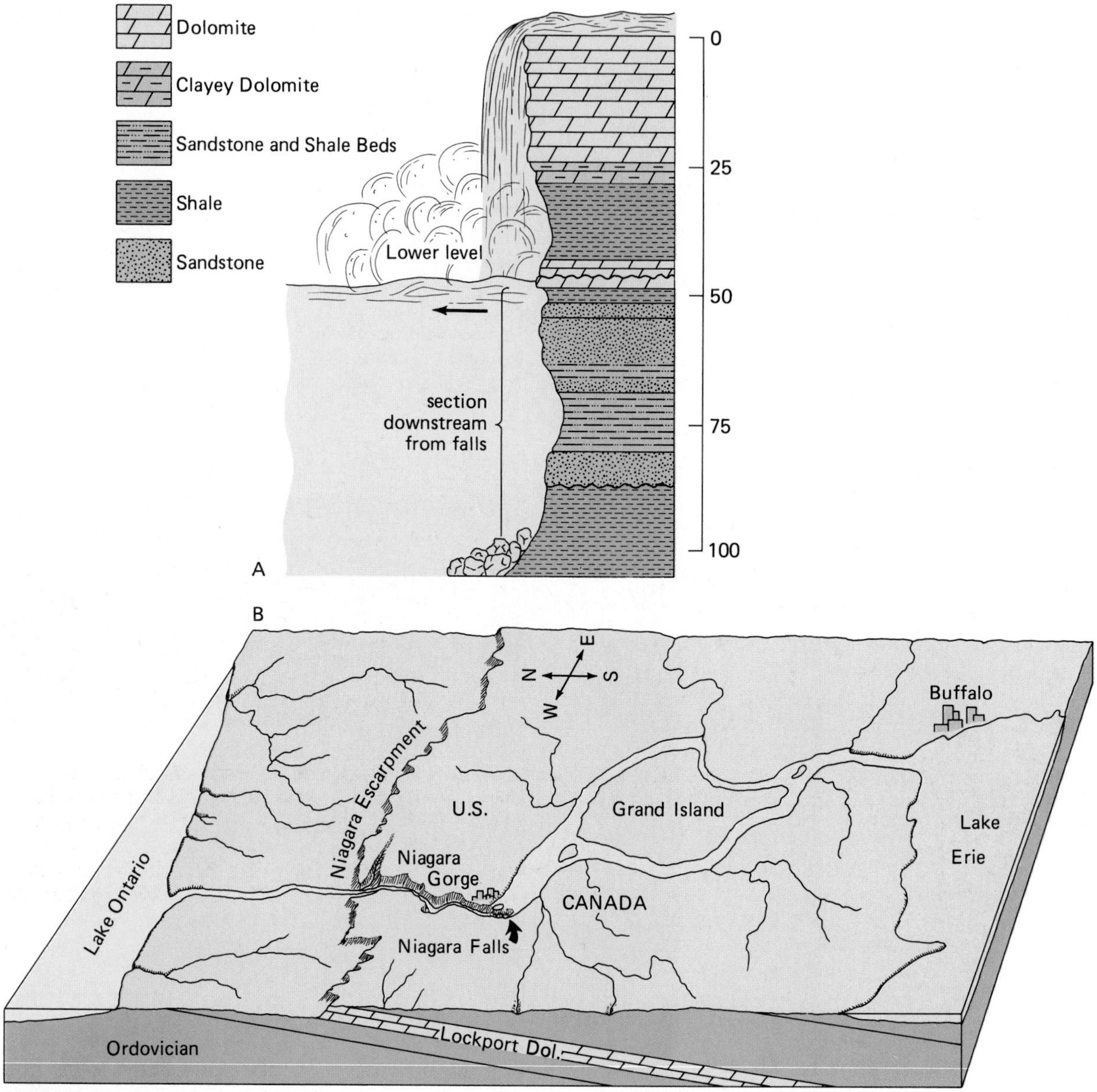

Figure 13–37 Block diagram and stratigraphic section of Niagara Falls. The Lockport Dolomite forms the resistant lip of the falls. The rocks dip gently to the south in this area, and where harder dolostone layers like the Lockport Dolomite intersect the surface, they form a line of bluffs known as the Niagaran Escarpment. (After E.T. Raisz.)

this line the "fall line." Water wheels located at waterfalls along the fall zone powered colonial flour mills.

Not all waterfalls are formed directly as a result of undermining by streams. In mountainous regions that have been sculpted by glaciers, waterfalls are common. Many of these formed when glacial erosion deepened the floor of major valleys far below the floor of entering tributary valleys. After the ice has melted, the tributary stream has no recourse but to plummet down the steeply excavated walls of the major valley.

Drainage Basins and Patterns

A river and its tributaries constitute a **drainage system**. Every such system drains an area called a **drainage basin** (Fig. 13–38), each of which is separated from other drainage basins by upland or ridge tracts termed **drainage divides**.

Drainage basins are separated by drainage divides.

The arrangement of streams within a drainage basin is determined by such factors as initial slopes, inequalities in rock resistance to erosion, whether strata are level or tilted, and even tectonic movements. The particular plan or design of the river system constitutes the **drainage pattern**. By the study of drainage patterns, it is often possible to infer some of the characteristics of the structural geology of an area as well as events in geologic history. For example, if the rocks underlying an area are essentially homogeneous or if they consist of flat-lying strata, the stream network will be characterized by an irregular branching of tributary streams. The overall drainage pattern is called **dendritic** (from the Greek dendrites: "tree-like") and rather resembles the branching of a tree (Fig. 13–39). By contrast, **trellis patterns** consist of a system of nearly parallel streams, usually aligned between the outcrops of folded rocks that are relatively resistant to erosion (Fig. 13–40). Patterns of this type are well displayed in the drainage systems of parts of the Appalachian Mountains where the drainage divides are formed by elongate ridges of resistant sandstones. Trellis drainage may also be controlled by aligned deposits of glacial debris or wind-blown sediment, but this pattern is less common in the United States.

In landscapes having trellis drainage patterns, the main streams often make nearly right-angle bends where they cross the drainage divides. In

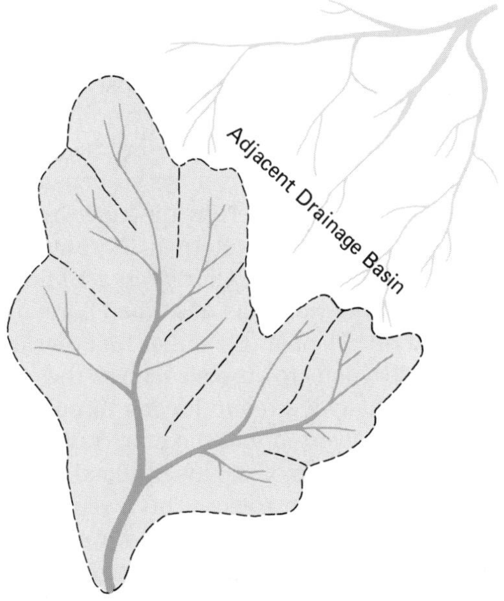

Figure 13–38 A small drainage basin. Dotted lines are divides. Refer now to Figure 13–1. Can you recognize the drainage divides in Figure 13–1?

Figure 13–39 Dendritic drainage, Republic of South Yemen at the southwestern corner of the Arabian Peninsula. This photograph was taken by astronauts on the space shuttle in 1984. The branching pattern of streams is termed dendritic because of its similarity to the way trees branch out to form finer twigs. (Courtesy of NASA.)

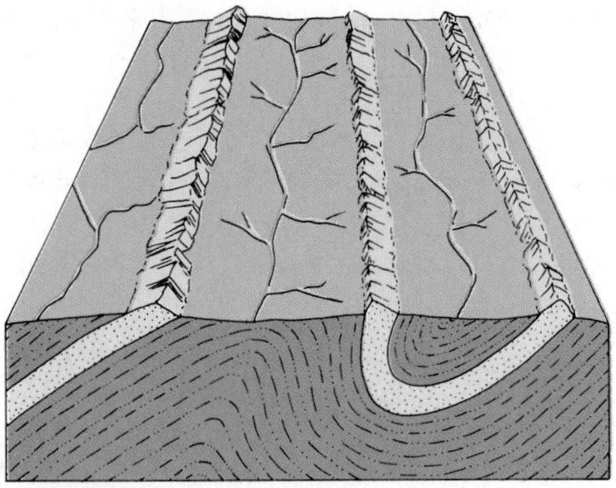

Figure 13–40 Trellis drainage pattern.

addition, tributary streams tend to enter main streams at approximate right angles. Two intriguing features that are occasionally found in regions of trellis drainage are windgaps and watergaps. **Watergaps** are gorgelike stream valleys cut directly through a well-defined mountain ridge. Most watergaps can be attributed to the erosional activities of a stream that established its course long ago in rocks having one type of structure and that gradually eroded its way down to an underlying sequence of rocks having quite different structure (Fig. 13–41; and see Fig. 13–17). The steam responsible for cutting the water gap is said to have been superposed (contraction of superimposed) on the

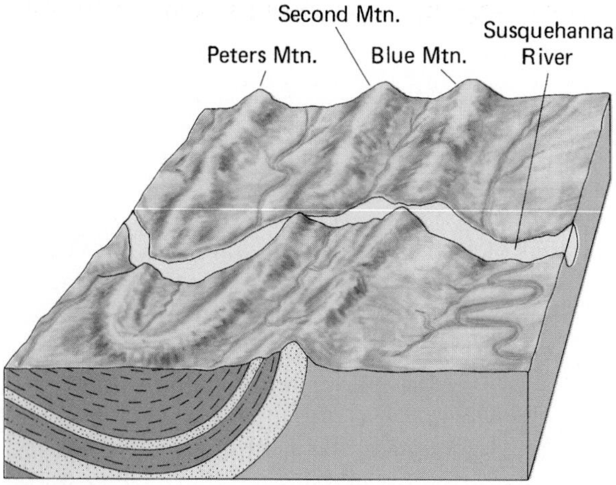

Figure 13–41 The Susquehanna River has cut a series of watergaps through resistant ridges in the Appalachian Mountains of southeastern Pennsylvania.

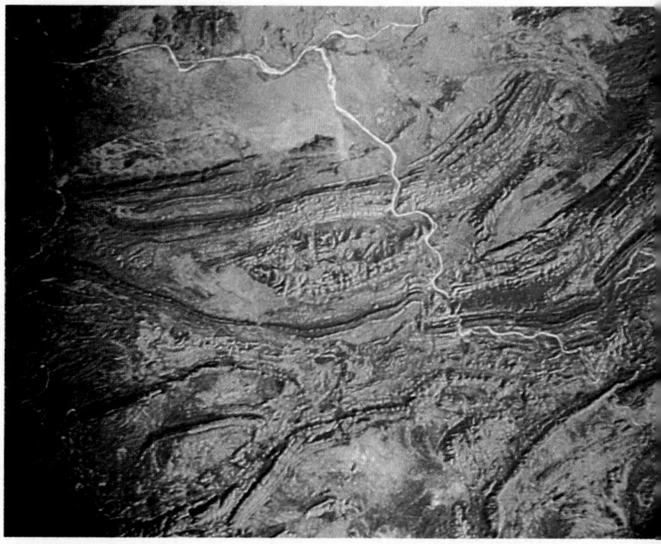

Figure 13–42 The major rivers in this area of the Macdonnell Ranges of Australia cut right across folded strata rather than being deflected into courses parallel to the ridges. From this pattern we deduce that the rivers probably had established their courses before reaching the erosional level where they encountered the folds. As the land was uplifted, they continued in their old courses, incising water gaps in resistant ridges of rock. (Space shuttle photograph courtesy of NASA.)

underlying structure. Because the stream has already established its course in the overlying alluvium, it is able to continue in this course cutting across resistant and nonresistant strata indiscriminately. Watergaps characterize orogenic regions where tectonic ridges rise across the paths of rivers (Fig. 13–42). If the river has sufficient energy to maintain its course across the ridge rising in its path, it may erode a watergap. Streams that have been able to maintain their direction of flow across active tectonic ridges as described here are called **antecedent**. The name signifies that they are older than (antecedent to) the structures they cut.

Occasionally a stream flowing through a watergap may be diverted into the more rapidly deepening channel of one of the longitudinal streams that flow across softer rocks. This event constitutes stream piracy and leaves the watergap without water. The notch cut into the ridge is now called a **windgap** (Fig. 13–43), although the action of wind has had nothing to do with its origin.

Figure 13–43 Windgap in Saddle Mountain, Grant County, West Virginia. (Courtesy of West Virginia Geological Survey, photo by G.S. Ratcliff.)

A third type of drainage pattern is that which develops around dome mountain uplifts or on the sides of volcanic cones. Because streams flow away from a high central area in all directions, the overall pattern is designated as **radial**. Drainage on exposed igneous stocks and laccoliths is also often radial in pattern (Fig. 13–44A).

Streams located around domal uplifts, where alternating layers of strata outcrop in a circular plan around the central mass, frequently develop an **annular drainage pattern** (Fig. 13–44B). As with trellis drainage, the pattern of channels is structurally controlled. Streams flow across the outcrop of softer rocks and between stream divides composed of more resistant rock.

Because streams tend to occupy belts of easily eroded rocks, ridges formed by outcrops of folded, resistant rock influence stream patterns.

Floods

Nearly every year the news media report a disastrous flood somewhere in the world. A flood is any relatively high streamflow that spills out of its channel onto the valley floor (Fig. 13–45). Everyone knows that floods develop whenever there is heavy runoff due to the persistence of intensity of rain, rapid melting of snow, or alteration of land surfaces that increases runoff. The channel of the steam is shaped primarily by its usual amount of discharge, and therefore when it receives an extraordinary volume of water, the surplus can only spill over the banks and flow out onto the floodplain. Floodplains, therefore, provide a place for the release of water the channel cannot hold. This

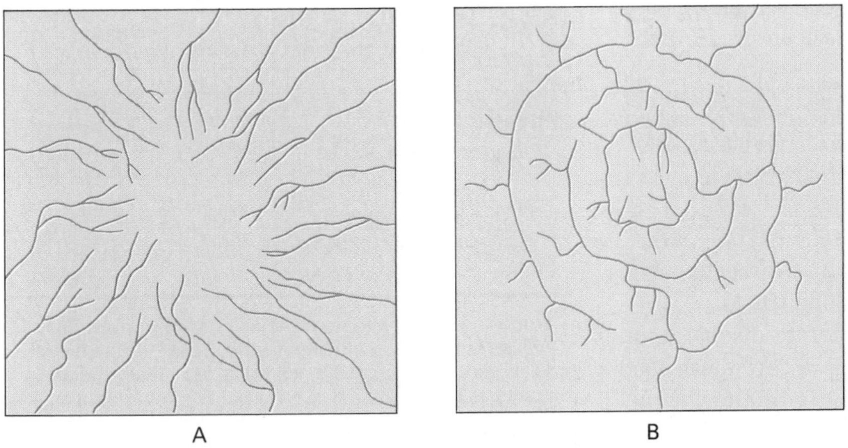

Figure 13–44 (A) Radial drainage pattern. (B) Annular drainage pattern.

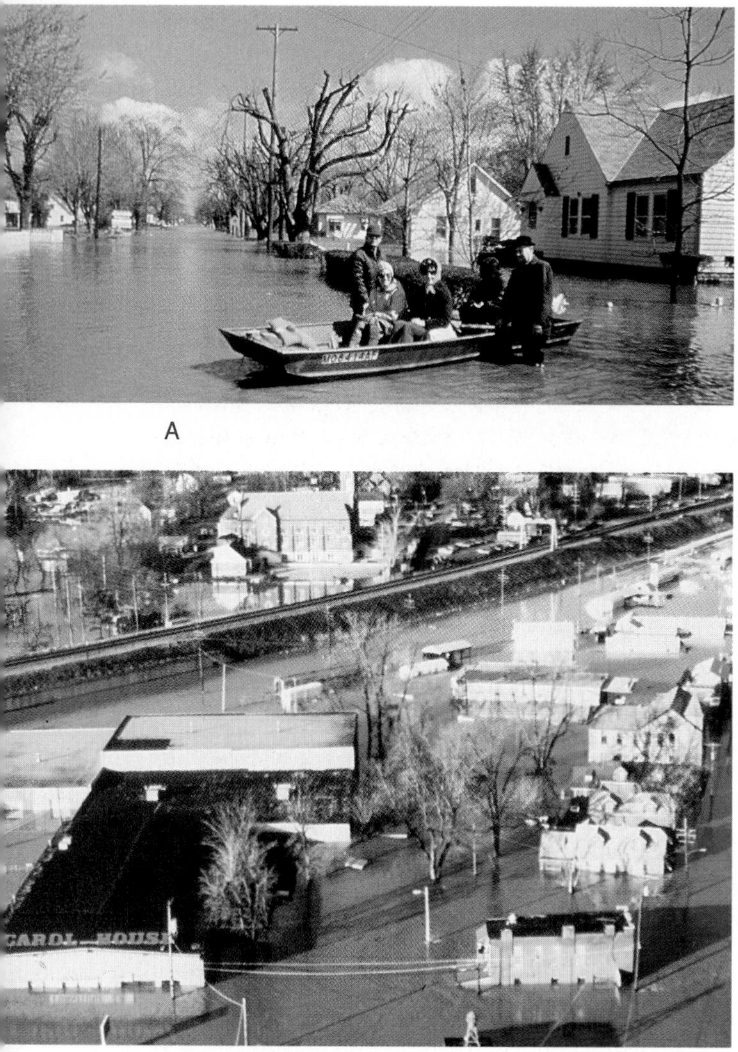

Figure 13–45 Floods are an abiding concern for people living in low-lying areas near major streams. (A) In 1973, flooding of the Missouri River inundated a major part of Alton, Illinois. (B) The Meramec River, a tributary of the Mississippi overflowed its banks in 1982, causing damage to homes and businesses in Times Beach, Missouri. (Courtesy U.S. Army Corps of Engineers.)

simple fact is often overlooked by overzealous and uninformed developers of housing tracts.

A river will have only a moderate discharge on most days of the year. Perhaps a few days each year, there may be enough water to fill the channel bank full but not sufficient to cause flooding. For most streams, actual flooding rarely occurs more frequently than about once every two or three years on the average. Streams do have relatively high temporary water storage capacity, and it takes a heavy rainfall or sequence of rains to cause flooding. In order to plan for the exceptional floods, scientists examine river flow charts for as many past years as there are records available. Using statistics, it is possible to calculate the probability of future flooding in specific streams. The method will not permit one to predict the precise date of a major flood but will indicate the probability of a flood of a certain size within a certain interval of years.

People have struggled with the problem of what to do about floods for centuries. It would seem the simple solution for avoiding damage from inundation, deposition of sediment, and erosion would simply be not to build on floodplains. Floodplains are difficult to leave alone, however. With each flood they are covered with a fresh layer of fertile sediment and are prized as agricultural lands. The level terrain makes farming easier and also provides convenient routes for highways, railroads, and utility lines. In order to ship goods by river barge, one must clearly have dock facilities on the banks of rivers. Various industrial developments and large communities are located on floodplains in order to have a ready source of water and natural system for disposal of waste.

Not only are industrial housing projects on

Table 13–1 Increase in Runoff Resulting from Urbanization in East Meadow Brook, Long Island, New York

Mean Annual Runoff (acre-feet)	*Year*	*Rainfall % Increase 1937–43*	*Runoff % Increase 1937–43*
920	1937–43		
1170	1943–51	0.6	27
2200	1952–58	5.7	140
2400	1959–62	7.0	270

Notice that, although each period is marked by an increase in rainfall, the percentage increase in runoff for each period is far greater than that of rainfall. (Data from Seaburn, G.E., U.S. Geological Survey Professional Paper 627-B, 1969.)

floodplains endangered by floods, but they may actually help to cause them. Buildings, roads, sidewalks, and paved parking lots prevent rainwater from seeping into the ground and making its way slowly to a stream. Instead, there is a rapid runoff of huge volumes of water, which streams cannot convey away rapidly enough to prevent flooding (Table 13–1). These effects can be illustrated by means of a chart called a hydrograph. The hydrograph depicts variation in stream flow with time. Figure 13–46 is a hydrograph depicting a period of *peak flow* and possible flooding and a low-water stage designated the *base flow.* Base flow is maintained largely by water flowing into the stream from underground. The interval between a time of heavy rainfall and peak flow is called the *lag time.* From Figure 13–46, it is apparent that rapid runoff resulting from too many roofs and too much pavement has sharply reduced the lag time and increased the amount of discharge. Conditions favoring flooding have been established.

We have seen that a stream's behavior at any time and place represents a composite of many variables. For this reason, artificial structures and engineering programs may provide only limited protection against flooding or perhaps shift the flood to localities that will not be as severely affected. Sometimes it is possible to protect parts of floodplains against the more frequent low-magnitude floods, although complete protection against

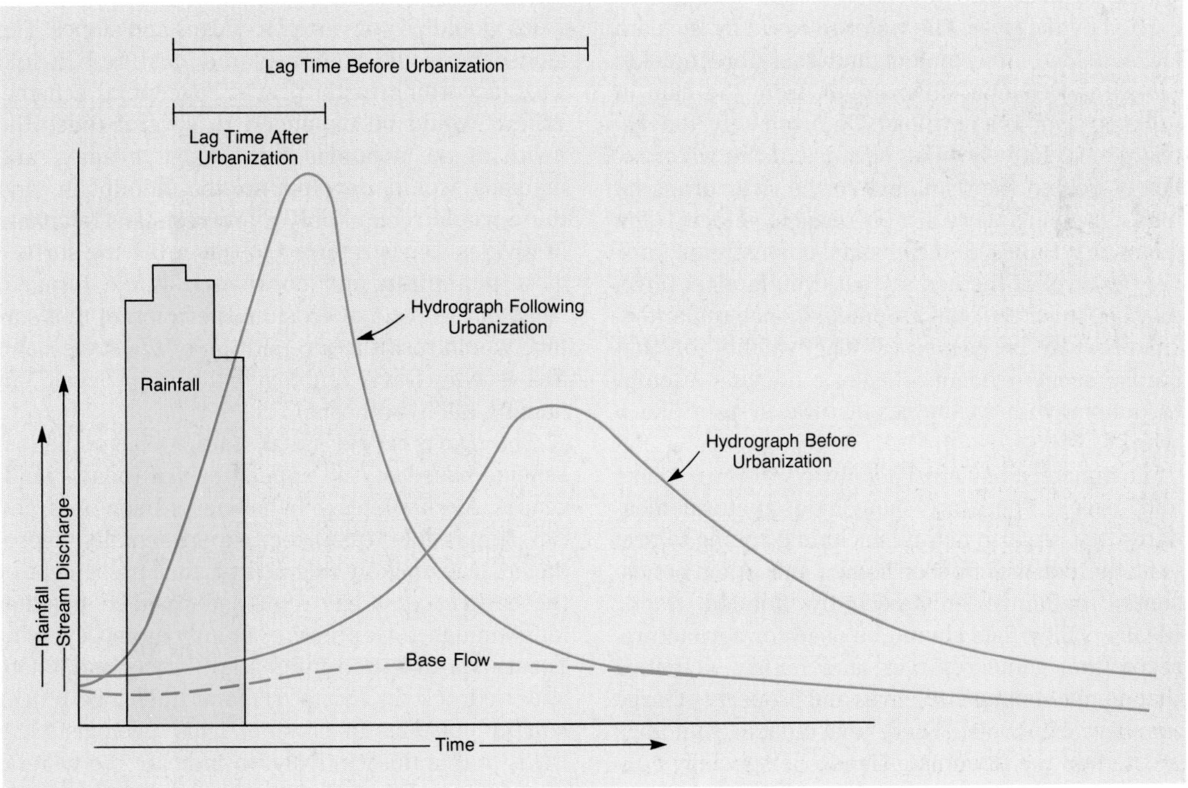

Figure 13–46 Two hypothetic hydrograph curves illustrating the effects of urbanization and lag time between heavy rainfall and peak discharge.

a 50- or 100-year flood would be a monumental and uneconomic feat of engineering. Such protection would be more costly than the financial resources of the public could bear. There are, however, practices that may reduce the loss of property and lives from flooding. For example, artificial levees and dikes may be emplaced to help hold the water in its channel. Before such structures are emplaced, however, they should be carefully evaluated, for they may contribute to flooding upstream. Channels may be artificially straightened so as to speed up the flow and move water out quickly before it can spill over the banks. Here too there is a danger, for by speeding the flow of water, locations downstream may be flooded.

In many flood-prone regions, dams have been constructed in order to store water and allow it to move downstream in a regulated manner. Dams, however, are not always as effective in flood control as we would like them to be. It is true that dams do relieve flooding by containing some of the water, but they also collect the sediment that was carried by the river. The water released by the dam lacks a load of sediment and therefore quickly erodes the channel downstream from the dam in order to acquire a new load. Unfortunately, the discharge is relatively small because of the water retained behind the dam, and so the river drops its load as soon as there is a decrease in velocity. The deposition builds up the channel downstream, raising the level of the bed so that even modest flows may spill over the banks. Someone once remarked, "nature can be perverse." The validity of that phrase seems evident whenever humans attempt to tamper with a complex natural system like a river.

Fortunately, one need not always resort to dams and levees. There are many nonstructural measures that are also helpful, including zoning to prevent the construction of homes and other nonessential buildings on frequently flooded tracts. Ideally, valley flats should be used for agriculture, recreation, game reserves, and human activities that do not endanger the lives and property of large numbers of people. Where unavoidable, buildings erected on the floodplain should be specially constructed to resist flood damage, and flood insurance should become a requirement of all who dwell on floodplains.

The Shapes of Landscapes

In the late 1890s, William Morris Davis, an eminent American geomorphologist, proposed a conceptual scheme of landscape changes in humid climates. Davis postulated that the principal factors in the development of a landscape are the *processes* of erosion (running water, moving ice, gravity) acting on land surfaces, the underlying *structure* of the bedrock, and the length of *time* over which the various processes have operated. He went on to propose that, beginning with a nearly flat initial surface (as for example, a newly emerged section of continental shelf), the landscape would pass through a cycle of changes designated "youthful," "mature," and "old age." Landscapes in youth would lack valley flats and would be separated by poorly drained interstream tracts composed of the initial surface (Fig. 13–47). As stream erosion and mass wasting progressed, valley flats would develop, drainage density would increase, the major stream would attain a graded condition, and divides would be converted to ridges and slopes. The landscape would be considered "mature." In "old age," according to the Davis conceptual scheme, valleys would be expanded to several times the width of the meander belt. Lakes, swamps, and marshes would develop on the floodplain, and there would remain only a few resistant remnants of divides. Davis referred to this low-lying surface as a **peneplain** and considered it the terminal stage in the cycle of erosion. Elevation of that surface would result in a repetition of the stages, and this is why Davis employed the term "cycle" in naming his hypothesis.

The Davis cycle is an entirely hypothetical scheme based on the appearance of today's landscapes. A complete cycle has never been observed (an impossible task) or experimentally reproduced. It is unlikely that any particular segment of the earth's crust has remained fixed in position long enough for an entire, uninterrupted cycle to occur. Nor are such influential factors as climate and bedrock likely to remain unchanged over lengthy episodes of erosion. Thus, geologists use Davis' terms descriptively, to indicate the general appearance of a landscape.

Davis used the terms "youth," "maturity," and "old age" to describe streams also. He defined

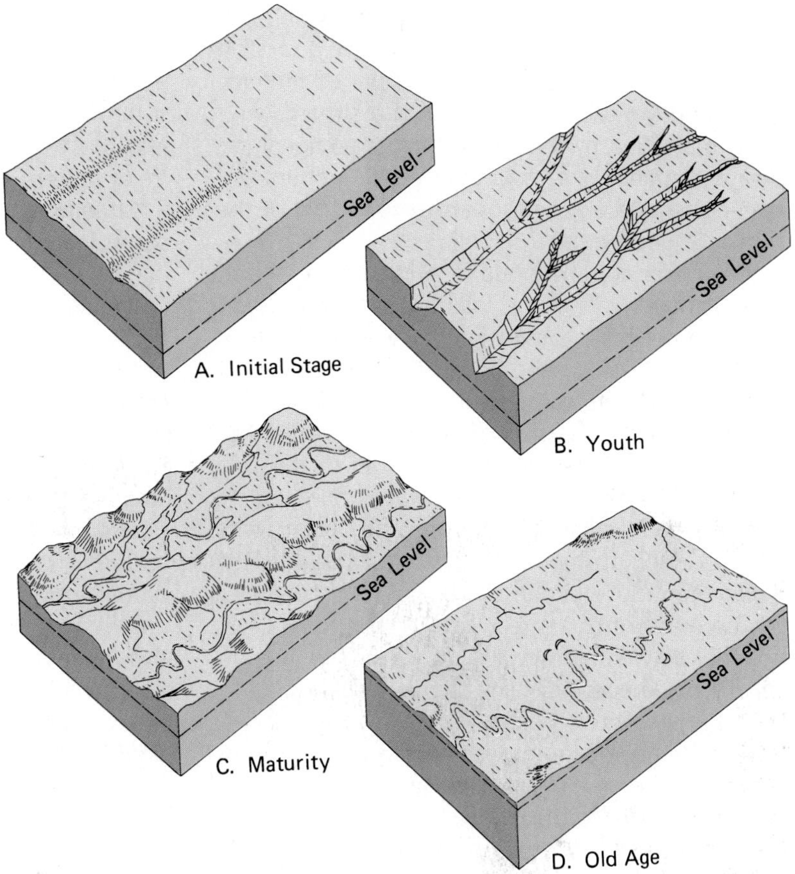

Figure 13–47 W.M. Davis' conceptual scheme for the evolution of landscape in a humid temperate climate.

"youthful streams" as those that had deep V-shaped valleys. "Mature streams" had flat broad valleys with floodplains at least as wide as the meander belt, and "old age streams" were slow-moving, widely meandering streams flowing over floodplains several times wider than the width of the meander belt. It should be remembered, however, that a single stream may have youthful characteristics in its headwaters and mature or old-age characteristics in its lower segments. Thus, the terms are again without consistent time or cyclic significance. Yet, they are useful and expedient in designating the shapes and characteristics of stream valleys.

Summary

Stream erosion has done more to develop the landscapes of our continents than any other group of geologic processes. Everywhere we see the ravines, valleys, canyons, ridges, and floodplains that result from the activity of streams. There are times when that activity is inspirational, as when streams topple over resistant ledges to form waterfalls. At other times the activity of streams may be catastrophic, as when floods sweep away buildings and undermine bridges. But most of the work of streams goes on largely unnoticed, as the flowing water sculpts valleys and transports the material that once filled them to the sea.

As is the case with mass wasting, gravity is the fundamental force that causes water to flow and is thus indirectly responsible for the work of streams. The component of gravity acting along the slope over which the stream moves is one factor that influences the river's velocity. *Velocity* in turn influences the amount and nature of the erosional and transportational work of a stream. Velocity multiplied by the stream's cross-sectional area determines *stream discharge* (the amount of water passing through a given cross-section of the river in a given time).

The material transported by a stream is referred to as its *load.* Part of the load is *dissolved material.* Another part, consisting mostly of clay and fine silt, is carried in the body of moving water and is termed the *suspended load.* The *saltation* and *tractive load* consists of larger material rolling, sliding, and skipping along the bottom of the channel. The total load that a stream can carry constitutes its *capacity.* The companion term competency refers to the maximal size particle the stream can move under a given set of conditions.

The gradient of a stream is the slope of its channel between any two specified points. It is readily obtained by dividing the horizontal distance between two points (in meters) along the course of the stream into the vertical difference in elevation (also in meters) between the same two points. A *graded stream* is one that has sufficient gradient to transport the material delivered to it by its tributaries, its own bank erosion, and mass wasting. It is a stream in equilibrium, in which any change in load, channel morphology, or discharge will result in changes tending to absorb the effects of those altered conditions. The gradients of graded streams tend to be steeper near their headwaters and more gently inclined toward the mouth. If one were to plot this changing slope along the full length of the stream, it would be apparent that graded streams have a concave-upward longitudinal profile.

Streams vary in their patterns of drainage, primarily according to the structure of underlying rock. If that rock is homogeneous or consists of flat-lying strata, the most common pattern will be *dendritic* and resemble the branching of veins in a leaf. If, however, the surface is underlain by a series of parallel, inclined, alternately weak and resistant rock layers, the tributary streams may come into the main stream at right angles, producing a *trellis pattern.*

Streams are responsible for many familiar landscape features. In mountainous regions, for example, one finds *alluvial fans* developed at places where streams emerge from mountain fronts and deposit their load of silt, sand, and gravel. Deltas may develop where rivers enter oceans or lakes. *Waterfalls* and *rapids* are found frequently in relatively narrow valleys, whereas the prominent features of broad valleys are *meanders, braided streams, natural levees,* and *stream terraces.* The *floodplain* is perhaps the most readily apparent feature of most streams. Floodplains are formed primarily by lateral planation and the downstream migration of the stream's bends or meanders. In this migration, the stream acts somewhat like a horizontal saw cutting into its banks on the outside of bends and depositing sediment on the inside of meanders.

People who live or work on floodplains are well aware of the fact that, at least once every two or three years for most streams, the discharge of the stream exceeds the storage capacity of its channel, and floods will occur. By studying the records from stream gauging stations over a sequence of years, scientists are able to determine the probability of a stream having a flood of a given magnitude within a specified number of years. However, it is not possible to predict the precise year in which a major flood will occur.

Rivers can be destructive, but their benefits far outweigh their dangers. They provide us with water for our homes and factories, assist in the disposal of waste, are inland waterways for the transportation of raw materials and manufactured products, and are important sources of hydroelectric power.

Terms to Understand

abrasion
alluvial fan
alluvium
annular drainage pattern
antecedent streams
base level
bed load
bottomset beds
braided channel pattern
capacity

competence
consequent stream
cutoff
delta
dendritic drainage pattern
discharge
distributaries
drainage basin
drainage divide
drainage pattern
drainage system
floodplains
foreset beds
graded stream
gradient
headward erosion
hydrologic cycle
impact
incised meander
lateral planation
load
longitudinal profile
meanders
natural levee
oxbow lake
peneplain
point-bar
pothole
radial drainage pattern
reach (of a stream)
saltation
solution
stream piracy (capture)
stream terrace
subsequent stream
superimposed stream
suspended load
suspension
temporary (local) base level
topset beds
traction
trellis drainage pattern
turbulent flow
ultimate base level
velocity
waterfall
watergap
windgap
Yazoo stream

Questions for Review

1. Calculate the discharge of a stream having a cross-sectional area of 120 square meters and a velocity of 2 meters per second. What would be the velocity of a stream with the discharge of 100 cubic meters per second and a cross-sectional area of 20 square meters?

2. Why is water velocity near the bed and banks of a stream generally lower than the velocity near the center of the stream?

3. What systematic changes in discharge, velocity, gradient, and channel shape occur in a river from its headwaters to its mouth?

4. How does a stream widen its valley?

5. Why is it that after a dam is constructed and the reservoir behind the dam is filled, the water that flows over the dam is clear, and this clear water tends to deepen the stream channel downstream from the dam?

6. What is the general configuration of the "longitudinal profile" of a stream? What is the cause of this configuration?

7. What geologic inference might one make on observing a stream that has incised meanders? For a stream that cuts across the structural "grain" of a region (that is, across ridges of resistant rock)?

8. What type drainage pattern usually develops on folded rocks? On horizontal rocks? What conditions are likely to result in a radial stream pattern?

9. What are the characteristics of the three stages of landscape evolution proposed by William M. Davis? What objections have been raised against the Davis scheme of landscape evolution?

10. What is a watergap? How might a watergap be converted into a windgap?

11. Why do many streams continue to flow even though there has been no precipitation in their drainage basins for weeks or months?

Supplemental Readings and References

Leopold, L.B., and Langbein, W.B., 1960. *A Primer on Water.* Washington, D.C., U.S. Dept. of the Interior, U.S. Gov't. Printing Office.

Leopold, L.B., Wolman, M.G., and Miller, J.P., 1964. *Fluvial Processes in Geomorphology.* San Francisco, W.H. Freeman & Co.

Morisawa, M., 1968. *Streams, Their Dynamics and Morphology.* New York, McGraw-Hill.

Ritter, D.F., 1974. *Process Geomorphology.* Dubuque, Ia., Wm. C. Brown & Co.

Schumm, S.A., (ed.), 1974. *River Morphology.* Stroudsburg, Pa., Dowden, Hutchinson & Ross.

Vitaliano, D.B., 1973. *Legends of the Earth.* Bloomington, In., Indiana University Press.

Underground Water

14

Groundwater emerging from a spring at the base of a hill. The groundwater moved along bedding planes and through the fracture system of the limestone that forms the cliff.

And Moses lifted up his hand, and with his rod he smote the rock twice: and the water came out abundantly: and the congregation drank and their beasts also.

Deuteronomy 6:11

Introduction

There has always been a sense of mystery associated with our thinking about water underground. Ancient peoples were certain that great rivers flowed unseen beneath their fields. The springs that formed where underground water flowed out onto the land were thought to be gifts of benevolent gods. Some believed these spring waters could cure the sick and give strength to the feeble. In desert communities, the survival of entire populations was dependent on success in finding underground water. The enormous importance of a supply of underground water to the ancient tribes of Israel is documented in the Bible by frequent references to the digging of wells and to the many battles fought to protect them. The well was the focal point of the community. Around it, information was transmitted and historic personal relationships were established.

Early Greek philosophers, including Homer, Thales, and Plato, believed that underground water was really derived from the oceans. According to their conception, water was transmitted inland in subterranean channels and then purified of salt and other contaminants as it ascended through the crevices and voids of rocks and soil. A far more accurate explanation for the origin of underground water was provided somewhat later in history by the Roman architect Vitruvius. Vitruvius correctly argued that an appreciable part of the water from rainfall percolates downward through soil and rock, flows along through whatever fractures and openings are available in the subsurface, and emerges at distant locations as springs. Vitruvius had a sound understanding of the hydrologic cycle. He recognized that the atmospheric return of water to the continents was more plausible than the underground return proposed by his Greek predecessors.

Of all the water that falls as rain or is derived from melting snow, about 32 per cent runs off at the surface as sheetwash and in streams. Approximately 65 per cent evaporates or is used by plants, and the small amount remaining infiltrates into the regolith and rocks beneath, percolating downward until there are no longer any air-filled openings, and then rising in level toward the surface as additional water is added. This underground water that nearly fills all openings in rock or sediment is called **groundwater** by geologists. It has been estimated that groundwater supplies about one-fifth of the total water needs of the United States.

Several factors influence the amount of water that will become part of the groundwater reservoir for a particular region. The more important of these factors are the amount of precipitation, the volume of openings in the subsurface materials, the slope of the land surface, and the nature of the vegetative cover at the surface. Once the water has percolated into the subsurface, it is capable of movement toward outlets such as wells and springs, can accomplish certain kinds of geologic work, can be cleansed of disease-causing organisms, and can contribute to the hot waters and vapors emitted by geysers and hot springs. In this chapter, we will examine the movements, benefits, hazards, and characteristics of groundwater. Initially, it will be necessary to understand the subsurface distribution of underground water and the factors that influence its movement. This understanding will provide a basis for a discussion of gravity and artesian springs, the formation of caverns, and the uses and misuses of underground water supplies.

Amount of precipitation, rock porosity, nature of the ground surface, and kind of vegetation are among the factors that determine the amount of water in groundwater reservoirs.

The Sources of Groundwater

Most of the water found underground is derived from rain or snow. Such water is called **meteoric**. Actually, only about 25 per cent of meteoric water finds its way into the underground water system. A larger part is carried away as runoff, evaporated, or utilized by plants. Of course, some meteoric water stored in lakes or flowing in streams may seep downward and become groundwater. In order for such seepage to occur, however, the river or lake bed must be above the level of the subsurface zone in which all openings are filled with water. Streams whose waters contribute to the groundwater supply are called **influent**. A larger number of streams receive part of their water from groundwater and are thus **effluent** with respect to groundwater. The water in many lakes, ponds, and bogs is also effluent in origin (Fig. 14–1).

Effluent streams derive part of their water from the underground supply, whereas influent streams contribute to groundwater by seepage.

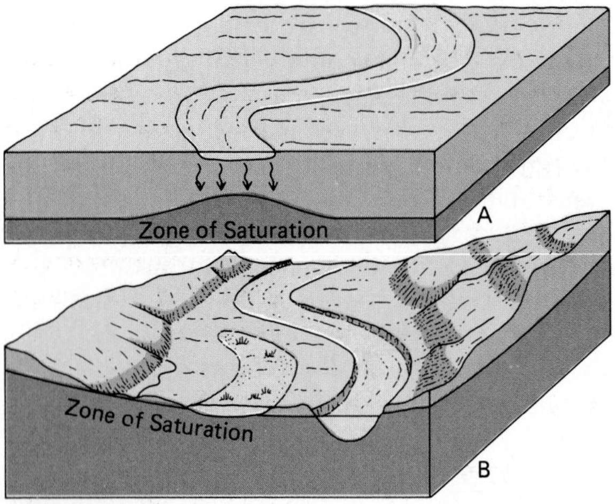

Figure 14–1 (A) An *influent stream* contributes stream water to the groundwater supply. (B) Streams that receive part of their water from underground sources are termed *effluent.*

Whenever sediment is deposited in water, water will necessarily fill all the open spaces between grains. Water trapped in sedimentary rocks in this way is termed **connate water**. Often connate water is salty, either because it was originally sea water or because of the mineral matter that has been dissolved in it over long intervals of geologic time. Petroleum engineers are quite familiar with connate water, for it is frequently encountered during well drilling operations.

Magmatic water is produced directly by igneous activity. Along with recirculated meteoric water, it is most visible to us as a component of the vapors emitted from volcanoes. Deeply buried magmatic bodies also contribute magmatic water to the surrounding rocks. It is likely that at least a part of the water emitted by some thermal springs is magmatic. In terms of its immediate usefulness to mankind, magmatic water is not as important as meteoric water. Historically, however, it is of immense importance, for the planet's entire hydrosphere was derived from a magmatic source.

Rock Properties Affecting Groundwater Supplies

Aquifers and Aquicludes

A body of rock or sediment through which water can move and that yields sufficient water for domestic or industrial use is termed an **aquifer**. Examples of aquifers include unconsolidated sands and gravels, sandstone strata, partly dissolved or cavernous limestones, and nearly any other rock type that is highly fractured (Fig. 14–2). Geologists employ the term **aquiclude** for rocks that do not permit passage of significant quantities of groundwater. By far the most common aquiclude is clay. Clay-rich rocks like shale may actually contain considerable amounts of water, but that water cannot flow freely through the microscopic pores between clay particles.

Aquifers are rocks that are able to hold and transmit water in usable quantities, whereas aquicludes are relatively impermeable.

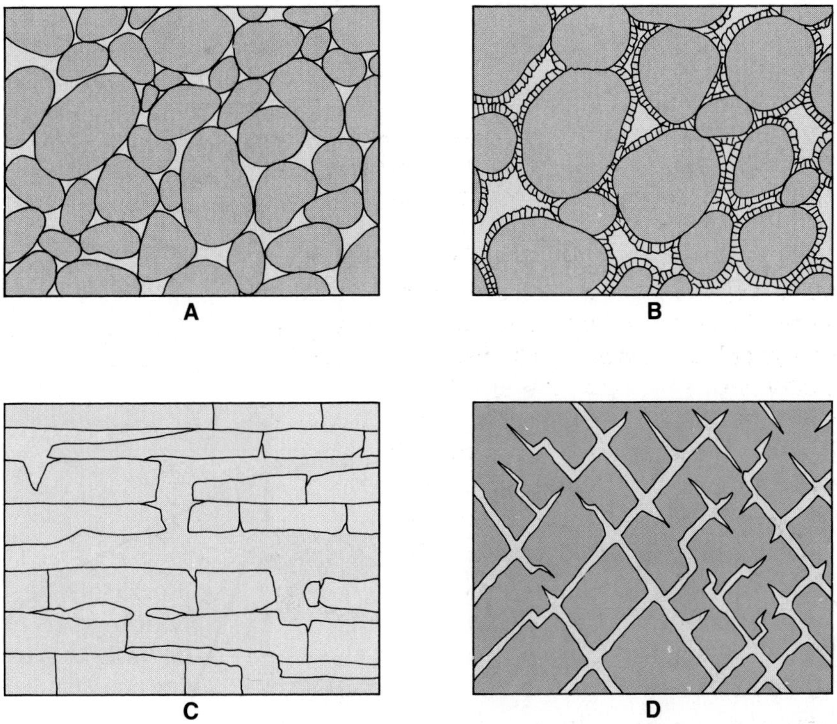

Figure 14–2 Differing modes of porosity. (A) Well-sorted sand having good porosity. (B) Porosity reduced by cement in sand converted to sandstone. (C) Rock rendered porous by dissolution. (D) Rock made porous by fracturing.

The quantity of water that can be obtained from wells drilled into aquifers depends in part on the amount and characteristics of the void spaces in the water-bearing rock, as well as the manner in which these voids are interconnected. The hydrologic terms used to describe these two attributes of aquifers are porosity and permeability.

Porosity

Porosity is defined as the total volume of open spaces, pores, or voids in a rock or sediment expressed as a percentage. The spaces may be fractures, cavities dissolved in rock by water, or simply openings fortuitously formed between grains as they were deposited. In a clastic sediment or sedimentary rock, porosity will be markedly influenced by sorting. As noted in Chapter 6, sorting is a measure of the uniformity of the sizes of particles in sediment or rock. In well-sorted sandstones, for example, the grains are mostly about the same size or fall within the narrow range of sand-sized particles. There are fewer of the tiniest clay and silt-sized particles to fill the spaces between the larger grains (Fig. 14–3). Poorly sorted sandstones contain a complete range of particles from clay-size to sand-size. Therefore, what might have been pore space in a rock having better sorting is occupied by fine particles in a poorly sorted rock. Therefore, poor sorting generally correlates with relatively lower porosity.

Another factor that influences the porosity of an aquifer is the manner in which the grains are packed together (Fig. 14–4). If that packing is relatively open, there will be a greater volume of space available for water storage. The shape of grains and the presence of cementing materials like calcium carbonate or silica can also influence porosity. One factor that does not directly affect porosity is grain size. Given an aggregate of larger spherical grains packed in open arrangement, and a second aggregate of smaller but similarly shaped and packed grains, both would have identical porosities. It is easy to convince oneself that pore volume between equal spheres is independent of their diameter if you were to examine Figure 14–5 with a magnifying glass. It would be apparent that magnifying the diameter of the spheres does not change the proportion of pore volume.

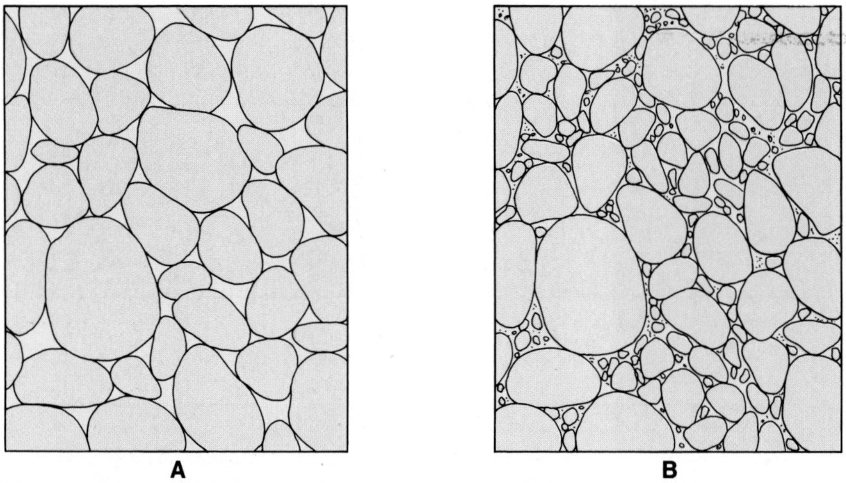

Figure 14–3 Because smaller particles occupy potential pore spaces, well-sorted sand (A) would generally have higher porosity than a sand having poor sorting (B).

The sorting that characterizes the grains in a clastic rock, and the way in which they are packed largely determines the porosity of the rock.

Rocks, of course, are not composed of perfectly spherical grains packed in the most open manner. As a result, they rarely achieve the theoretical maximum porosity. In general, a porosity of 20 per cent is considered large. A rock having between 5 and 20 per cent porosity would be considered medium, and porosities less than 5 per cent would be considered small. Representative porosities for various rocks and loose sediment are given in Table 14–1. Unconsolidated sediments, because they are commonly more loosely packed (having undergone less compaction) and lack cements, generally have higher porosity than do lithified clastic sedimentary rocks.

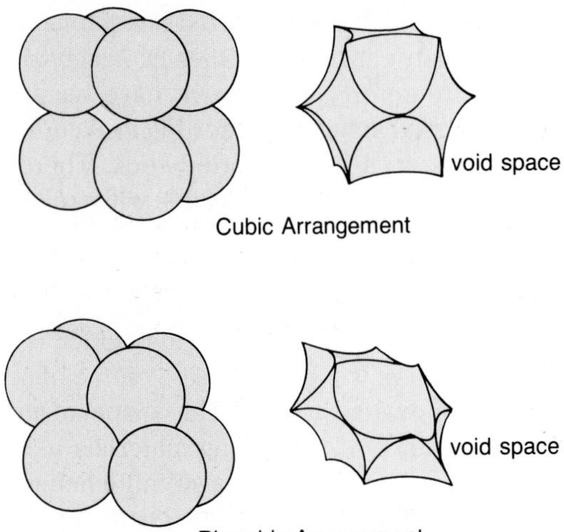

Figure 14–4 Illustration of the effect of packing on void space. The cubic arrangement of grains is the most open and provides the greatest porosity. In addition, the minimal cross-sectional area of the pore (shown in brown) is greater in cubic packing. In natural materials like sand, the grains would not be perfect spheres, they would not be of identical size, nor would they be so precisely arranged. (From Graton, L.C., and Fraser, H.J., *Systematic packing of spheres with particular relation to porosity and permeability, J. Geol.* 43:(8), 1935.)

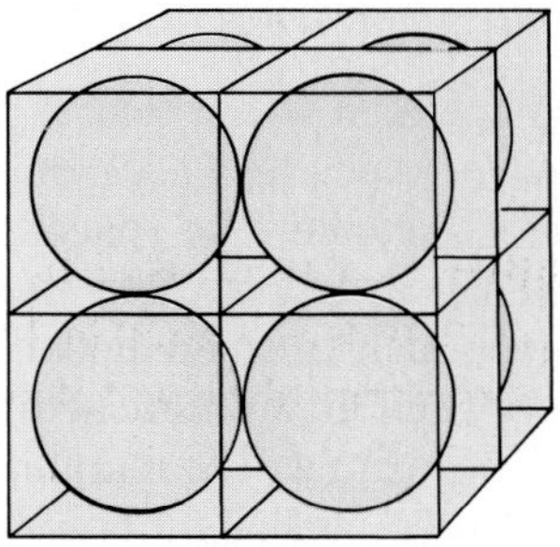

Figure 14–5 Each of the eight spheres packed in cubic arrangement occupies 52 per cent of the volume of the cube in which it rests. Porosity is therefore 48 per cent. Because of variation in particle size, packing, and particle shape, natural materials will not attain so great a porosity.

Table 14–1 General Porosity Ranges for Various Unconsolidated Sediments and Sedimentary Rocks

Sediment or rock	Porosity %
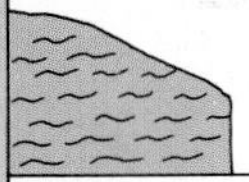 Clay	45-55
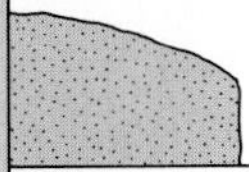 Sand	35-40
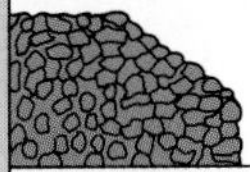 Gravel	30-40
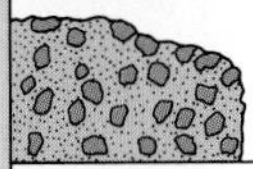 Sand and Gravel	25-35
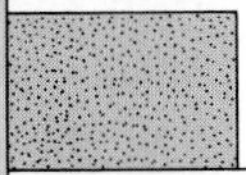 Sandstone	10-20
 Shale	1-10
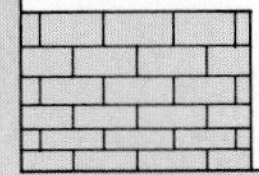 Limestone (non-cavernous)	1-10

For clastic rocks these values may vary because of cementation and packing, and for limestones they will vary if solution cavities and fractures occur. (After Meinser, O.E., U.S. Geol. Surv. Water Supply Paper 489, 1959.)

Permeability

Permeability is a term used by geologists to express the relative ease with which water moves through the spaces or interstices in a rock or sediment. These spaces must be interconnected if a rock is to have permeability. It is possible to measure permeability by subjecting a piece of rock to a given water pressure and recording the amount of water that passes through the rock in a measured interval of time. When this is done for materials having a range of grain sizes, it is apparent that the flow rate is greatest in coarser materials. Unlike porosity, permeability increases as the grain size increases. The reason is that the size of voids or interstices also increases in a sediment composed of larger grains. Grain surfaces retard flow, and there is more total surface area to do this in fine-grained sediment than in coarse-grained material. In fine silts and clays, the film of water adhering to each particle is in continuous contact with the film of adjacent grains, and there is less low-friction "throat area" for the uninhibited passage of fluid (Fig. 14–6). Rocks composed of clay may have considerably more porosity than sands, but because of their smaller interstices and the consequently higher frictional resistance to flow, clayey materials have much lower permeability than sands. Unconsolidated and well-sorted gravels, as well as cavernous limestones, are among the most permeable of earth materials.

In general, the finer the particles in a clastic sedimentary rock, the lower will be that rock's permeability.

The Occurrence of Water Under Ground

Underground Water Zones

Unless trapped above impermeable layers of rock, meteoric water will percolate steadily downward into the earth under the pull of gravity. Eventually, however, the water will reach a depth at which the great pressures of overburden have closed most of the open spaces and the rock has become impermeable. Continued additions of water will then fill all pore spaces above the impermeable zone, and the surface of the saturated zone will rise toward ground level as more water is added. The situation resembles what would occur if a tub were filled with gravel and then a shower of water added (Fig. 14–7). The water would percolate down to the bottom of the tub and then gradually rise as it filled up the spaces between the gravel. If the shower of water were stopped before the tub was filled, one could reach down through the gravel and find the level of the free water surface. In nature, that surface could be located by digging a hole into the ground until one reached a depth where water began to flow into the hole.

Figure 14–6 Conceptual example of the effect of grain size on permeability. (A) Rocks composed of large grains are likely also to have large openings between grains and less total surface area to cause frictional drag on fluids. (B) In finer-grained material, the low-friction throat areas are smaller and surface area relatively greater than in coarser material.

After a period of time, the surface of the water in the well would stand at the level of the free water surface in the rock. The subsurface zone below that level is appropriately called the **zone of saturation** (Fig. 14–8). Above the zone of saturation is the **vadose zone** or **zone of aeration**, so called because some of the pores are filled with air as the meteoric water infiltrates downward. If additional wells are dug nearby and the elevation of the free water surface is recorded in each well, then it would be possible to use that information to map the underground surface at the top of the zone of saturation. This upper surface of the saturated zone is called the **water table.**

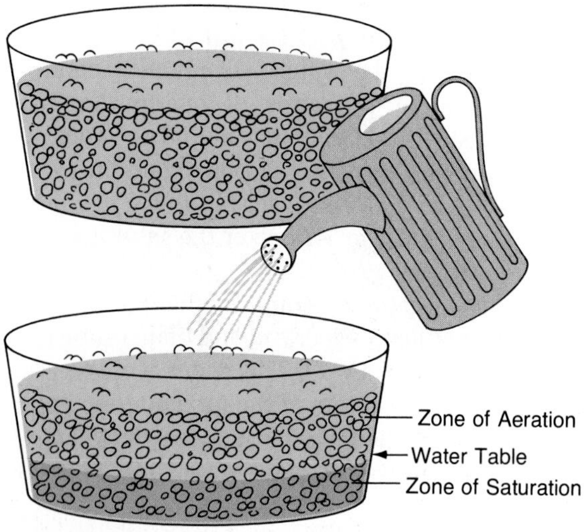

Figure 14–7 A simple experiment that indicates the manner in which the *zone of saturation* and *water table* develop.

One may define the water table as the upper surface of the zone of saturation.

Just above the water table, moisture may be pulled up into the vadose zone as a result of capillary attraction. This irregular zone of water is called the **capillary fringe**. In general, the capillary fringe is thicker in finer-grained materials than coarse clastic sediment. One can observe capillary attraction when a glass tube is inserted into a glass of water. Water will appear to oppose gravity and move upward in the tube (Fig. 14–9). Capillary attraction is the result of surface tension, or the tendency of a water surface to act like a stretched rubber membrane. Surface tension is caused by the mutual attraction of water molecules for one another. In a tubelike crevice or pore, water molecules are attracted laterally to the minerals adjacent to the pore and downward to other molecules, so that a membranelike concave surface is produced. The attempt by surface tension to flatten the concavity tends to draw the water up the tube. As soon as the weight of the raised column of

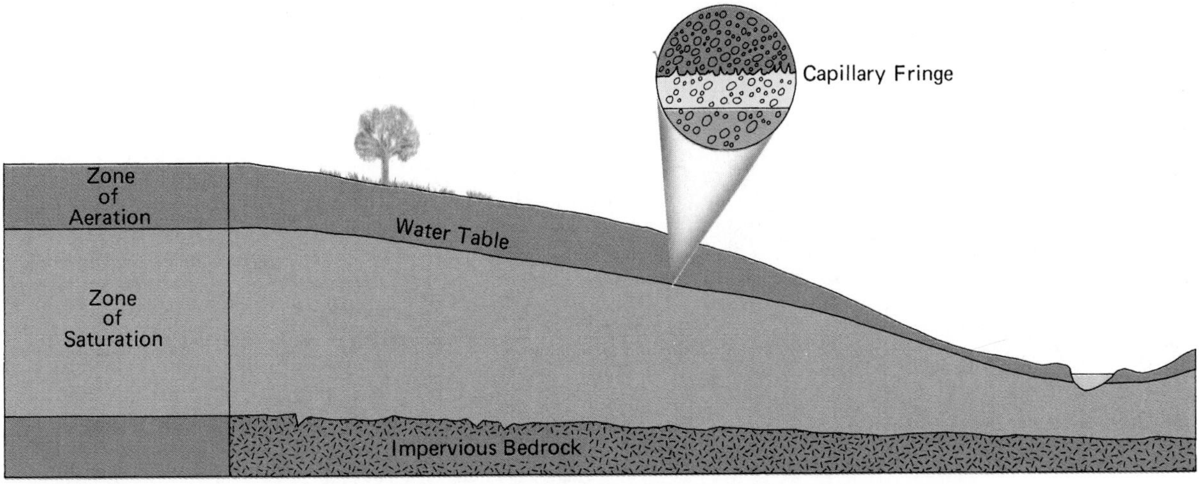

Figure 14–8 The water table in a humid region underlain by homogeneous sediment with impervious bedrock at depth. All openings below the water table are completely filled with water. In dry periods, the water table may drop, whereas in wet periods it is likely to rise.

water equals the force resulting from surface tension, water will cease to rise in the tube.

The Configuration of the Water Table

The water table is not a static surface but will vary in elevation and slope according to the configuration of the overlying topography, the permeability of the rocks and sediments, and the amount and frequency of rainfall. If a hilly terrain is underlain by homogeneous permeable earth materials, and if rainfall is relatively frequent, the configuration of the water table will roughly approximate the surface contours of hills and valleys (Fig. 14–10). In the event rain would fall continuously for a time, the zone of saturation would gradually extend upward to the surface of the ground, and water would flow into the lower-lying areas. On the other hand, if there were no rain for a long time, the water table would slowly flatten as the water in the saturated zone percolated downward and flowed laterally toward nearby stream valleys.

In humid regions, the time between periods of rainfall is usually not long enough to allow the water table to flatten completely. Also, water falling as rain over an entire hilly area does not infiltrate rapidly and rush to adjacent stream channels. In most aquifers, groundwater can move only slowly through the tortuous interstitial spaces, and before it can reach an outlet, more rain falls, causing a "pile up" of water in the divides farthest from

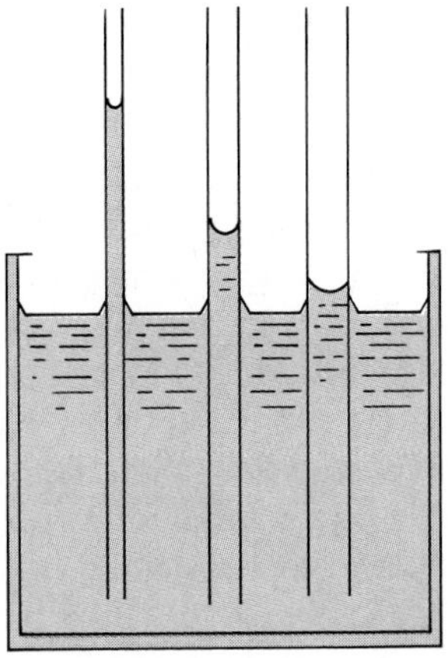

Figure 14–9 Capillarity can be demonstrated by suspending glass tubes in water. Water rises in the glass tube because the concave surface layer of molecules, in contracting, draws the water upward with it by molecular attraction. The height of rise varies inversely with the diameter of the tube. It is for this reason that the capillary fringe is thicker in fine sand than in a coarse gravel.

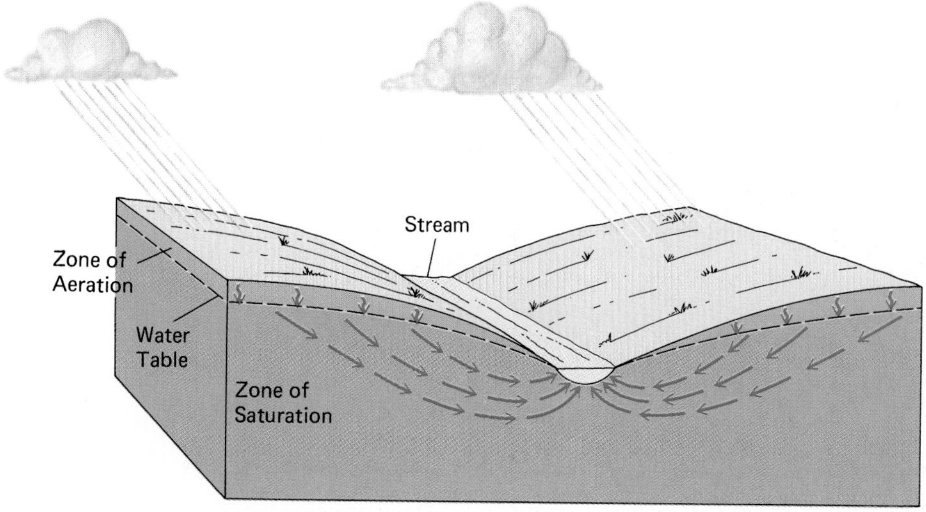

Figure 14–10 In a hilly area underlain by uniformly permeable material, the configuration of the water table will be a subdued reflection of surface topography, provided rainfall is frequent. The flow of groundwater through such permeable homogeneous material is indicated by the arrows.

the streams. It is for this reason that the water table rises under hills and descends beneath valleys so as to take the form of a subdued reflection of surface topography. It is a "subdued" reflection because the vertical distance from ground surface to the water table is greater along the divides than along the valleys.

In humid regions, the water table shows a similarity to surface topography in that it is higher beneath divides and lowest beneath valleys.

Although most areas are underlain by a body of groundwater having only one water table, there are also some places where localized impermeable beds occur at a shallow depth below the surface and prevent the further downward percolation of water toward the main water table. Water may collect on top of the impermeable strata to produce a local secondary water table above the main one. Such features are appropriately called **perched water tables** (Fig. 14–11). Glaciated areas and areas where there are frequent intertonguings of sandstones and shales are likely to have perched water tables. The quantity of water available from the limited saturated zone beneath a perched water table may be considerably less than beneath the main water table. When large quantities of groundwater are needed, it is usually advisable to drill wells through the perched water table and penetrate the main water table.

Groundwater Movement

Patterns of Flow

The energy required for the movement of water molecules through rock or sediment is supplied by gravity. Water responds to the pull of gravity by attempting "to seek its own level." This is why groundwater percolates from higher parts of the water table to lower parts and often out onto the surface as springs. Part of the groundwater beneath divides flows directly down the slope of the water table. An even greater part, however, moves slowly toward a nearby stream along curved paths of flow (see Fig. 14–10). Along these paths, water may be carried deep into bedrock before ascending toward the stream channel. These flow patterns replenish and replace the deeper realms of groundwater and prevent their stagnation. Water moving in this manner will dissolve and transport soluble matter and provide cements for formerly unconsolidated sediments.

The curved flow paths for groundwater are apparently the result of the fluid attempting to move toward points of least hydraulic head. Along any horizontal line, such as A-A′ in Figure 14–10, the water is under greater hydraulic head beneath a divide than beneath a stream, and so it moves toward the stream.

On reaching the zone of saturation, water movement occurs along curved paths toward topographically low areas.

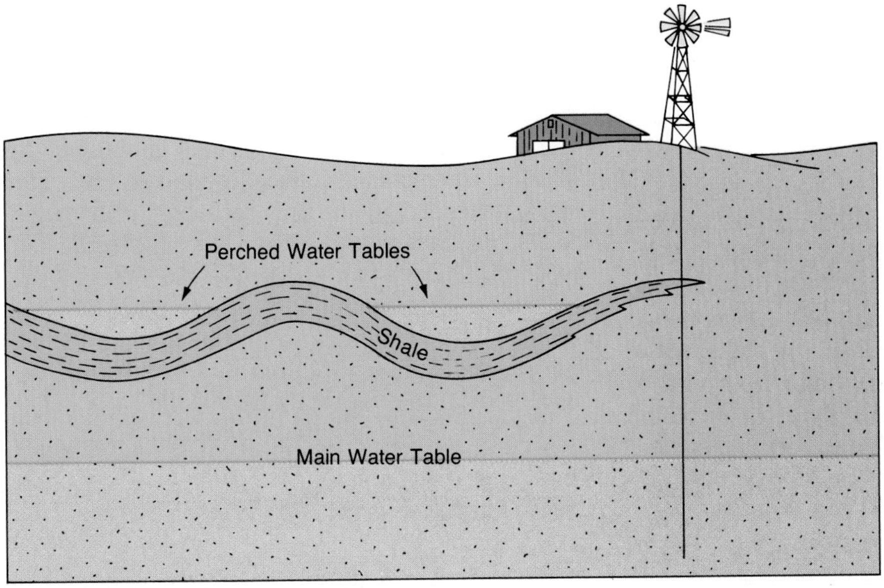

Figure 14–11 Perched water tables developed above a folded and pinched-out shale stratum.

Rates of Flow

Two factors that influence the rate at which water in the zone of saturation flows are the permeability of the aquifer and the slope of the water table. **Hydraulic gradient** is the term used to express the slope of the water table. Rather like the gradient of a stream, the hydraulic gradient is equal to the vertical distance between two points on the water table (h) divided by the horizontal distance between these two points (ℓ). The vertical measurement is termed the **hydraulic head**. One can quickly determine the hydraulic head by subtracting the elevation of the free-standing water surface in valley wells from similar measurements obtained in wells dug high on divides. For example, if the water stood at an elevation of 550 meters above sea level in an upland well, and at 500 meters above sea level in a nearby valley well, then the hydraulic head (h) would be 50 meters. If the wells were 1000 meters apart, then the hydraulic gradient would be 50 meters divided by 1000 meters or 0.05 (Fig. 14–12).

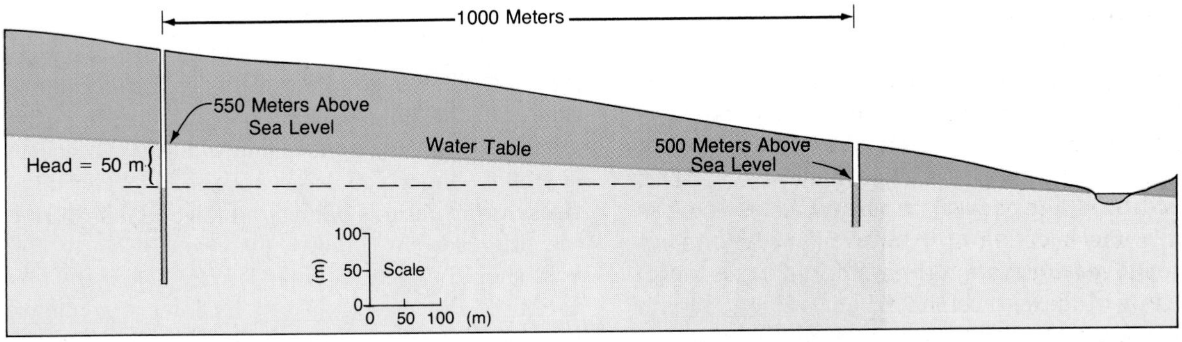

Figure 14–12 The hydraulic gradient between two points on the water table 1000 meters apart and having a 50-meter difference in elevation would be 50/1000 or 0.05.

The greater the hydraulic gradient and permeability, the greater will be the rate at which water in the zone of saturation will flow toward an outlet.

The relation between rate of groundwater flow, hydraulic gradient, and permeability was worked out in the 1860s by Henry Darcy, an engineer for the city of Dijon, France. Darcy made observations of the height of water in wells and attempted to relate these observations to the permeability of the aquifer and the movement of water from well to well. He also conducted laboratory experiments on the flow of water through a pipe filled with sand of various grain sizes. He observed that the flow rate was fastest for the most permeable (coarsest) sediment. Further, he was able to show that, for an aquifer of a certain permeability, the rate of flow from one location to another was directly proportional to the hydraulic head and inversely proportional to the horizontal distance the water travels. Thus, as the slope of the water table increases, so also does the rate of flow. In addition, the direction of maximal rate of flow will be in the direction of the steepest slope of the water table, that is, the direction of greatest hydraulic gradient. This important relationship is known as *Darcy's Law*.

Darcy demonstrated that, for a given aquifer, the rate of groundwater flow from one place to another is directly proportional to the drop in vertical elevation between the two places and inversely proportional to the horizontal distance the water travels.

In most aquifers, the velocity of groundwater flow rarely exceeds 1 or 2 meters per day, although the range of velocities recorded have included rates as fast as 100 meters per day and as slow as only a few cm per year. Several methods have been used to obtain velocity measurements. Some involve the injection of dyes or radioactive indicators in wells higher on the water table and the detection of these tracers in distant wells. In another interesting technique, wells having metal casings are wired together to form an electrical circuit. A highly conductive soluble compound is poured into the upland well, which on arrival at the downslope well causes a short circuit between the metal well casing and an electrode. The short circuit is automatically recorded by a clocklike device. The velocity of groundwater movement is obtained by dividing the distance traveled by the groundwater (with its tracer) by the time required for the transfer of the trace material from the first to the second well. Thus, if two wells were 400 meters apart, and the tracer required 200 days to reach the second well, groundwater would be moving toward the second well at the rapid rate of 2 meters per day.

Wells and Springs

It is always a delight and somehow surprising to see water flowing from a ledge of rock or spouting from a pipe sunk deep into the earth. These are, however, rather ordinary occurrences of groundwater emerging at the earth's surface as springs and wells. In general, wells and springs are classified into two groups. Those that are fed by water flowing through unconfined aquifers are called **ordinary wells** and **springs**. If, however, the groundwater is confined to an aquifer by impermeable rock layers, the groundwater system is called **artesian**.

Ordinary Wells

As indicated earlier, if a hole is dug to a depth below the water table and encounters permeable materials at that depth, groundwater will flow into the hole. The hole has become a water well. The presence of permeable sediment or rock is important, for without good permeability water will not flow into the hole rapidly enough to replace that which is withdrawn. Also important is the necessity of digging a well deep enough. The water table rises during periods of high rainfall and falls during the drier season (or drought years). Clearly, the well should penetrate considerably below the elevation of the water table at its lowest predicted elevation. There is another reason for drilling a well to a depth appreciably below the water table. When water is pumped from a well, a conical depression in the water table forms in the area surrounding the bore hole. This **cone of depression** (Fig. 14–13) may eventually reach to the intake level of the well, causing it to go "dry." Much to the consternation of neighboring users of groundwater, the decline in the water table around large cones of depression may also cause nearby wells to become dry. In the past, exceptionally wide and deep cones of depression developed around wells that supply the large quantities of water needed for irrigation and industry, and attempts to reestablish former levels of groundwater have been unsuccessful, even when pumping was discontinued.

Ordinary or Gravity Springs

A spring is a natural flow of groundwater that has emerged onto the surface of the earth. The quantity of water emerging in this way may be only a trickle, or it may be a gushing flow that may proceed over the surface as a stream. As indicated in Figure 14–14, ordinary or **gravity springs** may develop at points where the water table intersects the ground surface along the sides of a valley. Springs may also be localized by impermeable strata, bedding planes, fractures, faults, or coverings of debris from slumping or landslides. Some springs are associated with perched water tables and emerge above impermeable layers of rock exposed along hillsides.

Artesian Wells and Springs

In some wells groundwater encountered at depth is under sufficient hydraulic pressure that it is able to rise above the level of the aquifer containing it and may even flow out at the surface. These wells and springs are given the name **artesian** after the Old Roman province of Artesium in France (now called Artois). An artesian system (Fig. 14–15) is the result of the simultaneous occurrence of three geologic conditions. First of all, there must be an inclined aquifer. One or more dipping layers of sandstone or other coarse sediment, layers of gravel in old alluvial fans, or cavernous limestones may be suitable inclined aquifers. Second, the aquifer must be enclosed by impermeable beds (aquicludes), which serve to prevent the escape of water from the aquifer to the surface. The third condition needed for an artesian system is that the elevation of the intake area is sufficiently high that enough hydraulic pressure will be developed to force the water in wells upward above the level of the aquifer. The pressure that forces the water to rise in the well is caused mainly by the weight of the water coming down toward the well site from the higher collecting area. If the artesian system were free-flowing, then water would gush upward at the well site to the level of water in the upland intake area. That surface to which water in an aquifer would rise by hydrostatic pressure is called the **potentiometric surface**. Water does

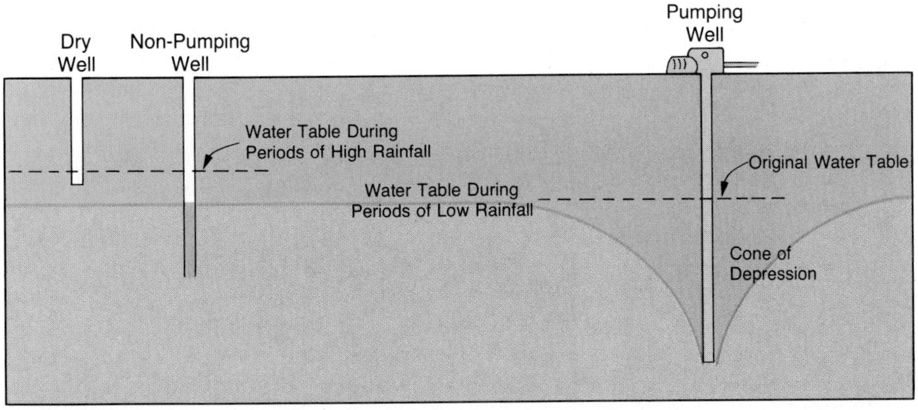

Figure 14–13 Because the water table may subside during intervals of drought, wells should penetrate deeply into the zone of saturation. Note the basin-like effect or cone of depression developed in the water table around the pumping well.

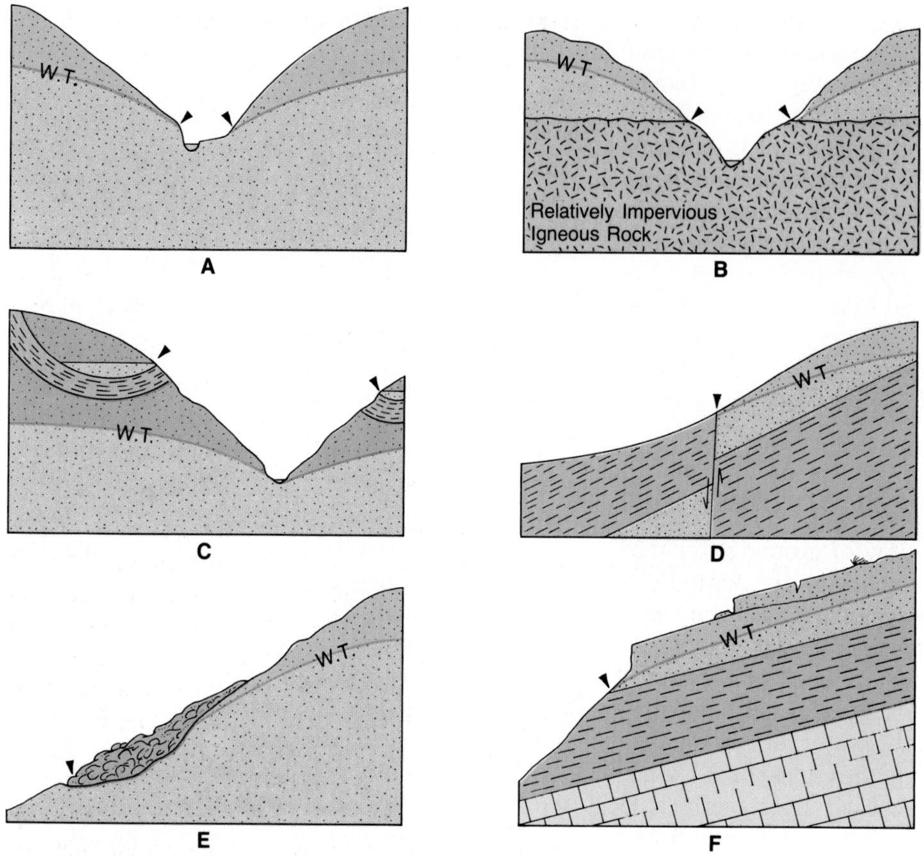

Figure 14–14 Ordinary springs may develop in a variety of geologic settings. (A) Springs developed at intersection of water table and valley slopes. (B) Springs formed at contact between impermeable crystalline rocks and relatively permeable overlying sediment. (C) Perched springs. (D) Spring localized by faulting. (E) Spring developed at toe of a landslide. (F) Perched spring formed at contact between permeable sandstone stratum and impermeable shale stratum.

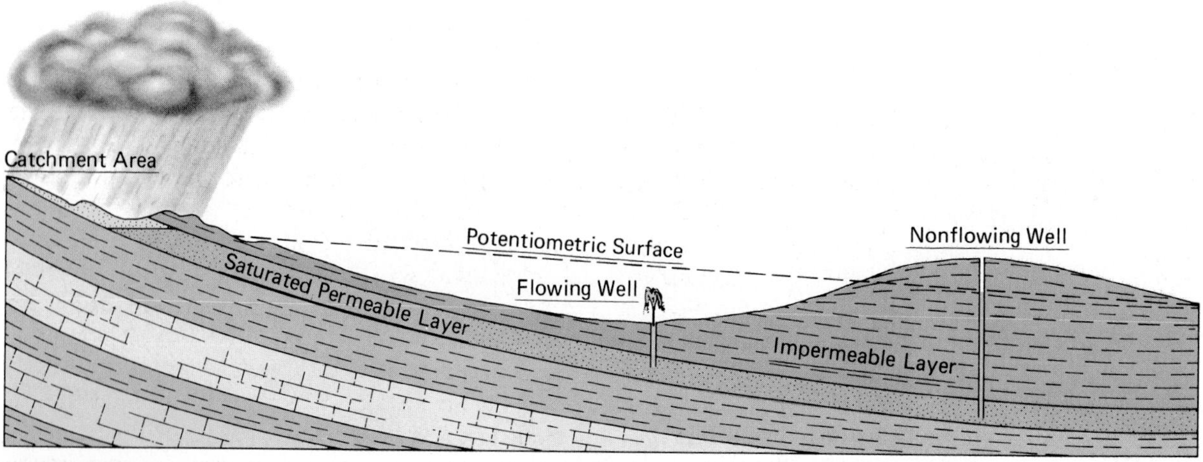

Figure 14–15 Typical conditions for the occurrence of an artesian well.

not rise to the potentiometric surface in natural artesian systems, however, because of the frictional loss of energy as it percolates through the aquifer. Nevertheless, to qualify as an artesian system, there must be sufficient head (and hydraulic pressure) to overcome friction.

An inclined aquifer with sufficient head to overcome friction and an aquiclude that will confine water within that aquifer are essentional requirements for an artesian system.

Perhaps the best known artesian system in the United States is that developed in the plains surrounding the Black Hills of South Dakota. Water enters the Dakota Sandstone where the formation is exposed in the outcrops that encircle the Black Hills. It then percolates through the sandstone and laterally outward beneath the plains. Overlying shale strata provide the impermeable cover. Half a century ago, hydraulic pressure in this confined system was sufficient to cause water from wells to gush a few meters into the air and even turn water wheels. There are now so many wells, however, that pressure is greatly reduced. The weight of overlying sediment has "squeezed" water toward wells but has not compensated for the lessening of head caused by withdrawal of water from over 10,000 wells in the region. Today, the yield in flowing wells is only a fraction of earlier amounts, and pumps must be employed on many wells from which water formerly flowed freely.

Beware of the Water Witch

Many people if asked, "Where shall I drill my well?" would receive the prompt reply, "Don't ask me, hire a water witch." A water witch, or dowser, is a person who claims to have the occult power of locating the site for a well by walking across the ground while holding a forked stick horizontally. When the butt end of the stick or "divining rod" is attracted downward, the well site has been located. It has been estimated that over 25,000 water witches are currently working in the United States. The majority are not charlatans, but sincere men and women of good character and benevolent intentions who sometimes do not even charge for their services. Indeed, they are often successful in finding what prove to be good sites for wells. The reason for their success may be related to their long-standing knowledge of groundwater occurrence in the local area, the ability to apply common sense geologic interpretations, the acquired knowledge they have about the water table being closer to the surface in valleys than on hilltops, and recognition that certain kinds of vegetation are clues to underlying aquifers. Of course, the principal reason dowsers find water in humid regions is that groundwater occurs ubiquitously below the surface, so that if a reasonably deep well is drilled, the probability of success is good.

There is, of course, no scientific basis for the alleged powers of the water witch. Studies conducted in Australia in the early 1900s indicated that there were twice as many dry holes among the well sites selected by dowsers than in those locations based on geologic information. The United States Geological Survey also examined the question in great detail. In 1917, O. E. Meinzer, speaking on behalf of the Survey, concluded, "It is difficult to see how for practical purposes the entire matter could be more thoroughly discredited, and it should be obvious to everyone that further tests by the United States Geological Survey of this so-called 'witching' for water, oil, or other minerals would be a misuse of public funds."

The Use and Misuse of Groundwater

Depletion

As is the case with most geologic resources, the supply of groundwater can be easily depleted. The groundwater that has been rapidly withdrawn during the past several decades required centuries to accumulate. In some regions, groundwater accumulated thousands of years ago at a time when precipitation was considerably greater than it is presently. In order to prevent wells from going dry and insure a supply of groundwater, communities have resorted to various techniques for artificially replenishing aquifers. In some instances, such artificial recharge involves water spreading, that is, covering flat surface areas with shallow ponds from which water can infiltrate into underlying aquifers. In other cases, water may be pumped down into the aquifers during seasons when surface water is readily available. Finally, after realizing the limitations of particular groundwater reservoirs, governments can enact regulations that sensibly limit the number of wells and the withdrawal rates so that all users can obtain a fair share of the water supply.

Contamination

Groundwater derived from sandstone or sandy sediments is famous for its purity. Contaminating particles adhere to the surfaces of the myriads of mineral grains and become trapped in interstitial spaces. Of course, there are limits to the filtering capacity of even sandy materials, and the ultimate purity of water from a well may depend on the volume of the pollutants being added to the underground reservoir and the distance over which the groundwater has traveled. Natural filtration is usually not effective for many dissolved chemical pollutants. Also, not all aquifers are composed of granular materials. Large open conduits such as those in cavernous limestones, basalts containing lava tunnels, and rocks with large joints provide little filtration.

Whenever sewage or chemical wastes are dumped at the earth's surface, there is a strong possibility that groundwater will become contaminated due to seepage of contaminants into the zone of saturation. This is particularly true if the area for dumping is underlain by highly permeable materials (Fig. 14–16). Not infrequently one discovers places where refuse has been dumped into pits that lead directly into cavernous conduits that supply water for nearby towns. Salty water from oil fields and gasoline from rusted storage tanks of neighborhood gasoline stations may similarly contaminate a community's primary source of water. Nitrates from fertilizers and pesticides used in agricultural fields may also contaminate groundwater supplies over vast areas.

A few decades ago, most people would have assumed that water falling as rain and seeping into porous rocks would be relatively free of contaminants. In recent years, however, it has become clear that rainwater in many regions has become seriously tainted even before it reaches the ground. When compounds such as sulfur dioxide, hydrogen sulfide, and nitrogen oxide are emitted by power plants, smelters, factories, and automobiles, they rise into the atmosphere and combine with water vapor to form sulfuric and nitric acid aerosols. These microscopic drops are carried by wind and suspended in the atmosphere until meteorologic conditions return them to earth as acidic rain and snow. **Acid rain** greatly increases rates of chemical weathering, and contributes significantly to the dissolution of limestones and other rocks. It dissolves important plant nutrients from soils, and is enormously damaging to the biology of lakes and streams. Acid rain seeping underground may contaminate groundwater supplies with traces of toxic metals, such as cadmium and mercury, that were originally emitted from factory smoke stacks. Some areas underlain by limestones are able to partially resist the assault of acid rain by chemically neutralizing it, but much of eastern Canada, Scandinavia, and New England, for example, are underlain by glacial materials or granitic rocks that do not neutralize acid rain. The solution to the acid-rain problem involves legislation limiting high sulfur emissions from industrial sources. Such limits are urgently needed, for as expressed by Canada's Minister of the Environment, "Acid rain is one of the most devastating forms of pollution imaginable, an insidious malaria of the biosphere."

Saltwater Intrusion

Along coasts and on small islands, two kinds of water compete for space in permeable sediments and rocks. Meteoric water derived from rain and snow infiltrates in the manner already described and gradually builds a zone saturated with fresh water. Saltwater from the adjacent ocean, however, also enters permeable layers, but because of its greater density, the brine tends to form a wedgelike zone beneath the fresh water. The density differences between the two kinds of water are such that a balanced state develops in which there are about 40 meters of fresh water beneath sea level for every 1 meter of elevation of the water table above sea level (Fig. 14–17). Thus, if the water table is 5 meters above sea level at a given location, fresh water may extend to a depth of 200 meters below sea level. If the water table, however,

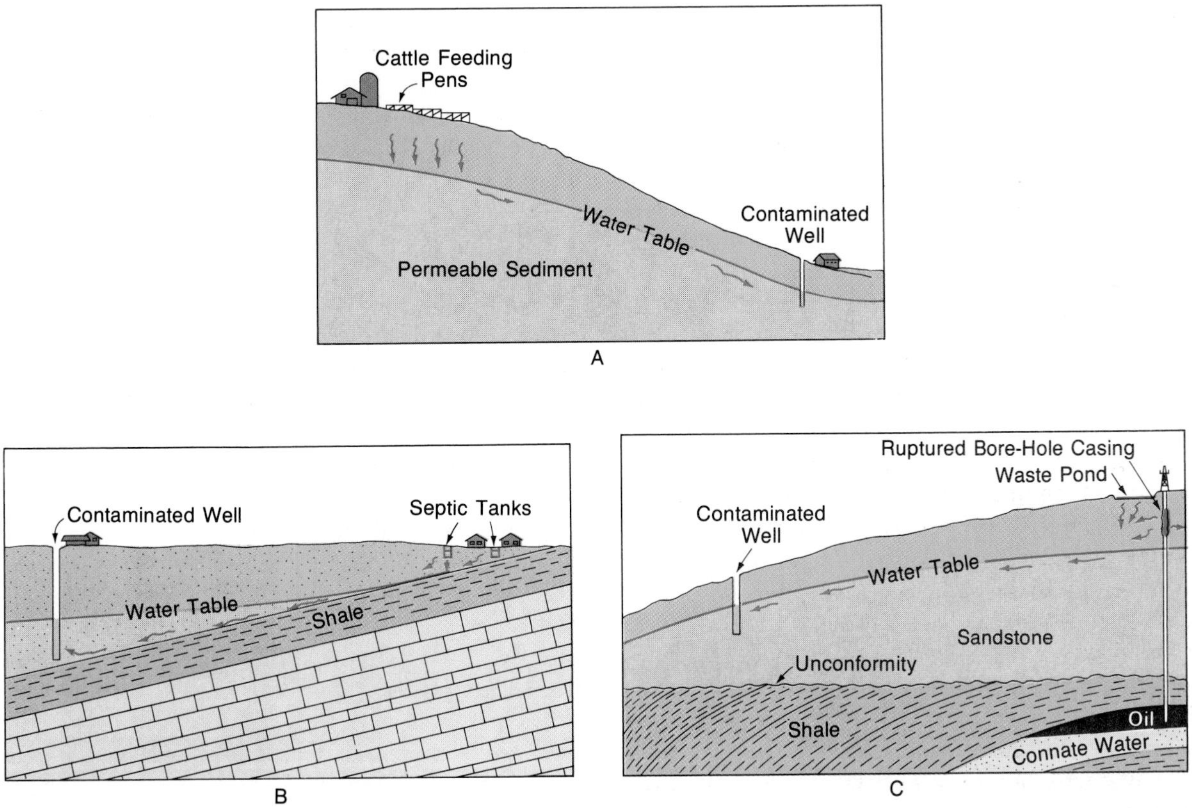

Figure 14–16 Well pollution. (A) Seepage of pollutants from feeding pens results in contamination of nearby wells located downslope. (B) Septic tanks provide the contamination that moves toward a well because of the inclination of an impermeable shale stratum. (C) Contamination of groundwater from a waste pond and ruptured casing of an oil well.

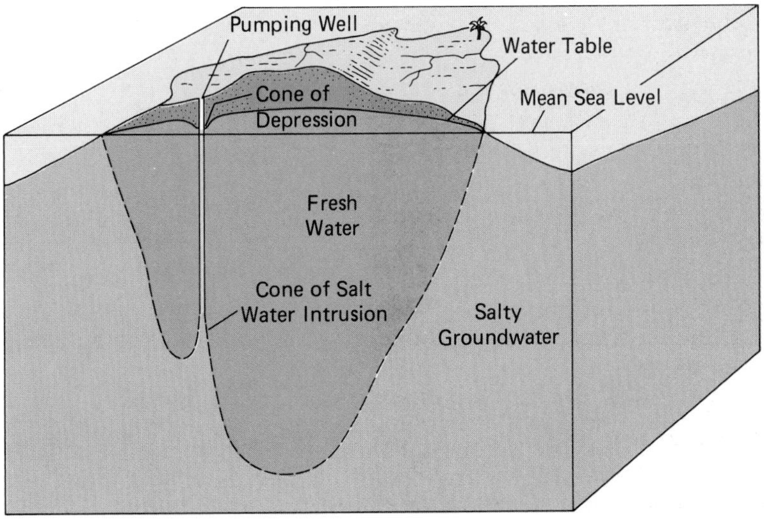

Figure 14–17 Relationship of fresh to salty groundwater on an island. Note that the fresh groundwater floats on the salty groundwater. The cone of salt water intrusion is produced by pumping. (The vertical scale of the area below sea level is one-half that of the area above sea level.)

is lowered by only 5 meters as a result of drawing too much water from wells, then the top of the saltwater zone may rise and contaminate wells.

One locality where saltwater incursions have been a major problem is on Long Island. The aquifers beneath Long Island consist of coarse highly permeable glacial sediments underlain by unconsolidated sediments that dip gently toward the Atlantic (Fig. 14–18). These beds receive their water from rainfall, and for a time they supplied all the water needs for the island. As the population grew, however, the number of wells put down increased. Withdrawal of groundwater became particularly rapid in the industrialized western part of the island. Inevitably, the freshwater-to-saltwater interface moved upward until saltwater entered the deeper wells. Today, water for much of the island must be supplied by pipes from the mainland.

Subsidence

An important environmental consequence of heavy withdrawal of groundwater (and also petroleum) is subsidence of the ground surface above

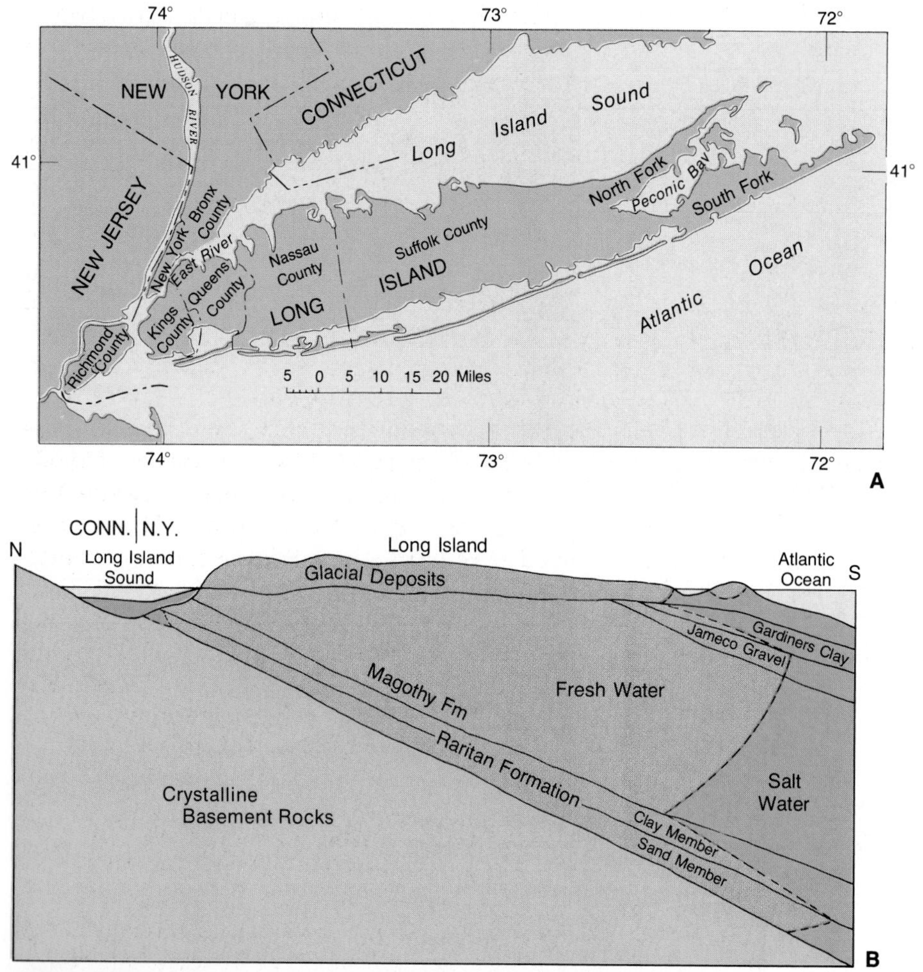

Figure 14–18 Map and north–south cross-section of Long Island showing general relationships of rock units and the freshwater-to-saltwater boundaries before development and rapid withdrawal of groundwater. (Adapted from Heath, R.C., *The Changing Pattern of Groundwater Development on Long Island, New York,* Washington, D.C., U.S. Geological Survey Circular, 524, 1966.)

the reservoir rocks. Fluid withdrawals from loose fine-grained materials lead to volume reductions and in artesian systems reduce the pressure with which the aquifer formerly resisted compression. Perhaps the most famous subsidence problems are found in Mexico City, where local sinking due to groundwater withdrawal has exceeded 7.5 meters. In places, the steel casings of wells now protrude above ground level and houses are entered through second-story windows. For some areas of the city, the rate of subsidence has been recently reduced by careful control of groundwater withdrawal.

Venice, Italy, is another city plagued with subsidence resulting from both groundwater depletion and the compaction of loose sediments beneath heavy structures. The city has subsided over 3 meters since it was founded 1500 years ago. Within the last few years, efforts to recharge the aquifers beneath the city and to regulate withdrawal of water from wells have reduced the rate of subsidence appreciably.

Advantages of a Groundwater System

One should not let the problems that are sometimes associated with dependency on groundwater supplies bias one's thinking about this valuable resource. The use of groundwater has far more advantages than disadvantages. Many a weary farmer on a hot summer day will agree that probably the best beverage on this planet is a cup of water freshly drawn from a well. In fact, groundwater has a justified reputation for its uniformly cool temperature and purity. Groundwater has other benefits even in areas where there is a supply of surface water. Often it can be reached within a few hundred meters of where it is to be used, and thus long pipelines from water purification centers are not needed. In its natural subsurface reservoirs, groundwater is protected from costly losses due to evaporation. Also, although the amount of water drawn from wells may be limited, the yield is generally more uniform than from small streams that may periodically become dry. As indicated in the preceding pages, groundwater can become contaminated, and so one cannot assume *a priori* that all well water is free of impurities. It is also true, as noted earlier, that aquifers composed of medium- or fine-grained sand or sandstone have a remarkable ability for trapping pathologic microorganisms and unpleasant suspended particles. Where care is taken to protect groundwater from contamination, it can usually be used with little or no expensive treatment.

Groundwater in Soluble Rocks

Except in those areas where water flows rapidly through caverns or lava tunnels, groundwater moves too slowly to cause appreciable mechanical erosion or to transport coarse particles. On a less perceptible scale, however, groundwater does accomplish geologic work by dissolving minerals, transporting the soluble matter, and precipitating that material as cements in granular sediments and as depositional features in caverns and other underground openings.

As described in the discussion of chemical weathering (Chapter 5), limestones, dolostones, marbles, and evaporites are particularly susceptible to solution by water that contains carbon dioxide. Limestone taken into solution may later be precipitated by either removing carbon dioxide (through agitation, photosynthesis, or heating) or by evaporating part of the water. Evaporation would increase the concentration of dissolved matter to the point where there would be more in solution than the diminished volume of water could retain. The "surplus" calcium carbonate would come out of solution and be deposited. The basic equation for the relation between water, carbon dioxide, and limestone is as follows:

$$\underset{\text{calcium carbonate}}{CaCO_3} + \underset{\text{water}}{H_2O} + \underset{\text{carbon dioxide}}{CO_2} \rightleftharpoons \underset{\text{calcium ion}}{Ca^{+2}} + \underset{\text{bicarbonate ions}}{2\,(HCO_3)^{-1}}$$

Notice that the addition of carbon dioxide to the left side of the equation provides the acidic condition that dissolves calcium carbonate. A loss of carbon dioxide from the bicarbonate on the right side results in precipitation of calcium carbonate.

Solution of limestones occurs at the most rapid rate near the surface. Here, recently fallen rainwater still retains the carbon dioxide absorbed from the atmosphere, and there are further additions of the potent gas from organic materials in soil. Also, the water itself is not yet saturated with lime and is capable of causing dissolution. The water flows along bedding planes, joints, and fractures and begins to widen them by solution. Surface collapse often occurs where large openings may be dissolved out of the limestone. Such depressions are called **sinkholes** (Figs. 14–19 and 14–20). Sinkholes may also develop where the ceiling of large solution cavities or parts of caverns collapse. Often sinkholes are connected to subterranean solution channels so that water is drained from them. Many sinkholes, however, become plugged with clay and debris, so that they form lakes and ponds. Although some sinkholes are as much as 1000 meters in diameter, on the average they are less than about 30 meters across. The progressive collapse of the roof of a cavern may produce a linear series of sinkholes, and these may coalesce to form a steep-walled trough called a **solution valley**. Solution valleys are recognized by their peculiar right-angle bends, which tend to follow the original joint pattern in the limestone bedrock. Here and there along a solution valley, the stream may pass beneath a segment of the original cavern roof that has not yet collapsed and thus form a natural bridge (Figs. 14–21 and 14–22). In other places, a considerable length of uncollapsed cavern may receive the flow of the surface stream. Such so-called *disappearing streams* may emerge elsewhere.

Caverns are, of course, the most intriguing of all solution features. Each year thousands of people visit caverns in order to observe the fascinating rock formations (Fig. 14–23) and examine organisms adapted to life in perpetual darkness. The development of caverns is almost inevitable in humid areas underlain by thick and extensive limestones. Caverns may be large or small, may have a simple or complex labyrinthic plan, may extend vertically as well as horizontally, and may have passages developed along multiple levels. Subterranean streams course through some caverns, whereas in other caverns streams are lacking. The rectilinear pattern of passages in some caverns indicates that they developed as water moved through an established system of joints (Fig. 14–24), whereas others appear to have no relationship to joints.

Most caverns are the result of dissolution of limestone, during which carbonic acid (formed by the combination of carbon dioxide with water) chemically reacts with the limestone.

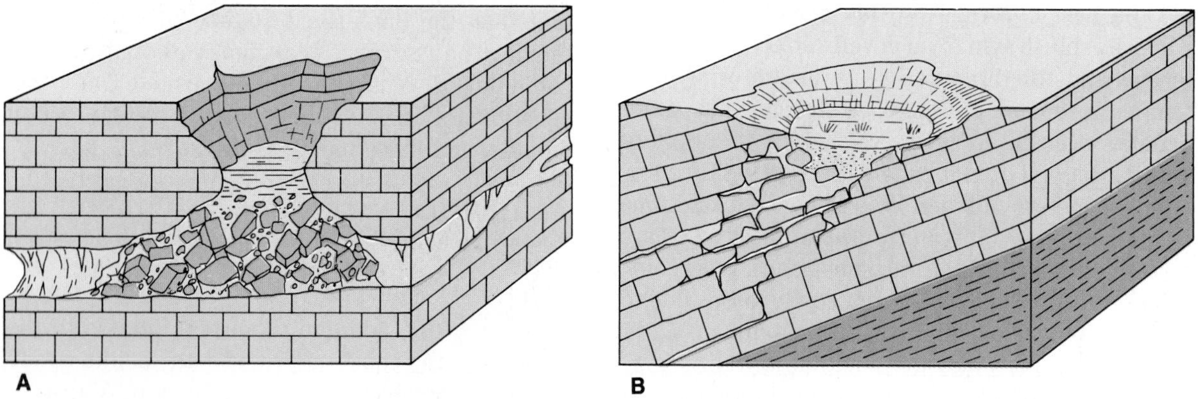

Figure 14–19 (A) A collapse sinkhole resulting from failure of the roof of a cavern. The sinkhole has been plugged with clay and rock debris, so that water accumulates within the depression. (B) Solution sinkholes are due to surface dissolution around some favorable point like a joint intersection or topographic depression.

Figure 14–20 This enormous sinkhole developed on May 8 and 9, 1981 in the town of Winter Park, Florida. The collapse destroyed buildings and part of the public swimming pool visible below. (Courtesy of the Florida Bureau of Geology.)

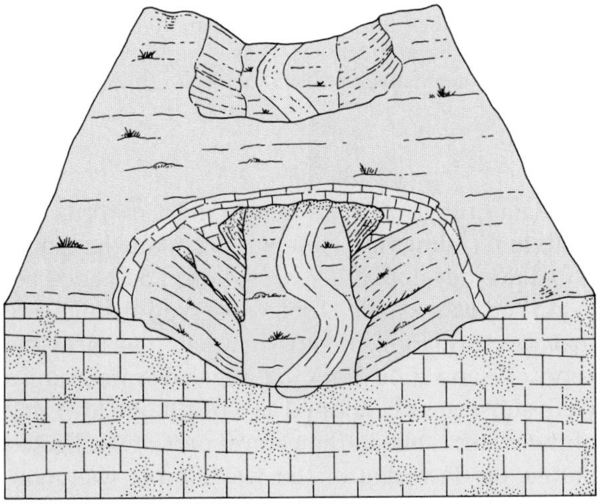

Figure 14–21 Collapse of a significant portion of the roof of a cavern has produced this *natural bridge* and *solution valley.*

Figure 14–22 This natural bridge, located about 30 miles northeast of Roanoke, Virginia, was formed by a combination of groundwater solution work and stream erosion. The bridge was originally surveyed by George Washington. (Courtesy of J. Brice.)

Figure 14–23 Calcium carbonate, deposited from groundwater dripping from the ceilings and walls of this large room in Luray Caverns, Virginia, has formed these dripstone features.

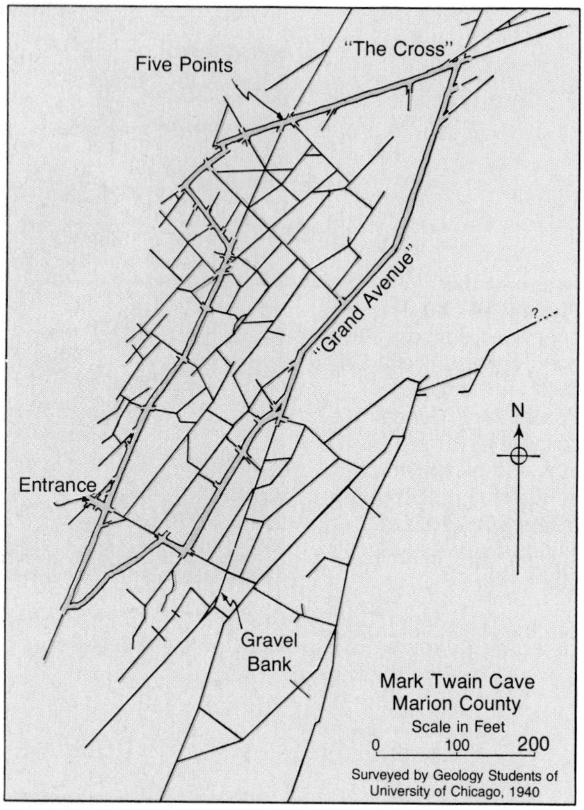

Figure 14–24 Mark Twain Cave. This cave, made famous because of its description in the novel *Tom Sawyer,* is located in Hannibal, Missouri. It is easy to see how Tom Sawyer and Becky Thatcher could have become lost in the labyrinthine passages, the locations of which are determined by three or more intersecting set of joints. (From Bretz, J.H., *Caves of Missouri,* Missouri Geological Survey publication, Vol. 39, 2nd sec., 1956.)

The depositional features of caverns are well known. There are the downward-extending iciclelike **stalactites**, as well as **stalagmites**, which grow in conic form upward from the floor of the cavern. The construction of a stalactite begins with a drop of lime-saturated water temporarily adhering to the cavern roof (Fig. 14–25). With the loss of carbon dioxide, calcium carbonate comes out of solution to form a thin layer around the drop. As more water flows down through the center and along the sides of the original layer, the stalactite grows by successive additions of calcium carbonate. Stalagmites are formed by drops of water that fall from the ceiling of the cavern and strike the floor, where both agitation and evaporation contribute to the build-up of calcium carbonate. **Columns** (Fig. 14–26) develop when stalactites and stalagmites eventually grow together. Features such as stalactites, stalagmites, and columns are composed of a rather dense banded variety of limestone called **cave travertine**.

Aside from their aesthetic value, features like stalactites help geologists unravel the history of cavern development. Clearly, such features must have formed when the cavern was above the water table and filled with air. Yet, in most caverns, the walls and ceilings are lined with crystals that could only have developed beneath water and hence beneath the water table. These observations indicate that the formation of most caverns began in the zone of saturation but that the cavernous areas were subsequently drained as a result of a lowering of the water table. In the central part of the United States, the lowering of the water table was most commonly in response to slight regional uplift, which raised the water-filled caverns slightly above the zone of saturation, causing a lowering of

Figure 14–25
Stalactites of a variety named helictites in an early stage of development in Castleguard Cavern, Castleguard, Alberta, Canada. What do you think may cause the horizontal extensions of these stalactites? Water required for the dissolution of limestone and formation of dripstone is partly derived from the melting of glacial ice that overlies the cavernous limestone. (From the film *Castleguard Caves*, Courtesy of the National Film Board of Canada.)

Figure 14–26 (A) Large symmetric column. (B) Cluster of stalagmites. (C) Stalactites in Carlsbad Cavern. (Courtesy of the U.S. National Park Service.)

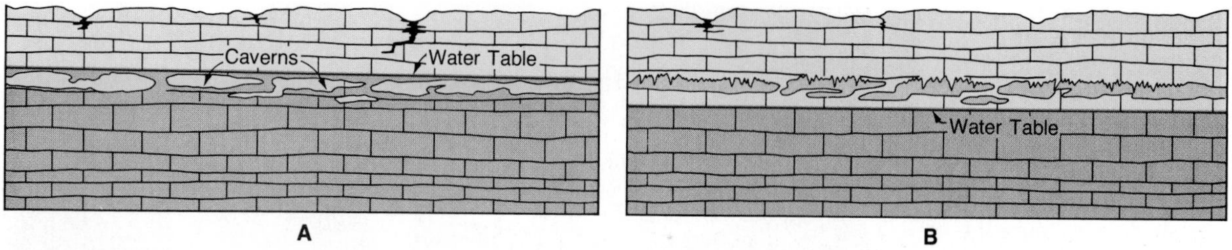

Figure 14–27 (A) Caverns formed below the water table and completely filled with water may subsequently be drained as the water table is lowered. (B) Relative lowering of the water table may result from regional uplift or an extended period of diminished precipitation.

the water table and draining of the caverns. Following uplift, stalactites and other cave features began to develop within the vacated subterranean chambers (Fig. 14–27).

People have been exploring caves throughout history. The ancient Assyrians, motivated by tales of gem minerals and precious metals to be found in caves, explored caves in the valley of the Tigris River. Their explorations were unsuccessful. Limestone caverns are not likely places in which to find such riches. The Roman philosopher Seneca left an account of cave exploration by some of his contemporaries who were also in search of mineral wealth. "These men," wrote Seneca, "dared to descend to a region where the whole of nature is reversed. The land hung above their heads and the winds whistled hollowly in the shadows. In the depths, frightful rivers led nowhere into the perpetual and alien night. After accomplishing so much, they now live in fear for having tempted the fires of hell."

Long before the classical Greek period, prehistoric people used caves as havens from inclement weather and shelter that could be defended against predatory animals and hostile tribes. The tools, arms, adornments, religious artifacts, and even the skeletons of these ancient cave dwellers are well known to archaeologists. Often paleontologists join the archaeologists in their subterranean digs, for caves may contain skeletal remains of Ice Age bears, pigs, sloths, and a host of other contemporaries of our early ancestors.

Without doubt, caves will continue to stimulate the imagination, not only of scientists, but of anyone likely to enjoy the mystery and excitement that comes as one moves cautiously through uncharted passageways. In the French Alps and Pyrenees, explorers have mapped cave segments more than a km from the surface and thrilled to the discovery of magnificent subterranean waterfalls and lakes. The less adventuresome can enjoy conducted tours through the better-known caverns of the world. Mammoth Cave in Kentucky is one such cavern that has an interconnected series of passages extending over a total distance of 34 km. In New Mexico, Carlsbad Cavern boasts a single chamber that is an astonishing 1200 meters long, 200 meters wide, and 100 meters high.

Karst Topography

Karst is a term applied to topography formed in regions of limestone or dolomite bedrock by the vigorous solution work of groundwater. One recognizes karst topography by the presence of large numbers of sinkholes, solution valleys, disappearing streams, and caverns (Fig. 14–28). The development of karst topography is enhanced by the presence of well-jointed carbonates (or evaporites) near the surface, abundant rainfall, and sufficient relief to insure continuous movement of groundwater that will carry away dissolved matter. The term karst comes from a limestone plateau in Yugoslavia where solution features are well developed. Similar topography can be found in Kentucky, Tennessee, Indiana, northern Florida, and Puerto Rico.

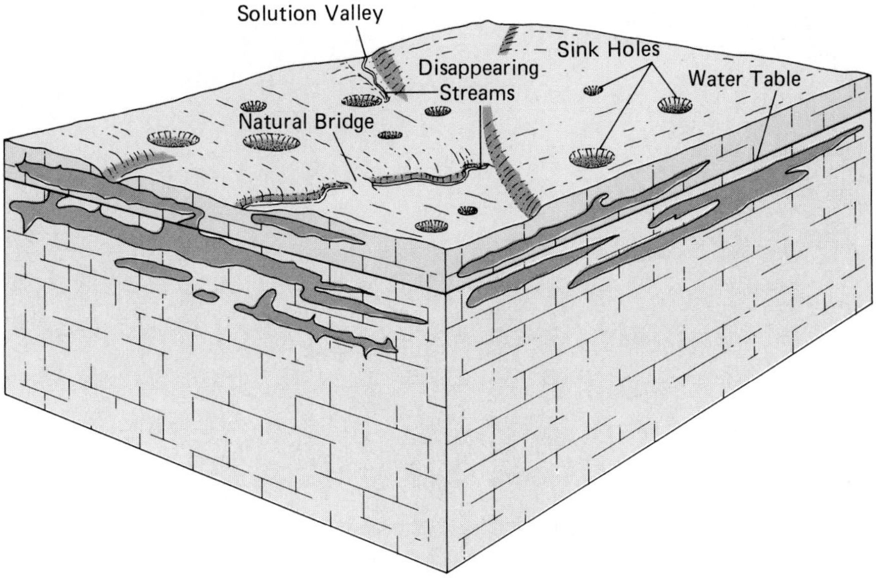

Figure 14–28 Karst topography. (Surface area of block is about 5 km^2.)

Hot Springs and Geysers

To qualify as a **hot spring** (Figs. 14–29 and 14–30) water emerging from the subsurface must be at least as warm as the body temperature of a human. Most hot springs are found in regions where there has been geologically recent igneous activity. In such regions, rocks immediately below the surface may still retain part of their original heat. Indeed, that heat may be retained for centuries because of the insulating effect of overlying layers. Groundwater that comes in contact with the hot rocks is

Figure 14–29 Mud hot springs, North Island, New Zealand. (Photograph by W.B. Hamilton, courtesy of the U.S. Geological Survey.)

Figure 14–30 Madison Hot Springs, Yellowstone National Park, Wyoming. The terraces are composed of calcium carbonate in the form of travertine. (Courtesy of L.E. Davis.)

heated and sometimes even mixed with hot magmatic water before finding its way to the surface and emerging as a hot spring.

Geysers (Fig. 14–31) are special kinds of hot springs that discharge intermittently in the form of a tall column or fountain of water and steam. The essential component of a geyser is one or more tubular conduits extending from the surface into a zone of hot rocks. Water that percolates downward through the conduits is heated above the boiling point but is temporarily prevented from boiling because of the weight of the overlying water column. Eventually, water at some level in the tube reaches a temperature where it boils at the higher pressure. Steam is rapidly generated and lifts the water above it until it flows out at the surface. The deeper superheated water, suddenly freed of the weight of overlying water, flashes into steam and blasts upward toward the surface. After the eruption has dissipated, water flows back down through the tube system, and is heated for another eruptive cycle. The cycles may be irregular or rather regular, as is the case for Old Faithful in Yellowstone Park. Until recently, this famous geyser erupted about every 65 minutes (with a range of 30 to 90 minutes), but presently the interval between eruptions is closer to 80 minutes.

In these times of concern over meeting our energy needs, increasing attention is being given to utilizing the steam produced in hot spring areas to generate power. In order to tap this resource, wells are drilled into moist hot rocks. The steam released into the well is fed into turbines that generate electricity. Energy produced in this way is relatively inexpensive, has no radiation hazards, and does not add pollutants to the atmosphere. Unfortunately, there are only a few sites around the world favorable for the development of geothermal power at reasonable costs. Even in these favored locations, corrosive solutions destroy pipes and equipment and may even endanger plant and animal life if not properly managed. One way to handle the disposal problem is to pump the waste water back underground where it will be heated for a second cycle of use.

Locations where hot igneous rocks exist close enough to the earth's surface to be used in generating geothermal power occur in California, Nevada, New Mexico, Wyoming, Utah, Montana, Italy, Alaska, New Zealand, Mexico, Iceland, Japan, and the U.S.S.R. The 80 wells of the geothermal facility at the geysers in Sonoma County, California, generate enough power to serve the electricity demands of at least a million people. Plans are underway to expand this facility. The geothermal plant at Lardarello, Italy, has been producing electricity from natural steam since 1904, and is thus the oldest facility of this type. In Reykjavik, Iceland, not only is geothermal steam used to generate electricity, but it is pumped into buildings for heating and used to warm greenhouses that produce tropical fruits and vegetables.

Figure 14–31 The geyser Old Faithful erupting in Yellowstone National Park, Wyoming.

Concretions and Geodes

Among the minor features associated with groundwater deposition are accumulations of mineral matter called **concretions**. Concretions may be ellipsoid, discoid, or variously shaped and may range in size from a few mm to 1 or 2 meters. Most concretions are composed of calcite or silica, although one also frequently finds hematite, limonite, siderite (iron carbonate), and pyrite concretions. Concretions developed in a rock after it has been deposited are usually the result of localized chemical precipitations of dissolved substances from groundwater. Often precipitation starts around a mineral or fossil particle that differs chemically from the enclosing rock. Concretions formed within sedimentary rocks in this way may retain relic laminations and textural features of the original rock. Other concretions are merely local concentrations of the cementing material in a clastic sedimentary rock and therefore are similar in composition to the host rock. Concretions may differ in composition from the rock in which they have formed. Chert nodules in limestone or dolostone are a common example.

Not all concretions develop in rocks after deposition. Some apparently form on the floors of lakes or seas as sediments are being deposited. Concretions that form in this way do not contain relic features of the rocks that enclose them.

Although the odd shapes of concretions are intriguing, most people find geodes more interesting because of their beauty. **Geodes** (Fig. 14–32) are hollow crystal-lined roughly spherical bodies varying from 1 or 2 cm to as many as 60 cm in diameter. They are most characteristic of certain limestone strata, although they are also known to occur in shale and basalt. The manner in which well-formed crystals grow inward toward the central cavity indicates that solution and subsequent precipitation of calcite or silica are important in the formation of geodes.

Figure 14–32 Cavity in a rock, which has become lined with quartz crystals. Such hollowed-out rocks are called *geodes.* (Courtesy of Ward's National Science Establishment, Inc., Rochester, N.Y.)

Summary

Approximately 20 per cent of all the water used in the United States is drawn from the subsurface by means of wells. This underground water that fills pores, cracks, and joints in rock and sediment is called *groundwater*. Most groundwater is derived from that portion of rainfall or melting snow that does not immediately flow to streams but infiltrates into the ground. The downward-infiltrating water eventually reaches rocks through which it cannot move and begins to fill all openings in overlying rocks so as to develop a zone of saturation. The upper surface of the *zone of saturation* is called the *water table*. Its configuration roughly parallels the land surface, and it rises or falls according to the amount of precipitation that infiltrates from above. In general, wells must be deeper than the level of the water table if they are to provide an adequate supply of water. Where the water table intersects the sides of valleys, *springs* may be formed.

Rocks and sediments below the surface of the earth that are capable of producing water from wells are termed *aquifers*. The quantity of water that can be obtained from an aquifer depends not only on the amount of rainwater that percolates into it but also on two fundamental properties of the aquifer itself. These properties are porosity and permeability. *Porosity* is expressed as the percentage of the rock or sediment that consists of openings. *Permeability* refers to the ability of a rock to transmit groundwater or other fluids through pores and interconnected openings.

The movement of groundwater below the water table in the zone of saturation is principally a slow seepage downward and laterally toward streams and lakes. Normally, the water follows curved paths of flow and moves at rates determined by permeability of the aquifer and the slope of the water table, that is, the hydraulic gradient. For most aquifers, the velocity of movement rarely exceeds 1 or 2 meters per day and may be as sluggish as a few cm per year.

We see groundwater at or near the earth's surface when it emerges in well borings or as springs. An *ordinary well* is simply a drill hole that extends downward below the water table, thereby allowing water to seep into the cylindric hole and be available for pumping. Such wells penetrate aquifers that are not confined by overlying relatively impermeable beds. Where aquifers are confined and dipping so that there is sufficient hydraulic head, water may rise in the well above the level of the aquifer and flow without pumping. Such wells penetrate *artesian groundwater systems*.

Groundwater is an important natural resource. It supplies many homes and factories with a conveniently nearby source of usually clean water, which is noted for its uniform temperature, and which is protected from the heavy evaporative losses experienced by surface water sources. Contamination of groundwater, however, does occur, and is especially prevalent in aquifers that have large conduit-like openings. One should be wary, for example, of groundwater drawn from cavernous limestones located downslope from garbage or sewage-disposal facilities. Users of groundwater should avoid rapid withdrawal of groundwater, as this may result in lowering of the water table to the depth of intake pipes of wells. In coastal regions, too rapid withdrawal can cause saltwater contamination. Actual *subsidence* of the ground surface is another possible consequence of heavy withdrawal of groundwater.

The geologic work of groundwater consists primarily of *solution of rocks* and minerals, the *transportation of the dissolved matter*, and the *precipitation of dissolved matter* when conditions are suitable. The solution work of groundwater is particularly evident in areas underlain by limestone, although rocks such as dolostone and gypsum are also susceptible. Usually, the solution process begins when groundwater containing carbon dioxide begins to move through bedding planes and joints, gradually widening them. In time, enough rock may be dissolved to produce large underground passages or caverns. As caverns are enlarged, some portions of their roofs may collapse, resulting in depressions at the surface called sinkholes. *Sinkholes* also may form independently of caverns by solution along intersecting joints at the bedrock surface. Depositon of the calcium carbonate dissolved in groundwater occurs whenever part of the water is lost as a result of evaporation, or there is a loss of carbon dioxide as a result of agitation, or a change in water temperature. *Stalactites* and *stalagmites* are among the many depositional features formed in caverns as a result of the work of groundwater.

Where hot rocks from geologically recent igneous activity exist near the earth's surface, groundwater may be heated and may flow to the surface to form *hot springs*. *Geysers* are an intermittent eruptive kind of hot spring. Their occurrence in Yellowstone National Park, Iceland, and New Zealand is well known. In some locations around the world, the steam from geysers and hot springs is being used to turn turbines that generate electricity.

Terms to Understand

acid rain
aquiclude
aquifer
artesian well
capillary fringe
cave travertine
caverns
columns
concretions
cone of depression
connate water
effluent streams
geodes
geyser
gravity spring
groundwater
hot spring
hydraulic gradient
hydraulic head
influent streams
karst topography
magmatic water (juvenile)
meteoric water
perched water table
permeability
porosity
potentiometric surface
sinkhole
solution valley
spring
stalactites
stalagmites
water table
well
zone of aeration (vadose zone)
zone of saturation

Questions for Review

1. What is the water table? What would be the effect on the slope and elevation of the water table of a period of excessive rainfall? Of a period of marked aridity?

2. A sandstone and a shale have identical porosities of 18 per cent. Which is most likely to be the most permeable and why?

3. In an obscure valley, Aaron Hatfield digs a well that penetrates 2 meters below the water table. In an adjacent valley, Jud McCoy has a well that penetrates 5 meters below the water table. Each summer during a two month dry spell, the Hatfield well goes dry (thus curtailing distilling operations). Provide a reason for Hatfield's well going dry?

4. What might you do to a sample of groundwater to prove that it contained substances in solution? What compound most frequently makes water "hard"?

5. Draw cross-sections illustrating the following:

a. Gravity spring.
b. Artesian groundwater system.
c. Zones of aeration and saturation.
d. Perched water table.

6. How does the functioning of a geyser differ from that of an artesian well?

7. What is the role of carbon dioxide in the solution of limestone and in the formation of such dripstone features as stalactites and stalagmites?

8. Why does the fountain produced by an uncapped artesian well not rise to the same elevation as the groundwater table in the catchment area?

9. What are the characteristics by which karst topography can be recognized?

10. Account for the observation that wells in coastal areas adjacent to the sea are often able to draw fresh water from a depth well below sea level. What might be the effect of rapid withdrawal of fresh water from such wells?

11. A liquid dye is poured into a garbage dump at 1:00 P.M. on January 1. The dye appears in a well located 1 km away on January 6 at 1:00 P.M. What is the velocity of groundwater flow in meters per hour between the dump and the well?

12. The elevation of the water table at well "A" is 200 meters above sea level and at well "B," located 1000 meters away, the elevation of the water table is at 180 meters above sea level. What is the head and hydraulic gradient between the two wells?

14. What is a "cone of depression"? Assuming uniform pumping, would the cone of depression be steeper in more permeable or less permeable aquifers? Why?

13. Describe the necessary natural conditions for an artesian groundwater system.

Supplemental Readings and References

Anderson, J., 1974. *Cave Exploring*. New York, Association Press.

Bouwer, H., 1978. *Groundwater Hydrology*. New York, McGraw-Hill.

Freeze, R.A., and Cherry, J.A., 1979. *Groundwater*. Englewood Cliffs, N.J., Prentice-Hall, Inc.

Heath, R.C., and Trainer, F.W., 1968. *Introduction to Ground Water Hydrology*. New York, John Wiley & Sons.

Jennings, J.N., 1971. *Karst*. Cambridge, Mass., M.I.T. Press.

Meinzer, O.E., 1923. The occurrence of groundwater in the United States. U.S. Geological Survey Water Supply Paper 489.

Sweeting, M.M., 1973. *Karst Landforms*. New York, Columbia University Press.

Walton, W.C., 1970. *Groundwater Resource Evaluation*. New York, McGraw-Hill Book Co.

Glaciers

15

Shivling Peak in the Himalayas is a spectacular example of a glacial horn. (Courtesy of D. Bhattacharyya.)

The ground of Europe, previously covered with tropical vegetation and inhabited by herds of great elephants . . . became suddenly buried under a vast expanse of ice, covering plains, lakes, seas, and plateaus alike.

Louis Agassiz, 1840. Etudes de les glaciers

Introduction

If one were to leave the main highways and explore the back roads of the northern United States, Canada, and northern Europe, it would be possible to find large boulders of granitic rock lying at random about the countryside. One would surmise that these boulders were far from home, for they lie on quite different soil and sedimentary rocks; and exposures of similar igneous rocks are not found nearby. Such displaced boulders and cobbles are called **erratics**. They give us cause to wonder about what gargantuan process could have transported and dumped them onto the hills and fields where they now rest.

Two centuries ago scientists would have explained erratics as having been carried by enormous currents of water and mud during the devastation accompanying the great biblical deluge. Indeed, these same scientists referred to the bouldery deposits as *diluvium*, a Latin word meaning "deluge." Charles Lyell, the eminent British geologist, was not entirely satisfied with this explanation of erratics. He noticed that they did not occur in the southern parts of Europe, that they were scratched and grooved like rocks scoured by alpine glaciers, and that they were not rounded as are rocks transported by water. Lyell argued, therefore, that the erratics were in some way associated with accumulations of snow and ice. He had observed large amounts of rock debris frozen into the icebergs that formed at the coastal terminations of glaciers. This observation led him to reason that similar boulder-laden icebergs had also once drifted across Europe at a time when the continent was inundated by the sea. As the icebergs melted, its boulders and cobbles were dropped randomly over the submerged landscape of Europe. Lyell proposed this imaginative hypothesis in 1833. Many of his colleagues found it hard to imagine the enormous rise in sea level required to float icebergs over Europe's high plateaus and mountains. Also, at some localities in Europe, the huge erratics lay directly on bedrock surfaces that appeared to have been scratched and polished by moving ice (Fig. 15–1). Some scientists, therefore, continued the search for an alternative theory.

An alternative theory was proposed in Switzerland, and its development began with studies made by a Swiss director of salt mines named Jean de Charpentier. Charpentier was fascinated by glaciers and explored them whenever he had the opportunity. Initially, he was not concerned with an explanation of erratics but rather used bouldery deposits as evidence that alpine glaciers once extended well beyond their present limits. In 1836, Charpentier brought a young friend and colleague, named Louis Agassiz, to his home in Bex, Switzerland, in order to show him the field evidence favoring his idea. Agassiz was a shrewd and discerning scientist. On every side he saw indications of the way glaciers moved, witnessed the way they had altered the landscape, and ob-

Figure 15–1 Fragment of limestone that has been scratched and smoothed by glacial action. (Courtesy of the Institute of Geological Sciences, London.)

served their distinctive markings and deposits. He not only became convinced that Charpentier was correct about the former greater extent of alpine glaciers but also was able to visualize how erratics were carried across the plains and lowlands of northern Europe. Erratics were transported, said the energetic Agassiz, by a blanket of ice so thick and so vast that nearly half of Europe was buried beneath it (Fig. 15–2). Although these new ideas were greeted initially with skepticism, by 1960 few scientists could dispute the evidence, and Agassiz' glacial theory was widely accepted.

There were, however, questions associated with the new theory that still had to be answered. What caused the build-up of the enormous continental glaciers? After having advanced over the northern parts of Europe and North America, why then did they subsequently recede? Was there cause for concern that the ice might return and devastate cities and farms? In order to formulate theories that might be useful in answering these questions, geologists went to the Alps, Alaska, Greenland, and Antarctica to learn what they could about how glaciers are formed, how they are able to flow, and the manner in which they so magnificently sculpt our most spectacular landscapes. In this chapter, we will review some of their findings.

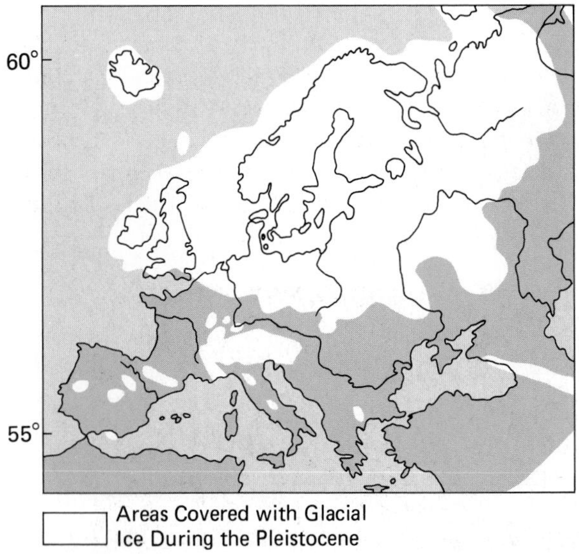

Figure 15–2 Areal extent of major glaciers in Europe during the Pleistocene Epoch.

Glaciers: Their Magnitude and Origin

A **glacier** is a large mass of on-land ice derived from crystallized snow, having definite limits, and showing evidence of present or past movement. At the present time, about 10 per cent of the earth's land area is covered by glaciers, and, as Agassiz correctly suggested, in the recent past that figure was about three times higher. Present-day coverage of glacial ice is sufficient to convince most geologists that we are presently living in a temporary warmer interval within an ice age whose end is not in sight. We would not have to convince an Eskimo of the validity of that concept, but a sunbather in Florida might be more skeptical. Perhaps that skepticism might be lessened on learning that glaciers cover about 14.5 million km^2 of our continents. Of this amount, 12.5 million km^2 are accounted for by the Antarctic ice sheet (Fig. 15–3) and 1.7 million km^2 by glaciers in Greenland. The remaining glaciated areas consist mostly of smaller ice caps on polar islands, glaciers along the perimeter of Alaska and Scandinavia, and the mountain glaciers of the Yukon, Alps, and Himalayas.

There are two conditions that are necessary for the growth and development of glaciers. Temperatures must be below freezing at least part of the time, and there must be snowfall in sufficient amounts to exceed losses that normally occur through melting and evaporation. These conditions are most prevalent in polar latitudes, high elevations, or both. Glaciers expand or contract depending on either seasonal or long-term climatic changes. Certainly during the climatic changes of the recent Ice Age, glaciers covered lowlands more extensively, and extended much lower into mountain valleys than they do today.

Glaciers and ice sheets grow when snow accumulation is greater than melting. They shrink when melting exceeds accumulation.

Because of the impressive ice sheet that covers Antarctica, one might guess that temperatures well below freezing are conducive to glacial develop-

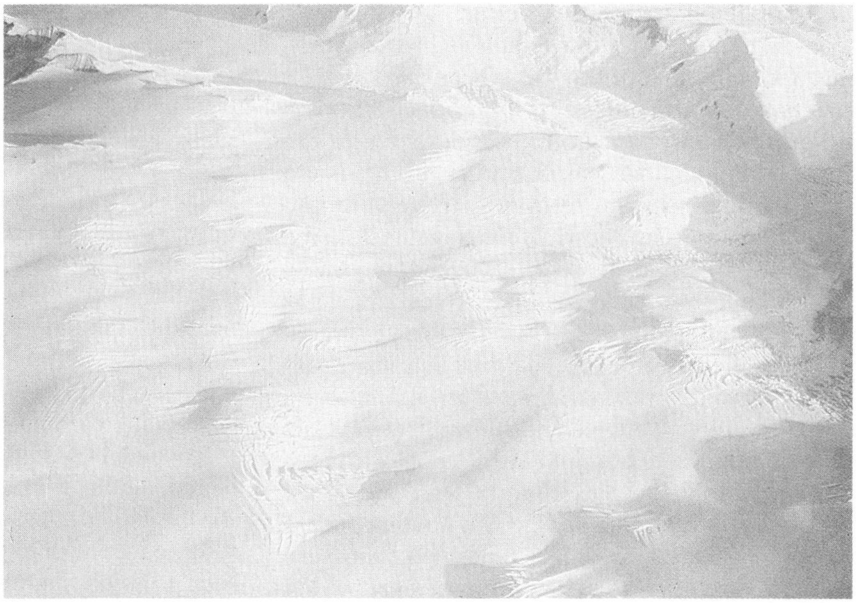

Figure 15–3 View of a small part of the Antarctic ice sheet near Victoria Land. (Courtesy of P.B. Larson and the U.S. Geological Survey.)

ment. Actually, extremely low temperatures are not necessary. In fact, glaciers are more likely to form when temperatures are only slightly below freezing, because under such conditions air can hold more moisture than at lower temperatures. Although snowfall is meager in the frigid Antarctic, the ice has been accumulating there for millions of years.

If the 24 million km^3 of ice covering various parts of the earth were to melt, sea level around the world would rise by about 100 meters. New York City, all of New Jersey, Florida, Louisiana, and vast areas of our eastern states and interior California would be covered by shallow seas. Landlocked midwesterners might travel relatively short distances to enjoy the surf. Of course, one should not hurry to purchase "ocean front" property in inland places like Missouri, for such melting and marine encroachments would take place very slowly over a period of several thousand years.

Ice from Snow

The first step in the formation of a glacier is the accumulation of snow as a permanent snow field. Flakes of freshly fallen snow are, of course, noted for their delicate hexagonal form and intricate growth patterns (Fig. 15–4). After the flakes lie at the surface for a period of time, or if they are bur-

Figure 15–4 Snow crystals show basic hexagonal pattern of branches. (Courtesy of National Oceanic and Atmospheric Administration.)

ied by additional snow, they begin to lose their delicate form as a result of sublimation, partial melting, and refreezing of the water near the center of the flake. (**Sublimation** is a process whereby water molecules pass from the solid state to the vapor state without passing through the liquid state.) The conversion to ice is enhanced by the weight of overlying snow. Eventually, most of the water once distributed as ice throughout the snowflake is concentrated as a granule. The material has now ceased to be snow and is in the form of granular ice called **firn**. The pellets of firn are really poorly formed hexagonal crystals or ice. Under the pressure of overlying ice and snow, these crystals begin a transformation into solid ice. Most of the air is gradually expelled from the mass as the ice pellets are packed tightly together. Melting occurs at contacts between grains (so-called "pressure points"), and the molecules of water are added elsewhere to growing ice crystals. As the process continues, adjacent crystals intergrow, and the mass becomes solid crystalline ice with a texture similar to, but coarser than, the texture of a coarse-grained gneiss. If snowfall is rapid, the transformation from snow to solid ice may occur in as little time as five years. On the frigid surfaces of Antarctica, the process may take thousands of years.

Snow is converted to ice by such processes as sublimation, partial melting and refreezing, and recrystallization. When enough ice has accumulated it will flow under the influence of gravity.

Kinds of Glaciers

One way glaciers are classified is based on their relation to surface topography. Great ice sheets, for example, may cover large parts of continents, burying mountains and plateaus indiscriminately. The size and shape of such glaciers is not dependent on the configuration of the ground surface. Glaciers that are confined by the mountain divides and valley sides constitute a second category of glaciers.

Glaciers Not Constrained by Topography

Ice sheets and **icecaps** are great domal masses of ice that differ primarily in size. Icecaps are generally smaller than ice sheets and rarely exceed 50,000 km^2. The ice in both types of glaciers flows outward from a central area of maximal ice accumulation much like thin pancake batter poured on a skillet. In cross-section, the maximal thickness is near the center of the ice sheet, and from that point the ice slopes away gently in all directions until it reaches the periphery where there is a marked steepening of slope.

Today icecaps are developed on Iceland, Baffin Island, and in the Canadian Arctic archipelago. As for the larger ice sheets, the only two that remain today cover most of Greenland and Antarctica. The Antarctic ice sheet (see Fig. 15–3) is over 3000 meters thick. Buried beneath this great blanket are hills, plateaus, and impressive mountain ranges (Fig. 15–5). Here and there mountains protrude through the glacial ice as isolated peaks called a **nunataks**. Around the margins of both Greenland and Antarctica, the moving ice reaches the ocean, where it may extend seaward as a floating ice shelf or break off as huge icebergs.

Glaciers Constrained by Topography

Glaciers that owe their size and shape to the characteristics of the landscape differ in many ways from icecaps and ice sheets. The most obvious difference is that they are smaller. In addition, they do not have the low domal shape of icecaps, and their direction of flow is not radial but rather down the existing slope of the valley. Typically, glaciers constrained by topography develop in high mountainous regions. They are initiated in lofty catchment areas where snowfall is sufficiently heavy and frequent to accumulate as a snowfield (Fig. 15–6). The snowfield is the nourishment area for the ice that moves down the slope as a **valley** or **alpine glacier** (Fig. 15–7). Although valley glaciers exist today that are over 100 km in length, most are less than 30 km long.

Where valley glaciers emerge from a mountain range, they spread out across the unconfined

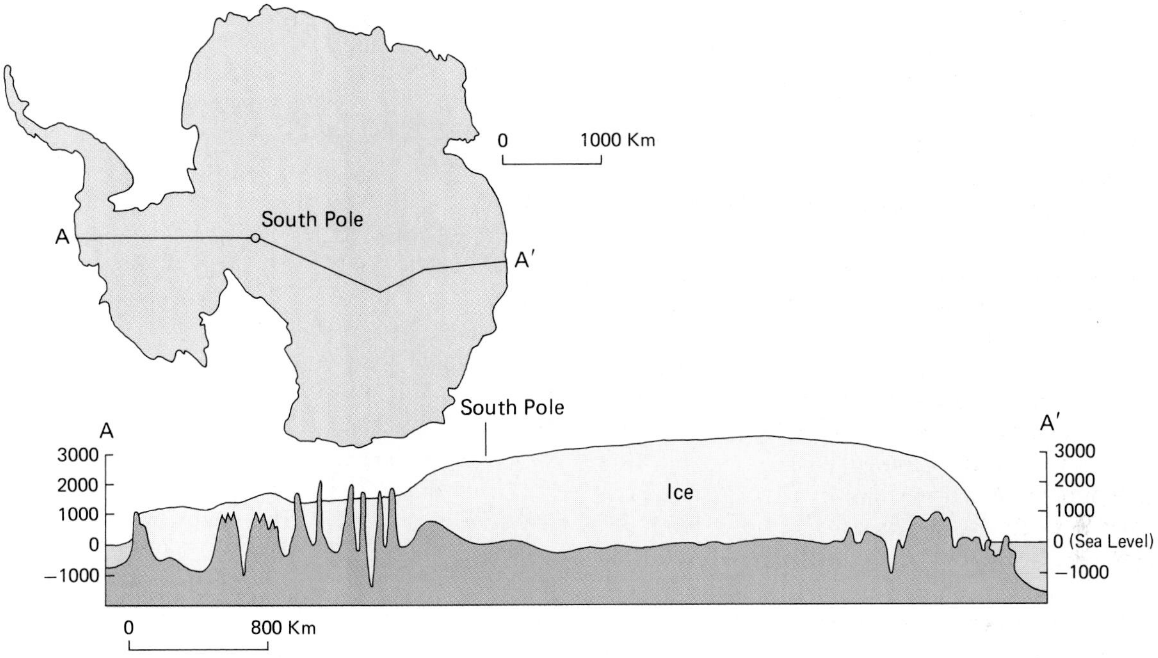

Figure 15–5 The Antarctic ice sheet in cross-section. Line of section A-A′ is indicated on index map. Vertical scale is exaggerated. (Adapted from Twidale, C.R., *Analysis of Landforms,* Sydney Australia and New York, John Wiley and Sons, Australia, Pty., Ltd., 1976).

Figure 15–6 Snowfield above several glaciers northeast of Sukkertøppen near the southwestern coast of Greenland. (Courtesy of M.S. Smith.)

Figure 15–7 Oblique aerial view of glaciers in the Fairweather Mountains of Alaska. Note the many medial flowlines. (Photograph by M.E. Dalechek, courtesy of the U.S. Geological Survey.)

mountain front as **piedmont glaciers**. Not uncommonly, several adjacent piedmont glaciers merge to form a compound piedmont glacier. Today, such glaciers are particularly numerous in Alaska. During the recent ice age, however, piedmont glaciers moved sluggishly across vast areas of the High Plains along the eastern margin of the Rocky Mountains.

Cold and Warm Glaciers

Overall form and relation to topography are not the only way glaciers are classified. Another scheme recognizes two categories of glaciers, according to their temperature. There are, for example, **polar** or **dry-base glaciers** in which the entire thickness of the ice is well below the pressure-melting temperature. The pressure-melting temperature is the temperature at which melting occurs under a given pressure. In general, when the pressure on ice is increased, the melting point of ice is decreased. We take advantage of this phenomenon when we squeeze snow into a snowball. The water melted in the snowball by applying pressure is refrozen when we release the pressure. In polar glaciers, the temperature of the ice even at the base is below the pressure-melting temperature. Therefore, melting does not occur, and the basal ice is usually frozen firmly to the rock beneath. Although some water has been detected beneath parts of the east Antarctica ice sheet, it is primarily a polar or "dry base" glacier. Even at great depth, the ice in Antarctica is colder than −23°C.

The second category in the thermal classification of glaciers can be termed **temperate** or **wet-base glaciers**. The ice in these glaciers is sufficiently close to the pressure-melting point that ice and water frequently coexist. Lubricated by the water, ice in temperate glaciers flows more readily, and erosional processes are enhanced. Many glaciers in the Pacific Northwest of North America, and the Alps are valley glaciers of the temperate kind.

It is sometimes impossible to designate a particular glacier as either polar dry-base or temperate wet-base. Therefore, the term **intermediate glacier** has been suggested for glaciers that alternate seasonally between the wet-base and dry-base condition or that vary in these characteristics in different parts of the ice mass.

Glacier Dynamics

The Glacial Balance

Glaciers that maintain their size and shape over a long period of time are said to be in balance. In such glaciers, the amount of ice that is added is balanced by the amount being lost through melting (ablation) and sublimation. In a valley glacier, the area in which there is a net gain in snow and ice usually is found in the upper reaches of the valley, in what is designated the **accumulation zone** (Fig. 15–8). Accumulation generally decreases toward the terminus (front) of the glacier until one passes into the area where there is a net loss of ice due primarily to melting. This lower end is called the **ablation zone**. From the snout, ablation decreases up-glacier. An imaginary line called the *line of equilibrium* separates the ablation and accumulation zones and itself represents that part of the glacier in which gains approximately balance losses. The line of equilibrium shifts along the glacier, depending on whether it is advancing or retreating. The terminus of a glacier will maintain itself at about the same position as long as the glacier is in equilibrium, that is, as long as net losses in the ablation zone are balanced by net gains in the accumulation zone. If, however, the glacier begins to acquire more ice than is lost through ablation, it will advance. Conversely, if the ablation rate exceeds the rate of accumulation, the glacier front will gradually recede. Even when a glacier is receding, however, the ice within it continues to move forward. Recession is caused by the terminus wasting away faster than the ice is advancing.

How Glaciers Move

We are accustomed to thinking of ice as rather brittle, unyielding material. When subjected to pressure of thick overlying layers of frozen waste, however, ice will deform or flow. In general, flow begins when the ice mass reaches a thickness of about 60 meters. No matter what kind of glacier, the prime mover that causes flow is the pull exerted on the ice by the force of gravity. As is the case for streams, valley glaciers are drawn down the slope of the valley floor by a component of gravity directed parallel to the slope. In contrast, ice sheets move forward largely because of the great weight near the centers of accumulation, which exerts pressure on the underlying ice, causing it to flow outward.

As suggested by the even pattern of ribbons of dark debris seen on valley glaciers (see Fig. 15–7), ice flows laminarly like slowly moving water. Because of friction with the bedrock floor, velocity of flow is least near the base of a valley glacier and increases from the base toward the top, except for

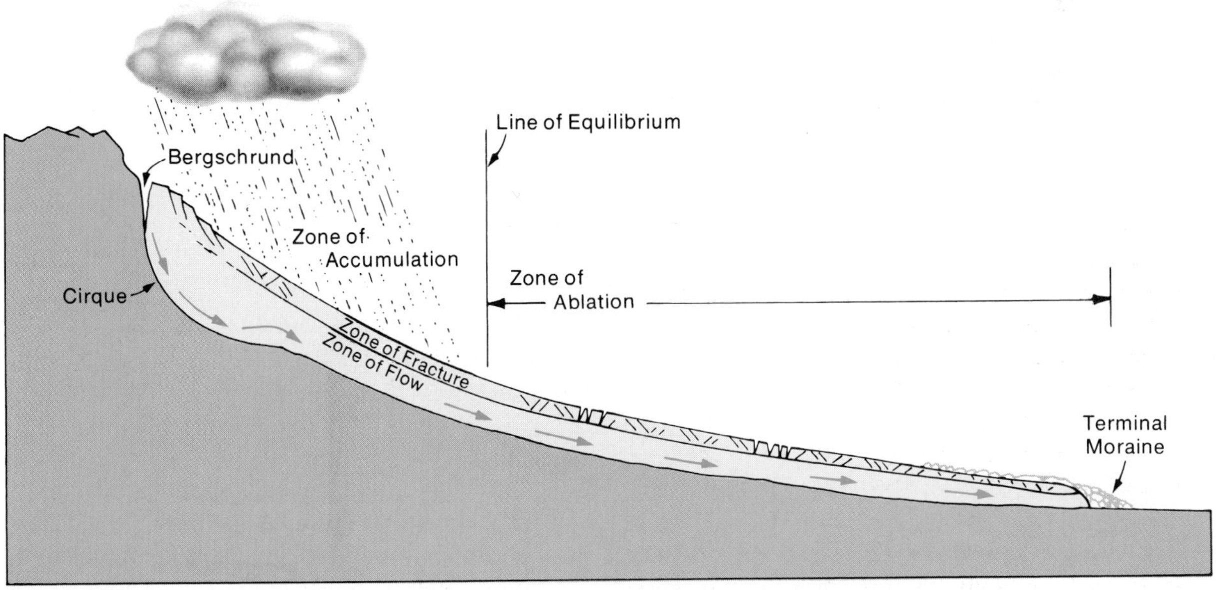

Figure 15–8 The *zones of accumulation* and *ablation,* and the *line of equilibrium* as ideally developed in a valley glacier.

the uppermost 60 meters. Ice in this top layer is under insufficient pressure to flow and is carried along as a brittle slab by the flowing ice below. This upper brittle portion of the glacier is designated the **zone of fracture** in order to distinguish it from the underlying **zone of flow**. As implied by its name, the zone of fracture is often characterized by chaotic fracturing and the development of deep crevasses. In plan view (Fig. 15–9), the center of a valley glacier moves along more rapidly than the flanks, where there is friction with the valley sides (Fig. 15–10). Agassiz and Charpentier were well aware of this phenomenon and had measured the surface velocity of the ice by placing stakes across the surface of a glacier and recording their progress from year to year. With the exception of sudden rapid movements called surges, the most rapid movement measured in alpine glaciers is about 75 meters per year.

The actual flowage of ice (see Fig. 15–10) in a glacier represents an adjustment to the pressures acting on the ice, and these pressures are in turn affected by the thickness of the mass and the slope of its surface. Meltwater trapped within the glacier and variations in temperature also influence how fast the ice moves.

Several mechanisms are involved in the movement of glaciers. What appears as plastic flowage is the result of mutual displacements of ice crystals relative to one another and displacements within crystals along sheets of atoms. Such displacements produce ice crystals that on microscopic examination appear to have been bent.

Other mechanisms for movement include granulation of one portion of the ice mass as it moves against another. Partial and temporary melting may also enhance the movement of the glacier, provided temperatures within the ice are not exceptionally low. For example, ice flowing across bedrock obstacles will experience higher pressure on the "upstream" side of the obstacle than on the "downstream" side. As a result, partial melting will occur on the "upstream" side and the meltwater will flow around the obstacle and freeze again on the "downstream" side. Elsewhere, water produced in this way (or by melting from other causes) may form thin layers of liquid that provide a low friction horizon along which overlying ice may slide at increased speed.

Not all of the movement in a glacier is the result of melt-freeze or plastic flow. Whenever stress on the ice mass exceeds its ability to respond with displacements within and between the crystals, rupture surfaces will develop in much the same way as do faults in rocks. Such rupturing and associated fracturing is mostly confined to the upper 30 to 60 meters of the glacier.

In addition to the movement of glaciers by actual flowage of the ice mass, glaciers progress by simply sliding along the underlying ground surface.

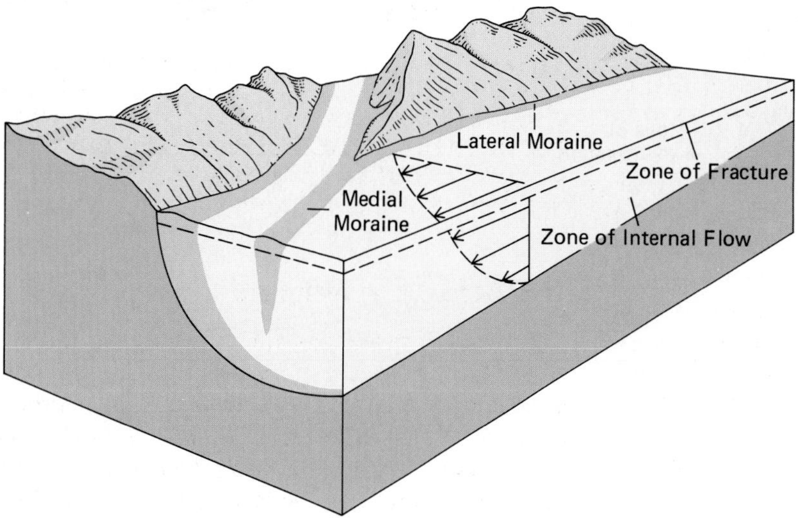

Figure 15–9 Forward motion in a glacier. The ice moves least rapidly near the contact with valley floor and sides because of friction.

Such movement is most prevalent in valley glaciers of temperate regions.

The internal flow of a glacier involves multiple displacements between and within ice crystals, granulation, and alternate partial melting and refreezing. External movement involves sliding of the ice on its bed.

The rates at which valley glaciers flow vary widely. Some move rapidly, others slowly, some erratically, and some with a constant rate throughout the year. A characteristic of many glaciers is a **surging behavior** in which a glacier having a slow and uniform flow periodically surges forward at higher velocity. Glaciers in the United States and the Himalayas are known to have sprinted forward on occasion at rates of over 300 meters per day. The movement is usually uneven and "jerky" and is accompanied by a shivering motion of the ground and loud cracking noises. Most surging in glaciers appears to be associated with the accumulation of excess water at the base of the ice mass. Such meltwater occurs in temperate glaciers whenever a thickening of the ice lowers the pressure-melting temperature sufficiently to produce a layer of water.

The Work of Glaciers

Because most of us live far away from regions undergoing active glaciation, we may fail to appreciate the awesome power of glaciers. Glacial sculpturing on a grand scale has greatly enhanced the beauty of the Rocky Mountains, the Alps, the Himalayas, and coastal regions of Alaska, Greenland, and Scandinavia. Across northern Europe and North America, glaciers of the past have scraped away topographic obstructions and molded hummocky lowlands. Vast amounts of fertile soils have been stripped from bedrock in some regions, whereas in others immense tonnages of rock debris have been spread across the land. The highly productive soils of the United States corn belt are partly composed of fertile soils transported into the region by continental glaciers.

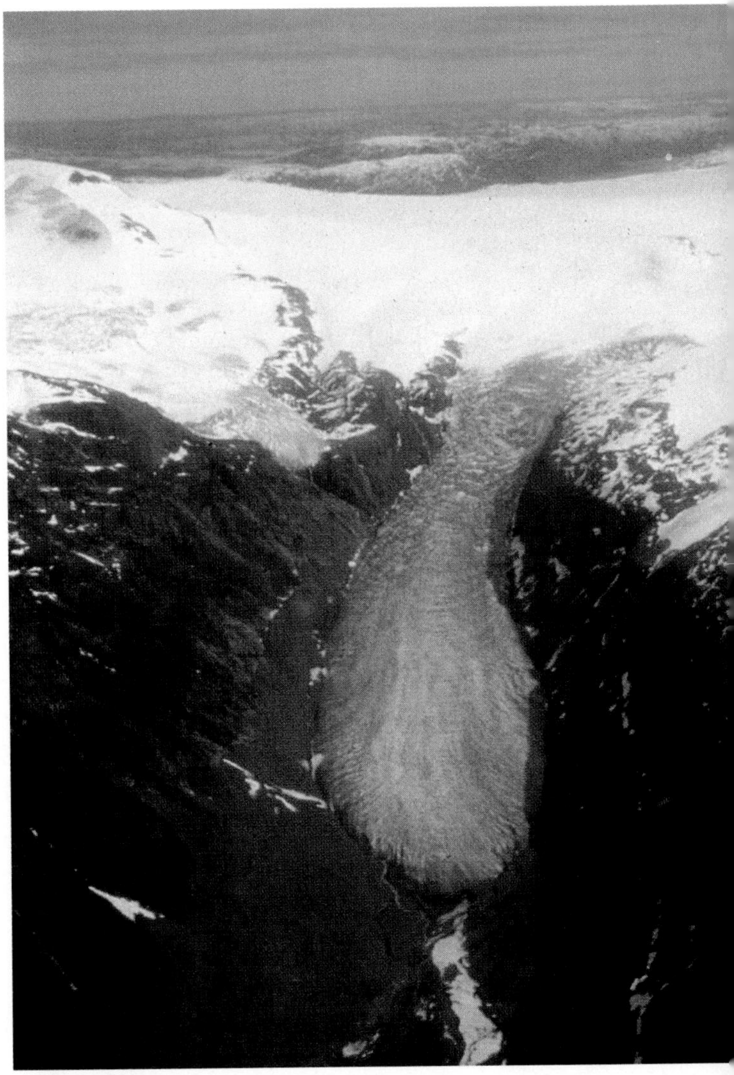

Figure 15–10 Glacier descending from a high-level snowfield. The lobate snout of the glacier results from a greater rate of flow at the center in comparison to the sides. The glacier depicted here is in southwestern Greenland between Sukkertøppen and Søndre Strømfjord. (Courtesy of M.S. Smith.)

Erosion by Glaciers

Erosion by glaciers is different from fluvial erosion. Flowing ice can carry more and larger rock tools with which to gouge and wear away bedrock than can rivers. Glaciers need not be moving as rapidly as water to accomplish their work, nor do they always flow over the lowest ground available.

The processes of glacial erosion include *abrasion*, *quarrying*, and *avalanching*. These processes vary greatly among glaciers, and a process that is highly effective in valley glaciers may be of less importance in an ice sheet.

Abrasion

Abrasion refers to the wearing, grinding, and rubbing away of solid rock by moving glacial ice armed with angular rock fragments. The debris within the ice acts somewhat like the teeth of a rasp in producing scratches and grooves in the rock over which the glacier moves. The scratches are termed **glacial striations** (Fig. 15–11). Boulders impinging on bedrock as the glacier inches forward may crack the bedrock into curved fractures called **chatter marks** (Fig. 15–12). The concave sides of these marks face the direction toward which the ice flows. Geologists attempting to ascertain movements of former glaciers are delighted when they find well-preserved glacial striations and chatter marks. In addition to scratches produced by rock fragments in the ice, pulverized rock (rock flour) along the sides and bottom of a glacier acts as a fine abrasive in producing polished bedrock surfaces. Regardless of whether bedrock is being scratched, gouged, or polished, the abrasion process is most effective if the glacier is actively moving and also if there is sufficient melting to continuously expose new rock tools along the sole of the glacier.

Figure 15–11 Glacial grooves and striations, Isle Royale National Park, Keweenaw County, Michigan. (Courtesy of the U.S. Geological Survey and N.K. Huber.)

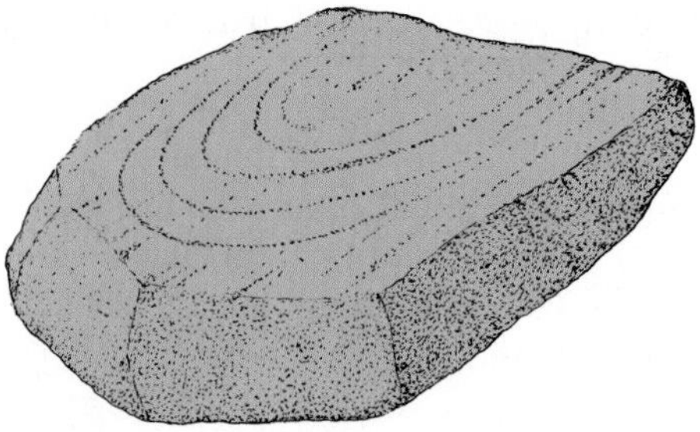

Figure 15–12 Chatter marks. The glacier moved from the lower left to the upper right. Long dimension of rock is 40 cm.

Glacial Quarrying

Glacial quarrying (glacial plucking) is the process by which a glacier extracts and carries away pieces of the bedrock over which it is moving. Quarrying is accomplished in several ways. The weight of a huge mass of ice pushing against an obstacle may itself be sufficient to dislodge rock masses provided the pressure required does not exceed that at which ice deforms. Also, the passing of the glacier over bedrock results in a pulling or tensional force on the ground surface, which may open incipient fractures and facilitate lifting pieces of rock into the mass of moving ice. Meltwater beneath the glacier seeps into fractures in the underlying bedrock and may then freeze solidly. The resulting expansion opens the fractures more widely and may cause new breaks that free chunks of bedrock from the parent mass. As movement continues, these blocks of rock, frozen solidly into the ice, are pulled out and carried away.

Avalanching

As used by glaciologists, the term **avalanching** is the sudden and rapid movement of snow and rock debris down mountain slopes and onto the surface of a valley glacier. The avalanching process is enhanced by the fact that abrasion and quarrying occur actively on the sides as well as the bottom of the glacier filling a valley. As a result, the valley sides are undercut and this leads to loss of support for loose material lying on the valley slopes. Inevitably, the unconsolidated debris slides or falls onto the ice surface and is then moved along as if on a conveyer belt. From a distance, the debris appears in the form of graceful dark bands, which parallel the valley sides (Fig. 15–13).

The work of glaciers includes erosion (by quarrying and abrasion), transport of the products of erosion and avalanching, and deposition of those products.

Erosional Features

Anyone who has visited mountains in Alaska (Fig. 15–14), the higher ranges of the Sierra Nevada, the Rocky Mountains, or the Alps is certain to be impressed by their rugged grandeur. Indeed, the effects of glacial erosion are most spectacular in mountains. On every side one finds magnificent U-shaped valleys, elegant waterfalls cascading from hanging valleys, truncated spurs, cirques, arêtes and cols, and placid tarns.

Valley glaciers, of course, occupy valleys eroded by streams. In mountainous regions, stream valleys characteristically have a V-shape cross-section caused by the fact that the stream itself can only cut within its channel and depends on mass wasting to establish the slope of the valley sides. In the glaciers that occupy those valleys, however, ice is in contact with, and can erode, rock at the base of the valley and well up onto the valley sides. As a result, glaciers sculpt a valley shape that has a parabolic cross-section and impressively steep walls (Fig. 15–15).

Glaciers not only change the cross-sectional appearance of the valleys they occupy but also straighten the valley by wearing away the bends. The result is a straight, fast, and direct route of escape for the ice. The movement of ice down the valley truncates divides that once existed between stream bends, so that steep cliffs or **truncated spurs** occur between tributary valleys.

Another distinctive feature of glaciated mountainous regions are **hanging valleys** (Fig. 15–16). They are the result of the fact that tributary glaciers erode at a slower rate than does the main

Figure 15–13 Medial flow bands of rock debris and fracturing of the ice surface on a glacier in the northwestern corner of British Columbia, Canada. Tulsequah Lake at the center receives water from the glacier and overflows catastrophically in late summer. (Courtesy of the U.S. Geological Survey and A. Post.)

Figure 15–14 The breeding grounds of glaciers. Angular, rugged mountains of a glaciated island on the east side of the Kenai Peninsula, Alaska. The coastline of the island shows evidence of about two meters of uplift that occurred during the Alaskan earthquake of 1964. (Courtesy of J. Brice.)

glacier. As a result, the floors of the tributary valleys meet the main valley high above the main valley's floor. After the glaciers have melted away, this difference in elevation between the tributary and main valley floor is dramatically exposed as a precipice over which waterfalls and cataracts may plummet (Fig. 15–17). Bridal Veil Falls in Yosemite National Park is an example.

Valley glaciers change stream valleys into less sinuous, U-shaped troughs with hanging tributary valleys.

At the head of a glaciated valley, one frequently finds a steep-walled recess, shaped like a half bowl and excavated primarily by glacial quarrying. These features are called **cirques** (Fig. 15–18). The walls of cirques may drop vertically for over 40 meters, and their floors are often gouged to a depth lower than the adjacent valley floor. A small lake called a **tarn** (Fig. 15–19) may occupy the depression within the floor of the cirque after the disappearance of ice it once contained. As the ice moves out of the cirque in association with the downslope movement of the entire glacier, an arcuate crevasse called a **bergschrund** develops next to the

Figure 15–15 Glaciated valley named Little Cottonwood Canyon, southeast of Salt Lake City, Utah.

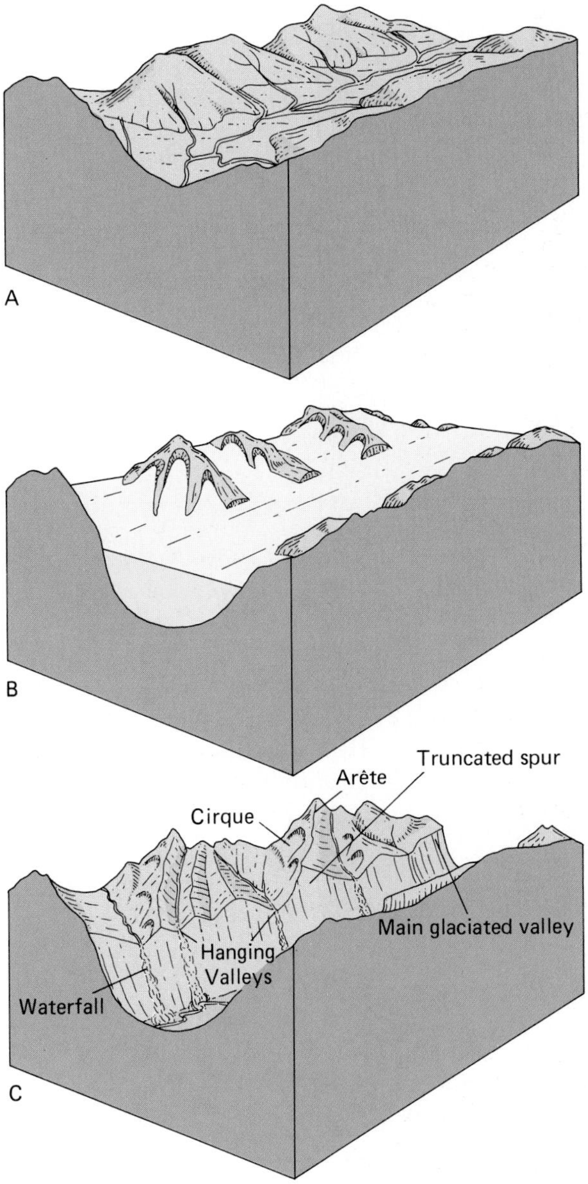

Figure 15–16 Landforms produced by glacial erosion in mountainous regions. (A) Before glaciation. (B) During glaciation. (C) After glaciation.

back wall. Meltwater flows into the bergschrund and freezes in bedrock fractures, thereby anchoring the ice to the rock. When the ice moves again, large pieces of rock are plucked from the mountainside, thus deepening and further sculpting the cirque.

Characteristically in glaciated mountainous regions one finds places where adjacent valley glaciers have sharply reduced the width of divides or where cirques on either side of a mountain are gradually eroded toward one another until only a knife-edge divide remains. Such sharp-edged precipitous ridges are known as **arêtes** (Fig. 15–20). Where three or more cirques come together, the spectacularly steep pyramidal peaks called **horns** are sculptured (see the photograph of Shivling Peak that opens this chapter). There are many horns in the Swiss Alps, but by far the most famous is the Matterhorn south of Zermatt (Fig. 15–21). Less spectacular features found on the floors of glaciated valleys are polished knobs of rock formed when the ice rode over a bedrock protuberance. In moving across the obstacle, ice smoothed the upstream side and roughened the lee side by quarrying the jointed rock (Fig. 15–22). These features are called **roches moutonnées** (rocks shaped like sheep).

In any catalogue of inspirational landforms, **fjords** (Fig. 15–23) would certainly rival horns in scenic splendor. Fjords are long, deep, steep-sided glacial valleys into which the sea extends following the disappearance of the glacier. They are sculptured by thick glaciers with steeply sloping surfaces, which flow down from the mountains into the sea and are able to erode their valleys to a depth below sea level. Today, the water in some fjords is even deeper than when they were filled with ice, because of the rise in sea level following the last ice age recession. In many of the famous Norwegian fjords, water depths exceed 1200 meters.

Erosional features produced by ice sheets are less spectacular than those caused by mountain glaciers. Ice sheets move bodily over entire ranges (see Fig. 15–5), and tend to smooth and bevel topographic irregularities. They do however, produce impressive glacial lineations and grooves in very dense igneous and metamorphic rocks. Grooves have been found in areas covered by the Pleistocene ice sheets that are over 2 meters deep and extend for distances of more than 100 meters. Such features provide striking evidence of the direction and abrasive powers of ice sheets.

Another way in which ice sheets erosionally modify landscapes is by scouring and deepening

Figure 15–17 Waterfall plunging from stream in hanging valley, Sogne Fjord, Norway. (Courtesy of the Norwegian Information Service.)

Figure 15–18 Large cirque (right of center) in the Cathedral Peaks of the Teton Range of northwestern Wyoming. (Courtesy of E. Moldovanyi.)

Figure 15–19 A tarn occupying the ice-gouged floor of a cirque in the San Juan Mountains of Colorado. (Courtesy of P. Carrara and the U.S. Geological Survey.)

topographic depressions. Preglacial stream valleys oriented parallel to the direction of the ice movement are particularly susceptible to this erosion. In such valleys, there are fewer barriers to movement, and the ice itself is thicker and hence more effective in erosion. After the ice has receded, the elongate basins fill with water to form picturesque lakes, notable examples of which are the Finger Lakes of New York.

Transportation by Glaciers

The work potential of glaciers differs from that of streams in that there is practically no limit to the size or amount of material that can be carried by a glacier as suspended load. The competency and capacity of glaciers are truly enormous. Indeed, parts of some glaciers contain more rock debris

Figure 15–20 *Arêtes* near Snowdon Summit, Carnavaron, northern Wales. (Courtesy of the Institute of Geological Sciences, London.)

Figure 15–21 The Matterhorn in the Swiss Alps. This magnificent peak is a classic example of a horn formed by glacial erosion. It is an erosional remnant of a much larger thrust sheet that originated in Italy and overrode Switzerland when Africa collided with southern Europe about 20 million years ago.

Figure 15–22 *Roches moutonnées,* Black Bay, Lake Athabasca, Saskatchewan, Canada. Direction of ice movement was from right to left. (Courtesy of Geological Survey of Canada.)

than ice itself. In general, the size of rock fragments carried by a glacier diminishes with increased distance of transport. This condition is a consequence of the crushing and wearing away of rock debris as the ice advances. Nevertheless, chunks of particularly durable rock like quartzite and granite may be transported for hundreds of kilometers without being destroyed. As an example, erratics whose origin was Ontario, Canada, were carried by Ice Age glaciers as much as 1000 km into Missouri, and boulders from Finland are now scattered across the German countryside.

Valley glaciers and ice sheets differ somewhat in the distribution of the load of rock debris they transport. Ice sheets, for example, carry very little load on the surface of the ice. Their surface is above the majority of mountain tops, and so landslides do not provide a surface load. In general, ice sheets also carry a smaller volume of debris per unit volume of ice than do valley glaciers, and that load is carried primarily near the bottom of the sheet or pushed along at the front.

In the case of valley glaciers, avalanching and mass wasting contribute to the accumulation of a significant amount of debris on top of the ice. The larger blocks of rock slowly work their way downward through the ice so that the load becomes generally coarser near the base of the glacier. Debris

Figure 15–23 Norway boasts the most beautiful fjords in the world. Sogne Fjord, north of Bergen, is a spectacular example. Note the hanging valley and waterfall in the distance. (Courtesy of Ian Duncan.)

from previous years may appear as superimposed layers in downstream cross-sections of the glacier. Rock on the ice that does not work its way downward is transported along the upper brittle layer in conveyer-belt fashion and is eventually dumped at the snout of the glacier.

Deposition by Glaciers

Glacial Sediment

Glaciers drop their load of coarse and fine sediment either directly from the moving ice mass or indirectly through meltwater. As one might expect, sediment deposited by streams of meltwater is likely to be better sorted than debris dropped directly from glacial ice. The general term for all deposits resulting from glaciers is **drift**. The term originated in the decade before the glacial theory was announced by Agassiz and alludes to the erroneous idea that all glacial sediment "drifted" to present locations within floating ice or during surging biblical floods. As used today, the term embraces all debris carried and deposited by glaciers, glacial meltwater, or even icebergs derived from glaciers.

Because of differences in origin, it is useful to distinguish two kinds of glacial deposits, namely stratified drift and till. **Till** (Figure 15–24) is poorly sorted rock and mineral fragments in a range of sizes and without a discernible arrangement of those particles. There is no trace of the kind of winnowing and sorting of particle sizes that characterizes sediments deposited by water, and therefore till is believed to have been dropped directly from a glacier without appreciable reworking by meltwater.

Till is a glacial deposit that has not been stratified or sorted by meltwater.

Stratified drift is glacial sediment deposited and reworked by water or by wind. Stratified drift deposited by meltwater streams is better sorted than till and often exhibits distinct sand and gravel layers. Cross-bedding may also be developed.

Stratified drift also accumulates in lakes. Many such deposits have distinctive lamination. Sediment accumulating during the warmer summer months when water is stirred by waves forms a layer that is somewhat lighter and coarser than a darker, thinner layer deposited during the winter when the lake is frozen over. The regularly repeated pairs of layers or **varves** record the years during which a particular glacial lake existed (Fig. 15–25). Each pair of light and dark bands or layers represents one year's accumulation. Counting the number of varves tells us how long the lake was there.

Figure 15–24 Glacial till in moraine of a valley glacier, Kenai Peninsula, Alaska. (Courtesy of the U.S. Geological Survey.)

The final category of stratified drift is deposited by wind. **Loess**, which is wind-blown dust, is the most common example. Loess is mostly associated with glaciation, and in many areas it appears to have been derived from the floodplains of meltwater-swollen streams.

Depositional Features Composed Primarily of Till

Moraine is probably the first term that comes to mind in a list of the depositional features of glaciers. It is a term used by eighteenth-century Swiss peasants for ridges of rock debris found near the margins of glaciers. Today, geologists define moraines as accumulations of till deposited directly by the action of glacial ice. The more easily recog-

Figure 15–25 Varved clays from glacial Lake Barlow, Ojibway, Ontario. The pocket knife at left center indicates scale. Contorted beds at base resulted from subaqueous slumping and flow. (Courtesy of the Geological Survey of Canada.)

nizable types are ground moraines, terminal moraines, and lateral moraines.

Ground moraine (Fig. 15–26) consists of a broad blanket of till in which there are not discernible ridges of glacial clastics. The material comprising ground moraine is deposited largely from the voluminous detritus carried near the bottom of the glacier, as well as sediment that has melted downward during ablation. Ground moraine may completely cover earlier topographic features and typically takes the form of broad undulating plains. Vast areas of the northern U.S. are mantled by the ground moraine of Ice Age continental glaciers.

Terminal moraines (Fig. 15–27) are accumulations of till along the lobate front of the glacier. They are produced by the conveyor-belt action of the moving ice, which brings rock fragments to the snout of the glacier even when it is in a state of equilibrium and not actively advancing. When the ice does progress forward, however, the glacier may push terminal morainal deposits forward in much the same way as a bulldozer might. If the amount of flow and melting in a glacier are in balance, the snout of the glacier will not advance, and the end moraine may become irregular and hummocky. Glaciers that retreat slowly and in increments may leave behind a succession of distinctive cresent-shaped terminal deposits called **recessional moraines** (Fig. 15–28). By contrast, when glaciers recede continuously, they deposit a layer of debris that is indistinguishable from ground moraines.

Rock debris that has fallen from mountainsides onto the margins of valley glaciers is carried along and eventually deposited as ribbons of till called **lateral moraines** (see Fig. 15–9). Long after glaciers have disappeared, these lateral moraines (Fig. 15–29) remain as ridges along margins of valleys. When glaciers from two adjoining valleys come together, neighboring lateral moraines also coalesce and form a **medial moraine** (Fig. 15–30).

Other features usually formed in till, and particularly characteristic of regions once covered by continental glaciers are drumlins. **Drumlins** are low, elliptic, egg-shaped hills resembling inverted teaspoons and having their long axes oriented parallel to the direction of glacial movement (Fig. 15–31). They are rarely more than about 50 meters in height, generally occur in large numbers within a given area, and have steeper slopes on the sides from which the glaciers came. They are left by a receding glacier and moulded by a subsequent one. Because of its role in American history, Bunker Hill in Boston is probably the most famous drumlin in the United States.

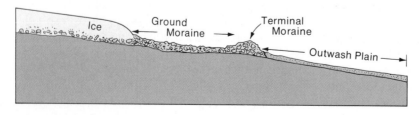

Figure 15–26 Terminal moraine, ground moraine, and outwash plain.

Figure 15–27 Terminal moraine at the terminus of Nisqually Glacier, south side of Mt. Ranier, Mt. Ranier National Park, Washington. A lateral moraine is also visible on the left side of the photograph. (Courtesy R.R. Crandell and the U.S. Geological Survey.)

Depositional Features Composed Primarily of Stratified Drift

Beyond the terminus of the glacier, one finds an apron of rock fragments, gravel, sand, and silty materials. This irregular blanket of debris constitutes the **outwash plain**. Its constituent particles are transported outward from the glacier by streams of meltwater, which are so choked with sediment that they frequently assume a braided pattern. In the outwash plain, material previously deposited as ground moraines may be reworked into deposits of stratified drift.

In Canada and Maine, one may sometimes encounter long sinuous ribbons of sand and gravel and wonder why someone went to all the trouble of preparing a raised railroad bed without ever adding the rails. These features, however, are definitely not railroad beds but rather winding ridges of stratified drift called **eskers**. Eskers mark the locations of former subglacial streams that flowed through the ice near the margins of the glacier.

Figure 15–28 Recessional moraines in the mid-ground of photograph were deposited by a small glacier occupying Bear Creek in the Yukon region of Alaska. Lingering snow patches in the furrows between crescentic ridges have inhibited lichen growth and created the light-colored swales. (Courtesy of the U.S. Geological Survey.)

A

B

Figure 15–29 Lateral moraines. (A) Athabasca Glacier, Jasper National Park, Canada. The terminus of the glacier is visible in the center of the photograph. Extending outward to the right is a lateral moraine deposited when the glacier extended farther down the valley. Note deposits of till in the foreground. (B) Short lateral moraines flanking a steeply descending glacier that is entering a fjord, southwestern coast of Greenland. (Photograph B courtesy of M.S. Smith.)

Whenever the velocity of such a stream was checked because of loss of water, rerouting, obstacles, or lack of sufficient gradient, the streams dropped their load of sediment within their tunnels. After the glacier had wasted away, there remained a winding ridge of glaciofluvial material (Fig. 15–32). Some eskers extend for over 10 km and reach heights of 15 to 30 meters.

The upper surface of a glacier is likely to hold numerous random depressions into which surface meltwater may wash sand and gravel. Also, along the margins of the glacier, small alluvial fans may be constructed out onto the surface of the ice. In either situation, when the ice melts, the accumulated sediment is lowered to the ground to form the irregular mounds called **kames** (Fig. 15–33).

Figure 15–30 Two lateral moraines coalesce to form the medial moraine of the Kaskawulsh Glacier, Kluane National Park, Yukon. (Courtesy of the Geological Survey of Canada.)

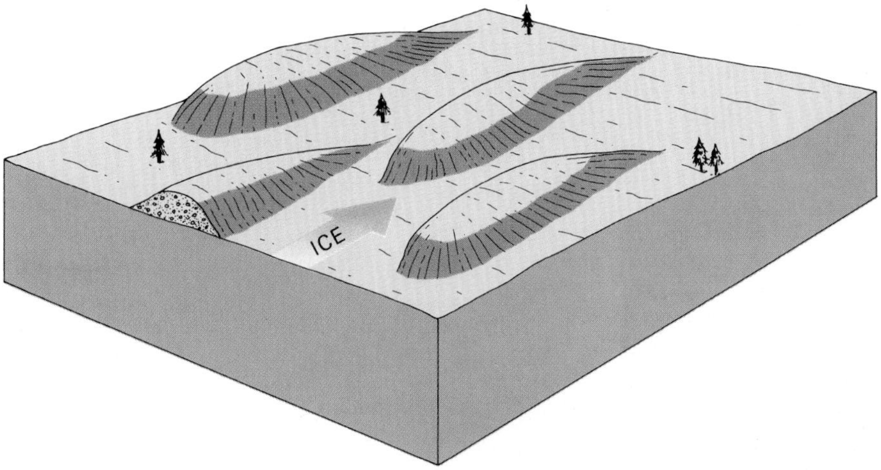

Figure 15–31 Drumlins. The steeper end is the ice-approach side. These drumlins are about 600 meters long.

Figure 15–32 An esker, resembling an abandoned railroad embankment, winds across part of the Snake River valley, Northwest Territories, Canada. (Courtesy of the Geological Survey of Canada.)

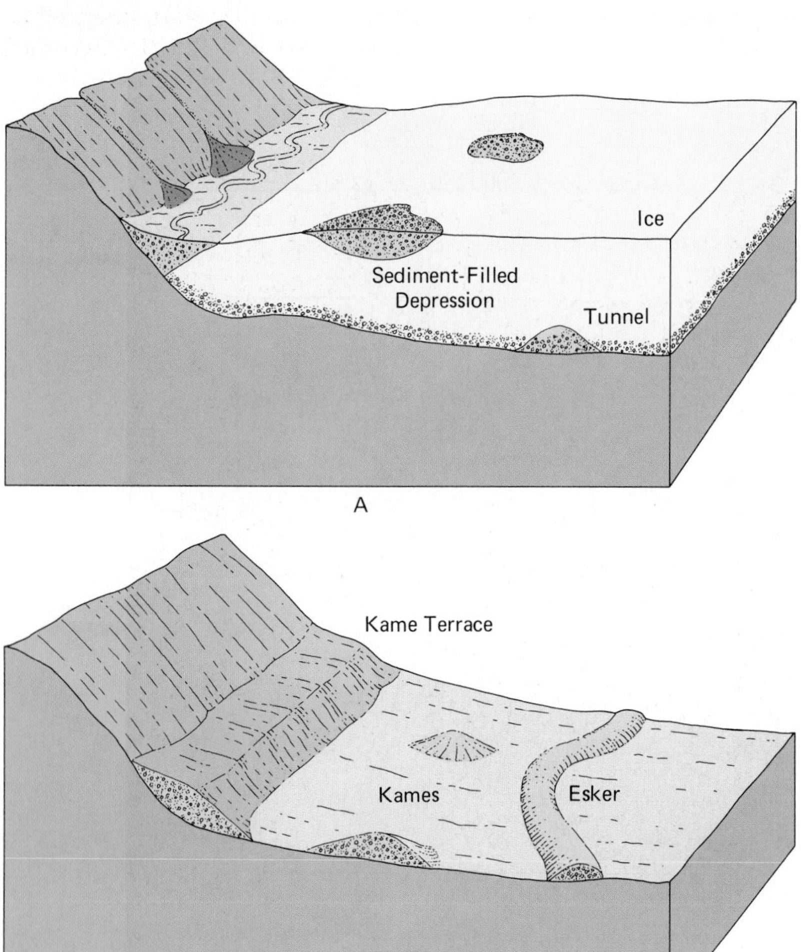

Figure 15–33 The formation of kames and eskers.

Kames may also form near the front of the glacier as sediment from meltwater is rapidly deposited.

In valley glaciers, streams of meltwater may develop between a valley side and the lateral margin of the glacier. Such streams also deposit sand and gravel, which when lowered to the bedrock floor after the glaciers have wasted away form raised embankments called **kame terraces** (see Fig. 15–33).

One kind of depositional feature that is not built up above the general ground level, but rather forms a depression, is called a **kettle** (Fig. 15–34). One can readily imagine that the load of debris carried by a glacier is not likely to be uniformly distributed throughout the ice. Here and there, a large mass of ice may exist that contains relatively less debris than adjoining areas. When the glacier melts away (Fig. 15–35), there will be less material lowered to the ground at the site of the lesser load. The result will be a small basin or kettle, which may subsequently be filled with water to form a **kettle lake**. Kettles may also develop from the melting of large blocks of ice that were buried in till. Where excessive rock material existed in scattered parts of the glacier, not kettles but **knobs** (Fig. 15–36) may form upon melting, thereby giving rise to kettle-and-knob topography.

The Pleistocene Ice Age

The development of vast continental ice sheets has occurred several times in the Precambrian and Paleozoic, but those ice sheets of the **Pleistocene Epoch** are the most recent and hence have the clearest records. You will recall from the discussion of the geologic time scale in Chapter 8 that the Pleistocene Epoch is the older of the two final

Figure 15–34 The ponds shown in this photograph are in depressions called kettles, made by the wasting away of a mass of glacial ice that had been buried in glacial debris. (Courtesy of the U.S. Geological Survey.)

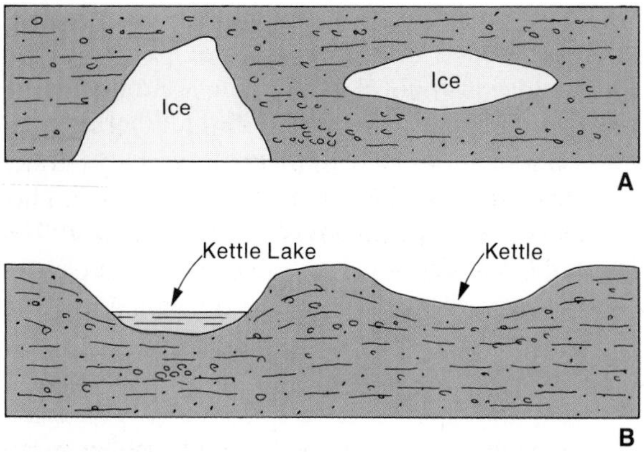

Figure 15–35 Formation of kettles. (A) Cross-section of drift containing buried masses of ice. (B) When the ice melts, depressions are formed at the surface by the lost volume of ice.

epochs of the Cenozoic Era. The Pleistocene is believed to have begun about 1.5 million years ago and ended 8000 or 10,000 years ago with the onset of the Holocene or Recent Epoch.

When the continental glaciers of the Pleistocene began their advance, they ultimately covered about 30 per cent of the earth's land area with over 40 million cubic miles of snow and ice. In North America alone, the ice sheet extended nearly 6400 km across the entire breadth of Canada (Figure 15–37). Such an extensive coverage of ice had profound effects, not only on the glaciated terrains themselves but also on lands far from the ice fronts. Climatic zones in the Northern Hemisphere were shifted southward, and arctic conditions prevailed across northern Europe and the United States. Mountains and uplands in Eurasia and North America were sculptured by spectacular mountain glaciers. While the snow and ice was accumulating in higher latitudes, rainfall increased in the lower latitudes, with generally beneficial effects on plant and animal life. Even as recently as 10,000 years ago, presently arid regions in north and eastern Africa were well watered, fertile, and populated by nomadic tribes.

Huge ice sheets covered North America and Europe during the multiple glaciations of the Pleistocene Epoch.

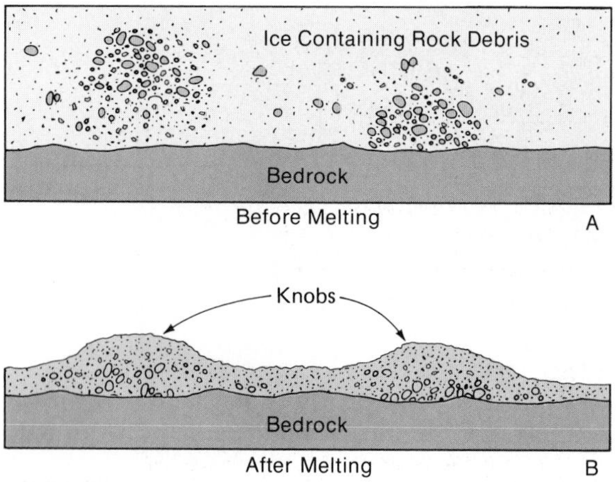

Figure 15–36 Formation of knobs by melting of ice in which there are local concentrations of debris.

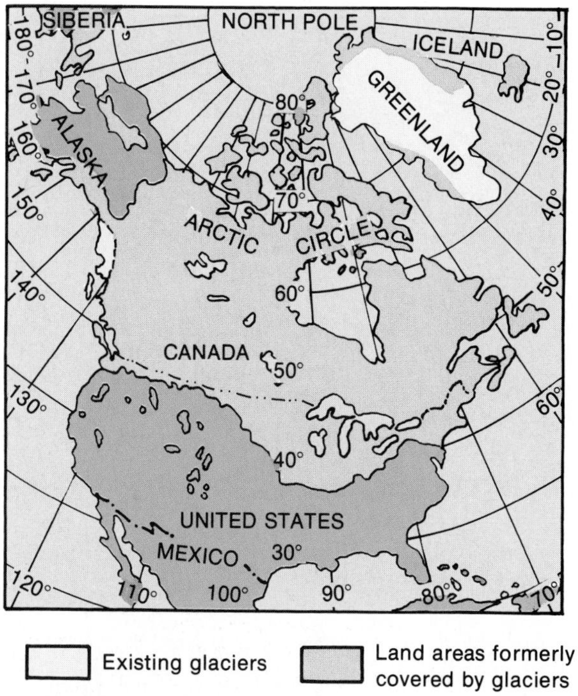

Figure 15–37 Maximal area of glacial coverage in North America. (From the U.S. Geological Survey Pamphlet, *The Great Ice Age,* U.S. Government Printing Office, 0-357-128, 1969.)

Pleistocene Sediments

Geologists studying the drift deposits of the United States and Europe have determined that during the Pleistocene Epoch there were at least four major advances of the ice separated by three warmer interglacial intervals. The deposits of each of these time intervals are called a stage, and each glacial and interglacial stage is named after a locality where deposits of that age are particularly well exposed for study (Fig. 15–38).

Terrestrial Deposits

Because they are characteristically unfossiliferous and generally similar-looking mixtures of coarse clastics, it is often difficult to say that a given blanket of drift is the equivalent of another at a different location. Yet such correlations are essential if geologists are to determine the extent and nature of the various glacial advances. Usually, a single criterion for correlation will not suffice, and several lines of evidence, including superposition, must be employed. One method is to examine closely the extent to which a sheet of drift has been dissected by streams. Older layers of drift, if they have been exposed since the withdrawal of the glacier, are likely to have a greater number of streams per unit area than younger layers. Of course, climate and the amount of precipitation also influence stream density and must be taken into account in the evaluation of this method.

The depth of oxidation and amount of chemical weathering in blankets of drift may also provide criteria useful in evaluating the relative ages of

NORTH AMERICA	ALPINE REGION	YEARS BEFORE PRESENT
		—10,000
WISCONSIN	Würm	
		—75,000
Sangamon	Riss-Würm	
		—125,000
ILLINOIAN	Riss	
		—265,000
Yarmouth	Mindel-Riss	
		—300,000
KANSAN	Mindel	
		—435,000
Aftonian	Günz-Mindel	
		—500,000
NEBRASKAN	Günz	
		—1,800,000
Pre-Nebraskan	Pre-Gunz	

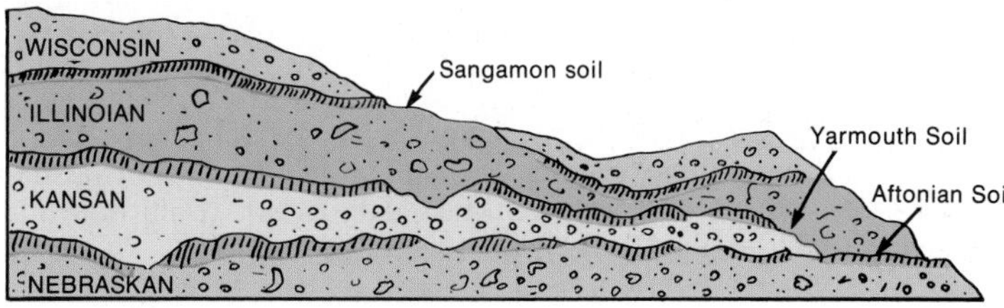

Figure 15–38
(A) Standard Pleistocene nomenclature for glacial and interglacial stages. (B) An idealized cross-section showing a succession of deposits of glacial stages and interglacially developed soils. (A section this complete would be a rare occurrence.)

deposits. Careful examination of fossil pollen grains in thick exposures of bog or lake deposits often clearly reflect fluctuations of climate and can be used to mark times of glacial advance and retreat. Varved clays deposited in lakes near the glaciers can sometimes be correlated to similar sediments of other lakes and in addition may provide an estimate of the time required for deposition of the entire thickness of lake deposits. However, the most accurate means of dating and correlating Pleistocene sediments is to extract pieces of wood, bone, or peat and determine their age by radiocarbon dating techniques. Unfortunately, even this method has its limitations, the most significant of which is the relatively short half-life (5570 years) of carbon 14. The method can be used only on material less than about 100,000 years old and thus is valid only for deposits of the Sangamon and Wisconsin glacial stages (see Fig. 15–38).

Deep-Sea Sediments

Deep-sea sediments can be very useful in Pleistocene geochronology. The deep ocean basins are sites of a relatively continuous sedimentary record; the column of sediments contain abundant fossil remains; and the deposits can be dated by relating them to paleomagnetic data and to radiometric isotopes having short half-lives.

Continuous sections of deep-sea sediments are obtained by means of piston-coring devices. These tools provide cores over 15 meters long. The cores are then subjected to various kinds of analyses. One approach is to use the oxygen isotope ratios of calcareous foraminifers within the cored sediment to determine the approximate amount of ocean water stored globally as glacial ice. That calculation would then provide an estimate of paleotemperatures. Because of its greater mass, less oxygen 18 is evaporated and precipitated as snow. Hence, during an ice age when ice is accumulating globally, there will be relatively greater percentages of oxygen 18 relative to oxygen 16 in sea water and in the shells of marine invertebrates. Plotted against depth in a deep-sea core, those isotopic ratios indicate variation in ice volume and temperature with time (Fig. 15–39). Indications of cooler conditions can then also be correlated with declines in worldwide sea level.

Another way to use deep-sea sediments in Pleistocene chronology is to plot at each level in the core the relative abundance of species of foraminifera that are known to be especially sensitive to temperature. For example, the tropical species

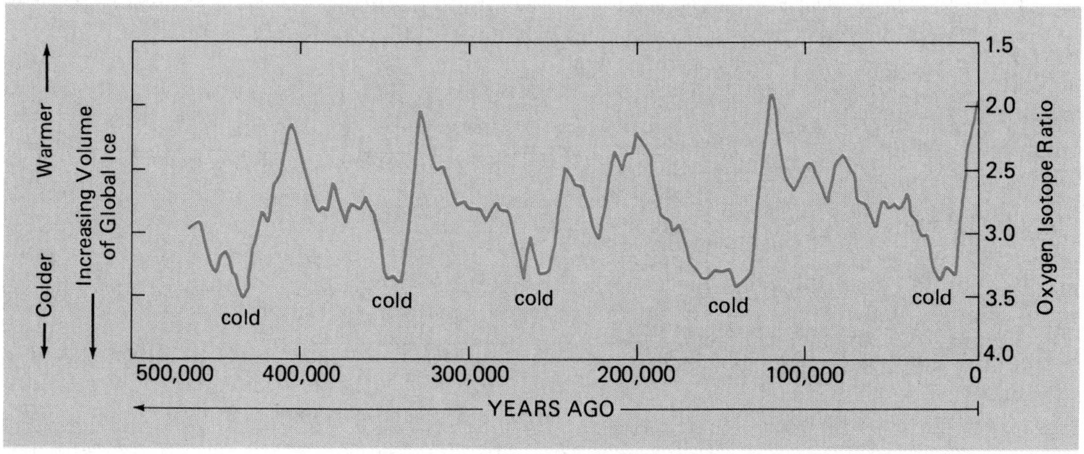

Figure 15–39 Curve reflecting variations in the global volume of ice (and, indirectly, paleotemperatures) during the past 500,000 years. Data from radiometric dating and isotope measurements of cores from the Indian Ocean. (Data from J.D. Hays and N.J. Shackleton. *Science* 194:1121–1132, 1976.)

Globorotalia menardii (Fig. 15–40) is alternately present or absent within Pleistocene cores from the equatorial Atlantic. The absence of the species in part of a core is taken to indicate an episode of cooler climates and glaciation. Carbon 14 dates in the upper part of such cores permit one to determine the rates of sedimentation, which is assumed to be uniform for the lower portions. Another foraminifera, *Globorotalia truncatulinoides*, also permits one to recognize alternate cooling and warming of the oceans. This species coils in a spiral with all the whorls visible on one side but only the last whorl visible on the other side. Individuals that coil to the right dominate in warmer water, whereas left-coiled individuals prefer colder water. During glacial advances, when ocean temperatures declined, the right-coiled populations of *G. truncatulinoides* in middle and low latitudes were replaced by populations in which left-coiling predominated. The record of such changes is clearly apparent in the deep-sea cores.

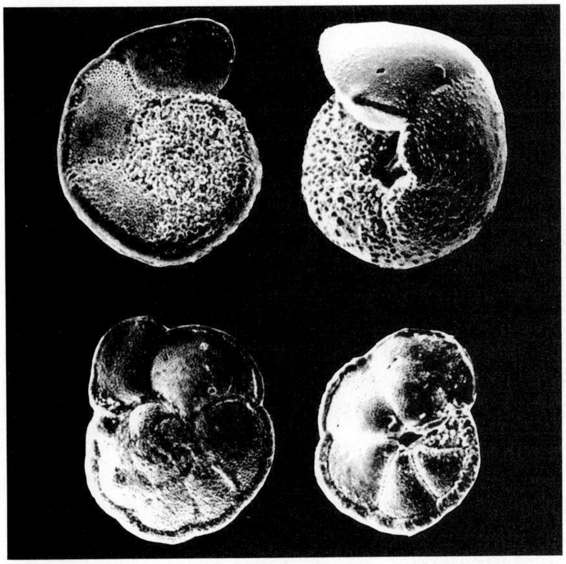

Figure 15–40 Two well-known species of planktonic foraminifers used in correlating deep-sea sediments. At the top are views of both sides of *Globorotalia truncatulinoides.* Below are views of opposite sides of *Globorotalia menardii.* (Magnification of all specimens ×50.)

The Effects of Pleistocene Glaciations

Earlier it was noted that during maximal glacial coverage, over 40 million km^3 of ice and snow lay on the continents, equivalent to a tremendous amount of water. Removal of that water from the oceans had a multitude of effects on the environment. It has been estimated that sea level may have dropped at least 75 meters during maximal ice coverage. Extensive tracts of the present continental shelves became dry land and were covered with forests and grasslands. The British Isles were joined to Europe, and a land bridge stretched from Siberia to Alaska. During interglacial stages, marine waters returned to the low coastal areas, drowning the flora and forcing terrestrial animals inland.

The glaciers had a direct impact on the erosion of lands and the creation of glacial landforms. The great weight of the ice depressed the crust of the earth over large parts of the glaciated area, in some places to a level of 200 to 300 meters below the preglacial position. With the removal of the last ice sheet, the down-warped areas of the crust gradually began to return to their former positions. This rebound is dramatically apparent in parts of the Baltic, the Arctic, and the Great Lakes region of North America, where former coastal features are now elevated high above former levels (Fig. 15–41).

As the great continental glaciers advanced, they obliterated old drainage channels and caused streams to erode new channels. These dislocations are especially evident in the north-central United States. Prior to the Ice Age, the northern segment of the Missouri River drained northward into Hudson Bay, and the northern part of the Ohio River flowed northeastward into the Gulf of St. Lawrence. A segment of the lower Ohio River drained into a preglacial stream named the Teays River. Today, geologists recognize the former location of the Teays by a thick linear trend of sands and gravels. Those parts of the Missouri and Ohio rivers that once flowed toward the north were turned aside by the ice sheet and forced to flow along the fringe of the glacier until they found a southward

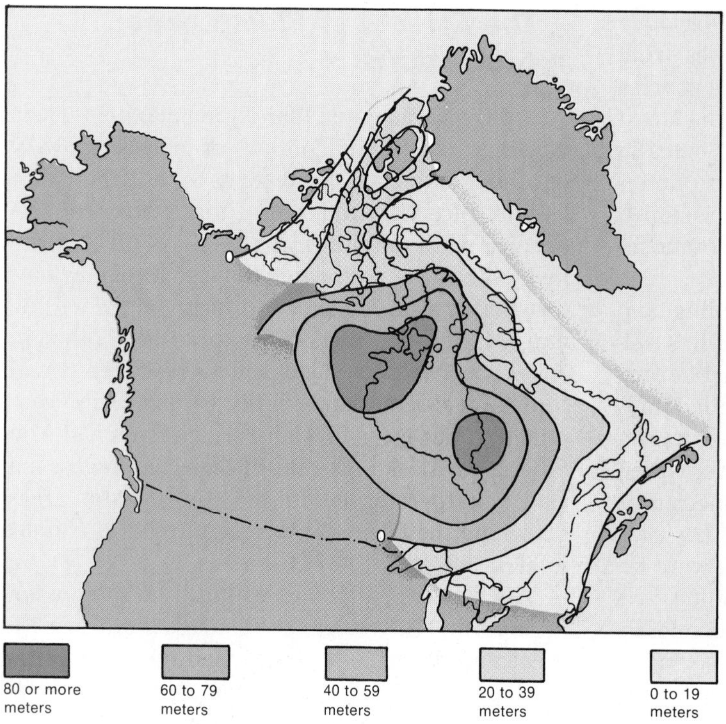

Figure 15–41
Postglacial uplift of North America determined by measuring elevation of marine sediments 6000 years old. (Simplified and adapted from Andrews, J.T., *The pattern and interpretation of restrained, postglacial and residual rebound in the area of Hudson Bay,* in *Earth Science Symposium on Hudson Bay, Ottawa, 1968,* Canadian Geological Survey Paper 68–53, p. 53, 1969.)

outlet. The present trend of the Missouri and Ohio rivers approximates the margin of the most southerly advance of the ice. The equilibrium of streams was also affected, for with glacial advances there was a lowering of sea level and thus an increase in stream gradients near coastlines. The reverse effect occurred with melting of the ice sheets.

Prior to the Pleistocene, there were no Great Lakes in North America. The present floors of these water bodies were lowlands. Glaciers moved into these lowlands and scoured them deeper. As the glaciers retreated, their meltwaters collected in the vacated depressions (Fig. 15–42). Niagara Falls, between Lake Erie and Lake Ontario, came into existence when the retreating ice of the Wisconsin glacial stage uncovered an escarpment formed by a southwardly tilted resistant stratum (see Fig. 13–37).

Another large system of ice-dammed lakes covered a vast area of North Dakota, Minnesota, Manitoba, and Saskatchewan. The largest of these lakes had been named Lake Agassiz. Today, rich wheatlands extend across what was once the floor of the lake. Other lakes developed during the Pleistocene, occupying basins that were not near ice sheets. These were formed as a consequence of the greatly increased precipitation and runoff that characterized regions south of the glaciers. Such water bodies are called pluvial lakes (Latin *pluvia*, rain). **Pluvial lakes** were particularly numerous in the northern part of the Basin and Range Province of North America, where faulting produced more than 140 closed basins. So-called pluvial intervals, when lakes were most extensive, were generally synchronous with glacial stages, while during interglacial stages many lakes shrank to small saline remnants or even dried out completely. Lake Bonneville in Utah was such a lake. It once covered over 50,000 km^2 and was as deep as 300 meters in some places. Parts of Lake Bonneville persist today as Great Salt Lake, Utah Lake, and Sevier Lake.

A spectacular event associated with Pleistocene lake formation occurred in the northwestern cor-

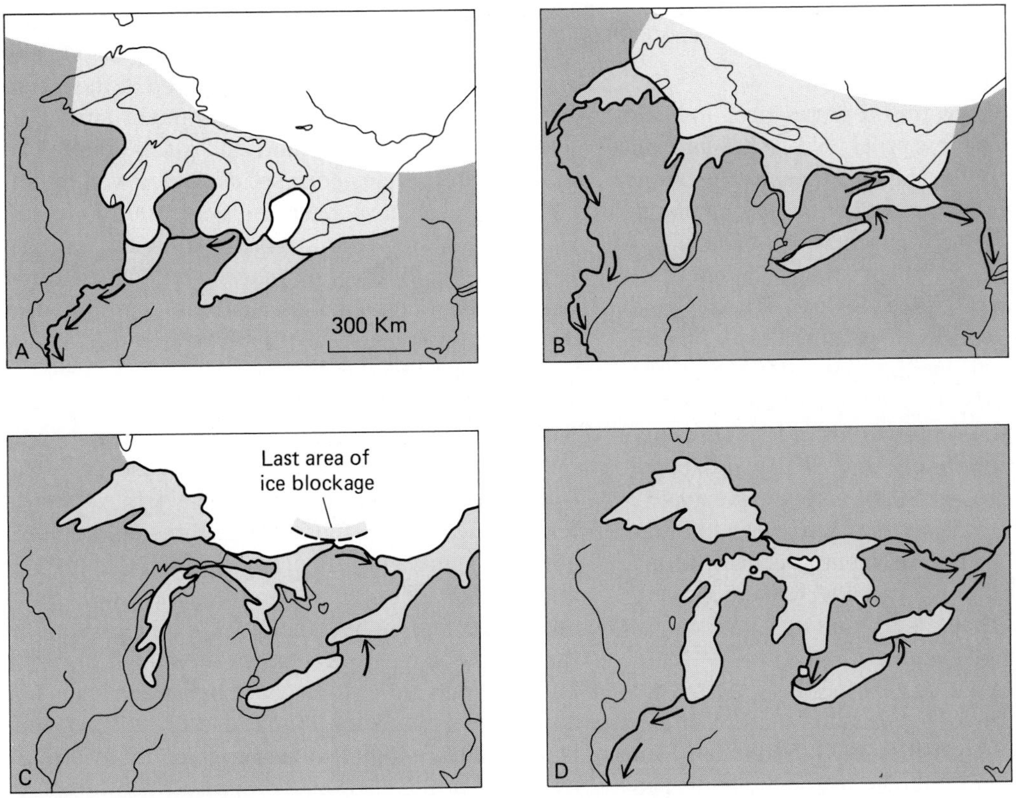

Figure 15–42 Four stages in the development of the Great Lakes as the ice of the last glacial advance moved away. (After Hough, J.L., *Geology of the Great Lakes,* Urbana, University of Illinois Press, 1958, Figs. 56, 69, 73, and 74.)

ner of the United States. Lobes of the southwardly advancing ice sheet repeatedly blocked the Clark Fork River, and the impounded water formed a long, narrow lake extending diagonally across part of western Montana. Called Lake Missoula, it contained an estimated 2000 km^3 of water. Ice is not a reliable material for holding back huge volumes of water, and on several occasions, the dam broke, releasing tremendous torrents of water onto the Columbia Plateau. Giant boulders were apparently transported in the rushing flood for miles. The landscape was chaotically modified over more than 38,000 km^2 of what is now appropriately termed the channeled scablands.

The glacial conditions of the Pleistocene also had an effect on soils. In many northern areas, fertile topsoil was stripped off the bedrock and transported to more southerly regions, which are now among the world's most productive farmlands. Because of the flow of dense cold air coming off the glaciers, winds were strong and persistent. Fine-grained glacial sediments that had been spread across outwash plains and floodplains were picked up and transported by the wind and then deposited as thick layers of loess.

Among the effects of Pleistocene glaciations were alternate lowering and raising of sea level, isostatic adjustments associated with the waxing and waning of ice sheets, alteration of stream gradients and patterns, and the development of pluvial lakes.

Cause of Pleistocene Climatic Conditions

The results of oxygen isotope research indicate that world climates grew progressively cooler from the Middle Cenozoic to the Pleistocene. The culmination of this trend was not a single sudden plunge into frigidity but rather an oscillation of glacial and interglacial stages. Any theory that adequately explains glaciation must consider not only the long-term decline in worldwide temperatures but the oscillations as well. Further, the theory must include reasons for the ideal combination of temperature and precipitation required for the buildup of continental glaciers. Although geologists, physicists, and meteorologists have been speculating about the cause of the Ice Age for over a hundred years, no single causative agent has been found. Indeed, it appears likely that Pleistocene climatic conditions came about as a result of several simultaneously occurring factors.

A widely accepted theory for the Ice Age was developed by the Yugoslavian mathematician Milutin Milankovitch (1879–1958). After 30 years of careful study, Milankovitch convincingly proposed that irregularities in the earth's movements and their influence on the amount of solar radiation received by the earth could account for glacial and interglacial stages of the Pleistocene. His calculations were based on three variables: the earth's axial tilt, precession, and orbital eccentricity. With regard to the first of these variables, Milankovitch recognized that the tilt of the earth's axis varies between about 22° and 24° over a period of around 41,000 years. This results in a corresponding variation in the seasonal length of days and in the amount of solar radiation received at higher latitudes. Precession, the second variable, refers to the way the axis of rotation moves slowly in a circle that is completed about every 26,000 years. The effect is equivalent to tilting a rapidly spinning top whose axis responds by describing a cone in space. The third variable is the eccentricity of the earth's orbital path around the sun, which over an interval of about 100,000 years varies by about 2 per cent. As a result of this variation, the earth is at times closer or farther away from the sun.

According to Milankovitch's calculations, the combination of the above variables periodically results in less solar radiation received at the top of the earth's atmosphere, and this might suffice to cause cooling and recurrent glaciations. The Milankovitch cycles correspond rather well to episodes of glaciation over the past 100,000 years. However, if the **Milankovitch effect** has been in existence for billions of years, why haven't there been Pleistocene-like glaciations continuously down through geologic time? Apparently, other factors must also be involved.

One such factor might be a variation in the amount of solar energy reflected from the earth back into space rather than being absorbed. The fraction of solar energy reflected back into space is termed the earth's **albedo**. At the present time, it is about 33 per cent. Theorists suggest that by the end of the Cenozoic, when continents were fully emergent, and hence highly reflective, temperatures may have been lowered enough for the Milankovitch effect to begin to operate. This is not an unreasonable suggestion, for only a 1 per cent lessening of retained solar energy could lead to as much as an 8°C drop in average surface temperatures. This would be sufficient to trigger a glacial advance if ample precipitation were available over continental areas. Still other geologists speculate that absorption of solar energy was hindered by cloud cover, volcanic ash, and dust in the atmosphere or fluctuations in carbon dioxide. A decrease in carbon dioxide content, for example, would cause a corresponding decrease in the warmth-gathering "greenhouse effect."

Although a unified theory for the cause of Pleistocene glaciations is not yet completed, it is likely that the Milankovitch effect, albedo, and atmospheric composition will be among the variables needed to account for the onset and fluctuations of glacial conditions.

Summary

Glaciers are flowing masses of land ice. The principal requirement for their development is the accumulation of snow in sufficient amounts to exceed losses that occur as a result of melting and evaporation. For example, glaciers in mountainous regions originate as *snowfields*, which, if melting is not excessive, may increase in thickness from year to year. *Snowflakes* falling in the area of the snowfield are transformed by partial melting and refreezing into small pellets of ice known as *firn*. As the blanket of firn thickens because of continuing snowfall, the grains of ice are pressed tightly together and metamorphosed into solid crystalline ice. When the mass of ice along with its overburden of firn and snow accumulates to a thickness of over about 60 meters, the lower levels of ice will begin to flow in response to gravity. Once this movement has begun, the snowfield has become a glacier.

Snowfields give rise to glaciers not only in mountainous areas but also on plains and plateaus. Of course, in glaciers unconfined by valley sides, the ice moves outward in all directions from the center of accumulation. Such glaciers are called either *ice caps* (if they cover less than 50,000 km^2) or ice sheets. Valley glaciers differ in that they are constrained by topography and flow predominantly in one general direction.

Along with mass wasting and such geologic agents as running water, wind, and waves, glaciers have great importance in the carving of landscapes. Valley glaciers are noted for their ability to scour U-shaped valleys from the formerly V-shaped stream valleys they occupy. *Fjords* are such glaciated valleys that have been cut deeply into bedrock where glaciers entered the sea and that were then filled by the sea when the ice receded.

Rock and sediment transported and deposited by glaciers are termed *stratified drift* if they show evidence of sorting and crude layering and *till* if they are unsorted and heterogeneous in appearance. In general, most deposits of stratified drift have been deposited by meltwater, whereas till is the product of deposition directly by ice. Materials deposited by glaciers may accumulate in variously shaped masses called moraines.

Although all glaciers are obviously cold, there is a difference in their temperatures at depth and the pressure-related temperature at which the ice within the glaciers will melt. For this reason, one may recognize cold *dry-base glaciers*, such as those in Antarctica, in which very little meltwater is associated with glacial movement, and warm *wet-base glaciers*, such as those in the Alps and Rocky Mountains, that contain both meltwater and ice simultaneously.

The behavior of a glacier is determined by the state of balance between the rate of accumulation of snow and ice, the rate of flow, and the rate of ablation or wastage. When these factors are in perfect equilibrium, the terminus of the glacier will neither advance nor retreat. If accumulation increases, the glacier will advance, whereas a decrease in accumulation or increase in ablation will result in a glacial recession.

Glaciers flow as a result of plastic deformation caused primarily by displacements within and between ice crystals and by granulation. Movement is slowest adjacent to the bedrock surface beneath the ice and increases in velocity upward to a point about 60 meters from the ice surface. Ice above this level is under insufficient pressure to deform plastically and is carried by the glacier as a brittle unit. As the glacier moves along, it incorporates large and small pieces of rock that comprise the tools by which abrasion is accomplished. Further erosion is achieved by quarrying, in which bedrock frozen into the base of the glacier is lifted out as the glacier moves forward. The falling and sliding of rock debris from mountainsides onto the surface of the ice supplement the glacier's load still further.

At various times in the geologic past, worldwide climates were cooler and glaciation more pervasive. One such ice age began within the period of time designated the Pleistocene Epoch. During the Pleistocene, vast ice sheets formed on the northern continents and alternately advanced and receded.

The effects of Pleistocene glaciation were profound. The ice, at times attaining thicknesses of 3000 meters, depressed the earth's crust, smoothed and rounded the topography of plains and plateaus, deranged drainage, and scoured out basins that became lakes. As the ice alternately accumulated and melted, sea level was caused to rise and fall, with corresponding effects on low-lying regions adjacent to the oceans.

Terms to Understand

ablation zone
abrasion (glacial)
accumulation zone
albedo
arête
avalanching
bergschrund
chatter marks
cirque
drift
drumlin
erratic
esker
firn
fjord
glacial striations
glacier
ground moraine
hanging valley
horn
icecap
ice sheet
intermediate glacier
kame
kame terrace
kettle
kettle lake
knob
lateral moraine
loess
medial moraine
Milankovitch effect
moraine
nunatak
outwash plain
piedmont glacier
Pleistocene Epoch
pluvial lakes
polar (dry-base) glacier
quarrying (glacial)
recessional moraine
roches moutonnées
stratified drift
sublimation
surging
tarn
temperate (wet-base) glacier
terminal moraine
till
truncated spur
valley glacier
varves
zone of flow
zone of fracture

Questions for Review

1. A valley glacier flows mainly because of the slope of the valley that it occupies, just as does a stream of water. Ice sheets, however, can move across level ground. Why does the ice sheet flow?

2. Under what conditions does the front of a glacier remain stationary, moving neither forward nor backward?

3. What is the "Milankovitch Effect?" Why is it unlikely that this effect alone could have been the cause of Pleistocene glaciations?

4. Why do glaciated valleys tend to have trough-like shapes (roughly semicircular in cross-section), whereas the valley of a mountain stream tends to be V-shaped?

5. In a 30-meter-long core of Pleistocene sediment from the ocean floor, how would you be able to distinguish sediments of a glacial stage from those of an interglacial stage?

6. What is the origin of the term *drift* and to what does it refer?

7. What kinds of evidence might one seek in the field in order to determine whether or not an ancient glacier once existed in a region? How can the direction of ice flow be inferred?

8. In general, how does the competence and capacity of a stream compare with the competence and capacity of a glacier?

9. How is granular snow developed from snowflakes? In what crystal system does ice crystallize? What evidence of crystal system is provided by snowflakes?

10. In what way are the following features developed: a *cirque*, a *fjord*, an *end moraine*, an *esker*?

11. What evidence indicates that glaciers move? Describe the velocity distribution within the glacier. How does movement in the upper 30 meters of the glacier differ from the movement at greater depths?

12. What evidence indicates that, within the last ¾ million years or so, four different ice sheets have at different times advanced across, and retreated from, the northern United States and that these advances and retreats were separated by long intervals of warmer climate?

13. What has been the effect on sea level of the advance and retreat of Pleistocene ice sheets? What role did continental glaciation play in the development of the Great Lakes?

14. What renowned Swiss naturalist championed the "glacial theory" of continental glaciation? Prior to his theory, how did scientists explain erratics and drift deposits?

15. What is the cause for the gradual rise in land elevations that has occurred within historic time around Hudson Bay, the Great Lakes, and the Baltic Sea?

Supplemental Readings and References

Baker, V.R., 1981. *Catastrophic Flooding: The Origin of the Channeled Scabland.* Stroudsburg, Pa., Dowden, Hutchingson, & Ross, Inc.

Embleton, C., and King, C.A.M., 1975. *Glacial Geomorphology*, 2nd ed. New York, John Wiley & Sons.

Ericson, D.B., and Wollin, G., 1964. *The Deep and the Past.* New York, Alfred R. Knopf, Inc.

Flint, R.F., 1971. *Glacial and Quaternary Geology.* New York, John Wiley & Sons.

Matsch, C.L., 1976. *North America and the Great Ice Age.* New York, McGraw-Hill.

Winds and Deserts

16

Dune fields of the Namibian coastal desert of the southwestern coast of Africa. The Atlantic Ocean is visible in the upper left corner of this photograph taken by astronauts on the space shuttle in 1984. The Namib Desert extends across a distance of 800 km (500 miles) and ranges from 40 to 140 km in width. The intricate pattern of large transverse sand dunes is caused mainly by dry westerly winds. Extending diagonally upward from the lower right corner is a dune-free tongue of gravel formed by flash floods draining from the barren, rocky hills on the right side of the photograph. (Courtesy of NASA.)

The wind grew stronger, whisked under stones, carried up straws and old leaves, and even little clods, marking its course as it sailed across the fields. The air and sky darkened, and through them the sun shone redly, and there was a raw sting in the air.

John Steinbeck, 1939 The Grapes of Wrath

Introduction

Unless one has had the misfortune to experience a great dust storm, it is difficult to realize the enormous capacity of wind to transport loose, dry, fine sediment. During such storms, dust blots out the light from the sun, filters into the tiniest crevice, buries crops, and even kills animals by suffocation. The smallest of the wind-borne particles are carried completely around the globe before finally settling to the ground. Indeed, it is likely that, because of the continuous and often invisible rain of tiny dust particles, every square kilometer of land contains at least some of the sediment from every other square kilometer.

The ability of the wind to transport particles of sediment interests geologists keenly, but it is not the most important function of wind. Wind picks up moisture from the seas and carries it over the lands. When that moisture condenses and falls as rain, it increases the discharge of the streams that help to carve our landscapes. Wind blowing across the ocean generates the waves that pound against the borders of continents and help to move sand along beaches. Winds bring us clouds, fogs, rains and snows, and are thus essential components of weather.

Although winds are important for all of the above reasons, they are not, of themselves, powerful erosional agents. Far greater geologic work is accomplished by running water and moving ice. Wind is however, capable of lifting and transporting relatively small particles of rock and mineral matter, if certain conditions exist. The wind must have sufficient velocity. The surface over which the wind blows must contain loose dry mineral particles, preferably in sizes from silt to fine sand. There must also be a lack of vegetative cover. Such conditions are found along some shorelines, on floodplains following periods of flooding, and in places where bare ground has been exposed by plow or bulldozer. Mostly, however, conditions favoring wind erosion occur in the warm deserts of the world (Fig. 16–1). In this chapter, we will discuss some of the goelogic features of such areas, describe the winds that blow over them, and examine the nature of the windblown materials deposited on them.

The Nature of Wind

Wind Direction

Wind is air in motion. Meteorologists would add that, in wind, the motion of air is predominantly horizontal. Such terms as updraft and downdraft are employed for vertical motions of air that do not qualify as wind in the meteorologic sense. Winds are named according to the direction from which they come. Thus, if winds blow from the east to the west, they are called east winds. Every sailor knows that the term **windward** refers to the direction from which the wind comes, and **leeward** the direction toward which it blows. One determines wind direction from the most venerable of all meteorologic instruments, the weather vane. Wind velocity is measured by any of several kinds of devices called **anemometers** (Fig. 16–2). Velocity is usually expressed in either km per hour or nautical miles per hour (knots).

Origin of Wind

The ultimate energy source for wind is the sun. When the sun warms one part of the earth more than another, the warm air expands and is displaced upward by cooler (hence, denser) air, which moves in along the earth's surface from adjacent regions. That movement is what we recognize as wind (Fig. 16–3). Because the air molecules are closer together in cool air, a column of cool air

Figure 16–1 The deserts of Tunisia contain vast dune fields, which are exceptionally difficult to traverse. Here a bulldozer is moving into position to prepare an access road for a geophysical survey. Note the well-developed eolian ripple marks on the surface of the dunes. (Courtesy of Western Geophysical, photograph by Volker Vagt.)

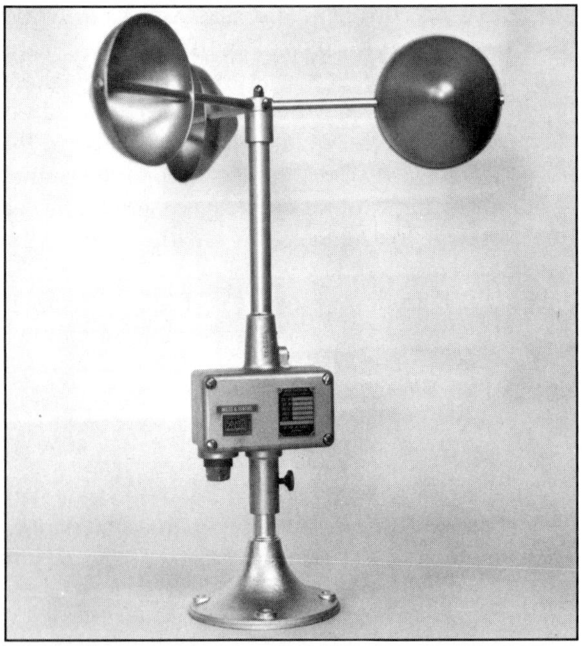

Figure 16–2 A three-cup anemometer. (Courtesy of NOAA.)

weighs more than a column of warm air of the same volume. As a result, a mass of cool air above the ground is appropriately called a *high-pressure area*, whereas warm air masses constitute *low-pressure areas*. Air moves as wind from high-pressure areas of descending cool air to low-pressure areas of rising warm air. Its direction of movement, however, is usually not directly from areas of high pressure to areas of low pressure because of the deflection of the wind caused by the earth's rotation.

One may construct a map showing the variation in atmospheric pressure across an area at any given time by connecting points of equal pressure with lines called **isobars**. Suppose, for example, the isobars on such a map extend north–south as shown in Figure 16–4. The high pressure to the east will cause the air to flow toward the west. The pressure difference between the two areas provides a **pressure gradient**. The pressure gradient on the east is about two units of pressure per 100 km, whereas the pressure gradient in the westward area is four units per 100 km. Because of the steeper gradient in the west, the winds are likely to be stronger.

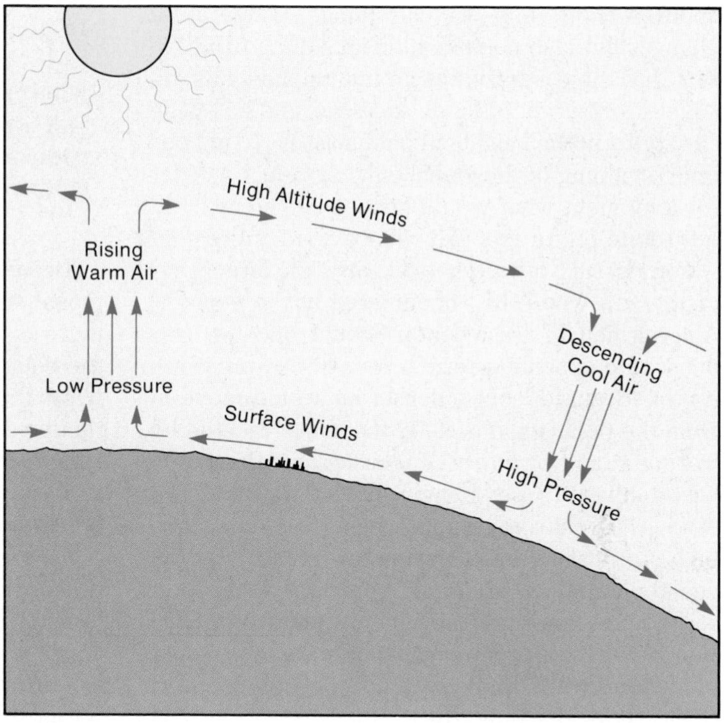

Figure 16–3 The generation of wind as a consequence of unequal heating.

Pressure differences from place to place are the principal forces driving winds.

Temperature is not the only factor affecting the pressures developed in highs and lows; moisture and altitude also influence the pressure gradient. In general, water vapor has a lower density than dry air and, when present, makes air lighter. This seems surprising because we often incorrectly think of air as "containing" water vapor. Actually air does not "contain" water vapor, water vapor *replaces* air. The replacing water vapor has a lower density than dry air. Thus, masses of moist air tend to rise until they reach cooler zones where condensation may occur.

The lessening of atmospheric pressure as one ascends to higher altitudes is a familiar phenomenon. Indeed, instruments that measure atmospheric pressure can be calibrated to indicate elevation above sea level. Atmospheric pressure is greater at lower altitudes because there is a taller column and greater mass of air above land near sea level.

The Global Circulation Pattern

The great wind systems operating across the surface of the earth are components of huge convection cells in the atmosphere. Global wind patterns are quite complex, for they represent re-

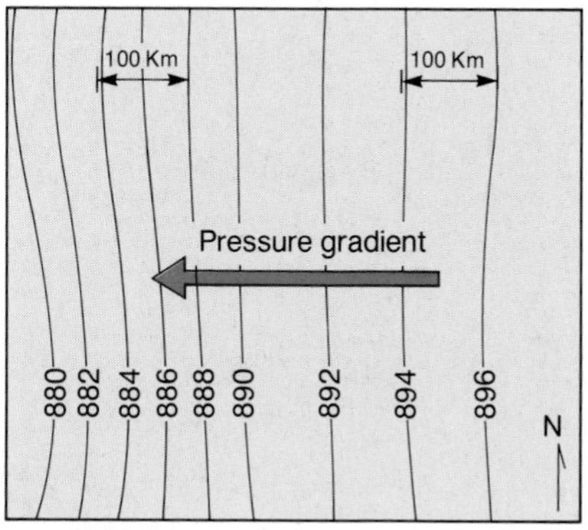

Figure 16–4 Map view of the flow of air from a high-pressure area to a low-pressure area. The lines connecting points of equal pressure are called isobars. In general, winds are relatively stronger where isobars are more closely spaced.

sponses not only to latitudinal temperature changes but also complications resulting from the fact that the earth rotates around an axis, has unequal distribution of lands and seas, and has wide variations in the heights of land masses. To aid our understanding of the global wind system, consider for a moment what would happen if the earth did not rotate on its axis and was covered entirely by water. On such a hypothetical earth, the direct rays of the sun would be encountered at the equator. Air warmed at the equator would become less dense and expand to form a zone of low pressure. As this occured, cooler denser air would move in from the polar regions to fill the space vacated by warm air, and a pattern of convection cells would be established resembling those shown in Figure 16–5. This relatively simple system would produce high pressure at the poles and low pressure at the equator, and the pressure differences would generate two large air-mass movements. In the Northern Hemisphere, the predominant surface winds would blow southward, while in the upper atmosphere, warm air would flow toward the North Pole. The situation would be reversed in the Southern Hemisphere.

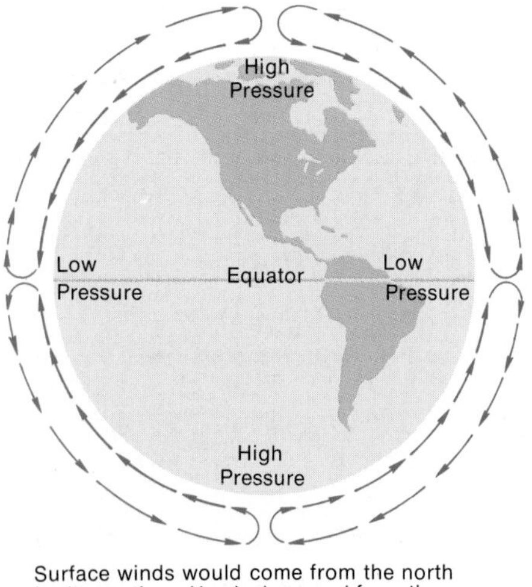
High Pressure
Low Pressure
Equator
Low Pressure
High Pressure
Surface winds would come from the north in the Northern Hemisphere and from the south in the Southern Hemisphere

Figure 16–5 Wind systems on a hypothetical nonrotating earth. (From Turk, J., and Turk, A., *Physical Science*, Philadelphia, W.B. Saunders Company, 1977.)

In general, wind patterns at lower latitudes resemble a giant convection cell wherein warm air rises over the equatorial zone, moves poleward at high altitudes, and returns to the equatorial zone at lower altitudes.

Because the earth does rotate, the simple atmospheric circulation described above is made more complex. The warm air masses that rise at the equator divide and move at high altitudes toward the poles. However, they do not move directly north or south, but in response to the earth's rotation, are deflected to the right in the Northern Hemisphere and to the left in the Southern Hemisphere.

This effect, which also causes similar deflections in ocean currents and other bodies that are moving freely with respect to the rotating solid earth, is called the **Coriolis effect**. To understand the Coriolis effect, imagine you are an observer far out in space watching a rocket that has been fired toward the south from the North Pole. From space, the rocket would indeed be seen to move in a straight line. If you were to plot its course across the surface of the solid earth, however, it would appear to curve gradually westward, because the earth is turning from west to east beneath the rocket as it speeds southward.

Air movement is subject to this very same effect. For example, as an air mass located in the Northern Hemisphere moves southward, the earth's rotation beneath the moving air mass will cause it to turn toward the west (that is, to the right). Conversely, an air mass moving northward in the Northern Hemisphere (like the prevailing westerlies in Figure 16–6), will be deflected eastward.

By the time air masses that had originated at the equator reach about 30° north and south latitude, they are traveling directly eastward. They have now cooled somewhat, and begin to descend, producing the subtropical high-pressure areas known as the **horse latitudes** (see Fig. 16–6). The winds in these regions are light and variable, rainfall is slight, and skies are generally clear. In this region, sailing ships were often becalmed. As water and food aboard ship became scarce, horses carried as cargo died and were thrown overboard. Their

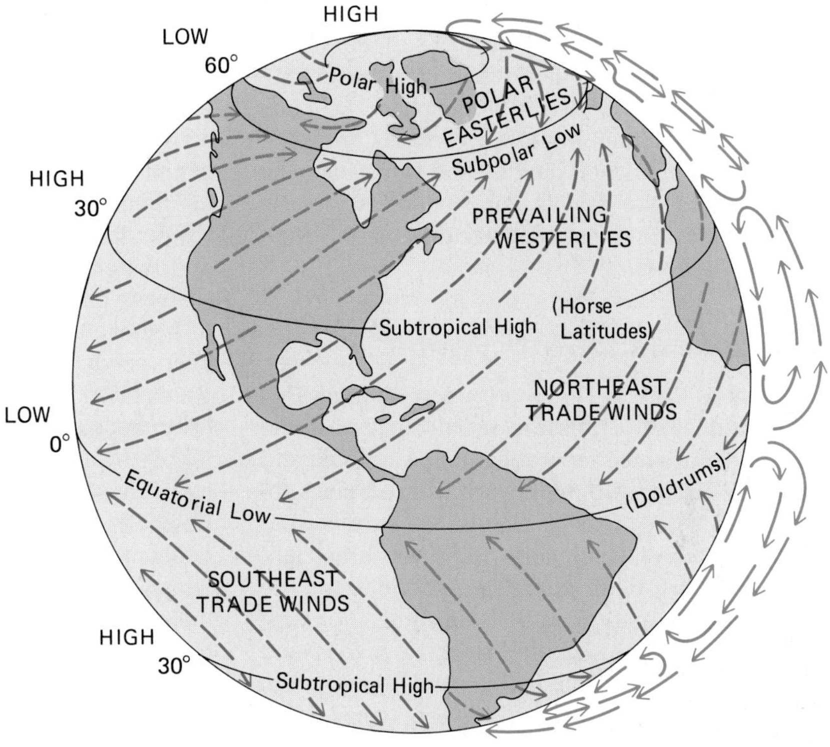

Figure 16–6 Wind and pressure systems of the earth.

bloated bodies seen floating on the sea gave the horse latitudes their odd name.

The air masses that descend at the horse latitudes return to the equator at low altitudes. Because of the earth's rotation, they veer and blow from the northeast in the Northern Hemisphere and southeast in the Southern Hemisphere. These are reliable winds and are called **trade winds** because they once provided the trade routes for merchant vessels. The cycle is complete when the air reaches the equator, where the convection cycle begins anew.

Poleward from the subtropical high-pressure regions are the wind belts known as the westerlies. In these regions, the prevailing wind directions are from the southwest in the Northern Hemisphere and from the northwest in southern regions. Subpolar low-pressure belts are developed at latitudes of about 60° north and south. Poleward of these areas, the air is cooled and compressed to form high-pressure zones. Cold air in this region sinks and moves generally southward in the Northern Hemisphere and northward in the Southern Hemisphere. Once again, the winds are deflected and are appropriately designated the polar easterlies. At latitudes of about 50° or 60° the polar easterlies encounter the northern margin of the belt of westerlies. The warmer westerlies are shunted above the colder polar air.

From this brief survey of atmospheric circulation, it is apparent that three factors determine the horizontal and vertical movements of air masses. Of predominant importance is the sun, which provides the energy needed to drive the system. A second and related factor is the change in temperature from the equator to the poles. Finally, currents of air generated by temperature differences are influenced by the Coriolis effect. The entire pattern of circulation acts to modulate the temperature differences on earth and carry excess heat from the equator toward the poles.

The temperature differences that are responsible for global air circulation are not the result of unequal heating of the atmosphere itself but rather of the earth's surface. Energy from the sun reaches the earth as short wavelength radiation—chiefly visible light and lesser amounts of ultraviolet and infrared light. Such short-wavelength radiation is absorbed only slightly by the atmosphere, so that most of it penetrates to the earth's surface, where it is absorbed as heat. The warm earth then radiates this heat back to the atmosphere as long-wavelength radiation. Long wavelengths are readily absorbed by carbon dioxide and water vapor in the air. Thus, the earth's atmosphere is directly heated by the continents and oceans and only indirectly by the sun.

Cyclones and Anticyclones

A **cyclone** (Fig. 16–7) is a roughly circular area of low atmospheric pressure, usually formed in the zone of disturbance separating masses of cold and warm air. As winds blow inward from all sides toward the center of the cyclones, the earth's rotation causes them to be deflected toward the right in the Northern Hemisphere (and to the left in the Southern Hemisphere). At the same time, the warm air in the cyclone's center rises to higher and cooler levels where clouds form and precipitation may occur. Hurricanes are severe tropical cyclones.

A high atmospheric pressure system is called an **anticyclone**. Circulation of air in an anticyclone is opposite to that in a cyclone. In the Northern Hemisphere, air moves downward through the center of the anticyclone and spirals outward with a clockwise deflection. The descending air of anticyclones tends to be cool and dry. As a result, anticyclones are generally associated with pleasant weather.

Local Winds

In addition to the earth's great prevailing winds, local winds may develop in particular areas whenever or wherever a pressure gradient is established. For example, people living along coastlines are familiar with the so-called land and sea breezes resulting from the unequal heat capacities of land and water. *Heat capacity* is the amount of heat (calories) needed to raise the temperature of 1 cm^3 of a substance by 1°C. (A calorie is the amount of heat needed to raise the temperature of 1 g of water by 1°C.) Water has one of the highest known heat capacities. Indeed, saltwater has a heat capacity of 0.9 calorie per gram, whereas the heat capacity of most common minerals is less than 0.2 calorie per gram. For this reason, sea water warms and cools much more slowly than solid ground. The air adjacent to a coastline heats quickly during the day under the action of the sun's rays and also cools rapidly during the night. In the daytime, air heated by contact with the warm soil rises and is replaced by cooler air flowing in from the sea. At night, these conditions are reversed. The land areas cool more quickly than the sea, and air flows from the cool land surface toward the water, which has retained its warmth (Fig. 16–8).

Similar pressure gradients are found in mountainous regions. On warm clear days, mountain slopes exposed to the sun's rays are heated. The air in contact with the slopes is warmed, expands, and flows up the flanks of the mountains. Because of the relatively low heat capacity of rock and soil, at

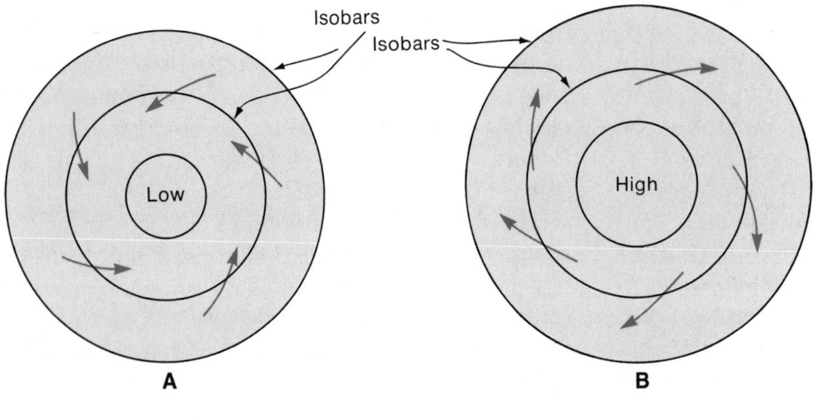

Figure 16–7 Typical distribution of wind in (A) a low-pressure cyclone and, (B) a high-pressure anticyclonic area in the Northern Hemisphere.

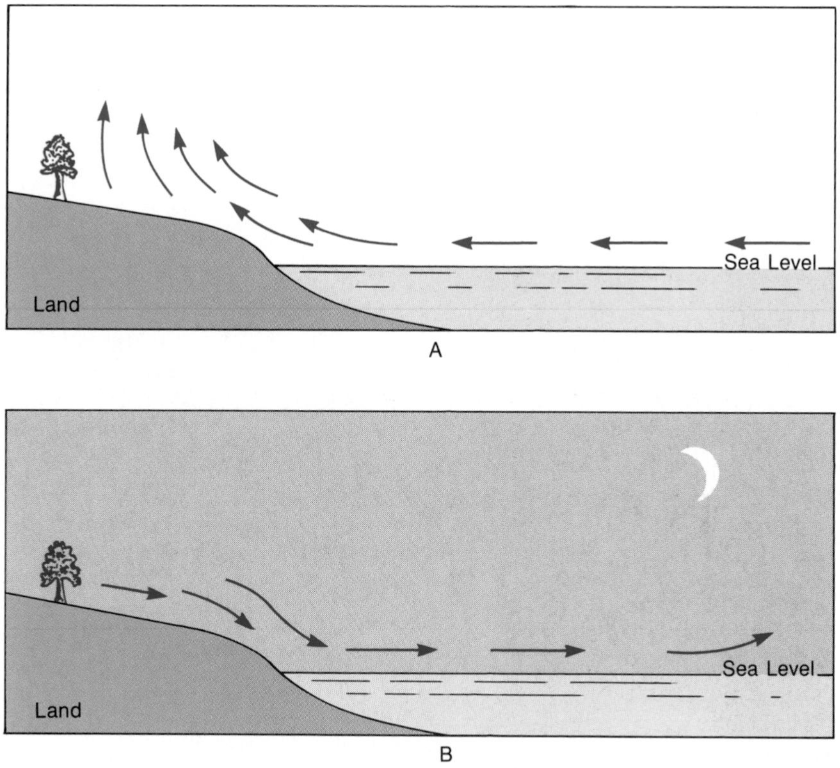

Figure 16–8 Land and sea breezes result from unequal heating of land and water.

night the mountain walls cool rapidly. Air in contact with the cooled surfaces is chilled, becomes denser, and flows down the mountains into the lowlands as a mountain breeze.

Monsoons

Monsoons are winds that are the result of seasonal temperature differences on continents relative to adjacent oceanic areas. Because of their enormous capacity for retaining heat, and continual circulation, oceans tend to maintain temperatures that vary only within relatively narrow limits. Continental areas are not capable of such heat retention and may become much cooler than nearby water bodies during the winter. At such times, there is a flow of cold dense air from the continents toward the oceans. The resulting winds are termed winter monsoons. In the summer, the land areas may become warmer than the adjacent ocean, and winds will blow inward from the oceans as summer monsoons (Fig. 16–9). As this water-saturated air rises over the land, it cools, and drops rain.

Wind Erosion

The effectiveness of wind in lifting and transporting fine soil particles was dramatically demonstrated in the western plains of the United States during the 1930s (Fig. 16–10). At that time, drought and crop failures had exposed millions of acres of loose soil to the ravages of gusting winds. A region from the Canadian prairie provinces into Texas became a huge dust bowl. Once-productive farms were stripped of their soils, and dust-weary farm families abandoned their homes and suffered the trauma of displacement so movingly described by John Steinbeck in his novel *The Grapes of Wrath.*

Today, as a result of better agricultural practices, full-scale dust-bowl conditions have been avoided. Nevertheless, an occasional dust storm still occurs, and causes a significant loss in valuable soil.

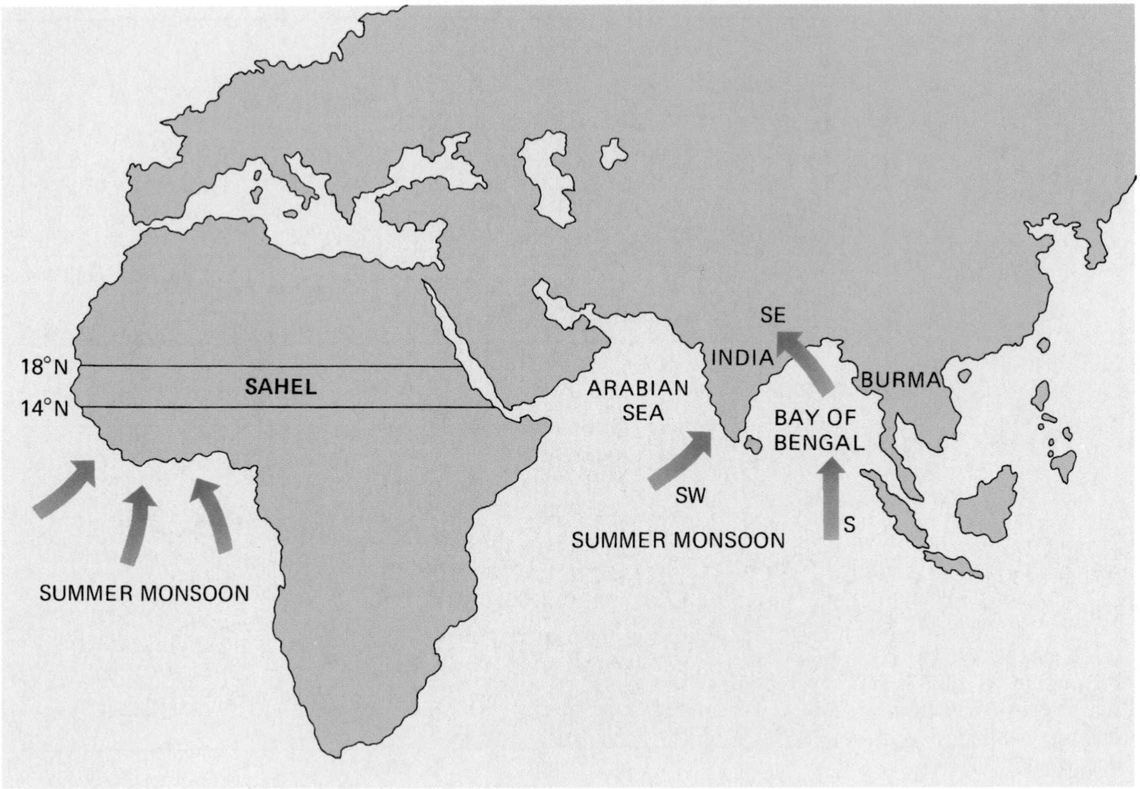

Figure 16–9 Locations having summer monsoons. The Sahel (Arabic, meaning "fringe") is a sub-Sahara region of Africa that is dependent on rainfall brought inland by monsoons. Famine threatens the Sahel whenever the monsoons fail to bring adequate moisture.

Transportation of Sand, Silt, and Clay

The flow of air that we call wind is in some ways similar to the flow of water. Like water, air tends to flow in a laminar fashion when moving very slowly but flows turbulently at higher velocities. Air, like water, is retarded by friction with the surface over which it passes, and—as is the case with water in channels—its lowest velocity is immediately above the solid surface over which it flows. Of course, there are also differences between the behavior of flowing air and flowing water. Flowing water is confined to channels and is driven by a component of the force of gravity that acts parallel to the surface across which it moves. Winds, on the other hand, are impelled by differences in air pressure resulting from unequal heating of the earth's surface. It is also readily apparent that air lacks the specific gravity and viscosity of water and is thus less effective as an agent of erosion. Moving air does have one advantage over water, however, in that sand grains carried in wind by saltation are not impeded as much as in the denser liquid medium. As a result, a skipping sand grain on impact can dislodge another grain more than six times its own size.

As noted earlier, wind can accomplish its work only if fine-grained, dry, unconsolidated, unvegetated materials exist on the ground surface (Fig. 16–11). As was the case with streams, sand and silt are picked up most readily. Such particles are large enough to provide a surface against which the wind can push and small enough that they are not too heavy to lift. The flattish clay particles tend to

Figure 16–10 This 1937 photograph of an abandoned farmstead in Oklahoma shows the disastrous effects of drought followed by wind erosion. (Courtesy of U.S.D.A.)

Figure 16–11 Dry conditions and a lack of vegetation, such as exist in this area of Death Valley, California, favor wind erosion. Death Valley also contains playas, pediments, and sand dunes characteristic of arid regions. Infrequent rainfall acting on the bare surface material has sculptured the badland topography shown here.

hug the ground closely and require extraordinary velocities before they are lifted into the air stream.

Once particles lying on the surface have been dislodged by wind, they may be transported above the ground as part of the wind's suspended load or moved by saltation and rolling as the bed load (Fig. 16–12). Generally, particles in the suspended load are less than about 0.15 mm in diameter. Often the smaller particles of the suspended load are buffeted high up into the atmosphere, where they are effectively segregated from other more earthbound materials. This helps to account for the striking absence of such fine sediment in the sandy deposits of deserts.

A wind's bed load consists mostly of particles in the 0.15- to 2.0-mm size range. Particles of this size will not stay in suspension in air but rather move along the surface of the ground by rolling and saltation. As was the case with transportation of sedimentary particles by streams, saltation is a pattern of movement whereby grains move forward in the direction of the air current by making low arcing jumps (Fig. 16–13). During saltation, grains may bounce off other grains on impact with the ground and frequently dislodge other grains so that they too begin the saltation movement downwind. Some additional transport occurs as myriads of impacting grains collide with grains on the ground, imparting a slow forward "creep" to these larger particles.

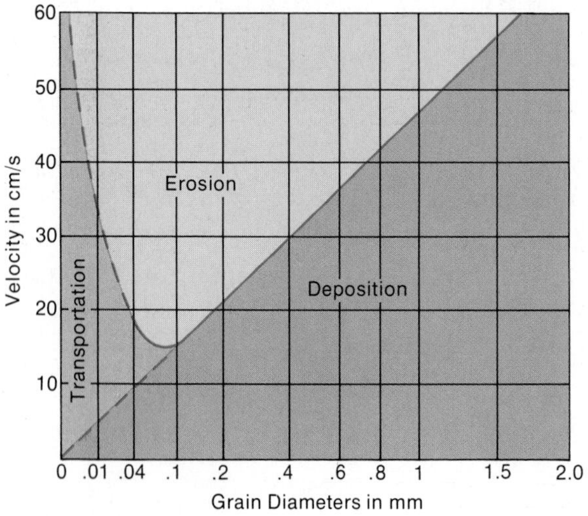

Figure 16–12 Erosion and transportation curves for wind. (Adapted from Bagnold, R.A., *The Physics of Blown Sand and Desert Dunes,* London, Methuen & Co., Ltd., 1941, p. 88.)

Saltating sand grains frequently rise to a height of 100 to 150 cm above the ground. During a sand storm, it is often possible to estimate the average maximal height of saltating grains by the clearly visible upper limit of the cloud of saltating grains. Even after the storms, the approximate height of saltation can be discerned on the sandblasted portion of the surface of telephone poles and other vertical objects. Because of the tremendous erosive powers of wind and sand, sheets of metal are sometimes placed around the base of the poles as protection.

The sand and other particles moved by wind are derived from a variety of sources. Ultimately, much wind-blown sediment is produced during the weathering of rocks or by the impact of grains of windblown sand on bare rock surfaces. In additon, fine clastics may be picked up by winds from alluvial fans, dry stream beds, floodplains, volcanic eruptions, glacial outwash deposits, and beaches.

Wind effectively sorts grains of sediment by depositing its bed load of larger rolling and saltating grains separately from its suspended load of finer particles.

Deflation and Abrasion

The erosion of rock and soil surfaces by the wind is accomplished by the two processes of deflation and abrasion. **Deflation** refers to the removal of loose and dry sediment as the wind blows across an unprotected surface (Fig. 16–14). Without deflation there would not be sand and dust storms. Deflation causes the development of **desert pavements** (Fig. 16–15) of pebbles and cobbles by selectively blowing away finer particles. Wherever patches of ground are unprotected by vegetation, deflation may sweep away surficial materials, producing shallow roughly circular depressions called **blowouts** (Fig. 16–16). Erosion of this kind can often be prevented by planting suitable vegetative cover and by the construction of wind breaks.

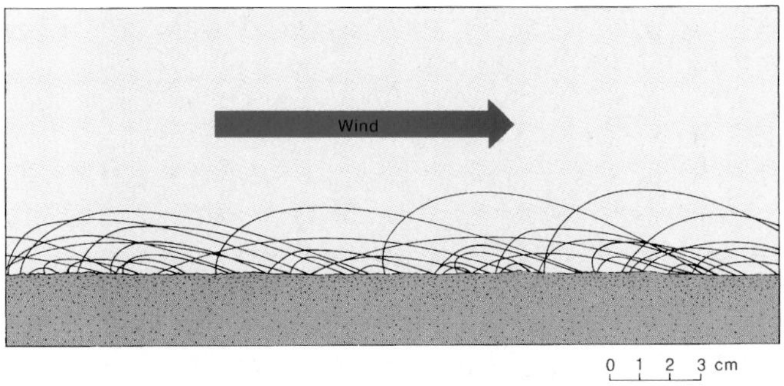

Figure 16–13 Saltation movement of sand particles over a surface composed of sand.

Abrasion, the second means by which wind accomplishes erosion, is the wearing away of rock or soil masses as they are repeatedly struck by myriads of wind-borne sand and silt grains. The process resembles sandblasting. As the grains forcefully strike against rock surfaces, they break grains, weaken cements so that particles fall free of the rock, or actually dislodge grains from the matrix. The sand grains freed from the parent rock then become additional tools in causing further abrasion. When examined closely, sand grains of this kind display small impact pits on their surfaces, which give them an overall "frosted" appearance. Frosted grains in ancient sandstones provide geologists with strong evidence that wind was the transporting and depositing agent.

Wind erosion is accomplished by abrasion and deflation. Erosion is facilitated if winds are strong and the ground surface contains loose, dry particles that lack plant cover.

If a large cobble projects into the wind in an area where abrasion is occurring, the windward side will be sandblasted to a smooth sloping surface. When discovered in desert areas, these peculiarly shaped stones (Fig. 16–17) appear to have been ground down on one or more sides. They are called **ventifacts** and when found in ancient sandstone formations, are indicators of wind action. If the cobble falls into a shallow pit excavated on its lee side by deflation, if it is turned by flash floods or frost action, or if there are two dominant wind directions, the ventifact may develop two smooth faces.

Deposition of Wind-Borne Sediment

Sooner or later, the particles carried by wind are deposited. The finest materials are distributed widely over the earth. Some fall into the oceans, some onto forested areas, and some onto fields where they are incorporated into soils. Rains wash the particles of dust from the air, and like other sediments, they are carried to the sea. The larger grains of wind-borne sediment are more readily

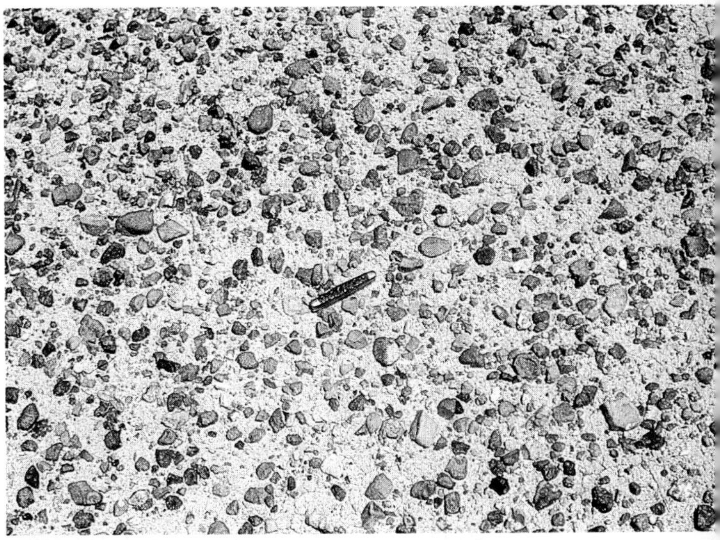

Figure 16–14 Lag gravel, a residual accumulation of pebbles and other coarse particles from which finer material has been removed by deflation. Further deflation of this surface would produce a continuous layer of pebbles and cobbles called desert pavement. (Courtesy of B. Stinchcomb.)

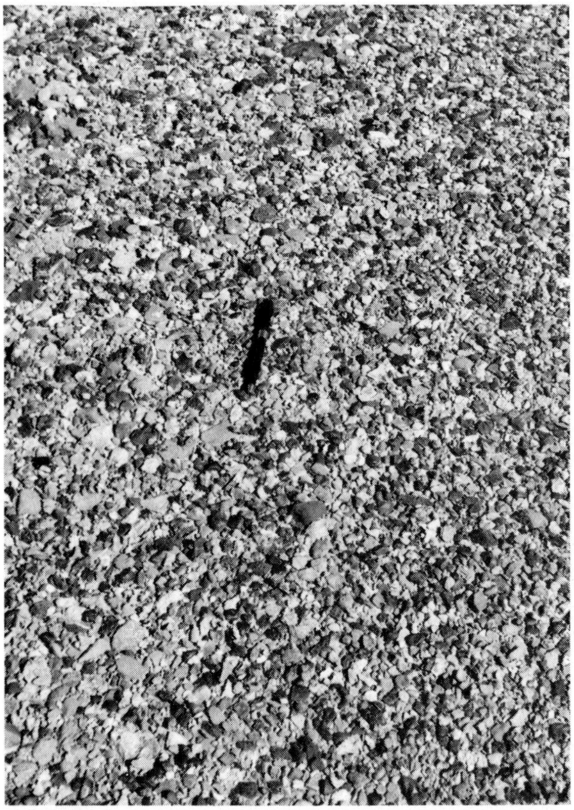

Figure 16–15 Desert pavement, Lake Mead National Recreation Area. Fountain pen (center) indicates scale. (Courtesy of John V. Bezy and the Southwest Parks and Monuments Association.)

visible to us, for they collect along shorelines and in deserts as dunes and blankets of sand. All windborne material, whether fine silt or sand, is described as **eolian**, a term derived from the name Aeolus, the Greek god who controlled the winds by releasing them at will from his cave.

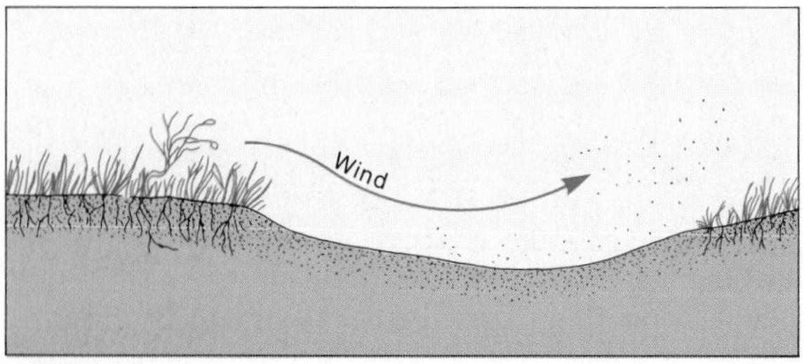

Figure 16–16 Blowout developing in area of patchy vegetative cover.

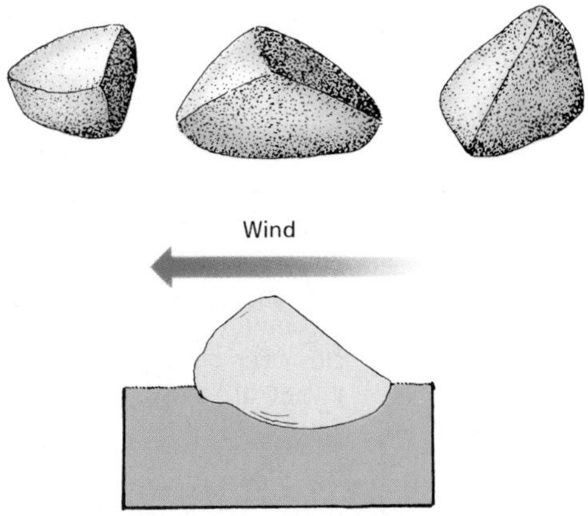

Figure 16–17 Ventifacts (top row). Pebble being shaped into ventifact by wind action (bottom.)

Loess

The term loess was first used to designate the tawny-colored unconsolidated silty deposits exposed along the bluffs overlooking the Rhine River in West Germany. Such deposits are also known in many other regions of the globe. In North America, loess deposits are several meters thick and blanket hundreds of square kilometers of terrain in the Mississippi Valley and Columbia Plateau (Fig. 16–18). Vast blankets of loess also cover the plains of northern China.

Geologists define **loess** as a soft, homogeneous, porous, unconsolidated, and unstratified deposit consisting predominantly of silt but also containing lesser amounts of clay and very fine sand. A striking characteristic of loess is the way in which exposures of the material break off in vertical slabs, leaving behind steep escarpments. In China and central Europe, there are a few places where people have carved their homes directly into such vertical exposures of loess.

Loess (pronounced "less") is wind-deposited silt, containing some clay and fine sand.

In regions where there is sufficient rainfall so that a layer of humus can develop, loess deposits develop into excellent soils. They are not only rich in minerals required for healthy plant growth but are permeable and easily tilled.

Several lines of evidence lend support to the belief that loess deposits are of eolian origin. Collections of dust from recent dust storms have been subjected to a statistical analysis of grain sizes and found to compare very closely with ancient loess deposits. The loess deposits themselves blanket hills, valleys, and slopes indiscriminately, just as one would expect in a sediment that has been deposited from the air. Thin vertical tubes, found abundantly in loess, are thought to represent the stems of plants that were buried by the falling dust. Fossil shells of air-breathing snails are commonplace in loess and indicate that the material was not deposited beneath a water body.

There appear to be several environments in which loess may originate. The blankets of loess in the central United States and Europe were deposited during the Ice Age beyond the margins of the great continental glaciers. Meltwater and debris from the ice sheets poured into nearby streams. Very likely, flooding was a frequent occurrence, and when the flood waters subsided, they left behind extensive tracts of fine fluvial sediments. Cold weather inhibited the development of a protective cover of vegetation on this sediment, so that gusty winds sweeping down from high pressure regions over the glacier were able to whip up clouds of fine sediments and deposit them widely across the Mississippi Valley and adjacent regions.

The extensive yellowish loess of northern China is made up of disintegrated rock debris carried by westerly winds from the great Gobi Desert. In places these loess deposits are an astonishing 300 meters thick. The reason for this huge thickness is the almost endless supply of sediment from the Gobi. In the summer, the Gobi is very hot and vegetation is sparse. As a result, the soil is exposed to wind action. A high rate of erosional activity also occurs during the very cold Gobi winters, when frost heaving and similar processes disturb the packing of the fine sediment and render it more susceptible to wind erosion. Old loess deposits in China are easily eroded and find their way readily into streams, coloring them yellow. Few people realize that some other name would have been used for the Yellow River and Yellow Sea were it not for the Gobi-derived loess deposits.

Figure 16–18 Loess exposed in a building excavation, St. Louis, Missouri. The loess extends down to the bench level and is underlain by clays of Pennsylvanian age.

Sand Deposits

Contemporary Dunes

Although eolian sand deposits may be blanket-like in form, they frequently accumulate as hills and ridges of sand called **dunes** (Fig. 16–19). Not

Figure 16–19 Sand dune in the Gran Desierto region of the Sonoran Desert, Mexico. The wind responsible for this dune blew from left to right. Note wind ripples on the windward side. (Courtesy of Jeff J. Plaut.)

all dunes are found in tropical deserts. They occur along the shorelines of oceans and large lakes and even on floodplains of streams in arid or semiarid regions. Even in arid regions, one finds dunes that derive their sand from the desert but actually accumulate just beyond the desert's perimeter.

Dunes may form wherever there is a decrease in the velocity of the wind that had been actively transporting sand grains. For example, anyone who has stood behind a tree during a wind storm is aware that the wind velocity in that protected area is much reduced. On any tract of land, one nearly always finds large and small obstacles that similarly offer a protected side for the accumulation of sand grains. The wind must sweep up and over such obstacles, and, in doing so, an area of low-velocity eddying air is left behind on the lee side as well as near the base on the windward side. These quieter areas are called **wind shadows**, and the boundary between the wind shadow and the overlying sweep of fast-moving air is termed a **surface of wind discontinuity** (Fig. 16–20). Sand grains that are carried into the lee wind shadow are deposited there, thus extending the obstacle downwind. Similarly, some of the grains swept over the top of the obstacle may fall through the surface of discontinuity and accumulate. Eddying air currents develop within the shadow zone and contribute to the process of dune formation by sweeping sand inward toward the central part of the wind shadow. Of course, once the dune begins to build, it provides an increasingly effective obstacle to wind. The steeper leeward slope of the dune is called the **slip face** because of the many small sand slides that occur there.

Dunes are mounds or ridges of sand deposited by the wind. They are localized by an obstacle that distorts the flow of air so as to cause deposition of sand grains.

The windward slope of a dune is less steeply inclined than the lee slope (Fig. 16–21). Wind directed along the windward slope picks up and moves sand grains up the gentler incline toward the crest of the dune. Thus, the windward slope is continually being beveled by the wind and rarely acquires a slope of more than about 10°. On reaching the crest of the dune, sand grains that have been moved up the gentler windward slope are left to cascade down the leeward slope, where they assume a natural angle of repose of between 30° and 40°.

The height to which sand accumulates on dunes is limited by the height at which wind velocity increases. The current of wind blowing over a dune is compressed into an increasingly smaller space

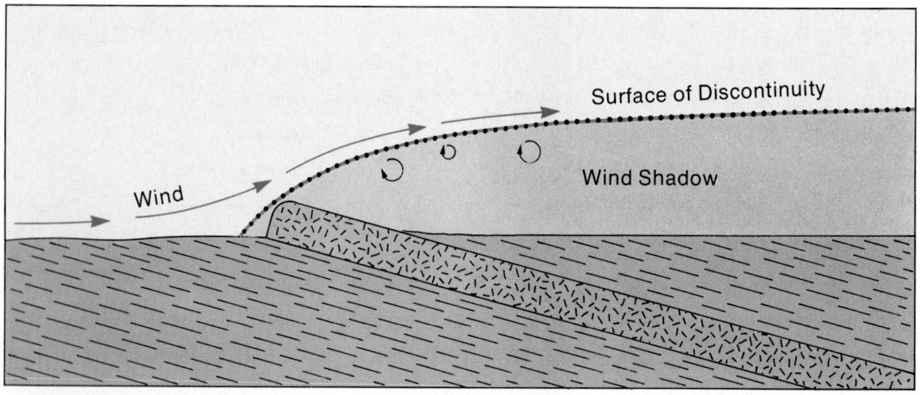

Figure 16–20 The air in front of an obstacle is divided by a surface of discontinuity—above which the air flows smoothly by and below which the air moves forward in eddies, and has a lower average forward velocity than the wind above the surface of disconformity. (Adapted from Bagnold, R.A., *The Physics of Blown Sand and Desert Dunes,* London, Methuen & Co., Ltd. 1941, p. 190.)

Figure 16–21 Aerial view of transverse sand dunes in Imperial Valley, California. Wind is blowing from the upper left side to the lower right of the area. Notice the more gentle windward slope of each dune and the more steeply inclined leeward slope. (Courtesy of the U.S. Geological Survey and J.R. Balsley.)

as it accelerates up the windward slope (Fig. 16–22). As more air is forced to rush through a smaller area, velocities increase to the point where grains blow off the top of the dune as fast as they are added. The height attained by a dune also depends on the overall wind speed as well as the size of the sand grains being deposited. In the Sahara Desert, dunes as tall as 100 meters are commonplace.

From the explanation of how dunes are formed, it is evident that they are not stationary features but rather experience more or less continuous migration. In some parts of the world, the migration of dunes poses a serious threat to villages and farms. Wind-breaks are constructed, and vegetation is planted to protect fields and property from the encroachment of sand.

Dunes usually occur in groups, either in roughly parallel migrating series or as irregular migrating complexes. Depending on the amount of sand available, the constancy of wind direction, and wind velocity, dunes may develop any of a variety of interesting and sometimes aesthetically pleasing shapes. Perhaps the most interesting forms are the **barchans** (Figs. 16–23 and 16–24). These are crescent-shaped dunes in which the "horns" are directed downwind. Barchans characteristically develop in regions where wind direction is rather constant and the supply of sand relatively limited. In fact, the surfaces of the ground adjacent to barchans may consist of bare hard rock, with no sand cover whatsoever. The "horns" on the barchans are a result of the wind sweeping around the sides of the accumulation of sand being less impeded and thus able to move sand somewhat more efficiently. As a result, the long horns of sand are extended to the leeward side. Crescent-shaped dunes may also form with their horns directed upwind.

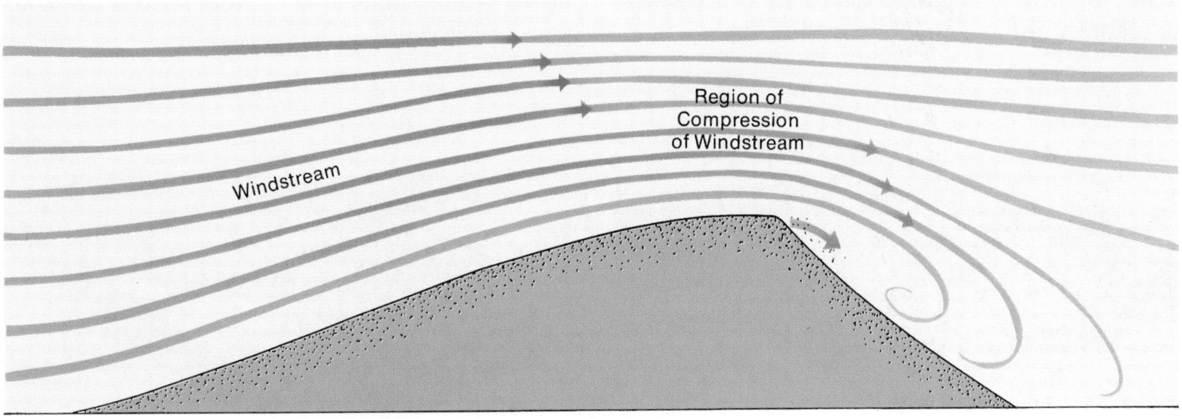

Figure 16–22 As the wind blowing over a dune is compressed into a smaller space, it increases in velocity and in the power to erode. A point is reached where the wind is able to blow grains off the top of the dune as quickly as they are deposited.

Figure 16–23 Barchan dunes, Egypt. (Courtesy of D. Bhattacharyya.)

Such accumulations are designated **parabolic dunes** and nearly always represent an accumulation of sand around a blowout (Fig. 16–25). Parabolic dunes are most abundant along shorelines.

Where winds are strong and there exists an abundant supply of sand, **transverse dunes** (see Fig. 16–21) with crests at right angles to the wind may develop. If such elongate accumulations are aligned parallel to prevailing wind directions, they are referred to as **longitudinal dunes** (Fig. 16–26). Longitudinal dunes are most prevalent in warm deserts where the wind blows strongly from a single direction and there is virtually no plant cover. In North Africa, such dunes are called **seifs**, an arabic word meaning "sword."

The shapes of dunes are largely controlled by wind strength, constancy, and direction, as well as sand supply and distribution of vegetation.

The sand in most dunes consists of quartz with lesser amounts of feldspar. At some localities, however, other minerals may predominate. For example, the dunes found in Bermuda are composed of calcite grains derived from the erosion of reefs and from accumulation of the shells of marine invertebrates. The brilliant snow-white dunes of the White Sands area of New Mexico are composed of gypsum.

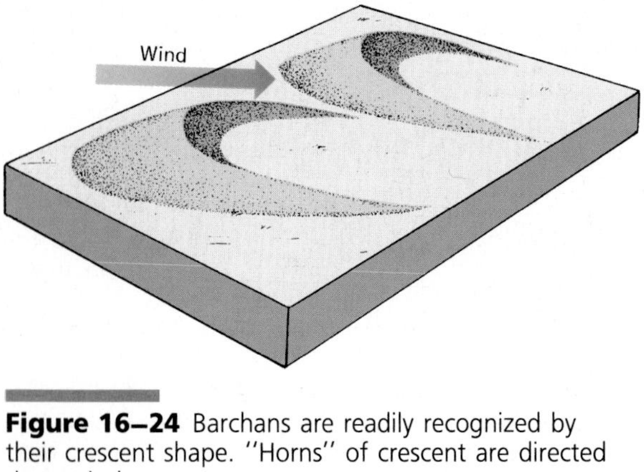

Figure 16–24 Barchans are readily recognized by their crescent shape. "Horns" of crescent are directed downwind.

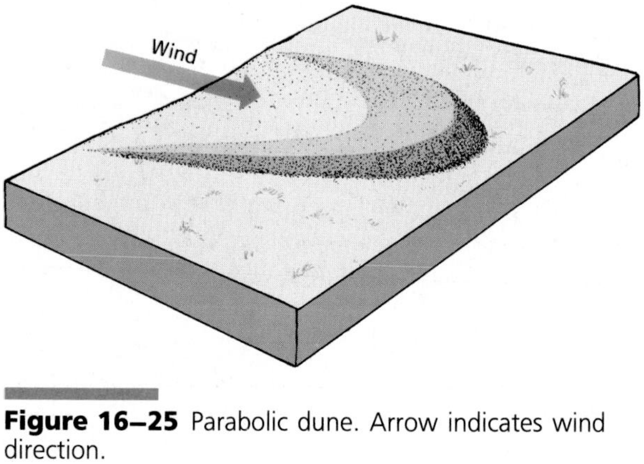

Figure 16–25 Parabolic dune. Arrow indicates wind direction.

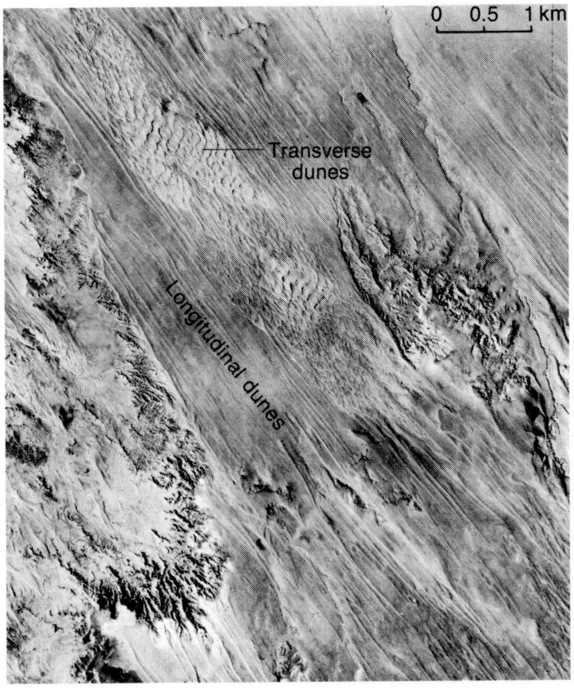

Figure 16–26 Longitudinal and transverse dunes, Coconino County, Arizona. Wind direction is toward southeast quadrant of photograph. (Courtesy of the U.S. Geological Survey.)

Eolian Cross-Stratification and Ancient Dunes

As grains are deposited on the lee slope of a dune, they periodically slide down in small avalanches in order to maintain the constant angle of repose. The result is a series of inclined layers of sand on the lee slope. These inclined layers are often developed at an angle to the beds in truncated underlying dunes and thus constitute eolian cross-stratification (Fig. 16–27).

The orientation of the inclined layers formed on the lee slopes of sand dunes is determined by the direction in which the wind blows. In general, these sloping foreset beds are inclined in the downwind direction. By applying this observation to ancient sandstone formations that exhibit eolian cross-bedding, it is sometimes possible to ascertain the probable directions of prevailing winds that existed hundreds of millions of years ago. For example, measurements made of the 200-million-year-old New Red Sandstone of Great Britain indicate it was deposited by winds that blew mostly from the east and northeast. By comparison with the present global wind patterns, the measurements confirm paleomagnetic evidence that, since the deposition of the New Red Sandstone, Britain and Europe have drifted northward and rotated clockwise by about 35°.

Unfortunately, it is not always easy to be certain that a particular sandstone is of eolian rather than subaqueous origin. One can find clues, however, that are useful in confirming an interpretation of eolian origin. Consider, for example, the Nugget Sandstone that crops out in Utah and Wyoming. In addition to interfering sets of cross-beds (which are often a response to variable wind directions), the Nugget contains ventifacts. Light flakey mica minerals are rare, and sorting excellent, just as one would expect in wind-deposited sediment. The sand grains of the Nugget are pitted as if by impact. Finally, rare finds of fossil plants and land reptiles have been discovered in the formation, thus attesting to its terrestrial origin.

The Nugget Sandstone is just one of many ancient formations containing well-preserved eolian cross-bedding. In Utah and surrounding states, geologists have been studying cross-stratification in the Navajo Formation for many years. The Coconino Sandstone displays magnificent eolian

Figure 16–27 Eolian cross-stratification in the Permian DeChelly Sandstone, Canyon DeChelly National Monument, Arizona. (Courtesy of E.O. McKee and the U.S. Geological Survey.)

cross-stratification in Arizona's Canyon de Chelly, and in England, the dunes of the Penrith Sandstone are famous among European geologists for their splendid eolian cross-stratification.

The dropping of sand grains over the crest of a dune onto the downwind slip face produces cross-stratification.

Ripple Marks

The surfaces of dunes and other sandy deposits, and also some of the bedding surfaces of eolian sandstones, are often characterized by subparallel ridges and hollows called **current ripple marks**. Current ripple marks may be formed either by water or wind. Wind ripples are a consequence of the saltation movement of sand. They will form spontaneously wherever there is an adequate supply of sand and the wind velocity is sufficient to erode and transport the sand-sized particles. After a few ripples have formed, additional ones develop quickly. Because of the average trajectory of incoming saltating grains, more sand will land on the windward side of the ripple than on the lee side (Fig. 16–28A). The windward side tends to be built up, while the lee side remains troughlike as a consequence of receiving too few grains. The ridge will steepen and the hollow deepen. A short distance downwind on the far edge of the trough, grains again impact with sufficient frequency to build a second ridge. This ridge in turn rises above the depression created by fewer grains impacting on its lee side. In this way ripple after ripple are formed. The spacing of ripples depends on the average length of the saltating jumps (Fig. 16–28B), which is in turn determined by the size of the sand grains and wind velocity. Maximal height of each ripple crest is reached when deposited grains intercept winds moving so fast that grains are blown off the crest as quickly as they are added.

Deserts

Characteristics of Deserts

Because deserts generally have only sparse soil-holding vegetation and also have dry soils, they are environments in which the effects of wind erosion are most noticeable. Geologists study such areas assiduously, for by recognizing present-day desert features in ancient sedimentary rocks, they can infer where desert conditions existed millions of years ago. It may come as a surprise to devotees of Hollywood films depicting the French Foreign Legion, but not all deserts are vast areas of drifting sands and searing temperatures. Indeed, only about one-fifth of the earth's total desert area is

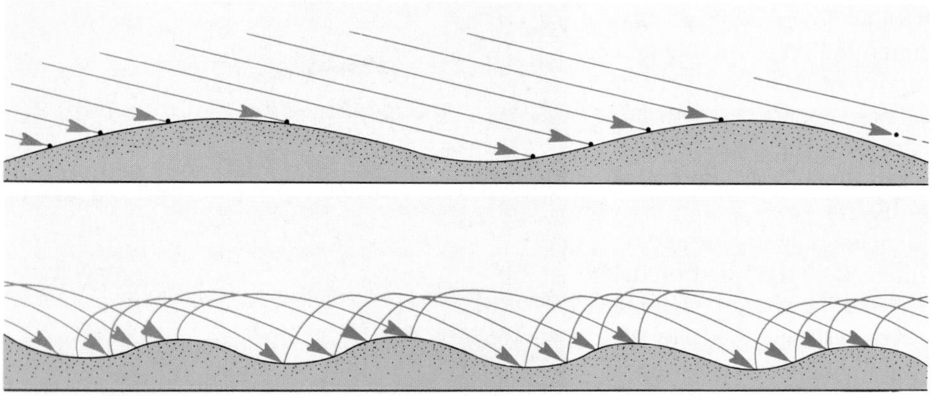

A

B

Figure 16–28 From cross-section A, it is apparent that the windward side of an uneven surface will receive a greater number of impacting grains than the lee side. Cross-section B suggests that there is a correspondence between the average distance between ripple crests and the average length of saltation paths. (After Bagnold, R.A., *The Physics of Blown Sand and Desert Dunes,* London Methuen and Co., Ltd., 1941, pp. 146, 150.)

covered with sand. Also, not all deserts are hot, for deserts also exist in frigid polar regions.

What all deserts do seem to have in common is a deficiency of vegetation so extreme that the region cannot support an appreciable human population. In polar regions, of course, the lack of vegetation results from the intense cold, but for most other deserts, the barren landscape is the result of an insufficiency of rainfall. In these warm deserts, annual rainfall does not exceed 25 cm, and the potential rate of evaporation exceeds that amount. Because the amount of rainfall is so low and evaporation so great, streams in warm desert regions rarely extend to the oceans. Vast floodplains are uncommon, and many streams are ephemeral. In addition to wind erosion, mass wasting, sheetwash, and rain-pelting are important erosional processes in deserts.

Low-Latitude Deserts

The majority of the world's great deserts are located in a belt between about 15° and 30° of latitude (Fig. 16–29). Here one finds the high-pressure belts and trade winds. These high-pressure belts (horse latitudes) are characterized by clear skies and dry cool air. As the cool air descends and is shunted obliquely toward the equator by the trade winds, it is warmed and thus tends to absorb moisture. Evaporation is favored over precipitation except where mountain ranges again deflect the winds to higher altitudes. The overall results of these globe-encircling belts of dry air are the world's most famous deserts. These include the Arabian and North African deserts (Fig. 16–30), the Kalahari of southwestern Africa, and the immense Victoria Desert of Australia.

Middle-Latitude Rain Shadow Deserts

Middle-latitude deserts are found beyond 30° of latitude to about 60°. Unlike the low-latitude deserts, they owe their existence to physiographic features near their locations. For example, many middle-latitude deserts are located on the lee sides of great mountain ranges. Winds blowing toward these mountains are shunted to high altitudes where they cool and drop the moisture they contain on the windward side of the range. As they continue across the mountains, they bring only dry

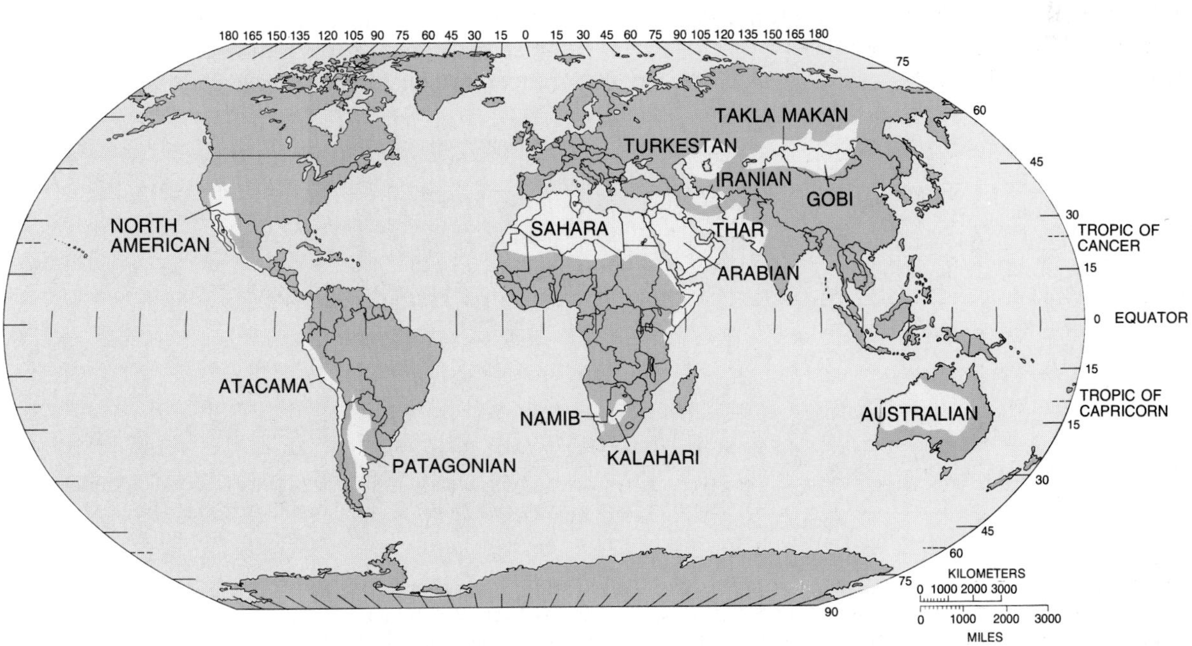

Figure 16–29 Deserts of the world. (From Navarra, J.G., *Atmosphere, Weather, and Climate,* Philadelphia, W.B. Saunders Company, 1979.)

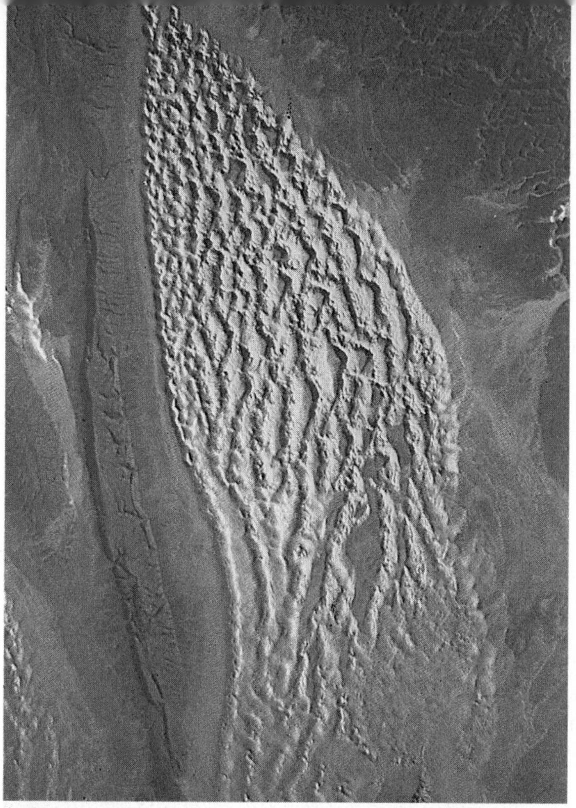

Figure 16–30 The Tifernine Dune Field of the Sahara Desert as viewed from the space shuttle in 1981. The dunes to the north (top) form long crescents (very large barchanes) with steep convex sides facing prevailing winds that blow from the west or southwest. To the south, the dunes are irregular, indicating no consistent wind direction there. The deep gullies were cut near the end of the Ice Age when rainfall was more abundant in this region. (Courtesy of NASA.)

air with high potential for evaporation to the regions downwind from the mountains. The desert areas of the Basin and Range Province are the result of such physiographic circumstances, as are several small deserts located along the east flank of the Andes in Argentina. It should be noted that these deserts actually owe their existence to plate tectonics. Plate convergence and subduction at the continental margin produced the mountains that resulted in the arid rain shadow.

Landforms of Deserts

Deserts are often harsh defiant regions that test the stamina of man and beast alike. Yet geologists freely relinquish studies of better watered terrain in order to solve a problem in desert geology, for in such places rocks lie naked at the surface and can be traced across the land for scores of miles.

Because they are not covered with a heavy mantle of weathered material and dense vegetation, landscape features of deserts have a stark ruggedness and clarity (Fig. 16–31). On every side one sees evidence of an insufficiency of water. As already noted, most desert streams are discontinuous and ephemeral. Channels may fill quickly after a sudden shower but are dry again in a few hours. For the most part, drainage systems are internal, and streams do not reach the ocean. The channels of most streams in deserts have rectangular cross-

Figure 16–31 A desert landscape in Canyon DeChelly National Monument, Arizona. The stream is termed an *ephemeral stream* because it is dry during most of the year and bears water only during and immediately after a rain. (Courtesy of the U.S. Geological Survey.)

Figure 16–32 Coalescing alluvial fans, Gulf of Suez area, Egypt. An ephemeral, braided stream struggles to transport the huge amount of debris supplied to it. (Courtesy of D. Bhattacharyya.)

sections as a result of caving of dry banks along vertical cracks, the intermittent nature of the flow, and the high proportion of sediment flushed down the channel as bed load. The often dry steep-walled channels are called **arroyos**.

Alluvial fans are often exceptionally well developed in desert regions that front on mountain ranges. As described in Chapter 13, alluvial fans are the result of rapid deposition of clastic sediment as a stream emerges from a mountain front and shifts radially back and forth across its former deposits. The alluvial fans of desert regions (Fig. 16–32) are noted for their symmetry. It is a characteristic of desert regions that as the fans along mountain fronts grow, they merge to form a large alluvial apron called a **bajada** (Fig. 16–33).

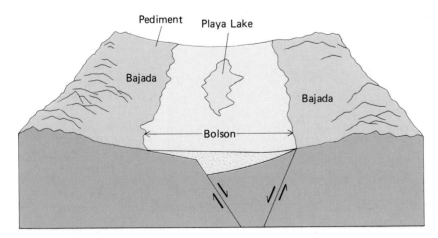

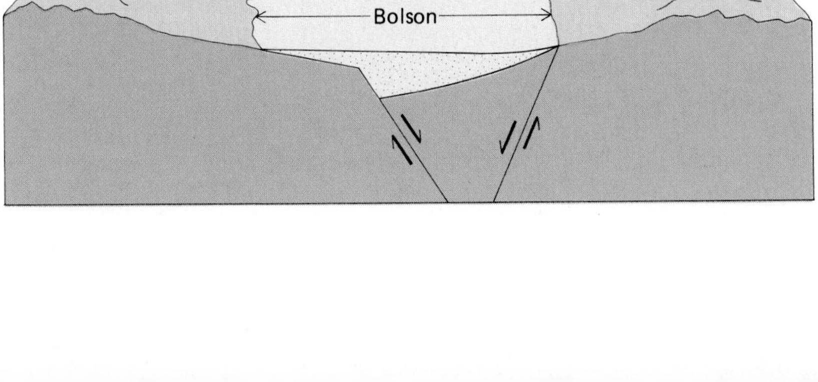

Figure 16–33 Playa lake formed on the bolson between adjacent pediments.

Figure 16–34 Playa lake with alluvial fans in the background, Death Valley National Monument, California. (Courtesy of L.E. Davis.)

Alluvial fans are ubiquitous features in the desert regions of the western United States and Mexico, where topographic basins are often fault-controlled and more or less surrounded by mountains. The basins, called **bolsons**, are great collecting areas for the erosional debris shed from the encircling ranges. Streams flow inward toward the center of the bolson.

During periods of rainfall, streams originating in the mountains may bring sufficient water into the center of the bolson, so that a temporary **playa lake** is formed (Fig. 16–34). Such lakes tend to lose their water rapidly as a result of evaporation and infiltration, and often only a salt encrusted former lake bed or **playa** remains (Fig. 16–35).

The mountains of arid regions are characteristically bordered by a sloping surface that extends downward toward the center of the bolson. The upper part of this slope is termed the **pediment** and is generally considered to have been formed primarily by erosion (see Figure 16–33). Bare rock is often exposed along the surface of the pediment, but it may also be thinly covered with alluvium. Pediments are slopes of transportation across

Figure 16–35 The dried bed of a playa lake is called simply a playa. This playa is located about 54 km west of Carson City, Nevada.

which streams, sheetwash, and mudflows carry sediment away from this mountain front. The history of pediment development begins with the uplift of a mountain range. Usually this uplift occurs along a fault system. For an initial period after uplift, alluvial fans form directly at the foot of the mountains. With time, the mountain front retreats and the smooth pediment begins to be leveled into the bedrock as indicated in Figure 16–36. Adjacent pediments may coalesce to form a pediplain.

Erosional processes in deserts differ from those in humid regions, principally because of sparse precipitation which results in a lack of vegetation and the development of internal drainage.

Landscape Evolution in Deserts

Because arid regions have interior drainage, their landscapes evolve somewhat differently than in humid regions. For example, in humid regions the plane toward which erosion progresses, namely sea level, is relatively constant. If there are no tectonic uplifts, the land area will be gradually and continuously reduced until it approaches sea level in elevation. By contrast, in deserts where streams empty into bolsons rather than the sea, the level toward which erosion progresses is the lowest part of the bolson. Also, as the bolson gradually fills, that level actually rises. Thus, with time the bolsons are raised and widened and the mountains reduced. Soon, the differences in elevation between mountains and basins become minimal, and geologic work decreases. Typically, after a bolson has been nearly filled with sediment, drainage will escape to another basin at a lower elevation. This causes renewed erosion of old alluvial fans and pediments.

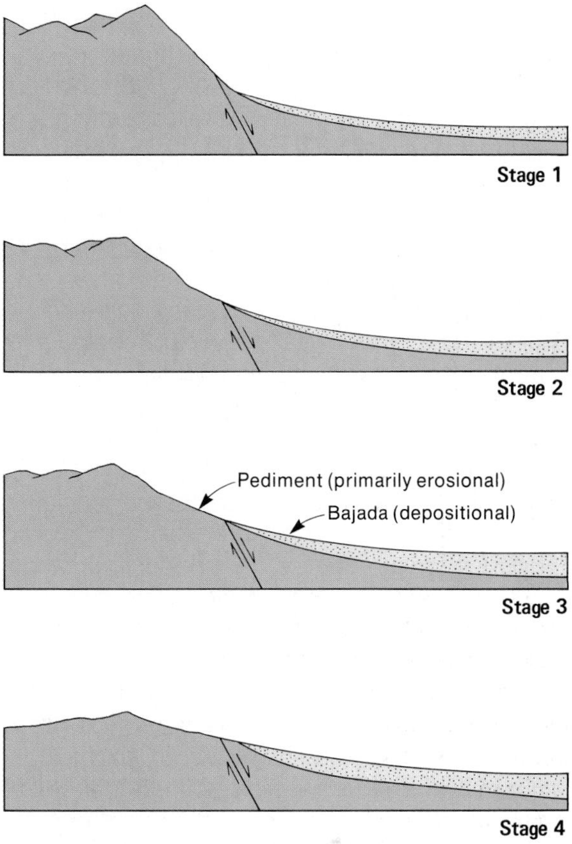

Figure 16–36 The development of a pediment and bajada.

Although the majority of deserts are characterized by internal drainage, there are a few notable exceptions. The Nile and the Colorado River, for example, cross desert regions and empty into the sea. These streams, however, originate outside deserts in areas having more rainfall. As a result, they have sufficient volume to maintain themselves as they cross the arid region despite losses due to evaporation and seepage.

Summary

Winds are movements of air masses over the continents and oceans. They are caused by unequal heating of the various parts of the earth's surface. For example, when an area of land is warmed by the sun's rays, the warmed ground radiates heat into the air. As the heated air expands, it becomes less dense, rises, and is replaced by cooler (hence denser) air from adjacent areas. In this way an air current is established. A column of warm air, being less dense than a column of cold air, does not exert as much pressure on the earth's surface. Areas overlain by warm air therefore constitute low-pressure areas, whereas areas of cold, dense, and hence heavy air are called *high-pressure areas*. Winds flow from high to *low-pressure areas*. Large roughly circular regions of low pressure and rising air are called *cyclones*, whereas *anticyclones* are characterized by high pressure and air that is descending.

Many winds are relatively local in origin and originate as a result of geographic or topographic conditions that lead to the establishment of local pressure gradients. There are also global patterns of air circulation that owe their existence to the greater amount of the sun's energy received at the equator, as contrasted to the poles, coupled with the effect of the earth's rotation.

There are many meteorologic reasons for studying winds, but geologists are most interested in the effectiveness of wind as an agent of transport and deposition. Air in motion does not have the density and weight of either water or ice and so is limited in the amount of geologic work it can accomplish. Even in deserts where dry soils and lack of plant cover enhance the opportunities for wind erosion, wind action does not sculpture major landforms. Rather, the abrasive blast of wind with its contained sand grains can only modify the features formed earlier by more powerful geologic agents. The effectiveness of wind in transporting fine sediments, however, must be considered an important geologic agent.

As is true for flowing water, wind velocity is lowest adjacent to the surface over which the wind moves. Also like flowing water, the kind of particles picked up most readily are within the size range of sand and silt. Clays tend to lie flat on the ground and ordinarily cannot be lifted into the air stream. At the other extreme, particles larger than sand are too heavy to be lifted by winds of ordinary velocities.

Whenever materials finer than sand are put into flight by wind, they are buffeted high into the atmosphere and carried away as *suspended load*. Normally this suspended load consists of particles finer than 0.15 mm in diameter. The fine material may be carried for great distances and is thus effectively separated from the sand grains moved along near ground level. This segregation of finer materials accounts for the absence of clay in the sandy deposits of many desert areas.

Bed load for wind is composed mostly of sand. These particles are transported by saltation or rolling along the ground. During saltation, they follow low-arcing trajectories that rarely exceed a meter or two in height. Upon striking the ground, the grain may temporarily remain, bounce back into the air, or dislodge another grain into the air stream.

Erosion by wind is accomplished by means of *deflation* and *abrasion*. Deflation is the actual lifting and blowing away of loose dry unprotected materials.

Abrasion (corrasion) is a form of natural sandblasting. Sand particles in saltation forcibly strike against rock surfaces and, in doing so, weaken bonding materials, free grains from their matrices, and even fracture grains.

Among the depositional features resulting from wind action are ripple marks and dunes. *Ripple marks* are formed more or less spontaneously as the result of bombardment of a sandy surface by saltating grains. Nearly always, this saltation will result in a slight unevenness of the sandy surface. The windward side of the uneven area will intercept more saltating grains than the lee side, and so a small ridge develops. The presence of one ridge or ripple creates a shadow zone, beyond which grains again land in greater numbers and form yet another ridge.

Dunes are depositional features that form as a result of the infall of sand in an area of weaker eddying currents. On their windward side, dunes develop a low slope as high-velocity winds sweep over the accumulated sand. The lee or backslope normally assumes the 35° angle of repose for dry sand. There are four types of dunes, depending on the supply of sand, wind velocity, and wind constancy. These are *barchan*, *parabolic*, *longitudinal*, and *transverse*.

Although interesting, sand dunes are not as important to human welfare as the wind-blown silty sediment known as *loess*. In the United States and Europe, particles constituting loess are thought to have been carried into suspension by winds blowing across aprons of glacial outwash or the broad floodplains of streams fed by glacial meltwater. In Asia, loess deposits represent the finer sediments carried by prevailing winds from arid regions.

Winds, of course, occur everywhere around the globe. The effects of wind action, however, are most noticeable in the world's great *deserts*. The landscapes of desert regions are generally more angular and bleak than those of humid regions where deep chemical weathering and a heavy covering of soil tends to round divides and reduce ruggedness. The drainage of deserts is internal, and streams flow into intermontane basins called *bolsons*. At the margins of bolsons, one finds the coalesced alluvial fans of *bajadas*, and, at the front of the mountains, erosionally leveled surfaces called *pediments*.

Terms to Understand

abrasion
anemometer
anticyclone
arroyo
bajada
barchan dune
blowout
bolson
coriolis effect
current ripple marks
cyclone
deflation
desert pavement
dune
eolian
horse latitudes
isobar
leeward
loess
longitudinal dune
monsoon
parabolic dune
pediment
playa
playa lake
pressure gradient
ripple mark
seif
slip face
surface of wind discontinuity
trade winds
transverse dune
ventifact
wind shadow
windward

Questions for Review

1. Explain how differences in temperature from place to place at the earth's surface can cause differences in pressure and how pressure differences can cause wind.

2. Why is wind erosion more effective in arid regions than in humid regions?

3. What are ventifacts and how are they produced?

4. Why do sand dunes rarely contain grains smaller than about 0.15 mm in diameter?

5. How do sand dunes form? Why do dunes have different shapes?

6. Why are the abrasional effects of winds limited to a height of a meter or so above the ground surface?

7. In what way is the lee face of a sand dune similar to the foreset beds of a delta?

8. Where uplifted horizontal-layered rocks of varying resistance to erosion are exposed in arid regions, the hard layers form steep cliffs and the softer layers more gentle slopes. Why is such cliff and slope topography more subdued in humid regions that have a similar geologic structure?

9. How does loess differ from residual soil? What evidence indicates loess was deposited by the wind?

10. What is a pediment? How does it differ in origin and composition from the alluvial fans that develop at the front of mountain ranges?

Supplemental Readings and References

Bagnold, R.A., 1941. *The Physics of Blown Sand and Desert Dunes*. London, Methuen & Co. (Repr. 1965, Halsted Press, New York).

Cooke, R.U., and Warren, A., 1973. *Geomorphology in Deserts*. Berkeley, Calif., University of California Press.

Glennie, K.W., 1970. *Desert Sedimentary Environments*. New York, Elsevier Publishing Co.

Goudie, A., and Wilkinson, J., 1977. *The Warm Desert Environment*. Cambridge, Cambridge University Press.

Ritter, D.F., 1978. *Process Geomorphology*. Dubuque, Ia., Wm. C. Brown Co.

The Ocean

17

Pacific Ocean along the Oregon Coast. Wave-cut bench exposed along shoreline. (Courtesy of L.E. Davis.)

Here in a protected environment covering almost three-quarters of the surface of the earth, the record of geologic events on and in the crust of the earth is likely to be preserved with minimum disturbance.

Tjeerd H. Van Andel 1968, Science

Introduction

There is a majesty and omnipotence about the sea. Poet and scientist alike are moved by its breadth and power, its ever-changing aspect, and its apparent timelessness. It is difficult not to be fascinated by the multitude of strange creatures that inhabit its depths, or intrigued by the deep chasms and spectacular mountains that exist in the darkness of its abyss. The ocean was the cradle of life, and in many ways it continues to nurture the inhabitants of this planet.

Programs to study the world ocean comprehensively are generally considered to have begun in 1872 with the scientific expedition of the naval ship *H.M.S. Challenger* (Fig. 17–1). The *Challenger*, a three-masted vessel with auxiliary steam power, was the research tool of a team of scientists charged by the British government to chart the ocean depths, measure movements of water masses, describe the sea's many creatures, and examine the ocean's chemistry and bottom deposits. In order to accomplish this formidable assignment, *H.M.S. Challenger* sailed over 110,000 km into all major ocean areas of the earth. During the 3.5 years of the expedition, enough data were collected to fill 50 heavy volumes. Over a century has elapsed since the *Challenger* expedition, yet scientists still refer frequently to the information provided in these reports.

The scientists aboard *H.M.S. Challenger* would have been astonished and envious of the sophisticated instrumentation now available for the study of the oceans. For example, they did not have the automated echo-sounding devices that are now regularly used in studying the topography of the ocean floors. Nor were they able to photograph the sea bottom, precisely measure subtle variations in magnetic and gravitational properties, or drill into the ocean floor and obtain core samples. These feats became possible with the tools aboard a modern research vessel, named the *Glomar Challenger* as an obvious tribute to its predecessor, *H.M.S. Challenger*. The *Glomar Challenger* (Fig. 17–2) was outfitted solely for the collection of fundamental knowledge about the seas. It has permitted scientists to know the age, composition, and relations of rocks and sediments on the sea floors, and verify concepts of sea floor spreading. Much of this information was derived from the vessel's capability for drilling into the ocean floor. In 1983, the *Glomar Challenger* was retired and was replaced two years later by the newly rechristened drillship, the *JOIDES Resolution* (Fig. 17–3).

Ocean Water

The Chemistry of Sea Water

The waters of the ocean are remarkably uniform in the kinds and proportions of dissolved elements. An average sample of sea water consists of about 35 parts per thousand dissolved matter (Tables 17–1 and 17–2). Of that dissolved matter, nearly 86 per

Figure 17–1 *H.M.S. Challenger* (From the Report of the Scientific Results of the Exploring Voyage of *H.M.S. Challenger* during the years 1873–1876, Narrative, Part II, 1885.)

Some of the elements in sea water are in the form of dissolved gases, the most important of which are carbon dioxide and oxygen. Carbon dioxide is essential to the growth of marine plants, and the amount of this gas often varies closely with the abundance of microscopic marine plants in any given area of the sea. By means of photosynthesis, plants replenish the supply of oxygen in sea water, and oxygen, of course, is essential to marine animal life.

Salinity is the measure of the ocean's saltiness. The principal elements contributing to the ocean's salinity are sodium and chlorine.

The ocean has a great capacity for absorbing carbon dioxide from the air, and this helps to regulate the amount of this gas in the earth's atmosphere. If the carbon dioxide content of the atmosphere rises, the rate at which it is dissolved in sea

A

B

Figure 17–2 (A) The *Glomar Challenger* lying at anchor in Okinawa Harbor in the Pacific. (B) "Roughnecks" preparing to add a length of drill pipe on the drill floor of the *Glomar Challenger.*

Figure 17–3 The *JOIDES Resolution,* successor to the *Glomar Challenger.* This oceanographic research vessel is designed for taking drill cores from the bottom of the floor of the ocean. The vessel is a floating oceanographic research center, with a seven-story laboratory stack that occupies 12,000 square feet. On-board facilities include laboratories for sedimentology, paleontology, geochemistry, and geophysics. The ship can suspend as much as 9000 meters of drill pipe to obtain core samples. (Courtesy of the Ocean Drilling Program, National Science Foundation.)

cent consists of sodium and chlorine. The other major constituents in order of decreasing abundance are magnesium, sulfate, calcium, potassium, and bicarbonate. Minute quantities of almost all the other naturally occurring elements are also present.

Table 17–1 Dissolved Solids in Ocean Water

Chemical Constituent	Content (parts per thousand)
Calcium (Ca)	0.419
Magnesium (Mg)	1.304
Sodium (Na)	10.710
Potassium (K)	0.390
Bicarbonate (HCO_3)	0.146
Sulfate (SO_4)	2.690
Chloride (Cl)	19.350
Bromide (Br)	0.070
Total dissolved solids (salinity)	35.079

From U.S. Geological Survey publication, *Why is the Ocean Salty?*

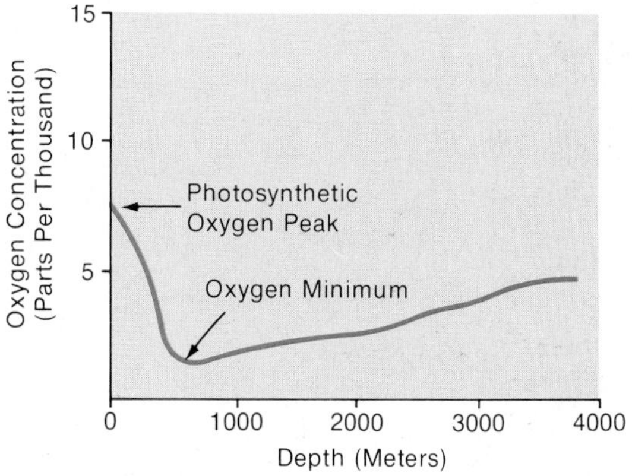

Figure 17–4 Variation of oxygen (O_2) in the North Central Pacific. (After McCormick, J.M., and Thiruvathukal, J.V., *Elements of Oceanography,* Philadelphia, W.B. Saunders Company, 1976.)

water also increases. Oxygen is also absorbed from the air above the oceans but in lesser amounts than carbon dioxide. Surface water tends to be richer in oxygen than deeper water because of its proximity to the air–water interface and also because marine plants flourish in the upper layers of water where there is abundant light for photosynthesis (Fig. 17–4). Beneath the surface layer, however, the amount of oxygen dissolved in the water is lower because of the consumption of the element by animals and bacteria, as well as in the reduction of organic waste as it sinks through the water column. Curiously, at depths greater than about 800 meters in many parts of the open ocean, oxygen concentrations increase somewhat, because of the transport of oxygen by deep currents from high latitudes where dense oxygen-rich water sinks and flows generally toward the equatorial regions (Fig. 17–5).

Table 17–2 Comparison of Percentage of Dissolved Solids in Ocean and River Water

Chemical Constituent	Percentage of Total Salt Content: Ocean Water	Percentage of Total Salt Content: River Water
Silica (SiO_2)	—	14.51
Iron (Fe)	—	0.74
Calcium (Ca)	1.19	16.62
Magnesium (Mg)	3.72	4.54
Sodium (Na)	30.53	6.98
Potassium (K)	1.11	2.55
Bicarbonate (HCO_3)	0.42	31.90
Sulfate (SO_4)	7.67	12.41
Chloride (Cl)	55.16	8.64
Nitrate (NO_3)	—	1.11
Bromide (Br)	0.20	—
Total	100.0	100.0

From U.S. Geological Survey publication, *Why is the Ocean Salty?*

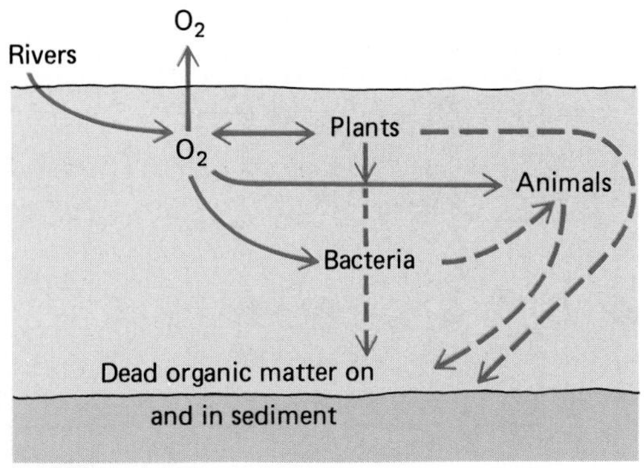

Figure 17–5 Oxygen cycle in the sea. Solid arrows indicate flow of elemental oxygen; dashed arrows indicate flow of oxygen in organic matter.

Physical Properties of Sea Water

Because of the dissolved salt in sea water, it differs from pure water in several physical properties. The density of pure water, for example, is 1 gram per cubic centimeter at 4°C. In constrast, the density of sea water increases with increasing salinity (Fig. 17–6) and is 2 to 3 per cent greater than the density of fresh water.

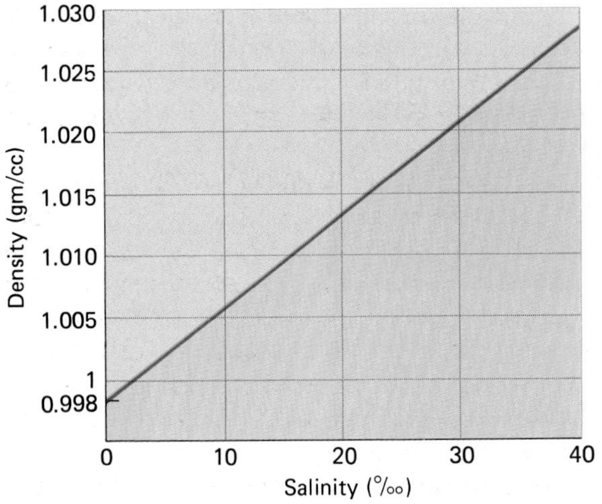

Figure 17–6 Variation in the density of ocean water with salinity, assuming constant temperature and atmospheric pressure. (From McCormick, J.M., and Thiruvathukal, J.V., *Elements of Oceanography,* Philadelphia, W.B. Saunders Company, 1976.)

The density of both sea and fresh water also changes slightly with variations in temperature. Once again, the changes are somewhat different in sea water than in fresh water. As fresh water is cooled to a temperature of 4°C, its density gradually increases. The dense water will sink to the bottom of a pond or lake and be continuously replaced by water from above until the entire water mass has reached 4°C. If fresh water is cooled further, it will expand and its density will decrease. The lighter water will rise to the surface where it freezes after the temperature has reached 0°C. By comparison, sea water of average salinity has a freezing temperature of about −2°C. Water much colder than −2°C is rarely encountered in even the high latitudes of the ocean. Unlike fresh water, the density of sea water increases all the way down to its freezing point (Fig. 17–7).

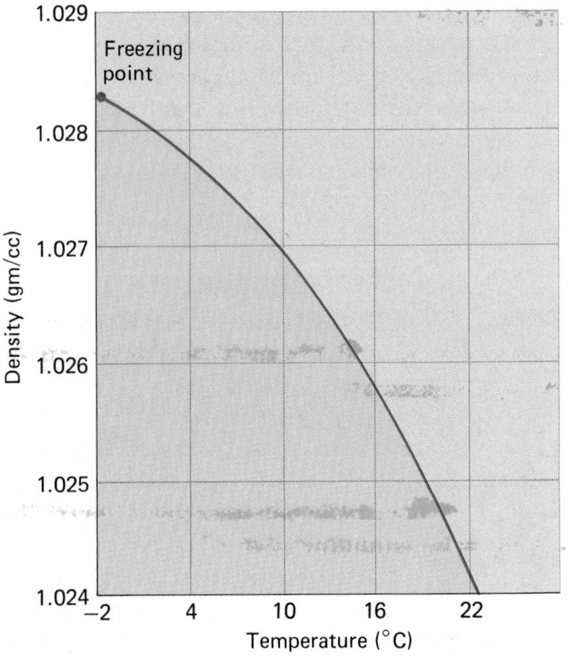

Figure 17–7 Variation in the density of ocean water with temperature at constant salinity and atmospheric pressure. (From McCormick, J.M., and Thiruvathukal, J.V., *Elements of Oceanography,* Philadelphia, W.B. Saunders Co., 1976.)

If temperature is lowered or salinity increased, ocean water will become more dense. This trend is reversed if temperature is raised and salinity reduced.

The temperature of the ocean varies with both latitude and depth and is strongly affected by oceanic circulation. Warmest water is found near the equator where temperatures average a balmy 28°C. Not unexpectedly, the coldest waters are found in polar regions where temperatures may reach −2°C. In the open ocean, the temperature of surface water changes very little with the seasons. This is because of the great volume of the oceans and the constant mixing that occurs. The ocean reacts very slowly to changes in air temperature and therefore serves as a heat regulator for the atmosphere.

In general, the temperature of ocean water decreases with depth. At moderate depths there is a layer of water in which the change of temperature with depth is at a maximum. This layer is called the **thermocline** (Fig. 17–8). Temperatures decrease only very slowly below the thermocline, until at a depth of 2000 meters the temperature remains virtually constant at 1° to 3°C.

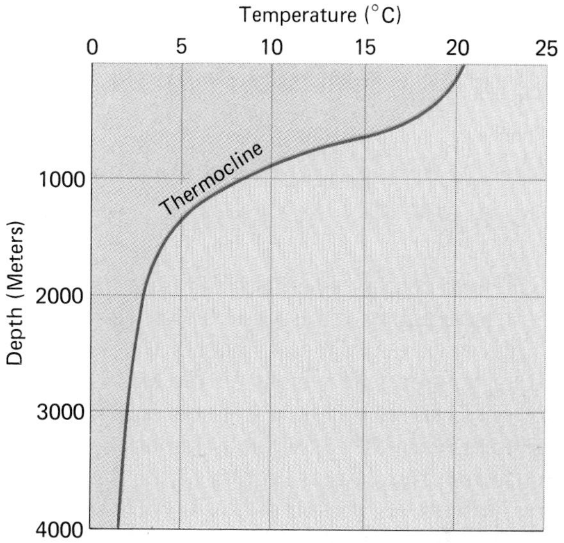

Figure 17–8 Warm water in the ocean is less dense than cold water and tends to stay at the surface, resulting in a layer of warm water over deeper cold water. The transition zone of rapid temperature change with depth is called the thermocline. (From McCormick, J.M., and Thiruvathukal, J.V., *Elements of Oceanography*, W.B. Saunders Company, 1976.)

The Depth and Breadth of the Ocean

As land dwellers, we sometimes need to be reminded that over 71 per cent of the earth's surface is covered by ocean. The ocean is not only broad but also deep. Although the average depth of the ocean is about 4000 meters, the floors of some deep-sea trenches are over 11,000 meters below sea level. The ocean depths are measured with the aid of **echo-sounding** devices. The method is based on knowledge of the speed with which sound travels in water. A short sharp signal or "ping" is emitted about every ¼ second by the research vessel as it travels along. Sound waves from the signal move through the water column and, upon reaching the sea floor, are reflected back to the ship (Fig. 17–9). The time interval between the emission of the sound and return of its echo from the sea floor is recorded, divided by 2 (because the sound makes a round trip), and multiplied by the speed of sound in ocean water (approximately 1500 meters per second). The results are automatically recorded on moving graph paper to provide a continuous profile of the sea floor above which the ship is traveling.

The principle involved in echo-sounding is also used to assist in the study of the sediments that blanket ocean floors. With a stronger signal, some of the energy can actually penetrate into the sea floor and is reflected off various layers of buried sediment and rock. This process is called **sub-bottom profiling**. In order to produce a signal strong enough to penetrate deep into the ocean floor, ships utilize an air gun or electric spark. The returned energy, which may be reflected from horizons as deep as 10 km below the ocean floor, is then detected by hydrophones, which are trailed behind the ship. Clearly, echo-sounding and bottom profiling are splendid tools for oceanographic research. As the vessel steams along, the moving graph paper reveals not only submarine canyons and mountains but an uninterrupted cross-section of what lies beneath the sea floor (Fig. 17–10).

The Earth's Surface Beneath the Oceans

Eighteenth-century scientists had little knowledge of the topography of the ocean floors. They lived at a time when depth measurements were made by letting down a lead weight on the end of a rope. Not only was this method time consuming, but in the open ocean it was virtually impossible to prevent error from lateral drifting of the weight, the ship, or both. As a result of these problems, only a limited number of soundings were made, except in bays and offshore areas where such information was vital for safe navigation. Oceanographers interpreted the few measurements available as indicating that the ocean floors were monotonous flat plains. With the advent of continuous topographic profiles from echo-sounding devices, it was shown that the ocean floors are as irregular as the surfaces of the continents. Beneath the waves lie canyons deeper than the Grand Canyon and mountain systems more magnificent than the Rockies.

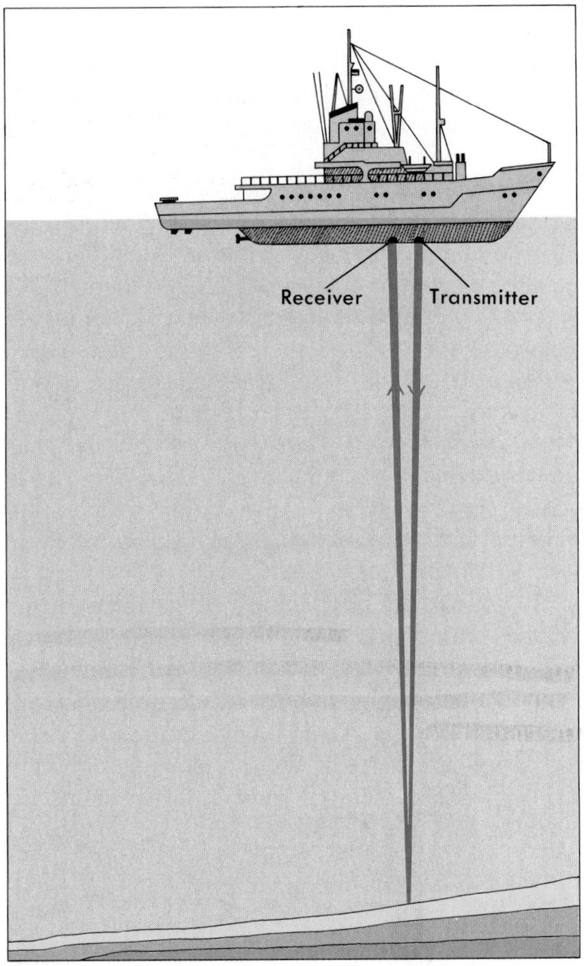

Figure 17–9 The principle of echo sounding. A transmitter sends a sound wave, which is reflected back to the surface by the ocean bottom and is picked up by a receiver. By knowing the total time involved and the speed of sound in the ocean (1500 meters per second), water depth can be determined. (From J.M. McCormick and J.V. Thiruvathukal., 1981. *Elements of Oceanography,* 2nd ed. Philadelphia, Saunders College Publishing.)

The Sea Floors near the Continents

Continental Shelves

As one departs on an ocean voyage from New York City, the first major oceanic topographic feature above which the ship travels is the continental shelf. The **continental shelves** are very gently sloping (an average of 0.1°) smooth surfaces that fringe the continents in widths that vary from only a few km to about 300 km and depths that range from low tide to about 200 meters (Fig. 17–11). In a geologic sense, these shallow areas are not part of the oceanic crust but resemble the continents in their structure and composition. They are, in fact, the submerged edges of the continents and have a readily apparent continuity with the coastal plains. The outer boundaries of the shelves are defined by a marked increase in slope to greater depths. The smoothness of parts of the continental shelves seems to have been produced in part by the action of waves and currents during the last Ice Age. At that time, sea level was periodically lowered by as much as 140 meters as a result of water's being locked in glacial ice. Waves and currents sweeping across the shelves shifted sediment into low places and generally leveled the surface.

For geologists specializing in the study of sedimentary rocks, the continental shelves hold great

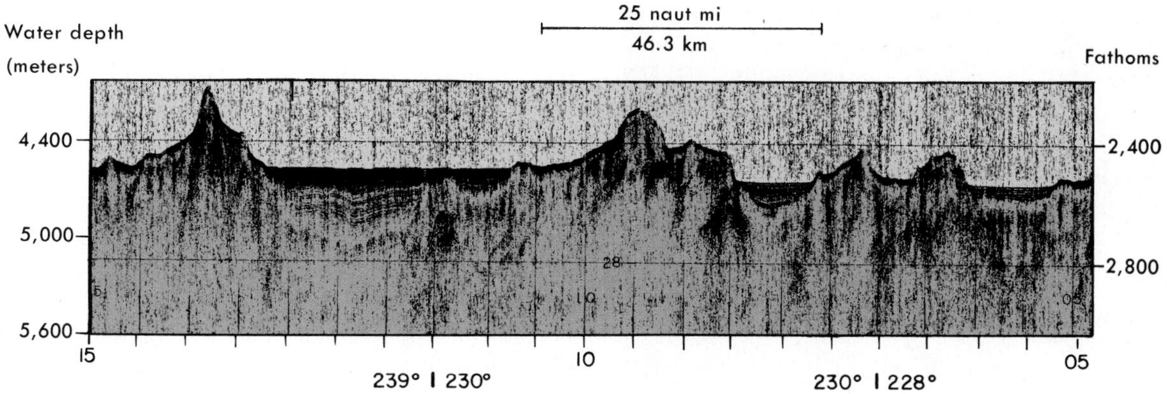

Figure 17–10 Continuous seismic profiles showing abyssal hills and abyssal plains along the edge of the Mid-Atlantic Ridge. (From Hayes, D.E., and Pimm, A.C., *Initial Reports of Deep Sea Drilling Project,* 1972. National Science Foundation publication, vol. 14, pp. 341–376.)

interest. All of the sediment eroded from the continents and carried to the sea in streams must ultimately cross the shelves or be deposited on them. Study of depositional patterns on the shelves has provided valuable insights into the origin of features in ancient sedimentary rocks now found high and dry on the continents. The shelves also have enormous biologic importance. Over most of their area, sunlight can penetrate all the way to the sea floor. Algae and other forms of plant life proliferate. Here one finds the "pastures of the sea" on which, directly or indirectly, a multitude of swimming and bottom-dwelling animals are dependent (Fig. 17–12).

Within the last few decades, we have become increasingly aware of the value of the continental shelves. Over 90 per cent of the world's seafood is captured in the waters of the shelves. Photographs of enormous offshore drilling rigs remind us daily of the oil and gas resources that lie beneath this part of the sea floor.

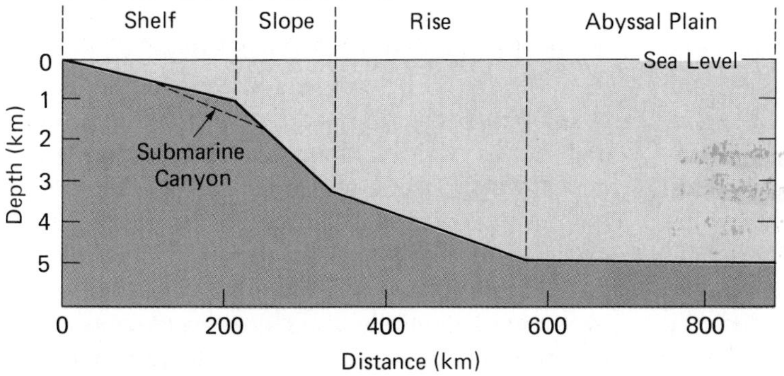

Figure 17–11 Vertical profile from a coastline outward into open ocean. The continental shelf is the relatively shallow part of the sea floor that extends from the coast to a line along which there is an abrupt increase in slope. In order to show the differences in the shelf, slope, and rise, a vertical exaggeration of nearly 200:1 is used (the vertical scale is 200 times that of the horizontal). Thus, on a true scale profile, differences in the elevations of the shelf, slope, and rise would not be nearly as apparent.

Figure 17–12 Marine life flourishes in many areas of the continental shelves, for light is required for the growth of marine plants, and plants are the basic components of the food chain that supports marine animals. (Courtesy of L.E. Davis.)

The Continental Slopes

Those areas of the ocean floor that extend from the seaward edge of the continental shelves down to the ocean deeps are named the **continental slopes**. Physiographic diagrams of the ocean floor are usually drawn with a large amount of vertical exaggeration, so that the continental slopes appear as steep escarpments. Actually, the inclination of the surface is only 3° to 6°. From the sharply defined upper boundary of the continental slope, the surface of the ocean floor drops to depths of 1400 to 3200 meters. At these depths the inclination of the ocean floor becomes gentler. The less pronounced slopes comprise the **continental rises**. Most of these areas of lesser slope and low relief appear to be wedgelike accumulations of sediment that have been transported across the shelves and deposited in the deep ocean at the base of the continental slopes.

Submarine Canyons and Valleys

Marine geologists conducting submarine topographic surveys of the continental shelves have long been aware of the large canyons and deep valleys on the continental shelves and slopes. Some of these features, like the Congo Canyon at the mouth of the Congo River, and the submarine canyon seaward of the mouth of the Hudson River, are clearly delineated on profiles obtained by echo-sounding. Other canyon-like features are now filled with sediment and have been detected by drilling and subbottom seismic profiling. The majority of such features on the continental shelves appear to have been formed by rivers at times when sea level was lower and broad tracts of the shelves were exposed to subaerial erosion. The best evidence for their fluvial origin is their continuity with valleys on land. Indeed, about 80 per cent of submarine valleys appear to be extensions of land valleys. Those that are not associated with land valleys are most often found at the outer edge of the shelf and on the continental slope. These canyons are inferred to have been eroded by submarine processes such as sand slides, slumping, and currents of sediment-laden water called turbidity currents.

The two processes for the origin of **submarine canyons** are not mutually exclusive. In many canyons (e.g., the Hudson River Canyon), the part closest to the coast had formed at low stands of sea level by stream erosion, but the distal parts were formed by submarine processes.

Water is said to be turbid when it contains an abundance of suspended particles, and a **turbidity current** is the movement of this dense mixture down a slope beneath clear or less turbid water (Fig. 17–13). The sediment in suspension causes the moving mass to have greater density than less turbid water. The heavy water mass flows close to the sloping ocean floor, picking up additional loose material as it races along. Turbidity currents can develop astonishing speed and are potent agents for the transport of sediment. A turbidity current initiated by an earthquake-triggered landslide off the Grand Banks of Newfoundland in 1929 crossed over and sequentially broke a series of trans-Atlantic telegraph cables. The time each cable was broken was automatically recorded, indicating the troublesome turbidity current was traveling 100 km per hour over the steeper parts of the continental slope and continued out into the Atlantic for well over 600 km (Fig. 17–14).

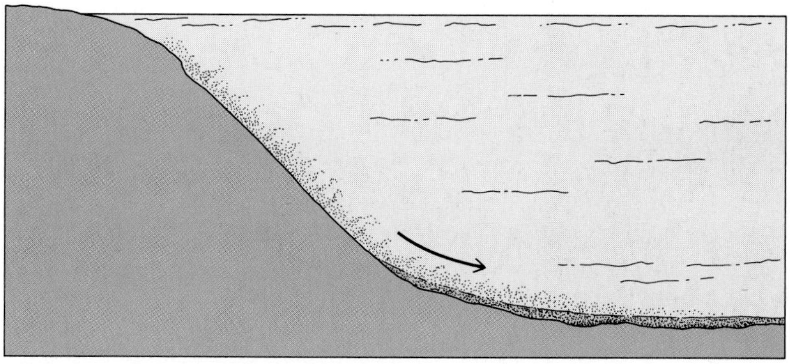

Figure 17–13 A turbidity current is a mass of moving water that is denser than surrounding water because of its content of suspended sediment. Turbidity currents flow along the slopes of ocean (or lake) floors because of their high density.

Submarine canyons formed by submarine processes are quite different from canyons and valleys formed subaerially on land. In most cases, their gradients are steeper, the canyon floors deepen consistently seaward, and the walls are more precipitous. Indeed, the height of the walls of submarine canyons may far exceed those of even the most magnificent of land canyons. The Grand Canyon of Colorado has a north wall that is about 1676 meters high, whereas the sides of the Great Bahama Submarine Canyon reach an astonishing height of 4400 meters above the canyon floor. Submarine canyons vary in length from only a few km to hundreds of km. The Bering Canyon north of the Bering Islands, for example, is 442 km long.

In those places where submarine canyons emerge onto the sea floor, the sediment carried down the canyons by turbidity currents may be deposited as **deep-sea fans** (Fig. 17–15) that resemble the alluvial fans found at the foot of mountains on land. Submarine canyons aligned with the mouths of major streams serve as conduits for huge amounts of land-derived sediment that produce the deep-sea fans.

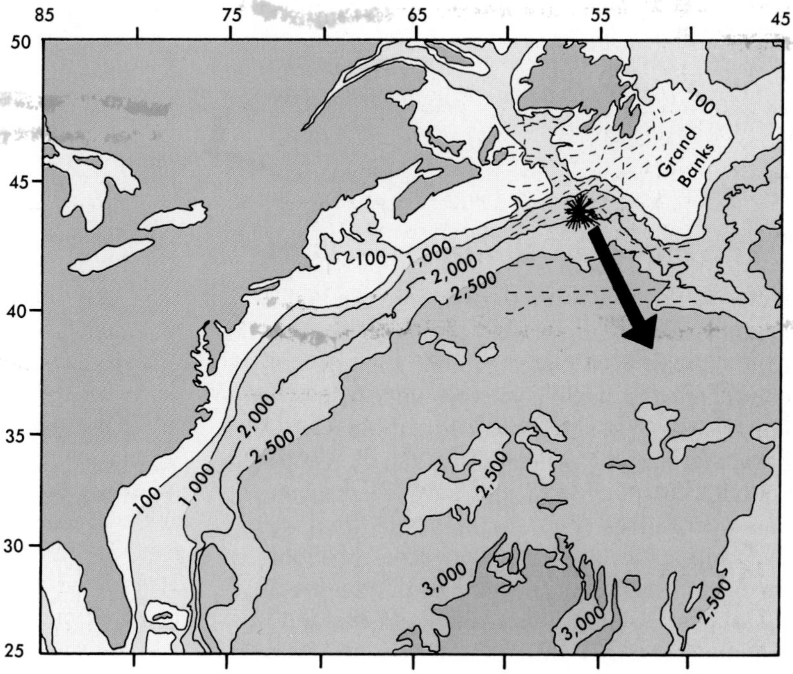

Figure 17–14 Breaking of the submarine telephone and telegraph cables after the 1929 Grand Banks earthquake near Newfoundland. The cables (dashed lines) were broken in a sequential manner after an earthquake triggered a turbidity current (indicated by the arrow). (After Heezen and Ewing, 1952, and the U.S. Naval Oceanographic Office, 1965. *Oceanographic Atlas of the North Atlantic Ocean,* Publication 700.)

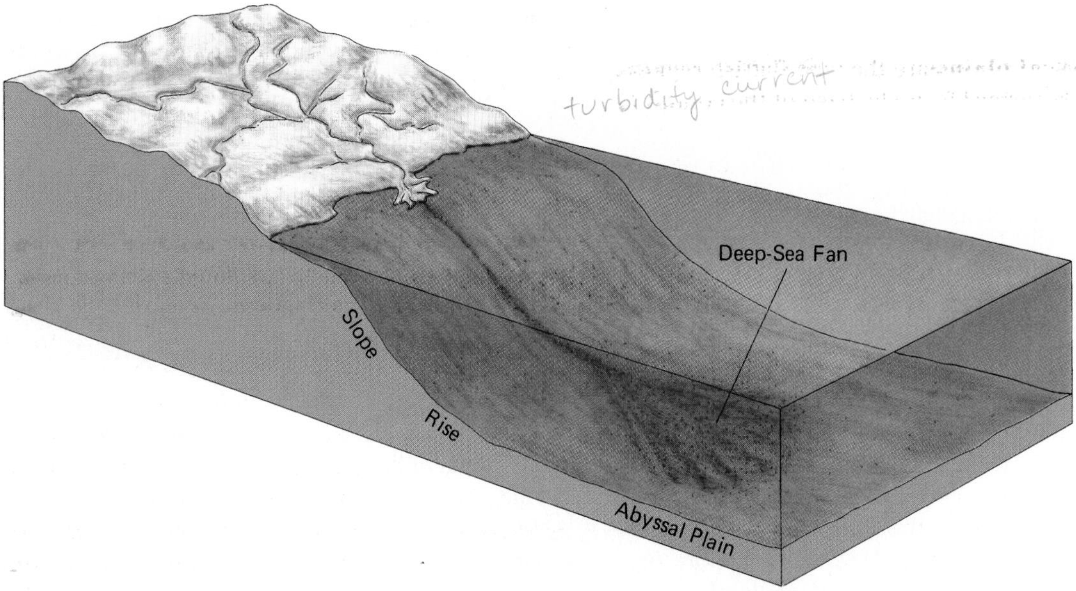

Figure 17–15 Deep-sea fan built of land-derived sediment emerging from the lower part of a submarine canyon. Such fans occur in association with such large rivers as the Amazon, Congo, Ganges, and Indus. (Vertical exaggeration 200:1.)

Deep sea fans at the base of submarine canyons are an exception to the generalization that most land-derived sediment carried to the ocean is deposited on the continental shelves.

The Floors of the Ocean Basins

The ocean basins are located between the continental rises. In general, they lie in the depth range from 4600 to 5500 meters. Unlike the crust of the continents, the ocean basins are underlain by igneous rocks of basaltic composition. Most of these basalts are rather deficient in alkali elements like sodium and potassium and have been given the name **tholeiites** (Fig. 17–16). There are a variety of topographic forms developed on the floors of the ocean basins, but the most distinctive are abyssal plains, abyssal hills, oceanic rises, and seamounts (see Fig. 1–12).

Figure 17–16 Pillow of tholeiitic lava from the summit area of a small oceanic volcano near the East Pacific Rise at about 12°N. Lat. The structure is termed a "trapdoor" lava pillow because liquid magma is withdrawn during growth through an opening and central cavity. The central cavity is visible on the top broken surface. (Courtesy of R. Batiza.)

Abyssal Plains

The **abyssal plains** are the vast flattish regions that extend seaward from the base of the continental rises. Their flatness can be attributed to sedimentation that has buried many of the irregularities that once existed on the ocean floor. Here and there, irregularities or extinct volcanoes interrupt the monotony of the abyssal plains. Such irregularities are called abyssal hills. They appear mostly in the areas where there has been insufficient sediment to cover them. Most of the sediment of abyssal plains consists either of very fine mineral particles derived from land or of the microscopic remains of marine organisms. In addition, an appreciable amount of the sediment on the abyssal plains appears to be fine particles from nearly spent turbidity currents.

Oceanic Rises, Seamounts, and Guyots

Oceanic rises (see Fig. 17–11) are elevated tracts of the sea floor that are hundreds of square miles in area, that stand a few hundred meters above the surrounding abyssal plains, and that are distinctly separated from continental margins or oceanic ridges. The surface of a rise may be either relatively smooth or quite irregular. Rises are found around Bermuda (the Bermuda Rise), as well as in the Pacific, Indian, and South Atlantic oceans. They are gentle upwarpings of the ocean floor, and should not be confused with features like the East Pacific Rise, which is part of the ocean ridge system and not a true oceanic rise.

Among the most remarkable features of the sea floor are rather symmetric isolated volcanic peaks having a height of 900 meters or more. They are called **seamounts** (Fig. 17–17) if they are conical and **guyots** if, in addition, they have flat tops. Seamounts are known on the continental rises but are much more numerous on the ocean basin floors along divergent plate boundaries where lavas are extruded. Others may have been formed in linear sequences as the ocean crust moved over "hot spots." Where the lava is built up above sea level, seamounts form islands such as those in the Hawaiian islands (see Fig. 4–35). Regardless of origin, there are more than 10,000 seamounts and guyots in the Pacific Ocean Basin.

Seamounts may originate along spreading centers, or they may form over local hot spots in the mantle.

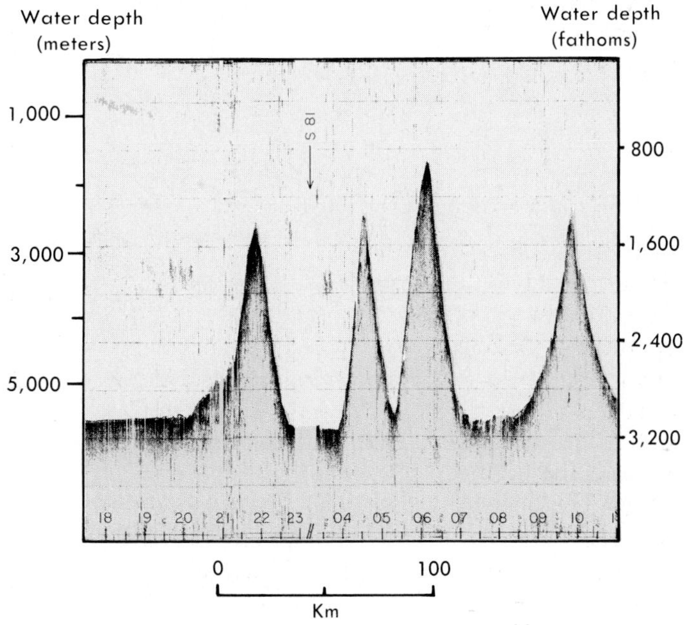

Figure 17–17
Continuous seismic profile across seamounts near the Mid-Atlantic Ridge. (From Hayes, D.E., and Pimm, A.C. *Initial Reports of the Deep Sea Drilling Project,* 1972. National Science Foundation, vol. 14, pp. 341–376.)

Guyots were named for the Swiss geographer Arnold Guyot (1807 to 1884). They are of special interest to geologists because of their remarkable flat tops. The most prevalent theories explain the flat tops as either drowned wave-beveled surfaces or coral reefs that had been built on the summits of barely submerged volcanoes. In support of these ideas, the top surfaces of many guyots are encrusted with the remains of ancient coral reefs, while a few others have yielded dredge samples of cobbles and pebbles that appear to have been shaped by wave action. Waves and coral reefs, however, are features that characterize the uppermost level of the sea. How then did the guyots reach the great depths at which they are now found? Either sea level rose to cover them, or the guyots were carried downward by subsidence of the oceanic crust.

Because the tops of some guyots lie 2000 to 3000 meters below sea level, eustatic rise in sea level cannot provide a general explanation for their submerged state. Rather, some guyots isostatically subsided simply because the oceanic crust has insufficient strength to support the weight of an entire volcano over long periods of time. Even more important, large numbers of guyots reached their present depths as a result of the gradual subsidence of oceanic crust when it cooled and was conveyed off the elevated flanks of the mid-oceanic ridges (Fig. 17–18).

Mid-Oceanic Ridges

Because of their impressive topography, distinctive structure, areal extent, and importance in global tectonics, **mid-oceanic ridges** deserve a prominent place in any classification of this planet's major geologic features (Fig. 17–19). Mid-oceanic ridges extend along 64,000 km of the sea floor and nearly equal the continents in total area (see Fig. 1–12). This world-encircling system can be traced down the middle of the Greenland Sea and the North and South Atlantic Oceans. It then swings eastward into the Indian Ocean basin and continues along between Australia and Antarctica to enter the Pacific Ocean. Once in the Pacific, the ridge is continued northward along the eastern side of the ocean basin, where it is called the East Pacific Rise.

The Mid-Atlantic Ridge is a widely studied segment of the ridge system and illustrates well many of the characteristics of mid-oceanic ridges. In its southward trend down the Atlantic Ocean basin, it parallels the margins of Africa and South America. Altogether, the ridge occupies about a third of the Atlantic Ocean basin (see Fig. 1–12) and rises over 3 km above the basin floors. Here and there some of its highest peaks project above sea level to form islands such as those of Iceland and the Azores.

In 1973 and 1974, the French-American Mid-Ocean Undersea Study (FAMOUS) undertook de-

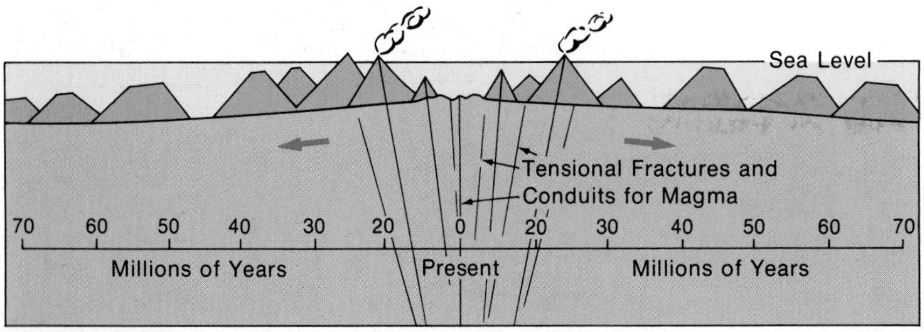

Figure 17–18 The history of guyots begins with volcanoes that originate near the axis of a mid-oceanic ridge or at a hot spot and progressively move away from their original location as they are conveyed by lithospheric plates. As the volcanoes are conveyed down the flank of the ridge or subside isostatically, their tops may be flattened by wave action, and they ultimately sink to great depths beneath the ocean surface.

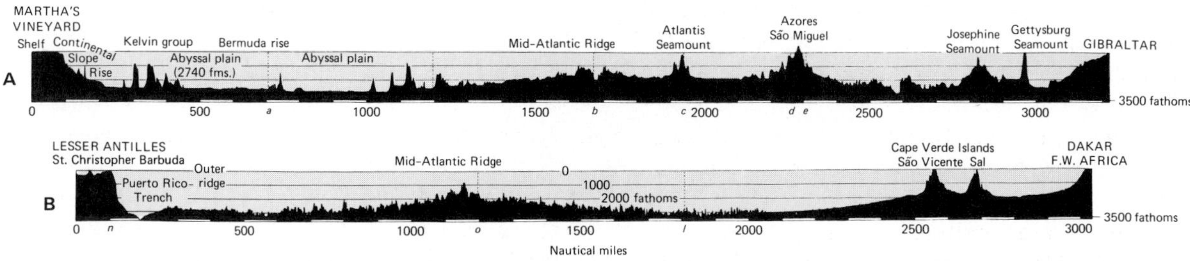

Figure 17–19 Bathymetric profiles of the Atlantic Ocean from (A) Martha's Vineyard, Massachusetts, to Gibraltar and (B) the island of Barbuda (about 100 km east of Puerto Rico) to Dakar, Africa. (From Heezen B.C. et al., 1959 *The floors of the ocean, pt. 1,* The North Atlantic, *Geol. Soc. Am. Special Paper* 65.)

tailed mapping of the Mid-Atlantic Ridge southwest of the Azores. Included in the results of this study were cross-sectional profiles across the ridge that clearly showed the presence of a central **rift valley** resembling a huge graben. This central rift was bounded on either side by normal faults, along which basaltic melts rose from deep zones to form dikes, sills, and submarine lava flows. Similar rift zones have been inferred from echo-sounding profiles along many other segments of the world ridge system. They appear to mark the location of divergent plate boundaries. As lithospheric plates move apart, the crust is subjected to tension, normal and transverse faults develop, and basaltic lava is able to rise (Fig. 17–20) and become incorporated into the trailing edges of the diverging plates. From this lava, the mid-oceanic ridges are constructed.

Deep-Sea Trenches

The deepest parts of the oceans are elongate, often arcuate troughs called **deep-sea trenches**. The floors of these great deeps lie as much as 11,000 meters below sea level. There is no light in the trenches, and pressures may exceed that at sea level by over 1000 times. Most deep-sea trenches are only 40 to 120 km in width, but a few are thousands of km in length. Trenches lie either along the oceanward side of volcanic island arcs, as exemplified by the Aleutian Trench, or adjacent to coastal ranges of continents, as is the case with the Peru–Chile Trench off the west coast of South America. The Mariana Trench (Fig. 17–21) in the North Pacific Ocean is particularly noteworthy for its record depth of 11,022 meters.

Deep-sea trenches and associated island arcs are vibrant with geologic activity. Volcanic eruptions and earthquakes are frequent events. When plotted on cross-sections of the trench and adjacent tracts, the foci of the earthquakes define a zone that slopes downward at an angle of 30° to 60° under the arcs and toward continental margins. Earlier we defined these trends as Benioff Seismic Zones. We also noted that strong negative gravity anomalies occur over deep-sea trenches. The Benioff Zones and negative anomalies indicate that, along the deep-sea trenches, the oceanic crust

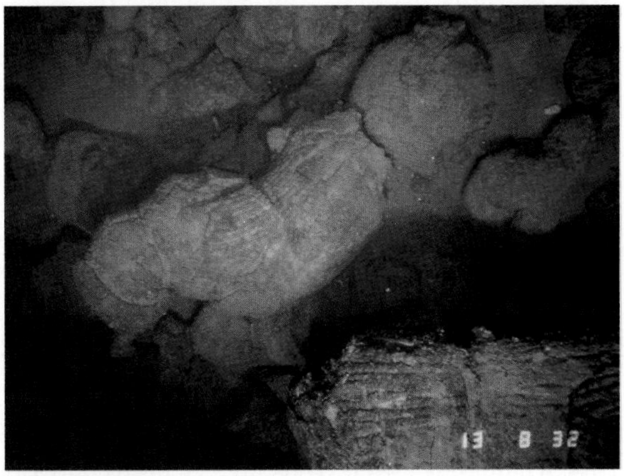

Figure 17–20 Pillow lavas photographed on the southern Juan de Fuca Ridge, Pacific Ocean floor. Photo covers 5 meters across. (Photograph by W.R. Normark, courtesy of the U.S. Geological Survey.)

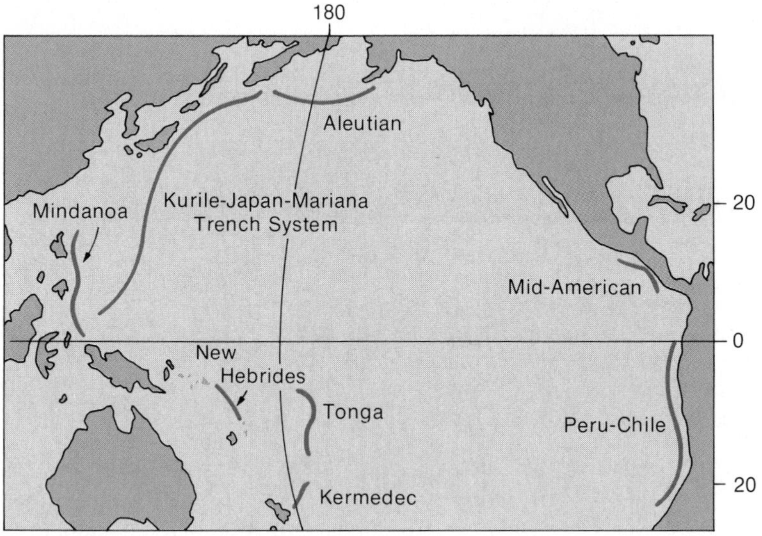

Figure 17–21 Major trenches of the Pacific Ocean.

is bent and carried downward into the upper mantle. The result on the ocean floor is the elongate depression of the deep-sea trench. As discussed in previous chapters, deep-sea trenches are ocean-floor manifestations of *subduction zones*. They are one of several major global features that form along the *leading edges* of tectonic plates.

Mid-oceanic ridges are zones along which magma rises from the mantle and joins the margins of diverging lithospheric plates. Deep-sea trenches are zones along which lithospheric plates plunge back into the mantle.

Deep-Sea Sediments

Deposits of the deep ocean basins were first examined by scientists aboard the *H.M.S. Challenger*, who found that there were two abundant kinds of sediment on the basin floors. In the deepest regions, the *Challenger* crew brought up samples of clay, which they called "red clay." Actually, abyssal clays are more of a brownish color, and so the term **brown clay** is now preferred. In the shallower regions of the ocean basins, various kinds of very fine organic deposits were raised onto the deck of the *Challenger*. Because of their fine texture and slippery feel while still wet, these materials were named **oozes**.

Inorganic Deep-Sea Deposits

The deep-sea sediment called brown clay consists of clay minerals and tiny particles of other silicates. For the most part, these particles originated on land and have been carried by winds and currents into the open oceans. Along with lesser quantities of volcanic and cosmic dust, these materials accumulate at the nearly imperceptible rate of 1 mm per 1000 years. Centuries elapse before a particle reaches the ocean floor, and so there is ample time for iron in the particles to be oxidized by even meager amounts of dissolved oxygen in the water column. Not all of the components of abyssal clays are derived from a distant source. A mineral named *phillipsite* grows in place on the ocean floor by the chemical alteration of particles of volcanic glass. Manganese nodules (Fig. 17–22) may also form in place on the sea floor. These curious spheric concretions are chemically precipitated from sea water as coatings of manganese dioxide around some "seed object" such as a fish tooth or rock fragment.

Occasionally, while examining the long cylindric cores of deep-sea sediments (Fig. 17–23), geologists are startled to discover a layer of sand or silt-sized particles. In most cases, the coarser material

Figure 17–22
Manganese nodule recovered from the floor of the Pacific Ocean, southeast of Hawaii, at a depth of 5500 meters. The nodule has been cut in half in order to show its internal structure and outside appearance. (Courtesy of the Institute of Geological Sciences, London.)

was brought to the site of deposition by turbidity currents. Relatively large pebbles and cobbles are also sometimes found in deep-sea tracts near the poles. This kind of larger debris has been dropped from drifting masses of melting glacial ice. It is especially prevalent on the floor of the ocean around Antarctica.

Organic Deep-Sea Deposits

There are science students who have been so impressed by the beauty of the skeletal remains found in oozes that they have devoted their lives to the study of these minute organic structures. The principal contributors to deep-sea oozes are protozoans, tiny mollusks, and algae. Depending on the composition of their hard parts, these organisms contribute to the formation of either calcareous or siliceous oozes. By definition, an ooze must consist of at least 30 per cent of organically derived material. Therefore, oozes are not found in nearshore areas because the influx of terrigenous sediment is so great that the organic contribution does not reach the 30 per cent level.

Calcareous Oozes

The principal constituents of calcareous oozes are the calcium carbonate coverings of foraminifers, coccoliths, and pteropods. **Foraminifers** (Fig. 17–24) are marine protozoans, whereas coccoliths (Fig. 17–25) are tiny calcite structures that encase the living cell of members of the algal family *Coccolithophoridae*. The term **Globigerina ooze** is frequently employed for a calcareous ooze composed of a mixture of coccoliths and tests of foraminifers. The "forams" in Globigerina ooze are species that, like *Globigerina*, are adapted to a floating (planktonic) existence. The tests are regularly vacated as part of the reproductive cycle of these organisms, and the abandoned shells settle to the sea floor. Only a very small percentage of the remains in oozes represent fatalities. The pteropods (Fig. 17–26) are a group of tiny marine gastropods (snails) that are adapted to a floating life. Their delicate shells are frequently found in Globigerina ooze, and locally they may be so abundant as to justify the name **pteropod ooze**.

Surface waters of the oceans are usually supersaturated with calcium carbonate. Foraminifers, coccolithophores, and other organisms extract this material in quantities that exceed by six times the amount of calcium carbonate contributed each year to the oceans from rivers. Initially, one might surmise that the biologic consumption of calcium carbonate would have depleted the supply long ago. There is, however, a recycling mechanism that helps to maintain the supply. If the tiny shells settle through a column of water no deeper than about 3000 meters, they will accumulate intact and form calcareous ooze (Fig. 17–27). Below that depth, however, the colder water holds more carbon diox-

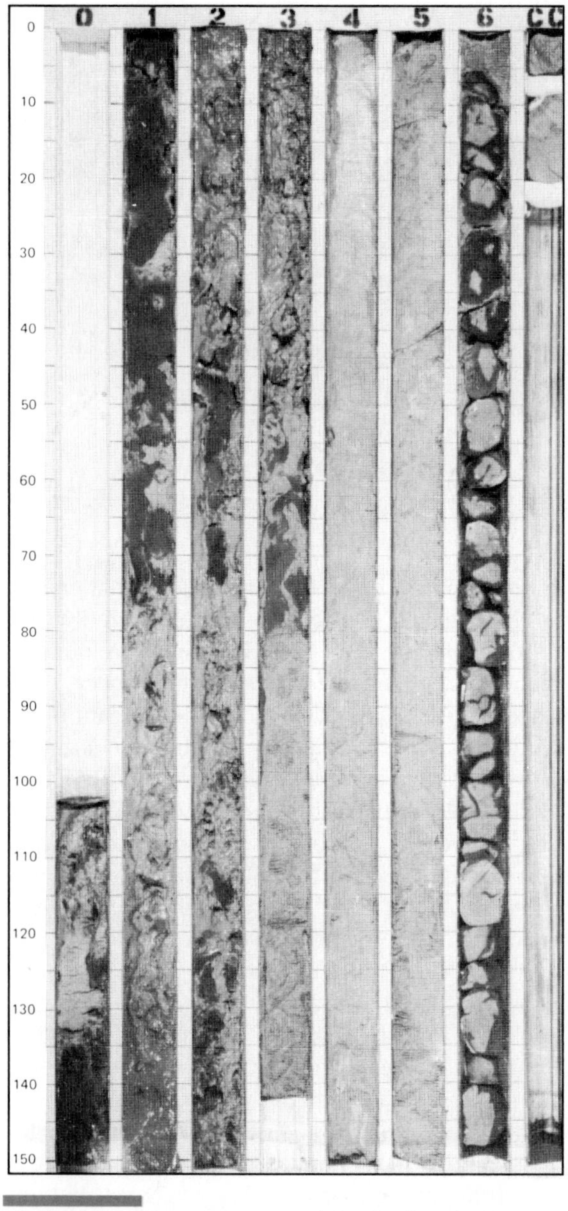

Figure 17–23 Core recovered from a location just east of the Mid-Atlantic Ridge, about 23°N, 43°W, by the *Glomar Challenger* as part of the Deep Sea Drilling Project. Sections 1–3 are disturbed sediment and oozes; sections 4–6 are largely calcareous oozes. Basalt begins at top of section 6. Scale is in cm. (Courtesy of the U.S. National Science Foundation, *Initial Reports of Deep Sea Drilling Project,* vol. XLV, Washington, 1972.)

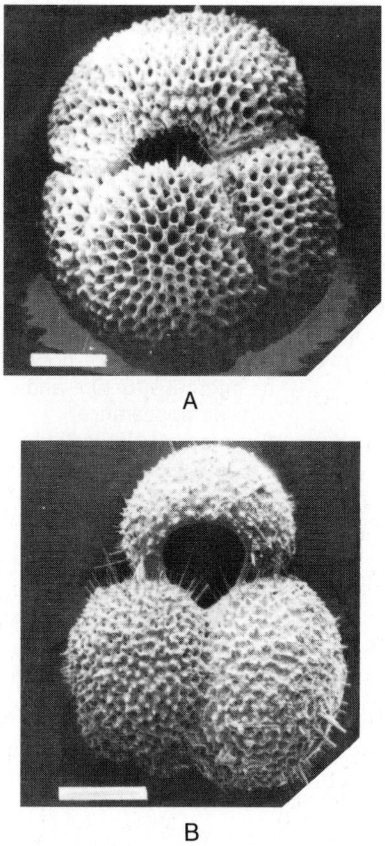

Figure 17–24 Present-day planktonic foraminifers. (A) *Globigerinoides conglobatus.* (B) *Globigerinoides rubra.* (Courtesy C.G. Adelseck, Jr., and W.H. Berger, Scripps Institution of Oceanography.)

ide, which increases its acidity. These deep waters are corrosive to the fragile shells of foraminifers and coccolithophores. At a depth of about 3700 meters, the supply of shell materials from above is approximately balanced by the amount being dissolved. As a result, calcareous oozes cannot accumulate. That depth at which calcium carbonate is dissolved as fast as it falls from above is the calcium carbonate compensation depth. The calcium carbonate that is dissolved below the calcium carbonate compensation depth is returned to the ocean to be used again by floating organisms near the surface. It is a well-balanced system, for if the organisms that utilize calcium carbonate in their shells were to increase in number, the depleted ocean water would respond by dissolving still more calcareous material on the ocean floor.

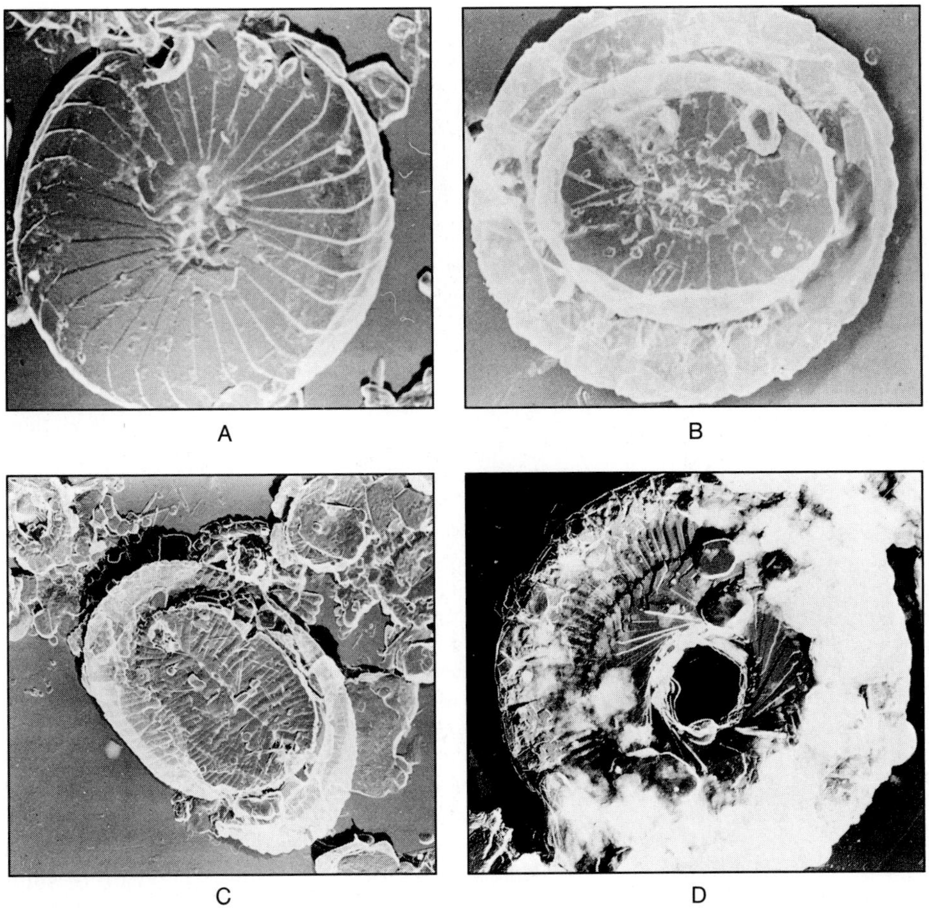

Figure 17–25 (A) *Coccolithus leptoporous* (distal side). (B) *Coccolithus leptoporous* (proximal side). (C) *Helicosphaera carteri.* (D) *Cyclococcolithus sp.*

The ability of cold deep water to dissolve calcareous materials accounts for the fact that calcareous oozes are most extensive on the shallower regions of ocean basins. Tall submarine mountains may have their summits, which reach shallower levels, capped by calcareous ooze. They resemble snow-capped peaks, with "snow lines" at about 3700 meters below sea level.

Siliceous Oozes

In order to be termed a siliceous ooze, a sediment must be composed of at least 30 per cent of organic silica. The two principal contributors to siliceous oozes are the unicellular **diatoms** (Fig. 17–28), which are algae, and the microscopic protozo-

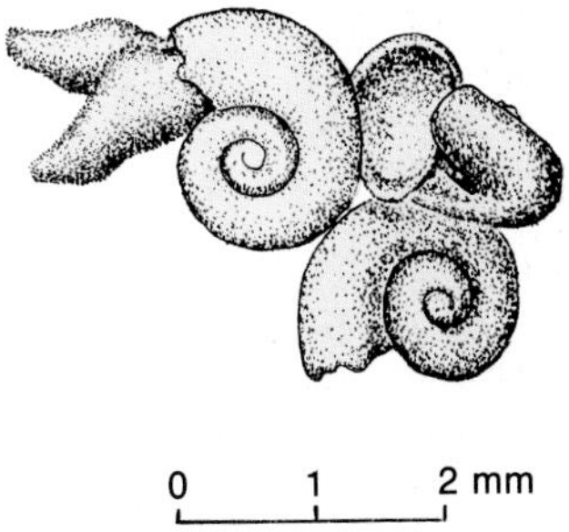

Figure 17–26 *Limacina,* a tiny swimming marine snail with foot modified into a pair of winglike lobes or fins, shown in specimen at left. At the right are two empty conches.

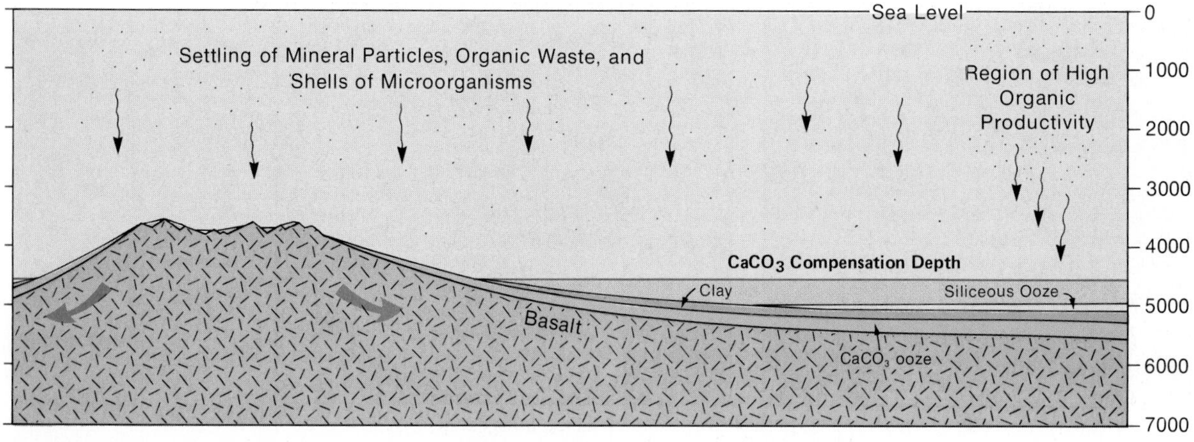

Figure 17–27 The calcium carbonate compensation depth is the depth below which particles of calcium carbonate from microorganisms are dissolved as fast as they descend through the water column. The compensation level varies at different locations in the world ocean. In this conceptual drawing, note that the calcium carbonate collects along parts of the mid-oceanic ridges that are above the compensation depth. The accumulated layer of calcareous ooze is then carried laterally as lithospheric plates diverge. When a given region of the ocean floor has reached depths in excess of the compensation depth, calcium carbonate is no longer deposited on the ocean floor, whereas clay particles and siliceous remains of radiolaria and diatoms may accumulate.

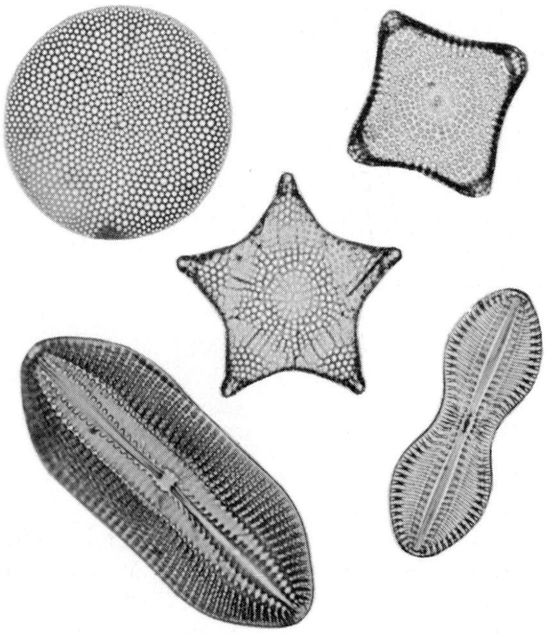

Figure 17–28 These modern diatoms were collected off the coast of Crete. (Photographs courtesy of Naja Mikkelsen, Scripps Institution of Oceanography.)

ans known as **radiolaria** (Fig. 17–29). Because the ocean is undersaturated at all depths with silica, microscopic siliceous shells are often dissolved before reaching the ocean floor. The rate of solution depends on the thickness of the skeletal structures and whether or not they have protective coatings of organic compounds. Siliceous oozes accumulate in colder or deeper oceanic regions where other sediments are lacking or in regions where an abundance of nutrients promotes high productivity of siliceous organisms (Fig. 17–30).

The principal contributors to calcareous oozes are foraminifers and coccolithophores. The principal contributors to siliceous oozes are diatoms and radiolaria.

Climatic History from Deep-Sea Sediment

A core of deep-sea sediment (see Fig. 17–23) can yield a bonanza of geologic information once it has been completely analyzed. Superposition tells

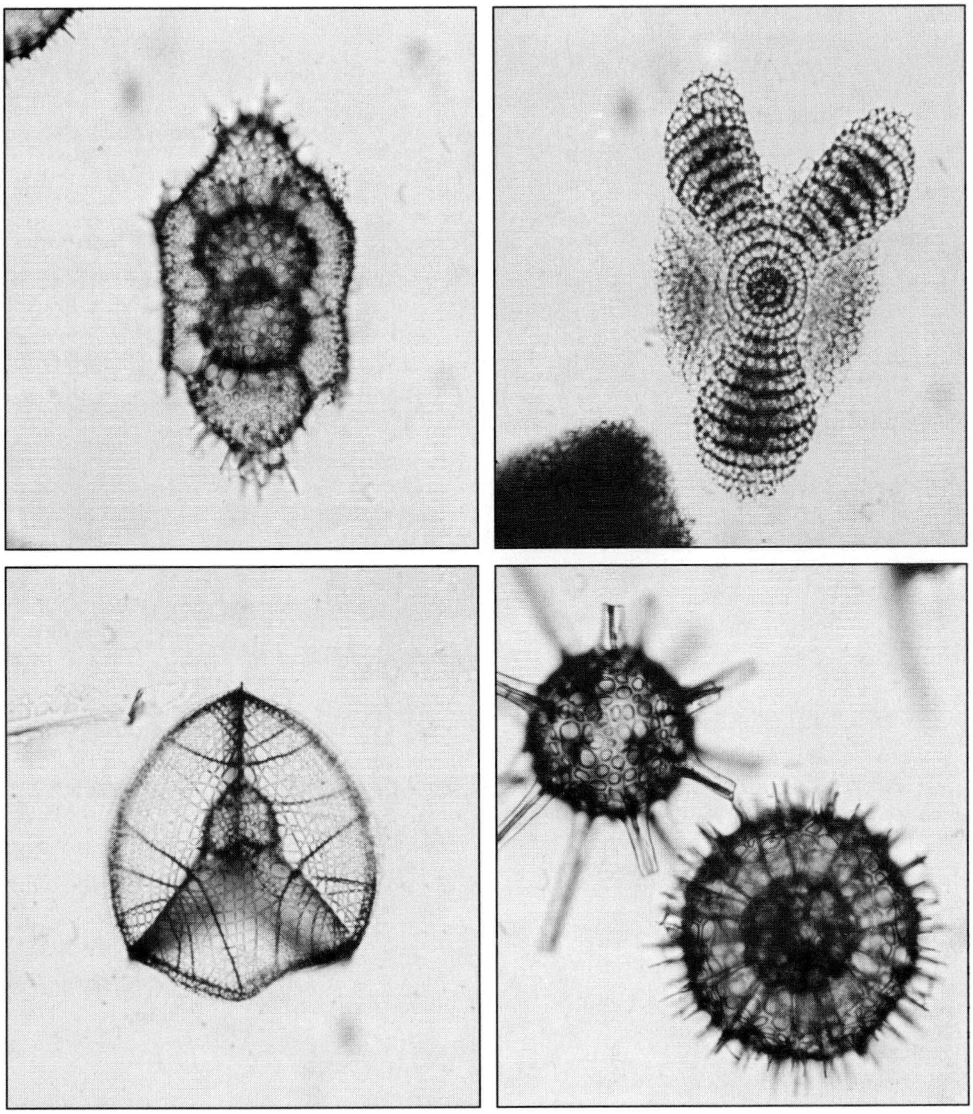

Figure 17–29 Quaternary radiolarians from the tropical Pacific Ocean. (Courtesy of Annika Sanfilippo, Scripps Institution of Oceanography.)

us that the lower layers of sediment are older and the higher layers consecutively younger. Using radioactive dating methods, scientists can determine the actual period of time during which the sediments in the core accumulated. Next the foraminifers, radiolaria, diatoms, or coccolithophores are carefully identified at each successively higher level in the core. Because many species of these organisms were able to live only within certain limits of water temperature, their fossil remains are indicators of ancient temperature. The layers might, for example, suggest an ancient cooling trend if warm-water species in the lower parts of the core gave way to cold-water species in higher layers. Even where good temperature-indicator organisms are absent, the calcareous shells may provide chemical clues to ancient temperatures. Scientists have discovered that the ratio of the iso-

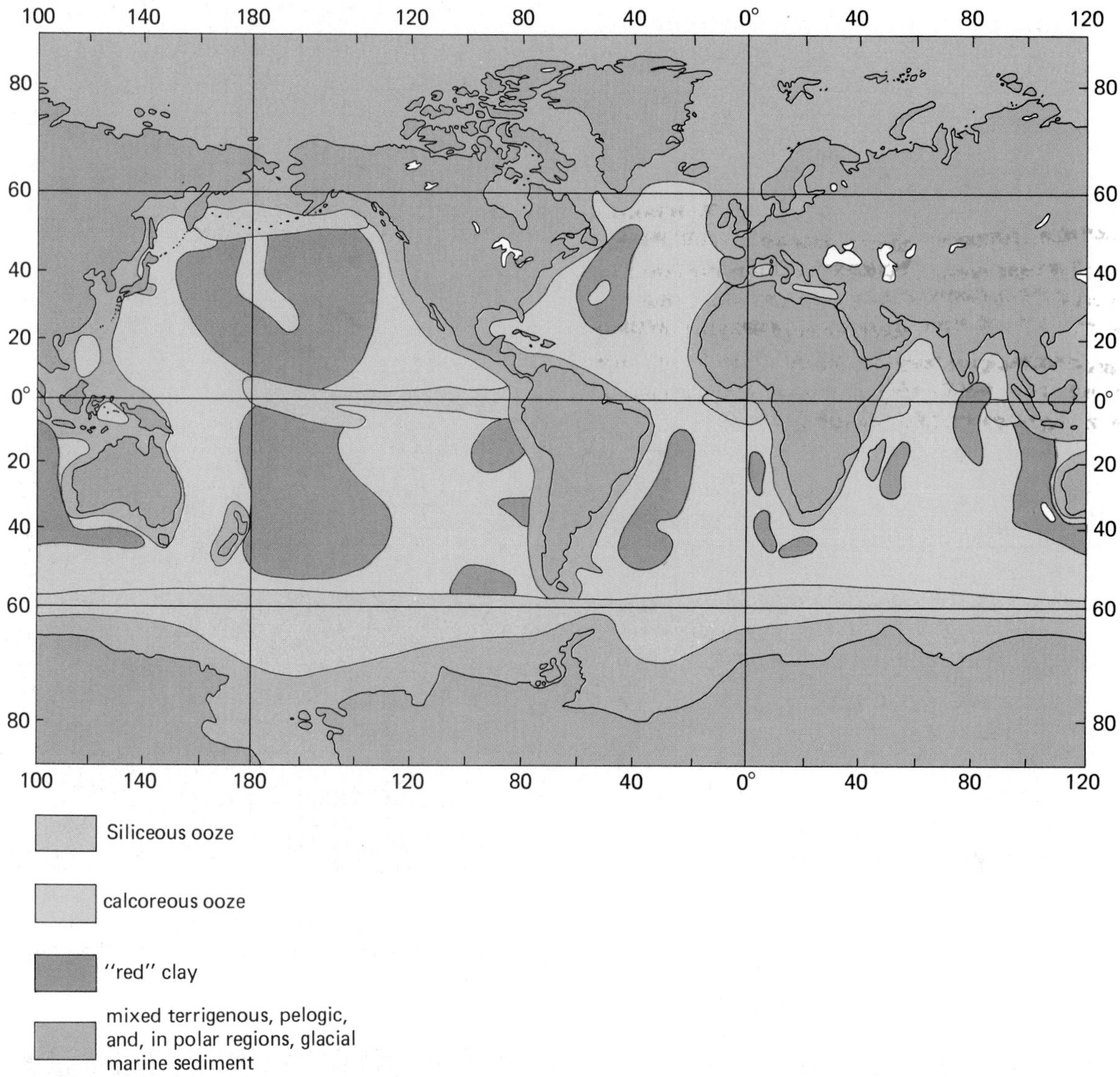

Figure 17–30 Distribution of sea floor sediment. (After Heezen, B.C., and Hollister, C.D., 1971. *The Face of the Deep,* New York, Oxford University Press.)

topes oxygen 16 and oxygen 18 in the calcium carbonate shells of marine creatures may reflect the enclosing water's temperature at the time the carbonate was being secreted.

Changes in climatic conditions through time can be reconstructed by study of core samples of sea floor sediment.

Coral Reefs and Atolls

Coral reefs are distinctive features of the marine environment and are common in the shallower warmer oceanic regions (Figs. 17–31, 32). The coral reefs growing today consist of a rigid framework composed of calcium carbonate structures built by vast colonies of tiny tentacled animals called *coral polyps* (Fig. 17–33). The coral reef, however, is not composed of coral structures alone, for encrusting calcareous algae and a variety of lime-secreting marine animals attach themselves to the framework of the reef and contribute to its strength and mass. Reefs are classified according to their proximity to shore. A **fringing reef** forms a narrow apron directly around the perimeter of an island. If there is a lagoon between the inner edge of the reef and the shoreline, the structure is termed a **barrier reef**. Finally, if there is no central island, but merely a circle of coral reefs and coralline islands surrounding a lagoon, the feature is called an **atoll**. All of these kinds of reef structures are related in origin. This fact was perceived long ago by Charles Darwin during his historic service aboard the naval survey ship *H.M.S. Beagle* in 1832.

Darwin had ample evidence that the coral reefs of the Pacific grew around the perimeter of volcanic islands. At a meeting of the Geological Society of London during the summer of 1837, he proposed that volcanic islands tend to sink slowly because of the inability of the ocean floor to support their great weight. Before this begins to happen, however, free-swimming larvae of coral animals reach newly formed volcanic islands, attach themselves to the shallow submerged areas around the island, and begin their development into adult polyps. Gradually, the coralline framework is constructed and other organisms add their own skeletal structures to the reef. The corals, however, are essential to the vitality of the reef. Darwin was aware that corals are rather fussy creatures in that they thrive only in shallow, well-lighted, relatively warm ocean waters. Depth of water is critical to the health of coral polyps. Thus, with the sinking of their volcanic island, the myriad of polyps might be doomed were it not for their astonishing efficiency in building their colonies upward so as to maintain an optimal habitat near the ocean surface. As the process of subsidence and growth continues, a barrier reef would develop, as shown in Figure 17–34. Eventually, the island would sink beneath the waves, leaving only a circle of coralline islands—the atoll. In Darwin's own words,

> *". . . as the land with the attached reefs subsides very gradually from the action of*

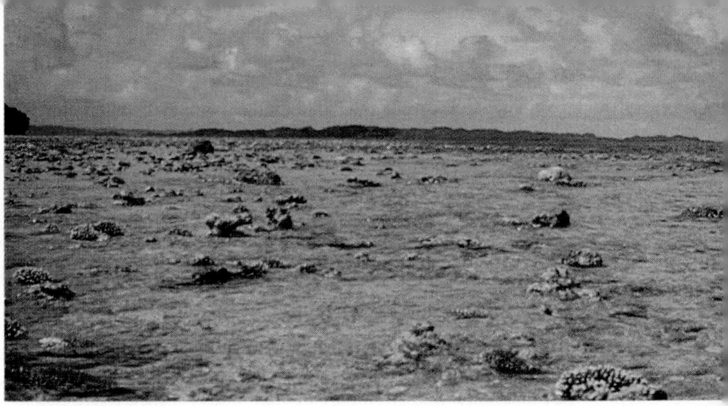

Figure 17–31 Reef exposed at low tide, Palau, West Carolina Islands, southwestern Pacific. (Courtesy of L.E. Davis.)

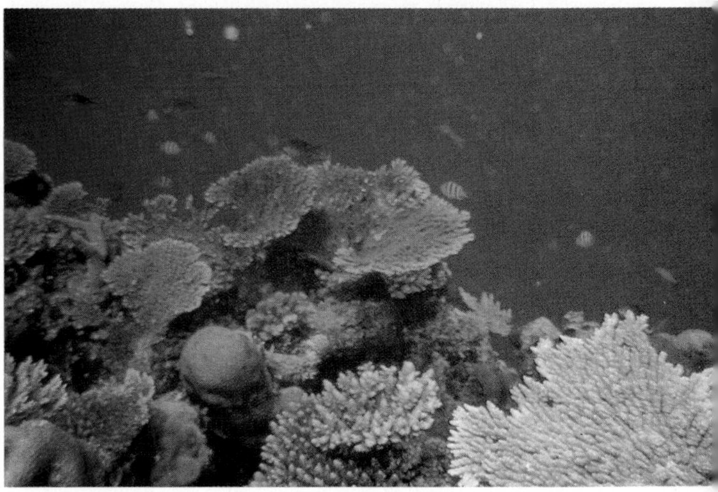

Figure 17–32 Looking down the steep front of a reef, Admiralty Islands, Papau, New Guinea. (Courtesy of L.E. Davis.)

Figure 17–33 Coral polyps of the star coral. (Courtesy of the American Museum of Natural History.)

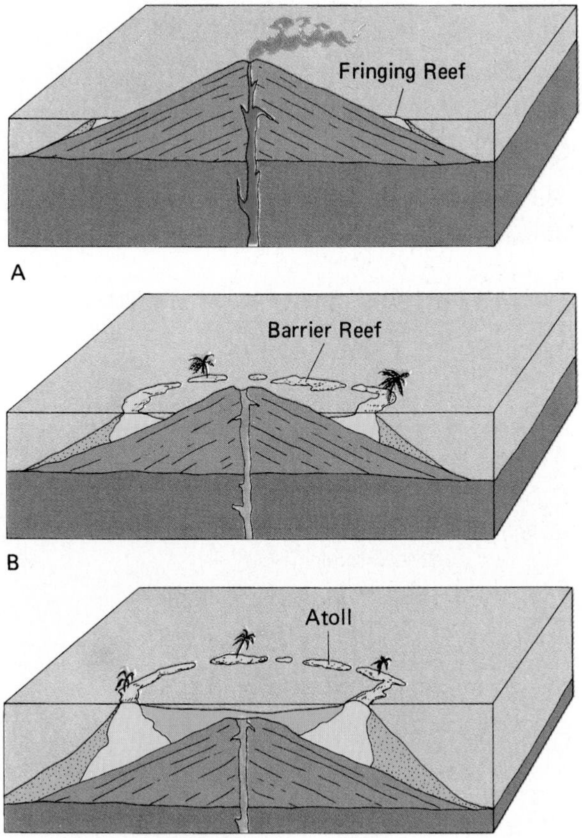

Figure 17–34 Three stages in the development of an atoll. (A) In initial stage, a fringing reef develops around the shoreline of a volcanic island. (B) The island begins to subside, and corals build upward in order to stay in their optimal shallow water life zone. The result is the development of a barrier reef backed by lagoons. (C) As subsidence continues, the original land area has become inundated, and a circle of reefs and coralline islands remains.

subterranean causes, the coral building polypi soon raise again their solid masses to the level of the water: but not so the land; every inch of which is irreclaimably gone: as the whole gradually sinks, the water gains foot by foot upon the shore until the last and highest peak is finally submerged."

Several deep borings into atolls have validated Darwin's theory. A boring at Eniwetok in the Pacific encountered the basaltic summit of the volcano at depths of about 1200 meters. Eocene reef structures were directly above these igneous rocks.

It is likely that the subsidence associated with the development of some atolls is related to sea floor spreading. Spreading begins along the crest of a mid-oceanic ridge, and as the plates move away from the ridge, they subside very slowly because of thermal contraction and isostatic adjustment. The subsidence is estimated at about 9 cm per year. Thus, a volcano formed at the crest of a mid-oceanic ridge is carried down the slight incline of the ocean plate until it is completely submerged. Coral growth that is able to keep pace with the subsidence will result in the formation of barrier reefs and atolls.

There is a possible alternative explanation for atoll formation. It is possible that corals might build fringing reefs around the perimeter of volcanic islands that had been erosionally truncated during a glacial period of low sea level. If the sea level subsequently rose (as it did when the great Pleistocene ice sheets began to melt), then the corals would strive to build the reef upward in order to stay at their optimal living depth. It seems reasonable that particular atolls may be primarily the result of subsidence of the host island, while others may have formed in response to a rise in sea level.

Atolls may form around the margin of a sinking island, or if sea level rises, a stable one.

Mineral Resources from the Ocean

Humans have extracted salt from sea water and caught fish from the oceans for thousands of years. Yet, only within the last few decades have we begun to appreciate the importance of the mineral resources in the sea. We now know more about the oceans, and because of advances in technology, it is now possible (if not always economically feasible) to extract resources from the ocean bottom.

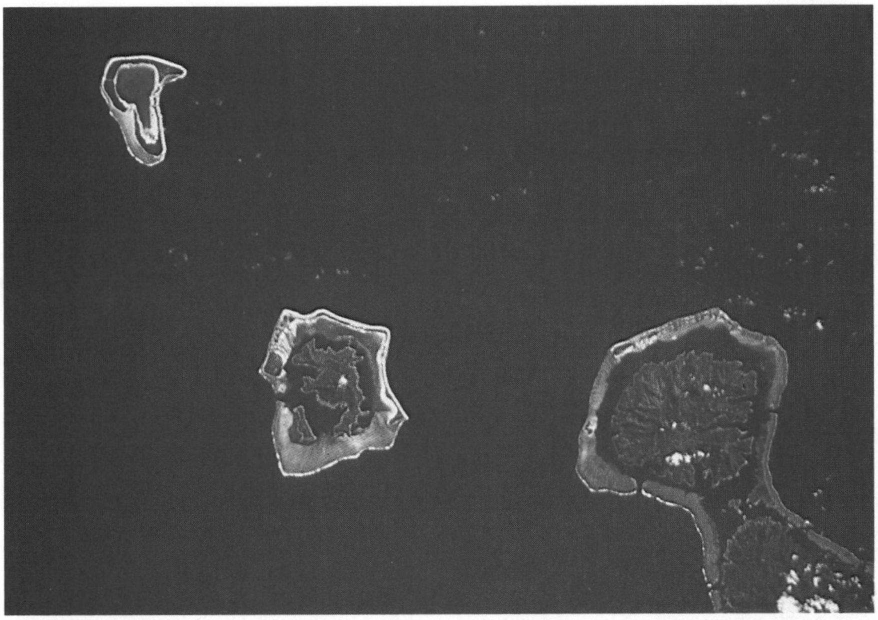

Figure 17–35 The Leeward Islands of the French Society Islands provide an illustration of the origin of coral reefs in deep oceans. At the bottom right, corals have constructed a fringing reef around the volcanic islands of Tahaa and Raiatea. Moving northward, the volcanoes are older and surrounded by barrier reefs. Bora Bora near the center of the photograph illustrates this phase of reef development. Tupai, visible in the top left of the photograph, is an atoll, and all that remains of the volcano around which the reef originally grew lies below the shallow floor of the central lagoon.

As geologists and economists study the increasing demands of a growing and highly industrialized society, they turn increasingly to the oceans as a potential source for many of our future mineral needs.

Petroleum and Natural Gas

Oil and gas are now produced on continental shelves adjacent to the coastlines of 25 nations (Fig. 17–36). These hydrocarbons constitute over 90 per cent of the total value of resources extracted from beneath the ocean floors and currently contribute approximately 20 per cent of the world's production. Presently, most offshore petroleum comes from water depths of less than 100 meters and from areas within 120 km of the coastline. Such areas of the continental shelves are sufficiently shallow so that drilling costs are not prohibitive. Further, the shallow shelves are commonly underlain by a thick sequence of sedimentary layers that include source beds rich in organic material as well as permeable and porous sandstone "reservoir rocks." The same kinds of oil-trapping structures found beneath the nearby coastal plains also occur beneath the ocean floor in offshore oil fields.

Saline Minerals

Although nearly every chemical element has been detected in sea water, only common salt (sodium chloride), magnesium, and bromine are now being commercially extracted. Of these three substances, magnesium is by far the most valuable, with annual world production worth about 75 million dollars. The ocean is the source of 70 per cent of the world's production of bromine, an element now used primarily as an antiknock compound for gasoline. Sediments beneath the sea floor in such

Figure 17–36 Drilling for oil on the continental shelf. Shown here is the semi-submersible drilling rig Zane Barnes. (Courtesy of Reading and Bates Drilling Co.)

places as the offshore area of the Gulf of Mexico include thick beds of gypsum (calcium sulfate), potassium salts, and common salt. These deposits were formed long ago by evaporation of sea water in basins having restricted circulation. Important magnesium salts were also deposited in such basins. Under the heavy pressure of a thick overburden of strata, rock salt and gypsum tend to flow plastically. In many instances in the Gulf Coast region of the United States and the Persian Gulf, the salt has squeezed upward along zones of relative weakness to form salt domes (see Fig. 6–51). Overlying younger sediments are faulted and folded, thereby producing favorable structures for the localization of petroleum. Elemental sulfur is also usually present in the cap rock of salt domes and can be extracted by the Frasch process, in which sulfur is melted by the injection of hot water into drill holes.

Manganese Oxide Nodules

Manganese nodules (see Fig. 17–22) are another important resource found on the ocean floor. These concretions occur near land in such areas as the Blake Plateau east of Florida and off the west coast of Baja California. The nodules are of interest not only because of their manganese content but also because they contain nickel, copper, and cobalt as well. Unfortunately, dredging for nodules is very costly, and it is not yet economically feasible to gather them. They are nevertheless an important resource that may be exploited within a decade or two.

Mineral-Rich Hot Brines

Geologists have long recognized that many ore bodies discovered on land were deposited in cracks and fissures by hot concentrated chemical solutions. Because the land deposits are frequently metallic sulfides, the depositing solutions were probably rich in sulfur. Recently, it was discovered that hot water rich in chlorides is also effective in leaching metals from rocks and depositing these metals as veins of ore. Ocean water, of course, contains ample chloride, so that all that is needed for this kind of ore genesis is sufficient heat and metal-bearing source rocks. High heat flow seems to be characteristic of the earth's rift zones (divergent plate boundaries). In such locations, fractures in basaltic bedrock permit sea water to enter the crust. There it is heated and circulated through cracks and fissures. The hot water is especially effective in dissolving metals, and as this enriched brine is returned to the cooler floor of the ocean, metals may be precipitated in fractured and porous bedrock. The process of ore deposition from hot brines has actually been discovered at three localities along the floor of the Red Sea. These Red Sea brines are so concentrated that they form "ponds" of heavy liquid in depressions on the ocean bottom.

There is another way that the hot brines may produce metalliferous deposits. When the hot brine with its dissolved metals makes its way to the surface, it eventually encounters cold sea water. Once chilled, the brine is a less effective solvent, and a reddish-brown precipitate is formed, consisting mainly of iron and manganese hydroxides, with minor amounts of cobalt, copper, and nickel. Except in the Red Sea, metal-rich muds are not mined today because of cost. In the future, however, they may be recovered by being converted to slurries and pumped to the ocean surface for processing.

Geologists are excited about hot sea-water brines. Not only do they provide evidence of how some ancient ore bodies developed, but their discovery has spurred efforts to find similar emplacements in ancient rift zones around the world. Many ore bodies in mountain ranges near coastlines may owe their origin to hot brine concentrates that have worked their way upward as melts following subduction along tectonic plate boundaries.

Metals dissolved in hot waters arising from submarine rift zones are precipitated on contact with cold water to form deposits rich in metals.

Summary

The world ocean covers 71 per cent of the earth's surface. This vast body of water has received careful scientific study for well over a century, beginning with the *Challenger* expedition of 1872. Scientists examining the chemistry of the oceans have found that ocean water contains about 3.5 per cent dissolved matter, of which sodium chloride is the most abundant. In general, the greater the amount of salt dissolved in sea water, and the colder its temperature, the denser it becomes. Differences in the density of masses of ocean water contribute to the circulation of the deeper zones of the world ocean. This circulation, along with such gases as oxygen and carbon dioxide, which are dissolved in sea water, are of vital importance to the maintenance of life in the seas.

The depths and configuration of the ocean floors are measured primarily by echo-sounding devices. These instruments determine depth by measuring the time required for a sonic signal to travel from a source of energy at the surface to the sea floor and back. Echo-sounding devices can provide a continuous profile of bottom topography along the traverse of the research vessel. The profiles have provided an extraordinary perspective of sea floor landscapes.

Major features of the ocean floors include the *continental shelves*, *continental slopes*, and *continental rises* that exist around the edges of land masses. Submarine canyons are found incised into the continental shelves and slopes. Other features of the ocean basins include *seamounts* and *guyots*. The most majestic elements of the ocean basins, however, are the *mid-oceanic ridges* and *deep-sea trenches*. Mid-oceanic ridges are submarine volcanic mountain systems formed along zones of crustal tension. They are spreading centers between tectonic plates. Deep-sea trenches are associated with zones of subduction, where one tectonic plate slides beneath an adjacent plate.

On the basis of their origin, the sediments of the ocean basins may be considered as either inorganic or organic. *Brown clays*, consisting of clay and other clay-size particles derived from the continents, are the predominant inorganic sediments of the deep sea. The organic materials consist of *siliceous oozes*, composed of the skeletal structures of diatoms and radiolaria, and *calcareous oozes* that consist of a myriad of tests of foraminifers and coverings of tiny plants of the family Coccolithophoridae.

Because they contribute to the formation of new land areas, coral reefs are also important oceanic features. Most coral reefs require the proliferation of colonial coelenterates, which build the general structure of the reef, thereby providing a framework on which other invertebrates and calcareous algae can attach themselves. Depending on their relationship to the host island or seamount, reefs are classified as *fringing reefs*, *barrier reefs*, or *atolls*. Barrier reefs and atolls form as a result of subsidence of the island around which coral growth is occurring. Certain atolls and barrier reefs may also develop as a result of eustatic rise in sea level.

The ocean contains quantities of mineral wealth. Some of this wealth is now being tapped by drilling and mining on the shallower sea floor in near-shore environments. The larger part of the ocean's metallic mineral resources cannot yet be extracted at a profit but may be worked at a future date. Petroleum and natural gas are currently being produced on the continental shelves of many countries and are clearly the most valuable resources presently obtained from beneath the sea floor. Sulfur is being produced from salt domes beneath the ocean bed and common salt, bromine, and magnesium from the water itself. In the future, manganese oxide nodules may be harvested from the ocean floors as well as metals precipitated from hot brines above submarine rift zones.

Terms to Understand

abyssal plain
atoll
barrier reef
brown clay
calcareous ooze
continental rise
continental shelf
continental slope
deep-sea fan
deep-sea trench
diatoms
echo-sounding
epicontinental marginal sea
foraminifer
fringing reef
Globigerina ooze
guyot
marginal platform
mid-oceanic ridge
oceanic rise
ooze (siliceous, calcareous)
pteropod ooze
radiolaria
rift valley
seamount
sub-bottom profiling
submarine canyon
thermocline
tholeiite
turbidity current

Questions for Review

1. Analyses of the water from major streams entering the oceans indicate a far greater calcium content than sodium (see Table 17–2). Why then is most of the dissolved salt in ocean water sodium chloride rather than calcium carbonate?

2. What is the effect on the density of ocean water if it is cooled? If it is evaporated? If it is mixed with rainwater?

3. How do scientists utilize their knowledge of the speed of sound in ocean water to determine submarine topography?

4. If one were to embark on an ocean voyage from New York to Lisbon, Portugal, above what major features of ocean floor would one travel?

5. What are turbidity currents? Why are they more effective as agents of submarine erosion and transportation than thermohaline currents?

6. In what way may such features as mid-oceanic ridges and deep-sea trenches be related to plate tectonics? With which of these features are Benioff Zones associated?

7. What differences in composition and texture are likely to exist between sediments recovered from the floor of the continental shelf in the Gulf of Mexico and sediments recovered from the abyssal plains of the ocean basins? What is the geologic explanation for these differences?

8. What is the origin of such ocean features as (1) atolls, (2) guyots, and (3) submarine canyons?

9. What characteristic of the living coral animal is important to any theory of the origin of atolls?

10. A "pinger" on an oceanic research vessel transmits a signal that is reflected off the ocean floor and returned to the ship in 1.8 seconds. What is the depth of the ocean floor beneath the ship?

11. What organisms are largely responsible for the formation of siliceous and calcareous oozes, respectively?

12. What is the most valuable resource currently being extracted from sedimentary rocks beneath the ocean floor?

13. What are deep-sea brines? How are they believed to originate? What is their relation to the origin of metallic mineral deposits?

Supplemental Readings and References

van Andel, T., 1977. *Tales of an Old Ocean*. New York, W.W. Norton Co., Inc.

Anikouchine, W.A., and Sternberg, R.W., 1973. *The World Ocean*. Englewood Cliffs, N.J., Prentice-Hall, Inc.

Corliss, W.R., 1970. *Mysteries Beneath the Sea*. New York, Thomas Crowell Co.

Davis, R.A., Jr., 1977. *Principles of Oceanography*. Reading, Mass. Addison-Wesley Co.

Kennett, J., 1982. *Marine Geology*. Englewood Cliffs, N.J., Prentice-Hall, Inc.

Thurman, H.V., 1987. *Essentials of Oceanography*. Columbus, Ohio, Merrill Publ. Co.

Tides, Waves, and Shorelines

18

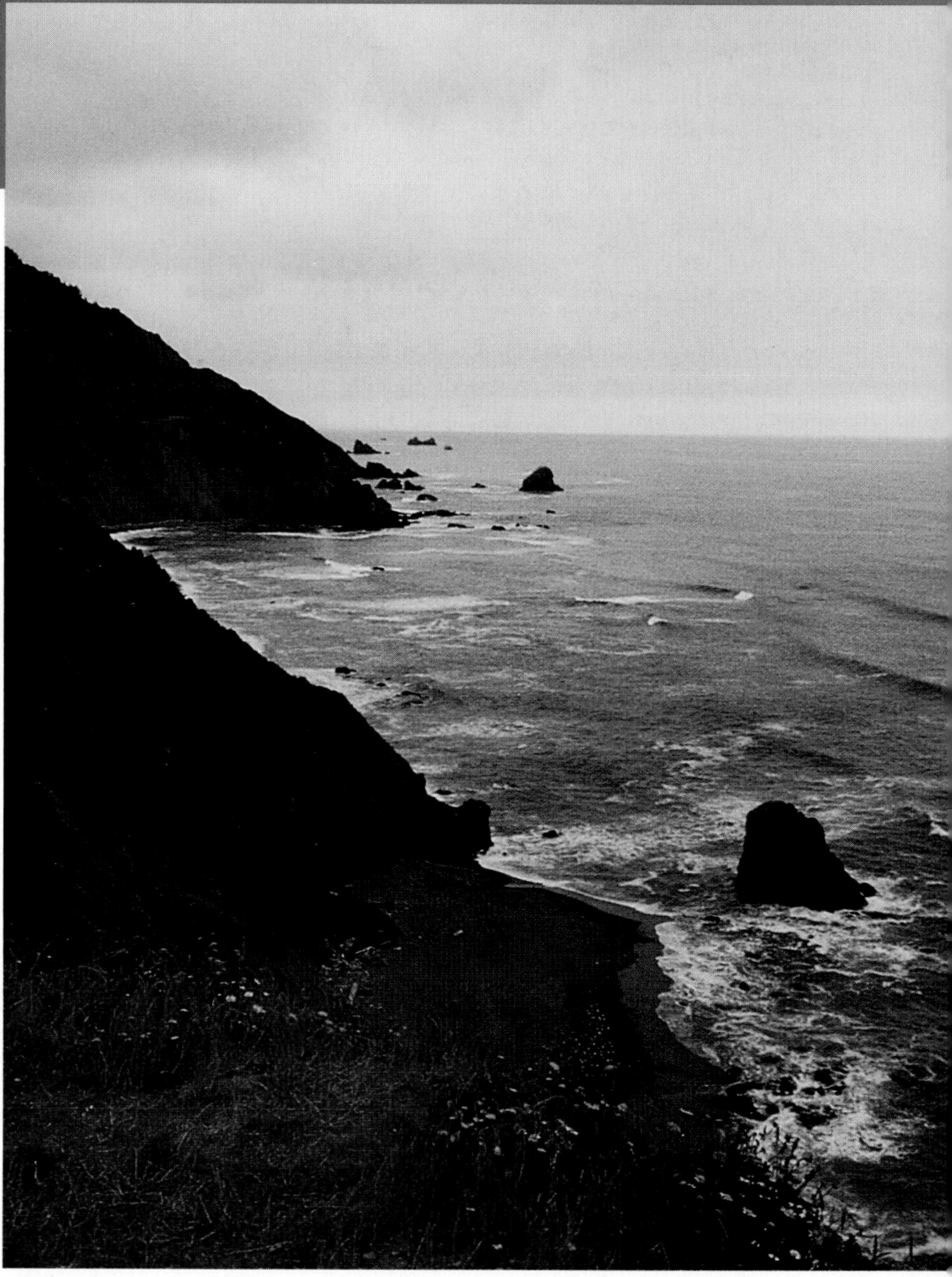

The Pacific Ocean and northern California shoreline south of Crescent City, California. The isolated masses of rocks separated from the coast by wave erosion are called stacks.

We stand on a rugged coast and watch the waves strike blow after blow with the relentless persistence of a trip hammer. The display of vast power is impressive, and some disintegration of the rocky walls proceeds before our eyes.

Sir Charles Lyell. Principles of Geology, 1860

Introduction

Only a few fortunate people have had an opportunity to see the ocean's depths or chart its currents. Most of us, however, have at least been able to visit the shore where we can observe the advance and retreat of the tides and the rhythmic breaking of waves along a stretch of beach. When we are in a meditative state of mind, the shore seems to evoke questions. What mighty force drives the tides? Why do they vary in size at different times and places? How do waves, which occur only at the surface of the sea, assist in the sculpture of tall sea cliffs? Why are there sandy beaches along some coastlines but only bare bedrock along others? And why are some coastlines straight, whereas others are punctuated with numerous bays and jutting promontories? These are a few of the questions we will address in this chapter.

Tides

Anyone who has spent a few hours at the seashore knows that the level of the ocean changes in a regular and predictable way twice in each span of approximately 24 hours. These changes are called tides. Their cause has intrigued curious people for thousands of years. Herodotus in 450 B.C. wrote about the tides he observed on the shores of the Mediterranean. Because both tides and moon followed a similar cyclical pattern, he concluded that they were in some way related. It was not until 1696, however, that the relationship of tides to the orbit of the moon could be adequately understood. In that year, Sir Isaac Newton presented his Law of Universal Gravitation.

The Forces That Cause Tides

Newton's **Law of Universal Gravitation** states that every body in the universe attracts every other body, and that the attraction is proportional to the product of the masses of the bodies and inversely proportional to the square of the distance between them. Thus, by increasing the mass of either or both bodies, gravitational attraction is increased. But if the distance between the two bodies is increased, gravitational attraction will be decreased.

The earth's tides are the result of the gravitational pull of the moon and the sun. The sun, although it has far more mass than the moon, plays a less important role in producing our tides because it is 390 times farther away from the earth. Twice each day, the attraction of the moon (and to a lesser degree the sun) acts to increase the depth of water in the ocean on opposite sides of the earth. Figure 18–1 depicts these two tidal bulges. Note that point **X** in the figure is closest to the moon. Therefore, matter at point **X** experiences the strongest gravitational pull from the moon, and a tidal bulge is raised. At point **Y**, which is farthest from the moon, gravitational attraction is weakest. Here, the moon tends to pull the entire earth away from matter (including ocean water) on the far side. The result is the second tidal bulge. The bulges at **X** and **Y** represent high tides, whereas low tides occur at locations midway between **X** and **Y** where water is being drawn away. As the earth rotates, any given location will experience two high tides and two low tides each day. (The solid earth behaves in a similar manner, but because of its rigidity, the bulges are extremely small.)

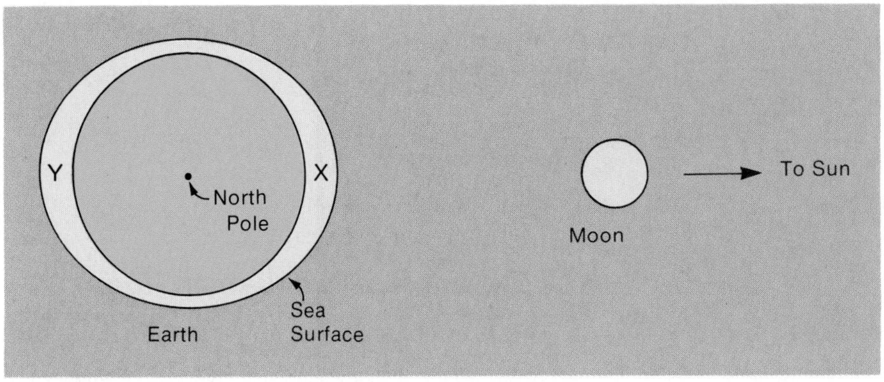

Figure 18–1 The tide-producing effect of the moon. Tidal bulges are exaggerated.

The gravitational attraction by the moon on the earth is the principal cause of tidal bulges. The sun also effects tides, but to a lesser degree because of its distance from the earth.

Behavior of Tides

As the earth rotates from west to east, tidal highs and lows appear to move generally westward. As described above, coastal locations experience two high tides within a 24-hour-and-50-minute interval or one high tide every 12 hours and 25 minutes. The reason the tides are not exactly 12 hours apart is because during the earth's daily rotation the moon has been moving forward in its orbit (Fig. 18–2). The earth, therefore, must turn for an extra 50 minutes to reach its previous day's position relative to the moon.

The two high tides experienced at certain coastal locations are usually of different magnitude because the orbit of the moon is at an angle to the plane of the earth's equator. Indeed, the moon's orbital path carries our satellite as high as 28.5° north and south of the equator. The tidal bulges occur on a line passing from the earth's center to the moon's center. As the earth rotates, a point along the coastline, at any given latitude, experiences first and second high tides of different magnitude except when the moon happens to be in the equatorial plane (Fig. 18–3).

Although the sun is not as important in causing tides as the moon, it nevertheless produces a tide that is two-fifths the magnitude of a lunar tide. At new and full moon, when the sun and moon are aligned with the earth (Fig. 18–4), the gravitational attraction of the sun reinforces that of the moon, and tides are at their maximal height. These unusually high bimonthly tides are termed **spring tides** (although they have no relationship to the season). At quarter moon, the sun and moon pull at right angles to each other so as to produce bimonthly weak tides called **neap tides**.

There are many other factors that complicate the earth's tidal patterns. The movement of tidal bulges around the earth is strongly influenced by the shape, depth, and interconnections of ocean basins. Tidal bulges may be diverted by continents and are affected by the Coriolis effect. As a result, tidal predictions for particular coastal locations require actual tidal observations (Fig. 18–5) over long periods of time.

It is difficult to measure the magnitude of tides in the open ocean, but observation stations on islands in the Pacific show that the water level seldom varies by more than about 1 meter in height. Along the irregular shorelines of continents, however, water movements may become concentrated by the configuration of the land and friction with the ocean bottom. The results are exceptionally high tides at some coastal locations. For example, in Nova Scotia and along the Brittany Coast of France, tides frequently exceed 15 meters. The

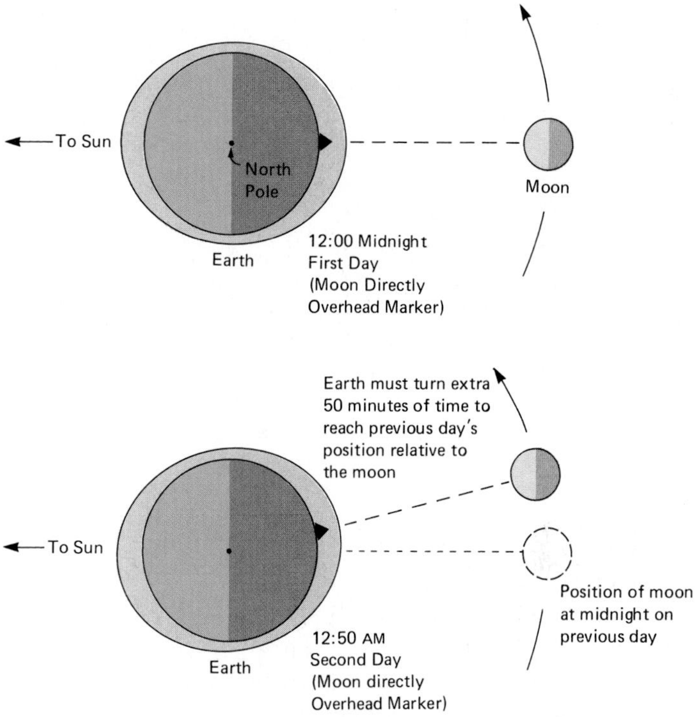

Figure 18–2 The moon passes a given location on the earth once every 24 hours and 50 minutes. Thus, the earth, after having completed its 24-hour rotation (the solar day), must continue to turn for 50 additional minutes to arrive at its previous day's position relative to the moon.

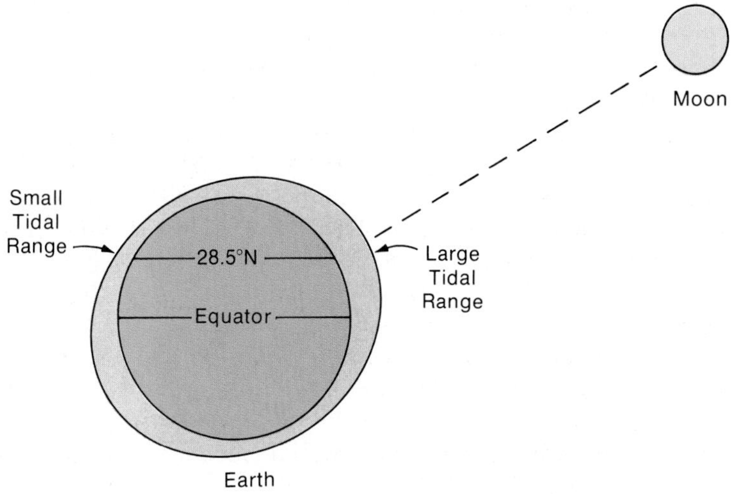

Figure 18–3 Because the center of the tidal bulges may lie at any angle from the equator to a maximum of 28.5°, tidal ranges along latitudes are not the same on opposite sides of the earth.

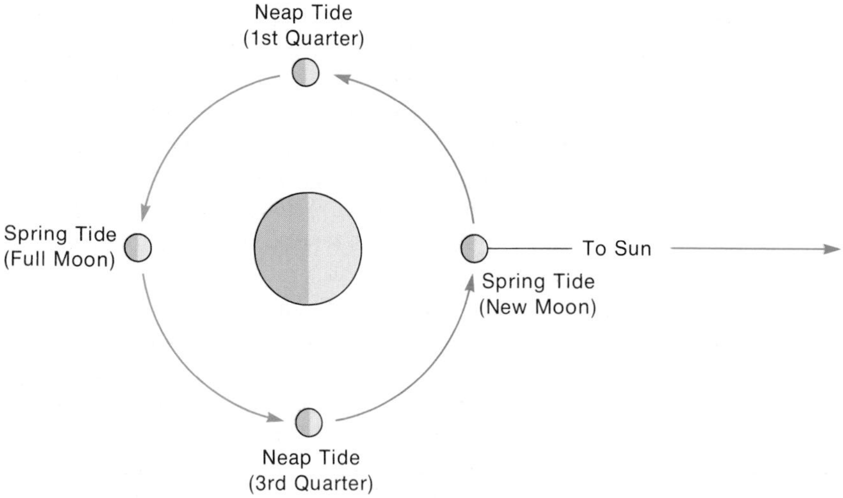

Figure 18–4 Position of the sun, earth, and moon at times of neap and spring tides.

maximal tidal range for most coastlines, however, is less than 3 meters.

Tidal bulges are unable to move unhindered around the earth because the continents obstruct their movement. Continents act as barriers behind which water piles up, and this piling-up accounts for higher tides along coasts as compared to open ocean.

Tidal currents are only minor agents of transportation and erosion. Nevertheless, in some shallow areas they have sufficient strength to transport sand and scour and erode the sea floor near the entrances to inlets and bays. At the mouths of some rivers, the advance of the tidal bulge may take the form of a rapidly flowing turbulent wave that moves up the river for miles before reversing its flow. Such tide-induced river waves are called **tidal bores**. Along the Amazon River, the tidal bore is often 5 meters in height and travels upstream at a speed of 22 km per hour.

Tidal Friction

As the earth rotates, there is a tendency for the two tidal bulges to be carried around with it. However, the moon's attraction prevents the earth from dragging the bulges very far. Thus, the two tidal crests tend to act as rather inefficient brake bands clamped on either side of the rotating earth and dragging back on the planet as it rotates. The bulges experience friction along shallower areas of the ocean bottom and are also retarded by the continents that stand in their way. These effects all contribute to a slight slowing of the earth's rota-

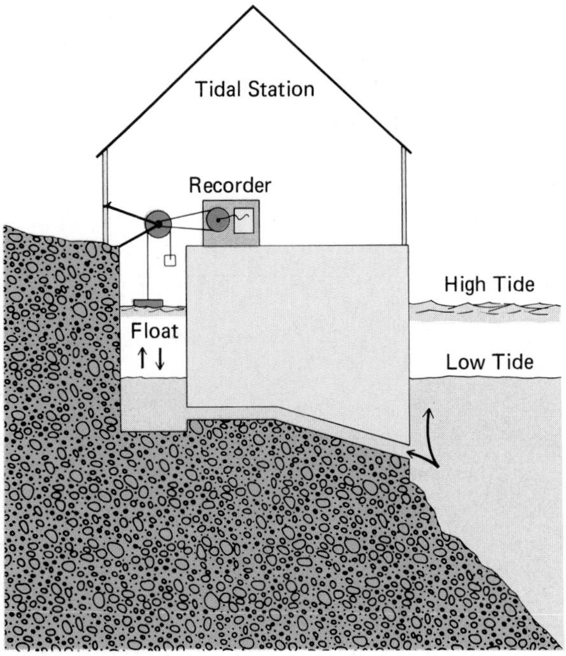

Figure 18–5 A tide gauge station. (From McCormick, J.M., and Thiruvathukal, J.V., *Elements of Oceanography*, 1976. Philadelphia, W.B. Saunders Company.)

tion. The slowdown means that days have been increasing in length through geologic time, and the number of days in the year have been decreasing. The rate of slowing, however, is very small. It has been calculated that the average length of a day has been extended by only one thousandth of a second in the past 100 years. The change is imperceptible in a human lifespan, but over hundreds of millions of years of earth history, that amount of slowing adds up. Earlier in geologic time, the planet rotated faster, and the moon was closer to the earth. Under these conditions, tides had greater magnitude and the interval of time between tides was shorter.

Evidence supporting the calculations that indicate the earth's rotation is slowing down was provided in the early 1960s by paleontologist John Wells. Wells, an authority on fossil corals, recognized that the fine growth lines on the exoskeletons of coral organisms might represent daily growth increments (Fig. 18–6). Thus, a coral might secrete one thin ridge of calcium carbonate each day. In addition to the fine daily growth lines, there are coarser monthly bands (presumably related to monthly breeding cycles during which carbonate secretion was inhibited) and still broader annual bands in corals living in areas of seasonal fluctuations. Wells counted the growth lines on several species of living corals and found that the count "hovers around 360 in the space of a year's growth." Proceeding next to fossil corals of successively older geologic age, Wells counted correspondingly larger numbers of growth lines in the yearly segment of the exoskeleton. For example, on corals that had been determined to be about 370 million years old, Wells found 398 daily growth lines in a yearly increment. This evidence suggesting that there were nearly 400 days in a year 370 million years ago correlated astonishingly well with the calculations provided by the astronomers. Geologists could now state with considerable confidence that long ago when the first land vertebrates began to appear, the days were shorter, but there were more of them in a year. Furthermore, once the relationship between number of days in the year and absolute geologic age had been plotted on a graph (Fig. 18–7), then it might be possible to derive the age of a stratum by counting the growth lines on the fossil corals collected from that stratum.

Largely because of tidal friction, the earth's rotation has been gradually slowing. Thus, in the geologic past there were fewer hours to the day and more days to the year.

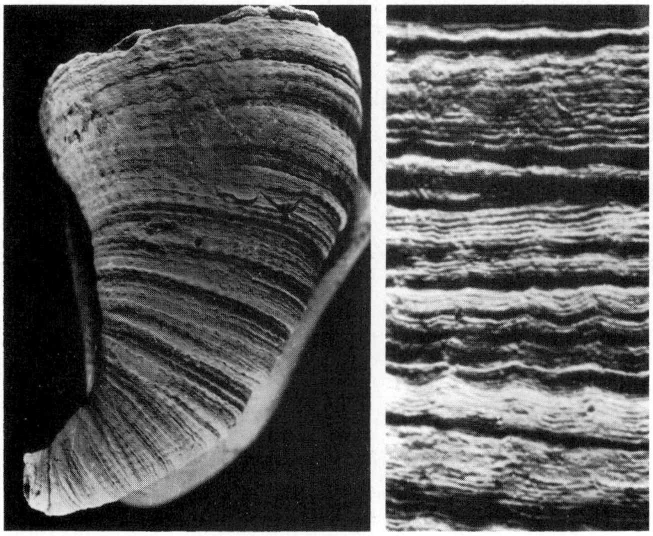

Figure 18–6 Growth banding displayed by a specimen of the extinct coral *Heliophyllum halli.* The finer lines may represent daily growth increments. Together with the annual bands, the growth increments can be used to estimate the number of days in a year at the time the animal lived. (Photograph courtesy of G.R. Clark, II.)

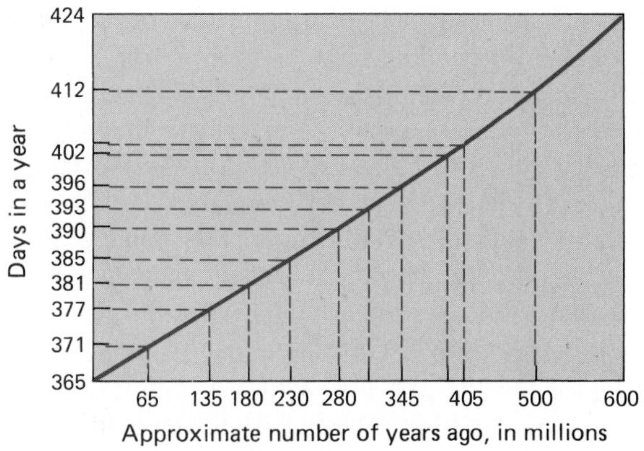

Figure 18–7 The changing length of the day through geologic time. (From J.W. Wells, 1963. *Nature* 197:948–950.)

Waves

The shores of the continents would not be as aesthetically interesting to us were it not for waves (Fig. 18–8). Waves contribute significantly to the enjoyment of bathers and make possible the sport of surfing. For geologists, waves are important because they accomplish more geologic work than either tides or surface currents. Waves are constantly at work changing the configurations of our shorelines. They transport sediment and contribute to the formation of other sediment-transporting currents. Waves comprise a great erosional mill for the reduction of large masses of rock to finer debris. From an environmental point of view, waves breaking at the sea surface bring vital carbon dioxide and oxygen into the water mass and provide a constant mixing of gases and nutrients in the upper layers of water. Someday, waves may have importance in producing electricity for use in lighting and heating our homes. It already is technically (but not economically) feasible to generate this power from waves.

Figure 18–8 Wave action has produced both depositional features (the beach in the foreground) and erosional features (the island-like stacks beyond the beach) along this section of the Oregon coast. (Courtesy of L.E. Davis.)

There are many different kinds of water waves. We have already discussed waves generated by earthquakes and volcanic eruptions (tsunami). Tides are actually also true waves characterized by great distances between crests and very low amplitude. The most important waves from a geologic point of view, however, are waves generated by wind. Wind waves occur in many sizes and shapes, ranging from a few centimeters in height to those with crests towering 30 meters above the wave trough. Waves are formed whenever the wind blows over the surface of the ocean and thereby transfers part of its energy through friction and pressure fluctuations to the water surface. Three factors control the size of waves. The first is the velocity of the wind, the second is the distance over which the wind has blown, and the third is the constancy or duration of time during which the wind has moved over the water. Waves frequently continue to travel hundreds of kilometers beyond the area in which they were generated. Such waves are termed **swells**.

Wave Terminology

Waves are described in terms of such properties as length, period, frequency, velocity, and height (Fig. 18–9). **Wave length** is simply the distance between two successive crests or troughs. The **period** of a wave is the time required for one full wave length to pass a fixed point. Thus, if it takes 10 seconds for two crests to pass a fixed buoy, the wave period is 10 seconds. The **frequency** of a wave is the number of waves passing an observation point in a given interval of time. **Wave velocity** is the speed at which a crest or trough travels

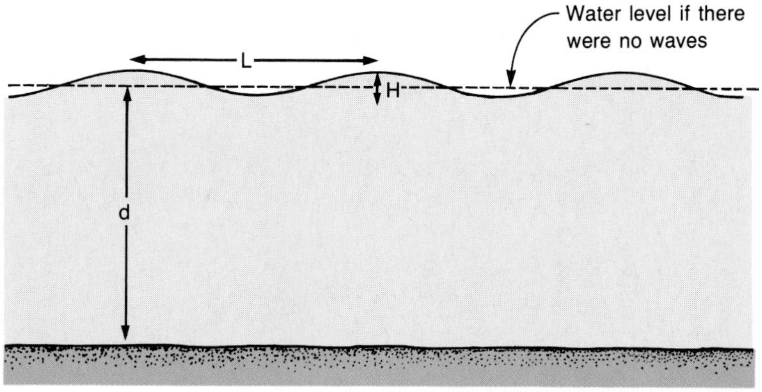

Figure 18–9 Ideal ocean waves. L = wave length, H = wave height, d = water depth. (From McCormick, J.M., and Thiruvathukal, J.V., *Elements of Oceanography,* 1976. Philadelphia, W.B. Saunders Company.)

and is equal to the wave length divided by the period. **Wave height**, of course, is the vertical distance between the top of a crest and bottom of the trough.

Waves in the Open Ocean

In deep water, the motion of waves is not affected by the sea floor, and so open ocean waves have different characteristics than waves in shallow water. A wave is called a deep-water wave if the ratio of water depth to wave length, that is, $\frac{d}{L}$, is at least one-half. The motion of waves in deep water is a bit more complex than one might surmise. For example, although the wave *form* appears to travel along, there is very little real transport of water. This seemingly contradictory fact is apparent to an observer watching a piece of driftwood floating on the ocean. The wood does not move appreciably forward with the wave but stays approximately in the same place as it rides up on the wave crests and down on the troughs.

While the wave *form* is moving across the ocean surface, water particles within the wave travel through a circular orbit and return approximately to their original positions. On the crest of a wave, all water particles are moving forward in the direction of wave travel. Particles move backward in the trough. In this manner, any given water particle makes one complete circuit in each wave period.

As indicated in Figure 18–10, the particle orbits exist not only at the surface but also at shallow depths. As we have seen, energy is imparted to a wave by wind at the surface, and this energy is progressively lost with depth. For this reason, the diameter of the particle orbits decreases with depth. At a depth of about one-half the wave length, motion is negligible. This depth is referred to as **wave base**. Scuba divers are usually not disturbed by small or medium-size waves because wave base is relatively near the ocean surface.

In deep water, waves have little or no effect on the sea floor because the sea floor is below wave base.

If it were somehow possible to "freeze" a wave at any given instant, and then draw lines between corresponding water particles in orbits from the surface downward as shown in Figure 18–10, then one would notice that the lines bend toward the crests and away from the troughs. In effect, the traveling wave is a series of convergences and divergences of water particles. Water, being incompressible, does not respond to these compressions and expansions by changing volume, as air might, but rather is forced to rise as a crest under converging water and to fall as a trough in divergent areas.

Far out to sea, waves usually do not exhibit a symmetric pattern of successive elongated crests and troughs (Fig. 18–11). Rather, the crests and troughs tend to have an irregularly elliptic form and uneven spacing. This irregularity is the result of waves from different source areas interfering with one another as well as fluctuations in the velocity and directions of the wind that generates the waves.

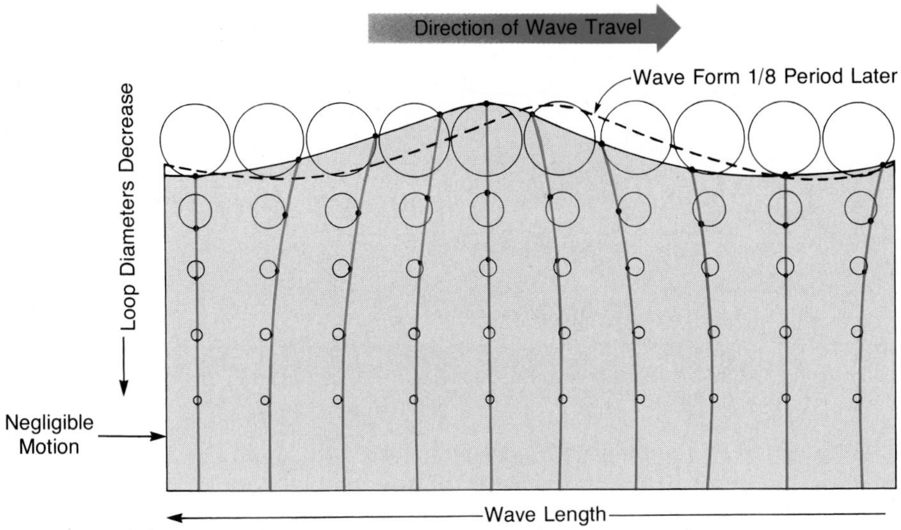

Figure 18–10 Generalized cross-section of wave showing orbital motion of water particles (black dots). Brown lines connect corresponding particles. (From Kuenen, Ph. H., *Marine Geology,* 1950. New York, John Wiley & Sons, Inc., p. 70.)

Waves in Shallow Water

As waves enter the shallow zone near the shoreline, their movement and form are significantly changed. Near a depth of one-half the wave length, the deepest orbits below the surface come into contact with the shallow sea bottom (Fig. 18–12). Here, actual lateral movement of water is initiated for the first time. Friction with the bottom causes the wave to be slowed. The still rapidly moving waves that follow tend to catch up with the wave that has been slowed, and in this way wave length is shortened. The effect rather reminds one of a pileup of automobiles that must slow down at a highway toll gate.

Figure 18–11 The complex wave pattern often developed on the open sea. (From McCormick, J.M., and Thiruvathukal, J.V., *Elements of Oceanography,* 2nd ed., 1981. Philadelphia, Saunders College Publishing; photograph courtesy of the U.S. Coast Guard.)

Also, in an attempt to accommodate the volume of water in the incoming waves, wave height increases. As these changes are taking place, the shapes of the water particle orbits are distorted from circular orbits to more oval configurations. The orbits near the bottom are eventually flattened, so that water particles move back and forth parallel to the sea floor. Near the surface, the water particle orbits are less distorted. As the wave enters still shallower water, the water particles at the interface between the ocean and the atmosphere are moving faster than the wave itself, with the result that the wave steepens and collapses as a line of breakers (Fig. 18–13).

Shallow-water waves collapse to form breakers when they encounter a water depth of 1.3 times their wave height. This observation was given careful attention by military commanders of amphibi-

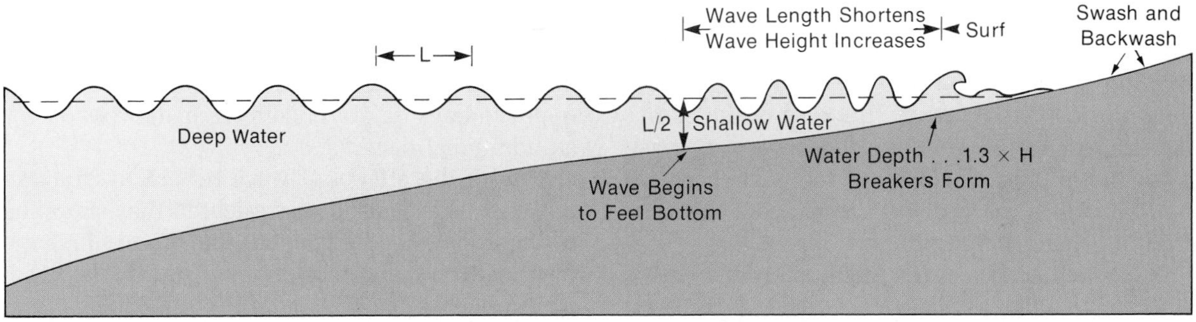

Figure 18–12 Profile of the near-shore zone indicating change in wave form as waves travel from deep water to shallow water to shore.

ous landings during World War II, for it provided a means of roughly estimating the nearshore depth of water in areas where detailed knowledge of bottom topography was lacking.

Most of the geologic work of waves is accomplished in the surf zone at depths of less than half the wave length.

Wave Refraction

There are many similarities in the behavior of water waves and light waves. Light waves, for example, are bent or refracted as they pass from one medium to another, as from air to water. Surface water waves are also bent, and this bending or refraction is caused by one part of a wave reaching shallow water before another part. If waves were

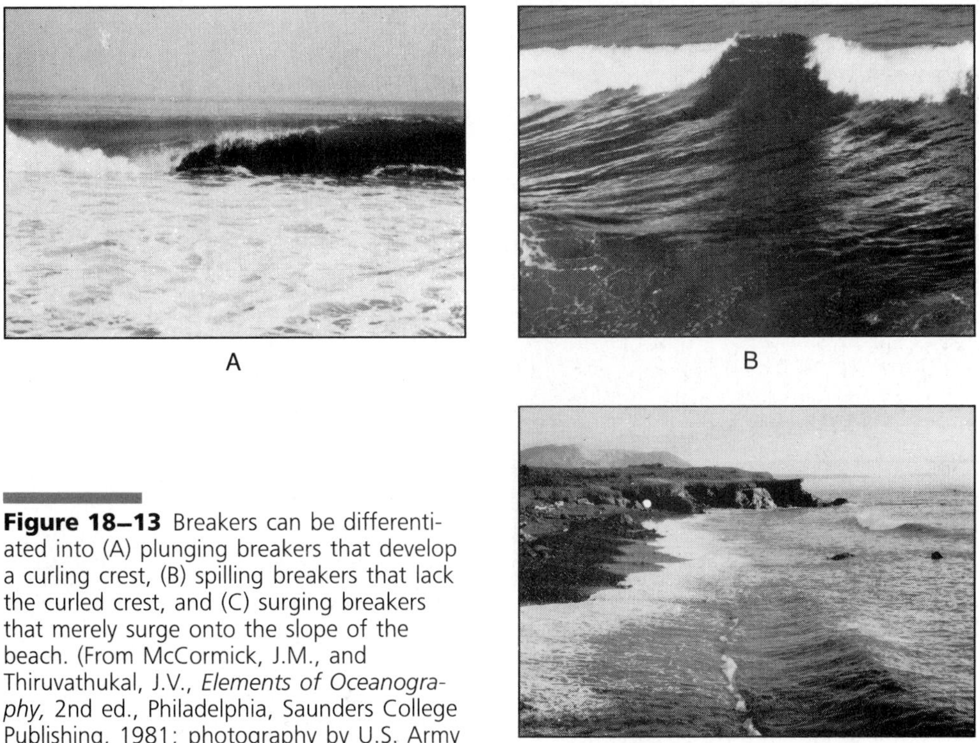

A

B

C

Figure 18–13 Breakers can be differentiated into (A) plunging breakers that develop a curling crest, (B) spilling breakers that lack the curled crest, and (C) surging breakers that merely surge onto the slope of the beach. (From McCormick, J.M., and Thiruvathukal, J.V., *Elements of Oceanography,* 2nd ed., Philadelphia, Saunders College Publishing, 1981; photography by U.S. Army Coastal Engineering Research Center.)

approaching a straight and uniformly sloping shoreline with crests parallel to the shore, then refraction would not be possible. Such conditions, however, occur only rarely. Because of the earth's spherical shape, as well as irregularities in coastlines and in the topography of the near-shore sea bottom, waves nearly always approach a shoreline from an angle, and are then refracted.

As waves approach a straight shoreline with a uniformly inclined sea floor, one end of the crest (or trough) will begin to "feel bottom" before the other (Fig. 18–14). The part of the wave that touches bottom first is slowed by friction, while the still unaffected more seaward part of the wave proceeds along at a faster speed. The result is that the wave crests and troughs swing around until they are within about 5° of being parallel to the shoreline. From above, one would observe the crest-to-crest distance diminishing and the crests forming a pattern of convex-landward curves.

If we were to take another helicopter flight over a very irregular shoreline, we would observe the refraction of wave crests into a pattern of curves that bend around jutting headlands and curve into intervening bays. This pattern is a manifestation of the fact that approaching waves "touch bottom" first adjacent to headlands where shallow water is initially encountered. Those parts of the wave front opposite bays do not encounter shallow water for an additional distance.

One of the effects of wave refraction along an irregular shoreline is to straighten that shoreline by vigorously eroding the headlands and filling the coves with transported sand and gravel. The headlands receive the first attack of the waves, and these wave segments have not dragged the bottom for any appreciable distance. Meanwhile, the remainder of the wave crests sweep into an arc so that their energy is dissipated over a wider area. The relationships are illustrated in Figure 18–15, which shows wave patterns as well as lines, called **orthogonals**, which are drawn perpendicular to wave fronts. Orthogonals are drawn with uniform spacing and are intended to indicate that equal amounts of energy are contained in the segments of the wave that are between two adjacent orthogonals. Thus, the amount of energy between any pair of orthogonals is constant regardless of refraction. Wherever orthogonals converge (as around headlands) there is a convergence of energy and a vigorous erosional attack. In the bays, orthogonals diverge, and the wave energy is spread widely. With lesser energy available, erosion is minimal, and waves are primarily at work distributing and depositing sand and silt derived from the headlands.

Figure 18–14 As the wave approaches this straight shoreline at an angle, the end of the wave at the right is the first to encounter the sea floor. That end is slowed, and a breaker begins to form that will move along the wave crest to the left. Atlantic coast near Rehoboth, Delaware.

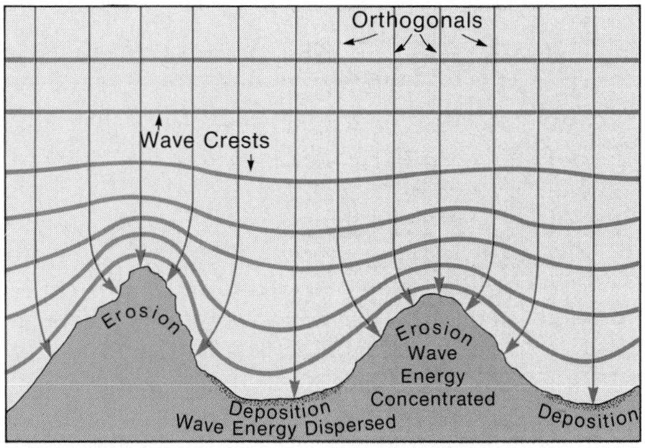

Figure 18–15 Wave refraction pattern for an irregular shoreline with a uniformly sloping bottom. Sea floor contours approximately parallel wave crests.

Wave refraction tends to straighten irregular shorelines by concentrating wave erosion on headlands and diminishing erosion in bays.

Along some stretches of straight coasts, one finds a pattern of concave and convex curves of waves somewhat resembling the refractions along more uneven coasts. Usually such a pattern is the result of irregularities in bottom topography. A submerged elevated area would retard waves and produce wave fronts with concave-landward crests (Fig. 18–16).

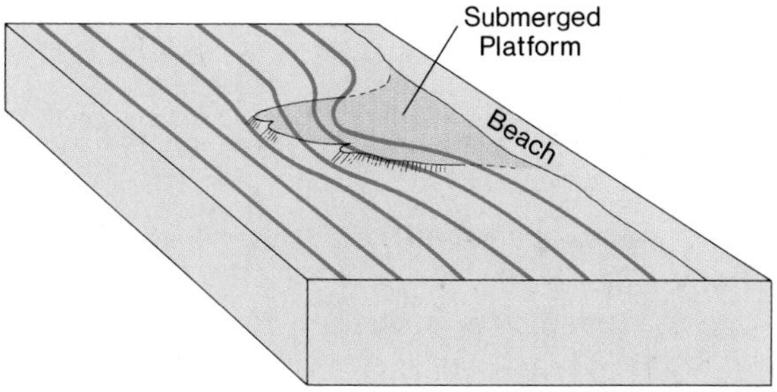

Figure 18–16 Wave refraction pattern for a straight coastline but with irregularity in bottom topography.

Near-Shore Currents

Anyone who has floated placidly on the sea on a balmy day has noticed how quickly one's position changes relative to some fixed point on the shore. Usually, we are being carried along by either longshore or rip currents. **Longshore currents** are nearshore drifts of water that parallel the shoreline. They are developed when waves approach a shoreline at an angle, so that there is a component of energy that causes water to move parallel to the beach and away from the acute angle made by the incoming waves and the shoreline (Fig. 18–17). The velocity of these currents generally increases with increases in the angle made by incoming waves and the beach, as well as increases in the slope angle of the beach, wave height, and wave period. Longshore currents are important agents in transporting sediment along a coastline.

Lifeguards at beaches sometimes caution vacationers about jetlike currents that flow out to sea (Fig. 18–18). Such narrow currents are called **rip currents**. They are formed when water brought to the shore by breakers escapes seaward. Often these currents are localized by depressions in the sea floor or breaks in offshore bars. Signs posted on beaches that warn swimmers of "dangerous undertow" are actually referring to these rip currents. Seen from above, they may be visible as long streams of foamy, turbid water extending approximately at right angle to the shore.

If, while swimming, you should be carried out to sea on a rip current, it would be unwise to exhaust yourself by attempting to swim back to shore against the current. Instead, swim parallel to the shore until out of the rip, and then swim directly to the beach.

Transport of Sediment Along the Shore

The transportation of sediment along a coastline is accomplished by longshore currents and by an interesting process known as **longshore drifting**. We may define longshore drifting as the irregular

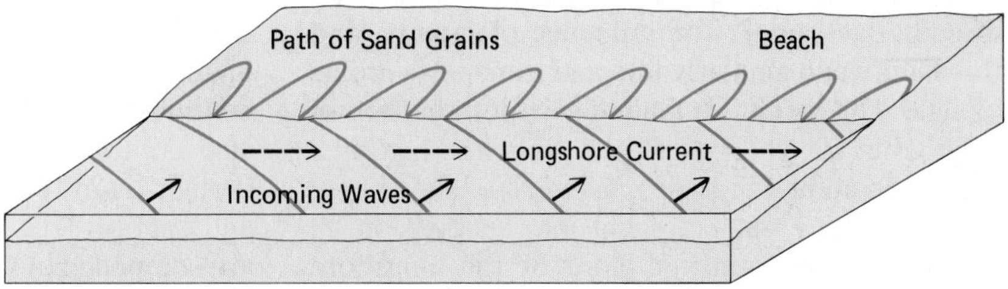

Figure 18–17 Longshore drift of sand along the beach face and in transport by longshore current.

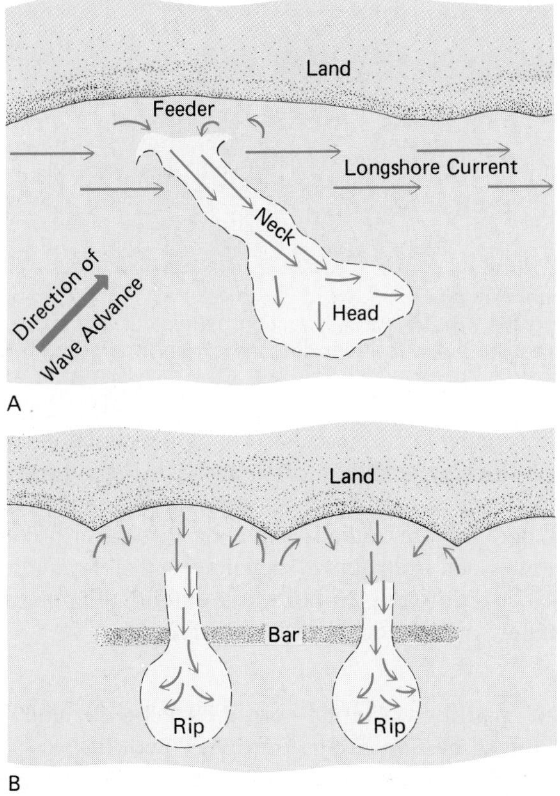

Figure 18–18 (A) Rip current formed along a coast by escape of water along a submarine topographic low. (B) Rip current caused by water piling up behind near-shore sandbars. (From McCormick, J.M., and Thiruvathukal. J.V., *Elements of Oceanography,* 2nd ed., 1981. Philadelphia, Saunders College Publishing.)

movement of sand and gravel along the shore, which results from the combined effects of incoming waves and the return flow of water from the beach. As indicated earlier, waves nearly always strike the shore at an angle and then continue along the same direction as they push up onto the beach. This incoming mass of turbulent water (**swash**) picks up particles of sand and carries them obliquely up the beach. The water then flows directly back under the influence of gravity, and this **backwash** similarly lifts and transports grains of sand. The result is a kind of sawtooth pattern of sand grain transport along the shoreline (see Fig. 18–17). In addition to this movement of sand along the beach face, sand brought into suspension by breaking waves is moved along by the longshore current parallel to the shoreline.

The direction of drifting may vary from time to time, but for most shorelines there is a cumulative movement determined by the prevailing wind direction. Longshore drifting and the carrying power of longshore currents are responsible for transport of sand over hundreds of miles of coastline. Because of these processes, many beaches exist where they might not exist otherwise. Submarine canyons, headlands, and artificial barriers may interrupt the transport of sand and produce barren stretches of coastline. Artificial barriers may also be emplaced to retard the transport of sediment and thereby preserve beaches. Groins, for example, are short walls built seaward so as to prevent loss of sand by longshore drift and to intercept moving sand (Fig. 18–19A). Breakwaters may also be emplaced to break the force of incoming waves and create an area of relative calm where sand is deposited to widen the beach (Fig. 18–19B).

Longshore drifting and longshore currents are able to transport enormous amounts of sand along shorelines.

The Beach

Millions of people who visit the beaches of the world will agree that beaches are indeed enjoyable and fascinating places. A **beach** can be rather simply defined as the zone of sediment that accumulates between the average low-water level and a landward change in topography (such as a sea cliff). Depending on the tectonic characteristics of a shoreline, beaches may extend continuously for hundreds of miles along a coast or they may be limited to protected bays and coves (Fig. 18–20). The materials that go into the building of beaches may be derived from either land or sea (e.g., reworked shell debris), but the far greater part is contributed by streams that drop their transported load of sediment when they reach the sea. Rock materials transported by glaciers, eroded from the continental margins, and transported by wind may also contribute to the formation of beaches. Although many beaches are composed of pebbles and cobbles (Fig. 18–21), the majority of beaches are composed of sand, and the most abundant mineral in most beach sands is quartz. Along granitic

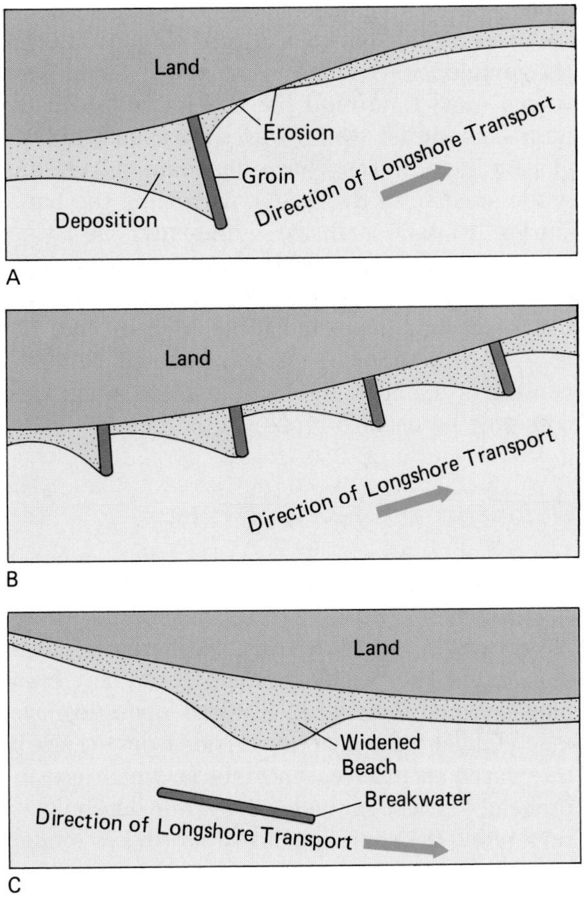

Figure 18–19 Groins and breakwaters are artificial structures designed to protect coastal areas. Groins are used to slow beach erosion, or to widen beaches by trapping sediment that was in longshore transport. (A) Unfortunately, although the groin may trap sediment on its updrift side, the area downdrift is deprived of sediment and may erode. (B) The response to such erosion is often the construction of additional groins, each emplaced to protect the beach or resorts or homes along the coast. (C) Breakwaters are built parallel to the shore to protect beaches and harbors from high waves. Because of the lessening of wave energy behind the breakwater, deposition occurs, widening the beach.

coastlines, feldspars also contribute to the composition of beach sand, whereas in semitropical or tropical regions, sand composed of broken and worn shell fragments may be prevalent (Fig. 18–22). Grains of black basalt and green olivine are common in the beaches of volcanic islands. Once it is deposited on a beach, sediment is subjected to repeated redistribution and transport by beach drifting and longshore currents.

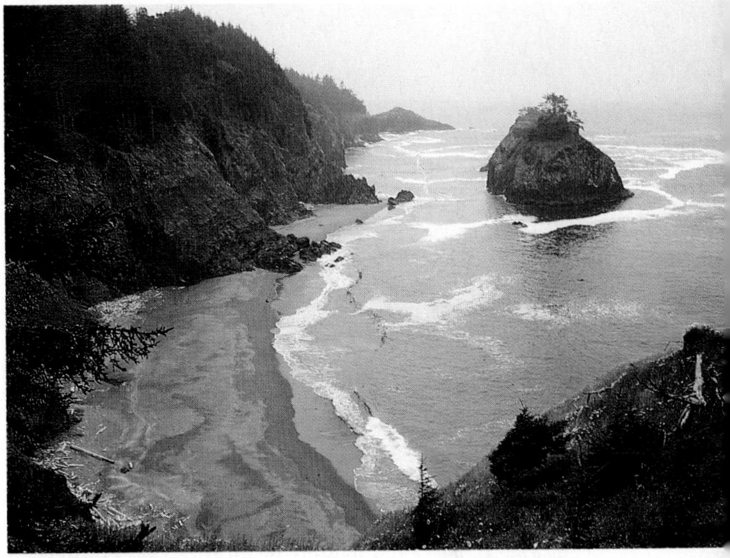

Figure 18–20 Beach developed in a small cove on the Oregon coast just north of Brookings, Oregon.

As seen in profile (Fig. 18–23), a beach can be divided into several subdivisions, although each of the subdivisions may not be present on every beach. Most beaches have a nearly flat upper part formed by deposition of material by waves. This level feature is called the **berm**, and a sloping lower part is called the beach face. The **beach face** is the area across which waves swash upslope across the beach and return as backwash. The

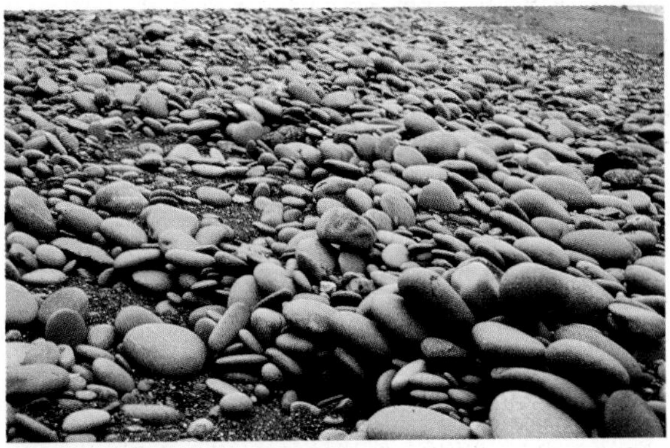

Figure 18–21 Beach face composed of cobbles and pebbles, and produced by vigorous wave action. (Courtesy of L.E. Davis.)

Figure 18–22 Beach sand composed of calcium carbonate skeletal debris, Andros Island, Bahama Islands.

slope of the beach face is largely determined by sediment particle size. In general, the larger the grain size, the steeper the slope. Very fine sand, for example, results in a beach face having only 1° or 2° slope, whereas coarse sand is found on beaches having a slope of about 9°. In addition to the berm and beach face, a **beach scarp** is sometimes also developed along some stretches of the shore. These features are cut at the lower margin of the beach face by large waves and generally do not exceed 2 meters in height.

Clearly, beaches are geologic features characterized by continuous change. They may be thinned or even swept away during times of stormy weather, or they may be built wider and thicker when seas are calm. Summer to winter changes occur in regions having seasonal climates. During the summer, waves tend to be long and low. Such waves wash sand onto the beach and widen the berm. In contrast, winter waves are usually tall and closely spaced. They have abundant energy and erode sand from the beach, narrowing the berm. During winter, sand-size grains may be swept away, leaving behind a beach face carpeted with pebbles and cobbles. The sand removed from the beach may be deposited in submerged bars offshore where the water is less agitated. With the coming of summer, the sand in these submerged bars may be used to replenish the beach.

Erosion of Shorelines

Waves are important agents in causing the erosion of coastlines (Fig. 18–24). As one might expect, large waves, such as those generated during storms, are exceptionally effective and may cause more change than smaller waves operating over much longer periods of time. Aside from the size of the waves themselves, the rate at which erosion proceeds is also influenced by the durability of rock along the shoreline, how that rock is jointed, the openness of the coast to attack, the slope of the sea floor adjacent to the coast, and the abundance and size of sedimentary particles that can be hurled by waves against the shore. The abrasive action of sand, gravel, and pebbles in surging water

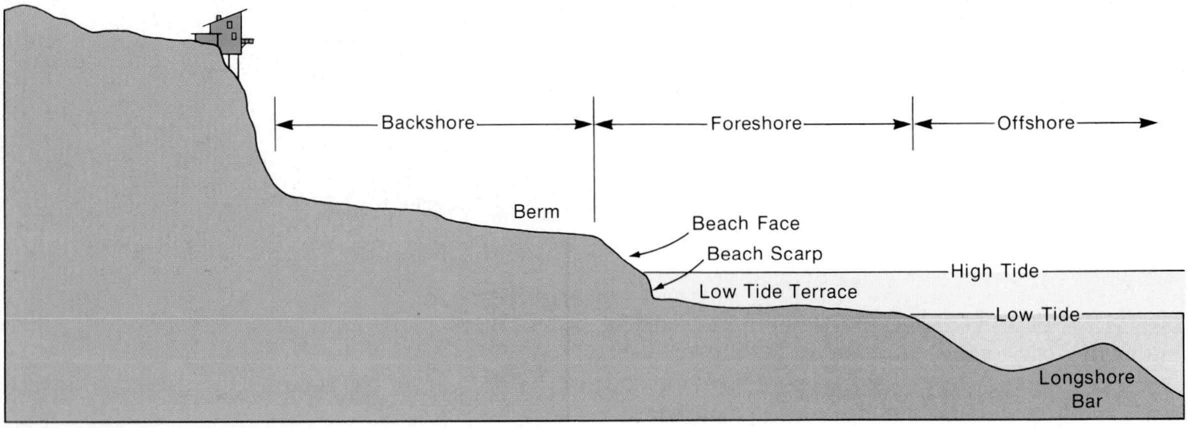

Figure 18–23 Major divisions of the beach and nearshore area.

effectively reduces rock particles to sizes that can be carried seaward in rip currents.

Rugged coastlines provide an impressive show of the effectiveness of wave impact in eroding rock masses (Fig. 18–25). The crash of incoming waves may exert pressures of over 6000 pounds per square foot. Air in joints and fractures is suddenly compressed and acts as a wedge in driving rocks apart. In the next moment, the water recedes, and the compressed air expands, often with explosive force. These mechanical stresses on the rocks are capable of reducing massive granitic headlands to piles of boulders and rubble.

There are many features of coastlines that are formed primarily by erosional activity (Fig. 18–26). Among these features are wave-cut cliffs, wave-cut benches, and an assortment of erosional remnants termed stacks, sea caves, and sea arches. **Wave-cut cliffs** are coastline escarpments resulting from the erosive powers of water that surges landward with each breaking wave. The action of the surf reminds one of a horizontal saw that repeatedly bites into the base of the cliff, sometimes chiseling out an indentation called a **notch** (Fig. 18–27A). Undermining by wave action causes periodic collapse of the face of the cliff (Fig. 18–27B), and the downfallen rubble is then further diminished by the surf. As the wave-cut cliff retreats before the onslaught of the waves, remnants may be left behind as **stacks** (Fig. 18–28). Elsewhere, waves deferentially erode along the more readily erodible joints and sculpture out **sea caves** and **sea arches** (Fig. 18–29). Extending seaward from the base of the cliff along shorelines where erosion predominates, a platform called a **wave-cut bench** (Fig. 18–30; see also Fig. 18–28) is produced by wave erosion. The erosionally beveled bedrock of the bench may be exposed, or it may be covered with beach sand and gravel.

Figure 18–24 Wave action is rapidly eroding the relatively soft sedimentary rocks exposed along this part of the northern coast of Puerto Rico. (Courtesy of W.H. Monroe and the U.S. Geological Survey.)

Figure 18–25 Waves contain astounding energy. They are able to break up huge masses of rock directly and by forcing water and air into rock fractures. A wave of moderate size, 10 feet high by 100 feet long, can exert a push of nearly 1700 pounds per square inch. Here we see waves breaking against the rugged coast of Mount Desert Island, Maine.

Depositional Features of Shorelines

In addition to beaches, there are several other depositional features that develop along coasts as a result of the activity of near-shore currents and waves. Some of these features are developed

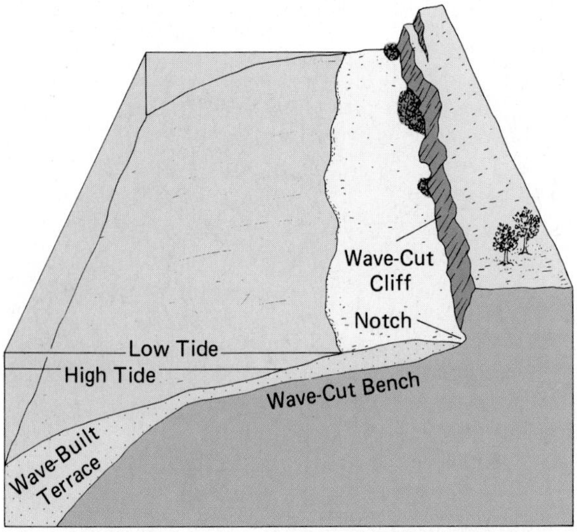

Figure 18–26 Erosional features of the shore profile are the wave-cut cliff, wave-cut bench, and notch. Depositional features include the beach and wave-built terrace.

below sea level, and others are built above sea level. Among the submerged features are wave-built terraces and longshore bars. **Wave-built terraces** (see Fig. 18–26) are simply accumulations of unconsolidated sediment swept back to sea from the beach environment or deposited by longshore currents. In most locations, the boundary between the beach and terrace is transitional. **Longshore bars** (see Fig. 18–23) are elongated, barely submerged, ridges of sand that parallel the shoreline. They are particularly characteristic of coastal areas in which the ocean floor slopes very gently seaward and which have an abundant supply of sand. Longshore bars migrate shoreward and increase in height during winter seasons when larger waves are prevalent. These trends are reversed during the less stormy summer season.

Several different coastal topographic features are built above sea level. They are distinguished by their form and position. For example, **spits** are ridges of sediment connected at one end to land and that extend into open water at the other end (Figs. 18–31 and 18–32). One can tell the direction of a longshore current by the direction in which the spit extends. Where spits partially extend across the mouth of a bay, wave action may cause them to curve landward (Fig. 18–33). The result is

A

B

C

Figure 18–27 (A) The notch formed by wave action at the base of a sea cliff near Capitola, California. (B) Collapse of the face of a sea cliff resulting from the undercutting that produced the notch. (C) Also at Capitola, the effect of such collapse on dwellings constructed too near the cliff. (Courtesy of R.E. Johnson.)

a feature called a **recurved spit** or **hook**. Spits may build completely across the bay opening so as to close off the harbor. In this way, they are transformed into **bay barriers**. When a bar built of sand or gravel extends from the mainland out to a nearby island or connects two islands, the resulting feature is termed a **tombolo**.

Barrier islands (Fig. 18–34) are elongated islands of sand extending parallel to a coast. These features are rather continuous along the Atlantic Coast of the United States as well as around Florida and along the Gulf of Mexico. Barrier islands may attain widths of several km and extend for over 100 km in length. Coney Island off the New York coast and Padre Island off Texas are good examples of barrier islands. The origin of these coastal features is rather complex. Some appear to have been formed simply as the result of the continuous extension of a spit along a coastline. The shoreward migration of offshore bars may also be responsible for having formed some barrier islands. One favored theory for the barrier islands along our eastern seaboard is that they formed during the rising of sea level with the last melting of glaciers about 10,000 years ago. The rise in sea level would have drowned low tracts of land behind coastal dunes and isolated the former dune areas as barrier islands offshore (Fig. 18–35).

Figure 18–28 A large sea stack (center background) and wave-cut bench (foreground) eroded in steeply dipping Carboniferous limestones, Worm's Head, Wales.

Figure 18–29 Sea arch eroded in a stack, Oregon coast. (Courtesy of L.E. Davis.)

Figure 18–30 Wave-cut bench eroded in resistant, dipping shales. Santa Cruz, California. The beds are dipping in a seaward direction. (Courtesy of R.E. Johnson.)

Kinds of Coasts

The marvelous variety of topographic features along different coastlines are, as we have noted, the result of many interacting variables. Coasts are dynamic geographic elements that are constantly changing as a result of erosion, deposition, and tectonic movements. Because of the changes and the variables, classifications of shorelines are difficult to formulate and should be considered as generalizations to which there are nearly always exceptions.

Drowned Coasts

The majority of the world's coastlines appear to have experienced varying amounts of submergence in relatively recent geologic time. The exceptions are coastal tracts that have been undergoing

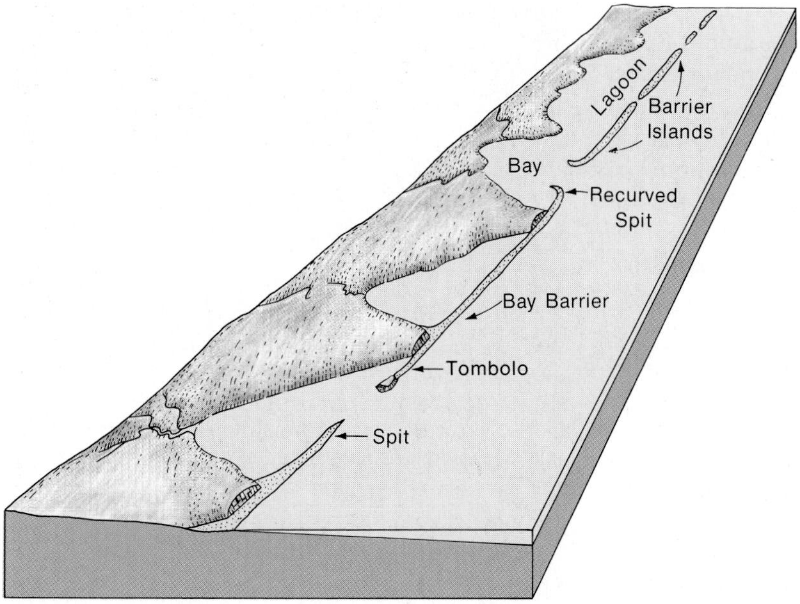

Figure 18–31 Depositional features along a coastline.

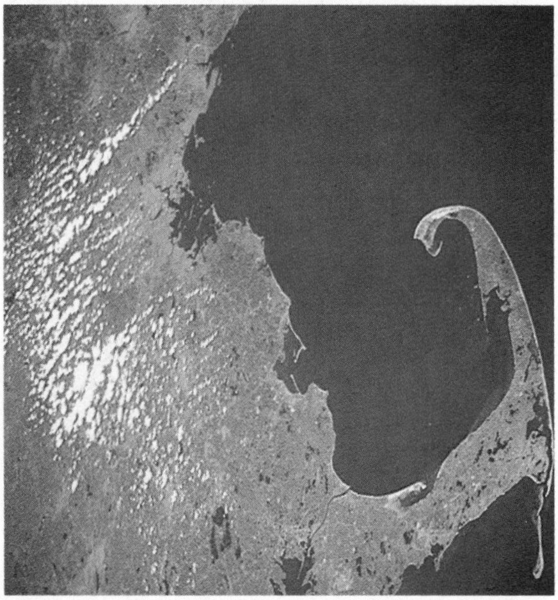

Figure 18–32 Wave refraction and longshore transport of sand derived largely from an earlier glacial moraine to the south produced the large spit and hook at the north end of Cape Cod, Massachusetts. This photograph was taken by the space shuttle, 260 km above the earth. (Courtesy of NASA.)

Figure 18–33 Northern California coast at mouth of the Klamath River. Spits have nearly closed off the bay to produce a bay barrier. Note landward curve of spit in lower left.

Figure 18–34 These resort hotels of Miami Beach, Florida, have been erected on a barrier island. A lagoon behind the barrier island is visible in the upper left. (Courtesy of the Florida Department of Commerce, Division of Tourism.)

active uplift. This does not necessarily mean that all of the inundated coasts subsided, for inundation can also result from rising sea level. Indeed, rising sea level appears to be the reason for most submerged coastlines today. A global rise in sea level began about 10,000 years ago as meltwater from continental ice sheets was poured back into the ocean.

Whether the result of subsiding land or rising sea level, drowned coastlines exhibit highly irregular outlines, somewhat resembling the outlines of oak leaves. The configuration results from the inundation of low-lying areas as the sea advances into river valleys and tributaries. The distinctive branching pattern is well developed in such areas as Delaware Bay (Delaware and New Jersey), Pamlico Sound (North Carolina), and Chesapeake Bay (Maryland and Virginia) (Fig. 18–36).

Fjords (Fig. 18–37) are characteristic of submerged coastlines in glaciated regions. They were developed by valley glaciers that excavated the lower ends of their valleys to depths well below

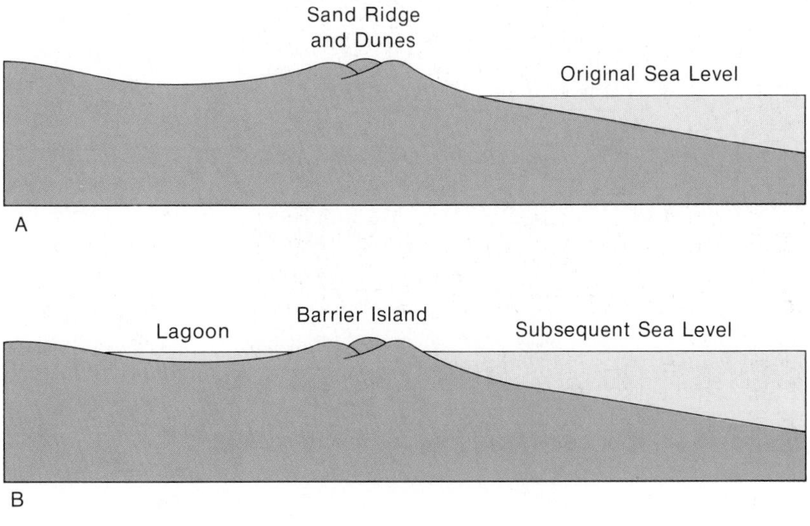

Figure 18–35 Development of a barrier island as a result of a rise in sea level.

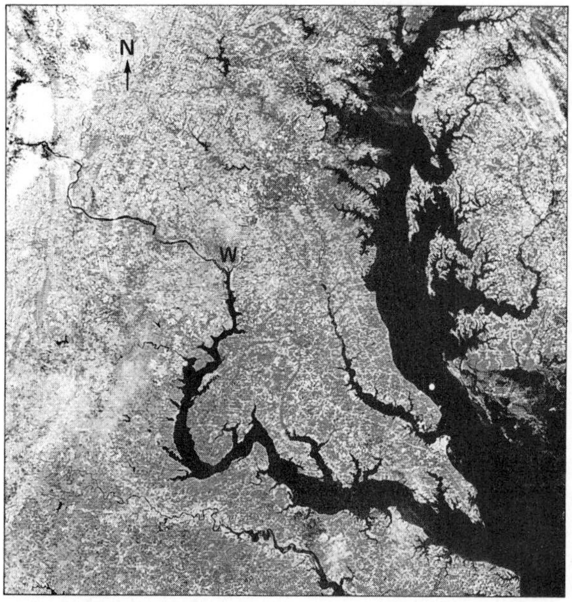

Figure 18–36 Earth Resource Technology Satellite photograph of Chesapeake Bay. The geologically recent increase in volume of water in the ocean with the melting of glaciers has caused an inundation of river valleys to produce bays such as this one. The city of Washington, D.C. is located at the "W" on the Potomac River. (Courtesy of NASA.)

Figure 18–37 Fjord near Juneau, Alaska (Courtesy of J.C. Brice.)

present sea level. The great depth of water in some fjords is the result not only of glacial action but also of the previously mentioned eustatic increase in sea level following the Ice Age. Fjord coasts have splendid protected harbors and some of the most spectacular scenery on earth. Although the fjords of Norway are probably the most famous, these features are also found along the coast of New Zealand, Greenland, Alaska, and British Columbia.

Deltaic Coasts

Deltaic coasts are those built seaward as a result of the deposition of sediment brought to the ocean by rivers. In order for a delta to form, the river must provide sediment at a rate faster than waves and near-shore currents can redistribute that sediment. For this reason, only rather large rivers are able to build and maintain deltas. If the coastline at the mouth of a river is relatively protected, and if tidal action is minimal (as in the Mediterranean and Gulf of Mexico), this will slow the removal of sediment by wave action and thereby improve the opportunities for the development of deltas. Deltas vary considerably in overall form. Some are truly in the form of the Greek letter delta (Δ), others are more broadly rounded, whereas others are called bird-foot deltas in reference to their finger-like extensions into the ocean (Fig. 18–38). The shape of a delta reflects the balance between sediment supply by the stream and the transport capability of ocean currents and waves. In the Mississippi Delta, sediment supply far exceeds the reworking capability of the marine environment, and sediment carried seaward in distributaries is deposited in the bird-foot pattern. In contrast, the sediment supply and reworking capabilities of the ocean are in much closer balance in the Niger Delta, and this balance accounts for its more uniform shape and nearly continuous perimeter of beaches and sand bars.

In the great marine deltas of the earth, the littoral zone has its maximal width. Here one finds shallow-water lagoons and saltwater marshes alternating with patches of land over a belt that may be 50 km wide. Not only is there a mixture of environments of sedimentation across the surface of

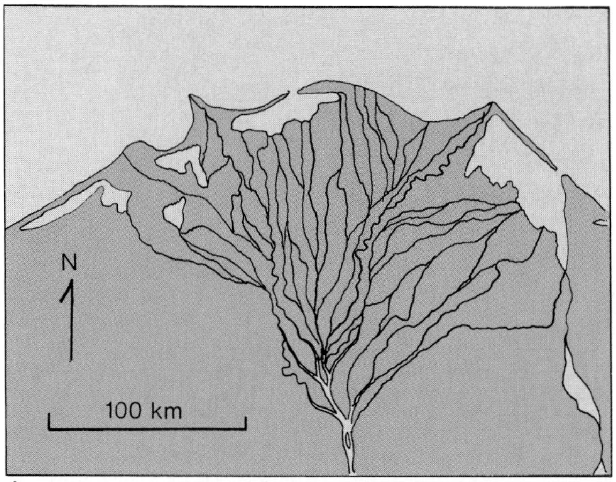

A

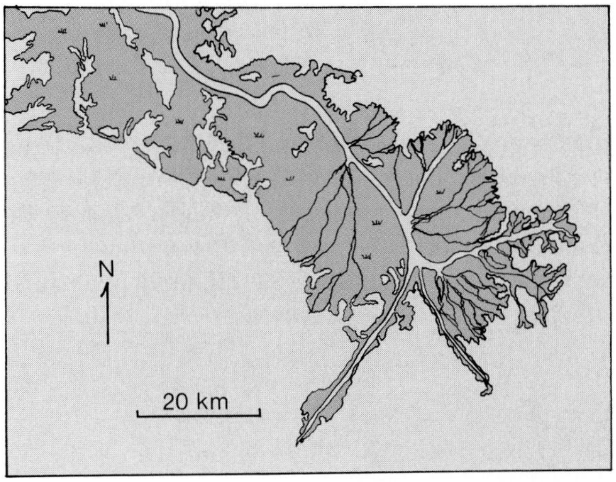

B

Figure 18–38 (A) The Nile Delta. The branching streams of this delta begin about 160 km inland at Cairo and fan out seaward over a great triangular area. (B) The most southern subdelta of the Mississippi Delta exhibits the bird's-foot pattern.

the delta, but vertical sections are characterized by an intertonguing of marine and nonmarine deposits caused by repeated shifting of the shoreline. Nearly always, sedimentation in deltas involves transportation of sands far out into the marine basins where they may become enclosed in organic rich muds. The sands are potential reservoir rocks for oil and gas, whereas the muds may provide the hydrocarbons and serve as impermeable barriers needed for oil entrapment. Clearly, ancient deltas are important because of the store of fossil fuels they may contain. They are no less important, however, because of the record they preserve of past conditions on earth.

Tectonically Active Erosional Coasts

Some of the most beautiful coasts in the world are those that have been subjected to periodic tectonic uplift and then vigorously eroded by marine agencies. Raised wave-cut cliffs and stairlike sequences of wave-cut benches (Fig. 18–39) are common along such coasts, and these features attest to the periodic uplift experienced by the unstable margin of the continent. Often very deep water exists immediately offshore from the mountainous coastal belt. Because of the rapid steepening of the sea floor, offshore bars are rarely developed. Tectonically active erosional coasts are particularly prevalent around the Pacific Ocean.

Coastlines Along Which Deposition Predominates

Another prevalent type of coastline is characterized by a broad shallow shelflike offshore zone lying adjacent to a low-lying coastal plain. Such coasts also have an abundant supply of sandy sediment. In such an environment, barrier islands and beaches are common and areally extensive. Some of the barrier islands may attain widths of several km and extend for great distances. Galveston (Fig. 18–40), Atlantic City, and Miami Beach are built on barrier islands. Usually the barriers are separated from the mainland by shallow lagoons. It is not uncommon for such lagoons to gradually silt up and thereby increase the size of continents.

Coasts Modified by Organisms

Yet another kind of coastline is shaped largely by the activity of organisms. As indicated in the previous chapter, coral reefs and calcareous algae abound in tropical areas and can add significantly to the dimensions of a coastal area. The largest of such structures is the 1600-km complex of reefs

Figure 18–39 Elevated marine terrace provides evidence for change in sea level at Rhossili Bay, Wales. (Courtesy of J.C. Brice.)

known as the Great Barrier Reef, which forms a gigantic reef barrier island and breakwater off the northeast coast of Australia. The Florida Keys are supported by reefs formed during an interglacial stage of the Ice Age when sea level was about 6 meters higher than at present. Similar organic coasts are prevalent around the islands of the southwestern Pacific. Besides corals, oysters, tube-secreting worms, calcareous algae, mangrove plants, and a host of other organisms contribute to the formation of coastlines modified by plants and animals.

Figure 18–40 Relatively straight barrier islands extend east–west immediately to the south of Trinity and Galveston bays, Texas Gulf Coast. Groinlike structures called jetties allow entrance to the bay. One can see sediment streaming out through the passage created by the jetties, and into the Gulf of Mexico where it is shunted westward by longshore currents. (Space shuttle photograph, courtesy of NASA.)

Summary

Tides are caused primarily by the gravitational attraction of the moon and, to a lesser extent, by the sun's gravitational pull. The earth develops two tidal bulges simultaneously. One of these is on the side facing the moon, and the other is on the opposite side of the earth. The bulge opposite the moon results primarily from the strong attraction of the moon for the water mass that is closest to it. Because gravity decreases with distance, the force acting on the body of the earth itself is greater than that acting on the ocean of the opposite hemisphere. Thus, the moon tends to pull the entire earth away from the opposite hemisphere's ocean. Because two opposite tidal bulges occur on earth, as the planet rotates, any given section of coastline will experience two high tides and two low tides each day. Friction between tidal bulges and the rotating solid earth has the effect of slowing the earth's rotation by a small amount. Because of this, early in the earth's history a complete rotation of the planet required less than 24 hours, and there were more days in a year.

Waves are caused by friction of wind as it passes over the water surface. The size of waves is determined by such factors as wind velocity, duration, and the distance over which the wind has blown. Waves are described according to their *wave length* (distance from crest to crest), *wave height* (vertical distance between crest and trough), and *wave period* (the time necessary for a crest to travel one wavelength). Water particles in the wave move in circular orbits that diminish with depth of water until little or no motion is evident at a depth equivalent to one-half the wavelength. As waves approach a coastline, they undergo a marked change in shape due to the effect of friction with the sea bottom. The waves become higher and more closely spaced, whereas the wave period does not change significantly. Eventually the unstable crest topples over and forms a line of breakers.

Most waves approach land obliquely. They "feel" the drag of the bottom first at that part of the wave that first attains a depth of one-half a wavelength. Because of the lessening of wave velocity at the part of the wave that first "feels bottom," the wave bends by refraction and breaks against the shore at an angle of about 5°. As sheets of water (swash) move obliquely upward on the beach, they carry along grains of sediment, which are subsequently carried directly back down the beach as the water returns in the backwash. This pattern of swash and backwash results in a net transport of sediment along the beach. The transport system is called *longshore drifting.*

Waves erode shorelines predominantly by *hydraulic action* and *abrasion,* both of which help to produce such erosional features as wave-cut benches, sea cliffs, stacks, sea caves, and sea arches. Refraction plays an important role in wave erosion by concentrating the energy of waves on headlands.

Waves are indirectly responsible for such inshore water movements as longshore currents and rip currents. *Longshore currents* are produced as waves strike a shore obliquely. The coast acts as a barrier, and water is deflected so as to flow parallel to the shore. *Rip currents* represent a return flow of water brought to the near-shore area by waves.

Sediment carried along a coastline by longshore currents contributes to the formation of *beaches,* as well as *longshore bars, wave-built terraces, spits, bay barriers, tombolos,* and *barrier islands.*

The physical appearance of any particular coastal area is determined by the size of waves, openness of the coast to wave action, nature of coastal bedrock, and changes in sea level of either a tectonic or eustatic nature. Coastlines having an irregular profile with many shallow inlets and estuaries are termed *drowned coastlines.* Deltaic coastlines often have a lobate or fanlike pattern and are developed where streams enter the sea and deposit their load, thereby extending the shoreline seaward. *Tectonically active coasts* are those in which recent tectonic uplifts have raised wave-formed features well above present sea level. Along marine depositional coasts, features formed by the deposition of sand and gravel predominate. Broad beaches, bay barriers, spits, and barrier islands are prevalent. Some coasts are built largely by the activity of calcium carbonate–secreting organisms. Coral reef coasts are the most important in this category.

Terms to Understand

backwash	Law of Universal Gravitation	sea arch	wave base
barrier island	longshore bar	sea cave	wave-built terrace
bay barrier	longshore current	spit	wave-cut bench
beach	longshore drifting	spring tide	wave-cut cliff
beach face	neap tide	stack	wave frequency
beach	notch	surging breakers	wave height
berm	orthogonal	swash	wave length
breaker	recurved spit (hook)	swell	wave period
centrifugal force	rip current	tidal bore	wave refraction
inertia		tombolo	wave velocity

Questions for Review

1. What is the relationship between the motion of water particles within a wave and the external form of a wave?

2. What is the relationship between wavelength and the water depth at which waves begin to slow down and form breakers?

3. What is the cause of wave refraction? What is the role of wave refraction in causing a general straightening of a formerly rugged and irregular coastline?

4. How might a coastline that has experienced recent uplift differ in appearance from one that has experienced recent subsidence?

5. Sketch an imaginary length of shoreline that exhibits beaches, spits, bay barriers, barrier islands, and tombolos. Label each feature.

6. Explain, using a diagram, the reason for neap and spring tides.

7. The velocity of waves in the open ocean can be determined from the formula

$$V = \frac{L}{T}$$

(where V is velocity, L is wavelength, and T is wave period). What is the velocity of a wave having a wavelength of 10 meters and a period of 10 seconds? What is the period of a wave having a wavelength of 12 meters and a velocity of 2 meters per second?

8. Some stretches of shoreline do not have beaches. Why might a shoreline lack beaches?

9. Far from the coast, the ocean surface does not have uniform waves with evenly spaced linear crests, but rather the surface has irregularly spaced elliptic crests and troughs. What is the explanation for this observation?

10. What changes in waves occur as they approach the shallow area of a coastline?

Supplemental Readings and References

Bascom, W., 1980. *Waves and Beaches,* 2nd ed. Garden City, N.Y., Doubleday.

Clancy, E.P., 1968. *The Tides, Pulses of the Earth.* Garden City, N.Y., Doubleday.

Davis, R.A., Jr., 1972. *Principles of Oceanography.* Reading, Mass., Addison-Wesley Co.

King, C.A.M., 1972. *Beaches and Coasts.* 2nd ed. New York, St. Martin's Press.

Shepard, F.P., and Wanless, H.R., 1971. *Our Changing Coastlines.* New York, McGraw-Hill Book Co.

The Earth's Geologic Resources

19

Copper mine at Bingham Canyon, Utah. This is a spectacular open-pit type of mining that recovers finely disseminated copper sulfides from the porphyry host rock. The ore contains about 1 per cent copper. The benches cut into the porphyry are about 15 to 20 meters high and 20 meters wide. (Courtesy of Eva Moldovanyi.)

A rising world population requiring an improved standard of living clashes with the oncoming realities of a planet of impoverished resources.

Loren Eiseley, 1970 The Invisible Pyramid, *Chas. Scribner's Sons*

Introduction

Since the time human hands first struck one piece of flint against another in an attempt to made a crude tool, humankind has had an intimate involvement with the earth's mineral resources. Our Ice Age ancestors were able to recognize in rocks the properties needed to make a sharp-edged scraper or spear point. It was not unusual for them to carry flint and chert with them when they moved to locations where such useful rocks were not available.

From stone, people turned to metals. At first the only metal to be fashioned into implements was copper, and this was not an altogether satisfactory material. It was soft, easily bent, and knives with copper blades would not keep a keen edge. Perhaps by fortunate accident, this deficiency was corrected when minerals containing both copper and tin were thrown into the fire together. The marriage of the two metals produced bronze, an alloy that when formed into tools would not bend or lose its edge as readily.

The next metal to be used by ancient people was iron. Actually, pieces of native iron were used even before anyone knew how to make a fire hot enough to melt the metal. The ancient Sumerians, for example, knew about iron from the meteorites they had found. Perhaps they even recognized the extraterrestrial origin of these objects, for the Sumerian name for meteorites means "metal from heaven." By about 1500 B.C., artisans among the Hittites were smelting iron, and 500 years later, metalsmiths in India were producing steel by adding small amounts of carbon (less than 1.7 per cent) to the melt of iron. From that time to the present, tools have improved, and with the improvements, factories and farms have been able to produce more than ever before. At first, however, technologic progress was slow. Then, early in the nineteenth century, two momentous inventions launched the world into an era of industrialization which continues today. One was the development of the blast furnace for producing large quantities of steel, and the other was the invention of the steam engine. Nations with the good fortune of having both abundant coal and iron became leaders in the production of manufactured goods and achieved wealth and high standards of living. For the people of many of these industrialized nations, relatively good living conditions have continued into the present century when petroleum replaced coal as the preferred fuel. Within recent decades, however, shortages of critical materials have developed. There is little doubt that these shortages, and those that will occur in the near future, are related to a rapidly increasing number of consumers.

Resources and the Population Explosion

The most important concept to be learned from this chapter is that our geological resources are finite and are presently being rapidly depleted, largely because of burgeoning global population. Populations of organisms, including human populations, tend to increase in geometric progression. Thus, if a woman has two children and both of these children in turn produce two children, and these offspring continue to reproduce in like manner, then the human species increases by the numbers 2, 4, 8, 32, 64, 256, and so on. Huge numbers are soon generated by this geometric growth in population. As an example, the earth's population in 1970 stood at 3.6 billion people. Projecting current rates of increase into the future, there will be 7. 5 billion people on earth in the year 2000 (Fig. 19–1). Only a decade into the 21st century, global population is likely to exceed 10 billion. Can our

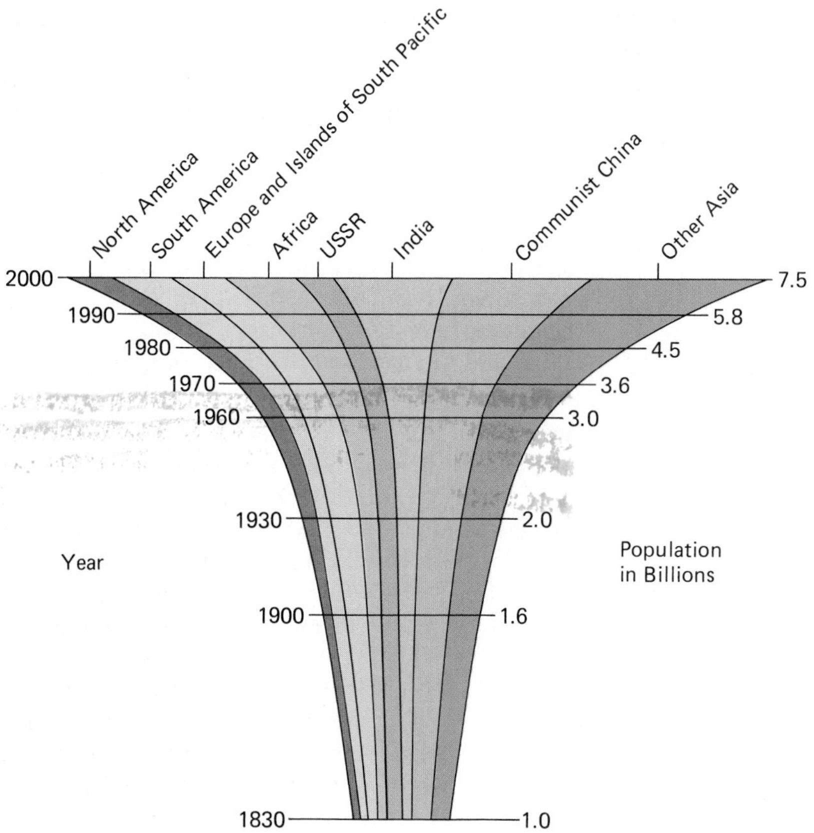

Figure 19–1 World population growth projected to the year 2000. (From U.S. Dept. of State information circular.)

planet sustain so large a population? With our present population of about 5.8 billion, there are already nations in dire need of food and the resources that can provide even a marginal standard of living. Even among many of the wealthier nations, there is evidence of a declining standard of living.

The problem of finding new reserves of nonrenewable resources for billions of humans cannot be solved by geologists alone, nor by scientists finding alternative materials for depleted resources, nor by technicians with skills in the recycling of metals and plastics. In the end, the problem can be solved only by having fewer consumers so that reserves will last longer.

The needs of the earth's rapidly increasing population cannot indefinitely be met by the earth's finite resources of fuels, minerals, water, and farmland.

Energy Resources

The major sources of fossil and mineral energy in the world today are crude oil, natural gas, coal, and uranium. Approximately 40 per cent of the energy consumed by nations comes from the burning of coal, another 40 per cent from oil, and about 10 per cent from gas. The remaining 10 per cent comes from hydrothermal, nuclear, geothermal, and solar facilities.

Coal

Long before petroleum came into use as a source of energy, coal was the traditional fuel (Fig. 19–2). It was mined by Europeans during the Middle Ages, used by some tribes of American Indians for warmth and to fire their pottery, and it fueled the industrial revolution. During the first half of the present century, coal supplied larger amounts of energy than any other material. With the advent of cleaner and more conveniently handled petroleum and natural gas, however, the use of coal in the United States diminished. Over the past three or four decades, coal has been used in the United States mostly for the smelting of iron and as a fuel in electric generating plants. Coal began to regain popularity in the 1970s because of the oil embargo and rising prices of gasoline and heating oil. Although coal is not without disadvantages, it is comforting to know that there is enough still remaining to help tide us over difficult times. Indeed, coal deposits in America are sufficient to provide at least twice the energy potentially available in all the Middle East oil fields.

The Origin of Coal

Coal is a brownish-black to black, carbon-rich combustible material that forms beds from a few centimeters to many meters in thickness and is interbedded with shales, sandstones, and limestones. It was formed long ago as the result of the accumulation of dead vegetation in an environment that favored the concentration of carbon. Swamps and bogs in mild temperate to equatorial climates are among the places that favored the initiation of coal-forming processes. Such environments have an abundance of plants. As the plants died and fell away, they were covered by water and soggy layers of additional rotting vegetation. Beneath this cover, oxidation decay was inhibited. Carbon in the dead plant tissue was prevented from combining with oxygen in the air and thereby being lost to the atmosphere. In this oxygen-deficient environment, degradation proceeded primarily by the action of anaerobic bacteria, which extracted oxygen from the plant tissue, released hydrogen (which is then recombined in various gases), and left behind a residue of carbon. By this time, the vegetal matter had been converted to **peat**, a water-soaked, porous brown mass of incompletely degraded plant remains containing about 50 per cent carbon. The coal-forming process continued as the peat was buried under still more plant debris as well as sand, silt, and clay. Beneath this heavy burden, water was squeezed out, and increasing temperature and pressure forced out volatile organic compounds such as methane (CH_4). The result was a further increase in the proportion of carbon. Given an abundance of time and continuing increases in pressure and temperature, the original mass passed through a series of stages leading to full conversion to coal.

Clearly, coal could not have developed in the past without an abundance of land plants. The fos-

Figure 19–2 Mining coal in West Virginia. Shown here is a remote-controlled continuous mining machine with a flooded bed scrubber for dust control. (Courtesy of Peabody Coal Co.)

sil record of land plants begins about 400 million years ago, and thus coal occurs in rocks of post-Devonian Age. The Pennsylvanian System is particularly well known for its abundance of coal.

Kinds of Coal

The kinds of coal formed by the processes described above vary according to the character of the vegetation and the amount of pressure brought to bear on the potential coal seam. Where the vegetation was only slightly compacted and therefore not subjected to sufficient pressure and heat, the soft porous "precoal" peat is formed. As already noted, peat is only partially altered, has a relatively low carbon content, and still contains as much as 30 per cent oxygen. Nevertheless, it is used as fuel in Scandinavia, parts of Holland, Ireland, and Scotland.

The further compaction and alteration of peat may result in a dark brown, relatively lightweight coal called **lignite**. Lignite contains about 70 per cent carbon and 20 per cent oxygen. With recurrent prospects of petroleum shortages, lignite is receiving increased attention because it can be used for the generation of gas and the production of synthetic gasoline.

The next rank of coal above lignite is **bituminous coal**. This coal is denser than lignite, contains 80 per cent carbon and only 10 per cent oxygen, and has a lustrous black color. Because of its higher carbon content, it burns hotter than lignite. Bituminous coal is plentiful and adaptable to many uses.

The highest ranking and most intensely compacted of all coal is **anthracite**. It is a jet-black, shiny coal that gives little smoke on burning and provides an exceptionally hot flame. For these reasons, it was once favored for domestic heating. Anthracite is actually metamorphosed bituminous coal in which the concentration of carbon reaches levels as high as 98 per cent.

Coal increases in carbon content from lignite through bituminous coal to anthracite coal.

Reserves and Uses

Geologists believe that they have already found most of the major coal basins of the world. It has been estimated that beds thick enough and close enough to the surface to be mined at a profit under current economic conditions constitute over 8 trillion tons, enough to supply world needs for about a century. In the United States, which is second only to the U.S.S.R. in coal reserves, thick seams of coal stretch from Pennsylvania to Alabama, with more in Indiana, Illinois, Missouri, Kansas, Oklahoma, Texas, and several Rocky Mountain states (Fig. 19–3).

Coal has a greater number of uses than most people realize. It is, of course, used in heating buildings and firing steam boilers that power electric generating turbines. In addition, the steel industry prepares coke from coal, which in turn is used to free metallic iron from its ore (Fig. 19–4). The process is summarized by the equation:

$$\underset{\text{iron ore (hematite)}}{2\,Fe_2O_3} + \underset{\text{coke}}{3C} \longrightarrow \underset{\text{iron}}{4Fe} + \underset{\text{carbon dioxide}}{3CO_2} \uparrow$$

Coal tar, a sticky black substance derived from coal, is used in the manufacture of a variety of products including dyes, sulfa drugs, insecticides, aspirin, and plastics. Finally, coal can be used as a raw material in the manufacture of synthetic oil and gas by means of processes called gasification.

Coal and the Environment

There were good reasons why Americans switched from coal to oil for their energy needs earlier in this century. Coal does not burn as cleanly as oil or gas, it usually requires expensive barge or rail transport, mining coal underground is expensive and dangerous, and without reclamation, surface mining can damage the landscape.

Attempts to solve these problems are underway, but at present totally acceptable solutions have not been found. For coal near the surface, surface mining (Fig. 19–5) permits the recovery of nearly all the coal (as contrasted to about 60 per cent recovery in underground mines). At the present time, in the 39 states where 95 per cent of America's coal is mined, laws have been enacted that protect against

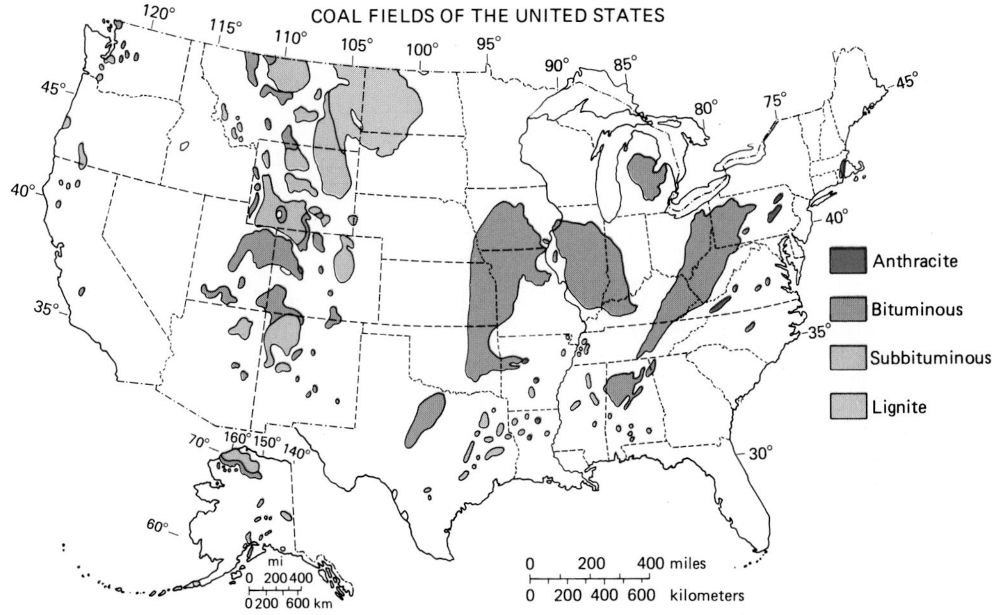

Figure 19–3 Location of major deposits of coal and lignite in the United States. (From U.S. Bureau of Mines Circular 8535.)

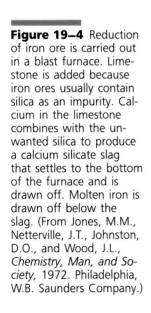

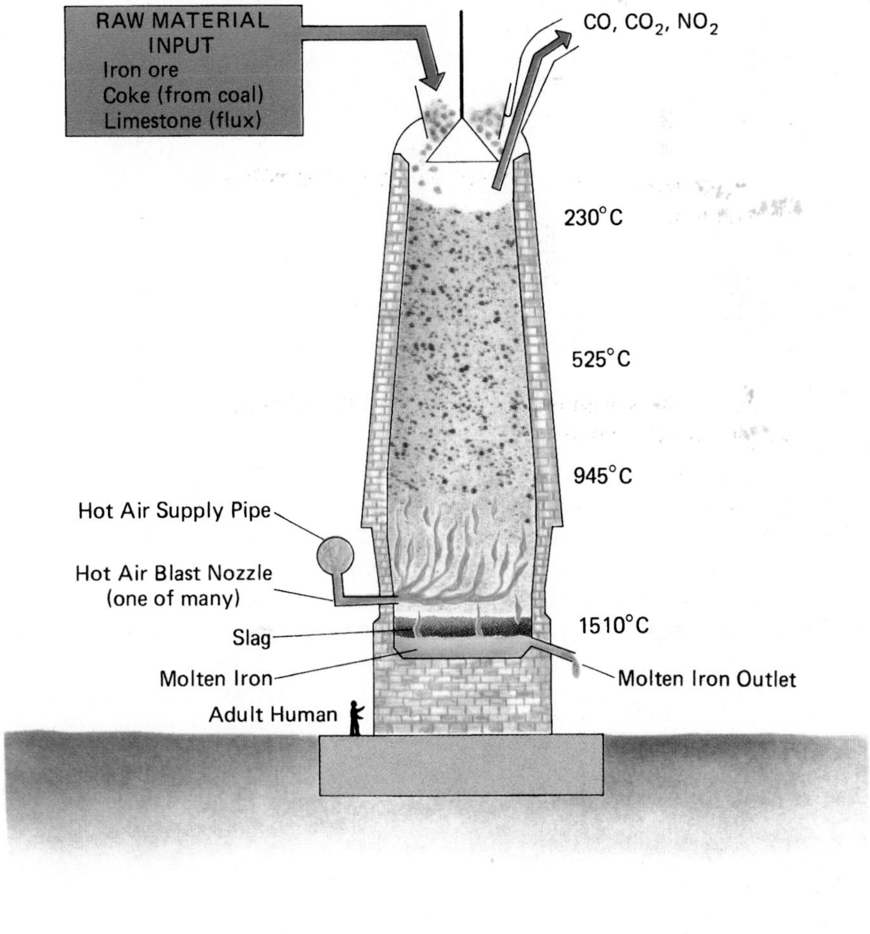

Figure 19–4 Reduction of iron ore is carried out in a blast furnace. Limestone is added because iron ores usually contain silica as an impurity. Calcium in the limestone combines with the unwanted silica to produce a calcium silicate slag that settles to the bottom of the furnace and is drawn off. Molten iron is drawn off below the slag. (From Jones, M.M., Netterville, J.T., Johnston, D.O., and Wood, J.L., *Chemistry, Man, and Society*, 1972. Philadelphia, W.B. Saunders Company.)

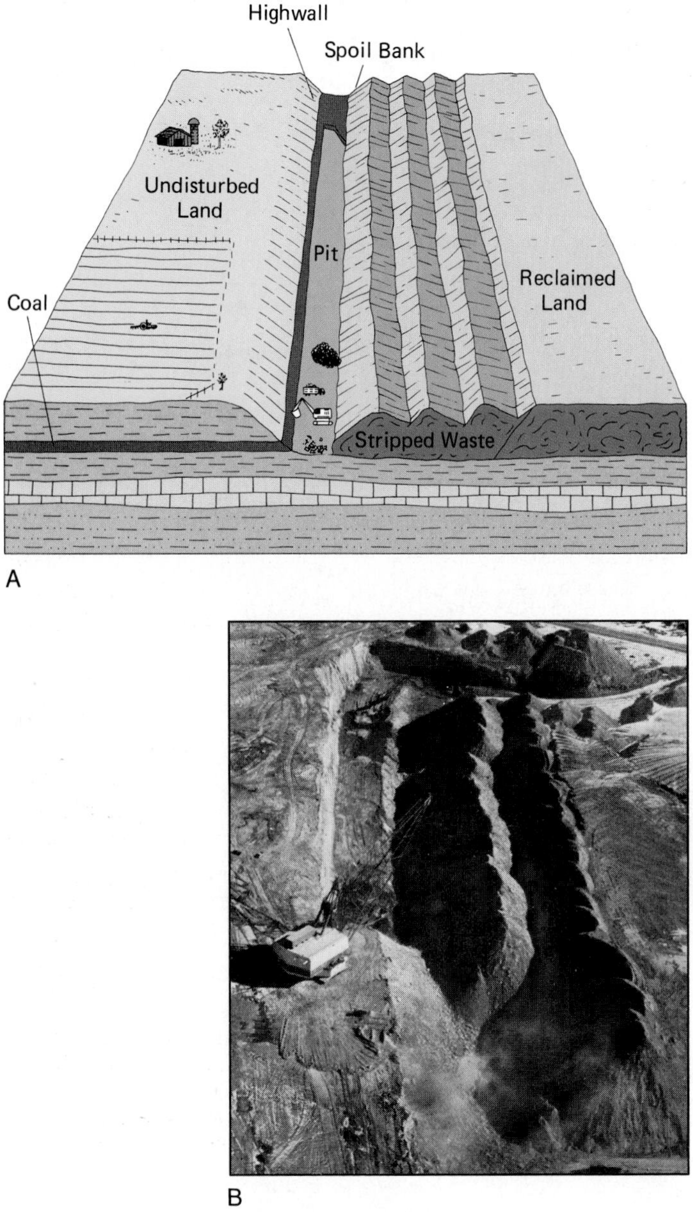

Figure 19–5 Surface mining. (A) Diagram illustrating terminology and mining method. (B) An aerial view of strip mine with dragline at work. (Courtesy of the U.S. Geological Survey.)

the environmental damage of surface or strip mining.

Gaseous emissions given off when coal is burned constitute another problem associated with its use. As with other fossil fuels, when coal is burned oxygen is consumed and carbon dioxide released.

$$\underset{\text{coal}}{C} + \underset{\text{oxygen}}{O_2} \xrightarrow{\text{fire}} \underset{\text{carbon dioxide}}{CO_2}$$

Carbon dioxide concentrations in the atmosphere have been rising steadily for the past 75 years, and there is concern that this added carbon dioxide may alter the earth's temperature balance and adversely affect life.

Coal contains other elements besides carbon, and when coal burns, these elements combine with oxygen to form particularly noxious gases. All types of coal, for example, contain sulfur. As the sulfur burns along with the coal it forms sulfur dioxide and sulfur trioxide. These gases can cause respiratory illness. Nitrogen is also present in coal and produces nitrogen oxide and nitrogen dioxide. The latter gas provides the smell and brown coloration of smog, and along with other pollutants, contributes to the formation of acid rain. Finally, even if coal were free of sulfur and nitrogen, incomplete burning produces carbon monoxide, a colorless and odorless lethal gas.

Petroleum and Natural Gas

For several decades, oil and natural gas have served America and many other nations as the most important source of energy. Today, as everyone knows, the supply is dwindling. Unless consumption is markedly showed by increasing our use of alternate energy sources, it is likely that over 90 per cent of the planet's oil will be depleted by about A.D. 2040. The time to begin the switch to other energy sources and conserve our remaining oil reserves is now at hand.

Petroleum and natural gas are today's primary sources of energy.

The Origin of Oil and Gas

Petroleum is a complex mixture of many chemical compounds called hydrocarbons because of their content of hydrogen and carbon. An example of a very simple hydrocarbon is methane (CH_4), the major component of natural gas. Both petroleum and natural gas are commonly found together in similar geologic environments. Although there is some debate about the precise way in which petroleum originates, most geologists would agree that the liquid was ultimately derived from the decay of protozoans and algae that proliferated in seas and inland water bodies and whose dead remains fell continuously on the basin floors (Fig. 19–6). Scientists also agree that a lack of oxygen probably enhanced the oil-forming process. If oxygen were available, it would combine with carbon in the decaying organism to form carbon dioxide, which would then be dispersed. In the oxygen-poor environment, carbon tends to accumulate, and oxygen in organic material is consumed by anaerobic bacteria. Burial, with attendant increases in pressure and heat, contributed to the transformation of the remains of aquatic life into oil.

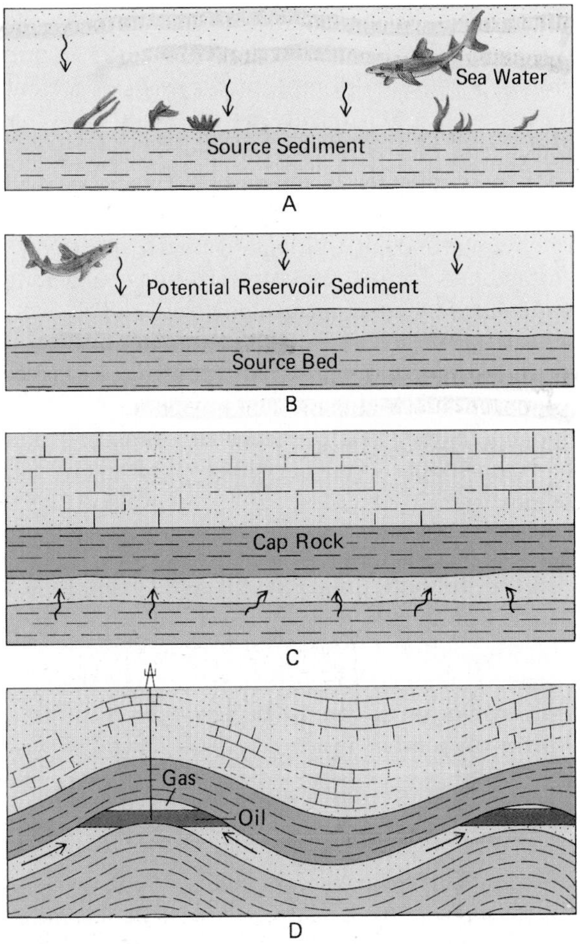

Figure 19–6 An ideal sequence of events that might lead to the accumulation of petroleum. (A) Marine area with accumulation of organic debris in sea floor sediment. (B) Deposition of sand above the organic-rich sediment. (C) Deposition of impermeable shale (and subsequently other overlying sedimentary layers). (D) Deformation to produce structures into which oil slowly migrates.

A layer of sediment in which exceptional quantities of organic material were once deposited is called a **source rock** by petroleum geologists. Usually it is a shale or siltstone that is colored dark gray or black by its abundance of carbon. From source rocks, particles of oil once flowed into other rocks, which, because of their porosity and permeability are called **reservoir rocks**. Sandstones, porous limestones, and the rocks of ancient buried reefs often quality as excellent reservoir rocks.

If there is water in the reservoir rock, oil will tend to continuously rise to the top of the saturated zone and may ultimately reach the earth's surface and be lost. This upward movement may be blocked by an impermeable layer or rock overlying the reservoir rock. The stratum that prevents the upward migration of petroleum and natural gas is appropriately dubbed the **cap rock** (a term also applied to the disk-like layer of evaporites and limestones that occur at the top of many salt domes.)

In the geologic past, petroleum formed from chemically decomposed organic matter deposited on the floors of ancient seas. The resulting fluids moved through the rocks until caught in geologic traps to form oil pools.

Oil Entrapment

Once petroleum has entered a reservoir rock, one might surmise that its journey from a suitable source bed would be ended. This is not the case, for the oil continues to migrate through the more porous rock until a change in rock structure or lithology prevents further migration. Usually, the agent that drives the oil along is water. Everyone knows that oil will float on top water. Most reservoir rocks are of marine origin and therefore contain water in their pore spaces. As noted above, oil makes its way upward through pore spaces until it is on top of the water-saturated zone within the reservoir rock. If that rock is tilted, the oil will continue to rise until stopped by some impermeable barrier. Below the barrier, additional petroleum continues to accumulate, and an oil "pool" is formed. Because gases within the crude are lighter than the oil itself, they tend to separate, rise, and accumulate at the top of the oil trap.

There are a great many different kinds of oil traps, although most can be placed within one of two categories. There are, for example, structural traps, which have been formed in folded or faulted strata (Fig. 19–7A and B). These include oil that has accumulated high up beneath the arch of anticlines and domes, as well as traps caused by faulting of reservoir rocks against updip impermeable layers. The other broad category of oil trap is stratigraphic in nature. They include traps formed at the updip edges of reservoir rocks where they thin to a feather-edge and are enclosed by shales (Fig. 19–7C) or where the reservoir layer is abruptly truncated and sealed at an unconformity (Fig. 19–7D). Often several different kinds of oil traps may exist within the same oil field, as, for example, around salt domes (see Fig. 6–51).

The geologist engaged in selecting a location for a prospective oil well must constantly keep in mind the many different factors that govern the chances of obtaining commercially profitable amounts of petroleum. There must be evidence that a formation suitable as a reservoir rock does exist at a depth that can be reached by drilling. The geologist will feel the chances of finding oil will be better if within the sequence of subsurface formations there are one or more strata that contain several per cent of organic matter and that may have served as source rocks. With the aid of drilling, surface mapping, and geophysical methods, he or she must ascertain the location of relatively impermeable cap rocks. Finally, and most important, the geologist must determine the probable existence of a favorable structure high in the reservoir rock and beneath the cap rock seal. If all these conditions for an oil pool appear to exist, drilling can begin (Fig. 19–8) and with it, an interval of anxiety and excitement for the geologist.

The requirements for an oil pool are a reservoir rock, cap rock, source rock, and a favorable geologic structure.

Finding Oil

The most direct evidence for the existence of underground accumulations of petroleum are *oil seeps*—places where oil actually oozes onto the surface of the ground. One of the greatest oil wells in history, the Cerro Azul No. 4 in Mexico, was drilled above an oil seep in 1916 and produced almost 60 million barrels of oil. Oil seeps, however,

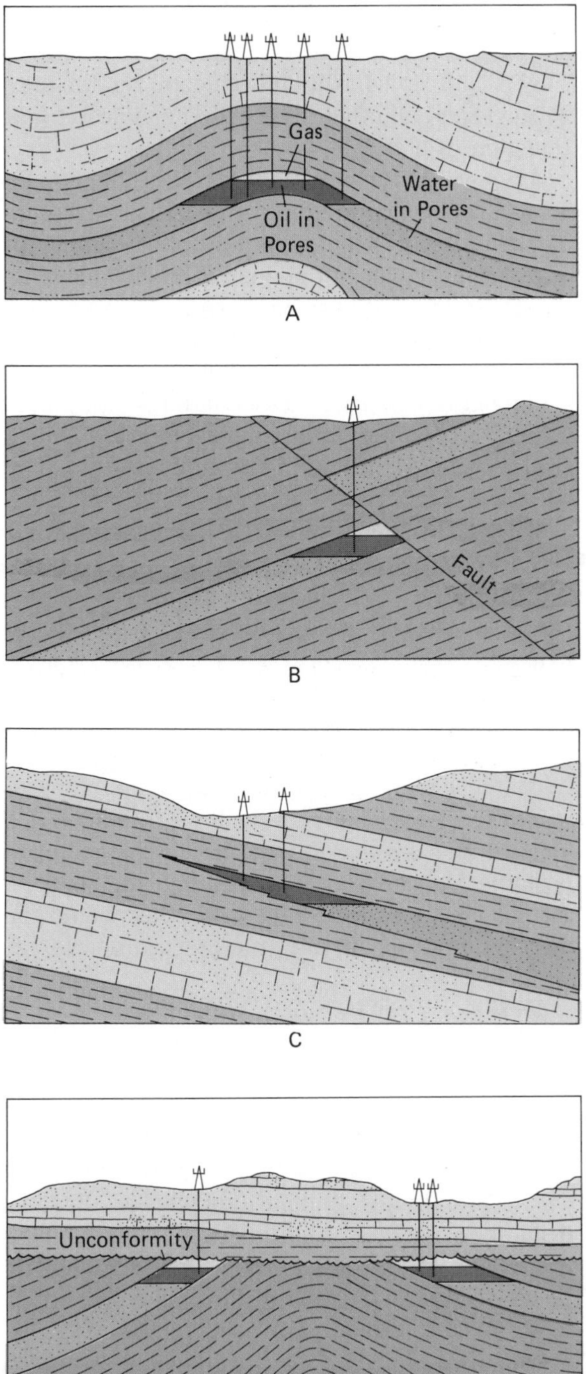

Figure 19–7 Idealized cross-sections depicting geologic conditions that may lead to the accumulation of petroleum and natural gas. (A) Anticlinal trap. (B) Fault trap. (C) Stratigraphic trap. (D) Entrapment beneath an unconformity.

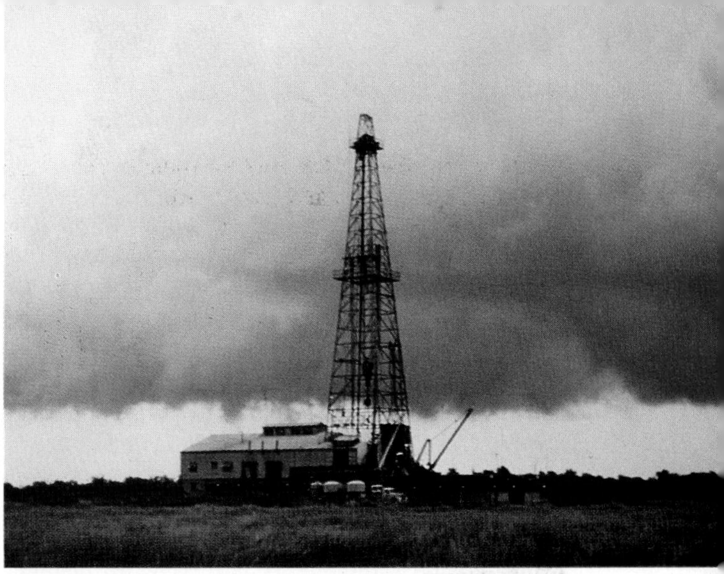

Figure 19–8 Drilling for oil in eastern Kansas.

are relatively uncommon, and so people in the business of finding oil in the late 1800s sought other means to look for it. Some of the more adventurous decided simply to drill into the earth and hope for good luck. As a result of this sort of random drilling in the Appalachians, it became evident that when oil was discovered it invariably had accumulated along the crest of an anticline (Fig. 19–9). Armed with this knowledge, geologists in the early 1900s devoted themselves to the systematic search for anticlines and similar structures.

Of course, geologic structures detected at the earth's surface may have no relationship to those

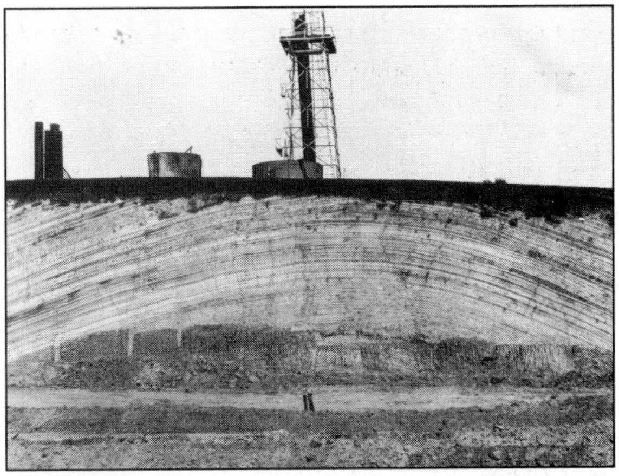

Figure 19–9 A 1931 photograph of a small anticline in Los Angeles, California. The oil well at the top of the photograph produced oil from a reservoir rock several hundred meters beneath this exposure. (Courtesy of the U.S. Geological Survey and M.N. Bramlette.)

deeper in the earth. Horizontal layers of rock at the surface may conceal well-developed structures that lie at depth beneath unconformities. The search for oil trapped in these deep hidden structures demanded geophysical methods of exploration.

Geophysical Methods. Geophysical methods of petroleum exploration involve measurements of gravity, magnetic, and seismic properties of rocks and the use of such measurements to detect geologic structure. *Gravity methods*, for example, employ an instrument called a **gravimeter**, which measures the pull of gravity on a weight suspended from a spring. Gravimeters can measure changes less than one ten-millionth of the total force of gravity. They are especially useful in detecting oil traps associated with salt domes. Because salt is a lighter material than the surrounding sediments, the gravimeter above a salt dome will record a slightly lesser gravitational force at the surface (Fig. 19–10). Alternatively, some anticlines involve uparching over relatively denser rocks, thus providing a higher-than-normal gravity reading at the surface. When new territories are opened for exploration, the geophysicists employed by oil companies are given the task of conducting gravity surveys. Gravity readings are taken at regularly spaced intervals, and from the data, gravity maps are prepared that show the distribution of gravity "highs" and "lows." Careful evaluation of the gravity maps and all other information may lead to the selection of a site for the test drilling.

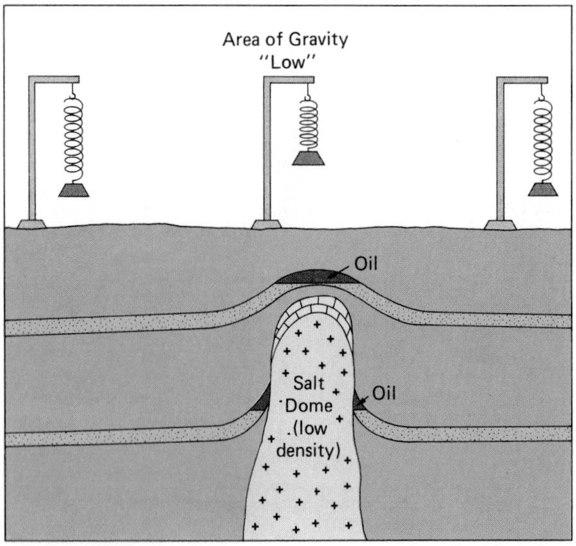

Figure 19–10 Conceptual diagram indicating how measurements of gravity can be used to locate salt domes.

Magnetic methods of exploration attempt to discern variations in the earth's magnetism that can be attributed to the configuration and composition of rocks in the upper 4 or 5 km of the earth's crust. Magnetite-bearing igneous rocks and some ore deposits can be detected by magnetometer surveys. In the search for favorable structures for oil accumulation, magnetometer studies have been used to locate deeply buried igneous ridges. Occasionally such ridges are associated with overlying anticlinal stratigraphic oil traps.

The most widely used of all geophysical methods for oil exploration is called **seismic reflection shooting** (Fig. 19–11). In this technique, seismic waves are sent into the earth either by detonating dynamite in shallow drill holes (Fig. 19–12) or by hydraulic vibrators. The latter method is more frequently used today because it is less disturbing to the environment. The depth to the stratum that reflects these waves is determined from the time required for the waves to travel down to that reflecting bed and back to the surface for detection by portable seismographs. Repeating the process along a predetermined traverse or seismic line permits one to ascertain the structural attitude of subsurface strata.

Well-Logging Methods. With every well that is drilled, whether oil is encountered or not, the geologist's knowledge of what lies beneath the ground has been increased. At the location of the well, he or she knows what strata are present, their lithology, their thickness, and their relation to strata above and below. Similar knowledge obtained from other wells is used to correlate strata and to prepare cross-sections of the subsurface geology. To construct these cross-sections one uses vertical records of well data called well-logs.

A **well-log** is a long scroll of paper that records one or more characteristics of the strata encountered at successive depths during the drilling of a well. There are many different kinds of well-logs. On some, information about lithology is recorded, whereas on others, fossil occurrences are noted.

Other logs record natural radioactivity, bore-hole diameter, or the electrical properties of the formations penetrated. Of all the instrumental logs, none is more useful than electrical logs. When the tool used for producing electric logs is raised out of the drill hole, it measures the natural electrical properties of the formations as well as their resistance to an induced flow of electricity. Rocks of different composition and fluid content (oil, fresh water, saltwater) provide distinct patterns of lines on the electric logs (Fig. 19–13). These patterns can be used not only to infer lithology and the nature of fluids in the rock but also to record formation thicknesses and boundaries.

Oil Shale

A variety of fine-grained laminated sedimentary rock of special interest called **oil shale** (Fig. 19–14) contains an oil-yielding organic compound called **kerogen**. Like coal and petroleum, oil shale originated in an oxygen-poor environment, and

Figure 19–11 Vibrator trucks transmit energy into the ground during a seismic survey in West Texas. Echoes from strata beneath the surface are picked up by sensitive receivers and processes by computer to indicate depth and attitude of subsurface formations. (Courtesy of Exxon.)

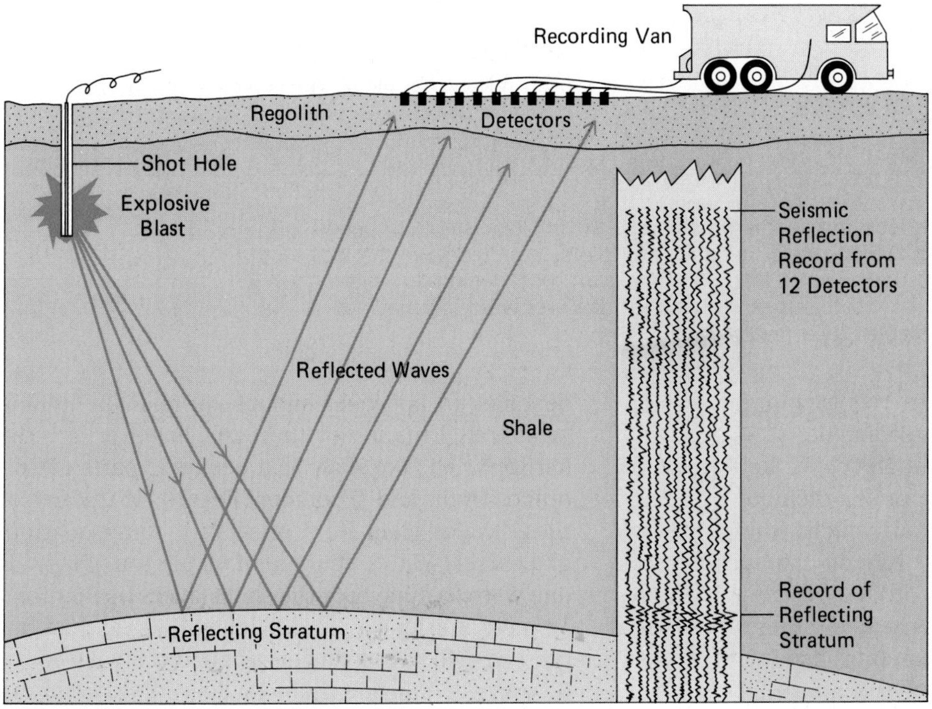

Figure 19–12 Seismic reflection shooting.

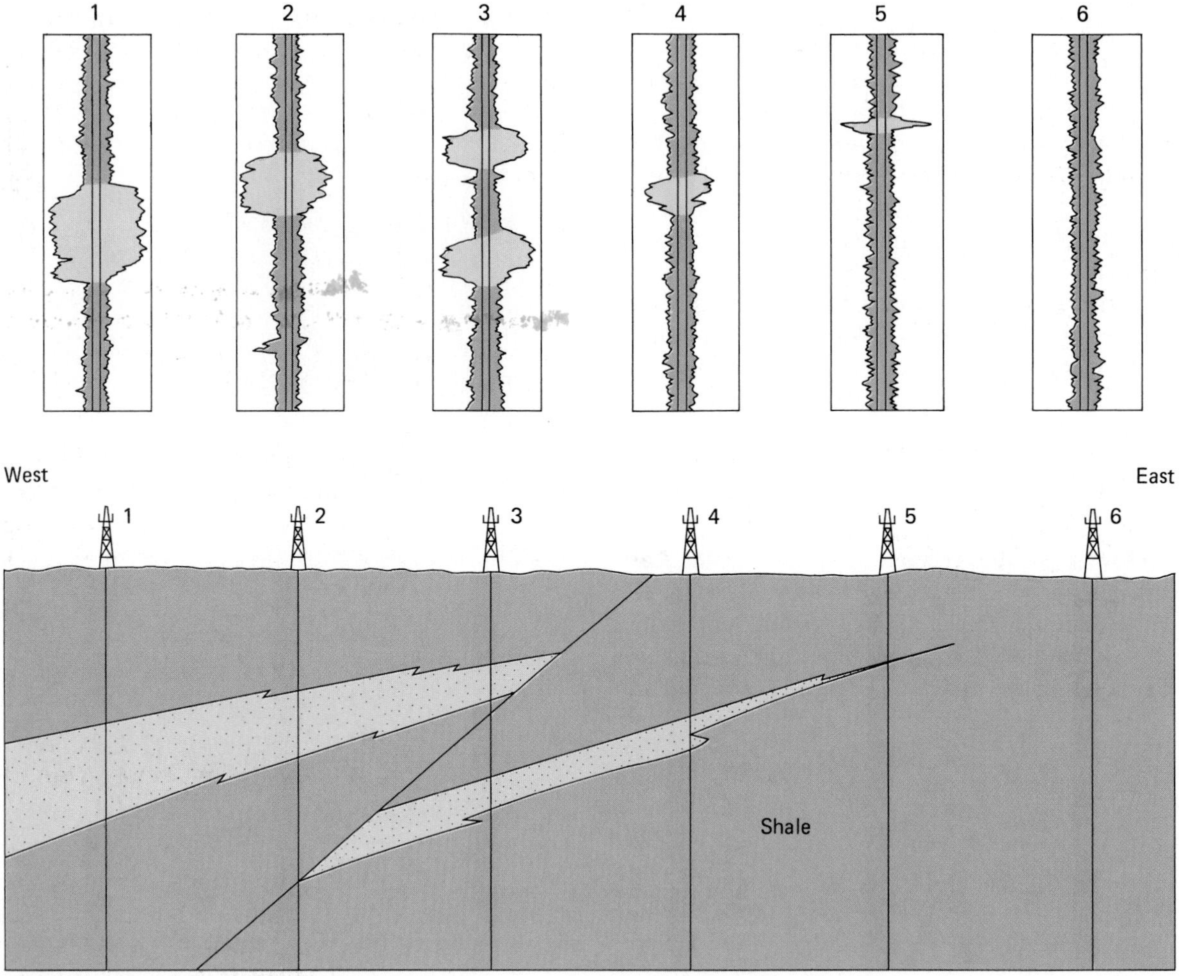

Figure 19–13 Six electric logs and an interpretation of the geologic structure penetrated by wells from which the logs were prepared. The prominent "kicks" in the logs record a petroliferous sandstone. From a study of the electric logs, geologists developed an interpretation of a petroliferous sandstone that dips westward, pinches out toward the east, and is disrupted by a reverse fault.

long periods of time were required for its development from organic-rich sediment.

When heated to about 480°C, the kerogen in oil shale vaporizes. The vapor can then be condensed and will form a thick oil, which, when enriched with hydrogen, can be refined into products in much the same way as ordinary crude oil. The retort in which the shale is heated rather resembles a giant pressure cooker that fuels itself with the very gases generated during heating.

Oil shale is quite common. It occurs in many places around the world, but the majority of the deposits do not yield enough oil to make mining profitable. This is not the case, however, for the kerogen-rich lake shale that occurs in parts of Colorado, Utah, and Wyoming (Fig. 19–15). Many of these shales yield from one-half to three-quarters of a barrel (42 U.S. gallons) of oil per ton of rock. If one were to mine only the shale layers thicker than 10 meters in these areas, they would yield an impressive 540 billion barrels of oil. If this oil were to be produced during the next decade, it would appreciably reduce the United States' dependency on foreign crude oil. Its production, however, requires

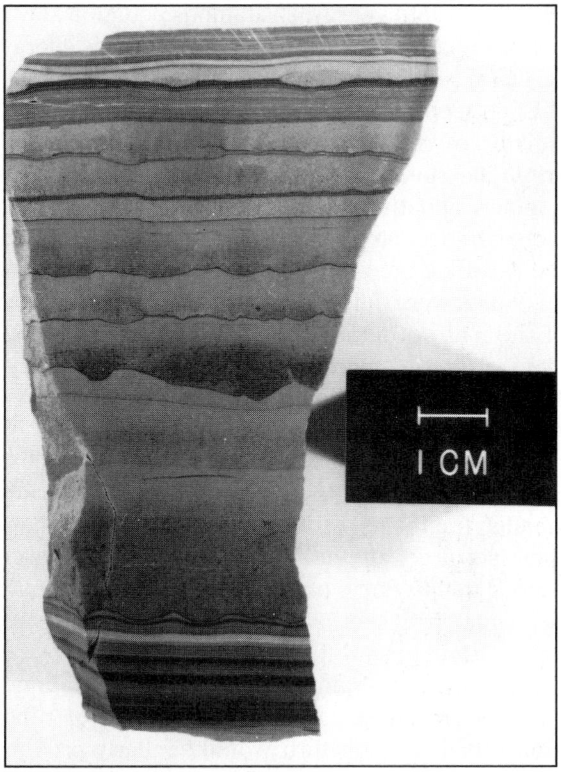

Figure 19–14 A piece of laminated oil shale from Mahogany Ledge, Rio Blanco County, Colorado. (Courtesy of the U.S. Geological Survey and D.A. Brobst.)

the use of vast amounts of water and also might result in harm to the environment. Thus, Americans face the question of how to balance the harmful effects resulting from the use of a resource with critical need for that resource.

Two environmental problems have prevented full-scale mining and retorting of oil shales during the 1970s. The first is a waste disposal problem. In the process of being crushed and retorted, oil shale expands to occupy about 30 per cent more volume than was present in the original rock. Engineers call this the "popcorn effect." Where can this great volume of light, dusty material be placed? The second problem has to do with air quality, for the processing of shale is likely to release considerable amounts of dust into the atmosphere. The shales crop out in arid regions, and there is at present little water available for use in processing the shale, holding down the dust, and assuring the success of revegetation measures. Until these problems are solved to the satisfaction of the American public, oil shale production will be slow. Perhaps this is a good thing, for we may need this oil for processed goods a century or two from now, when the internal combustion engine may be as rare as is the horse and buggy today.

Tar Sands Another natural material having the potential for yielding large amounts of petroleum is tar sand. **Tar sands** are porous sands filled with heavy relatively solid asphaltic hydrocarbons. Most of these tarlike materials are residues remaining in reservoir rocks after lighter hydrogen-

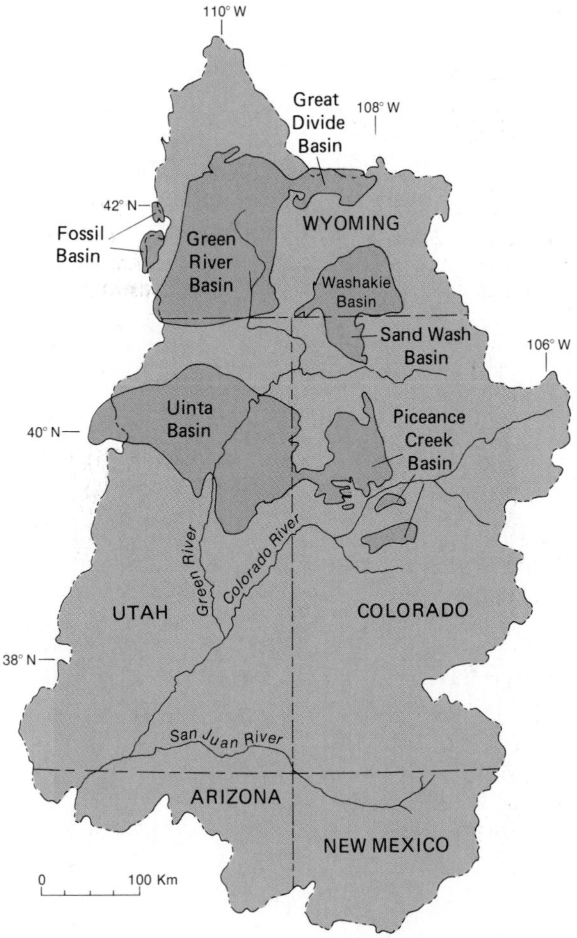

Figure 19–15 Principal oil shale deposits of Colorado, Utah, and Wyoming. Map boundary is the Upper Colorado River drainage basin. (Simplified from Rickert, D.A., Ulman, W.J., and Hampton, E.R., (eds.) 1979. *Synthetic Fuels Development,* U.S. Geologic Survey publication.)

rich crude oils escaped. The asphaltic material (called *bitumen*) may be driven out of the sandstone with steam or hot water. Hydrogen is then added to produce an oil that can be processed in the same way as ordinary liquid crude oils.

The largest of the world's tar sand deposits occur in northern Alberta along the Athabasca River (Fig. 19–16). Tar sand layers in this area are over 60 meters thick and lie near the surface over a total area of about 50,000 km^2. They contain an estimated potential yield of 600 billion barrels of oil.

Oil shale and tar sands are potential sources for petroleum when present supplies near depletion.

Nuclear Fuels

Energy can be defined as the capacity to do work. Early in the present century, Albert Einstein wrote that energy could be converted into matter and matter into energy. His famous equation showing this relation was $E = mc^2$, in which E is energy, m mass, and c^2 the speed of light squared. Einstein's theory indicates that tiny bits of matter are capable of releasing tremendous amounts of energy. The speed of light, after all, is a very large number, and the square of that number is even larger. Hence, when multiplied by even a small m, the resulting amount of energy will be enormous. Scientists were quick to realize that, because most of the mass of an atom is in the nucleus, the nucleus also contained most of the potential energy of the atom. How might it be possible to release that energy?

In the late 1930s it was discovered that atoms of uranium, when bombarded by neutrons, split into smaller fragments of less mass, and at the same time set free an enormous amount of energy. Among the products of the reaction were neutrons, the same particles that were responsible for causing the atom to split. The neutrons produced by the splitting of one uranium atom might then, under suitable conditions, cause the splitting of other atoms, and these in turn would similarly produce further neutrons to spread the splitting reactions through the mass of uranium. This splitting of an atom into smaller fragments with the release of energy is called **fission** (Fig. 19–17). Note that fission is not the same as radioactive decay (dis-

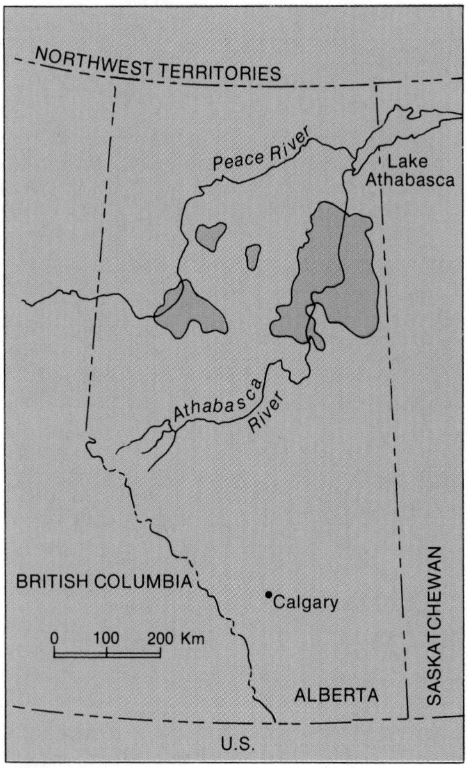

Figure 19–16 Location of the tar sand deposits of Alberta, Canada.

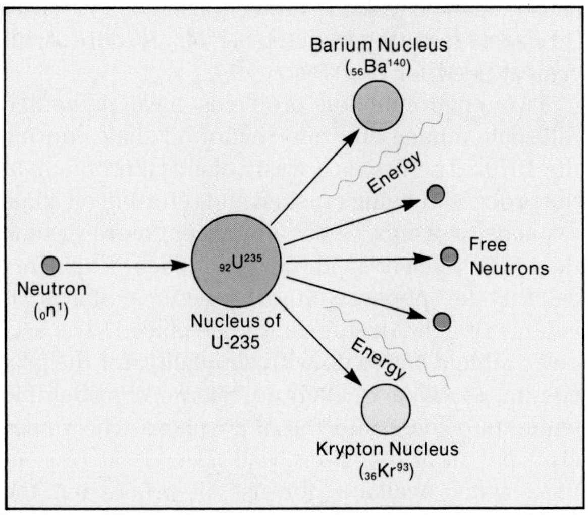

Figure 19–17 Typical fission of a uranium 235 nucleus.

cussed in Chapter 8), which is the spontaneous nuclear disintegration of certain radioactive isotopes.

Energy from Fission

Having discovered a way to split an atom, scientists next devised a method for producing an ongoing series of fissions. In their experiments, they allowed the neutrons given off by split atoms to collide with other atoms, causing additional splitting in a sustained chain reaction.

The device in which these reactions were allowed to occur was termed a uranium pile (Fig. 19–18). The pile had within it rods of fissionable uranium, retractable rods of neutron-absorbing materials to control the rate of fission, and graphite packing to slow the neutrons to a velocity most conducive for causing fission. As illustrated in Figure 19–19, heat from the reactor can be used to produce steam. The steam in turn propels the blades of turbines, and the turbines provide the power to operate the electric generators in nuclear power plants (Fig. 19–20).

Fission, which involves the splitting of atomic nuclei, is the only type of nuclear power currently in use.

Early in the experimentation that preceded the development of the uranium pile, it was discovered that not all isotopes of uranium split under neutron bombardment. Pure natural uranium, for example, consists mainly of two isotopes, uranium 235, which is fissionable, and uranium 238, which is not. Less than 0.72 per cent of all atoms in natural uranium are of the fissionable uranium 235 variety. Furthermore, exploration for uranium over the past three decades indicates that the supply of fissionable uranium 235 may only last a few more decades. Thus, other methods for utilizing nuclear energy will have to be devised for the future. One method now being studied would attempt to breed fissionable atoms from nonfissionable materials in a so-called **breeder reactor**. These reactors would bring the neutrons of decaying uranium 235 into the nuclei of uranium 238 atoms so as to convert them to fissionable plutonium 239. Similarly, nonfissionable thorium 232 might be converted into fissionable uranium 236. It is known that the conversions can be accomplished at a rate that produces more fissionable material than is consumed. With breeder reactors, energy from fission might be extended over several centuries, because lower-grade ores could be utilized and more energy would be derived from any given amount of ore. Breeder reactors, however, also pose dangers. They produce large quantities of radioactive plutonium, and the disposal of a radioactive waste that requires six to ten centuries to be rendered biologically safe is of great concern to every human being.

Disposal of Nuclear Waste

Perhaps the most difficult question associated with the use of fission power is what to do with nuclear wastes. When radioactive elements un-

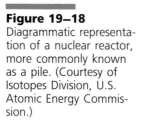

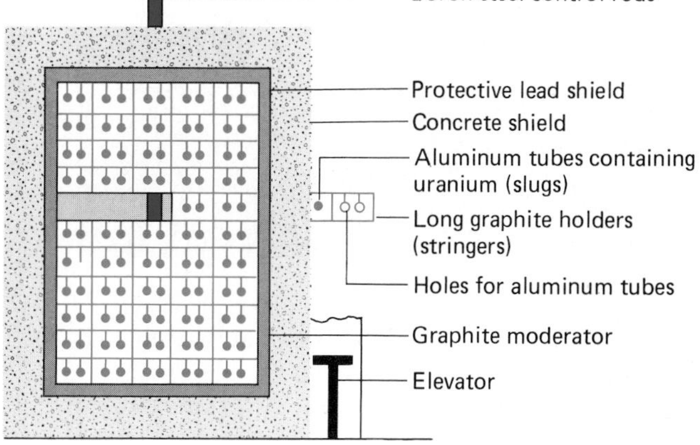

Figure 19–18
Diagrammatic representation of a nuclear reactor, more commonly known as a pile. (Courtesy of Isotopes Division, U.S. Atomic Energy Commission.)

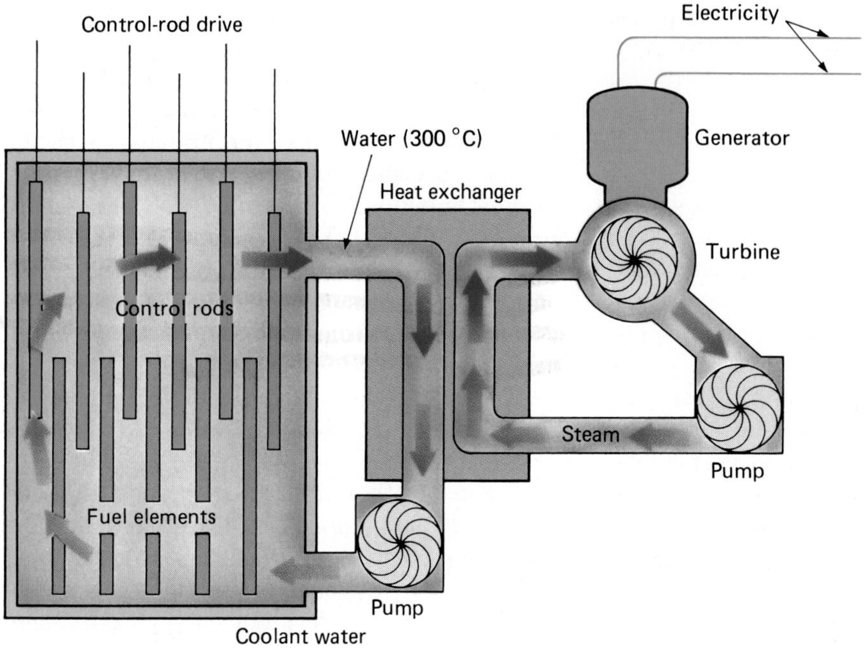

Figure 19–19 The production of electricity in a nuclear power plant. (From Johnston, D.O., Netterville, J.T., Woods, J.L., and Jones, M.M., 1987. *Chemistry and the Environment,* Philadelphia, W.B. Saunders Company.)

Figure 19–20 Nuclear power plant located in Callaway County, Missouri. Cooling tower is at the right and domed reactor containment building is at the center. (Courtesy of Union Electric Co.)

dergo fission, they produce products that are themselves unstable and emit radiation for periods of as long as 250,000 years. Strontium 90, which is in nuclear waste, tends to concentrate in parts of the body of humans and other animals. As in the substitution of ions in minerals, strontium 90 can substitute for calcium in bone. Once this happens, the presence of the isotope in the body can lead to the development of cancers and other illnesses.

As noted above, these hazardous wastes must be stored in places where they will be locked out of the biologic environment for immense spans of time. Presently, scientists are considering storage in an underground repository to be constructed at Yucca Mountain, Nevada. If the project is shown to be feasible, storage chambers for radioactive waste will be excavated in layers of rhyolite ash-flow tuff that lie 300 m beneath the surface of the ground. Abandoned salt mines have also been proposed as repositories. Thick beds of salt are relatively impervious to water. Fractures in salt tend to be self-mending, and salt provides about as good a shield against radioactivity as does concrete. On the other hand, pockets of brine (saltwater) have been found in these underground mines that might jeopardize retrieval of the waste if the storage chamber were flooded. Contaminated groundwater might find its way into mine shafts or wells.

Finding Uranium

There are many different minerals that contain uranium, but only a few occur in deposits sufficiently rich to be mined. The most important is a heavy, black, shiny, noncrystalline oxide of uranium called pitchblende. Uraninite is another uranium oxide that occurs in association with pitchblende but is crystalline. Both uraninite and pitchblende occur in veins and are often formed during magmatic and hydrothermal activity. Another abundant uranium mineral (containing also potassium and vanadium) is carnotite (Fig. 19–21). This canary-yellow mineral is often found as a cement in sandstones. The uranium in carnotite was probably derived from the weathering or hydrothermal alteration of older igneous rocks and was subsequently carried by solutions into nearby clastic sediments.

Exploration for uranium need not entail the use of complex geophysical methods. As is the case with the search for other forms of ores, geologist first examine multitudes of geologic reports and maps to determine areas that have the kinds of rocks and structures most likely to contain uranium ores. Next, exploration crews go out into the field, armed with a device called a scintillometer

Figure 19–22 Radiation detecting instruments. (A) Geiger counter. (B) Scintillometer. (C) Gamma-ray spectrometer. (From Olson, J.C., Butler, A.P., Jr., Finch, W.I., Fischer, R.P., and Nash, J.T., 1978. *Nuclear Energy Resources,* U.S. Geological Survey pamphlet.)

(Fig. 19–22). This instrument is capable of detecting radioactive elements and even distinguishing between some varieties. Scintillometer readings are made along a prearranged survey pattern. Where readings indicate appreciable subsurface natural radioactivity, more detailed geologic work is undertaken. Ultimately, the prospects may be drilled so that geologists can better ascertain the quality and dimensions of the deposit.

The richest deposits of uranium ore in the United States are found in Mesozoic rocks of the Colorado Plateau. Canada's ores are found principally in the Great Bear Lake region, Northwest Territories, in the Blind River district north of Lake Huron, and in northern Saskatchewan at Cluff Lake and Key Lake. As is the case with the generation of electricity in nuclear-power plants, the mining and milling of uranium ore are hazardous occupations. Workers are in constant danger of illness from radiation.

Figure 19–21 Uranium ore minerals. The two yellow specimens on the left are carnotite. The large dark mineral is pitchblende.

Fusion: Hope for the Future

Imagine a source of energy that is cheap and abundant, produces no hazardous waste, and does not jeopardize the health of humans nor degrade the environment. Such an ideal source of power is theoretically possible through a process known as **fusion**. Fusion is the combining of lighter nuclei to form a single heavier one. It is the nuclear process that provides the sun's heat. The commercial fusion reactor, which at present is only a theoretic possibility, would combine nuclei of the heavy hydrogen isotopes deuterium (or tritium, which is made from deuterium) to produce helium, which has slightly less mass than the two heavy hydrogen nuclei. The "missing" amount of mass is converted into energy in the form of heat. The heat, in turn, could be used to generate electricity in much the same way as existing fission reactors. The fuel for fusion presents no problem, for the oceans can provide an almost limitless supply. Indeed, calculations indicate that the amount of deuterium in only 1 km^3 of sea water could provide as much energy as all of the world's present oil reserves.

There are, however, formidable technologic problems that must be solved before fusion can be used to supply our homes and factories with power. The fusion reaction requires a temperature enormously higher than that on the surface of the sun. What can be used to contain the deuterium as such high temperatures? Presently, scientists are experimenting with the containment of the deuterium and tritium at high temperatures within a strong magnetic field. With this method, thermonuclear fusion has already been accomplished in experimental models (Fig. 19–23). Another method still in small-scale experimental stage attempts to heat deuterium and tritium with intense bursts of laser light. The laser's energy, focused on the atoms contained within microscopic pellets, heats them to 50 million °C. Under these conditions, the walls of the pellet collapse inward, compressing the atoms. Deuterium atoms, moving rapidly in response to the heat and in close contact because of the compression, collide with each other and for a brief moment may fuse. In that brief moment, a helium nucleus, a neutron, and abundant energy are produced.

The technology for fusion is still in its infancy. If fusion can be accomplished on a commercial scale—and many are skeptical that it can be—then it is probable that nuclear plants generating power from sustained fusion reactions will still not be ready until sometime in the middle of the next century. As suggested by Glen Seaborg, former head of the Atomic Energy Commission, fusion is "the most difficult scientific-technological problem ever undertaken by mankind." Nevertheless, scientists from a dozen countries are bending every effort to make fusion work, for if they are successful, it will provide safe and unlimited energy for the entire world.

Figure 19–23 The Princeton Plasma Physic Laboratory Tokomak Fusion Test Reactor for producing experimental fusion reactions. (Courtesy of the Princeton Plasma Physics Laboratory.)

Energy from Sun, Wind, and Water

Energy derived from coal, oil, and minerals containing radioactive elements is contained within earth materials and hence is the direct concern of geology. There are, however, other sources of energy that do not originate at the mine or oil well. Most important of these is hydroelectric power from streams (Fig. 19–24). It is conceivable that streams, if fully developed, could supply about 15 per cent of our present electrical needs. Of course, such an extensive development of hydroelectric power would bring with it problems of silting of artificial lakes behind dams, flooding of valuable farmland, disruption of the stream ecology, and the breeding of disease organisms in lakes and canals associated with the dams.

Water power, which can supply only a small part of our energy needs, has the advantage of being clean, renewable, and relatively inexpensive.

Still other sources of energy also exist and are presently being evaluated or used mainly as supplements to existing sources. These include the geothermal energy discussed in Chapter 4, as well as solar energy, wind power, tide and wave power, energy from burning trash, and energy derived from fuels artificially produced from growing plants. Of this group, solar energy appears to hold the most promise. In many parts of the world, energy from the sun is already being used to operate solar furnaces and heat systems for buildings. Futuristic proposals have been made for placing solar energy collectors in space and beaming the energy down to earth in microwave form. Presently, however, efficient utilization of solar energy is hampered by lack of sufficiently efficient storage devices, of cheap solar collectors, and by the variability in the energy received from the sun at various locations. Also, an enormous amount of space is required for the large number of solar collectors needed to generate the electrical needs for large cities.

Wind power involves the use of specially designed windmills (Fig. 19–25) to generate electricity and works well where winds are strong and persistent. Unfortunately, in most land areas, winds are rather unsteady at times, and hence large energy storage facilities are needed to supply electricity during slack periods. Wind power is certainly technologically feasible but presently not economically practical on a large scale.

Figure 19–24 Hoover Dam on the Colorado River in southern Nevada. The dam is 220 meters high and holds back the waters of Lake Mead, which is visible in the background. In addition to providing electric power and recreational opportunities, water held in Lake Mead is used to irrigate thousands of acres of farm lands. (Courtesy of U.S. Bureau of Reclamation, photo by E. Hertzog.)

Figure 19–25 These three 55-kilowatt wind turbines provide electric power to a large shopping center near Copenhagen, Denmark. (Courtesy of Bent Sørensen.)

Tidal power is another energy source that has a proven but limited potential. At present there are about 50 sites around the world with enough variation between high and low tide levels to harness the energy in surging tidal waters to drive turbines. Even if fully developed, it is likely that tides could only supply about 2 per cent of the world's electrical needs.

We know that energy derived from tides, wind, and water will not supply the world's energy needs. Nevertheless, it is important that these sources be fully utilized. They can be of immense importance to local and regional economies, and they lessen utilization of nonrenewable and pollution-causing fuels.

Energy sources that are supplements to fossil fuels and nuclear power include water power, solar energy, geothermal power, wind power, and tidal power.

Our concern about having enough for the future is a mandate for conservation. Today, large amounts of energy are wasted or badly used. It is true that many of us now drive smaller cars, yet we still mostly ride one to a car and tend to avoid public transportation. We transport cargo by air when trains and trucks would be more efficient. Electricity is used in smelting where direct burning of coal would do as well. We use aluminum cans but know full well that aluminum refining requires huge expenditures of electric power. We recognize the need for more energy-efficient machines, better electrical conductors, and an energy priority system but do little to make the needed changes. We know that the oil used in generating electricity may someday be needed to make petrochemicals and clothing and to fuel our automobiles, yet oil continues to be burned in this way. In short, we are not making adequate progress in our attempts to conserve the energy resources still remaining.

Metallic Ore Deposits

Ores and Economics

A metal **ore** is a natural mineral occurrence that is close enough to the earth's surface and sufficiently concentrated to permit the extraction of one or more metals at a profit. It is evident from this definition that the term embraces both geologic and economic considerations. Whether a mineral deposit can be mined at a profit depends on such factors as its depth, shape, location, and size, the ease or difficulty with which the metal can be extracted from the host rock, distances over which raw materials would have to be transported, costs of labor, the efficiency of the operator, and the current market value of the product.

As is true for fossil fuels, ore deposits are nonrenewable, potentially exhaustible, and irreplaceable materials having only a sporadic occurrence around the globe. At the same time, they are the basis for the wealth and power of nations and for the well-being of their citizens. It is already evident that the world supply of many readily available ore minerals is approaching exhaustion. As a result, mining companies are forced to process lower-grade ores and dig deeper to find them. In order to do this they must have abundant inexpensive energy at a time when there is an energy shortage.

Thus, the price needed to extract a ton of iron increases, and so also does the price of the automobiles made from that iron. Ultimately, these events are likely to result in lower standards of living even among the highly industrialized nations. To slow the approach of such more difficult times, it is necessary that resources be conserved, that we increase the efficiency of our scrap-metal recovery, develop substitutes for vital metals that are used in nonvital products, improve our recovery and exploration methods, and if possible, find cheap sources of energy to reduce the cost of extracting lower-grade ores.

Most geologists prefer to reserve the term *ore* for materials from which metals can be extracted economically with present technology. Thus, limestone used for cement manufacture or sandstone used to make glass would not be called ore. The metallic ores include a great variety of minerals, the most abundant of which are listed in Table 19–1. As is evident from the table, metallic ores include sulphides, oxides, and silicates as well as uncombined or native metals. The ore minerals yield iron, aluminum, copper, lead, mercury, tin, zinc, and such alloying metals as chromium, cobalt, manganese, molybdenum, nickel, and tungsten, which are essential to the construction of jet airplanes, spacecraft, skyscrapers, and computers, as well as the many other products of our technologic ages.

Table 19–1 List of Selected Common Metals and Ore Minerals

Metal	*Ore Mineral*	*Composition*
Aluminum	Bauxite	$Al_2O_3 \cdot 2\ H_2O$
Antimony	Stibnite	Sb_2S_3
Bismuth	Bismuthite	Bi_2S_3
Chromium	Chromite	Fe_2CrO_4
Cobalt	Cobaltite	$CoAsS_2$
Copper	Native copper	Cu
	Bornite	Cu_5FeS_4
	Chalcopyrite	$CuFeS_2$
	Malachite	$CuCO_3 \cdot Cu(OH)_2$
	Cuprite	Cu_2O
Gold	Native gold	Au
Iron	Hematite	Fe_2O_3
	"Limonite"	$Fe_2O_3 \cdot H_2O$
	Magnetite	Fe_3O_4
Lead	Galena	PbS
Manganese	Pyrolusite	MnO_2
Mercury	Cinnibar	HgS
Molybdenum	Molybdenite	MoS_2
Nickel	Pentlandite	$(NiFe)_9S_8$
Silver	Native silver	Ag
Tin	Cassiterite	SnO_2
Titanium	Rutile	TiO_2
	Ilmenite	$FeTiO_3$
Tungsten	Scheelite	$CaWO_4$
Uranium	Uraninite	UO_2
Zinc	Sphalerite	ZnS

To qualify as an ore, a mineral deposit must contain a valuable metal in sufficient concentration to make it profitable to mine.

The Origin of Ores

Minerals have been concentrated into ores in many different ways. Some concentrations appear to have been clearly the result of processes associated with the slow cooling and crystallization of magmas. Other concentrations show evidence of having been formed by metamorphic processes operating around the boundaries of ancient igneous intrusions. Hot water, with its ability to dissolve a host of materials, has risen from great depth in the past and, on cooling, deposited metallic treasures in veins and fissures. Even such a mundane but pervasive process as weathering has produced ores by leaching away worthless components of a rock and leaving behind a residue of valuable oxides of iron and aluminum.

Ores from Magmas

As described in Chapter 3, magmas are silicate melts from which common silicate minerals are formed. As they undergo slow crystallization, chemical reactions among the melt, early-formed crystals, and adjacent solid rock are constantly occurring. These reactions occasionally result in the formation of valuable ore minerals that may be concentrated as a result of magmatic processes. For example, while still in the molten condition, chromium and iron may combine with chemically bound oxygen to form chromite. Chromite crystals are heavy and may therefore sink to the bottom of the magma chamber and accumulate as a concentrate. The process has been called **early crystal settling**. In addition to chromium, early crystal settling has provided rich deposits of magnetite (iron ore), nickel, and platinum-rich pyroxenes.

At some stage in the solidification of a magma, it exists as a "mush" of liquid and crystals. If such a partly crystallized magma were to be squeezed by forces associated with earth movements, part of the liquid might be forced out of the chamber and injected into small cracks and fissures. The liquid may contain a concentration of metals such as iron and titanium, which then crystallize in the fractures as magnetite and ilmenite. This magmatic process is called **filter pressing** (Fig. 19–26) because it resembles the action of an antique wine press in which grapes are placed in a wooden barrel perforated with small holes and pressure is applied by a piston from above. The grape juice is forced out through the holes, while the less valuable pulp remains in the barrel. The world's largest iron ore bodies, located at Kiruna, Sweden, were formed by filter pressing.

Although the concentration of valuable ore minerals along a specific zone or vein facilitates economical removal, some materials have such extraordinary value that they may be thinly distributed throughout a rock mass, and the entire rock can still be worked at a profit. Ores of this nature are called **disseminated magmatic deposits**. Diamonds (see Fig. 2–44) are among the exceptionally valuable minerals mined from disseminated deposits. Diamond ores yield only about ½ to 1 carat (100 to 200 mg) for every 3 tons of rock mined, and yet mining is still profitable.

The host rock for diamonds is a dark, dense iron and magnesium silicate called kimberlite. This material was intruded into narrow pipelike vents from great depths, possibly even the upper mantle. The extreme pressures that exist at such depths are required for the formation of diamond crystals from carbon in the kimberlite melt.

Some occurrences of ores have been termed **late magmatic deposits** because they show evidence of having formed late in the cooling history of the magma. In these deposits, the metal-bearing residual liquids are injected into still unsolidified zones within the magma rather than into the surrounding country rock. In other deposits, it is evident that the ores are not injected at all, but that they simply crystallized in pockets and layers that were still unsolidified.

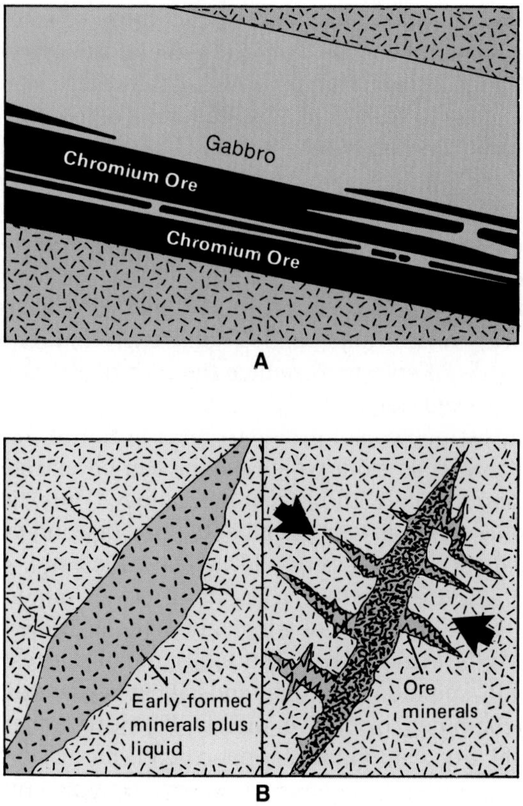

Figure 19–26 (A) Zones of chromite ore in a dipping gabbro sill. The ore has been concentrated as a result of settling of early-formed crystals. (B) Two stages in the formation of an ore body by filter pressing. The drawing at the left shows early crystallization of ore minerals in the melt. In the adjacent drawing, liquid has been squeezed into the joint and fracture system of the country rock. The liquid that has entered joints and fractures crystallizes to form valuable ore minerals, whereas the early-formed valueless minerals are left behind.

In Chapter 3 we examined the relation of igneous bodies to subduction zones at the boundaries of tectonic plates. Later in Chapter 17, we further noted the existence of rich deposits of metalliferous sediment associated with hot brines in basins within zones of divergence like the Red Sea. Is it possible that such sediments, rich in gold and silver as well as sulfides, oxides, and carbonates of iron, manganese, zinc, and copper, may somehow be clues to the origin of ancient ore bodies now known on the continents? Many geologists believe a relationship does exist and embrace a new hypothesis that provides an explanation for many magmatic ore deposits. According to these new ideas, metals are concentrated in marine sediments along diverging plate boundaries by hot chemically active solutions, which percolate through fractures in the crust. As these solutions rise and cool, they deposit part of their metal as carbonate and sulfide veins in the recently extruded basaltic oceanic rock and part in the sediment that lies above the basalt. As divergence continues, these deposits become part of the trailing edges of the two opposing plates and move slowly away on their journey toward distant subduction zones (Fig. 19–27).

On arrival at the subduction zone, both the metal-bearing rock and sediment from the mid-oceanic ridge and ferromagnesian nodules and low-grade concentrates of metals deposited on the sea floor as the plate was traveling descend toward the interior. As the plate is subducted, temperature and pressure increase, causing various mineral components to melt and move upward. Mostly, these are the silicate minerals found in igneous rocks associated with subduction zones. In addition, however, the metals are mobilized at various depths and migrate into neighboring rock masses. It seems likely that such events have been going on at least as long ago as Early Precambrian. If so, plate tectonics provides an interesting model for explaining the existence of large concentrations of valuable metals of specific age and extent in the orogenic belts around the earth.

Ores Resulting from Contact Metamorphism

When a magma works its way upward into overlying older country rock, a zone of contact metamorphism forms at or near the boundary. Along this zone, hot emanations from the magma chemically interact with the country rock to form assemblages of high-temperature minerals. If, by chance, the compositions of the emanations and invaded rock are suitable, valuable ore minerals may be developed. In general, the invaded rocks most susceptible for ore genesis by contact metamorphism are impure limestones and dolostones. Some of the iron deposits of the Urals, copper deposits in Arizona and Utah, zinc deposits in new Mexico and Ontario, and graphite deposits in the Adirondacks are of contact metamorphic origin.

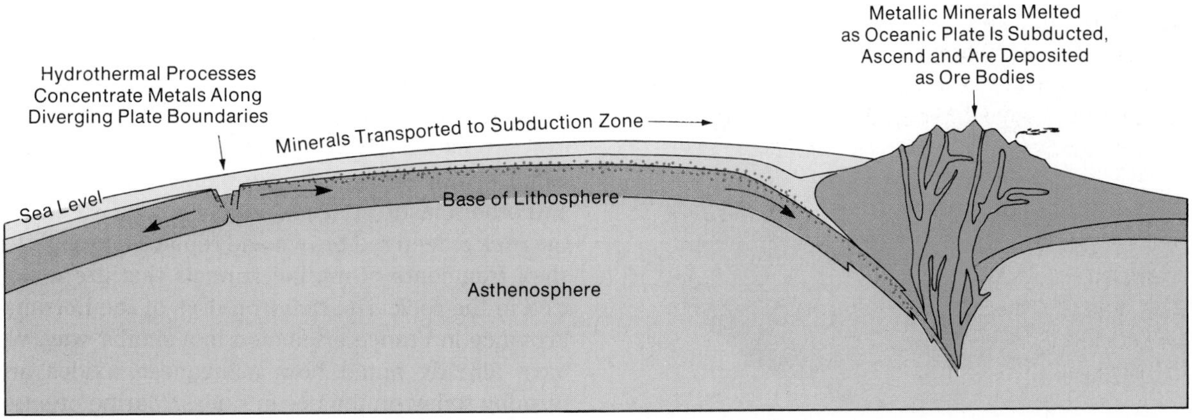

Figure 19–27 Conceptual diagram of the manner in which plate boundaries may influence the accumulation of mineral deposits. Hydrothermal processes emplace metallic minerals in both oceanic crust and marine sediments. These deposits are conveyed to subduction zones where they are melted and ultimately emplaced into a variety of magmatic and hydrothermal deposits.

Hydrothermal Deposits

Hydrothermal ore deposits have crystallized from hot watery solutions. These solutions are hot because of the high temperatures inherent to the depths where they originated, or they may have derived their heat from magmatic bodies. In either case, as they penetrate into the cracks and crevices of rocks with which they may be chemically reactive, and which may be cooler and subject to lower pressures, the valuable metals they hold in solution are precipitated, fill voids, and form veins. Some of the world's great gold deposits were formed in this way, as well as copper deposits of Butte, Montana, lead deposits of Idaho, zinc and lead deposits of Missouri (Fig. 19–28) and Oklahoma, and silver deposits of the Comstock Lode in Nevada.

Figure 19–28 Cubic crystals of galena from a lead mine near Joplin, Missouri.

Not all hydrothermal deposits occur in veins. As the hot steamy solutions move through rock, the ore minerals may precipitate out in disseminated form. Perhaps the best examples of deposits of this kind are the so-called **porphyry copper deposits** found in the Solomon Islands and in Utah (Fig. 19–29). Hydrothermal solutions deposited copper as small grains scattered throughout porphyritic intrusives. Porphyry coppers are not high-grade ores. Typically, they contain only 1 to 4 per cent copper minerals and require careful management in order for mining operations to be profitable.

Hydrothermal ores are precipitated in veins or disseminated throughout a rock by the action of circulating hot water arising from magmas or areas of high heat flow.

Ores Concentrated by Sedimentary Processes

Many valuable deposits of both metallic and nonmetallic minerals have been concentrated as a result of ordinary sedimentary processes. We are already familiar with the manner in which clay minerals (for ceramic use) may be deposited in quiet areas of seas or lakes, or salts of sodium, potassium, and boron may be precipitated from saline lakes and evaporating lagoons. Iron dissolved from rocks and carried to the sea by streams may also be deposited from sea water. The iron ore mined from the Clinton Formation in Alabama was deposited as iron oxide around öolites and other clastic grains of sediment. It has colored the rock a deep red-brown and replaces the calcite-shell fragments of marine animals that are fossilized in the rock. The rich iron ores of the Lorraine Province in France originated in a similar way. We have already noted how manganese oxides are forming today on the ocean floors. Marine precipitation of manganese in the geologic past is responsible for the rich manganese deposits of the U.S.S.R.

Some ore minerals, because of their exceptional weight, are effectively concentrated by waves, currents, or wind action. Such accumulations of heavy

Figure 19–29 Porphyry copper deposit being mined at Butte, Montana, by the open-pit method. The truck in the foreground carries 100 tons of ore. See the chapter-opening illustration as well. (Courtesy of J.C. Brice.)

mineral particles are called **placer deposits**. For example, pure gold (Fig. 19–30) has a density of 19 g per cm^3, whereas quartz has a density of only 2.65 g per cm^3. In a current of water carrying both of these minerals, the lighter quartz is likely to be transported selectively and the heavy gold particles dropped from suspension whenever there is a local slowing of velocity, as often occurs behind an obstruction. Of course, for a placer deposit to form there must be a primary deposit of the ore material nearby. Weathering, erosion, and mass wasting serve to bring the valuable minerals into the stream or onto a beach where current and wave action concentrate the heavy ore minerals.

Placer deposits are famous in the history of mining. A placer deposit found at Sutter's sawmill in 1848 set off the great gold rush of California. The discovery of placer gold in the Klondike triggered the wild 1890s Alaskan gold rush.

Gold is not the only material concentrated in placer deposits. The black placer deposits of the Oregon Coast are concentrates of chromite, and tin occurs as placer ore in Indonesia. Diamonds are concentrated by wave action along some beaches in southwestern Africa, and the thorium-bearing mineral monazite occurs as placer concentrates at a few places along the Malabar Coast of India.

Placers are current-deposited accumulations of gold or other heavy minerals.

Ores Formed by Secondary Enrichment

Secondary enrichment is a term used to describe the chemical processes whereby a deposit may become enriched by the addition of mineral matter from percolating solution. Consider, for example, an area where low-grade porphyry copper is exposed at the earth's surface. Over long intervals of time, meteoric water near the surface percolates down through openings in the rock, dissolves sulphides such as pyrite (FeS_2), and produces sulfuric acid and residual iron oxides. The acidic waters also dissolve primary copper in the porphyry. This upper leached and oxidized zone is developed within the zone of aeration. In the field it can be recognized by the rusty brown color imparted to it by iron oxides. As the copper-bearing solutions continue to trickle downward, they eventually reach the zone of saturation where oxygen is deficient. In this zone, which is just below the water table, copper in solution is precipitated as copper sulfides (Fig. 19–31), thus secondarily enriching the primary occurrences of copper in the porphyry. In the important Utah Copper Mine of Bingham Canyon, the upper part of the copper-bearing igneous rock has been enriched along a zone that is nearly 30 meters thick. over $1\frac{1}{4}$ billion tons of ore have already been taken from this mine. The 600-meter-deep stair-stepped excavation is one of the greatest holes ever dug by human beings.

Figure 19–30 Typical placer gold from the North Trinity River of Trinity County, California. (Courtesy of the U.S. Geological Survey and P.E. Holt.)

Residual Ores Formed by Weathering and Leaching

Ore deposits formed when weathering processes cause the removal of nonvaluable rock materials and leave behind a residuum of valuable minerals are called **residual concentrations**. For such deposits to develop, there must, of course, be a source rock that contains a valuable substance that is resistant to chemical attack (so that it won't be dissolved and carried away), there must be abundant water to carry off the worthless more soluble components, and the topography must be level so that ore minerals are not eroded and carried away. Residual concentrations may require millions of years of slow chemical attack. A good example of an ore mineral formed as a residual concentrate is **bauxite**, the hydrous aluminum oxide that is an ore of aluminum. The aluminum in bauxite is derived from feldspars in primary source rocks or even from the clays present in impure limestones. It is technically possible to extract aluminum directly from clay minerals or feldspars, but the process is too expensive and presently uneconomical. Weathering in a warm humid climate is necessary to break down the silicate minerals into hydrous aluminum oxides (bauxite) from which the metal, now concentrated and rendered easier to process, can be extracted. Residual concentrations of bauxite are widespread but tend to be concentrated in the tropics. Even the bauxite deposits of France and Arkansas were formed when those locations were experiencing tropical climatic conditions. In addition to the U.S. and European occurrences of bauxite, ores are presently being mined in Jamaica, Surinam, Guyana, Australia, and Africa.

Aluminum is not the only metal concentrated by weathering. The banded iron ores of the Lake Su-

Figure 19–31 Copper minerals found in ore bodies formed by secondary enrichment. The mineral on the left with the gold metallic color is chalcopyrite. The dark blue mineral at the right is bornite. The center specimen contains both calcopyrite ($CuFeS_2$) and bornite (Cu_5FeS_4).

perior region owe their high metal content to deep chemical weathering of iron-bearing silicate minerals and to the slow leaching away of enough silica to leave behind bands of rock in which iron is concentrated. The iron ores of the Labrador Trough in North America, Cerro Bolivar in Venezuela, Krivov Rog in the Ukraine, and Minas Gerais in Brazil have iron similarly concentrated by the leaching.

Nonmetallic Minerals for Chemistry

Salts, nitrates, borates, phosphates, compounds of sulphur, potash, coal, and oil are among the nonmetallic earth materials utilized in large quantities by the chemical industries. Ordinary salt or halite is, of course, the most familiar of this group of nonmetallics (see Fig. 2–4). Along with potash sulfates, large halite deposits have formed in the geologic past by evaporation of saline waters in lakes or epicontinental seaways. Layers of salt formed in this way and subsequently deeply buried by the accumulated strata of long geologic periods may become mobilized by the great pressure and flow upward to form salt domes. Bedded salt deposits and salt from these domal structures have been mined for many years. In addition, salt is also obtained for commercial uses directly from sea water by evaporation.

Another mineral resource derived from ancient evaporative basins are potash minerals. Potash is a rather loosely used term for carbonates, oxides, and hydroxides of potassium. Potash minerals are extensively used in fertilizers. They are effective in speeding plant growth, accelerating sugar and starch formation in the plant, and improving the overall quality of the crops.

Nearly always, commercial fertilizers contain nitrogen and phosphorus in addition to potash. The nitrogen is obtained from soda niter ($NaNO_3$), the principal deposits of which are in Chile. It can also be gathered directly from the air by a nitrogen fixation method developed by German chemists during World War I. Most commercial deposits of phosphates are obtained from marine sedimentary rocks in which the element has been concentrated (Fig. 19–32). Some of the phosphorus was deposited in the rock as a direct chemical precipitate from sea water, whereas additional amounts are of organic origin and derived from the remains of marine plants and animals, nearly all of which contain phosphorus in their tissues.

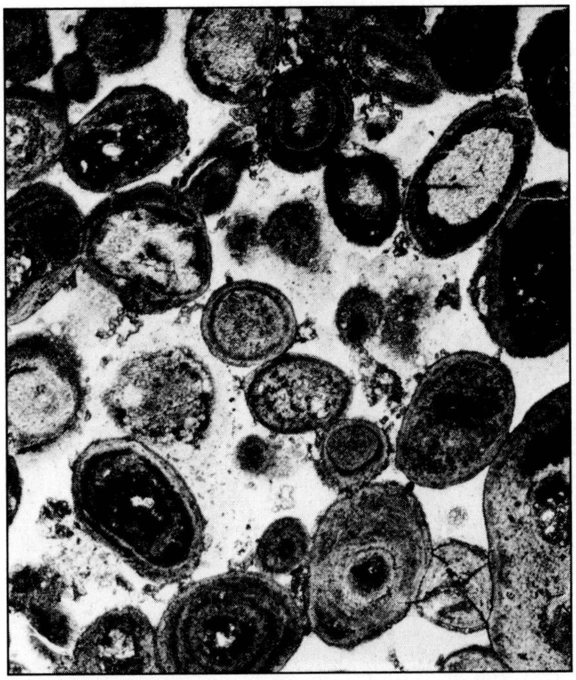

Figure 19–32 Pelletal, oölitic phosphate rock. The phosphatic pellets and oöids are cemented by microcrystalline quartz. This sample was obtained from an outcrop of phosphate rock in Madison County, Montana. (Courtesy of the U.S. Geological Survey and E.R. Cressman.)

Sulfur is a common nonmetallic resource employed in a variety of ways. Much is used to manufacture sulphuric acid, which in turn is used to manufacture fertilizers and as a chemical reagent in hundreds of industrial processes. These include the processing of wood pulp needed in the manufacture of paper, rubber, and explosives. Most sulfur is obtained from the cap rock of salt domes. The method of extraction is known as the Frasch process (Fig. 19–33). In the Frasch method, superheated water is pumped down the outermost of three concentric pipes. The hot water melts the sulfur, which is then converted to a light froth by air pumped down the second pipe. Water and air pressure then force the sulfur froth to the surface where it is collected in large vats.

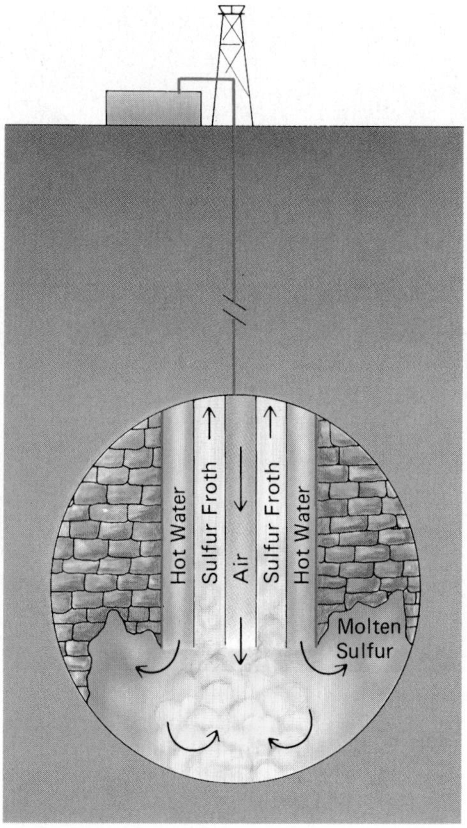

Figure 19–33 Diagram of sulfur mining by the Frasch process, which uses melting and pressure. Three concentric pipes are directed into the sulfur deposit. Compressed air and hot water are sent down two of the pipes, and molten sulfur is forced up the third pipe.

Building Materials

Building Stone

The construction of buildings, roads, canals, and bridges involves the use of such enormous amounts of rock materials that humans must be considered as agents of erosion on earth. Fortunately, most mineral building materials are abundant, easily extractable, and relatively inexpensive. Various rocks are suitable for use in construction. Granite, sandstone, limestone, and marble are often cut at the quarry into blocks of prescribed dimensions and used for specific parts of structures. Slate, because of its characteristic cleavage, is widely used in roofing, stair treads, flooring, and blackboards. Thin slabs of sparkling mica schist are overlapped to form a durable roof for cottages and barns throughout the Swiss Alps. All across the land, a discerning eye will recognize in the houses of a town the same rocks that are exposed in a nearby bluff or quarry, for by using local supplies of construction materials, builders can avoid excessive transportation expenses.

Clay Products

Long before humans had learned how to cut rock for construction, they had discovered how to make bricks out of clay. Indeed, the art of brickmaking began in the days of the Babylonian empire. Clay has two properties that make it eminently useful in making bricks and other ceramic products such as tiles and drainpipes. It is plastic when wet and can therefore be formed into the required shapes, and once it has been fired, it is hard and resistant to water and weathering. Clay deposits vary in origin. Some are residual accumulations that form "in place" on feldspar-rich rocks in warm humid climates. Other clay accumulations have been transported and occur as sedimentary clay beds.

Cement

Everyone knows that cut stone and bricks are not all that is needed in the construction of buildings. One must also have cement to bind the rock or brick into a solid wall or to be combined with sand, gravel, or crushed rock to form concrete. The most widely used cement today was originally developed by an English engineer named Joseph Aspidin in 1824. Because it resembled Portland stone, a limestone frequently used in English buildings, he named his product Portland cement. Cement is manufactured by mixing limestone with appropriate amounts of shale or clay. The mixture is first ground and then heated to a temperature of about 1480°C. During heating, carbon dioxide is expelled from the $CaCO_3$ (forming quicklime, CaO) and the mixture is formed into pelletlike clinkers. When cooled, the clinkers are ground to a fine powder that slowly hardens when mixed with water.

Gypsum Products

Gypsum (see Fig. 6–50) is a soft form of calcium sulfate containing water in its crystal structure and designated by the formula $CaSO_4 \cdot 2H_2O$. It is a relatively abundant mineral commonly associated with evaporite sequences. Because of its softness and solubility, it is not used as a building stone. When, however, gypsum is heated to 177°C, about 75 per cent of its contained water is driven off, and it changes into a new material, $2CaSO_4 \cdot H_2O$, called *plaster of Paris*. The name is indeed derived from quarries near Paris, which provide gypsum for the manufacture of a particularly high-quality plaster. In the plaster of Paris used on the interior walls of buildings, water and sand are added. As the wet mixture begins to congeal, tiny crystals of gypsum form in the plaster and interlock to form a firm solid. The plaster of Paris mixture need not be applied directly to walls but can be mounted on paper backings to produce wallboard (see Fig. 6–50B).

Figure 19–34 Outcrop of St. Peter Sandstone (massive rock beneath overhanging beds of dolostone). The St. Peter Sandstone is widely used in the manufacture of plate glass.

Quartz Sands for Glass

When one thinks of products made of glass, the first items that come to mind are light bulbs, bottles, ovenware, and spectacles. We often overlook the many uses to which glass is put in construction. Glass windows and doors are essential components of modern architecture. Translucent blocks of glass provide walls that allow light to penetrate into rooms. Glass is spun into thin strands and formed into mats of "glass-wool" and other forms of glass-fiber insulation.

The making of glass is an ancient art. The Egyptians melted sand and ashes together over 4000 years ago and produced glass for ornaments and vases. Today, glassmakers mix the quartz grains from quartz sandstones with quicklime (CaO), sodium carbonate, and borax in order to be able to melt the silica at a temperature lower than the 1713°C required to melt pure quartz.

Sand that is used in glass manufacturing must be of high purity. Normally, this means that the raw material consists of over 95 per cent quartz. Impurities, even in small amounts, can impart undesirable colors and murkiness to finished glass. Devo-

Figure 19–35 Veins of asbestos in serpentinite, Thetford, Quebec. (Courtesy of K. Schultz.)

Figure 19–36 Asbestos. (Courtesy of the Institute of Geological Sciences, London.)

nian deposits in West Virginia and Pennsylvania and the Ordovician St. Peter Sandstone of Missouri and Illinois (Fig. 19–34) are among the most notable glass sands in the United States.

Asbestos

Asbestos is the name applied to a group of minerals that have long thin flexible crystals. Asbestos occurs in veins that criss-cross through the metamorphic rock serpentinite (Fig. 19–35). Serpentinite forms by metamorphism of the basic igneous rock peridotite. The use of asbestos fibers (Fig. 19–36) in making fireproof cloth, electrical insulation, and brake linings is well known. In the building industry, it is used as a heat insulator around pipes and as a filler in asphaltic tiles, Portland cement products, and shingles. Its use in materials that will fray or "dust" should be guarded against, because the tiny fibers can get into human lungs and the digestive tracts and cause cancers to develop. Currently, asbestos materials are being systematically removed from buildings.

Summary

In 1971, when the crew of the Apollo 14 mission to the moon had returned to earth, astronaut Edgar Dean Mitchell described his appreciation of the earth as follows: "It is so incredibly impressive," he said, "when you look back at our planet from out there in space and realize so forcibly that it's a closed system—that we don't have unlimited resources, . . .'' Millions of earthbound humans, viewing earth from space on their television screens, were also more aware that the earth's resources were finite and could not support uncontrolled population growth indefinitely.

The principal energy resources derived from earth materials are coal, natural gas, crude oil, and minerals containing fissionable radioactive elements. The first of these, coal, is the result of the accumulation of plant remains under conditions that favor the conversion of plant tissues into deposits rich in carbon.

Petroleum is also of organic origin, although the contributing life forms were not large plants but myriads of tiny, often unicellular, protozoans and algae. Oil derived from these organisms formed in *source rocks* such as black shales, then migrated into more permeable *reservoir rocks* like sandstones, and ultimately accumulated in favorable geologic structures beneath impermeable *cap rocks*.

Although crude oil is one principal source of liquid hydrocarbons today, oil can also be obtained from *tar sands* located in Canada that are currently being mined and processed. In addition, when the environmental problems associated with the use of *oil shale* are solved, this material can be used to extend the period of availability of hydrocarbons for fuel and processed goods such as synthetic textiles, synthetic rubber, and plastics.

Nuclear power presently supplies only a small percentage of the total energy used in the United States. The danger of leakage of radioactive materials and problems of disposal of radioactive wastes have slowed the development of nuclear power plants. Nevertheless, plants continue to be built as a result of pressures resulting from decline in the availability of domestic crude oil and increased costs of fossil fuels. Present nuclear reactors use uranium 235 as their principal fuel. The nuclei of uranium 235 atoms are caused to split, or fission, into two smaller nuclei and at the same time give off abundant heat, which is used to produce steam and drive electric generating turbines. Because the world supply of uranium 235 is limited, attempts are currently underway to develop reactors that convert isotopes like uranium 238, which will not fission, into fissionable plutonium 239. These newer so-called *breeder reactors* will provide much more energy from a given amount of uranium ore than is now possible.

Fusion is a nuclear process in which two nuclei of a light element like hydrogen are combined to yield a new and larger nucleus. During fusion, a small amount of mass is converted to energy in the form of heat. Fusion requires enormously high temperatures, and there are presently formidable technologic problems that must be solved if it is ever to become commercially feasible.

Energy resources are intrinsically linked to the metallic and nonmetallic resources on which we depend for the products of our civilization. Enormous amounts of energy are needed to extract ores and manufacture products from the refined metals and chemicals. As easily obtainable mineral resources are exhausted, ever-increasing amounts of energy will be necessary to mine and process deeper and lower-grade raw materials.

Ores are natural mineral occurrences that are sufficiently concentrated to permit the profitable extraction of one or more metals. Concentration may have resulted from any of a variety of geologic processes, including early crystal settling in magmas, filter pressing, late magmatic crystallization, and contact metamorphic reactions.

Magmas characteristically emit large amounts of hot watery solutions, which penetrate into overlying rocks, and on cooling, precipitate their dissolved minerals. The results are *hydrothermal deposits*. Many gold, silver, zinc, and copper deposits are of hydrothermal origin.

Not all metallic minerals are associated with igneous activity. Iron and manganese oxides may be formed by direct precipitation onto the floors of lakes and oceans. Gold may be eroded from exposures of igneous rocks, transported by streams, reworked by currents and waves, and deposited as *placer deposits*. Percolating waters are known to have leached metallic elements form surface rocks, carried them down to zones near the groundwater table, and precipitated them as secondarily enriched ore deposits. There are also many examples of ores of iron and aluminum (bauxite) that have been concentrated residually when undesirable components of the host rock were dissolved and wasted away.

The nonmetallic earth materials are no less important to our industry, technology, and well-being than are the metallic minerals. The nonmetallics include mineral fertilizers such as potash and phosphates, sulphur for a variety of chemical uses, asbestos, which is used as a thermal and electrical insulator, salt for our tables and chemicals, limestone for cement, and clean quartz sand for the manufacture of glass.

Terms to Understand

anthracite coal
asbestos
bauxite
bituminous coal
breeder reactor
cap rock
disseminated magmatic deposit
early crystal settling
filter pressing
fission
fusion
gravimeter
hydrothermal deposit
kerogen
late magmatic deposit
lignite
oil shale
ore
peat
placer deposit
porphyry copper deposit
reservoir rock
residual concentration
secondary enrichment
seismic reflection shooting
source rock
tar sands
well-log

Questions for Review

1. What is the most common iron ore mineral? What is the role of carbon in the smelting of iron?
2. What are the various kinds of coal and on what basis are they distinguished? What environmental conditions are conducive to the origin of coal?
3. What are the essential natural requirements for the occurrence of an oil trap?
4. What environmental hazards are associated with the use as fuel of each of the following materials: coal, petroleum, uranium, and oil shale?
5. What is a common ore mineral for each of the following metals: lead, tin, copper, zinc, mercury, aluminum, titanium, chromium, and nickel?
6. What factors other than the percentage of metal in rock, determine whether or not a metal can be mined at a profit?
7. Which process of ore concentration in magmas is most dependent on gravity and the specific gravity of the ore minerals?
8. What are porphyry copper deposits? How did they obtain their copper content?
9. What are placer deposits? What metallic resources are concentrated as placer deposits?
10. Aluminum is a common constituent of the feldspars in igneous rocks. Why do we not obtain this metal directly from feldspathic igneous rocks rather then the weathered products of those rocks?
11. What are the constituents of Portland cement? How is this important material manufactured?
12. Why is it necessary to add such things as quicklime, borax, and sodium carbonate to quartz sand in order to make glass?
13. Describe the phenomenon that makes nuclear power possible.

Supplemental Readings and References

Craig, J.R., Vaughan, D.J., and Skinner, B.J., 1988. *Resources of the Earth*. Englewood Cliffs, N.J., Prentice-Hall Inc.

Flawn, P.T., 1966. *Mineral Resources*. Chicago, Rand McNally & Co.

Park, C.F., Jr., 1975. *Earthbound*. San Francisco, Freeman, Cooper & Co.

Park, C.F., Jr., and MacDiarmid, R.A., 1975. *Ore Deposits*. 3rd ed. San Francisco, W.H. Freeman & Co.

Skinner, B.J., 1986. *Earth Resources*. Englewood Cliffs, N.J., Prentice-Hall Inc.

Planetary Geology

20

Martian canyonlands. (A) High-resolution view of intricate cliffs, mesas, and canyons, (B) Oblique view of cliffs and landslides in the Valles Marineris. (Courtesy of the U.S. Geological Survey and NASA.)

Give me matter and motion and I will construct the universe.

René Descartes, 1640

Introduction

Long ago primitive people of the Ice Age watched the night sky and wondered about the meaning of the stars and moon. Since that time, we humans have continued to be fascinated by news of what exists in space beyond our own planet. The drive to learn and to explore has fueled our technologic inventiveness and carried us into the present age of planetary exploration (Fig. 20–1). Humans have landed their devices on the surface of Mars and held the soil of the moon in their hands. Less than a century ago, these feats would have been regarded as the fantastic dreams of science-fiction writers.

With the space age still in its infancy, it appears certain that exciting new discoveries lie ahead. So that we may better understand tomorrow's news from space, it is important to examine some of the perspectives and insights already provided by artificial satellites, telescopes, and spacecraft. Such information is not inappropriate for a book about geology, for the earth's history is intrinsically tied to the history of the solar system and, indeed, the entire universe.

Planet Earth

Compared with the other planets in our solar system, the earth has many unique features. No other planet contains so much water. The earth is also notable for its nitrogen–oxygen atmosphere. It is not so close to the sun as to be seared by intense heat, nor so far away as to be locked in deep cold. Most important, life has developed and has been hospitably sustained on earth.

Although our planet is in some ways special, it is not at all a planetary oddity. The elements of which it is composed occur also in other planets, although in different proportions. The earth has a general similarity in size and density (Table 20–1) to the other three inner planets (Mercury, Venus, and Mars), and it shares a general similarity in the directions of movements with most of the other planets. In fact, interpreting the data and photographs gathered recently from the earth's neighboring planets is achieved largely by applying geologic knowledge that we have about the earth.

Like its neighbors in the solar system, the earth rotates on its axis, and this movement has caused it to assume the shape of a ball slightly flattened at the poles. As a result, the distance from the center of the earth to the poles (6357 km) is 21 km less than the distance from the earth's center to the equator. The planet's circumference is about 40,000 km (25,000 miles).

Figure 20–1 A human visitor stands on the surface of the moon. (Courtesy of NASA.)

Table 20–1 Physical and Rotational Characteristics of the Planets

Name	Equatorial km	Radius ÷ Earth's	Mass ÷ Earth's	Mean Density (gm/cm^3)	Oblateness	Surface Gravity (Earth = 1)	Sidereal Rotation Period	Inclination of Equator to Orbit
Mercury	2,439	0.3824	0.0553	5.43	0.0	0.378	58.646^{d}	0°
Venus	6,052	0.9489	0.8150	5.24	0.0	0.894	243.01^{d}R*	177.3
Earth	6,378.140	1	1	5.515	0.0034	1	$23^{h}56^{h}04.1.^{s}$	23.45
Mars	3,397.2	0.5326	0.1074	3.93	0.005	0.379	$24^{h}37^{m}22.662^{5}$	25.19
Jupiter	71,398	11.194	317.89	1.36	0.061	2.54	$9^{h}50^{m}$to $> 9^{h}55^{m}$	3.12
Saturn	60,000	9.41	95.17	0.71	0.109	1.07	$10^{h}30.9^{m}$	26.73
Uranus	26,145	4.1	14.56	1.30	0.03	0.8	12^{h} to 24^{h}	97.86
Neptune	24,300	3.8	17.24	1.8	0.03	1.2	$18^{h} \pm 10^{m}$	29.56
Pluto	1,500–1,800	0.4	0.02	0.5–0.8	?	~0.03	$6^{d}9^{h}17^{m}$	118?

*R signifies retrograde rotation.

The masses and diameters are the values recommended by the International Astronomical Union in 1976, except for a 1977 value for the radius of Uranus. Densities and surface gravities were calculated from these values.

Source: Pasachoff, J.M., 1985. *Contemporary Astronomy*, 3rd ed., Philadelphia, Saunders College Publishing.

The Earth's Place in Space

The earth is one of nine planets that revolve around a rather average star, our sun. The planets, in order of increasing distance from the sun, are Mercury, Venus, Earth, Mars, Jupiter, Saturn, Uranus, Neptune, and Pluto. A belt of asteroids orbits the sun in the region between the paths of Mars and Jupiter. This grouping of planets and asteroids around the sun constitutes our solar system (Fig. 20–2). Probably ours is not the only such system in the universe. It is true that the planets of other systems are too distant to be detected directly, but we infer their existence because of the wobbles their orbital movements and gravitational pull cause in the stars they circle.

Our solar system is a small part of a much larger aggregate of stars, planets, dust, and gases called a **galaxy**. Galaxies are also numerous in the universe, and their constituents are arranged into several different general forms: tightly packed elliptic galaxies, irregular galaxies, and the more familiar discoidal spiral galaxies with their glowing central bulge and great curving arms.

The galaxy in which our solar system is located is called the **Milky Way Galaxy** because, when we

Figure 20–2 The solar system. (From Turk, J., and Turk, A.P., 1987. *Physical Science*, Philadelphia, W.B. Saunders Company.)

look in a direction parallel to the plane of the galaxy, we see a great milky haze of light. The Milky Way Galaxy (Fig. 20–3) is of the spiral category. It rotates slowly in space, completing one rotation about every 240 million years. Our sun is located about two-thirds of the distance (or 26,000 light years) outward from the center of the galaxy.

The Sun

The largest and for many reasons the most important member of our solar system is the sun. Although only a modest star in comparison with others in our galaxy, it nevertheless has a mass 333,000 times that of the earth and is 109 times larger in diameter. It is composed mostly of hydrogen (about 73 per cent) and helium (about 25 per cent). Remaining heavier elements exist as gases in the hot interior of the star, where temperatures exceed 15 million°C.

The part of the sun that we are able to observe directly is called the solar atmosphere. There are three general regions of the solar atmosphere, each having transitional boundaries with adjacent regions. The uppermost region is called the photosphere (Fig. 20–4). The light that reaches us here on earth emanates from the photosphere. Immediately above the photosphere is the chromosphere. One can see the hot hydrogen that forms the chromosphere as a reddish glow visible during a solar eclipse. The chromosphere merges gradually into the outermost atmospheric region known as the corona. This seething tenuous shroud of fiery gases extends millions of km outward from the sun.

Although an enormous amount of solar energy is intercepted by the earth, our planet is neither frozen nor roasted but is able to maintain a range of temperatures roughly between −50° and +60°C. The maintenance of this vital temperature range is made possible by three factors. First, because of

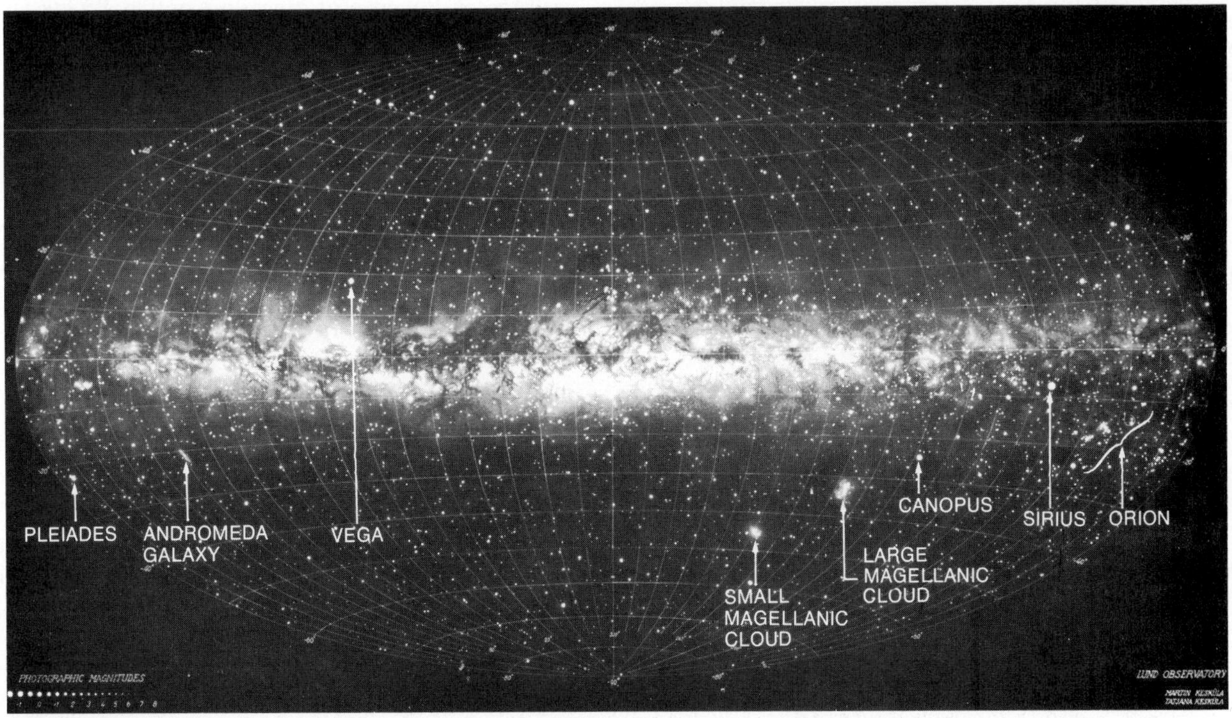

Figure 20–3 A drawing of the Milky Way, made under the supervision of Knut Lundmark at the Lund Observatory in Sweden; 7000 stars plus the Milky Way are shown in this panorama, which is in coordinates such that the Milky Way falls along the equator. (Courtesy of Lund Observatory, Sweden; from Pasachoff, J.M., 1989. *Contemporary Astronomy,* 4th ed., Philadelphia, Saunders College Publishing.)

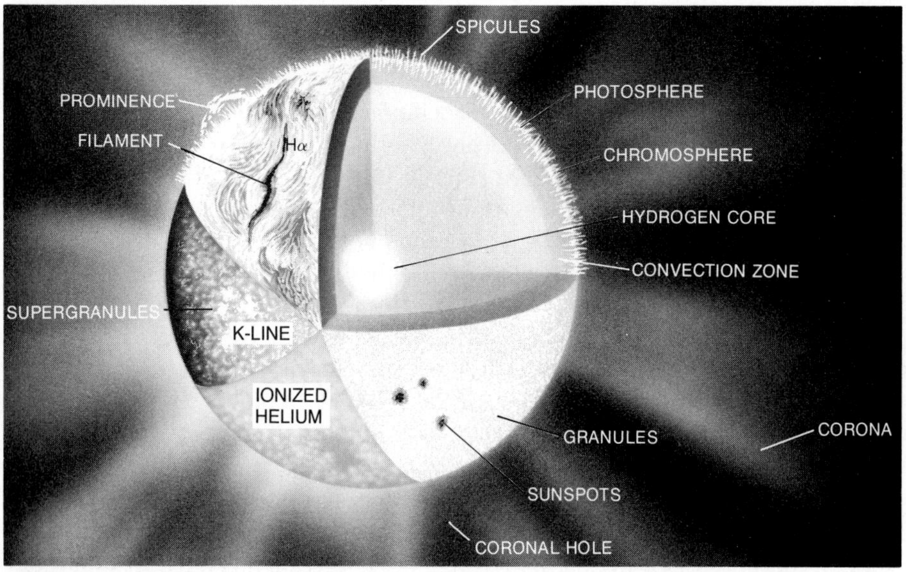

Figure 20–4 Parts of the solar atmosphere and interior. (From Pasachoff, J.M., 1989. *Contemporary Astronomy,* 4th ed., Philadelphia, Saunders College Publishing.)

rotation, the earth receives energy from the sun on one hemisphere, while it returns heat to space over its entire surface. Second, some of the incoming radiation is reflected off the atmosphere, dust, and clouds and is directed back into space without ever reaching ground level. Finally, a part of the intercepted radiation is absorbed by the atmosphere and radiated back into space without warming the earth's surface.

There are other forms of radiation besides visible sunlight that are intercepted by the earth (Fig. 20–5). Three per cent of the incoming rays are ultraviolet. Because of their destructive effects on life, it is fortunate that most of this short-wavelength radiation is absorbed in the ozone layer of the atmosphere. Much of the infrared radiation is absorbed by water vapor and carbon dioxide, causing the lower zones of the atmosphere to be warmed.

The energy that maintains the sun as a great glowing sphere of gases is derived from a continuous thermonuclear reaction called **fusion**. In the fusion process, four hydrogen nuclei combine in a series of steps to form helium. Helium has slightly

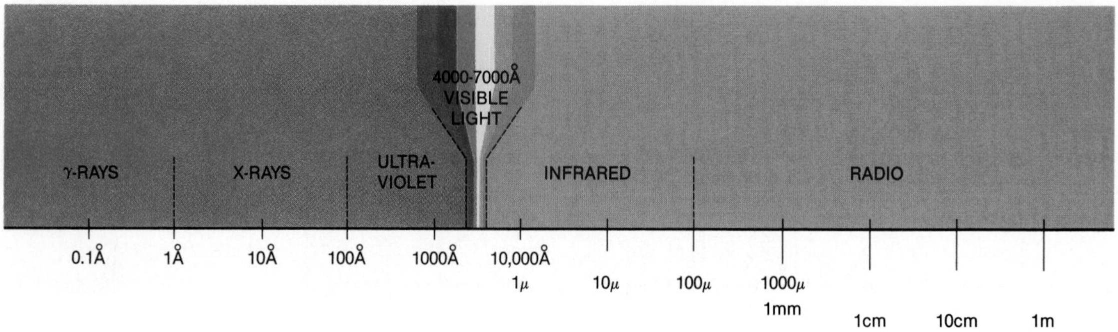

Figure 20–5 The electromagnetic spectrum. In the scale, an angstrom (A) is 10^{-8} cm (0.000 000 01 cm) and a micron (μ) is 10^{-4} cm (0.0001). (From Pasachoff, J.M., 1989. *Contemporary Astronomy,* 4th ed., Philadelphia, Saunders College Publishing.)

less mass than the hydrogen nuclei from which it was derived, and that "missing mass" is converted into energy in the form of heat (Fig. 20–6). Each second, the sun transmits an incredible amount of energy, and this energy is essential for life on earth. It is the ultimate force behind the many geologic processes that sculpture the earth's surface. For example, the sun's rays aid in the evaporation of surface waters, which in turn results in clouds that provide the precipitation required for erosion. Along with the earth's rotation, the sun's radiation results in winds and ocean currents. Some scientists believe that protracted variations in the heat received from the sun may trigger episodes of continental glaciation or may reduce lush forests to barren wastelands. Even primitive people knew the sun's importance and recognized it as the fountainhead and sustainer of life.

The Birth of the Solar System

Any account of the origin of the solar system is obliged to conform to certain basic constraints within the system. These include dynamic constraints relating to the movements of planets, compositional constraints, and age constraints.

Dynamic Constraints

The planets of our solar system all revolve around the sun in the same direction and lie approximately in the plane of the sun's rotation. Thus, the solar system is like a disk in shape. The direction taken by the planets as they revolve around the sun is counterclockwise (called **prograde**) when viewed from a hypothetical point in space above the sun's north pole. The direction of rotation of the planets on their axes is also counterclockwise, with the exception of Venus and Uranus. With few exceptions, satellites mimic the movements of the planets they orbit.

Compositional Constraints

As indicated by their differing densities (see Table 20–1), as well as by earth, moon, and meteorite samples, the planets differ in composition. Mercury, Venus, Earth, and Mars have mean densities of 5.4, 5.2, 5.5, and 3.9 respectively. These four planets constitute the inner or **terrestrial planets**. They are small, dense, rocky, and rich in metals. Jupiter, Saturn, Uranus, and Neptune have lesser densities of 1.34, 0.69, 1.27, and 1.58 respectively. They are termed the outer or **Jovian planets**. Like the sun, the compositions of Jupiter and Saturn are dominated by hydrogen and helium.

The planets in our solar system fall into two groups. There are the small, stony inner planets, and the large, gaseous outer planets.

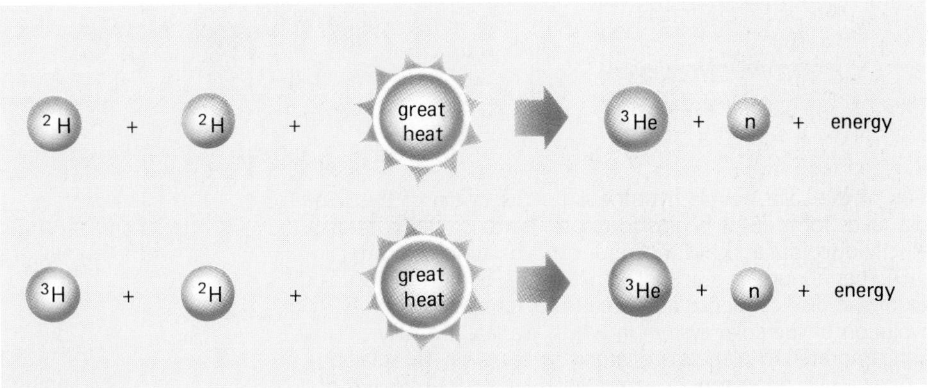

Figure 20–6 Examples of two fusion reactions (n = neutron).

Age Constraints

As indicated in Chapter 8, there are rocks on the surface of the earth that have been present for at least 4.2 billion years, and samples from the moon that are 4.4 billion years old. Meteorites that have not been melted after they were originally formed (melting would reset their radioactive clocks) yield ages of about 4.5 billion years. This age agrees with theoretical calculations of the age of the sun.

The Solar Nebula Hypothesis

When all the dynamic, compositional, and age constraints are taken into account, the favored hypothesis is one that concludes that the solar system was born about 4½ billion years ago from a rotating cloud of hot gases called the **solar nebula**. Immanuel Kant first suggested the idea in 1755. About 40 years later, Pierre LaPlace fully developed Kant's idea, and in recent years it has been improved with the rich flow of information emanating from the space program.

The **nebula hypothesis** begins with contraction of the nebula in response to gravitational forces. When a slowly rotating nebula contracts, its rotation speed increases as a consequence of the law of conservation of angular momentum. (This law accounts for the ability of a figure skater to spin faster when she pulls in her arms.) At the same time that this contraction and increase in speed of rotation was occurring, the nebula was heated by its own gravitational energy. As temperatures rose, any solids that may have been present were vaporized in the swirling mass of hot gases. Eventually, the nebula collapsed into disk form (Fig. 20–7). At the center of the disk, continued collapse led to self-sustaining nuclear reactions that ultimately produced the sun. The nebula could now be termed a solar nebula.

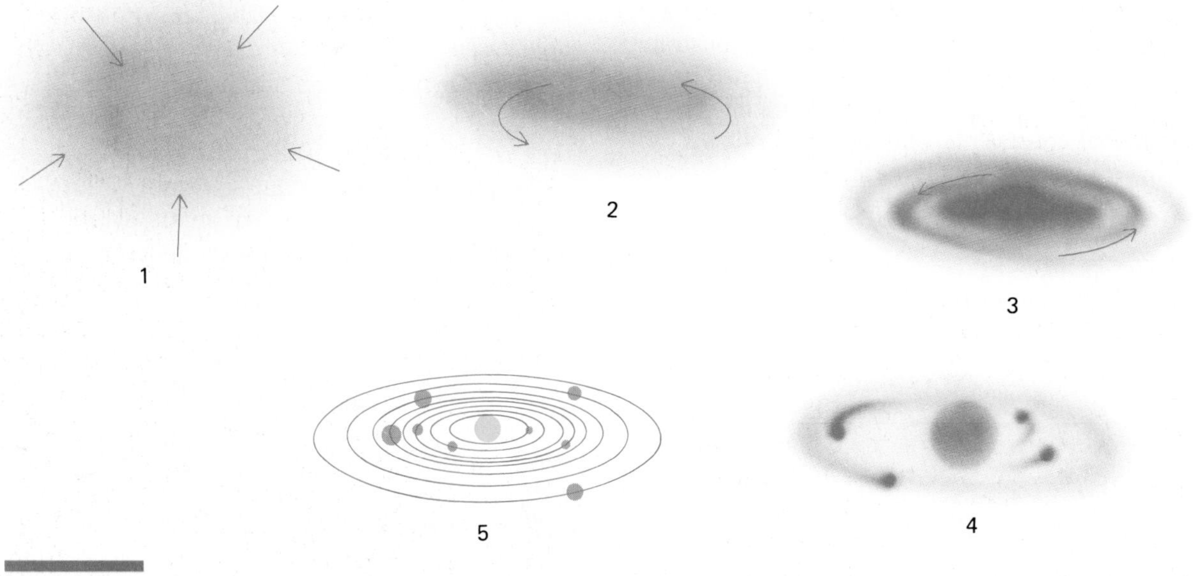

Figure 20–7 An illustration of the solar nebula hypothesis for the origin of the solar system. (A) The solar nebula takes form. (B) It begins to rotate, flattens, and contracts. (C) The nebula is now distinctly discoidal and has a central concentration of matter that will become the sun. (D) Thermonuclear reactions begin in the primordial sun as material in the thinner area of the disk condense and accrete to form protoplanets. (E) The late stage in the evolution of the solar system in which planets are fully formed, and materials not incorporated in planets are largely swept away by solar wind and radiation. (From Abell, G.O., Morrison, D., and Wolff, S.C., 1988. *Realm of the Universe.* Philadelphia, Saunders College Publishing.)

The sun was formed when atoms in the center of the nebula became sufficiently compacted as to cause fusion to occur.

The sun kept the inner parts of the system hot, but with no further gravitational energy to sustain high temperatures, the outer realms of the disk began to cool. Areas farthest from the sun cooled progressively more than those closer, just as today the outer planets are coldest. As cooling progressed, gases and vapors within the cloud were able to combine chemically and form compounds, and the compounds condensed into liquid droplets or solid grains of ice or mineral matter. Ultimately these particles combined into increasingly larger aggregates until much of the solid material was in the form of bodies a few kilometers to tens of kilometers in diameter. Some were large enough to gravitationally sweep up their neighbors and gradually assume the dimensions of planets. Meanwhile, the sun was emitting a stream of radiation (sometimes called **solar wind**) that drove enormous quantities of lighter elements out toward the outer reaches of the system. Planets nearest the sun lost huge amounts of lighter matter but were able to retain heavier elements. This ability accounts for the smaller masses but greater densities of the terrestrial planets relative to those of the Jovian planets, which retain their considerable volumes of hydrogen, helium, and light hydrogen compounds.

Asteroids and Comets

Asteroids are chunks of rocky material orbiting the sun that are smaller than planets but usually larger than meteors. They are concentrated in the so-called asteroid belt, which lies between the orbits of Mars and Jupiter. The asteroid belt may represent an area in the solar system where a planet failed to develop. The origin of comets is far more speculative. Astronomers using conventional methods to study the spectra of comets found indications that they are composed of frozen methane, ammonia, and water in which particles of heavier elements may be embedded, somewhat like sand grains in a snowball. It appears that comets are composed in part of the flux driven spaceward by solar wind, or they may be made of matter that once was too far out on the periphery of the nebula to have been drawn into the evolving planets. They may represent the initial condensates of the solar nebula.

Some comets approach sufficiently close to the earth to be within range of our present space technology. The March 1986 close approach of the comet Halley (Fig. 20–8), for example, was observed by a flotilla of six spacecraft. Instruments aboard the spacecraft made direct measurements of the comet's dust and gas and unexpectedly discovered that most of the dust was in the form of

Figure 20–8 Comet Halley photographed from Australia in December 1986. This color picture is produced from three separate exposures taken through individual filters, which accounts for the background pattern. (Courtesy of Anglo-Australian Observatory and Abell, G.O., Morrison, D., and Wolff, S.C., 1988. *Realm of the Universe.* Philadelphia, Saunders College Publishing.)

small particles of carbon and hydrocarbon compounds, rather than silica, as was once suspected.

The solar nebula hypothesis will certainly be modified and improved in the next several years. Many questions relating to the role of gravitation and the distribution of certain of the heavy elements remain to be answered. As the concept now stands, it is probably the most widely favored of hypotheses for solar system origin. Large telescopes have revealed the existence of true nebulae between stars, and some of these great swirls of gas and dust appear to be forming new stars. Planetary systems like our own may be forming even now. The hypothesis also accounts for the known spacing of planets, their directions of rotation and revolution, and the distribution of angular momentum within the system.

For geologists, the solar nebula hypothesis represents a beginning on which to build the history of the earth. One may begin with a solid earth that originated by the collection of particles derived from a cloud of interstellar matter and view the solar system as having formed from rather ordinary cosmic processes. The formation of the earth was not a unique occurrence.

Meteorites as Samples of the Solar System

Because meteorites are masses of mineral or rock that have reached the earth from space, they are of great importance to scientists interested in the origin and history of planets. They are the only objects from the universe beyond the moon and the earth that can currently be held in hand and scrutinized in the laboratory. This research indicates that meteorites are fragments of asteroids formed when asteroids collided in space and were shattered. Many of these fragments were once part of meteors (Fig. 20–9), those rapidly moving, luminous bodies we refer to as "shooting stars." Most of these particles burn up when they enter the earth's atmosphere, but if a portion survives its fall through the atmosphere and crashes into the earth, it is then termed a **meteorite**. About 500 meteorites as large as or larger than a baseball reach the earth's surface each year. Weathering and erosion on the earth have erased most of the craters made by their impacts, but there remain about 70 partially preserved craters or clusters of craters.

Figure 20–9 A meteor crossing the field of view of the Palomar telescope while it was taking a 15-minute exposure of the Comet Kobayashi-Berger-Milon, August 1975. (Hale Observatory photography/John Huchra; from Pasachoff, J.M., 1989. *Contemporary Astronomy,* 4th ed., Philadelphia, Saunders College Publishing.)

Figure 20–10 Aerial view of Meteor Crater, formed by the impact of a large meteorite about 50,000 years ago. Conconino County, Arizona. (Courtesy of W.B. Hamilton.)

The best preserved meteor crater in the world is Meteor Crater in Arizona (Figure 20–10). The impact that produced Meteor Crater 50,000 years ago is estimated to have released energy equivalent to a 20-megaton nuclear bomb. In the past much larger impacts have occurred. Meteor showers or the impacts of asteroids may even have contributed to episodes of mass extinctions on earth. There is evidence, for example, that an asteroid with a diameter of 10 km or greater struck the earth about 66 million years ago, lifting millions of tons of dust into the atmosphere. The dust blocked the sun's rays so that darkness and cold prevailed for several weeks, or possibly even several months. These effects could certainly have been catastrophic for life on earth, and have been proposed as a major cause for the extinction of the dinosaurs.

Meteorites can be classified according to their composition as ordinary chondrites, carbonaceous chondrites, achondrites, irons, and stony-irons (Fig. 20–11). The most abundant of these types are the **ordinary chondrites** (Fig. 20–12), which are crystalline stony bodies composed of high-temperature ferromagnesian minerals. Ordinary chondrites can be dated by the uranium–lead, strontium–rubidium, and potassium–argon methods, and they are found to be about 4600 million years old. Many ordinary and carbonaceous chondrites contain spherical bodies called **chondrules** (see Fig. 20–12). Their spherical shape suggests that chondrules solidified from molten droplets splashed into space during collisions of objects swirling about in the solar nebula.

Less abundant than the ordinary chondrites are the **carbonaceous chondrites** (Fig. 20–13). In these meteorites, one finds the same abundance of metallic elements as in ordinary chondrites, with the addition of about 5 per cent of organic compounds, including inorganically produced amino acids. Suspended in the blackish earthy matrix of carbonaceous chondrites are both chondrules and

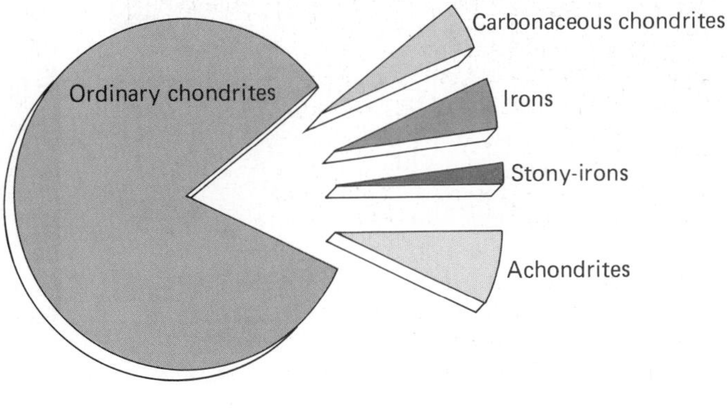

Figure 20–11 The major categories and proportions of meteorites.

Figure 20–12 Stony meteorite of the type termed a chondrite. Chondrites contain small, rounded particles called chondrules composed of ferromagnesian minerals. In most cases, chondrules are only about 2 or 3 mm in diameter, but the one visible in the lower right area of this specimen is approximately 2 cm in diameter. (Washington University collection.)

Figure 20–13 Small carbonaceous chondrite. Scale divisions are in millimeters. This meteorite fell near Murray, Kentucky, in 1950. (Washington University collection.)

irregular pieces of high-temperature minerals that may have condensed from a cooling vapor.

One of the most interesting findings about carbonaceous chondrites is that they contain nonvolatile elements in approximately the same proportion as those that occur in the sun. Indeed, if it were possible to extract some solar material, cool it down, and condense it, the condensate would be chemically similar to carbonaceous chondrites. This similarity suggests that carbonaceous chondrites are samples of primitive planetary material that formed when the sun formed and since that time have never been remelted.

A small percentage of stony meteorites do not contain chondrules and therefore are termed **achondrites** (Fig. 20–14). Most of these have compositions similar to terrestrial basalts, and many exhibit angular broken grains. These fragmental textures indicate that achondrites may be the products of collisions of larger bodies.

The **iron meteorites** (Fig. 20–15) comprise about 6 per cent of all observed falls of meteorites. Most iron meteorites are intergrowths of two varieties of nickel-iron alloy. The large size of the crystals and metallurgic calculations indicate that some iron-nickel meteorites cooled as slowly as 1°C per million years. Such a slow rate of cooling would be possible only in objects at least as large as asteroids. Thus the history of the iron meteor-

Figure 20–14 Meteorite of the achondrite variety, found in Sioux County, Nebraska. (Washington University collection.)

ites probably involved a long episode of slow crystallization within a parent body sufficiently large to provide insulation for the hot interior, followed by violent disruption of that body by a collision to produce the meteorites.

The least abundant of all the categories of meteorites are the **stony-irons** (Fig. 20–16). As indicated by their name, they are composed of silicate minerals and iron-nickel metal in about equal amounts. The stony-iron meteorites are generally considered to represent fragments from the interfaces between the silicate and metal portions of asteroidal bodies. In some cases, they may have originated at the boundary between an iron core and the silicate mantle of a planetary body destroyed by collision.

The principal kinds of meteorites are ordinary chondrites, achondrites, iron, and stony iron.

A

B

Figure 20–15 (A) Collecting an iron meteorite on the Antarctic ice. (B) Cut and polished iron meteorite found in Mohave County, Arizona, showing the interesting pattern of interlocking crystals, called the Widmanstatten pattern. (A, courtesy of NASA; B, from the Washington University collection.)

The Moon

Moon Features

As satellites go, our moon is large in relation to the size of its parent planet. It has a diameter of over one-fourth that of the earth, and its density of about 3.3 g/cm^3 is the same as that of the upper part of the earth's mantle. Gravity on the moon is insufficient to maintain an atmosphere. The moon and the earth revolve around their mutual center of gravity as a sort of "double planet." The moon orbits the earth and rotates on its axis at the same rate. As a result, we always see the same side of the moon and must depend on transmissions from space vehicles for images of the far side (Fig. 20–17). These photographs have revealed a densely cratered surface interrupted only by few and relatively small level areas.

Telescopic observation of the near side of the moon was first made in 1609 by a rather irascible professor of mathematics at the University of Padua. The mathematician, of course, was Galileo Galilei. With the use of his primitive telescope, Galileo was able to see that the surface of the moon was not smooth but "uneven, rough, full of cavities and prominences. . . ." He recognized "small

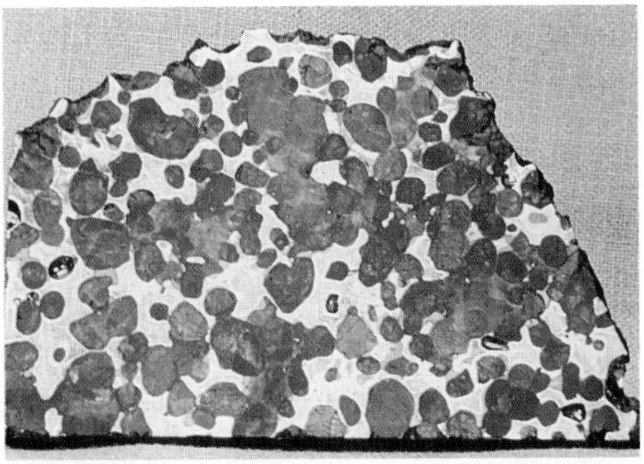

Figure 20–16 Stony-iron meteorite. The light-colored areas consist of iron-nickel, and the darker minerals are mostly olivine. (Washington University collection.)

spots," which we now know are craters, and dark patches that children imagine to be the eyes and mouth of the "man in the moon." These dark patches lie within large basins. Galileo named them **maria** ("seas") and incorrectly suggested they were filled with water. Today, we know the dark areas have been flooded, not with water, but with basaltic lavas. In many maria, the lava has been contained within depressions that are hundreds of km across and 10 to 20 km deep. These depressions are termed **mare basins** (Figs. 20–18 and 20–19). Mare basins had to have developed prior to the extrusion of the dark lavas that form the maria, for these lavas are contained within the basins.

Figure 20–17 The far side of the moon as photographed by Apollo 16. (Courtesy of NASA.)

The moon's maria are vast basins initially formed by meteorite impact and subsequently filled by lava flows.

Galileo also provided the name for the lighter-hued rougher terrains of the moon. He called these regions terrae, although modern space scientists refer to them as **lunar highlands**. The highlands are the oldest parts of the moon, having already formed long before the earliest mare lavas appeared. The highlands are heavily cratered regions reminiscent of a battlefield following a heavy artillery barrage. They provide stark evidence of the moon's early episodes of intense meteoritic bombardment.

On the moon, as on the earth, craters are roughly circular steep-sided basins normally

Figure 20–18 The near side of the moon. In order to show the whole moon but still show the detail that does not show up well at full moon, the Lick Observatory has put together this composite of first and third quarters. Note the dark maria and the lighter heavily cratered highlands. Two young craters, Copernicus and Kepler, can be seen to have rays of light (that is, lighter than the background) material emanating from them. This is a ground-based view of the moon, and the satellite is shown inverted as it would be when viewed through a telescope. (Lick Observatory photograph; from Pasachoff, J.M., 1989. *Contemporary Astronomy,* 4th ed. Philadelphia, Saunders College Publishing.)

Figure 20–19 A view of part of the Mare Tranquillitatis (Sea of Tranquility), where astronauts aboard the Apollo 11 spacecraft made their landing. (Courtesy of NASA.)

caused by either volcanic activity or meteoritic impact (Fig. 20–20). Although a very few lunar craters show questionable evidence of volcanic origin, the great majority are clearly the result of meteoritic impact. Unlike the earth, the moon has an insufficient atmosphere to burn up approaching meteorites, and thus the frequency of impact is high. Lunar geologists are able to determine the relative ages of certain areas on the moon's surface by the density of craters. Younger areas have fewer craters than do older ones. Another aid to recognizing differences in the age of lunar features is provided by the rays of material splashed radially outward from the crater at the time of impact. Younger rays and other impact features cross and partially cover older features. Light-colored impact rays composed of finely crushed rock are exceptionally well developed in the large lunar crater known as Copernicus (Fig. 20–21). This familiar feature is about 90 km across and is rimmed by concentric ridges of hummocky material blasted out by the impact of the crater-forming body. Not all craters have rays as spectacular as those around Copernicus. Impact rays are sometimes obliterated by the rather continuous rain of tiny meteorites that strike the moon. For this reason, younger craters may still be adorned with impact rays, whereas around older craters these features have been lost.

There is ample evidence of former igneous activity on the moon (see Fig. 4–45). We have already noted the extensive lava floods of mare basins. In addition to these flood basalts, a few dome-shaped volcanoes measuring about 10 km in basal diameter have been observed. About 1 per cent of the lunar craters may have resulted from volcanic activity. Meandering channelways occur on the surface of our satellite. Most of these are thought to have been produced by the turbulent flow of lava or by the collapse of lava tunnels, although the evidence is not conclusive. These winding features are distinctly different from straight channels, which probably result from faulting. Tall volcanic peaks are not found on the moon. Perhaps this is because lunar lavas, like lavas associated with shield volcanoes on the earth, are highly fluid and spread laterally rather than piling up around a vent.

Are lunar volcanoes still active today? In 1958, Russian astronomers observed evidence of gas

Figure 20–20 A view of the heavily cratered far side of the moon, photographed by the crew of the Apollo 11 spacecraft. (Courtesy of NASA.)

emission from the crater Alphonsus, and on two occasions in 1963, observers at the Lowell Observatory saw glowing red spots in and near the crater Aristarchus.

Moon Rocks

Direct examination of samples returned from the moon (Fig. 20–22), as well as instrumental data gathered by orbiting spacecraft, indicate that there are three main kinds of rocks on the moon. One of these forms the floor of the smooth mare regions, whereas the other two are characteristic of the lunar highlands. The mare rocks resemble basalt, a finely crystalline dark-colored igneous rock that was described in Chapter 4. Plagioclase, pyroxene, olivine, and ilmenite are the principal minerals in basalt.

The rock types that occur in the lunar highlands differ from the mare basalts in both texture and composition. Texturally they are decidedly coarser (Fig. 20–23), and compositionally they are richer in feldspar. Most of them are aggregates of crushed rock mixed with material that appears to have been molten. Although several compositional varieties of highland rocks have been identified, in general they resemble the family of rocks on earth known as gabbros.

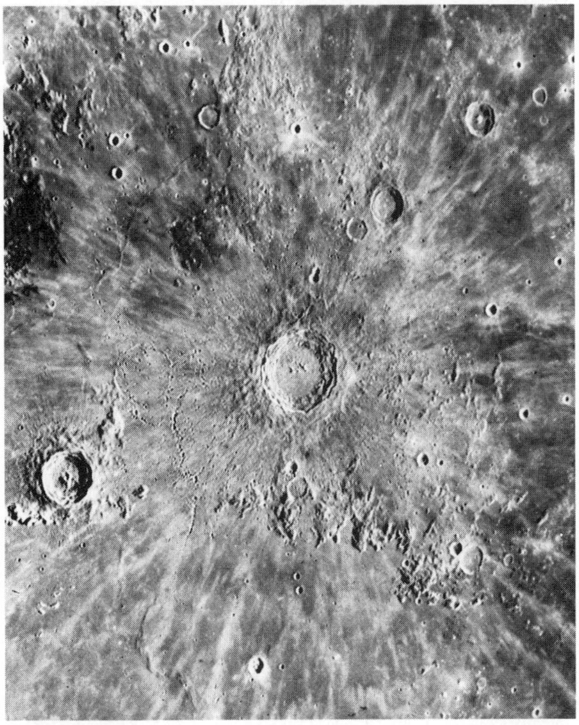

Figure 20–21 The crater Copernicus, as seen from earth-based telescope, has rays of lighter colored material emanating from it. This light material was ejected radially when the meteorite that formed the crater struck the moon's surface. (From Pasachoff, J.M., 1989. *Contemporary Astronomy,* 4th ed., Philadelphia, Saunders College Publishing.)

The samples brought to earth from the moon were all obtained from the blanket of rock fragments and dust that covers the moon's surface. Within this loose material, called **regolith**, it is possible to find fragments of rocks of all sizes, ranging from microscopic grains to huge boulders. Some of the boulders and cobbles consist of aggregates of smaller rock fragments welded together to form lunar breccias.

At Tranquillity Base, site of the first moon landing (1969), the regolith was estimated to be between 3 and 4 meters thick. It is believed to have formed as a result of meteorite impacts, each of which would dislodge a mass of debris many hundred times greater than its own mass.

Radiometric dating of lunar rocks has been of great importance in an understanding of the sequence of events in the moon's history. Most mare basalts are between 3.2 and 3.8 billion years old and were extruded during the great lava floods that produced the mare basins (Fig. 20–24). For the most part, the lunar breccias were formed during impact events that occurred prior to these floodings. The oldest rocks on the moon were taken from the lunar highlands, and they have been found to be about 4.6 billion years old. This age represents the time that meteoritic impacts formed the rugged highland terrain and crushed and melted the rocks.

The lunar highlands are remnants of ancient crust originating more than 4 billion years ago.

Figure 20–22 Moon rocks. At the top is a photograph of a lunar basalt from the Sea of Tranquillity. This mare basalt was collected by Apollo 15 astronauts. The lava solidified so rapidly that bubbles formed by escaping gases were trapped and formed vesicles. At the center is a nonvesicular lunar basalt, and the bottom is a photomicrograph of a thin section of lunar basalt composed of interlocking grains of augite (larger crystals), plagioclase (the zoned crystals), and olivine (center cluster of small crystals). The cubes are 2.5 cm square and indicate size. The area of the thin section is 3.7 mm. (Courtesy of NASA.)

Figure 20–23 Lunar rock from the lunar highlands. This coarsely crystalline rock is composed of approxi mately equal amounts of olivine (yellow grains) and plagioclase. (Courtesy of NASA.)

Samples of moon rocks now being studied in the United States have been found to retain weak magnetism imparted to the rocks at the time they solidified. This indicates the existence of a once liquid but now solidified core and that the moon had a significant magnetic field between about 4.2 and 3.2 billion years ago. The moon has no general magnetic field today.

Lunar Processes

In Chapter 11, we described earth processes capable of raising mountain ranges, as well as such opposing processes as erosion and weathering that are constantly at work wearing away the lands. Geologic processes also operate on the moon, but

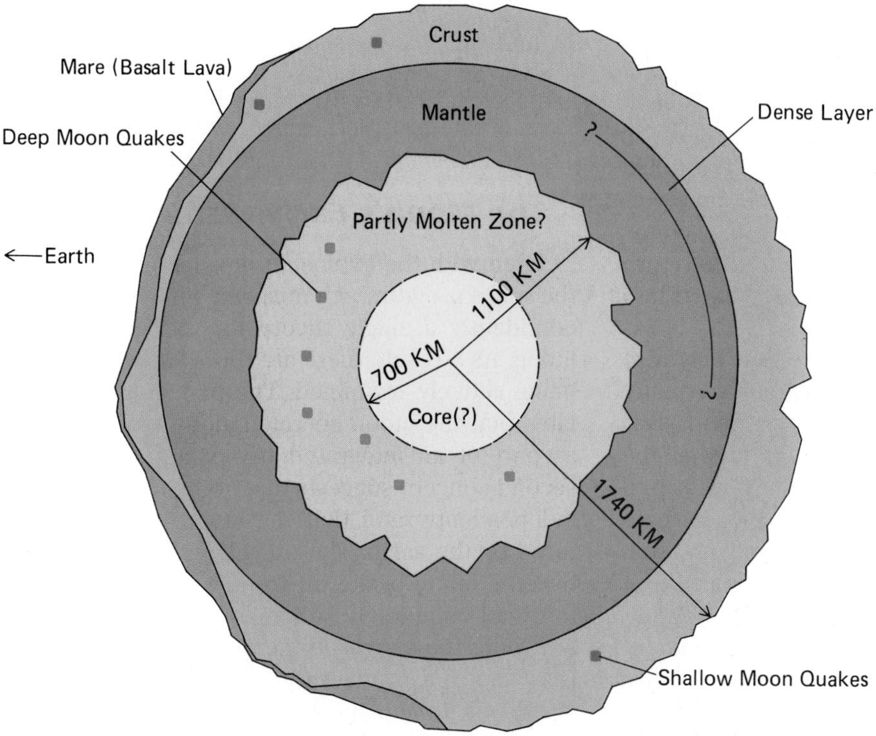

Figure 20–24 Cross-section through the moon showing an outer crust, an inner mantle, and an innermost zone that may still be partly molten. Within the innermost zone there may be a small metallic core (dashed circle.) The earth-facing side is relatively smooth and shows larger accumulations of the mare basin lavas. The far side is more rugged and has almost no lava. The drawing is not to scale, and the ruggedness of the far side is exaggerated. Data for the drawing were provided by A.M. Dainty, N.R. Goins, and M.N. Toksoz. (Courtesy of NASA. From French, B.M., *What's New on the Moon?)*

they are slower and of a different kind. The moon is a desolate world in which there is no air or surface water. Whenever gases or liquids escape to the lunar surface, they are dissipated into space because the moon lacks sufficient gravitational attraction to hold them. There can be no stream-cut canyons and valleys, no glacial deposits, and no accumulations of wave-sorted sand. Rocks brought back to earth from the moon seem bright and newly formed and lack the discoloration common to weathered earth rocks. Yet erosion does occur on the moon, primarily through the continuous bombardment of the lunar surface by large and small meteorites. The larger meteorites, of course, produce craters and rough terrain, but it is the rain of small meteorites and **micrometeorites** (those with diameters of less than 1 mm) that subtly alter the moon's surface and are important in the formation of the lunar soil. The upper or exposed surfaces of lunar rocks brought back to earth from the moon were riddled with tiny glass-lined pits produced by the impact of fast-moving micrometeorites. Geologists call these tiny depressions microcraters or **zap pits**. Most are less than .01 mm in diameter (Fig. 20–25). The micrometeorite barrage continues today, as was evident on examination of the Apollo spacecraft after its return to earth. Ten tiny microcraters were chipped into its windows.

Temperature variations on the moon may also contribute to the production of loose surface materials. Lunar day and night are each about two weeks long. During daytime, temperatures at the surface rise to 134°C, whereas at night they drop to −170°C. Although not yet proven, the alternate expansion with heating and contraction with cooling weakens grain boundaries and could cause fracturing.

Although gravity on the moon is only ⅙ that on earth, it is nevertheless capable of moving lunar materials from high to low places. Whenever loose material on slopes is unable to resist the pull of gravity, it will break away and produce masses of slumped debris. Steplike slump masses are recognized as a common feature along the rims of some lunar craters. Elsewhere, photographs reveal tracks in lunar soils that record the rolling and sliding of large boulders down slopes.

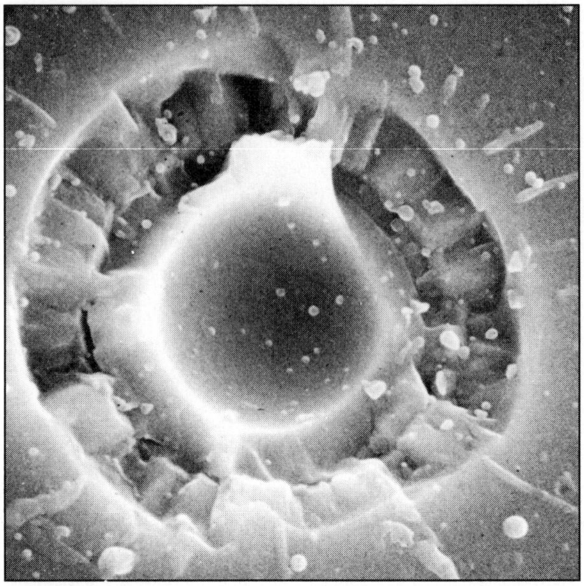

Figure 20–25 Photograph of a *zap pit* or microcrater. The tiny pit is only 20 μ (20/1000 mm) in diameter and was photographed with a scanning electron microscope. (Photo by F. Horz, courtesy of G.J. Taylor.)

The Moon's Early History

Even with the explosion of scientific data from the space program, scientists are unable to support confidently a single theory for the origin of the moon. As a result, there are three hypotheses now being actively examined. The first of these stipulates that the moon accreted along with the earth as part of an integrated two-planet system. The second concept suggests that both bodies accreted independently and that the moon was then captured by the earth. The third hypothesis proposes that the moon broke off from the earth before either had completely solidified.

If both the earth and the moon formed together from the same part of the solar nebula, then they should have similar composition (as indicated in part by specific gravity). However, the average specific gravity of the earth is about 5.5, whereas that of the moon is only 3.3. This difference, together with the absence of a magnetic field on the moon, indicates that it lacks a large iron core. These considerations have caused skepticism about the double-planet concept for the origin of the moon.

The compositional difficulties seem to be resolved in the second hypothesis, in which the moon is considered to have formed in a distant part of the solar system but was subsequently captured when it approached the earth's gravitational field. Unfortunately, there are some severe problems relating to the mechanics of such a capture, and long-term measurements of the moon's orbit required for verification of the concept are presently lacking.

In 1898, George Darwin, the son of Charles Darwin, speculated that the moon was composed of a large mass of material that spun off the earth at an early time when its rotation was much faster than now. The earth's centrifugal effect, together with the sun's tide-raising ability, first caused a huge tidal bulge on the earth. The bulge then separated to form the moon. When the moon's composition was found to be much like that of the earth's mantle, scientists speculated that the development of the earth had first proceeded to the point at which the earth's core of iron and other heavy elements had formed. After these heavy elements (which are substantially depleted in the moon's composition) had settled inward, a great bulge of mantle material formed, then pinched off, and began to orbit in widening circles around the earth. This interesting theory will continue to be examined over the next several years. It cannot be accepted as yet, largely because of calculations indicating that if the moon spun off the mantle, it would subsequently be drawn back again. Geophysicists are also uncertain that there would be sufficient force available to lift the mass of the moon from the earth and throw it into orbit.

As some scientists continue to debate the possibility that the moon was derived from the earth's mantle, a large number of investigators have come to favor the hypothesis that the moon, along with the other solar system bodies, was created about 4.6 billion years ago by the accumulation of matter and smaller bodies from the original dust cloud. Theoretically, the infall of small objects released so much energy that the outer part of the moon became molten to a depth of several hundred km. At this time, a melt of lighter minerals (such as feldspars) floated to the surface of this vast ocean of molten rock and formed a crust. Heavier minerals like pyroxene, olivine, and ilmenite settled to form an underlying layer about three times thicker than the crust. Then, about 200 or 300 million years later, great quantities of molten rock rose through fractures in the crust and hardened on the surface.

Next, the moon's primordial surface was subjected to a meteoritic barrage that lasted nearly 1.5 billion years and that produced the scarred and cratered landscapes of the lunar highlands. The internal heat generated by the steady decay of such radioactive minerals as uranium and thorium began to melt portions of the moon again, especially at rather shallow depths of 100 to 250 km. The results of this melting event were massive extrusions of basaltic lava that spread over lower parts of the lunar surface during the period from about 3.8 to 3.1 billion years ago. The moon was much quieter after the last of these eruptions. Meteorite impacts continued to sculpture its surface, sometimes forming spectacular craters like Copernicus (see Fig. 20–21).

The Earth's Neighboring Inner Planets

Mercury

Mercury has the distinction of being the smallest and swiftest of the inner planets. It revolves rapidly, making a complete journey around the sun every 88 earth days. Mercury makes a complete rotation on its axis every 59.65 days, which means that the planet rotates three times while encircling the sun twice.

Mariner 10, launched in 1973, provided our first close look at Mercury (Fig. 20–26). Following its launching, Mariner 10 flew by Venus and, using the gravity of that planet to deflect its trajectory, went on to Mercury. Indeed, the spacecraft's orbit permitted visits over Mercury on three occasions. Photographs transmitted back to earth revealed a planet with a moonlike surface. Densely cratered terrains as well as smooth areas resembling maria were quite apparent. Some of the craters exhibited rays of light-colored impact debris, just as do craters on the moon. Long linear scarps were discerned, and these are thought to be fracture zones

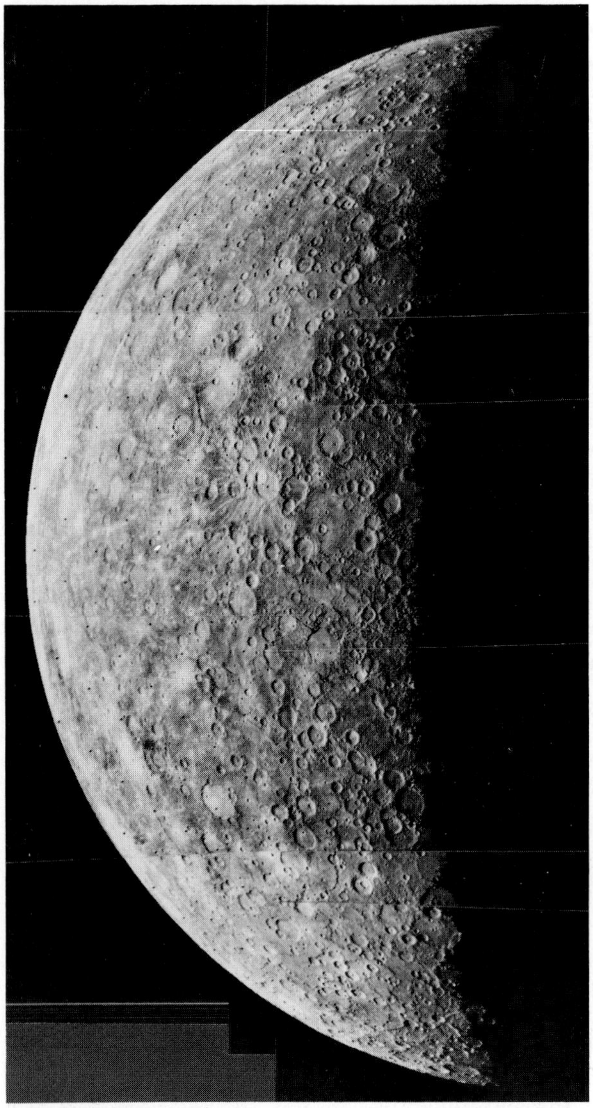

Figure 20–26 Photomosaic of Mercury as photographed by the Mariner spacecraft. (Courtesy of NASA.)

along which crustal adjustments occurred. Perhaps tidal forces, or shrinkage of the planet as it cooled, may have produced these elongate scarps.

Mercury has a weak magnetic field, a lightweight crust covered with fine dust, and an iron core that comprises nearly 75 per cent of its radius. The existence of the iron core is inferred from Mercury's high density, its magnetic field, and other evidence suggesting that the planet experienced melting and differentiation.

Mercury, like our moon, has a heavily cratered surface and lacks an atmosphere.

Venus

Although Venus is similar to the earth in size and shape, its surface conditions are quite different from those on earth. Indeed, our planet is a paradise compared with Venus. Maximal surface temperature on our nearest planetary neighbor is a searing 475°C. The planet is shrouded in a thick layer of clouds that extend 40 km above the surface (Fig. 20–27). Carbon dioxide is the principal gas in the atmosphere of Venus. The large amount of carbon dioxide serves as an insulating blanket in trapping solar energy by the greenhouse effect (the heating of an atmosphere by the absorption of infrared energy remitted by the planet as it receives light energy in the visible range from the sun). Surface atmospheric pressure on Venus is 90 times greater than it is on earth. The yellowish appearance of the planet as seen from the earth is the

Figure 20–27 The clouds of Venus as photographed in ultraviolet light by Pioneer Venus spacecraft. In visible light no structure at all can be discerned (Courtesy of NASA.)

result of droplets of sulfuric acid in the planet's clouds.

Venus rotates through one complete turn every 243 days. Its direction of rotation is retrograde with respect to the stars, that is, opposite to the paths of revolution of the planets around the sun. Its period of revolution around the sun is 225 days. From its size and density, planetary scientists infer an interior rather like that of the earth. Unlike the earth, however, Venus has no oceans. It is too close to the sun for water vapor to condense and form oceans. Without oceans in which carbon dioxide can be combined with calcium and magnesium to form carbonate rocks, there is no apparent mechanism to remove carbon dioxide from the atmosphere. The gas persists and is held in the atmosphere by gravity.

Because of the dense cloud cover, direct photography of Venus' hard surface from orbiting spacecraft is not possible. The American Pioneer Venus Orbiter, however, was equipped with cloud-piercing ground-scanning radar and with this device was able to map the entire surface of the planet. These maps reveal that over 60 per cent of Venus consists of gently rolling hills. Over most of this terrain, elevation varies only about 1000 meters between high and low points. There are, however, far more dramatic topographic features on Venus. Mountains as high as Mount Everest have been identified, as well as spectacular rift (fault) valleys. One of these valleys extends for a distance of over 1400 km, and is 250 km wide and nearly 5 km deep. Another striking topographic feature is a magnificent plateau (named Ishtar Terra after the Assyrian goddess of love and war), which is higher than the earth's Tibetan Plateau and extends across an area as large as the U.S. (Fig. 20–28).

In August of 1990, the U.S. spacecraft Magellan will be in orbit around Venus. Because of its highly sophisticated radar system, it will transmit images of Venusian landscapes unprecedented in their detail and clarity.

The first close look at the ground surface of Venus was provided in 1975, when the Russian robot spacecraft Venera 9 and Venera 10 succeeded in landing on the planet. In spite of the crushing atmospheric pressure and scorching temperatures (sufficient to melt tin), these spacecraft survived for about an hour. Even more successful, however, were the landings of Venera 13 and 14 in March of 1982. Like their predecessors, the two robots separated from their mother ships and drifted under parachute through the corrosive clouds to touch down on a mountainous region called Phoebe just below the Venusian equator. Venera 13 survived 2 full hours, whereas its sister craft stopped functioning in less than an hour. The electronic eyes on both landers began working immediately and transmitted to earth remarkable photographs of a landscape strewn with rust-

Figure 20–28 An artist's conception of Ishtar Terra, a magnificent plateau about the size of the United States in area and higher than the Tibetan plateau. The reconstruction is based on measurements taken by the Pioneer Venus Orbiter spacecraft. (Courtesy of NASA.)

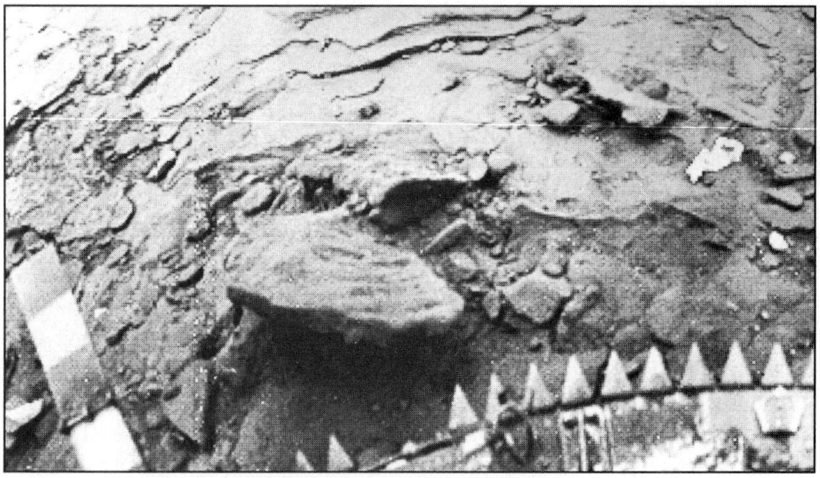

Figure 20–29 A view of the surface of Venus taken by Venera 14. The rocks are interpreted to be flow-layered basalts. The photograph was given by the U.S.S.R. to scientists at the U.S. Geological Survey in appreciation of help in choosing the landing sites for Venera 13 and 14.

colored slabby rocks and rubble (Fig. 20–29). In addition to accomplishing their photographic mission, the landers were able to drill out a few cm of Venusian soil and determine its composition. The analyses indicated the samples were basalt. Indeed, they had an uncanny resemblance to the basalts extruded from fissures along mid-oceanic ridges on earth. Many scientists believe that this discovery, along with the observation of great rift valleys, indicates that Venus, like the earth, is still a very dynamic planet with internal and surficial changes occurring constantly.

Venus is about the same size and density as the earth. Its atmosphere is largely carbon dioxide, and its surface temperature is about 475°C.

Mars

General Characteristics

The planet Mars (Fig. 20–30) has always been of special interest to humans because of speculation that some form of life, however humble and microscopic, may exist or at one time may have existed there. It is a planet that is only a little farther from the sun than is our own; it has an atmosphere that includes clouds; it has developed white polar caps; and it has seasons and a richly varied landscape. That landscape (Fig. 20–31) has been splendidly revealed to us by the Mariner and Viking missions. It includes magnificent craters, colossal volcanic peaks, deep gorges, sinuous channels, and an extensive fracture system. Evidence of wind erosion is clearly seen on photographs of the Martian surface, and there are indications of the work of ice as

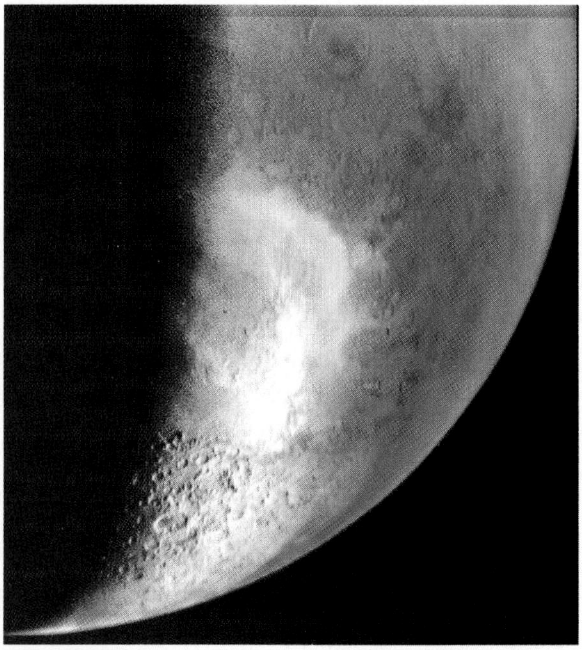

Figure 20–30 Planet Mars as photographed from the approaching Viking spacecraft in 1976. (Courtesy of NASA.)

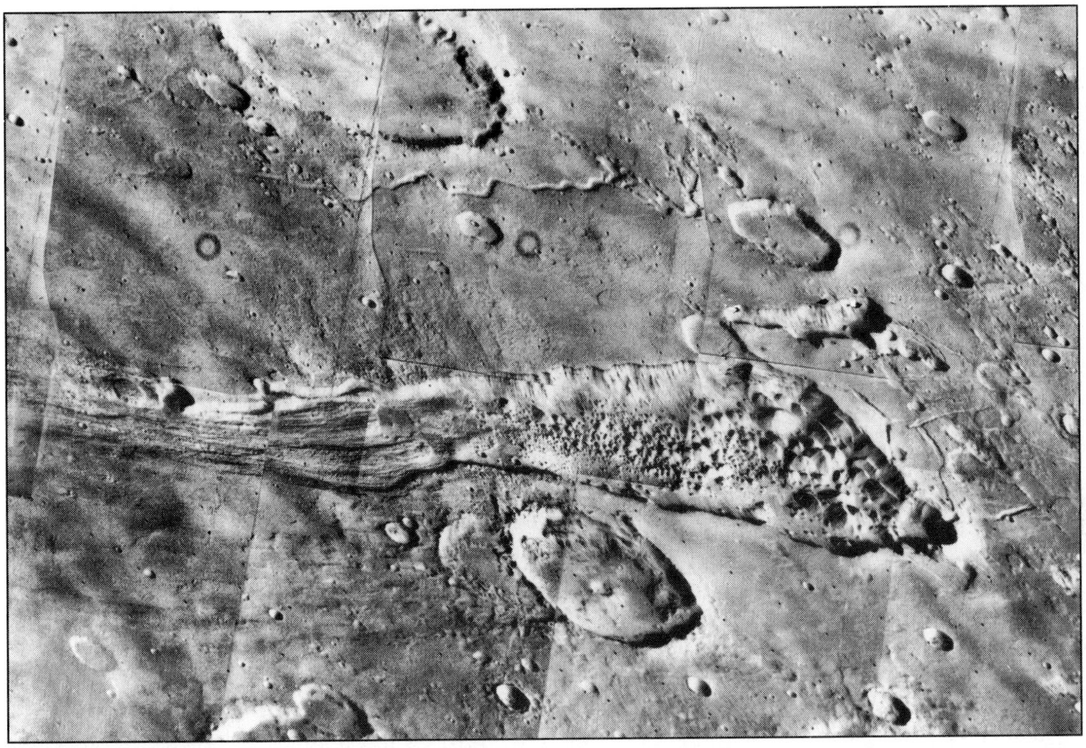

Figure 20–31 Martian surface as revealed in a mosaic of photographs taken by Viking I on July 3, 1976. The large valley in the center was probably caused by downfaulting, possibly in association with the melting of subsurface ice. The hummocky topography on the valley floor may have also resulted from this process. (Courtesy of NASA.)

well. The planet has a diameter of only about 50 per cent of that of the earth and has only 10 per cent of the mass of the earth.

A year on Mars is 687 earth days long, and the planet's axial tilt is similar to that of the earth. Carbon dioxide is the principal gas in the Martian atmosphere, although small amounts of nitrogen, oxygen, and carbon monoxide are also present. The atmosphere is nearly 150 times thinner than the earth's atmosphere. The planet has virtually no greenhouse effect, and temperatures vary at the equator from about +21° to −85°C.

Mars has two moons. Their names are *Phobos* Fig. 20–32), meaning "fear," and *Deimos*, meaning "panic." They are large chunks of rock about 27 and 15 km across, respectively.

The somewhat lower density of Mars, compared with that of other inner planets, and its lack of a detectable magnetic field suggest that its iron may be largely scattered throughout the planet. This interpretation is strengthened by the color photographs transmitted from the Viking landers. Those photographs show the rusty red and orange hues typically developed on earth by limonite and hematite. Indeed, the sky often takes on a pinkish hue, probably because of the rust-colored dust. Apparently, iron at the Martian surface has reacted with water vapor and oxygen to provide the color that so intrigued early planetary observers.

Martian Landscape Features

As a result of the work accomplished by the Mariner and Viking orbiters, nearly the entire surface of Mars had been photographed by 1976. Crisp high-resolution photographs revealed craters, dunes, volcanoes, and canyons in unsurpassed clarity. On a grand scale, the planet's physiography

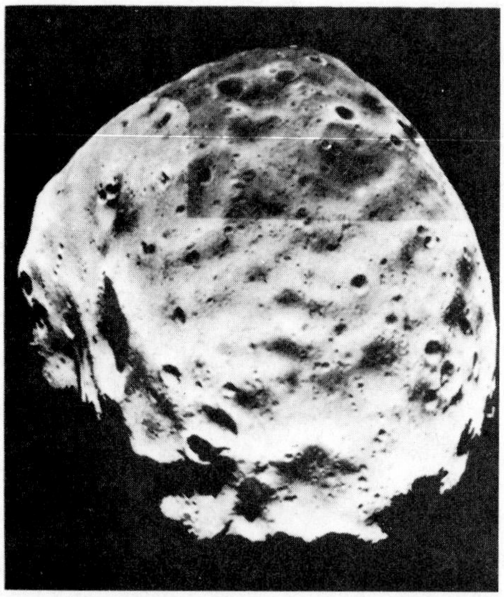

Figure 20–32 Photomosaic of Mars' inner satellite, Phobos. The photographs incorporated into this mosaic were obtained when Viking Orbiter 1 approached to within 480 km of Phobos in 1977. Phobos is about 21 km across and 19 km from top to bottom. North is at the top. (Courtesy of NASA.)

is divisible into two rather different halves. The southern hemisphere is mostly heavily cratered rough terrain. This surface is inferred to be older than the smoother only lightly cratered surface of the northern hemisphere.

The Viking spacecraft that reached Mars in the summer of 1976 each contained two parts, namely, a lander and an orbiter. The landers, of course, provided scientists with the first ground view of Mars. The photographs transmitted from the landers to the orbiters and thence to the earth show a sandy-looking terrain littered with cobbles and boulders (Fig. 20–33). Here and there, undisturbed bedrock was visible. Many of the rocks strewn about were clearly of volcanic origin. Some had spherical holes made by gas bubbles, whereas others were finely crystalline like many igneous rocks on earth. The devices on the lander that analyzed the soil indicated that the fine loose material was rich in iron, magnesium, and calcium. Very likely, it was derived from parent rocks rich in ferromagnesian minerals.

Craters and Dunes

The density of craters in the southern half of Mars is similar to the crater density in the lunar highlands. Most of these craters were probably excavated between 4 and 4.5 billion years ago by the same torrential barrage of meteorites that pelted the lunar highlands.

Photographs clearly show row upon row of dunes on the floors of some of the larger craters. If the thinness of the Martian atmosphere is taken into account, then the size and spacing of these dunes can be used to estimate the velocity of the winds that formed them. Calculations indicate that the dune-forming winds blew at speeds of up to 100 meters per second.

Dust storms on Mars are prodigious events. As Mariner 9 began its encirclement of the planet, a dust storm was in progress, and useful photographs of the Martian surface could not be transmitted until the dust subsided. Viking orbiters encountered similar problems. The dust storms are not driven by excessively violent winds, however. One of the Viking landers recorded a gust of 120 km/hour, but average wind velocities were much lower. Dust on Mars is sometimes swirled upward into towering tornado-like columns that, on a smaller scale, we would call "dust devils." Unlike those on earth, Martian dust devils loom a spectacular 6 km above the parched terrain.

Volcanic Features

Although most of the craters on Mars have been formed by the impact of meteorites, a smaller number are certainly volcanic. Easily the most spectacular of the volcanic craters is Olympus Mons (see Fig. 4–44). This gigantic feature is more than 500 km across at its base and rises 30 km above the surrounding terrain. It is equal in volume to the total mass of all the lava extruded in the entire Hawaiian Island chain. Around its summit are lava flows and the narrow rilles interpreted as old lava channels or possibly collapsed lava tubes. Flood lavas are also evident, not only in the smoother northern hemisphere but also among the craters of the rugged southern half of Mars.

Channels and Canyons

There is so very little water today in the Martian atmosphere that if all of it were to be condensed in one place, it would probably provide only enough water to fill a community swimming pool. Yet the surface of the planet is dissected with channels and canyons that grow wider and deeper in the downslope direction. Some of these channels have braided patterns, sinuous form, and tributary branches such as are characteristically developed by streams on earth (Fig. 20–34).

These features suggest that flowing water was an important geologic agent at some earlier time in Martian history. There may once have been a considerable reserve of water frozen in the subsurface as a kind of permafrost. Permafrost is ground that remains below the freezing temperature and contains ice throughout the year. Many investigators theorize that the larger channels, some of which are over 5 km wide and 500 km long, were excavated by water gushing out onto the surface as the subsurface layer melted. Large tongues of debris closely resembling mudflows may also have received their water from such a process. Many channels on Mars dwarf our own Grand Canyon in size and, in order to form, would have required torrential floods so spectacular as to be hard to visualize by earth standards.

Figure 20–33 Viking Lander photograph of the surface of Mars. The device on the left is a part of the Lander that was purposely discarded. For scale, it is about 20 cm long. Note the two trenches made by the lander's arm in taking samples of Martian soil. This photograph was created by combining low-resolution color images with high-resolution black-and-white images so as to show the Martian surface as it would appear on earth. The computer color balancing was done by M.A. Dale-Bannister. (Courtesy of NASA.)

Ice Caps

Long before spacecraft landed on Mars, observers using telescopes noticed the planet had ice caps that grew and shrank with the seasons. Initially, scientists inferred that the ice was frozen carbon dioxide (rather like commercial dry ice). Data transmitted by Viking orbiters, however, indicated that the residual northern ice cap was actually made of water ice (although the seasonal ice that forms in the fall and dissipates in the spring is confirmed as carbon dioxide).

Mars has a variety of landscapes, including great canyons, sinuous valleys, lava flows, polar snow fields, dunes, and craters.

Martian History

Like the earth, Mars was transformed from a protoplanet by accretion of a multitude of smaller objects. This accretion was followed by an episode of differentiation during which the once homogeneous body became partitioned into masses of different chemical composition and physical properties. Very likely, the differentiation process involved melting, so that fluids could move from one region of the planet to another.

During the initial billion years or so of Martian history, the planet experienced heavy bombardment of meteorites and asteroids, with the result that the exposed crust became densely cratered. The craters have been modified by relatively recent wind erosion. The crust was disturbed, not

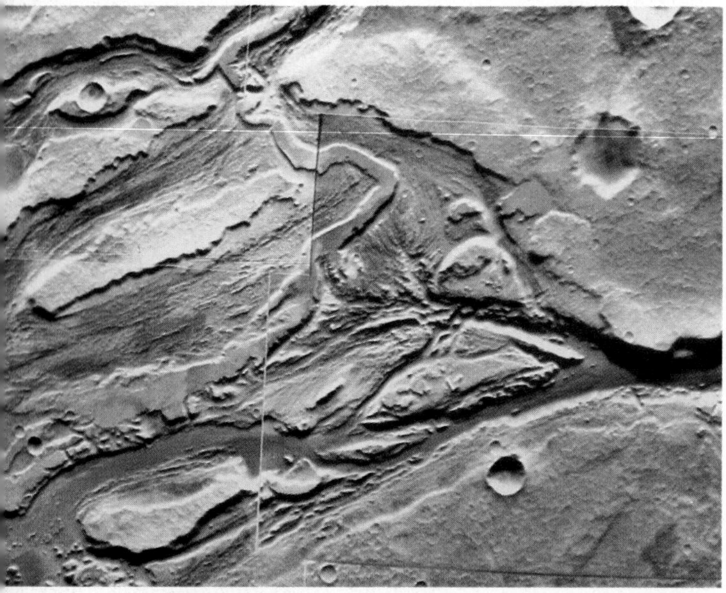

Figure 20–34 A large channel system named Mangala Vallis on Mars. The system, which now resembles dry river beds on earth, flowed into the lower plains of the northern Martian hemisphere. Note the streamlined islands and intricate connections between channels. (Courtesy of NASA.)

only by cratering, but also by the faulting and fracturing that accompanied volume changes in the mantle.

It is likely that a dense atmosphere formed on Mars as a result of volcanic outgassing (removal of embedded gas by heating) and that the production of water and carbon dioxide far exceeded its loss into space. The planet was warmer at that time, perhaps because of a greenhouse effect. For a time, running water was an important geologic agent, as evidenced by the sinuous furrows cut into the cratered terrain. In certain regions of the planet, a small decline in surface temperatures may have caused the water contained in surface materials to freeze, thus trapping still liquid subsurface water in pockets and channels. With local melting, the subsurface water may have gushed upward and eroded large channels while simultaneously causing entire areas of surficial material to collapse in mud-flows.

Eventually, the supply of atmospheric water and carbon dioxide became so depleted that the planet began to cool. Remaining carbon dioxide and water migrated to polar caps and to subsurface reservoirs, where it has remained.

Because Mars was once warmer, had more water, and possessed the elemental raw materials thought to be required for the development of organisms, scientists have long speculated on the possibility of finding primitive forms of life on the planet. For this reason, a series of experiments aboard the Viking lander was designed to search for signs of present or former life. The results were disappointing. Not a single microorganism or even a trace of organic molecules could be detected in the Martian soil at the landing site. However, it is still too early to rule out the possibility that continuing research and exploration of other sites in the future may provide evidence of organisms.

The Outer Planets

Beyond the orbit of Mars lies the ring of asteroids, and beyond the asteroids are the orbits of Jupiter, Saturn, Uranus, Neptune, and Pluto. The first four of these planets are also called the Jovian (for Jupiter) planets. They are similar to one another in having large size and relatively low densities. Pluto is not a Jovian planet. It is so far away and small that it is difficult to observe. As a result, there is considerable uncertainty about its origin.

Jupiter, Saturn, Uranus, and Neptune are the giant outer planets of our solar system. Because of their great size and mass, they have sufficient gravity to retain an atmosphere of light gases, including hydrogen, helium, and compounds of hydrogen such as methane and ammonia.

Jupiter

Jupiter, named for the leader of the ancient Roman gods, is the largest planet in our solar system. The diameter of this huge planet is 11 times greater than the earth's diameter, and its volume exceeds that of all the other planets combined. Its mass is 318 times that of the earth, yet it is only ¼ as dense. Jupiter spins rapidly on its axis, making one full rotation in slightly less than 10 earth hours.

This rapid rotation results in the formation of the encircling colored atmospheric bands for which the planet is famous (Fig. 20–35). Hydrogen, helium, and lesser quantities of ammonia and methane are the predominant gases in Jupiter's atmosphere. On the basis of experiments with these gases, some investigators believe that the reddish and orange colors in the atmosphere are derived from sulphur or phosphorus-based compounds. It is impossible to discern the surface of the solid interior of Jupiter with optical telescopes. The atmosphere, which is several hundred km thick, passes gradually to liquid and eventually to solid matter. The interior of the giant planet may consist of highly compressed hydrogen, possibly surrounding a rocky core. Jupiter has at least 16 satellites. The large planet with its many moons rather resembles a miniature solar system.

We have learned about Jupiter not only from earth-based telescopes but from close encounters by unmanned spacecraft. The most successful of the robot explorations occurred in 1979 when the two Voyager spacecraft swept past Jupiter and transmitted splendid images of the planet's brightly colored atmosphere, the tempestuous Great Red Spot (see Fig. 20–35), five of Jupiter's moons (Fig. 20–36), and a ring of debris less than 0.6 km thick that circles the planet about 55,000 km above the tops of the clouds. The discovery of the ring makes Jupiter the third planet in the solar system known to possess such a feature (Saturn and Uranus are the other planets known to have rings).

Jupiter's Great Red Spot, which is twice the size of the earth, is an intriguing feature known to astronomers for centuries. As clearly revealed by the Voyager mission, this many-hued disturbance is a gargantuan atmospheric storm that extends deep into the cloud cover and turns in a complete counterclockwise revolution every six earth days.

In addition to the Great Red Spot, devices aboard the Voyager spacecraft radioed back the distinct images of great bolts of lightning, whose occurrence had only been suspected before the flyby. Another discovery was an auroral display far brighter than any northern lights ever seen on earth. Voyager also confirmed the presence of an immense magnetic field surrounding Jupiter (Fig. 20–37). First detected by the Pioneer 10 spacecraft in 1973, the magnetic field extends out-

A

B

Figure 20–35 (A) Jupiter, photographed by Voyager 1 spacecraft from a distance of 33,000,000 km. (B) A close-up of Jupiter's famous Great Red Spot. Note the turbulence to the west (left) of the Great Red Spot. (Courtesy of NASA.)

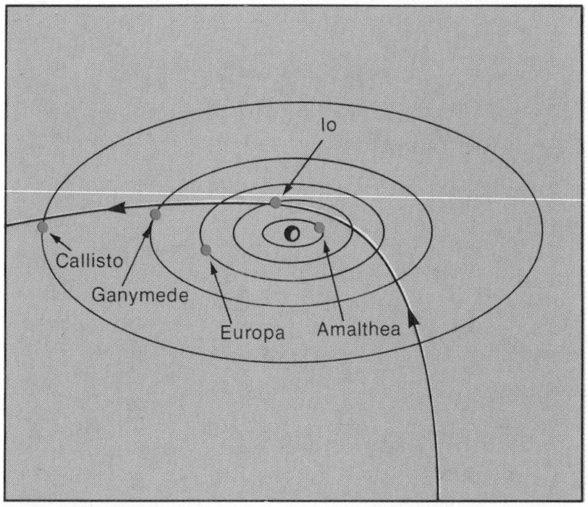

Figure 20–36 Voyager 1's flight path allowed scientists to obtain close-range photographs of five of Jupiter's 13 satellites. Each is shown at its closest point to Voyager 1's outbound flight away from Jupiter. The spacecraft approached within 280,000 km of Jupiter. (Courtesy of NASA.)

ward from the planet for 16 million km, and encompasses all of the larger satellites. The presence of the field indicates that Jupiter has a magnetic core resulting from fluid movements within its spinning interior.

The two Voyager spacecraft were not launched at the same time. Voyager 2 reached Jupiter four months after its companion spacecraft, and was able, therefore, to fill in many of the gaps in information. Voyager 2 also provided even better images of Jupiter's larger moons than had been transmitted by its sister ship. These larger satellites are called the Galilean moons to commemorate their discoverer. Individually, they are named *Callisto*, *Ganymede*, *Europa*, and *Io*. Each has characteristics quite different from those of its neighbors. Callisto (Fig. 20–38), the outermost of the Gallilean moons, is riddled with craters resulting from meteorite impact. On Callisto's icy dirt-laden surface, Voyager cameras detected huge concentric rings

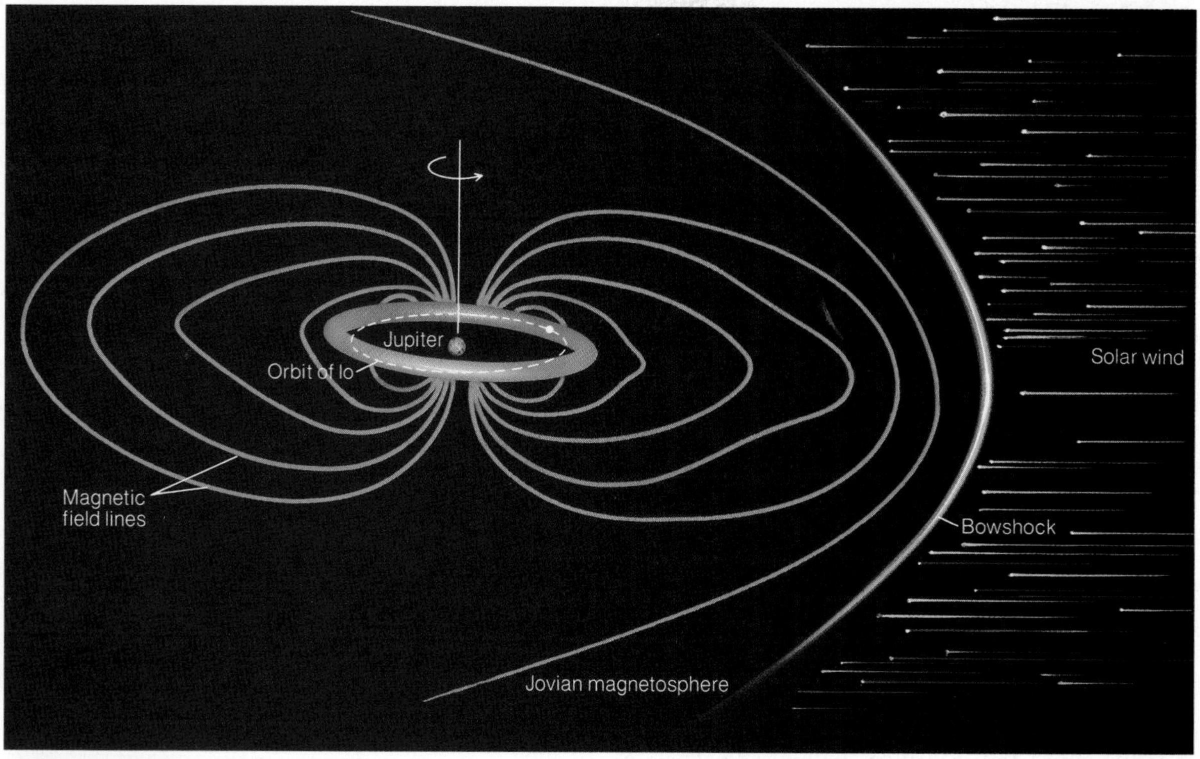

Figure 20–37 The magnetosphere of Jupiter as revealed by the Pioneer and Voyager spacecraft explorations. The blue zone consists of hot ionized gases (plasma). (From Abell, G.O., Morrison, D., and Wolff, S.C., 1988. *Realm of the Universe,* Philadelphia, Saunders College Publishing.)

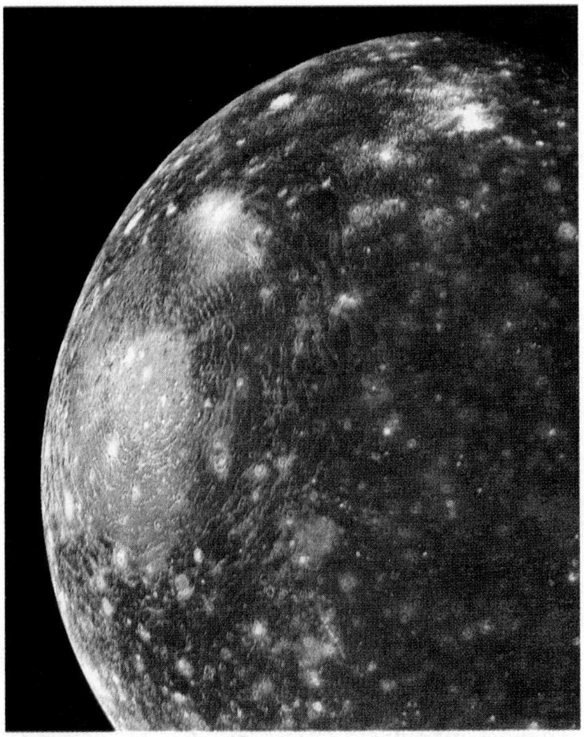

Figure 20–38 Callisto, the outermost of the Galilean moons, and the most heavily cratered body in our solar system. (Courtesy of NASA.)

Figure 20–39 Ganymede, Jupiter's largest satellite, as photographed by Voyager 1 from 216 million km. (Courtesy of NASA.)

that outline extraordinarily broad impact basins. Callisto's nearest neighbor is Ganymede (Fig. 20–39), the largest of Jupiter's satellites. Ganymede also shows the effects of meteorite bombardment in its dark cratered terrain and individual craters that have bright streaks or rays that fan out from the crater rim like huge splash marks. Sinuous ridges have been detected on Ganymede's surface, as well as criss-crossing fractures suggesting that fault movements have occurred on the satellite.

Europa (Fig. 20–40) is the brightest of the Galilean moons. It appears to have a thin surface crust of ice that may rest on somewhat softer ice or water. The surface of Europa is marked by huge fractures and ridges, but few impact craters have been detected. This suggests that Europa's surface is being continuously renewed, perhaps by the development and slow migration of glacierlike masses of ice.

When images of Io first reached scientists, they immediately compared the appearance of that

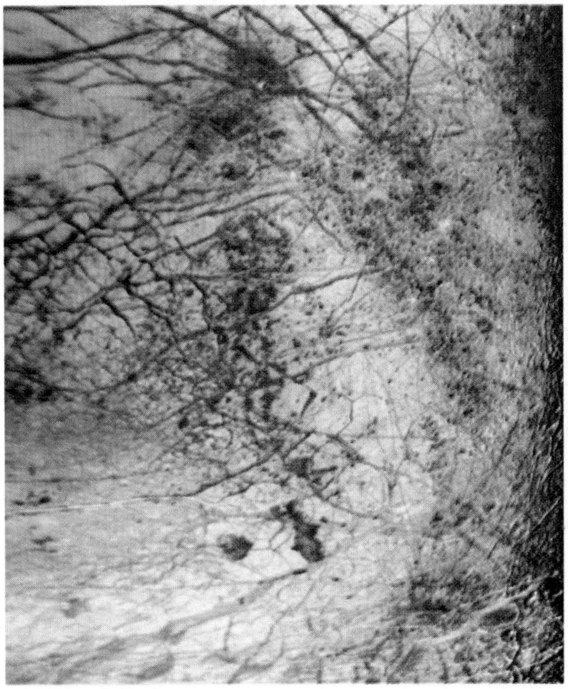

Figure 20–40 A view of the surface of Europa. The satellite's surface is composed of ice crossed by a complex system of cracks and low ridges. (Courtesy of NASA.)

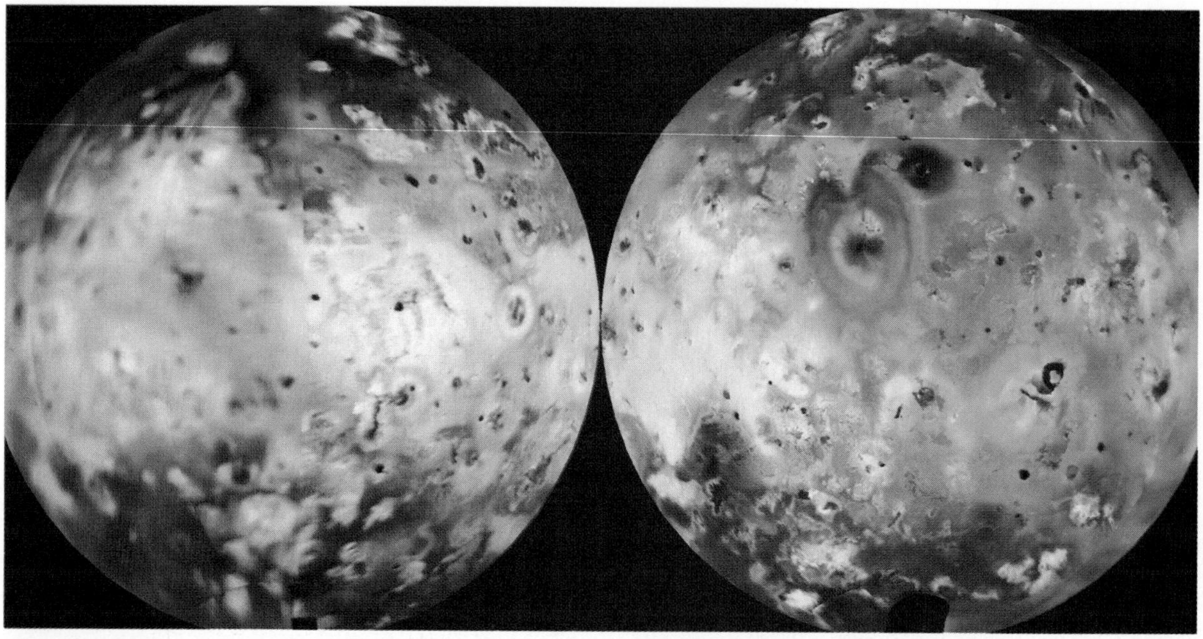

Figure 20–41 The satellite Io. The relatively smooth surface of Io and its active volcanoes suggest it may be Jupiter's youngest satellite. Large amounts of sulfur and sulfur dioxide frost are thought to be present on the surface. (Courtesy of NASA.)

moon's surface with a pizza (Fig. 20–41). The mottled yellow-orange color of the satellite is distinctive and is probably derived from the large amounts of frozen sulfur dioxide or amorphous silica in surface materials. Io's surface is thought to be young, for it is relatively smooth and scarred by only a few impact craters. Geologists were especially interested in the discovery of active volcanoes on Io (Fig. 20–42). Altogether, eight volcanoes have been detected, some with plumes extending over 300 km above the surface of the satellite. Sufficient sulfur dioxide is being emitted by the volcanoes to form a ring of ionized sulfur and oxygen atoms around Jupiter near Io's orbital path.

Voyager 2 also passed near Amalthea, Jupiter's innermost satellite. The spacecraft's camera showed that Amalthea was a reddish colored nonspherical satellite about 165 km long and 150 km wide. It is without doubt the most irregularly shaped satellite known in our solar system.

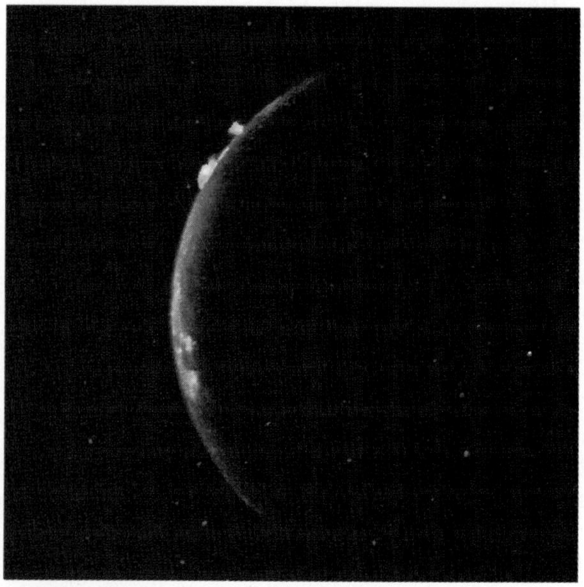

Figure 20–42 Voyager 2 photograph of Io showing two active volcanoes. (Courtesy of NASA.)

Saturn

Saturn's density is only about 70 per cent that of water. Its mass, however, is equivalent to 95 earth masses. Measurements of Saturn's size and density suggest that the planet may have a central core of heavy elements (probably mostly iron) surrounded by an outer core of hot compressed volatiles such as methane, ammonia, and water. These two core regions, however, constitute only a small part of the total volume of the planet, in which hydrogen and helium are overwhelmingly the predominant elements.

As was true for Jupiter, our knowledge of Saturn has been greatly improved as a result of data collected by unmanned spacecraft (Fig. 20–43). In particular, the electronic eyes of Voyager 2 in August of 1981 provided truly spectacular views of this second largest planet, its rings, and its moons. The data transmitted back to earth helped planetary scientists confirm that Saturn has a magnetic field, trapped radiation belts, and an internal source of heat.

The general appearance of Saturn's atmosphere is similar to that of Jupiter. The alternating light and dark bands and swirls of gas, however, are partially veiled by a haze layer above the denser clouds. Temperatures near the cloud tops range from 86°K (−305°F) to 92°K (−294°F) with the coolest temperatures occurring near the center of the equatorial zone. The planet is well illuminated, for even its dark face receives a substantial amount of light reflected from the rings. On this dark side, Voyager 2 detected radio emissions rather like those given off during lightning discharges.

The Rings of Saturn

There is no more beautiful sight in the solar system than the rings of Saturn (Fig. 20–44). With his primitive telescope, Galileo was the first to recognize these rings, but he was unable to discern them clearly. Fifty years later, in 1655, Huygens was able to differentiate three concentric rings, the brightest and broadest of which was in the center. There are, however, far more than three rings around Saturn. As a result of pictures taken by Pioneer 11, and the more recent and truly dazzling photographs taken by Voyager 1 and Voyager 2, we now know that the once neatly defined grouping of three rings is really a complex system of thousands of rings and rings within rings. Saturn's rings are composed of billions of orbiting particles of dust and ash, as well as many larger fragments measuring tens of meters in diameter. All revolve around Saturn in the approximate plane of the equator. Saturn's 17 or more moons also revolve in this plane. The entire ring system has an outside diameter of about 275,000 km. Although the area of the

Figure 20–43 Saturn and its rings as photographed by Voyager 2 from a distance of 18 million km. (Courtesy of NASA.)

Figure 20–44 Computer-enhanced photograph of Saturn's rings as photographed by Voyager 2. The exaggerated color shows the reflection and transmission of different wavelengths of light by particles of different sizes in various parts of the ring system. (Courtesy of NASA.)

rings is truly enormous, most of the material is concentrated within a thickness of only about 16 km. Thus, the entire system reminds one of a phonograph record with the rings corresponding to the record's many tiny grooves. Scientists are uncertain about the origin of Saturn's rings. Some suggest that the particles in the rings are remnants of an exploded satellite. Others argue that the rings are a remnant of the particle cloud from which satellites form.

Saturn's rings are composed of billions of separate particles orbiting in the planet's equatorial plane.

Saturn's Moons

Largely as a result of the Voyager missions, we know that Saturn ranks first among the planets in the number of its satellites (Fig. 20–45). Altogether, over 17 confirmed moons have been recognized. Some have been formally named, whereas others bear only numbers. The largest of the named satellites is *Titan*. Titan is about 5120 km in diameter and has a relatively low density about twice that of water ice. The satellite's surface is obscured by thick dense haze. Titan has a nitrogen-rich atmosphere, and lakes of liquid nitrogen may exist at the poles where surface temperatures are an estimated 90°K (−300°F). The more important

Figure 20–45 Saturn and some of its moons are shown in this montage compiled from Voyager 1 and Voyager 2 photographs. The satellites shown clockwise from upper left are Titan, Iapetus, Tethys, Mimas, Enceladus, Dione, and Rhea. (Courtesy of NASA.)

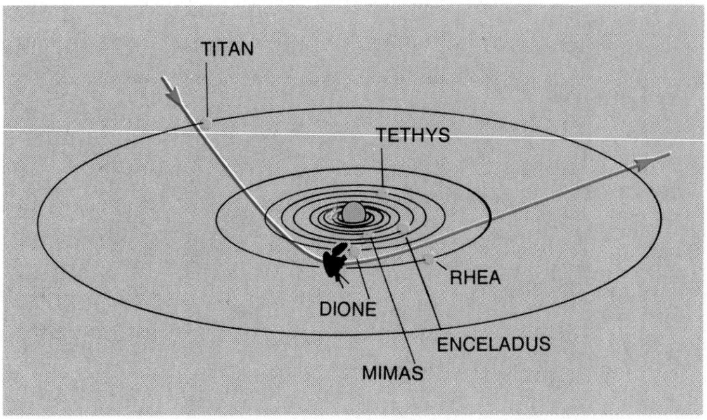

Figure 20–46 Voyager 1 approached within 124,000 km (77,000 miles) of Saturn's cloud tops. Six of the satellites that were photographed are shown in their approximate positions at closest approach by the spacecraft. (Courtesy of NASA.)

inner satellites of Saturn are named *Mimas, Enceladus, Tethys, Dione,* and *Rhea* (Fig. 20–46). Their densities and surface brightness suggest that they too are composed mainly of water ice. The five satellites range in size from the 390-km diameter of Mimas to the 1530-km diameter of Rhea. All are densely cratered. The outer satellites of Saturn include tiny *Phoebe*, potato-shaped *Hyperion*, and *Iapetus*.

Uranus

Beyond the orbit of Saturn lies the planet Uranus (Fig. 20–47). Like Saturn and Jupiter, Uranus also has rings. They have been photographed from earth with infrared film. Uranus has a low density (1.2 g/cm^3) and a frigid surface temperature of about −185°C. Five moons circle Uranus. They are named *Ariel, Miranda, Oberon, Titania,* and *Umbriel.* Thus far, earth-based instruments have only been able to detect hydrogen and methane on Uranus, although helium may also be present. The methane may be responsible for the planet's greenish hue. Perhaps the most distinctive characteristic of Uranus is the orientation of its axis of rotation, which lies nearly in the plane of its orbit. Because it is upside down, the planet's direction of rotation is the reverse of that of all other planets except Venus. This retrograde direction is also characteristic of the orbits of the five satellites around Uranus.

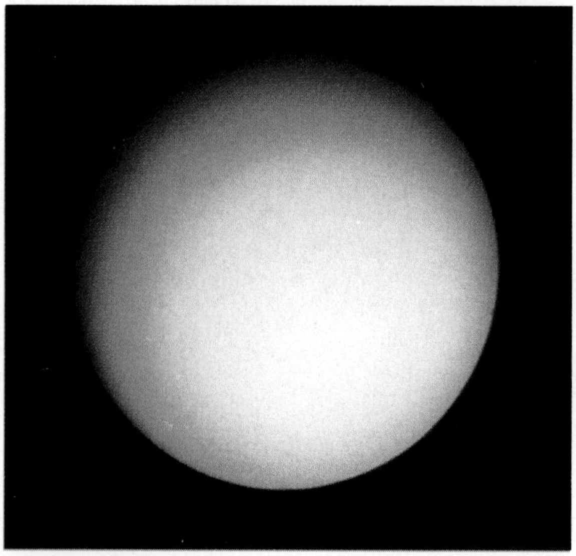

Figure 20–47 Uranus as photographed from Voyager 2 on January 17, 1986, from a distance of 5.7 million miles. (Courtesy of NASA.)

Neptune

Because they are similar in size and other physical properties, Neptune and Uranus are often called the twin planets. Through the telescope, Neptune, like Uranus, appears as a greenish orb. Neptune's density of 1.7 g/cm^3 is just slightly greater than the density of Uranus. Neptune's atmosphere may also resemble that of Uranus. Both hydrogen and helium have been detected spectroscopically, and probably methane is also present. Two moons, named *Triton* and *Nereid,* circle the planet. Triton is one of the four largest moons in the solar system (the other three being Ganymede, Titan, and Callisto). Its surface may contain ice made of frozen methane, and it may even have a thin atmosphere of methane gas. Recently, astronomers have found evidence that Neptune may have a third moon and possibly even a ring system.

Pluto

Far out on the outer limits of our solar system lies the orbit of Pluto. It is so small and so distant that less is known about it than about the other planets. We do observe that Pluto's orbit is tilted an unusual 17° to the ecliptic plane and that its orbit crosses that of Neptune. As a result, Pluto is actually closer to the sun than Neptune during certain periods (as during the current period from 1979 to 1999). Because of the high inclination of Pluto's orbit, however, there is no danger of a collision with Neptune. Its small size and orbital peculiarities suggest that Pluto may not be a true planet but rather a former satellite of Neptune that escaped the gravitational pull of that planet and that now occupies a separate orbit. Evidence for this theory, however, is not conclusive.

In 1978, scientists of the U.S. Naval Observatory found that Pluto has a moon. They named it *Charon*, after the mythologic boatman who ferried the souls of the dead across the river Stynx to Hades. Some scientists believe the existence of Pluto's moon makes the "escaped satellite" theory for Pluto's origin unlikely. Others argue that Pluto and Charon may have been ejected from the Neptunian system as a satellite-pair.

Because of its unusually small size, Pluto probably had a different origin than the four giant outer planets.

The Future of the Solar System

Anyone interested in the history of the earth is naturally led to a consideration of its future. Clearly, all that has happened, is happening, and will happen to this planet is inevitably linked to the history and destiny of the sun. The sun is a star, and like other stars, it will run its course from birth to death. By observing other stars that are at various stages in their histories, astronomers have pieced together a story of what happens when these bodies begin to deplete their store of nuclear fuel. Our star has been "burning" for about 5 billion years. In this span of time, it has consumed about half of its nuclear fuel and so has about another 5 billion years of life remaining.

What will "doomsday" for the sun and earth be like? Most astronomers believe that once the hydrogen in the sun's core has been largely converted to helium by fusion, the core will begin to contract. As the core contracts, the outer layers of the sun will expand and cool, causing the sun to puff up into an enormous sphere and to turn blood-red. Our sun will have become a "red giant." As the great ball of rarefied gases continues to expand, it will engulf and vaporize Mercury and Venus.

What will happen to the earth can be readily imagined. The searing heat will boil away the oceans and heat the crust even to the melting point of some surface rocks. This purgatorial state of affairs may continue for a billion years or so, until the sun finally dissipates its energy. The dying sun may then begin to shed its outer layers, leaving behind a dense hot "white dwarf" star about the size of the earth. Such white dwarf stars have exhausted their supply of nuclear fuels and remain incandescent only because of their residual internal heat. Eventually, even this light will fade as the sun becomes a "black dwarf," so called because it is now too cool to radiate visible light. The scorched and frazzled earth will continue as a barren frigid sphere unable to free itself from the gravitational grasp of the dead sun.

Summary

The planet earth is one of a group of nine planets revolving around a modest-sized star, our sun. The planets and sun form our solar system, which, together with myriads of other stars, planets, dust, and gas, make up the Milky Way Galaxy. Our sun provides energy to the earth, which drives geologic processes and which has ultimately determined the character of terrestrial life. The sun is composed mostly of hydrogen and helium. The helium in the core results from the fusion of hydrogen nuclei with the conversion of a part of the hydrogen into solar energy.

Theories for the origin of the solar system must conform to the known facts of the system's distribution of mass as well as the orbital and rotational characteristics of the planets. Currently, the most widely held theory is that the planets originate by condensation and accretion from a rotating nebula of gases and dust particles.

Meteorites are important to planetary geologists because they are truly extraterrestrial samples of the solar system. They can be classified as *ordinary chondrites, carbonaceous chondrites, achondrites, irons,* and *stony-irons* and are believed to be fragments resulting mostly from collisions of one or more small planetary bodies.

The moon, which has been studied by telescope for centuries, was first visited by humans in July 1969. Its general features include the relatively smooth dark *mare basins* and the rugged *lunar highlands.* Samples of the loose regolith on the moon have been brought back to earth. They include basaltic rocks from the maria and specimens of anorthosite from the highlands. Study of the age and distribution of these rocks suggests that lunar history began about 4.6 billion years ago, when the moon had developed its surficial skin of anorthositic igneous rock. This surface was then flooded in various regions by the basalts of the maria about 1 billion years later.

The planets of our solar system can be divided into two main groups. Those resembling the earth are the *inner (or terrestrial) planets.* The outer planets resemble Jupiter and are termed *Jovian planets.* The terrestrial planets tend to be small, dense, and rocky whereas the Jovian planets are large and have many satellites and dense thick atmospheres.

The terrestrial planet closest to the sun is Mercury. It is the smallest of the inner planets and is very moonlike in appearance. Next in the progression outward from the sun is Venus. Venus is about the size of the earth. It has a dense carbon dioxide atmosphere, which traps heat from the sun. The surface temperature of Venus is higher than that of any other planet in the solar system. Its rocky surface is enshrouded by a cloud cover too dense to see through except by radar.

Mars is of interest because of its great variety of surface features. It has a cratered topography reminiscent of the moon's and at the same time has features that appear to be the result of processes similar to those operating on earth. Dunes, stratified sediments, and sinuous rills suggest erosion by wind, water, and ice. Mars has a thin carbon dioxide atmosphere and polar caps composed of water ice and frozen carbon dioxide.

Among the outer planets, Jupiter is known for its tremendous size. It is larger than all other planets combined, has a massive hydrogen, helium, methane, and ammonia atmosphere, and rotates so rapidly as to form beautiful arrays of colored cloud bands. The planet has an internal heat supply. Its strong magnetic field suggests the presence of a metallic core.

Saturn is rather similar to Jupiter although smaller. Saturn's most distinctive feature is the thin complex ring system that encircles its equator like the brim of a straw hat. At least 17 moons accompany Saturn on its revolution around the sun.

Uranus and Neptune are similar in size and general characteristics. Both have atmospheres rich in hydrogen compounds, and Uranus appears to have a small ring system. Beyond Neptune lies Pluto. Pluto is roughly the size of the earth's moon and is thus the smallest planet in the solar system.

The earth's environment and history are inexorably tied to the history of the sun. Our sun will continue to shine with a hot yellow light for about another 5 billion years as its hydrogen is slowly converted to helium. When the hydrogen fuel nears exhaustion and helium becomes the principal material in the solar core, the sun will expand to become a red giant, and its heat will sear the earth's surface. After this phase of its history, the sun's emission of energy will gradually decline until it becomes negligible.

Terms to Understand

achondrite
carbonaceous chondrites
chondrite
chondrule
fusion
galaxy
iron meteorites
Jovian planet
lunar highlands
mare basins
maria (sing. mare)

meteor	Milky Way Galaxy	solar nebula	stony meteorite
meteorite	prograde	solar nebula hypothesis	stony-iron meteorite
micrometeorites	regolith	solar wind	terrestrial planet

Questions for Review

1. What are the names of the planets in our solar system? In what galaxy is our solar system located?

2. Given the amount of solar radiation intercepted by the earth, why is the earth's surface not much hotter than it is?

3. Compare Mercury, Mars, and the moon. What do they have in common, and how do they differ?

4. What regularities or uniformities exist in the solar system that indicate that it is truly a coherent system?

5. In general, how do the terrestrial planets differ from the Jovian planets?

6. The surface of Io is noted for its bright yellow and orange colors. What is the cause of this coloration?

7. What are the principal kinds of meteorites that have been found on the earth? In what way are they thought to have originated? What are chondrites?

8. What evidence indicates that, unlike the earth, the moon does not have an iron-nickel core?

9. How do the rocks of the lunar maria and lunar highland differ in composition and age?

10. What is the derivation of the reddish coloration of the planet Mars?

11. How does Mars differ from the earth in density, internal structure, and average range of equatorial temperatures?

12. What is the origin of the craters on the earth's moon?

13. How do lunar geologists determine the relative ages of craters on the moon?

14. What is lunar regolith? How is it thought to have originated?

Supplemental Readings and References

Abell, G.O., Morrison, D., and Wolff, S.C., 1988. *Realm of the Universe.* 4th ed. Philadelphia, Saunders College Publishing.

Cadogan, P., 1981. *The Moon: Our Sister Planet.* New York, Cambridge University Press.

Carr, M.H., 1981. *The Surface of Mars.* New Haven, Yale University Press.

Hartmann, W.K., 1983. *Moons and Planets: An Introduction to Planetary Science.* 2nd ed. Belmont, Calif., Wadsworth Publishing Co., Inc.

Hartmann, W.K., 1988. *Astronomy: The Cosmic Journey.* 4th ed. Belmont, Calif., Wadsworth Publishing Co., Inc.

Murray, B., Malin, M.C., and Greeley, R., 1981. *Earth-like Planets.* San Francisco, W.H. Freeman & Co.

Skinner, B.J. (ed.), 1981. *The Solar System and Its Strange Objects: Readings from American Scientist.* Los Altos, California, Wm. Kaufmann, Inc.

Appendix A Topographic Maps

Among the graphic tools employed by geologists, none are more widely used than topographic maps. Topographic maps show the size, shape, and distribution of features of the earth's surface, and from them one can learn much about the nature of rocks that lie beneath the surface. They are used as base maps for a variety of geologic investigations.

The three major categories of features depicted on topographic maps are *relief* (hills, valleys, plains, cliffs, and the like), *drainage* (streams, seas, lakes, swamps, and canals), and *culture* (the works of humans, including towns, cities, roads, railroads, boundaries, and names). Maps published by the U.S. Geological Survey all bear a key or legend on which symbols are defined and additional items, such as tints for forests or urban areas, are explained.

Relief

Contour lines connecting points of equal altitude are the standard method for portraying relief on a topographic map. These lines, printed in brown color, are an effective device for showing accurately a third dimension on a flat sheet of paper (Figs. A–1 and A–2). With practice, it is possible to readily visualize hills and valleys from the pattern of contour lines. On some topographic maps, landscape features are even more easily recognized because of the use of shading. Such maps, called *shaded-relief topographic maps*, are tinted so as to simulate in color the appearance of sunlight and shadows on the landscape and thereby provide the illusion of a three-dimensional land surface.

The vertical distance between any two contour lines is termed the *contour interval*. A satisfactory contour interval is one that depicts the landscape features adequately without the need to space contour lines so closely that reading the map becomes difficult. In Figure A–1, the contour interval is 10 feet. As a further aid to readability, every fourth or fifth contour is printed somewhat heavier and labeled with the elevation. These lines constitute *index contours*. The steepness of slopes can be ascertained on a contour map by the spacing of contour lines. A simple rule is that the closer the spacing the steeper the slope. Contour lines bend or loop upstream in valleys and can thus be used to determine direction of stream-flow. In becoming acquainted with contour maps, it is also useful to keep in mind that the ground surface must rise away from stream channels.

Map Scale

The scale of a map depicts the relation of a given distance on a map to the actual distance on the surface of the earth. Two widely used types of scales are *fractional* and *graphic*. The fractional scale is usually printed on the bottom margin of the map as a ratio, such as 1:24,000 or 1:125,000. The meaning of a scale of 1:24,000 is that 1 unit of measurement on the map represents 24,000 of the same units on the earth's surface. A graphic scale (Fig. A–3) consists of a line that has been divided into units representing length, such as km or fractions of km. Graphic scales can be used directly on the map to determine distances.

Direction

Compass directions on a standard U.S. Geological Survey topographic map are indicated by the meridians

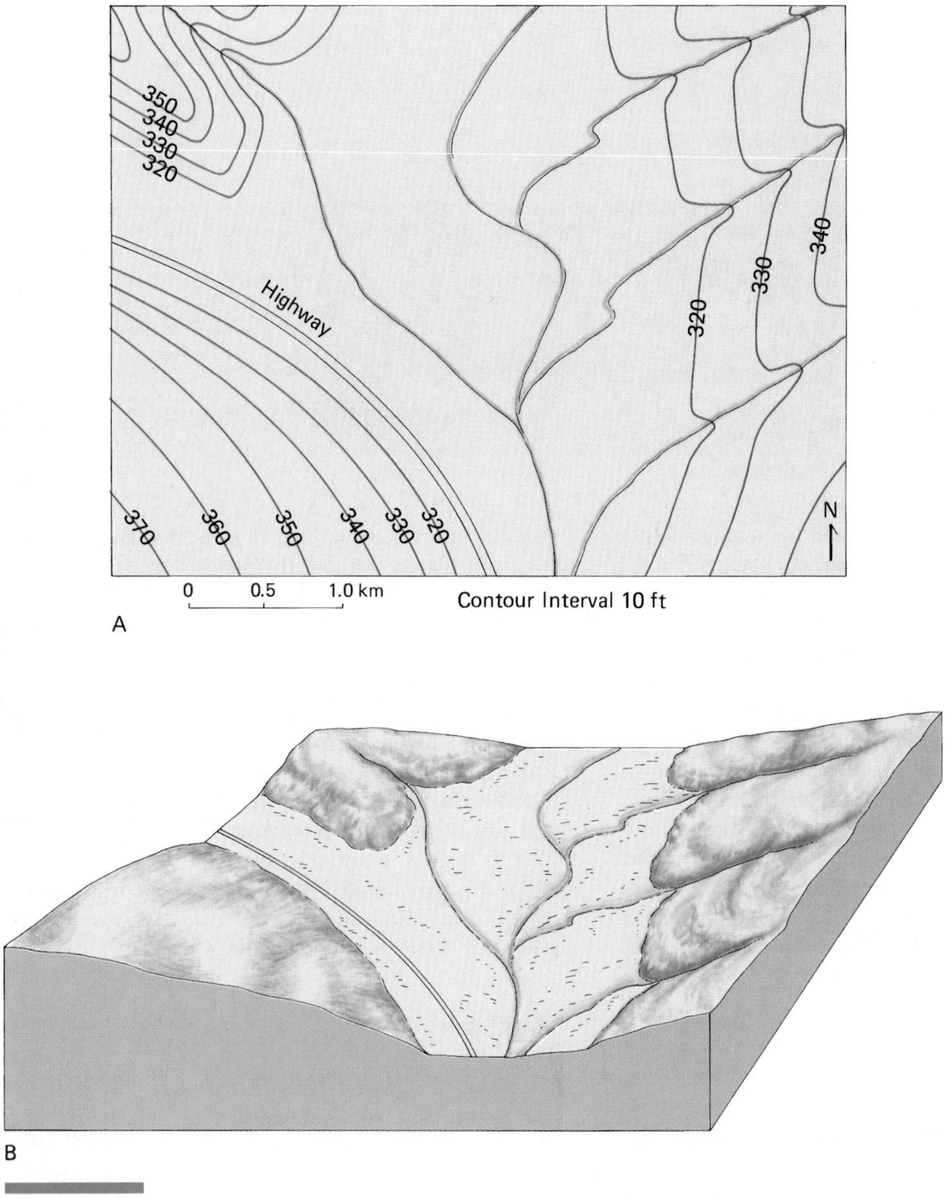

Figure A–1 Contour map (A) of the area shown in perspective (B) illustrating how contour lines express the form, slope, and altitude of various aspects of a landscape.

and parallels of latitude. By convention, the top of the map is north, and the right margin, consequently, is east. An important additional aid to orientation is the small diagram on the lower border of the map, which indicates true north, magnetic north, and the angle between the two, known as the *magnetic declination* (Fig. A–4). Magnetic declination is expressed in degrees and changes slowly. Corrections of magnetic declination can be made by referring to isogonic charts published each year by the U.S. Coast and Geodetic Survey.

Location

If you noticed some interesting feature like an ox-bow lake, sinkhole, cirque, or campsite on a topographic map, how would you describe its location accurately?

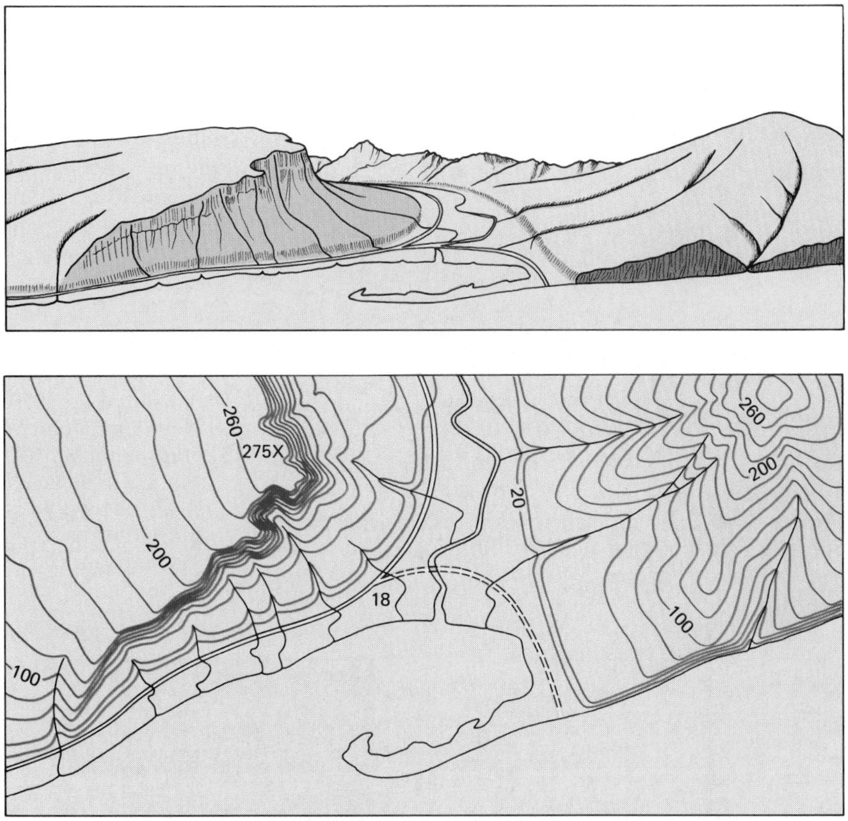

Figure A–2 Sketch of a coastal area and the topographic map of the same area. Contour interval is 20 feet. (From the circular *Topographic Maps,* U.S. Geological Survey.)

Usually one would employ one of the two principal aids to determining location imprinted on U.S. Geological Survey topographic maps. One of these aids is the global grid of *parallels of latitude* (east-west lines parallel to the equator) and *meridians of longitude* (true north-south lines that converge at the pole). Zero degrees of latitude corresponds to the equator, and 90°N to the North Pole. By agreement, the zero circle of longitude is placed so as to cross the former site of the Royal Greenwich Observatory in England. Longitude is measured in degrees east or west of the meridian that crosses Greenwich.

Another location method widely used in the U.S. because of its practicality is the *township and range system.* In this scheme, areas of the earth's surface are systematically divided according to a grid that is based on carefully surveyed east–west lines called base lines and north–south lines called principal me-

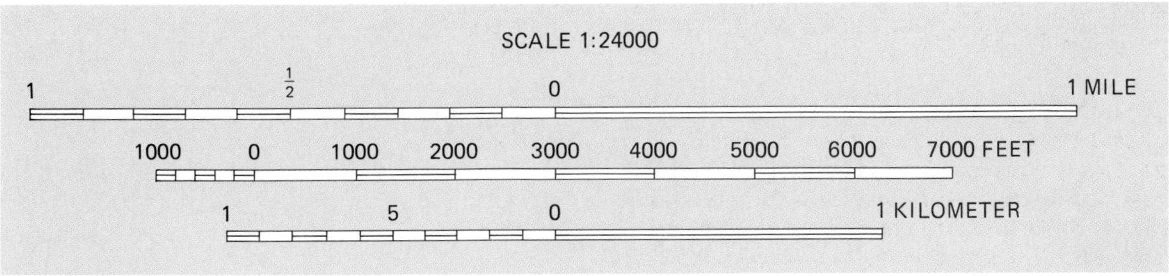

Figure A–3 The fractional and graphic scales as provided on the bottom margin of U.S. Geological Survey topographic maps.

Figure A–4 Direction diagram from a U.S. Geological Survey topographic map.

ridians. The basic unit of the grid is a square of land measuring 6 miles on a side and called a *township*. Vertical rows of townships, called ranges, are laid off east and west of the Principal Meridian. Each township is divided into 36 smaller squares, measuring 1 mile on a side and called sections. These sections contain 640 acres and may be subdivided according to fraction and geographic position as shown in Figure A–5.

References

Dickinson, G.C., 1979. *Maps and Air Photographs*. New York, John Wiley & Sons.

Thompson, M.M., 1979. *Maps for America*. United States Geological Survey. Topographic Maps. Pamphlet obtainable without charge on request to the Map Information Office, U.S. Geological Survey, Washington, D.C. 20242.

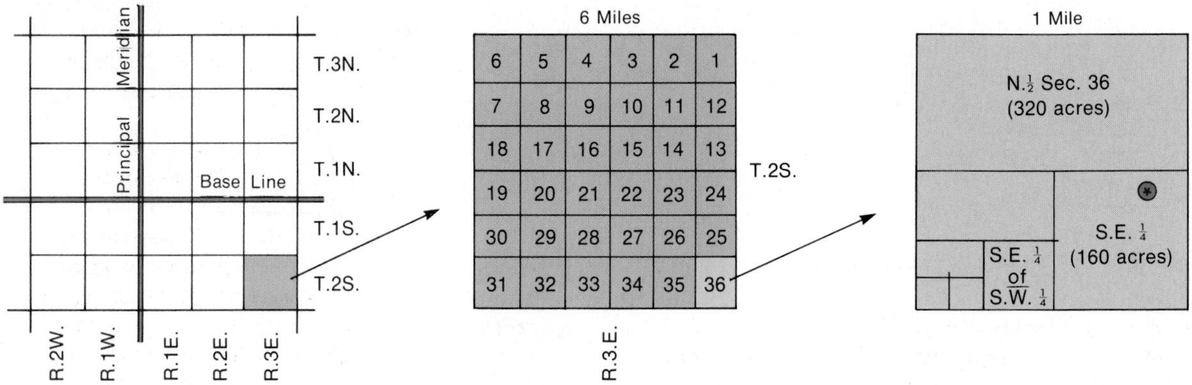

A. Numbering of townships from a base line and principal meridian

B. Numbering of sections in a township.

C. Scheme by which sections are subdivided.

⊛ Example, description of the location indicated by the circled star is N.E. ¼ of S.E. ¼ of Sec. 36, T.2S., R.3E.

Figure A–5 The township, range, and section system for defining areas of land and locations in the United States. A rather similar system is used in parts of Canada.

Appendix B Physical Properties of Minerals

The minerals listed in the following tables have been selected on the basis of their general abundance at the earth's surface or their special interest. They are arranged alphabetically so as to facilitate quick reference. The physical properties most often used in mineral identification include streak, color, luster, hardness, cleavage, fracture, general form, and specific gravity. In addition, some minerals can be recognized by distinctive taste, feel, odor, magnetism, or reaction to acid.

The **streak** of a mineral is the color of its powder and can be obtained by firmly scraping the mineral across a hard white porcelain streak plate. A mineral's streak is often less variable than its overall color. **Luster** refers to the appearance of a mineral in reflected light. The two major categories of luster are metallic and nonmetallic. Nonmetallic minerals can be further grouped into those having adamantine (as in diamond), vitreous (bright, shiny), resinous, waxy, pearly, greasy, or just dull lusters. **Hardness** is an especially useful aid to identifying minerals in hand specimen. Hardness refers to the resistance of a mineral to scratching. One does not necessarily require the minerals that comprise Mohs' Hardness Scale (see Table 2–4) in order to use this property. It is often more convenient to use one's fingernail (hardness of 2–2½), a copper penny (hardness 3), a good knife blade (hardness 5½), or a streak plate (hardness 7). If you are able, for example, to scratch a mineral with your fingernail, then the mineral's hardness would be less than 2. If the mineral will scratch a copper penny but won't scratch a knife blade, its hardness would be in the range from 3 to 5½.

As described in Chapter 2, **cleavage** refers to the tendency of certain minerals to break in preferred directions along smooth planes. Cleavage surfaces are usually parallel to a real or possible crystal face. Minerals vary in the number and perfection of cleavage planes. Frequently this property cannot be distinguished on the many tiny grains of minerals that ordinarily occur as aggregates.

For the identification of the relatively few exceptionally common minerals, it is often useful to start by observing if the mineral has a metallic or nonmetallic luster. If metallic, your choice may be among such minerals as galena (distinctive high specific gravity), pyrite (distinctive color and streak), magnetite (attracted to magnet), and hematite (distinctive streak). For nonmetallic minerals, consider first those that have a dark color and are harder than a steel knife blade. If good cleavage is evident, you may have a pyroxene or amphibole. If cleavage is absent or poor, the mineral may be olivine or garnet. Many other possibilities exist as well. For the softer dark-colored nonmetallic specimens, one may find the identification among such minerals as biotite, sphalerite, chlorite, goethite, or, again, hematite. Feldspars can often be recognized by the light coloration, hardness in excess of 5, nonmetallic luster, and prominently good cleavage. Similarly colored quartz minerals will lack this cleavage. Other light-colored minerals such as halite, gypsum, calcite, and muscovite can be identified by their softness and often well-developed cleavage planes.

Appendix B Common Minerals and Their Properties

Mineral	Chemical Composition	Form, Cleavage, Fracture	Usual Color	Hard-ness	Streak	Specific Gravity	Other Properties
Actinolite	$Ca_2(MgFe)_5Si_8O_{22}(OH)_2$	Slender crystals, radiating, fibrous	Blackish-green to light green	5–6	Pale green	3.2–3.6	Vitreous luster. Common in green schists.
Agate (onyx)	SiO_2	Always massive and banded	Banded, red white, pink, brown, green	7	White	2.6	Variety of quartz. No cleavage, tough, translucent; waxy luster.
Alabaster (gypsum)	$CaSO_4 \cdot 2\ H_2O$	Granular and massive	White, gray, pink	1–2.5	White	2.2–2.4	Pearly to dull luster.
Albite	$NaAlSi_3O_8$ (sodic plagioclase)	Good cleavage in 2 directions, nearly 90°	White, gray	6–6.5	White	2.6	May show fine striations (twinning lines) on cleavage faces.
Almandite (garnet)	$Fe_3Al_2(SiO_4)_3$	No cleavage, crystals 12- or 24-sided	Deep red	6.5–7.5	White	4.2	Vitreous to resinous luster. Metamorphic index mineral.
Amethyst	SiO_2	No cleavage, uneven fracture	Purple	7	White	2.6	Vitreous luster. Hexagonal crystals or massive crystalline.
Andalusite	Al_2SiO_5	Usually in square prisms	White, gray, green, violet, brown	6–7.5	White	3.1	Crystals may show dark interior cross.
Anhydrite	$CaSO_4$	Granular masses, crystals with 2 good cleavage directions	White, gray, blue-gray	3–3.5	White	2.9–3	Brittle; resembles marble but acid has no effect.
Anorthite	$CaAl_2Si_2O_8$ (calcic plagioclase)	Good cleavage in 2 directions at 94°	Colorless, white, gray, green	6	White	2.8	May show fine striations (twinning lines) on cleavage faces.
Apatite	$Ca_5(OH,F,Cl)(PO_4)_3$ (calcium fluorphosphate)	Massive, granular	Green, brown, red	4.5–5	Pale red-brown	3.1	Crystals may have a partly melted appearance, glassy
Asbestos (var. chrysotile)	$H_4Mg_3Si_2O_9$	Fibrous	White to pale olive-green	1–2.5	White	2.6–2.8	Pearly to greasy luster. Flexible, easily separated fibers.
Augite (a pyroxene)	$Ca(Mg,Fe,Al)(Al,Si_2O_6)$	Short stubby crystals have 4 or 8 sides in cross-section	Blackish-green to light green	5.6	Pale green	5–6	Vitreous, distinguished from hornblende by the 87° angle between cleavage faces.
Azurite	$Cu_3(CO_3)_2(OH)_2$ (copper carbonate)	Varied, may have fibrous crystals	Azure blue	4	Pale blue	3.8	Vitreous to earthy, effervesces with HCl, an ore of copper.
Barite	$BaSO_4$	Tabular crystals, 3 cleavages, crystal-line masses	White	2.5–3.5	White	4.5	Vitreous to pearly luster. Easily determined by high specific gravity.
Bauxite	Hydrous aluminum oxides; not a true mineral	Earthy masses	Reddish to brown	1.5–3.5	Pale reddish-brown	2.5	Dull luster, claylike masses with small round concretions.
Beryl	$Be_3Al_2(SiO_3)_6$	Uneven fracture, hexagonal crystals	Green, yellow, blue, pink	7.5–8.0	White	2.6–2.8	Vitreous luster. Gem variety is aquamarine.

Appendix B (Continued)

Mineral	Chemical Composition	Form, Cleavage, Fracture	Usual Color	Hardness	Streak	Specific Gravity	Other Properties
Biotite	$K(Mg,Fe)_3AlSi_3O_{10}(OH)_2$	Perfect cleavage into thin sheets	Black, brown, green	2.2–2.5	White, gray	2.7–3.1	Vitreous luster, divides readily into thin flexible sheets.
Bornite	Cu_5FeS_4	Compact, massive	Purplish	3	Gray-black	4.9–5.4	Metallic luster, brittle, an ore of copper.
Calcite	$CaCO_3$	Perfect cleavage into rhombs	Usually white, but may be variously tinted	3	White	2.7	Transparent to opaque. Rapid effervescence with HCl
Carnotite	$K_2(UO_2)_2(VO_4)_2$	Earthy, powdery	Canary yellow	Very soft	Pale yellow	4	Soft, earthy. An ore of uranium.
Cassiterite	SnO_2	Massive or as sand-size grains	Black	6–7	Dark brown	7	Submetallic, heavy, hard grains.
Chalcedony	SiO_2	Fractures uneven or conchoidal	White, gray, blue	7	White	2.6	Waxy luster, tough, translucent, variety of quartz.
Chalcocite	Cu_2S	Usually massive	Dark lead gray	2.5–3	Lead gray	5.5–5.8	Metallic luster, often with bluish tarnish.
Chalcopyrite	$CuFeS_2$	Uneven fracture	Brass yellow	3.5–4.5	Greenish-black	4.2	Metallic luster, softer than pyrite.
Chlorite	Hydrous ferromagnesian aluminum silicate	Perfect cleavage as fine scales	Green	2.0–2.5	Gray, white, pale green	2.8	Pearly to vitreous luster. Low-grade metamorphic mineral.
Chromite	$FeCr_2O_4$	Massive, granular, compact	Black	5.5	Dark brown	4.4	Metallic to submetallic luster. Ore of chromite occurring in serpentine.
Cinnibar	HgS	Compact, granular masses	Scarlet red to red-brown	2.5	Scarlet red	8	Color and streak distinctive. Ore of mercury.
Clay (clay minerals)	Hydrous aluminum silicates, with Ca, Na, K, Fe, Mg	Soft, compact, earthy masses	White, but tinted by impurities	1–2	White	2.2–2.6	Greasy feel, adheres to tongue, earthy odor when moist.
Corundum	Al_2O_3	Short, six-sided barrel-shaped crystals	Gray, light blue, and other colors	9	None	3.9–4.1	Ruby and sapphire are corundum varieties. Hardness is distinctive.
Diamond	C	Octahedral crystals	Colorless or with pale tints	10	None	3.5	Adamantine luster. Hardness is distinctive.
Diopside (a pyroxene)	$CaMg(Si_2O_6)$	Usually short thick prisms; may be granular	White to light green	5–6	White to greenish	3.2–3.6	Vitreous luster. A contact metamorphic mineral in carbonates.
Dolomite	$CaMg(CO_3)_2$	Cleaves into rhombs; granular masses	White, pink, gray, brown	3.5–4	White to pale gray	3.9–4.2	Effervesces slightly in cold dilute HCl.

Appendix B (Continued)

Mineral	Chemical Composition	Form, Cleavage, Fracture	Usual Color	Hardness	Streak	Specific Gravity	Other Properties
Enstatite (a pyroxene)	$MgSiO_3$	Cleavage good at 87° and 93°; usually massive	Pale green, brown, gray, or yellowish	5.5	White	3.2–3.5	Vitreous luster. Common in mafic igneous rocks.
Epidote	Hydrous Ca, Al, Fe, silicate	Usually granular masses; also as slender prisms	Yellow-green, olive-green, to nearly black	6.7	Pale yellow to white	3.3	Vitreous luster; occurs as contact metamorphic mineral in carbonates.
Fluorite	CaF_2	Octahedral and also cubic crystals	White, yellow, green, purple	4	White	3.2	Cleaves easily, vitreous, transparent to translucent.
Galena	PbS	Perfect cubic cleavage	Lead or silver gray	2.5	Gray	7.6	Metallic luster. Ore of lead.
Glauconite	Hydrous silicate of K and Fe	Occurs as grains or granular masses	Green	1–2	Greenish-white	2.2	Dull luster. Common constituent of "greensands."
Goethite	$HFeO_2$	Massive, in fibrous aggregates or foliated	Yellow-brown to dark brown	5–5.5	Yellow-brown	3.3–4.4	Adamantine to dull. An iron ore.
Graphite	C	Foliated, scaly, or earthy masses	Steel gray to black	1–2	Gray or black	2.2	Feels greasy; marks paper.
Gypsum	$CaSO_4 \cdot 2H_2O$	Tabular crystals, fibrous, or granular	White, pearly	1–2.5	White	2.2–2.4	Thin sheets (selenite), fibrous (satinspar), massive (alabaster).
Halite	NaCl	Granular masses, perfect cubic crystals	White, also pale colors and gray	2.5–3	White	2.2	Pearly luster, salty taste, soluble in water.
Hematite	Fe_2O_3	Granular, massive, or earthy	Brownish-red	2.5	Dark red	2.5–5	Often earthy, dull appearance.
Hornblende (an amphibole)	Complex Ca, Mg, Fe, Al, silicate	Elongate crystals	Dark shades of green	5–6	Pale green	3.2	Crystals six-sided with 124° between cleavage faces.
Jadeite (a pyroxene)	$NaAl(Si_2O_6)$	Compact fibrous aggregates	Green	6.5–7	White, pale green	3.3	Vitreous luster.
Jasper	SiO_2	Fracture uneven to conchoidal	Red, yellow-brown	7	White	2.6	Dull to waxy luster.
Kyanite	Al_2SiO_5	Good cleavage in one direction; bladed aggregates	White, pale blue, gray, green	4–7	White	3.6	Vitreous to pearly; crystals may have blue interiors.
Labradorite	A mixture of $CaAl_2Si_2O_8$ and $NaAlSi_3O_8$ (60:40)	Cleavage perfect in two directions	Dark gray to grayish-white	6	White	2.7	Vitreous, fine striations on one cleavage face, play of colors.
"Limonite" (amorphous hydrous iron oxide, not a true mineral)	$2Fe_2O_3 \cdot 3H_2O$	Earthy fracture	Brown or black	1.5–4	Brownish-yellow	3.6	Earthy masses that resemble clay.

Appendix B (Continued)

Mineral	Chemical Composition	Form, Cleavage, Fracture	Usual Color	Hardness	Streak	Specific Gravity	Other Properties
Magnetite	Fe_3O_4	Uneven fracture, granular masses	Iron black	5.5	Iron black	5.2	Metallic luster. Strongly magnetic.
Malachite	$CuCO_3 \cdot Cu(OH)_2$	Uneven splintery fracture	Bright green, dark green	3.5–4	Emerald green	4	Effervesces with HCl. Associated with azurite.
Marcasite	FeS_2	Uneven fracture, arrow-shaped crystals	Pale brass yellow	6	Black	4.9	Metallic luster (never in cubes as in pyrite).
Muscovite	$KAl_3Si_3O_{10}(OH)_2$	Perfect cleavage into thin sheets	Colorless if thin	2–2.5	White	2.7–3	Vitreous or pearly; flexible and elastic; splits easily.
Olivine	$(MgFe)_2SiO_4$	Uneven fracture, often in granular masses	Various shades of green	6.5–7	White	3.2–3.3	Vitreous, glassy luster. Common in basalt; gem variety peridot.
Opal (a mineraloid)	$SiO_2 \cdot nH_2O$	Conchoidal fracture, amorphous, massive	White and various colors	5.5–6.5	White	2.1	Vitreous, greasy, pearly luster. May show a play of colors.
Orthoclase	$KAlSi_3O_8$	Good cleavage in two directions at 90°	White, pink, red, yellow-green, gray	6	White	2.6	Vitreous to pearly, a common associate of quartz in granite.
Pyrite ("fools gold")	FeS_2	Uneven fracture cubes with striated faces, octahedrons	Pale brass yellow (lighter than chalcopyrite	6–6.5	Greenish-black	5	Metallic luster, brittle, very common.
Quartz	SiO_2	No cleavage, massive and as six-sided crystals	Colorless, white, or tinted any color by impurities	7	White	2.6	Includes rock crystal, rose and milky quartz, amethyst, smoky quartz, etc.
Rutile	TiO_2	Prismatic cleavage, uneven fracture	Reddish-brown to black	6.7	Light brown to light gray	4.3	Adamantine to submetallic, An ore of titanium.
Serpentine	$Mg_3Si_2O_5(OH)_4$	Uneven, often splintery fracture	Light and dark green, yellow	2.5	White	2.5	Waxy luster, smooth feel, brittle.
Siderite	$FeCO_3$	Good rhombohedral cleavage	Brown	3.5–4	Pale yellow or yellow-brown	3.8	Vitreous luster, brittle, cleavage faces often curved.
Sillimanite	Al_2SiO_5	Good cleavage in one direction, fibrous	Brown, light green, white	6–7	White	3.23	Vitreous luster. Index to high-grade metamorphism.
Specularite (var. hematite)	Fe_2O_3	Cleavage absent or micaceous	Dark steel gray	2.5–6.5	Red or reddish-brown	4.4–5.3	Metallic luster. Bright sparkling scales.

Appendix B (Continued)

Mineral	Chemical Composition	Form, Cleavage, Fracture	Usual Color	Hardness	Streak	Specific Gravity	Other Properties
Sphalerite	ZnS	Perfect cleavage in 6 directions at 120°	Shades of brown and red	3.5	Reddish-brown	4	Resinous luster, may occur with galena, pyrite. Ore of zinc.
Spinel (a mineral group)	$MgAl_2O_4$	No cleavage, rare octahedral crystals	Black, dark green, or various colors	7.5–8.0	White	3.5–4.1	Vitreous luster. Hardness is distinctive.
Staurolite	$Fe^{+2}Al_5Si_2O_{12}(OH)$	Fair cleavage in one direction	Dark brown to nearly black	7	White to grayish	3.6–3.8	Subvitreous to resinous. Crystals sometimes twinned as crosses.
Stibnite	Sb_2S_3	Perfect in one direction	Dark lead gray	2	Gray or black	4.5	Metallic luster; slender prismatic crystals.
Talc (soapstone)	$Mg_3Si_4O_{10}(OH)_2$	Perfect in one direction	Green, white, gray	1–1.5	White	1–2.5	Greasy feel; occurs in foliated masses.
Topaz	$Al_2SiO_4(F,OH)_2$	Cleavage good in one direction; conchoidal fracture	Colorless, white, pale tints of blue, pink	8	Colorless	3.5–3.6	Vitreous luster.
Tourmaline	Complex boron aluminum silicate with Na, Ca, F, Fe, Li	Poor cleavage, uneven fracture; striated crystals	Black, brown, green, pink	7–7.5	White to gray	4.4–4.8	Vitreous, slightly resinous. A gemstone.
Tremolite (an amphibole)	$Ca_2Mg_5(Si_8O_{22})(OH)_2$	Perfect cleavage at an angle of 124°	White, light gray, light green	5.6	White	2.9–3.1	Vitreous to silky. Crystals usually long-bladed or short and stout.
Uraninite	UO_2 with small amounts of Pb, Ra, Th, Y, N, He, and A	Massive or botyroidal; no cleavage	Brownish-black	5–6	Brownish-black	6.4–9.7	Submetallic luster to greasy or dull.
Wollastonite	$CaSiO_3$	Usually in fibrous masses of elongate crystals	White, pink, gray, colorless	4.5–5.0	White	2.8–2.9	Vitreous luster. A mineral of contact metamorphism of limestone.
Zircon	$Zr(SiO_4)$	Cleavage poor, but often well-formed tetragonal crystals	Colorless, gray, green, pink, bluish	7.5	White	4.7	Adamantine luster.

Appendix C The Periodic Table, Atomic Weights, and Atomic Numbers

Scientists generally group or organize things that are similar in order to better understand them and determine how they relate to one another. An important step in the organization of elements was made in 1869 by Dimitri Mendeleev. Mendeleev showed that, when elements are arranged in order of their increasing atomic weights, their physical and chemical properties tend to be repeated in cycles. The arrangement of elements was depicted on a chart called the Periodic Table of Elements, a modern version of which is shown as Table C–1. In the periodic table, each box contains the symbol of the element and its atomic number. Except for two long sequences set apart at the bottom, the elements appear in the increasing order of their atomic numbers. The vertical columns are called *groups* and contain elements with similar properties. For example, in column VIIA, one finds fluorine, chlorine, bromine, and iodine. All of these elements are colored, highly reactive, and share other similarities. Fluorine, however, is chemically the most active, chlorine somewhat less active, bromine still less active, and iodine the least active of the four. Except for hydrogen, all of the elements in group IA are soft, shiny metals and very reactive chemically. The horizontal rows on the table are called *periods* and contain sequences of elements have electron configurations that vary in characteristic patterns. It is thus apparent that the periodic table shows relationships among elements rather well and certainly better than an arbitrary listing. Alphabetical lists, however, also provide a useful reference (Table C–2).

Table C–1 Periodic Table of the Elements

Key: 26 / Fe / 55.847
- Atomic number (Z)
- Element symbol
- Atomic mass of naturally occurring isotopic mixture; for radioactive elements, numbers in parentheses are mass numbers of most stable isotopes

IA	IIA	IIIB	IVB	VB	VIB	VIIB	VIII	VIII	VIII	IB	IIB	IIA	IVA	VA	VIA	VIIA	O
1 H 1.0079																1 H 1.0079	2 He 4.00260
3 Li 6.941	4 Be 9.01218											5 B 10.81	6 C 12.011	7 N 14.0067	8 O 15.9994	9 F 18.998403	10 Ne 20.179
11 Na 22.98977	12 Mg 24.305											13 Al 26.98154	14 Si 28.0855	15 P 30.97376	16 S 32.06	17 Cl 35.453	18 Ar 39.948
19 K 39.0983	20 Ca 40.08	21 Sc 44.9559	22 Ti 47.90	23 V 50.9415	24 Cr 51.996	25 Mn 54.9380	26 Fe 55.847	27 Co 58.9332	28 Ni 58.70	29 Cu 63.546	30 Zn 65.38	31 Ga 69.72	32 Ge 72.59	33 As 74.9216	34 Se 78.96	35 Br 79.904	36 Kr 83.80
37 Rb 85.4678	38 Sr 87.62	39 Y 88.9059	40 Zr 91.22	41 Nb 92.9064	42 Mo 95.94	43 Tc (98)	44 Ru 101.07	45 Rh 102.9055	46 Pd 106.4	47 Ag 107.868	48 Cd 112.41	49 In 114.82	50 Sn 118.69	51 Sb 121.75	52 Te 127.60	53 I 126.9045	54 Xe 131.30
55 Cs 132.9054	56 Ba 137.33	57 *La 138.9055	72 Hf 178.49	73 Ta 180.9479	74 W 183.85	75 Re 186.207	76 Os 190.2	77 Ir 192.22	78 Pt 195.09	79 Au 196.9665	80 Hg 200.59	81 Tl 204.37	82 Pb 207.2	83 Bi 208.9804	84 Po (209)	85 At (210)	86 Rn (222)
87 Fr (223)	88 Ra 226.0254	89 †Ac 227.0278	104 Unq (261)	105 Unp (262)	106 Unh (263)	107 Uns		109									

*Lathanide Series

58 Ce 140.12	59 Pr 140.9077	60 Nd 144.24	61 Pm (145)	62 Sm 150.4	63 Eu 151.96	64 Gd 157.25	65 Tb 158.9254	66 Dy 162.50	67 Ho 164.9304	68 Er 167.26	69 Tm 168.9342	70 Yb 173.04	71 Lu 174.967

†Actinide Series

90 Th 232.0381	91 Pa 231.0359	92 U 238.029	93 Np 237.0482	94 Pu (244)	95 Am (243)	96 Cm (247)	97 Bk (247)	98 Cf (251)	99 Es (252)	100 Fm (257)	101 Md (258)	102 No (259)	103 Lr (260)

Note: Atomic masses shown here are 1977 IUPAC values.

Table C–2 International Table of Atomic Weights

Based on the assigned relative atomic mass of $^{12}C = 12$.

The following values apply to elements as they exist in materials of terrestrial origin and to certain artificial elements. When used with the footnotes, they are reliable to ±1 in the last digit, or ±3 if that digit is in small type.

	Symbol	Atomic Number	Atomic Weight		Symbol	Atomic Number	Atomic Weight		Symbol	Atomic Number	Atomic Weight
Actinium	Ac	89		Hafnium	Hf	72	178.4_9	Promethium	Pm	61	
Aluminum	Al	13	26.98154[a]	Helium	He	2	4.00260[b,c]	Protactinium	Pa	91	231.0359[a,f]
Americium	Am	95		Holmium	Ho	67	164.9304[a]	Radium	Ra	88	226.0254[f,g]
Antimony	Sb	51	121.7_5	Hydrogen	H	1	1.0079[b,d]	Radon	Rn	86	
Argon	Ar	18	39.94_8[b,c,d,g]	Indium	In	49	114.82	Rhenium	Re	75	186.207
Arsenic	As	33	74.9216[a]	Iodine	I	53	126.9045[a]	Rhodium	Rh	45	102.9055[a]
Astatine	At	85		Iridium	Ir	77	192.2_2	Rubidium	Rb	37	85.467_8[c]
Barium	Ba	56	137.3_4	Iron	Fe	26	55.84_7	Ruthenium	Ru	44	101.0_7
Berkelium	Bk	97		Krypton	Kr	36	83.80	Samarium	Sm	62	150.4
Beryllium	Be	4	9.01218[a]	Lanthanum	La	57	138.905_5[b]	Scandium	Sc	21	44.9559[a]
Bismuth	Bi	83	208.9804[a]	Lawrencium	Lr	103		Selenium	Se	34	78.9_6
Boron	B	5	10.81[c,d,e]	Lead	Pb	82	207.2[d,g]	Silicon	Si	14	28.08_6[d]
Bromine	Br	35	79.904[c]	Lithium	Li	3	6.94_1[c,d,e,g]	Silver	Ag	47	107.868[c]
Cadmium	Cd	48	112.40	Lutetium	Lu	71	174.97	Sodium	Na	11	22.98977[a]
Calcium	Ca	20	40.08[g]	Magnesium	Mg	12	24.305[c,g]	Strontium	Sr	38	87.62[g]
Californium	Cf	98		Manganese	Mn	25	54.9380[a]	Sulfur	S	16	32.06[d]
Carbon	C	6	12.011[b,d]	Mendelevium	Md	101		Tantalum	Ta	73	180.947_9[b]
Cerium	Ce	58	140.12	Mercury	Hg	80	200.5_9	Technetium	Tc	43	98.9062[f]
Cesium	Cs	55	132.9054[a]	Molybdenum	Mo	42	95.9_4	Tellurium	Te	52	127.6_0
Chlorine	Cl	17	35.453[c]	Neodymium	Nd	60	144.2_4	Terbium	Tb	65	158.9254[a]
Chromium	Cr	24	51.996[c]	Neon	Ne	10	20.17_9[c]	Thallium	Tl	81	204.3_7
Cobalt	Co	27	58.9332[a]	Neptunium	Np	93	237.0482[f]	Thorium	Th	90	232.0381[f]
Copper	Cu	29	63.54[c,d]	Nickel	Ni	28	58.70	Thulium	Tm	69	168.9342[a]
Curium	Cm	96		Niobium	Nb	41	92.9064[a]	Tin	Sn	50	118.6_9
Dysprosium	Dy	66	162.5_0	Nitrogen	N	7	14.0067[b,c]	Titanium	Ti	22	47.9_0
Einsteinium	Es	99		Nobelium	No	102		Tungsten	W	74	183.8_5
Erbium	Er	68	167.2_6	Osmium	Os	76	190.2	Uranium	U	92	238.029[b,c,e]
Europium	Eu	63	151.96	Oxygen	O	8	15.999_4[b,c,d]	Vanadium	V	23	50.941_4[b,c]
Fermium	Fm	100		Palladium	Pd	46	106.4	Wolfram			(see Tungsten)
Fluorine	F	9	18.99840[a]	Phosphorus	P	15	30.97376[a]	Xenon	Xe	54	131.30
Francium	Fr	87		Platinum	Pt	78	195.0_9	Ytterbium	Yb	70	173.0_4
Gadolinium	Gd	64	157.2_5	Plutonium	Pu	94		Yttrium	Y	39	88.9059[a]
Gallium	Ga	31	69.72	Polonium	Po	84		Zinc	Zn	30	65.38
Germanium	Ge	32	72.5_9	Potassium	K	19	39.09_8	Zirconium	Zr	40	91.22
Gold	Au	79	196.9665[a]	Praseodymium	Pr	59	140.9077[a]				

[a] Mononuclidic element.

[b] Element with one predominant isotope (about 99 to 100 percent abundance).

[c] Element for which the atomic weight is based on calibrated measurements by comparisons with synthetic mixtures of known isotopic composition.

[d] Element for which known variation in isotopic abundance in terrestrial samples limits the precision of the atomic weight given.

[e] Element for which users are cautioned against the possibility of large variations in atomic weight due to inadvertent or undisclosed artificial isotopic separation in commercially available materials.

[f] Most commonly available long-lived isotope.

[g] In some geological specimens this element has a highly anomalous isotopic composition, corresponding to an atomic weight significantly different from that given.

Appendix D Measurement Systems

Most of us are familiar with the English system of measure in which units of length include the inch, foot, yard, and mile, and units of mass include ounces, pounds, and tons. Unfortunately, such units are inconvenient for conversion and arithmetic computation. Scientists prefer the International System of Units (abbreviated SI), which is based on metric units. In the metric system, the meter is the fundamental unit of length. It was once based on the length of a standard platinum bar but is now defined in terms of wavelengths of light. A meter is about 1.1 yards in length. It is divided into 100 divisions called centimeters (cm), each about 0.4 inch long. Table D–1 provides a convenient method for converting from the English to the metric system for the more commonly used units. The names of the metric units and their numerical values are indicated in Table D–2.

The three temperature scales in general use are the Kelvin (K) scale (water freezes at 273 K and boils at 373 K), the Celsius (C) scale (water freezes at 0°C and boils at 100°C), and the Fahrenheit (F) scale (water freezes at 32°F and boils at 212°F). The relation between the Celsius and Fahrenheit scales is depicted in Table D–3.

The SI unit for temperature is the Kelvin. The size of a Kelvin degree (one Kelvin) is the same as the size of a Celsius degree. One can change from Celsius degrees to Kelvins simply by adding 273 to the Celsius temperature.

Table D–1 Convenient Conversion Factors

To Convert from	*To*	*Multiply by**
Centimeters	Feet	0.0328 ft/cm
	Inches	0.394 in/cm
	Meters	0.01 m/cm
	Microns (micrometers)	10000 μm/cm
	Miles (statute)	6.214×10^{-6} mi/cm
	Millimeters	10 mm/cm
Feet	Centimeters	30.48 cm/ft
	Inches	12 in/ft
	Meters	0.3048 m/ft
	Microns (micrometers)	304800 μm/ft
	Miles (statute)	0.000189 mi/ft
	Yards	0.3333 yd/ft
Kilometers	Miles	0.6214 mi/km
Gallons (U.S., liq.)	Cu. centimeters	3785 cm^3/gal
	Cu. feet	0.133 ft^3/gal
	Liters	3.785 ℓ/gal
	Quarts (U.S., liq.)	4 qt/gal
Grams	Kilograms	0.001 kg/g
	Micrograms	1×10^6 μg/g
	Ounces (avdp.)	0.03527 oz/g
	Pounds (avdp.)	0.002205 lb/g
Inches	Centimeters	2.54 cm/in
	Feet	0.0833 ft/in
	Meters	0.0254 m/in
	Yards	0.0278 yd/in
Kilograms	Ounces (avdp.)	35.27 oz/kg
	Pounds (avdp.)	2.205 lb/kg
Meters	Centimeters	100 cm/m
	Feet	3.2808 ft/m
	Inches	39.37 in/m
	Kilometers	0.001 km/m
	Miles (statute)	0.0006214 mi/m
	Millimeters	1000 mm/m
	Yards	1.0936 yd/m
Miles (statute)	Centimeters	160934 cm/mi
	Feet	5280 ft/mi
	Inches	63360 in/mi
	Kilometers	1.609 km/mi
	Meters	1609 m/mi
	Yards	1760 yd/mi
Ounces (avdp.)	Grams	28.3 g/oz
	Pounds (avdp.)	0.0625 lb/oz
Pounds (avdp.)	Grams	453.6 g/lb
	Kilograms	0.454 kg/lb
	Ounces (avdp.)	16 oz/lb

*Values are generally given to three or four significant figures.

(From Turk, J., and Turk, A., *Physical Science*, 1977, Philadelphia, W.B. Saunders Company.)

Table D–2 Metric Units, Prefixes, and Scientific Notation

Metric units:

length	meter (m)
volume	liter (l)
mass	gram (gm or g)
time	second (sec or s)

Prefixes for use with basic units of metric system:

Prefix	Symbol	Power	Equivalent
tera	T	10^{12} = 1,000,000,000,000	Trillion
giga	G	10^{9} = 1,000,000,000	Billion
mega	M	10^{6} = 1,000,000	Million
kilo	k	10^{3} = 1,000,*	Thousand
hecto	h	10^{2} = 100	Hundred
deca	da	10^{1} = 10	Ten
— — —	—	10^{0} = 1	One
deci	d	10^{-1} = .1	Tenth
centi	c	10^{-2} = .01	Hundredth
milli	m	10^{-3} = .001	Thousandth
micro	μ	10^{-6} = .000001	Millionth
nano	n	10^{-9} = .000000001	Billionth
pico	p	10^{-12} = .000000000001	Trillionth

*Example: 1000 meters = 1 kilometer = 1 km

Scientific notation

In scientific notation, a number is usually written as a figure between 1 and 9.99 multiplied by a power of 10. The power of 10 is called the exponent, and it indicates the number of places the decimal point must be moved to restore the number. The decimal point is moved to the right if the exponent is positive, and to the left if it is negative. For example, 150,000,000 is 1.5 × 100,000,000, which in scientific notation would be written 1.5×10^{8}. The figure 0.00000576 would be written 5.76×10^{-6}.

D–3 Relation between the Fahrenheit, Celsius, and Kelvin Temperature Scales

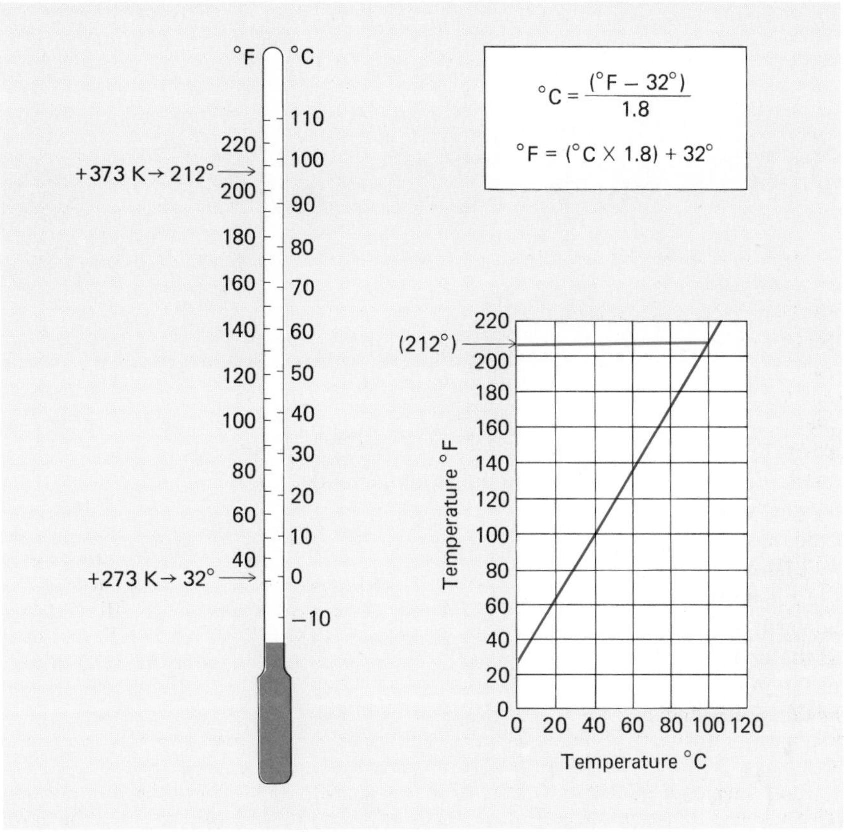

A third temperature scale, widely used in astronomy, is the *Kelvin scale.* It starts with the coldest possible temperature, −273°C and uses the same size degrees as does the Celsius scale. The Kelvin scale degrees are called *kelvins* and are abbreviated K (not °K). In converting Celsius to Kelvin, −273°C would be 0 kelvin, 0°C would be +273 K, and 100°C would be +373 K.

Appendix E Earth, Moon, and Sun Data

Volume of the earth = 1,083,230 km^3
Volume of the oceans = 1.4×10^9 km^3 (1.4 billion cubic kilometers)
Total mass of seawater = 1.4×10^{24} g
Average depth of the ocean = 3,800 meters
Average elevation of land = 840 meters
Equatorial radius of the earth = 6,378 km
Polar radius of the earth = 6356 km
Average density of the earth = 5.5 g/c^3
Average density of the moon = 3.3 g/c^3
Average density of the sun = 1.4 g/c^3
Mass of the earth = 5.976×10^{27} g
Mass of the moon = 7.347×10^{25} g
Mass of the sun = 1.971×10^{32} g
Gravitational acceleration of the earth = 980 $cm/second^2$ = 9.8 $meters/second^2$
Mean distance to moon from earth = 384,393 km (238,860 mi)
Mean distance to sun from earth = 149,450,000 km (92,900,000 mi)
Average thickness of earth's continental crust = 35 km
Average thickness of earth's oceanic crust = 8 km
Average thickness of earth's mantle = 2883 km
Mean density of earth's continental crust = 2.8 g/c^3
Mean density of the earth's oceanic crust = 2.9 g/c^3
Mean density of earth's mantle = 4.5 g/c^3

Appendix F Rock Symbols

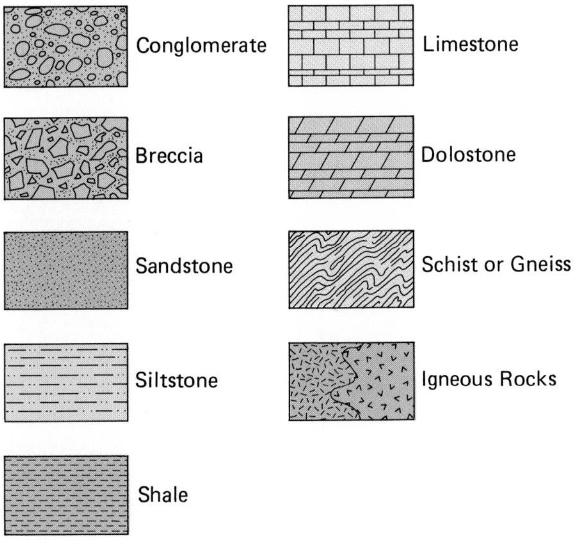

The standard geologic symbols used by geologists and in diagrams in this book are indicated above.

Glossary

Aa. Basaltic lava having a jagged, rough, clinkery surface.

Ablation. The combined processes of melting and evaporation by which a glacier wastes. The ablation zone is the lower margin of a glacier where loss of water exceeds additions.

Abyssal plains. Nearly level areas of the oceanic depths below about 2000 meters.

Achondrite. A type of stony meteorite that lacks the small rounded mineral bodies called chondrules.

Actualism. The concept that the laws of nature are invariant and responsible for changes that have occurred on earth through time.

Albedo. The ratio of the amount of incoming radiation reflected by a natural surface (ice, snow, water, clouds) to the amount incident on that surface, commonly expressed as a percentage.

Alluvial fan. A cone-shaped deposit of unconsolidated, poorly sorted sediments made by a stream where it passes from an area of steep gradient to lower gradient.

Alluvium. Unconsolidated poorly sorted detrital sediments ranging from clay to gravel sizes and characteristically fluvial in origin.

Alpha particle. A particle, equivalent to the nucleus of a helium atom, emitted from an atomic nucleus during radioactive decay.

Amphiboles. Ferromagnesian minerals characterized by good prismatic cleavage in two directions intersecting at 56° and 124° and including hornblende and actinolite.

Amygdules. Gas cavity or vesicle in an igneous rock that has become filled with a secondary mineral such as calcite or quartz.

Andesite. An extrusive igneous rock approximately intermediate in chemical composition between basalt and rhyolite. Andesites are composed of about 75 per cent plagioclase and lack quartz or orthoclase.

Andesite line. A line drawn on a map of the western hemisphere that separates the region of basaltic extrusive rocks of the Pacific from the more andesitic rocks of the circum-Pacific.

Angle of repose. The maximal angle, measured in degrees, at which material such as loose rock, sand, or silt remains stable.

Angular unconformity. An unconformity in which the older strata dip at a different (usually steeper) angle than younger strata.

Anion. An ion that carries a negative charge.

Annular drainage pattern. Streams and their tributaries that form a ringlike pattern because their development is controlled by the circular outcrop of strata associated with structural domes and basins.

Antecedent stream. A stream whose valley antedates the structures across which it flows and that has been able to maintain its course largely unaffected by uplift, folding, or faulting.

Anthracite coal. Hard black lustrous coal containing a high percentage (80 to 90 per cent) of fixed carbon and usually less than 5 per cent volatile matter.

Anticline. Geologic structure in which strata are bent into an upfold or arch and in which older rocks are found toward the center of curvature.

Anticyclone. Atmospheric circulation pattern having a generally clockwise rotation in the Northern hemisphere and counterclockwise rotation in the Southern Hemisphere.

Aphanitic. A textural term used for rocks in which the crystalline components are too small to be recognized with the unaided eye.

Aquiclude. A rock unit that does not transmit water fast enough to provide an appreciable supply for a spring or well.

Aquifer. A formation that, because of good porosity and permeability, is able to transmit water in sufficient quantity to supply springs and wells.

Aragonite. Orthorhombic variety of calcium carbonate.

Arkose. A sandstone in which more than 25 per cent of the grains are feldspar.

Arête. A narrow jagged ridge developed by glacial erosion along the divide separating valley glaciers.

Arroyo. The channel of an intermittent or ephemeral stream

Artesian well. A well in which the groundwater tapped has sufficient hydrostatic head to rise above the level of its aquifer.

Asbestos. General term applied to minerals of the amphibole and serpentine families that are characterized by a fibrous habit.

Asthenosphere. Zone within the earth between 50 and 250 km, below which seismic waves travel at much reduced speeds, possibly because of less rigidity in the rocks. Presumably, convective flow of material may occur in the asthenosphere, and plates of the lithosphere move over the asthenosphere.

Asymmetric fold. A fold (anticline or syncline) in which the axial plane is inclined, and the two limbs dip unequally in opposite directions.

Atmosphere. The envelope of air surrounding the earth and bound to it by the earth's gravitational attraction.

Atoll. An island or circle of islands surrounding a central lagoon.

Atom. The smallest divisible unit retaining the characteristics of a specific chemical element.

Atomic mass (atomic weight). A quantity essentially equivalent to the number of neutrons plus the number of protons in an atomic nucleus.

Atomic number. The number of protons in the nuclei of atoms of a particular element.

Attitude (of strata). The relation of some directional feature of a rock, such as a bedding plane or joint, to a horizontal plane. Attitude is usually recorded as a statement of strike and dip.

Avalanche. A large mass of snow and ice moving in fast motion down a mountainside. Debris avalanches may be composed of rock waste and lack snow or ice.

Axial plane (of a fold). The plane that divides a fold as symmetrically as possible.

Bajada. A relatively flat surface formed adjacent to a mountain front by the coalescence of a series of alluvial fans.

Barrier reef. A coral reef separated from the coast of the mainland or an island by a lagoon.

Base level. The limiting lower level to which stream erosion can proceed. Ultimate base level is equivalent to sea level.

Basalt. A dark-colored, finely crystalline, generally extrusive igneous rock composed predominantly of ferromagnesian minerals and plagioclase.

Basin. Structurally, a circular or elliptic downwarp of strata with younger beds at the center. In a depositional context, a basin refers to a depressed region that serves as a catchment area for sediments.

Batholith. A very large (at least 100 km^2 in area) intrusive igneous body of irregular shape.

Bathyal. Term that refers to the benthic environment from 200 to 2000 meters in depth.

Bauxite. A mixture of hydrous aluminum oxides and hydroxides commonly formed as a residual clay in tropical and subtropical regions. Bauxite is an ore of aluminum.

Bay barriers. A ridge of sand or gravel that is built up across the mouth of a bay and thus blocks entry into that bay.

Beach. The wave-washed gently sloping accumulation of clastic sediment along a shore. The beach extends from the outermost breakers to the landward limit of wave action.

Benioff seismic zone. An inclined zone along which frequent earthquakes occur and that marks the location of the plunging forward edge of a subducting lithospheric plate.

Benthic environment. Refers to all of the bottom environment from the shoreline to the deepest areas.

Benthos. The organisms that live at the bottom of the ocean, either stationary, attached, or able to move by crawling, burrowing, or swimming near the ocean floor.

Bergschrund. A deep crevasse located at the head of a valley glacier and separating the headwall of a cirque from the ice mass.

Berm. A terracelike portion of a beach or backshore commonly formed by storm waves and composed of wave-transported sand and gravel.

Beta particle. A charged particle, essentially equivalent to an electron, emitted from an atomic nucleus during radioactive disintegration.

Bioclasts. Detrital particles of sediment composed of fragments of the skeletons of marine invertebrates or calcareous algae.

Biofacies. The biologic aspect of a stratigraphic unit that differs discernibly from that of adjacent units. Biofacies are recognized by their fossil content.

Biogenic. A term pertaining to a deposit that originated as a result of physiologic activities of organisms.

Bituminous coal. Compact gray-black brittle coal having 50 to 80 per cent fixed carbon.

Blowout. A basin or shallow pit excavated by wind erosion.

Body seismic waves. Seismic waves that, like P- and S-waves, travel through the "body" of a medium rather than along a free surface.

Bolson. In arid regions, a closed basin or depression more or less rimmed by mountains and characterized by drainage directed inward toward the basin's center.

Bottomset beds. Layers of sediment deposited in an ocean or lake beyond the advancing margin of a delta and eventually covered by the delta.

Breakers. Waves that break along the shore or on encountering a shallowing area such as a reef.

Breccia. A clastic rock composed predominantly of angular fragments of granule size or larger.

Breeder reactor. An atomic reactor capable of producing more fissionable material than it consumes.

Caldera. A large rather circular steep-sided volcanic depression commonly at the summit of a volcano and containing volcanic vents.

Caliche. Calcium carbonate precipitated as surface or near-surface crusts by the evaporation of moisture in the pore spaces of soils.

Capacity (of a stream). Term used to express the maximal amount of solid and dissolved material that a stream can carry under a given set of conditions.

Capillary fringe. That zone immediately above the water table along which water rises in void spaces by capillarity.

Carbonation. The chemical addition of carbon dioxide materials during weathering.

Carbonate spar. A variety of limestone composed of a mosaic of calcite crystals.

Cataclastic metamorphism. Metamorphism involving deformation of rocks by shattering without appreciable chemical reconstitution.

Cation. A positively charged ion.

Cave travertine. Calcium carbonate deposited from solution in limestone caverns, often in the form of stalactites and stalagmites.

Centrifugal force. The apparent outward force experienced by an object moving in a circular path. Centrifugal force is a manifestation of inertia—the tendency for moving things to travel in straight lines.

Chalk. A white soft aphanitic variety of limestone composed largely of the calcium carbonate skeletal remains of marine planktonic organisms.

Chemical element. A substance all of whose atoms have the same atomic number.

Chemical precipitation. The formation of solid particles in a solution as a result of changes in composition, temperature, pressure, or evaporation.

Chemical weathering. The combination of chemical reactions that act on rocks to cause their decomposition. During chemical weathering, preexisting minerals may be dissolved or converted to new minerals more stable under conditions at the earth's surface.

Chatter marks. Scare made in bedrock as debris carried along at the base of a glacier chip into the underlying surface.

Chert. A dense hard sedimentary rock or mineral composed of submicrocrystalline quartz. Unless colored by impurities, chert is white, as opposed to flint, which is gray or black.

Chromosphere. The portion of the sun's atmosphere that lies between the photosphere and the corona.

Cinder cone. A steeply sloping conical hill composed of pyroclastics built up around a volcanic vent.

Cirque. A steep-walled recess in the side of a mountains sculptured by glacial erosion.

Clastic. Consisting of broken fragments of rock, mineral, or skeletal material.

Cleavage. The tendency of certain minerals to split in preferred directions along planes parallel to real or possible crystal faces.

Coal. A combustible earth material containing at least 70 per cent by volume of carbonaceous matter.

Coccoliths. Tiny discoidal calcareous platelets secreted by marine planktonic golden-brown algae known as coccolithophorids.

Columnar joints. Joints that break igneous rocks roughly into long six-sided columns. Columnar jointing is particularly characteristic of basalt sills and flows.

Competence (of a stream). Term used to express the largest particle that a stream can transport under a given set of conditions.

Concordant pluton. An intrusive igneous body having contacts parallel to the stratification of foliation of the preexisting intruded rock.

Cone of depression. Conical depression of the water table in the area surrounding a well in which water is being withdrawn.

Conglomerate. Coarse-grained clastic sedimentary rock composed of rounded rock and mineral fragments larger than 2 mm in diameter.

Connate water. Water trapped in void spaces in a sedimentary rock at the time it was deposited.

Conodonts. Small conical or variously shaped fossils composed of calcium phosphate.

Consequent stream. Streams whose courses are a consequence of an original or preexisting slope.

Continental rise. The ocean floor beyond the base of the continental slope, generally with lower gradient than the continental slope.

Continental shelves. The shelflike gently sloping area extending from the shore of a continent seaward to a line marked by an abrupt increase in slope.

Continental slope. The submerged region of steep slope extending from the seaward edge of the continental shelf down to the upper margin of the continental rise.

Core (of earth). The earth's innermost zone, the outer boundary of which is at a depth of about 2900 km.

Coriolis effect. The tendency for a body moving on the surface of the earth to be deflected to the right in the Northern Hemisphere and the left in the Southern Hemisphere as a result of the earth's rotation.

Corona (of sun). The outermost region of the sun, characterized by temperatures of millions of degrees.

Corrasion. Mechanical erosion accomplished by moving ice, flowing water, or wind as they and the rock particles they contain strike or move against solid rock.

Correlation (stratigraphic). The matching of rock units from different areas in order to determine their age equivalence.

Cosmic rays. Extremely high energy particles, mostly protons, that move through the galaxy and frequently strike the earth's atmosphere.

Country rock. The preexisting rocks that are penetrated by, and that surround, igneous intrusive bodies.

Covalent bonding. Chemical linkage between two atoms produced by sharing electrons in the region between the atoms.

Craton. That portion of a continent that is composed of very ancient crystalline rocks and that has been tectonically stable for several hundred million years.

Creep. The slow downhill movement of regolith, which results from continuous rearrangement of particles. Creep, measured in centimeters per year, is the slowest form of mass wasting.

Cross-bedding (cross-stratification). Beds or laminations arranged at an oblique angle to the main bedding.

Crust. The outer layer of the earth extending from the solid surface to Mohorovičić discontinuity.

Cryoturbation. Mixing and sorting of regolith as a result of alternate thawing and freezing.

Cryptozoic Eon. The span of geologic time that preceded the Cambrian Period.

Cutoff. A relatively new stream channel formed when the stream cuts through the neck of a meander.

Cyclone. Circular wind systems that move in a counterclockwise pattern in the Northern Hemisphere and a clockwise pattern in the Southern Hemisphere.

Darcy's Law. A law that governs the relation between the velocity of percolation of groundwater, the permeability of the aquifer, and the hydraulic gradient. Darcy's Law can be expressed as $V = kh/\ell$, in which V is velocity, k an experimentally determined constant for the water bearing material, h the head, the ℓ the length or distance over which the water moves.

Deep-sea fans. Submerged fan-shaped deposit of clastic sediment often located at the seaward margins of submarine canyons.

Deep-sea trench. A deep narrow trough in the ocean floor inferred to mark the line along which an oceanic tectonic plate is undergoing subduction.

Deflation. The removal of fine-grained sediment from a surface by wind.

Delta. An accumulation of alluvial sediments deposited at the mouth of a stream where it enters a sea or lake.

Dendritic drainage patterns. A drainage pattern that branches irregularly, rather like the pattern of branching in trees.

Density. The mass per unit volume of a substance expressed in grams per cubic centimeter (g/c^3).

Desert pavement. A blanketlike residual concentration of wind-eroded closely packed rock fragments or gravel, which remains at the surface after wind has removed finer particles.

Diatom. Unicellular microscopic plant commonly having a siliceous case or frustule.

Diatreme. A volcanic vent that has been filled with angular rock fragments resulting from explosive eruption.

Dike. A discordant tabular body of igneous rock that cuts across preexisting stratification of structures in the country rock.

Diorite. A usually intrusive igneous rock composed primarily of sodic plagioclase, hornblende, biotite, and/or pyroxene.

Dip. The angle formed by an inclined layer of rock or other planar feature with a horizontal plane.

Discharge (of stream). The amount of water passing through a given cross-section of a stream in a given time.

Discordant pluton. A pluton that cuts across the bedding, structure, or foliation of the country rock it intrudes.

Disintegration. (See *Mechanical weathering.*)

Disseminated magmatic deposit. Ore deposit in which economically valuable minerals are scattered throughout the host rock.

Distributaries. The downstream branches of a stream, as seen in a river that divides upon reaching a delta.

Divergent tectonic plate boundaries. The boundary along which tectonic plates move apart to permit upwelling and the formation of new lithosphere.

Dolostone. A sedimentary rock composed largely of the mineral dolomite, $CaMg(CO_3)_2$.

Dome. A rather symmetric uparching of layered rocks so that they dip about equally away from a center point.

Doppler shift. The change in wavelength that results when the source of waves and the observer are moving relative to each other.

Drainage basin. The area or region that contributes water to a specified stream.

Drainage density. The cumulative length of all the channels in a drainage system divided by the total surface area of the drainage basin.

Drainage divide. The upland tract or ridge that separates adjacent streams or drainage basins.

Drainage system. A given stream and all of its tributaries.

Drift. Earth materials such as clay, sand, gravel, or boulders deposited as a result of glacial activity.

Drumlin. A streamlined hill usually composed of till and having its long axis in the direction of glacial movement.

Dynamothermal metamorphism. (See *Regional metamorphism.*)

Earth flow. The sluggish erratic flow of clayey or silty regolith down relatively gentle slopes.

Echo-sounding. Geophysical method for determining depth of water by measuring the time required for a sound signal to travel to bottom and return.

Ekman spiral. The theoretic representation of the effect of a steady wind blowing across an ocean of uniform viscosity and unlimited depth and breadth, such that the surface layer of water would move at an angle of 45° to the right of the wind direction in the Northern Hemisphere and 45° to the left in the Southern Hemisphere.

Elastic limit. The maximal amount of stress to which an earth material can be subjected before it will begin to deform permanently by fracture or flow.

Elastic rebound. The abrupt release of elastic strain that has slowly accumulated in a rock mass.

Electron. A component of an atom that exists outside the nucleus, has very low mass, and carries one unit of negative charge.

Entrenched meanders. Meanders deeply incised into bedrock as a result of regional uplift.

Ephemeral stream. A stream or stream segment that derives its discharge almost entirely from precipitation and is thus often dry between periods of rainfall.

Epicenter. A point of the earth's surface directly above the focus (true center) of an earthquake.

Epicontinental marginal seas. Depressed submerged areas of the continental margins having greater topographic irregularity than the continental shelves.

Epoch. The chronologic subdivision of a geologic *period.* Rocks deposited during an epoch constitute the series for that epoch.

Erratic. A rock fragment that has been transported from a distant source area, usually by glacial or floating ice, and that rests on bedrock of a different kind.

Esker. A sinuous ridge composed of stratified drift.

Eugeosyncline (Eugeocline). The oceanward and more unstable portion of a geosyncline, characterized by great thickness of poorly sorted clastic sediment, siliceous sediment, and volcanics.

Eustatic. Worldwide simultaneous changes in sea level, such as may result from a change in the volume of continental glaciers.

Evaporites. Sediments precipitated from a water solution as a result of evaporation. Evaporite minerals include anhydrite gypsum ($CaSO_4$) and halite (NaCl).

Exfoliation. The breaking off of successive thin outer shells of a rock mass in response to weathering.

Facies. A particular aspect of one part of a rock body that is distinct from adjacent parts.

Fault breccia. A breccia, the angular rock fragments of which were produced by crushing between the walls of a fault.

Fault plane. A generally planar fault surface.

Feldspars. A group of abundant rock-forming aluminosilicate minerals containing sodium, calcium, or potassium and having framework atomic structure. Feldspars are the most common of any mineral group, constituting an estimated 60 per cent of the earth's crust.

Ferromagnesian minerals. Generally dark-colored silicate minerals containing iron and magnesium and including olivine and pyroxene.

Fetch. The continuous area of water across which wind blows in a constant direction and generates waves.

Filter pressing. The process whereby the liquid portion of a partially crystallized magma is squeezed out when the magma is compressed by earth movements.

Firn. Snow that has been partially consolidated by successive thawing and freezing but has not yet been altered into compact glacial ice.

Fissility. The property of rocks, including shale, which causes them to split into thin slabs parallel to bedding.

Fission, nuclear. The splitting of an atomic nucleus.

Fjord. A narrow arm of the sea representing a former stream valley that had been further eroded by a glacier. Fjords are recognized by their steep walls uneven bottom profile, and streams that enter as waterfalls and rapids.

Flood basalts (plateau basalts). Layered lava flows that issued from fissures and that extend over an entire region as a sequence of flat or nearly flat layers.

Flood plain. The lowland that borders a stream, is composed of sediments deposited by the stream, and that is dry except when the steam overflows its banks during flood stages.

Focus (of an earthquake). The true center of an earthquake and the point at which rupture occurs and strain energy is converted to elastic wave energy.

Fold. A bend or flexure in layered rocks.

Foliation. A textural feature especially characteristic of metamorphic rocks in which laminae develop by growth or realignment of minerals into roughly parallel orientation.

Footwall. The mass of rock that lies beneath an inclined fault.

Foraminifers. Mostly marine and microscopic protozoans that commonly secrete skeletons composed of calcium carbonate.

Foreset beds. The inclined layers deposited along the forward slope of a body of sediment that is advancing, as in a delta or a dune.

Formation. A mappable lithologically distinct body of rock having recognizable contacts with adjacent rock units.

Fractional crystallization. The separation of components of a cooling magma by sequential formation of particular mineral crystals at progressively lower temperatures.

Fringing reef. Coral reef that is attached to and thus forms a fringe around an island.

Fumaroles. Vents open at the earth's surface that emit volcanic gases, steam and hot water.

Fusion. The joining of nuclei of atoms into heavier nuclei.

Gabbro. A coarsely crystalline (phaneritic) igneous rock having the same mineralogic composition as basalt.

Galaxy. An aggregate of stars and planets separated from other such aggregates by distances greater than those between member stars.

Galilean moons. The four brightest moons of Jupiter (lo, Europa, Ganymede, and Callisto).

Gamma rays. Very high frequency short-wavelength electromagnetic waves.

Geologic map. A map that depicts the surface distribution of the rock types that crop out over a given area and that shows structural features and age relationships of rock masses.

Geologic period. A subdivision of an era.

Geologic system. A time-rock unit representing the rock deposited or emplaced during a geologic period.

Geosyncline. A great linear basin in which immense thicknesses of sediment accumulate over an extended period of time and which may ultimately be compressed to form mountain systems.

Geothermal gradient. The change in temperature of the earth with depth.

Geyser. A hot spring or fountain from which heated water or steam is emitted intermittently.

Glacial surging. Abrupt and sometimes catastrophic forward movement of a glacier.

Glacier. A large mass of ice, having definite lateral limits, that flows under the influence of gravity.

Glacioeolian. Pertaining to sediments deposited by glaciers and then recycled as a result of wind action.

Glaciofluvial. Pertaining to the deposits of streams flowing from glaciers.

Glaciolacustrine. Pertaining to the deposits of lakes that owe their origin to glaciation.

Globigerina ooze. A widespread pelagic calcereous ooze in which at least 30 per cent of the sediment consists of tests (shells) of planktonic foraminifers, including species of the genus *Globigerina.*

Glossopteris flora. An assemblage of fossil plants prevalent in rocks of Late Paleozoic and Early Triassic age in Southern Hemisphere continents.

Gondwanaland. The great Permo-Carboniferous Southern Hemisphere continent, comprising the present continents of South Africa, South America, India, Australia, Africa-Arabia, and Antarctica.

Graben. A troughlike down-faulted mass bounded on either side by normal faults.

Graded stream. A stream in a state of equilibrium so that any change in load, velocity, or discharge will result in a displacement of equilibrium to absorb the effects of the changes.

Gradient (of stream). The slope of a stream channel between any two specified points.

Granitization. The formation of granite from previously existing rocks by metamorphic processes. Granitization does not require that the origin of granite be directly from a silicate melt.

Gravity settling. The gravitational settling out of early formed minerals in a magma.

Graywacke. A poorly sorted sandstone, composed of angular quartz, chert, and feldspar grains as well as mica, set in a clayey matrix and often exhibiting graded bedding.

Greenschist facies. Metamorphic rocks developed under conditions of relatively low temperature and pressure and usually including greenschists, which owe their color to dominant chlorite minerals.

Greenstone. Metamorphosed basic igneous rocks, which owe their color to the presence of chlorite, epidote, and hornblende.

Groundwater. Underground water that is in the zone of saturation.

Ground moraine. Glacial debris dropped from a glacier onto the ground surface over which the glacier has moved.

Guide or Index fossils. Fossilized organisms noted for their wide geographic distribution and narrow stratigraphic range. Such fossils are especially useful in determining the age and correlation of strata.

Gutenberg discontinuity. The surface separating the mantle of the earth from the core below. The Gutenberg discontinuity lies at about 2900 km below the earth's surface.

Guyot. A flat-topped seamount.

Gyre. A circular system of ocean currents of regional extent, which rotates in circular motion in each of the major ocean basins.

Hadal. Very deep zones of the ocean; generally greater than 6500 meters in depth.

Half-life. The time in which one-half of an original amount of a radioactive species decays to daughter products.

Hanging valley. A tributary valley whose floor at the point of junction with the main valley lies above the main valley floor.

Hanging wall. The rock mass that lies above an inclined fault.

Headward erosion. The lengthening of a gulley or valley in an upslope direction by water that flows into it at its upper end.

Hinge (of a fold). The line along which a fold has its maximal curvature.

Horn. A steep-sided pyramidal peak formed by glaciers that have eroded cirques on three or more sides.

Horst. An elongate block of crustal rock that is bounded by normal faults and stands conspicuously above adjacent downfaulted blocks.

Hot spot. A relatively localized area near the top of the mantle from which basaltic lava is fed upward. If the hot spot remains fixed in position as a lithospheric place moves over it, the result may be a chain of volcanoes.

Humus. Organic matter in soil that is so thoroughly decayed that one cannot ascertain the nature of the parent organisms.

Hydration. A process whereby water is absorbed by a mineral and incorporated into the weathering product.

Hydraulic gradient. The slope of the water table expressed as h/ℓ, where h is the head (difference in elevation between two points on the water table) and ℓ is the distance between these two points.

Hydraulic head. As related to groundwater, the difference in elevation between any two points on the water table.

Hydrolysis. An important chemical process in weathering of minerals whereby a compound is split into two parts, with the hydrogen from water uniting with one part of the compound to make an acid, leaving the hydroxide ion; or the hydroxide ion uniting with one part to form a base, leaving the hydrogen ion.

Hydrosphere. Generally, the water portion of the earth (oceans, rivers, groundwater) as opposed to the solid part.

Hydrothermal or deposit. One minerals precipitated in rock from hot watery solutions.

Icecap. A great domal or platelike ice mass generally less than 50,000 km^2 in area and unconstrained by topographic features.

Igneous rocks. Rocks formed by the cooling and solidification of magma or lava.

Ignimbrites. Volcanic rocks formed by the welding together of explosively ejected hot partly molten particles.

Immature sandstone. A poorly sorted sandstone that contains abundant, relatively unstable, often angular constituents. Graywacke is an example of an immature sandstone.

Incised meanders. (See *Entrenched meanders.*)

Intermittent stream. Streams that are dry part of the year, that contain water at other times of the year, and that may also receive water from the groundwater reservoir.

Inertial. The property of a body that causes it to remain at rest or at constant velocity unless forced to change.

Insolation. In general, solar radiation received at the earth's surface.

Intraclasts. Nonskeletal particles of sand size or larger that have been broken from nearby consolidated carbonate sediment and incorporated into younger deposits.

Ion. An atom or group of atoms having a positive or negative electric charge because of having gained or lost electrons.

Ionic bonding. Chemical bonding by transfer of electrons to provide an attraction between oppositely charged ions.

Isobar. In meteorology, a line connecting points of equal pressure along some given reference surface such as sea level.

Isograd. In studies of regional metamorphism, a line connecting points of equal metamorphic intensity as indicated by metamorphic index minerals.

Isoseismal line. A line connecting points of equal intensity of earthquake shock.

Isostasy. A condition of flotational equilibrium between adjacent parts of the lighter crust on denser underlying layers.

Isotope. An atom having the same number of nuclear protons as another atoms but with a different number of neutrons.

Joint. A fracture in a rock along which little or no movement has occurred.

Jovian planets. The giant planets Jupiter, Saturn, Uranus, and Pluto.

Juvenile water. Water of volcanic or magmatic origin added to the underground and surface water supply.

Kame. A mound or ridge of gravel deposited in contact with glacial ice.

Kame terrace. A body of stratified drift, with a terracelike configuration, that was deposited between a wasting glacier and the confining valley wall.

Karst topography. Topography developed by solution of carbonate bedrock and recognizable by an abundance of sinkholes, caverns, and underground drainage.

Kerogen. The waxy bituminous substance in oil shales that yields oil when appropriately processed.

Kettle. A depression in glacial drift usually created by the melting of a stagnant block of ice that was enclosed by sediment.

Klippe. The remnant of an overthrust sheet, presently isolated from the main thrust sheet by erosional loss of the connecting portion.

Laccolith. An intrusive igneous body that has domed upward the strata that it has intruded.

Lahars. A debris flow composed largely of volcanic ash, breccia, and boulders mixed with rain water or water derived from a failed natural reservoir.

Laminar flow (of water). Smoothly flowing water in which the paths taken by individual water molecules are relatively straight or gently curved and parallel to the paths of neighboring molecules.

Lateral moraine. A ridgelike accumulation of till derived from valley sides and deposited along the margins of a glacier.

Laterite. A soil that forms in humid tropical regions and is characterized by its abundance of iron and aluminum hydroxides.

Laurasia. A hypothetical former supercontinent composed of land masses presently comprising Europe, Asia, Greenland, and North America.

Lava. Molten rock that has reached the earth's surface and is extruded in the course of vent and fissure eruptions.

Lava tunnels. A cavernous opening formed where lava has drained out from an area beneath a hardened surface crust.

Leeward. The downwind side of a dune or other obstacle in the path of wind.

Lignite. A brownish-black soft coal formed by the burial of peat.

Limestone. Sedimentary rock composed mainly of calcium carbonate.

Lineation (in metamorphic rocks). A parallel arrangement of prismatic or needlelike crystals in a metamorphic rock.

Liquefaction. The process of water being added to unconsolidated sediment during an earthquake so that the sediment loses all strength and flows as a thick mush.

Lithographic limestone. A very fine-grained uniformly textured limestone.

Lithosphere. The outer shell of the earth, which lies above the asthenosphere and comprises both the crust and uppermost mantle above the low-velocity zone. The term also refers to the solid earth in the tripartite division of the earth into atmosphere, hydrosphere, and lithosphere.

Lithostatic pressure. The equidimensional pressure in the lithosphere, which is due to the weight of overlying rock.

Littoral zone. The marine environment located between high- and low-tide levels.

Loess. A soft porous buff-colored accumulation of mostly silt-size wind-deposited particles.

Longitudinal dune. A parallel relatively straight dune having length many times greater than width.

Longshore bars. Sand ridges that are barely submerged and extend parallel to a shoreline.

Longshore current. Ocean or lake current that flows parallel to a shore and is most commonly generated by the oblique approach of waves.

Lopolith. A large concordant igneous intrusion in which the intruded strata generally dip inward toward a central point.

Love surface waves. A surface seismic wave that vibrates in a direction that is transverse to the direction of wave travel and that lacks a vertical component of vibration.

Low-velocity zone. An interior zone of the earth, immediately below the uppermost mantle, in which seismic wave velocities sharply decrease. Theoretically, the low-velocity zone may be the plastic layer over which lithospheric plates move.

Mafic. Term used to describe dark-colored or basic minerals and rocks that contain abundant iron and magnesium and relatively low amounts of silicon.

Mantle. An interior zone of the earth that lies above the core and beneath the crust.

Mare. One of the smooth areas on the moon or on some other planets.

Mascons. Concentrations of mass under the surface of the moon, discovered from their gravitational effect on spacecraft that have encircled the moon.

Mass spectrometer. An instrument for separating ions of different mass but equal charge and measuring their relative quantities.

Mass wasting. Those processes by which earth materials are moved downslope by gravity.

Mechanical weathering. The process by which rock is physically broken into smaller pieces. Also termed disintegration.

Medial moraine. A single elongate ridge composed of till and formed by the confluence of two adjacent lateral moraines.

Mélange. A body of intricately folded, faulted, and severely metamorphosed rocks deformed at convergent plate boundaries. An example is the Franciscan sequence of California.

Mesosphere. The lower part of the earth's mantle.

Metamorphic aureole. The zone of metamorphic alteration that surrounds an igneous intrusion.

Metamorphic facies. Metamorphic rocks that, by their mineral composition, can be inferred to have formed under a particular range of temperature and pressure conditions.

Metamorphism. The transformation of preexisting rocks into new types by the action of heat, pressure, and chemical solutions.

Meteor. A track of light in the sky from rock or dust burning in the earth's atmosphere.

Meteoric water. Water derived from the atmosphere.

Meteorite. An interplanetary piece of rock that has impacted on the earth, moon, or another planet.

Micrite. Mechanically deposited and lithified lime mud having particles in the size range from 1 to 3 microns.

Micrometeorites. Meteorites that are generally less than 1 mm in diameter.

Mid-oceanic ridge. A major median mountain range with rugged topography and a central rift zone that extends through the North and South Atlantic, Indian, and South Pacific Oceans, and that delineates the belt along which tectonic plates move apart and new oceanic crust is created.

Migmatite. A rock having both igneous and metamorphic properties, probably formed by intense metamorphism without actual melting.

Mineral. A natural chemical element or compound having a limited range in chemical composition, distinctive properties, and form that reflects its atomic structure.

Miogeosyncline (miogeocline). The inner side of a geosyncline characterized by thick accumulations of shale, carbonates, and sandstones but generally lacking in volcanics.

Mohorovičić discontinuity. The seismic discontinuity that marks the boundary between the earth's crust and mantle.

Molecule. The fundamental particle of a compound, composed of a group of atoms joined by chemical bonds, which displays the properties of that compound.

Moraine. An accumulation of till carried on a glacier or left behind by a glacier.

Mudflow. A relatively rapid flow of regolith that has the consistency of mud and behaves as a viscous fluid.

Mylonite. A fine-grained lithified fault breccia developed by the grinding action associated with movement along major faults.

Nappe. A large mass of rock that has been moved a considerable distance over underlying formation either by thrust-faulting, recumbent folding, or both.

Natural levee. An alluvial ridge built up on either side of a stream channel by deposition from water that spilled from the channel during times of flooding.

Nekton. Aquatic organisms that are able to direct their own movements by swimming, as opposed to those that drift with currents.

Neutron. An electrically neutral particle of matter existing along with protons in the atomic nucleus of all elements except the mass 1 isotope of hydrogen.

Norite. A hypersthene-rich variety of gabbro (hypersthene is a member of the pyroxene group).

Normal fault. A fault in which the hanging wall has been depressed relative to the footwall.

Nuée ardente. A highly heated mass of gas-charged lava ejected from the side of a volcano's summit and traveling swiftly down the flanks of the volcano.

Nunatak. A mountain peak that projects above the surface of an ice field.

Oblique-slip fault. A fault in which the direction of movement is between the strike and direction of dip.

Obsidian. Volcanic glass.

Oceanic crust. The type of crust that forms the floor of the oceans. It is about 5 km thick, has a density of about 3.0 g/cm^3, and has a basaltic composition.

Oceanic zone. The realm of the open ocean beyond the edge of the continental shelf.

Oil shale. A variety of shale that on heating will yield gaseous hydrocarbons.

Oöids. Spherical carbonate particles, mostly of sand size, that are constructed of concentric calcium carbonate laminae.

Ooze. A fine-grained deposit of the deep ocean floors which contains more than 30 per cent skeletal remains of marine organisms.

Ore. A general term for a metalliferous natural deposit that can be worked at a profit.

Orogeny. The deformational crustal process by which great mountain systems are formed.

Outwash plain. The area at the front of a glacier on which glaciofluvial sediments are deposited.

Overturned fold. A fold in which one limb has been "overturned," that is, rotated through more than 90°.

Oxbow lake. A horseshoe-shaped lake formed in an abandoned meander.

Oxidation. The process by which oxygen combines with a different element or compound.

Pahoehoe. A basaltic lava with a ropy surface texture.

Paleomagnetism. The earth's former magnetic field and magnetic properties.

Pangaea. In Alfred Wegner's theory of continental drift, the former supercontinent that included all of the present major continental masses.

Panthalassa. The great universal ocean that surrounded the supercontinent Pangaea prior to its breakup.

Parabolic dunes. A dune that in plain view has the shape of a parabola, with the concave side facing toward the wind.

Partial melting. The process by which a rock mass subjected to high temperatures and high pressure is partly melted and the liquid component is moved to another location where it may serve as a source for rocks of different composition.

Peat. The dark brown or black residue produced by alteration of plant debris in a swampy, boggy, or marshy environment.

Pedalfer. A soil formed in regions of ample rainfall and characterized by its abundance of aluminum and iron.

Pediment. A relatively smooth, gently sloping, primarily erosional surface formed at the front of a mountain range and often covered with a thin veneer of alluvium.

Pedocal. A calcium-rich soil that generally develops in arid or semi-arid regions having prairie or desert scrub vegetation.

Pegmatite. Extremely coarsely crystalline igneous rock.

Pelagic environment. In general, pertaining to the waters of the open ocean, as distinct from the ocean bottom environment. Includes the entire mass of water in the neritic and oceanic environments.

Peneplain. A low-lying relatively smooth to gently undulatory regional erosional surface that has developed after a long episode of erosion.

Perched water table. Groundwater held above the main water table by an impermeable stratum or other natural barrier.

Peridotite. A phaneritic equigranular dark-colored igneous rock containing abundant olivine or pyroxene or hornblende, but lacking quartz and feldspar.

Permeability. The ability of a rock to transmit fluids.

Phaneritic. Pertaining to a rock texture in which constituent mineral grains or crystals are large enough to be seen with the unaided eye.

Phanerozoic Eon. An eon of geologic time during which the earth was populated by a diversity and abundance of organisms. The Phanerozoic includes the Paleozoic, Mesozoic, and Cenozoic Eras.

Phenocrysts. In igneous rocks, large crystals that occur scattered through an otherwise more finely crystalline rock.

Photosphere. The region of a star from which most of its light radiates.

Photosynthesis. Process by which plants synthesize chemical compounds by utilizing the radiant energy of the sun.

Phytoplankton. Floating or drifting forms of aquatic plant life such as diatoms, dinoflagellates, coccolithophores, and sargassum weed.

Pillow lavas. Spheroidal or billowy masses of lava formed by extrusion beneath a body of water.

Placer deposit. Sediment rich in a valuable mineral or metal that has been concentrated because of its high specific gravity.

Plate tectonics. Large-scale movements and interactions of the many rigid plates of the earth's lithosphere.

Playa. The low-lying flat center area of an undrained desert basin.

Pleistocene Epoch. A subdivision of geologic time that includes the last glacial age.

Pluton. A generally deep-seated intrusive igneous body.

Pluvial lake. A lake formed during a former rainy climate in an area not associated with glaciation.

Podzol. An ashy gray bleached soil, deficient in iron and lime and formed under cool moist conditions.

Point-bar deposits. Sand and gravel deposited on the inner margins of meander bends where stream velocity is lower.

Porphyritic. A textural term for igneous rocks in which large crystals (phenocrysts) are enclosed in a matrix of much more finely crystalline minerals or glass.

Porphyroblasts. A large crystal, in a metamorphic rock, that is enclosed in a matrix of finer-grained minerals.

Porosity. The percentage of the total volume of a rock that consists of void spaces.

Pothole. A small basin eroded into the rock of a stream bed by the abrasive work of sand and gravel transported by the stream.

Primary seismic waves. A compressional seismic wave in which vibration occurs parallel to the direction of propagation.

Proton. A fundamental particle that is found in the nuclei of all atoms and that has a positive electric charge and mass similar to that of a neutron.

Pumice. A very light "frothy" glassy lava.

Pyroclastics. Volcanic rock that has been fragmented during ejection.

Pyroxene. A group of iron- and magnesium-rich silicate minerals having single-chain tetrahedral structure.

Quarrying (glacial). The process by which pieces of jointed or fractured bedrock are lifted out of place by glaciers.

Radial drainage pattern. Stream pattern in which streams radiate outward from an elevated central area rather like spokes on a wheel.

Radiolaria. Marine protozoans that build complex internal siliceous skeletons.

Rayleigh surface waves. A seismic wave propagated along the earth's surface and involved in the phenomenon of ground roll during an earthquake.

Reach (of a stream). A longitudinal segment of a stream between any two specified points.

Recumbent fold. A fold that has been so severely overturned that the axial plane is horizontal.

Regional metamorphism. Metamorphism that occurs across a vast area (region) and results from deep burial or the high pressures and temperatures associated with deformation and mountain-building.

Regolith. The loose material overlying bedrock. On planets other than earth, the fragmented and powdered surface material formed by repeated meteoritic bombardment.

Remanent magnetism. Permanent magnetism induced in an igneous rock at the time it cooled through the Curie point or induced in a sedimentary rock at the time of deposition by the rotation that occurs during settling of magnetic crystals so as to become aligned with the earth's magnetic field.

Reverse fault. A fault formed by compression in which the hanging wall appears to have moved up relative to the footwall.

Rhyolite. A generally aphanitic or porphyritic igneous rock having a composition similar to granite.

Rift valley. A valley whose development was controlled by faults. Also applied to deep fracture zones extending along the crest of mid-oceanic ridges.

Rip current. A strong narrow current representing the seaward return of water piled up near shore by waves.

Ripple marks. An alternating sequence of small troughs and ridges produced on the surface of fine clastic deposits by wind or water movements.

Roches moutonnées. Large rounded boulders that have been overrun by a glacier.

Rock glaciers. Glacierlike masses of unsorted rock waste found near or above the timberline in high mountains.

Rock-slide. The rapid free sliding along a sloping joint or bedding plane of recently loosened segments of bedrock.

Saltwater intrusion. A phenomenon usually associated with groundwater in which salt encroaches on the space occupied by fresh water, often as a result of too rapid withdrawal of fresh water from wells.

Saltation. The process by which particles of sediment are repeatedly lifted by currents, carried for a short distance, and than dropped.

Sandstone. A sedimentary rock composed of detrital grains ranging in size form 1/16 to 2 mm in diameter.

Scoria. Highly vesicular, usually basic, dark-colored, and cinderlike congealed volcanic ejecta.

Sea arch. A natural arch produced in the side of a headland by wave erosion.

Sea floor spreading. The process by which new oceanic crust is produced along mid-oceanic ridges (divergence zones) and slowly conveyed away from the ridges.

Seamounts. A submerged isolated mountain, commonly of volcanic origin, rising conspicuously above the ocean floor.

Secondary enrichment. The enrichment of a relatively low-grade ore body by solutions bearing metals derived from decomposition of nearby rocks.

Secondary seismic waves. A seismic wave in which the direction of vibration of wave energy is perpendicular to the direction of wave propagation.

Sediment. Any earth material carried and deposited by streams, winds, or glaciers, or precipitated from solution, or moved by gravity. Sediment may also include deposits of organic origin, such as exist in algal or coral reefs.

Sedimentary rocks. Rocks formed from accumulations of sediment.

Seiche. A large wave of oscillation set up in an enclosed basin, like a bay or lake, by the motion of an earthquake.

Seif. Term used to describe a large longitudinal dune, many of which exceed 100 km in length.

Seismograph. An instrument used to measure motions of the earth's surface caused by seismic waves.

Shale. A laminated or fissil fine-grained detrital sedimentary rock composed primarily of silt and clay.

Sheeting. The development in massive igneous rocks of closely spaced joints roughly parallel to the ground surface.

Shields. Broadly convex regional expanses of exposed or thinly covered Precambrian igneous and metamorphic rocks.

Shield volcano. A broadly convex (like a Roman shield) volcanic cone built of relatively low-viscosity lava flows.

Silica tetrahedron. A fundamental arrangement of silicon and oxygen ions in silicate minerals, consisting of a centrally located silicon ion linked to four oxygen ions placed symmetrically in tetrahedral arrangement around the silicon.

Sill. A tabular intrusion of magma that lies more or less parallel to preexisting stratification.

Sinkhole. A depression or small basin found in areas of karst topography and caused by solution of limestone bedrock near the surface or by collapse of a cavern's roof.

Sinuous rilles. Term used to describe the winding channels detected on the surface of Mars and believed to have been formed by erosion from flowing water.

Slaty cleavage. That type of foliation, characteristic of slates, resulting from a parallel alignment of flaky minerals.

Slump. A form of slide motion in which a mass of rock or regolith slips along a concave upward curved surface, while simultaneously rotating backward on a horizontal axis.

Soil. Weathered earth material that will support the growth of rooted plants.

Solifluction. The slow flowage over permafrost of soil and rock debris that is water-saturated and that may be subjected to alternate freezing and thawing.

Solution valley. A valley formed by the dissolving of limestone, gypsum, or salt at the surface, or by the collapse of the roof of subsurface solution channels.

Sorting. A measure of the uniformity of particle sizes in a sediment or sedimentary rock.

Spatter cone. A cone composed of lava that was ejected as globs of molten rock around the perimeter of a volcanic vent.

Specific gravity. The weight of a given volume of a mineral (or substance) divided by the weight of an equal volume of water at 4°C.

Speleologist. Someone engaged in the study of caverns and associated features.

Spheroidal weathering. The concentric spalling off of layers of rock to produce rounded boulders.

Spit. A sandy bar projecting into a body of water from a coast.

Stacks. Irregular columnlike masses of rock standing a short distance from a coastline and formed when waves are able to remove segments of a promontory.

Stalagmites. Pillars and protuberances of calcium carbonate built upward from a cavern floor by solutions dripping from the ceiling.

Stalactites. Icicle-like deposits of calcium carbonate suspended from the roof of a cavern.

Stock. A roughly circular intrusive mass of rock covering less than 100 km^2 in map area and having nearly vertical contacts with country rock.

Stoping. A process whereby rising magma engulfs blocks of country rock that then sink into the magma and are assimilated.

Strain. Changes in the shape of a body as a result of forces that are applied to that body.

Stratified drift. Deposits of glacial sediment that exhibit sorting and crude stratification suggesting transport and deposition by meltwater.

Strato-volcano. A usually large volcanic cone constructed of alternate layers of pyroclastics and congealed lava.

Stratum. A single coherent layer of sedimentary rock.

Stress. The force per unit area applied to a rock or other material.

Strike. A direction indicated by the intersection of an inclined bed or structure and a horizontal plane.

Subduction zone. An inclined planar zone, recognized by its high frequency of earthquakes, that is thought to locate the descending leading edge of a lithospheric plate.

Submarine canyons. A steep-sided underwater canyon cut into the continental shelf or slope.

Subsequent stream. A stream having a course determined by lithologic or structural variations in the bedrock over which it flows.

Surface currents (oceanic). Slow drifts of ocean water set in motion as a result of prevailing winds that blow across the water surface.

Suspended load. The sediment of relatively small particle size that remains suspended in a body of flowing water because the settling velocity is exceeded by the upward velocities of turbulent eddies.

Swell. A broad low-amplitude wind-generated wave that has traveled a considerable distance from the area where is was originally developed.

Symmetric fold. A fold having limbs that dip in opposite directions but at the same angle.

Syncline. A downfold in which the strata of the limbs dip toward the axis and which, on erosional truncations, exhibits younger beds along the axis.

Talus. A pile of rock fragments that has accumulated at the base of a cliff or ridge.

Tectonic. Pertaining to the broader large-scale structural features of the earth and how they originated.

Terminal moraine. A ridgelike accumulation of till that forms around the snout of a glacier.

Thermohaline oceanic currents. Deep-water flow of water controlled by the temperature and salinity of the water.

Thixotropic clays. Clayey wet sediments that become unstable and fluid when disturbed by earthquake or artificial vibrations.

Tholeiites. Basalts composed largely of plagioclase, pyroxenes, and iron oxide minerals occurring as phenocrysts in a glassy matrix. Olivine is usually absent.

Thrust fault. A reverse fault having a fault plane inclined at a low angle from the horizontal plane.

Tidal bore. A tide-generated wave that advances up a river or estuary and usually has a distinct or abrupt front.

Till. Unsorted unstratified glacial debris deposited directly from from glacial ice.

Tillite. A sedimentary rock composed of lithified till.

Time-distance graph. A graph on which the travel time of seismic waves is plotted against distance.

Time-rock unit. Rocks formed during a particular segment of geologic time. The Triassic System, for examples, is the time-rock for the segment of time known as the Triassic Period.

Tombolo. A sand or gravel bar that connects an island to the mainland.

Topset beds. Sediment deposited in horizontal layers over the foreset beds of a delta.

Traction. The combined processes of rolling and sliding by which mineral and rock fragments are transported along the stream bottom.

Transform fault. A strike-slip fault characteristic of mid-oceanic ridges and along which ridges are offset.

Transverse dune. A dune extending transverse to the direction of the dominant winds, having a steeper leeward than windward slope, and most characteristic of desert areas having scanty vegetation.

Travertine. Porous masses of calcium carbonate, usually deposited by groundwater in caverns or as incrustations around springs.

Trellis drainage pattern. Stream pattern in which tributary streams enter main streams nearly at right angles because of constraints imposed by folded and beveled rocks having differing resistance to erosion.

Tsunami. A high-velocity sea wave of long period that has been generated by a submarine earthquake, volcanic eruption, or massive landslide.

Turbidity current. A current of turbid water that, because of its content of suspended sediment, is denser than surrounding water and flows down submarine slopes.

Turbulent flow (of water). Type of flow in which water particles move in irregular swirls and eddies.

Unconformity. A surface of erosion or nondeposition that represents a gap in the geologic record and separates younger strata from older strata or rocks.

Uniformitarianism. A general principle that suggests that the past history of the earth can be interpreted in accordance with our knowledge of natural processes still operating today.

Unloading. The removal of great masses of rock, usually as the result of long episodes of erosion.

Varve. A thin sedimentary layer or pair of contrasting layers that represent the depositional record of a single year.

Ventifact. A fragment of rock that has been faceted by the erosive action of wind-driven particles of sediment.

Volatiles. Generally pertaining to products other than water vapor, given off by a material as gas or vapor.

Water gap. A pass in a topographic ridge through which a stream is flowing.

Water table. The upper surface of the zone of saturation.

Wave base. The water depth below which wave action is insufficient to disturb sediment on the sea floor.

Wave-built terrace. A level or gently sloping embankment build seaward from a beach by wave deposition.

Wave-cut bench. A gently sloping surface formed by wave erosion and extending seaward from the base of a wave-cut cliff.

Wave-cut cliff. A cliff produced as a result of landward erosion by incoming waves along a coast.

Wave frequency. The number of wavelengths that passed a fixed point per unit of time.

Wave height. The vertical distance between a wave's trough and the preceding crest.

Wavelength. The horizontal distance between two successive wave crests or wave troughs.

Wave period. The amount of time that elapses as two successive crests or troughs pass a fixed point.

Wave velocity. The speed at which a wave progresses.

Welded tuff. A rock composed of ash-size pyroclastic materials welded together while hot.

Windgap. A pass in a topographic ridge through which a stream formerly flowed.

Windward. The direction from which a wind is blowing.

Xenolith. A piece of country rock that has been engulfed, but not yet assimilated, by an invading magma.

Yazoo stream. A tributary stream that, because of the natural levee on the main stream, flows parallel to the larger stream until it can join it through a pass in the levee.

Zone of aeration. The zone above the water table in which interstices are partly filled with air and water is seeping toward the water table.

Zone of saturation. The zone beneath the water table in which all interstices are filled with groundwater.

Zooplankton. Floating or drifting forms of aquatic animal life.

Index

In this index, definitions are indicated by boldface page numbers. Italicized page numbers indicate illustrations. A small t following a page number indicates a table. The index includes formation names in the correlation charts. Terms in the glossary are not included.

188
95
96
279

.93
3 | 275
270
9

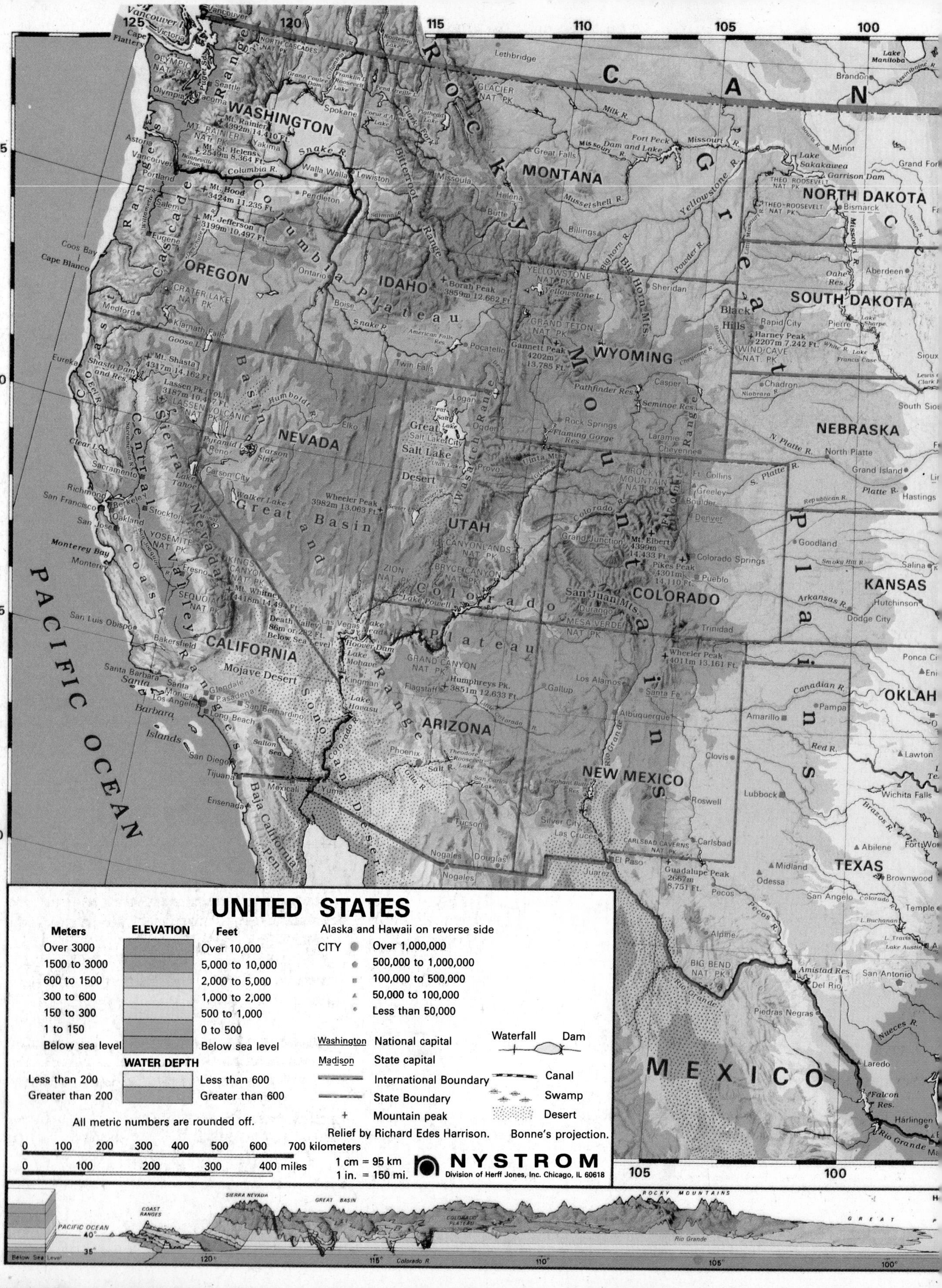
UNITED STATES
Alaska and Hawaii on reverse side
ELEVATION
Meters
Feet
Over 3000
Over 10,000
1500 to 3000
5,000 to 10,000
600 to 1500
2,000 to 5,000
300 to 600
1,000 to 2,000
150 to 300
500 to 1,000
1 to 150
0 to 500
Below sea level
Below sea level
WATER DEPTH
Less than 200
Less than 600
Greater than 200
Greater than 600
All metric numbers are rounded off.
CITY
Over 1,000,000
500,000 to 1,000,000
100,000 to 500,000
50,000 to 100,000
Less than 50,000
Washington National capital
Madison State capital
International Boundary
State Boundary
Mountain peak
Waterfall
Dam
Canal
Swamp
Desert
Relief by Richard Edes Harrison.
Bonne's projection.
0 100 200 300 400 500 600 700 kilometers
0 100 200 300 400 miles
1 cm = 95 km
1 in. = 150 mi.
NYSTROM
Division of Herff Jones, Inc. Chicago, IL 60618
PACIFIC OCEAN
CANADA
MEXICO
WASHINGTON
OREGON
CALIFORNIA
NEVADA
IDAHO
MONTANA
WYOMING
UTAH
COLORADO
ARIZONA
NEW MEXICO
NORTH DAKOTA
SOUTH DAKOTA
NEBRASKA
KANSAS
TEXAS
Rocky Mountains
Great Plains
Columbia Plateau
Great Basin
Colorado Plateau
Cascade Range
Coast Ranges
Sierra Nevada
Central Valley
Sonoran Desert
Mojave Desert
Great Salt Lake
Salt Lake Desert
Mt. Rainier 4392m 14,410 Ft.
Mt. Hood 3424m 11,235 Ft.
Mt. Shasta 4317m 14,162 Ft.
Mt. Whitney 4418m 14,494 Ft.
Borah Peak 3859m 12,662 Ft.
Gannett Peak 4202m 13,785 Ft.
Mt. Elbert 4399m 14,433 Ft.
Pikes Peak 4301m 14,110 Ft.
Wheeler Peak 3982m 13,063 Ft.
Wheeler Peak 4011m 13,161 Ft.
Humphreys Pk. 3851m 12,633 Ft.
Harney Peak 2207m 7,242 Ft.
Guadalupe Peak 2667m 8,751 Ft.
Death Valley 86m or 282 Ft. Below Sea Level
125
120
115
110
105
100
45
40
35
30
PACIFIC OCEAN
COAST RANGES
SIERRA NEVADA
GREAT BASIN
COLORADO PLATEAU
ROCKY MOUNTAINS
GREAT
Rio Grande
Colorado R.
Below Sea Level